CHAPTER 7

Loans, debts, 285, 289, 290
Credit card purchases, 285, 291
Financing equipment purchases, 286, 287, 292
Real estate taxes, 287
Certificates of deposit, 294, 296, 314
Cost of inflation, 304
Investment, 305
Sinking funds, 307
Installment loans, 308
Mortgage, 312, 314
Trust funds, 314

CHAPTER 8

Buying preferences, 320
Stock performance, 324
Rainfall data, 324
Academic grades, 325
Product survey, 328
Waiting times in line, 334
Weather forecasting, 339
Options in cars, 344
ESP, 347
Protein molecules, 352
Car accidents and drinking, 360
Quality control, 362, 366
Diabetes tests, 362, 388
Opinion polls, 376
Demographic spread, urban/nonurban, 377
Social-economic data, 379
Teenage smoking, 380
Multiple choice tests, 380
Behavior patterns, 381
Lawyers' assessment of clients, 385
Supplier reliability in manufacturing, 386
Reliability of pregnancy test kits, 388

CHAPTER 9

Quality control, 395
Manufacturing, 399
Length of phone calls, 408
Payoffs on roulette, 409
Insurance premiums, 410
Food price futures, 410
Oil exploration, 413
Genetic theory of plants, 418
Opinion polls, 419
Drug company claims, testing products, 419
Test taking, 419
Lifetime of lightbulbs, 438
Sampling from production line, 438
College enrollments, 447
Voter confidence, 455
Reliability of a production process, 456

CHAPTER 10

Behavioral patterns, 464
Brand switching, 465
Population transition, 469
Queuing theory, 469
Genetics, 471
Learning processes, 474
Heredity, 475, 478
Shopping habits, 476
Waiting lines, 477
Machine performance, 484
Distribution of inventory, 484
Morra, 492
Advertising payoffs, 496
Investment options, 498
New bank customer development, 498
Irrigation/rainfall mixed strategy, 506
Union contract negotiations, 510
Insecticide spraying, 510

second edition

Applied Finite Mathematics

Dalton R. Hunkins
Saint Bonaventure University

Larry R. Mugridge
Kutztown University of Pennsylvania

Prindle, Weber & Schmidt
Boston, Massachusetts

PWS PUBLISHERS

Prindle, Weber & Schmidt • ✦ • Duxbury Press • ♦ • PWS Engineering • ⬥
Statler Office Building • 20 Park Plaza • Boston, Massachusetts 02116

Dedicated to our wives

Jeanette and Sandy

for their patience and understanding during the many long hours, late nights, early mornings, and missed meals while this text was being brought to fruition.

Library of Congress Cataloging in Publication Data

Hunkins, Dalton R.
    Applied Finite Mathematics.

    Bibliography: p.
    Includes index.
    1. Mathematics—1981- . I. Mugridge, Larry R.
    II. Title
QA39.2.H859 1985   510   84-20591
ISBN 0-87150-861-3

ISBN 0-87150-861-3

This text was composed in Times Roman and Eurostyle by Polyglot PTE Ltd. and printed and bound by the Maple-Vail Book Manufacturing Group. Designed by Amato Prudente and production work by Comprehensive Graphics. Cover designed by Susan London.

Printed in the United States of America
10 9 8 7 6 5 4 3 2 1 – 88 87 86 85

Preface

Applied Finite Mathematics is written to present important mathematical tools and applications to students with a wide variety of majors. It is our belief that they will be better prepared to understand the use of quantitative mathematical concepts in their careers in such areas as business, economics, social sciences, and life sciences, as well as other courses that are part of their major courses of study.

FEATURES

It is assumed that the average student has a high school algebra background. However, for those who need it, a brief review is included in sections 1.1 through 1.3. A diagnostic test is provided at the beginning of Chapter 1 for students who wish to check their own level of proficiency.

The stated objectives continue to be met with the second edition. A wide variety of *real world* applications are used to introduce virtually all new concepts throughout the text. Examples, exercises, and applications are the key to understanding. A particular effort is made to relate exercises to examples, with several challenging problems placed towards the end of most exercise sets. The even-numbered exercises are paired with the odd-numbered ones, with the answers to the latter group given in Appendix C.

NEW ADDITIONS

From suggestions by users of the first edition, this edition is enhanced with the following:

• the exercise sets being reworked and expanded to over 1500 exercises;

• the inclusion of a section on logic;

• the discussion of systems and linear forms of two variables put into a separate chapter;

• the discussion of standard minimization problems in linear programming using duality;

• more examples that demonstrate the use of a computer program in the solution process;

• the inclusion of additional applications of Markov chains.

SUBJECT MATTER CHAPTER BY CHAPTER

Chapter 1 provides a review of concepts from high school algebra that are required in the remainder of the text. A section on logic is also included in this chapter.

Linear functions with applications has been moved to chapter 2. Chapter 2 also contains a new section on solving systems of two linear equations in two unknowns. This new section and the last section of the chapter on *word problems* will help the student prepare for systems in more than two unknowns and linear programming problems.

Chapters 3 and 6 cover linear programming problems from the graphical method through the simplex method. The chapter on the simplex method, chapter 6, has been expanded to include a section on solving standard minimization problems using duality. The coverage of chapter 3 may be postponed to follow chapter 5 and before chapter 6.

Systems of linear equations in more than two unknowns and matrix algebra are covered in chapters 4 and 5. Chapter 5 also includes a linear production model and lines of best fit. Lastly, a discussion on theoretical aspects on the number of solutions to a system of linear equations has been moved to the last section of chapter 5.

Chapter 7 develops material on simple and compound interest, bank discount, annuities, and other applications of the mathematics of finance. A greater emphasis has been made in this chapter on the use of a calculator in solving problems.

Chapter 8 deals with basic probability concepts, including a discussion of sets, Venn diagrams, and counting techniques. The section on counting principles has been divided into two sections. Chapter 8 could be followed with chapter 10, Markov chains and game theory, giving additional applications of probability theory.

Chapter 9 covers basic statistics and provides applications to such areas as insurance, production control, voting models, and medicine. The concepts of descriptive statistics are related to probability through random variables.

The applications of Markov chains and matrix games are combined into chapter 10. The applications of Markov chains has been expanded to include also a genetics model and queueing theory.

Chapter 11, Computers and BASIC Language Programming, is included because of the increasingly important role of computers in problem solving. Students using this chapter should be able to use, read, and/or make simple modifications in BASIC programs.

SUGGESTIONS FOR COURSE STUDY

The following graph illustrates the chapter dependencies. Starting with chapter 1, the instructor can choose a sequence of chapters that would be appropriate for his/her course. For example, a one-semester course on linear mathematics could start with chapter 1 (as review) and cover chapters 2 through 6, in the order 2, 3, 4, 5, and 6 or in the order 2, 4, 5, 3, and 6. A one-semester course consisting of some of the linear mathematics in the text and probability theory could cover chapters 1, 2, 4, 5, 8, and 10. Chapter 7, Mathematics of Finance, and/or chapter 11, Computers and BASIC Language Programming, could be included at any point in a course where the instructor would find it to be appropriate.

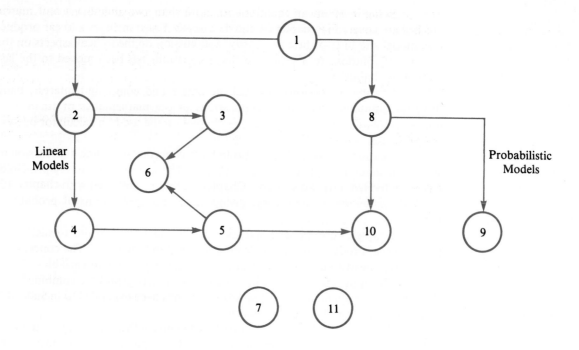

ACKNOWLEDGMENTS

We are grateful to a number of people who were helpful with valuable suggestions. The following reviewers helped bring the first edition to fruition.

Ronald D. Baker, University of Delaware; Barbara J. Bulmahn, Indiana University–Purdue University; William E. Coppage, Wright State University; Sam Councilman, California State University, Long Beach; David Cusick, Marshall University; John J. Dinkel, The Pennsylvania State University; William G. Frederick, Indiana University–Purdue University; Alan Gleit, University of Massachusetts, Amherst; Rose C. Hamm, The College of Charleston; Robert Hathway, Illinois State University; T.L. Herdman, Virginia Polytechnic Institute & State University; J.G. Horne Jr., Virginia Polytechnic Institute & State University; Edward L. Keller, California State University, Hayward; David E. Kullman, Miami University; William Margulies, California State University, Long Beach; Hal G. Moore, Brigham Young University; Peter J. Nicholls, Northern Illinois University; and W. Wiley Williams, University of Louisville.

The following reviewers of the first edition made many valuable suggestions for the improvements that appear in the second edition.

Richard Cutler, Villanova University; Stephen Gendler, Clarion State University; Patricia Hickey, Baylor University; J.R. Ingraham, The Citadel; Raymond Maruca, Delaware County Community College; and W. Wiley Williams, University of Louisville.

Last, we like to thank the editorial staff of Prindle, Weber & Schmidt, especially production editor Susan London, for their patience and understanding.

Dalton R. Hunkins
Larry R. Mugridge

Contents

1 Getting Started   1

Pretest   2
Answers to Pretest   5
1.1   Equations and Inequalities   8
1.2   Coordinate Systems: Graphing a Linear Equation   14
1.3   Writing the Equation of a Line   27
1.4   Logic   32
Summary of Terms   47
Review Exercises   47

2 Applications of Linear Equations   51

2.1   Linear Functions   52
2.2   Applications of Linear Functions to Business   59
2.3   Solving Systems of Two Equations in Two Unknowns   65
2.4   Applications of Systems of Two Equations   71
Summary of Terms   82
Review Exercises   83

3 Linear Programming: Graphical Methods 85

3.1 Graphing Linear Inequalities 85
3.2 Graphical Solution of Linear Programming Problems 98
Summary of Terms 115
Review Exercises 116

4 Systems of Linear Equations 118

4.1 Linear Forms and Systems of Linear Equations 118
4.2 Augmented Matrices and Gaussian Elimination 127
4.3 Gauss–Jordan Reduction 140
Summary of Terms 158
Review Exercises 158

5 Matrix Algebra with Applications 161

5.1 Matrix Operations 161
5.2 Row-Column Product 168
5.3 Matrix Multiplication 174
5.4 Lines of Best Fit 183
5.5 Square Matrices and Inverses 192
5.6 Linear Production Models 204
5.7 The Number of Solutions of a Linear System 213
Summary of Terms 216
Review Exercises 217

6 Linear Programming: The Simplex Method   **219**

6.1   Introducing Slack Variables   220
6.2   Solving Standard Maximization Problems   233
6.3   Standard Minimization Problems: Duality   250
6.4   Linear Programming Problems in General: The Two-Phase
        Method and Computer Solutions   265
Summary of Terms   279
Review Exercises   280

7 Mathematics of Finance   **283**

7.1   Simple Interest and Bank Discount   283
7.2   Compound Interest   293
7.3   Annuities   304
Summary of Terms and Formulas   315
Review Exercises   317

8 Probability   **319**

8.1   What Is Probability?   319
8.2   Properties of Probability Models   331
8.3   Equiprobable Models: The Principle of Counting
        and Permutations   341
8.4   Equiprobable Models: Combinations   352
8.5   Conditional Probability   359

8.6 Sequence of Experiments   372
8.7 Bayes's Formula   382
Summary of Terms and Formulas   389
Review Exercises   390

9 Statistics   394

9.1 Random Variables   394
9.2 Expected Value   403
9.3 Bernoulli Trials: Binomial Random Variables   414
9.4 Variance and Standard Deviation   423
9.5 Normal Random Variables: Normal Curve   428
9.6 Normal Approximation of a Binomial Random Variable   442
9.7 Hypothesis Testing   450
Summary of Terms and Formulas   458
Review Exercises   460

10 Decision Making: Markov Chains and Matrix Games   462

10.1 Matrix Models of Markov Chains   463
10.2 Regular Markov Chains and Applications   478
10.3 Matrix Games   491
10.4 Two-Person Zero-Sum Games with Mixed Strategies   499
Summary of Terms and Formulas   511
Review Exercises   512

11 Computers and BASIC Language Programming 516

11.1 Variables, Constants, and the LET Statement 517
11.2 Running a Program and More BASIC Statements 523
11.3 Control Structures: IF-THEN Statements and FOR-NEXT
    Loops 532
11.4 Arrays and MAT Statements 540
Summary of Terms 549
Review Exercises 549

Appendices 553

Appendix A: Real Numbers 555
Appendix B: Tables 559
    I. Binomial Random Variable X 559
    II. Standard Normal Curve 564
    III. Compound Interest and Annunity 565
Appendix C: Answers to Odd-Numbered Exercises 580

Index 635

CHAPTER 1

Getting Started

Introduction

The concepts of solving equations and inequalities, graphing linear equations, and writing the equation of a line make up the content of the first sections of this chapter. Section 1.4 deals with some basic concepts of the logic of compound statements and valid arguments. The main part of the text develops from your understanding of these basic ideas and the fundamental properties of the real numbers, a review of which is given here and in appendix A. Some of you may already have sufficient grasp of the material presented in appendix A and the first three sections of this chapter and may wish, therefore, to proceed directly to section 1.4. However, to determine if you need to spend some time reviewing the material of the first three sections, it is recommended that you take the provided pretest that appears first.

The pretest is divided into three parts, corresponding to the first three sections of the chapter.

- PART 1   Equations and Inequalities
- PART 2   Coordinate Systems: Graphing a Linear Equation
- PART 3   Writing the Equation of a Line

The answers immediately follow the pretest.

PRETEST

PART 1   Equations and Inequalities

For exercises 1–8 solve the equation for x.

1. $3x + 5 = 26$

2. $\dfrac{4}{3}x = 16$

3. $\dfrac{9}{2}x + \dfrac{1}{2} = 3x - \dfrac{5}{8}$

4. $1.3x + 1.2 = 0.5x + 5.6$

5. $2(x + 1) - 7 = 10$

6. $\dfrac{1}{3}(9x - 45) = -3x - 1$

7. $8(x + 3) - 11 = 5(2x - 5)$

8. $4(x - 1) - 3(x + 2) = 7(x + 2) - 6$

For exercises 9–12 solve the inequalities for x and graph the solution set on the number line.

9. $5x - 13 \leq 22$

10. $8x - 1 > 6x + 7$

11. $-2(2x + 1) - 6 > 4(x - 1)$

12. $0.4x - 9.6 \leq 1.2(4x + 3)$

For exercises 13–18 let x represent the number of people waiting in line at a checkout counter of a supermarket. Symbolize the given statement with an inequality.

13. There are fewer than 8 people waiting in line.
14. There are at least 8 people waiting in line.
15. There are more than 8 people waiting in line.
16. There are no more than 8 people waiting in line.
17. There are no fewer than 8 people waiting in line.
18. There are at most 8 people waiting in line.

PART 2   Coordinate Systems: Graphing a Linear Equation

For exercises 1–6 plot the points in the coordinate plane.

1. $(3, 8)$

2. $(0, -5)$

3. $(-4, 8)$

4. $(-4, -6)$

5. $\left(\dfrac{5}{2}, 0\right)$

6. $\left(\dfrac{5}{2}, -6\right)$

Exercises 7–12 refer to figure 1.1.

7. What is the ordinate of P_1?

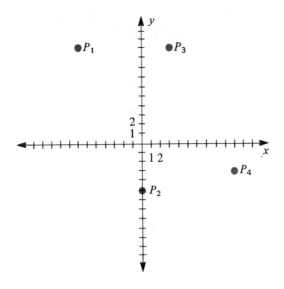

FIGURE 1.1

8. What is the ordinate of P_2?

9. What is the abscissa of P_3?

10. What is the abscissa of P_4?

11. What are the coordinates of the point P_2?

12. What are the coordinates of the point P_3?

For exercises 13–16 describe where the point (x, y) would be located in the coordinate plane.

13. $x < 0$ and $y > 0$ 14. $x = 0$ and $y = 0$

15. $x > 0$ and $y = 0$ 16. $x \geq 0$ and $y \geq 0$

For exercises 17–22 find the slope of the line passing through the two given points.

17. $P(4, 6)$ and $Q(6, 9)$ 18. $P(2, 9)$ and $Q(7, 4)$

19. $P(-1, 9)$ and $Q(5, -15)$ 20. $P(4, -3)$ and $Q(4, 7)$

21. $P\left(\dfrac{5}{4}, 11\right)$ and $Q\left(\dfrac{6}{7}, 11\right)$ 22. $P\left(\dfrac{1}{2}, 2\right)$ and $Q\left(3, \dfrac{7}{2}\right)$

For exercises 23 and 24 determine whether or not the three points lie on the same line.

23. $P(-2, 7)$, $Q(0, 1)$,   and $R(4, -13)$

24. $P(-4, 2)$, $Q(0, -10)$,   and $R(-2, -4)$

25. Determine the pair (x, y) that satisfies the equation $5x - 4y = 40$ when:

    a. $x = 0$ b. $y = 0$ c. $x = -5$ d. $y = 5$

26. Determine the pair (x, y) that satisfies the equation $\frac{4}{5}x + \frac{3}{2}y = 1$ when:

    a. $y = 0$ b. $x = 1$ c. $y = \frac{2}{3}$ d. $x = -10$

For each of the equations in exercises 27–30 find the slope m and the y-intercept b, and draw the graph of the equation.

27. $y = -5x + 7$ 28. $y + 5 = 0$

29. $5x - 4y = 40$ 30. $\frac{4}{5}x + \frac{3}{2}y = 1$

PART 3   Writing the Equation of a Line

For exercises 1–4 write the equation of the line for the given slope m and the given y-intercept b.

1. $m = \frac{6}{7}$ and $b = 0$ 2. $m = 3$ and $b = -1$

3. $m = -7$ and $b = \frac{9}{13}$ 4. $m = \frac{11}{8}$ and $b = -15$

For exercises 5–8 write the equation of the line for the given slope m and the given point P.

5. $m = -2$ and $P(0, 28)$ 6. $m = \frac{1}{3}$ and $P(9, 18)$

7. $m = 4$ and $P\left(\frac{7}{8}, -\frac{3}{2}\right)$ 8. $m = 0.25$ and $P(3.6, 0.3)$

For exercises 9–14 write the equation of the line passing through the two given points.

9. $P(0, 0)$ and $Q(2, 14)$ 10. $P(1, 9)$ and $Q(3, 15)$

11. $P(4, 6)$ and $Q(6, 9)$ 12. $P(1, -5)$ and $Q(-3, -5)$

13. $P(-2, 7)$ and $Q(2, -3)$ 14. $P\left(\frac{5}{3}, -4\right)$ and $Q\left(\frac{5}{3}, 6\right)$

ANSWERS TO PRETEST

PART 1

1. $x = 7$

2. $x = 12$

3. $x = -\dfrac{3}{4}$

4. $x = 5.5$

5. $x = \dfrac{15}{2}$

6. $x = \dfrac{7}{3}$

7. $x = 19$

8. $x = -3$

9. $x \leq 7$. For the graph see figure 1.2.

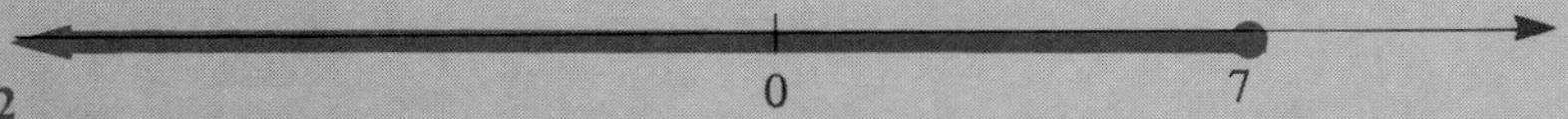

FIGURE 1.2

10. $x > 4$. For the graph see figure 1.3.

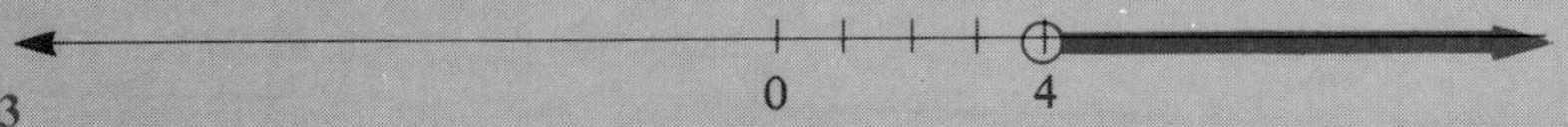

FIGURE 1.3

11. $x < -\dfrac{1}{2}$. For the graph see figure 1.4.

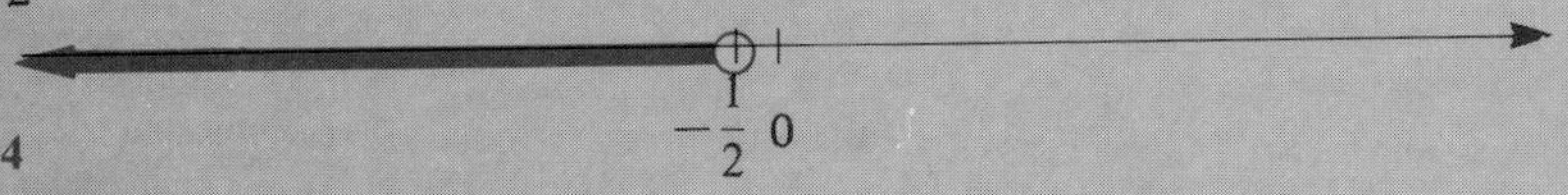

FIGURE 1.4

12. $x \geq -3$. For the graph see figure 1.5.

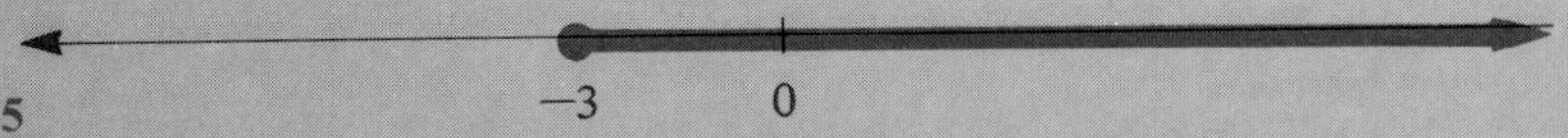

FIGURE 1.5

13. $x < 8$

14. $x \geq 8$

15. $x > 8$

16. $x \leq 8$

17. $x \geq 8$

18. $x \leq 8$

PART 2

For exercises 1–6, see figure 1.6

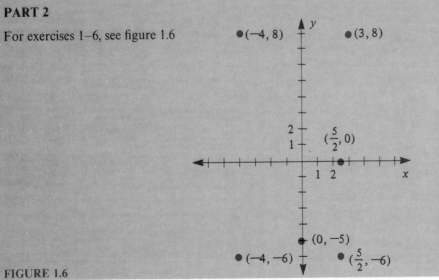

FIGURE 1.6

7. 10

8. −5

9. 3

10. 10

11. (0, −5)

12. (3, 10)

13. quadrant II

14. origin

15. positive *x*-axis

16. quadrant I, positive *x*-axis, positive *y*-axis, or origin

17. $\dfrac{3}{2}$

18. −1

19. −4

20. not defined

21. 0

22. $\dfrac{3}{5}$

23. not on same line

24. on the same line

25. a. (0, −10)

b. (8, 0)

c. $\left(-5, -16\dfrac{1}{4}\right)$

d. (12, 5)

26. a. $\left(\dfrac{5}{4}, 0\right)$

b. $\left(1, \dfrac{2}{15}\right)$

c. $\left(0, \dfrac{2}{3}\right)$

d. (−10, 6)

27. $m = -5$; $b = 7$. For the graph see figure 1.7.

28. $m = 0$; $b = -5$. For the graph see figure 1.8.

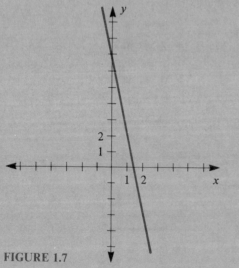

FIGURE 1.7

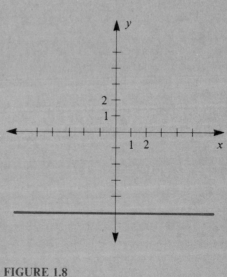

FIGURE 1.8

29. $m = \dfrac{5}{4}$; $b = -10$. For the graph see figure 1.9.

30. $m = -\dfrac{8}{15}$; $b = \dfrac{2}{3}$. For the graph see figure 1.10.

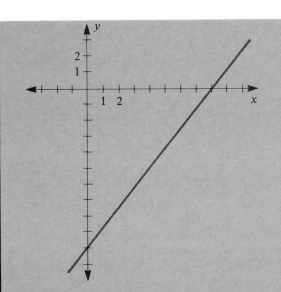

FIGURE 1.9

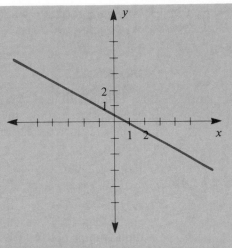

FIGURE 1.10

PART 3

1. $y = \dfrac{6}{7}x$

2. $y = 3x - 1$

3. $y = -7x + \dfrac{9}{13}$

4. $y = \dfrac{11}{8}x - 15$

5. $y = -2x + 28$

6. $y = \dfrac{1}{3}x + 15$

7. $y = 4x - 5$

8. $y = 0.25x - 0.6$

9. $y = 7x$

10. $y = 3x + 6$

11. $y = \dfrac{3}{2}x$

12. $y = -5$

13. $y = -\dfrac{5}{2}x + 2$

14. $x = \dfrac{5}{3}$

1.1 Equations and Inequalities

Equations occur frequently in applications. For example, Al, who owns the Pizza Place, figures that if he makes x pizzas, his cost (in dollars) is $1.20x + 98$. He also knows that his income from selling x pizzas is $4x$. Al wants to find the number of pizzas he must sell in order to break even; that is, he wants to determine x so that his cost equals his income. Therefore, setting cost equal to income, he arrives at the equation

$$1.20x + 98 = 4x.$$

By using rules of equations, he finds that $x = 35$. Therefore, Al must make and sell 35 pizzas to break even.

In chapter 2 we consider how to write expressions such as Al's cost equation. But for now, let us review the properties and rules used to arrive at Al's answer.

A **solution** to an equation containing one variable x is a number that can be substituted for x such that the statement is true. For example, the solution to the equation

$$x + 3 = 7$$

is clearly

$$x = 4$$

since

$$4 + 3 = 7.$$

As another example, the solution to

$$2x = -12$$

is

$$x = -6$$

since

$$2(-6) = -12.$$

For simple equations like the preceding ones, we can give the solution from our knowledge of adding and multiplying real numbers. However, if we cannot see the solution to an equation immediately, we can derive simpler **equivalent equations** from the original equation. An equivalent equation is obtained by rewriting expressions using properties of real numbers and the following rules of equality.

Rules of Equality

■ 1. We can add or subtract the same quantity on both sides of an equation. That is,

$$a = b \text{ is equivalent to } a + c = b + c$$

and

$$a = b \text{ is equivalent to } a - c = b - c.$$

2. We can multiply or divide both sides of an equation by the same nonzero quantity. That is,

$$a = b \text{ is equivalent to } ac = bc, \text{ when } c \neq 0$$

and

$$a = b \text{ is equivalent to } \frac{a}{c} = \frac{b}{c}, \text{ when } c \neq 0. \quad ■$$

In the following examples we solve equations by deriving equivalent equations.

EXAMPLES

1.
$$5x + 5 = 3x + 19$$
$$5x + (-3x) + 5 + (-5) = 3x + (-3x) + 19 + (-5) \quad \text{(rule 1)}$$
$$2x = 14$$
$$x = 7$$

2.
$$2(x + 1) = 3(x + 6) - 4$$
$$2x + 2 = 3x + 18 - 4$$
$$-x = 12$$
$$x = -12$$

3.
$$0.7x + 0.35 = 0.65x - 1$$
$$70x + 35 = 65x - 100$$
$$5x = -135$$
$$\frac{5x}{5} = \frac{-135}{5} \quad \text{(rule 2)}$$
$$x = -27$$

4.
$$2(3x - 5) = 15 - \frac{1}{2}(3x - 4)$$
$$6x - 10 = 15 - \frac{3}{2}x + 2$$
$$\frac{15}{2}x = 27$$
$$x = \frac{2}{15}(27) = \frac{18}{5}$$

5.
$$0.1x - 0.3[2x - (x + 0.1)] = -0.8x + 0.33$$
$$0.1x - 0.6x + 0.3(x + 0.1) = -0.8x + 0.33$$
$$-0.2x + 0.03 = -0.8x + 0.33$$
$$0.6x = 0.3$$
$$x = 0.5$$

The next two examples illustrate how equations can be used to solve word problems.

6. Let x be the number of electronic slot machines produced on production line A. Suppose that a second production line B can produce three-fourths as many slot machines as line A. Determine x if the total output from both lines is 280 slot machines.

   If x slot machines are produced on line A, then $\frac{3}{4}x$ slot machines are produced on line B. Thus, the total output from both lines is represented by

   $$x + \frac{3}{4}x.$$

   Since the total output is 280, we have

   $$x + \frac{3}{4}x = 280.$$

   Solving for x, we obtain

   $$\frac{7}{4}x = 280$$

   $$x = 160.$$

7. The developer of the Buckeye Industrial Park plans to divide 572 acres of land into 8-acre and 12-acre plots. If there are to be three times as many 12-acre plots as 8-acre plots, find the number of each type of plot that can be in the industrial park if the entire 572 acres are developed.

   Let x denote the number of 8-acre plots; then $3x$ is the number of 12-acre plots. The amount of land needed for the 8-acre plots is $8x$ acres. Similarly, the amount of land needed for the 12-acre plots is $12(3x) = 36x$ acres. The total acreage of land needed is then given by

   $$8x + 36x.$$

   Since there are 572 acres to be used,

   $$8x + 36x = 572.$$

   Solving for x, we obtain

   $$x = 13.$$

   Therefore, the developer should plan 13 eight-acre plots and $3(13) = 39$ twelve-acre plots.

Suppose Al wants to find x, the number of pizzas he should make so as to make a profit. That is, he wants his income to be more than his cost. In this case, Al would find x so that

$$4x > 1.20x + 98.$$

Al would now use rules of inequalities to obtain the solutions

$$x > 35.$$

We consider applications of inequalities in chapters 3 and 6. But for now, let us see how to solve inequalities.

For an inequality containing one variable x, the **solution set** is the set of values for x that satisfies the inequality. We say that the inequality has been **solved** when the solution set is obtained. To help solve inequalities, we can derive **equivalent inequalities** by using the properties of real numbers and the following rules of inequalities.

Rules of Inequalities

■ 1. If we add or subtract the same quantity on both sides of an inequality, the direction (sense) of the inequality does not change. For example,

$$a \leq b \quad \text{is equivalent to} \quad a + c \leq b + c$$

$$2 \leq 3 \quad \text{is equivalent to} \quad 2 + 1 \leq 3 + 1$$

and

$$a > b \quad \text{is equivalent to} \quad a - c > b - c$$

$$4 > 2 \quad \text{is equivalent to} \quad 4 - 2 > 2 - 2.$$

2. If we multiply or divide on both sides of an inequality by the same positive quantity, the direction (sense) of the inequality does not change. For example,

$$a \geq b \quad \text{is equivalent to} \quad ac \geq bc, \text{when } c > 0$$

$$2 \geq 1 \quad \text{is equivalent to} \quad 3 \cdot 2 \geq 3 \cdot 1$$

and

$$a < b \quad \text{is equivalent to} \quad \frac{a}{c} < \frac{b}{c}, \text{when } c > 0$$

$$3 < 5 \quad \text{is equivalent to} \quad \frac{3}{2} < \frac{5}{2}.$$

3. If we multiply or divide on both sides of an inequality by the same negative quantity, the direction (sense) of the inequality changes. For example,

$$a \geq b \quad \text{is equivalent to} \quad ac \leq bc, \text{when } c < 0$$

$$2 \geq 1 \quad \text{is equivalent to} \quad (-1)2 \leq (-1)1$$

and

$$a \leq b \quad \text{is equivalent to} \quad \frac{a}{c} \geq \frac{b}{c}, \text{when } c < 0$$

$$7 \leq 9 \quad \text{is equivalent to} \quad \frac{7}{-3} \geq \frac{9}{-3}. \quad ■$$

In the following examples we derive equivalent inequalities to solve the original inequality and graph the solution set on a number line.

EXAMPLES

8.
$$2x + 3 > 15$$
$$2x > 12$$
$$x > 6$$

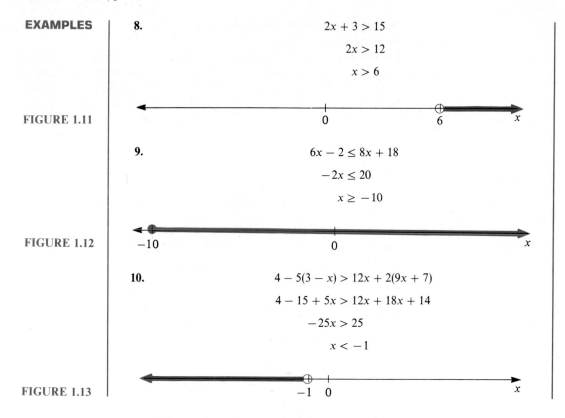

FIGURE 1.11

9.
$$6x - 2 \leq 8x + 18$$
$$-2x \leq 20$$
$$x \geq -10$$

FIGURE 1.12

10.
$$4 - 5(3 - x) > 12x + 2(9x + 7)$$
$$4 - 15 + 5x > 12x + 18x + 14$$
$$-25x > 25$$
$$x < -1$$

FIGURE 1.13

In dealing with applications in later chapters, we will find statements that express inequalities but are not given explicitly as greater than or less than. For example, suppose x represents the number of people waiting in line. Then

a. there are at most 10 people waiting in line   means $x \leq 10$

b. there are no more than 10 people waiting in line   means $x \leq 10$

c. there are at least 10 people waiting in line   means $x \geq 10$

d. there are no fewer than 10 people waiting in line   means $x \geq 10$

EXERCISES 1.1

For exercises 1–8 state whether or not the pairs of equations are equivalent. If a pair of equations is equivalent, show how the second equation can be derived from the first.

1. $2x - 4 = 23$ and $2x = 19$

2. $5x + 4 = 1$ and $5x = -3$

3. $9x - 8 = 5x + 13$ and $4x = 21$

4. $2x + 6 = -4x + 10$ and $6x = 16$

5. $8 - 3(x + 1) = 2x - 17$ and $5 - 3x = 2x - 17$

6. $4x + 2(3x - 4) = 10$ and $10x - 4 = 10$

7. $2 - 6(2x + 1) = 4(1 - 3x) + 8$ and $24x = 8$

8. $-4(x - 1) + 5(x + 3) = -3x + 2$ and $4x = -17$

Solve each of the equations given in exercises 9–24.

9. $9x = 36$ 10. $-2x = 68$ 11. $-3x = -32$

12. $\frac{2}{3}x = 16$ 13. $\frac{3}{4}x = -6$ 14. $\frac{5}{7}x = \frac{15}{14}$

15. $4x + 15 = 11$ 16. $4 - 3(x + 1) = 10$

17. $3x + \frac{1}{7} = \frac{9}{2}x - \frac{1}{14}$ 18. $\frac{1}{2}(4x - 14) = -3x - 2$

19. $3(2x - 5) + (3x - 13) = 17$

20. $3(-2x + 5) - (4x - 11) = 15$

21. $0.7x - 0.5 = 0.2x + 0.1$

22. $3.4x + 0.4(120 - 7x) = -0.6x + 42$

23. $7[4(2x + 1) - 3(2x + 1)] = 8(2x + 1) + 9$

24. $7x + 5[(4 - x) - 2x] = 2(5 - 3x) - 5x$

25. Find the number of pizzas (x) that Al must sell in order to break even if setting the cost equal to the income gives the equation

$$1.40x + 252 = 5.60x.$$

26. Let x be the number of microcomputers assembled on production line A. Suppose that line B can assemble four times as many microcomputers as A. Determine x if the total output from both lines is 1200 microcomputers.

27. An investor wants to divide $8000 between an income fund and a growth fund. Suppose the investor wants to put into the growth fund two-thirds of what he invests in the income fund. Find the amount put into each investment if the entire $8000 is invested.

28. Sally must buy new cabinets for the storage room at her company. She plans to buy two types of cabinets. Type A costs $54 each and type B costs $80 each. Sally wants twice as many type A cabinets as type B. If she has a total of $2820 to spend, how many of each type can she buy?

For statements in exercises 29–34, answer true or false.

29. $-3 > 5$ 30. $-2 \le 0$ 31. $\frac{-2}{12} \le -\frac{1}{6}$

32. $\frac{5}{7} < \frac{4}{9}$ 33. $-\frac{3}{11} \le -\frac{9}{26}$ 34. $-1.4 \ge -2.5$

For exercises 35–42 graph the solution set on the real number line for each of the inequalities.

35. $x \leq 5$

36. $x > -3$

37. $x \geq -1$

38. $x > 6$

39. $-4 < x \leq 8$

40. $0 < x < 5$

41. $-3.5 \leq x < -1.5$

42. $1 \leq x \leq 7$

For exercises 43–48 state whether or not the pair of inequalities is equivalent. If a pair of inequalities is equivalent, show how the second inequality can be derived from the first.

43. $-2x \leq -10$ and $x \leq 5$

44. $22 > 4x$ and $x < \dfrac{11}{2}$

45. $3x + 2 < 5x - 4$ and $-2x > -6$

46. $9x + 13 \geq 15x - 5$ and $x \geq 3$

47. $-1.2x - 4 < 0.8x + 6$ and $2x > -10$

48. $-\dfrac{2}{3}x + 1 \geq \dfrac{4}{5}x - 2$ and $10x - 15 \leq -12x + 30$

For exercises 49–56 solve each of the inequalities and graph the solution set on the real number line.

49. $-3x \leq 15$

50. $7x > -21$

51. $-\dfrac{21}{10} \geq -\dfrac{3}{5}x$

52. $-\dfrac{7}{6}x < \dfrac{35}{4}$

53. $7x - 12 < 5x + 4$

54. $2.3x + 20 \geq -1.7x - 4$

55. $-3(x - 2) - 8 > 4(x + 3)$

56. $-4(x - 4) + 4 \leq 2(3x + 17) - 3x$

For exercises 57–68 let x and y represent two real numbers. Symbolize the given statement in terms of an inequality.

57. y is non-negative

58. x is positive

59. x is non-positive

60. y is negative

61. x is at least 10

62. x is no more than 20

63. y is no less than 1

64. y is at most 15

65. x is no greater than y

66. y is at least as large as x

67. y is no smaller than x

68. x is at least twice as large as y

1.2 Coordinate Systems: Graphing a Linear Equation

There is a one-to-one correspondence between points on a line and the set of real numbers. (To review the real number line, refer to appendix A.) By using two number lines, we can exhibit a one-to-one correspondence between points in a geometric plane and the set of ordered pairs of real numbers. For an **ordered pair** (a, b) of real numbers, $(a, b) \neq (b, a)$ unless $a = b$.

Displaying or plotting ordered pairs is an important means of exhibiting data visually. Often a picture can show more readily what is happening. For instance, if monthly sales figures are plotted as ordered pairs with the first number of the pair representing the month and the second number representing the sales (in thousands of dollars), we can see at a glance the trends in sales over a period of time (see figure 1.14). Let us give the details for exhibiting data as shown in figure 1.14.

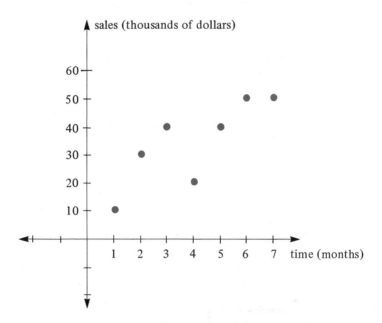

FIGURE 1.14

First, we draw two number lines so that they are perpendicular and intersect at their origins. The horizontal number line is called the **x-axis** and the vertical number line is called the **y-axis**. Together these two **coordinate axes** form a **Cartesian coordinate system** or a **rectangular coordinate system**. A geometric plane together with a coordinate system is called a **coordinate plane**. (See figure 1.15.)

The point 0 of intersection of the two coordinate axes is called the **origin** of the system and corresponds to the pair $(0, 0)$. The x-axis is set up so that points on the axis with positive coordinates lie to the right of the origin and the y-axis is set up so that points on this axis with positive coordinates lie above the origin. The coordinate axes of the system in figure 1.15 have the same **scale**; that is, the two points on each axis labeled with the number 1 are the same distance from the origin. In general the coordinate axes need not have the same scale.

An ordered pair (a, b) of real numbers identifies a unique point P in the plane. To **plot** (locate) this point P, we first locate the point on the x-axis with coordinate a and draw a line through that point perpendicular to the x-axis. Similarly we draw a perpendicular line through the point on the y-axis with coordinate b. The intersection of these two lines locates P (see figure 1.16). We can say that the first number of the pair (a, b) tells us how much to move horizontally

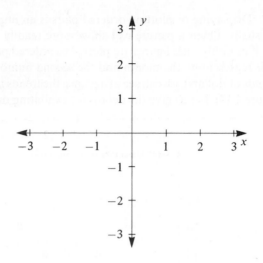

FIGURE 1.15

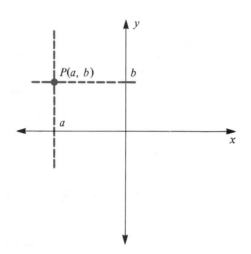

FIGURE 1.16

(right if $a > 0$, left if $a < 0$, and none if $a = 0$), and the second number tells us how much to move vertically (up if $b > 0$, down if $b < 0$, and none if $b = 0$).

We can reverse the process and identify with each point P in a coordinate plane an ordered pair (a, b). We say that the ordered pair (a, b) gives the **coordinates** of the point P. The first coordinate a of (a, b) is called the **x-coordinate** or **abscissa** of P and the second coordinate b of (a, b) is called the **y-coordinate** or **ordinate** of P. In giving the coordinates of a point, we can write either $P = (a, b)$ or $P(a, b)$.

EXAMPLE 1

The five points P, Q, R, S, and T with respective coordinates $(5, 2), (-4, 3), (-2, -2), (3, -4)$, and $(5, 0)$ are given in figure 1.17.

Note that a coordinate system divides the plane into four regions called **quadrants**. These quadrants are numbered as in figure 1.18. All points in

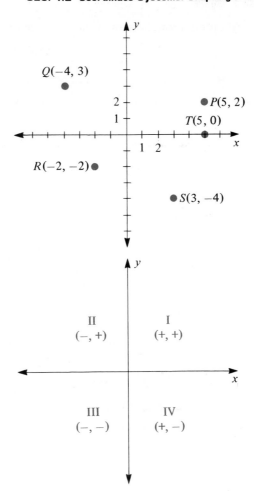

FIGURE 1.17

FIGURE 1.18

quadrant I have the property that both coordinates are positive. All points in quadrant II have the property that the abscissa is negative and the ordinate is positive. Analogous statements can be made about the points in quadrants III and IV.

An important use of a coordinate system is **graphing** an equation in x and y. We will consider here only equations of a special kind.

Let us return to the businessman Al who owns the Pizza Place. If he sells x pizzas, Al knows that his profit y (in dollars) is given by the formula

$$y = 2.80x - 98.$$

For example, if he sells 100 pizzas, then his profit is $182. This can be seen by setting $x = 100$ into the formula and obtaining

$$y = 2.80(100) - 98$$

$$= 182.$$

Al's profit formula is an example of a **linear equation**, as it can be written

$$-2.80x + y = 98.$$

Definition ■ An equation in the variables x and y of the form

$$Ax + By = C$$

where A, B, and C are constants and A and B are not both zero, is called a **linear equation in x and y**, or simply a **linear equation**. ■

Another example of a linear equation is

$$-2x + y = 1. \tag{1}$$

Equation (1) also can be written in the equivalent form

$$y = 2x + 1.$$

Definition ■ The **graph** of a linear equation $Ax + By = C$ is the set of points in the Cartesian plane with coordinates (x, y) that satisfy the equation. ■

For example, consider the linear equation given in (1). Coordinates (x, y) that satisfy $-2x + y = 1$ can be generated by setting one of the variables equal to any constant and solving for the other variable. For example, setting $x = 0$ and solving for y, we have $y = 1$; thus, $(0, 1)$ belongs to the graph of equation (1). Setting $y = 3$ and solving for x, we have $x = 1$; thus, $(1, 3)$ also belongs to the graph of equation (1). Similarly, it can be shown that $(2, 5)$, $(-1, -1)$, and $(-2, -3)$ are other points belonging to the graph. These points are plotted in figure 1.19.

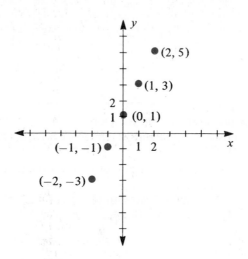

FIGURE 1.19

The graph of a linear equation contains infinitely many points. We have plotted only five points of the graph of equation (1). These points suggest that the graph is a straight line. This observation is true for any linear equation and is stated in the following theorem.

Theorem 1   ■ The graph of a linear equation
$$Ax + By = C$$
is a straight line.   ■

Knowing this it is easy to draw the graph because of the following statement from geometry.

Two points uniquely determine a line in the plane.

In particular, we can obtain the graph of a linear equation by

1. determining two ordered pairs that satisfy the equation,
2. plotting the two points, and
3. drawing a straight line through the two points.

EXAMPLES

2.
a. Consider the equation $y = \frac{4}{3}x - 2$. When $x = 0$, $y = -2$ and when $x = 3$, $y = 2$. Therefore two points that satisfy this equation are $(0, -2)$ and $(3, 2)$. These two points are plotted in a coordinate system and the line is drawn through them. (See figure 1.20a.)

b. Consider the equation $2x + 3y - 9 = 0$. Substituting $x = 0$ into the equation, we find $y = 3$. Thus $(0, 3)$ is a point in the graph. Similarly, substituting $y = 0$ into the equation and solving for x, we have $x = \frac{9}{2}$, and $(\frac{9}{2}, 0)$ is a second point in the graph. These two points are plotted in a coordinate system and the line is drawn through them. (See figure 1.20b.)

c. Consider the equation $y = 4$. Any point P in its graph has coordinates of the form $(x, 4)$. Thus the graph of $y = 4$ is a horizontal line located 4 units above the x-axis. (See figure 1.20c.)

d. Consider the equation $x = -2$. Any point P in its graph has coordinates of the form $(-2, y)$. Therefore the graph of $x = -2$ is a vertical line located 2 units to the left of the y-axis. (See figure 1.20d.)

3. Graph Al's profit formula
$$y = 2.80x - 98$$
where y is the profit (in dollars) obtained by selling x pizzas.

The profit formula is a linear equation; its graph is a straight line. We determine two **data points** by letting x be two convenient values. For example, when $x = 0$, $y = -98$ and when $x = 100$, $y = 182$. Plotting the two data points $(0, -98)$ and $(100, 182)$, we then draw the desired line (see figure 1.21). Notice that since x represents the number of pizzas sold, $x \geq 0$ and, therefore, we delete that part of the line in quadrant III.

4. A manufacturing company finds that the cost C (in thousands of dollars) is given by
$$C = 0.50x + 25$$
where x is the number of units produced. Draw the graph of this cost equation.

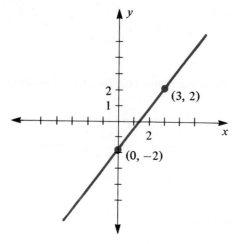

(a) Graph of $y = \frac{4}{3}x - 2$

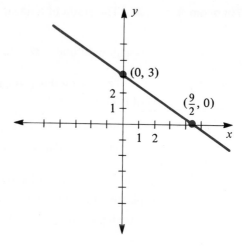

(b) Graph of $2x + 3y - 1 = 0$

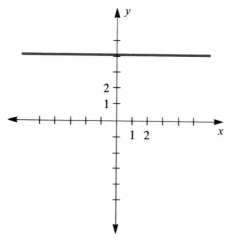

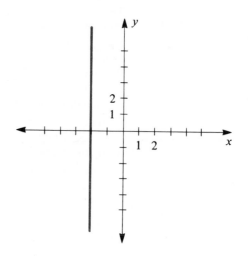

FIGURE 1.20   (c) Graph of $y = 4$

(d) Graph of $x = -2$

Since the cost equation is linear, we need to determine only the cost for each of two values of x and plot these points. For example, when $x = 0$, $C = 25$ and when $x = 10, C = 30$. The two data points $(0, 25)$ and $(10, 30)$ and the line determined by them are shown in figure 1.22.

Since the graph of a linear equation is a straight line, there is a close relationship between linear equations and straight lines. Therefore we now investigate some properties of straight lines that can be used in our study of applications of linear equations.

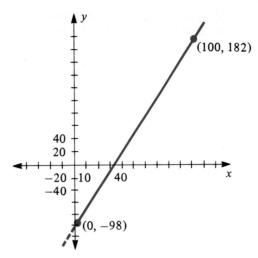

FIGURE 1.21

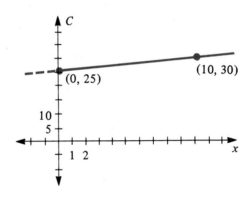

FIGURE 1.22

Consider a straight line ℓ in a coordinate plane and two points P_1 and Q_1 on the line, as shown in figure 1.23. We observe that if a point moves along the line ℓ from P_1 to Q_1, it must move up and to the right. The ratio of the amount of change in the vertical direction to the amount of change in the horizontal direction is the slope of the line ℓ. Knowing the slope of a straight line enables us to predict data. For example, if we know how much sales increase with advertising expenditure, we can predict how much we would need to spend on advertising in order to reach a particular sales goal. Let us consider the formal meaning of slope.

Definition

■ Let ℓ be a line in a coordinate plane and let P_1 and Q_1 with coordinates (x_1, y_1) and (x_2, y_2), respectively, be any two distinct points on ℓ. The **slope** m of ℓ is given by

$$m = \frac{y_2 - y_1}{x_2 - x_1} \qquad x_1 \neq x_2. \tag{2}$$

If $x_1 = x_2$, the line ℓ containing P_1 and Q_1 is a vertical line; the slope of a vertical line is not defined. ■

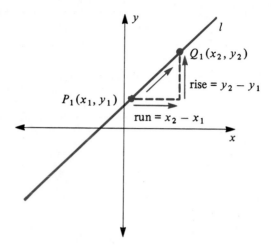

FIGURE 1.23

The difference $y_2 - y_1$ measures the vertical change, called the **rise**, and the difference $x_2 - x_1$ measures the horizontal change, called the **run**, when moving along the line from P_1 to Q_1 (see figure 1.23). Thus, the slope of a line measures the **ratio of rise to run**.

In using formula (2), it does not matter which of the two points is taken first since

$$\frac{y_1 - y_2}{x_1 - x_2} = \frac{y_2 - y_1}{x_2 - x_1}.$$

Furthermore it does not matter which two points we pick on the line to compute the slope—we will always obtain the same number m.

EXAMPLE 5

Find the slopes of the lines ℓ_1, ℓ_2, and ℓ_3 given in figure 1.24 using the points

$$P_1 = (1, 1) \qquad \text{and} \quad Q_1 = (2, 3) \qquad \text{on } \ell_1$$
$$P_2 = (-3, 4) \qquad \text{and} \quad Q_2 = (4, 4) \qquad \text{on } \ell_2$$
$$P_3 = (-2, -1) \quad \text{and} \quad Q_3 = (4, -3) \quad \text{on } \ell_3.$$

Let m_1, m_2, and m_3 denote the slopes of ℓ_1, ℓ_2, and ℓ_3, respectively.

By formula (2)   *By rise-to-run*

up 2

$$m_1 = \frac{3 - 1}{2 - 1} = 2 \qquad\qquad (1, 1) \qquad\qquad (2, 3)$$

right 1

$$m_1 = \frac{2}{1} = 2$$

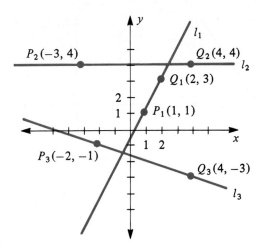

FIGURE 1.24

$$m_2 = \frac{4 - 4}{4 - (-3)} = 0$$

up 0

$(-3, 4)$ $(4, 4)$

right 7

$$m_2 = \frac{0}{7} = 0$$

down 2

$$m_3 = \frac{-3 - (-1)}{4 - (-2)} = -\frac{1}{3}$$

$(-2, -1)$ $(4, -3)$

right 6

$$m_3 = \frac{-2}{6} = -\frac{1}{3}$$

Note that from example 5 we have the information in table 1.1. See figure 1.25.

The concept of slope also gives us a method to determine if three points are on the same line.

TABLE 1.1

Line	Slope	Slant of line from left to right (see figure 1.25)	Information about a point $P(x, y)$ tracing the line
ℓ_1	positive	upward	as x increases in value, y increases in value
ℓ_2	zero	horizontal	as x increases in value, y remains the same
ℓ_3	negative	downward	as x increases in value, y decreases in value

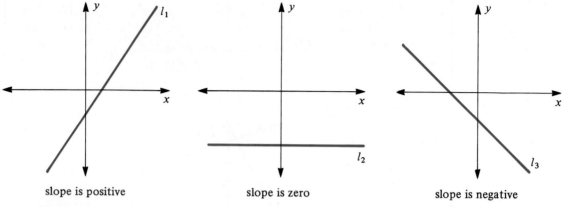

slope is positive   slope is zero   slope is negative

FIGURE 1.25

Theorem 2   ■ Let P, Q, and R be three distinct points in the coordinate plane.

1. If the slope of the line between P and Q is equal to the slope of the line between Q and R, then the three points are on the same line.

2. If the slope of the line between P and Q is not equal to the slope of the line between Q and R, then the three points are not on the same line. ■

EXAMPLE 6

a. Determine if $P(-1, 3)$, $Q(0, -1)$, and $R(5, -21)$ are on the same line.
The slope between P and Q is

$$m_1 = \frac{-1-3}{0-(-1)} = -4.$$

The slope between Q and R is

$$m_2 = \frac{-21-(-1)}{5-0} = -4.$$

Since $m_1 = m_2$, by part 1 of theorem 2, we conclude that P, Q, and R are on the same line.

b. Let us determine if $P(1, -3)$, $Q(-1, -2)$, and $R(3, -1)$ are on the same line.
The slope between P and Q is

$$m_1 = \frac{-2-(-3)}{-1-1} = -\frac{1}{2}.$$

The slope between Q and R is

$$m_2 = \frac{-1-(-2)}{3-(-1)} = \frac{1}{4}.$$

Since $m_1 \neq m_2$, by part 2 of theorem 2 we conclude that P, Q, and R are not on the same line.

The concept of slope is important in helping us to understand linear equations in applications. Some of these applications will be developed in chapter 2. In the meantime, in section 1.3 we consider techniques for writing the equation of a line.

EXERCISES 1.2

Plot the points given in exercises 1–6 in the coordinate plane.

1. $P_1 = (-5, 4)$ 2. $P_2 = (2, 6)$ 3. $P_3 = (0, 3)$
4. $P_4 = (5, -4)$ 5. $P_5 = (-5, -6)$ 6. $P_6 = (-5, 0)$

Exercises 7–14 refer to figure 1.26.

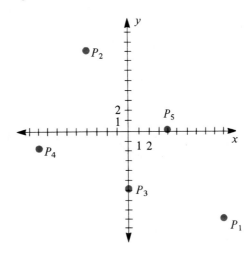

FIGURE 1.26

7. What is the ordinate of P_1?
8. What is the ordinate of P_2?
9. What is the abscissa of P_3?
10. What is the ordinate of P_4?
11. What are the coordinates of P_1?
12. What are the coordinates of P_2?
13. What are the coordinates of P_3?
14. What are the coordinates of P_5?

For exercises 15–22 describe where in the coordinate plane the point (x, y) would be located.

15. $x \leq 0$ 16. $y \geq 0$ 17. $x = 0$
18. $y = 0$ 19. $x = 0$ and $y \leq 0$ 20. $x < 0$ and $y = 0$
21. $x < 0$ and $y > 0$ 22. $x > 0$ and $y > 0$

23. Determine the pair (x, y) that satisfies the equation $5x + 3y = 12$ when:

    a. $x = 0$ b. $y = 0$ c. $x = -3$ d. $y = 11$

24. Determine the pair (x, y) that satisfies the equation $7x - 2y = 18$ when:

    a. $y = -2$ b. $x = 0$ c. $x = 5$ d. $y = \dfrac{7}{4}$

25. Determine the pair (x, y) that satisfies the equation $\frac{3}{4}x + \frac{2}{3}y = 1$ when:

    a. $y = \dfrac{3}{2}$ b. $x = 16$ c. $x = \dfrac{1}{9}$ d. $y = 0$

26. Determine the pair (x, y) that satisfies the equation $\frac{1}{7}x + \frac{1}{5}y = 2$ when:

    a. $x = 5$ b. $y = \dfrac{25}{3}$ c. $y = 0$ d. $y = \dfrac{49}{5}$

Draw the graph of each of the equations given in exercises 27–32 and also determine the slope of the line.

27. $y = 3x - 4$ 28. $y = \dfrac{1}{2}x + 5$ 29. $7x - 2y = 18$

30. $y = 6$ 31. $\dfrac{3}{4}x + \dfrac{2}{3}y = 1$ 32. $\dfrac{1}{7}x + \dfrac{1}{5}y = 2$

Find the slope, if it is defined, of the line passing through the two points given in each of the exercises 33–38.

33. $P(2, 3)$ and $Q(6, 7)$ 34. $P(-5, 3)$ and $Q(1, -9)$

35. $P(15, -2)$ and $Q(15, 7)$ 36. $P(11, 20)$ and $Q(8, 13)$

37. $P\left(8, \dfrac{1}{2}\right)$ and $Q\left(6, \dfrac{7}{4}\right)$ 38. $P\left(-4, \dfrac{4}{7}\right)$ and $Q\left(-1, \dfrac{20}{35}\right)$

Exercises 39–42 refer to figure 1.27. State whether the slope of the line is positive, negative, zero, or undefined.

39. l_1 40. l_2 41. l_3 42. l_4

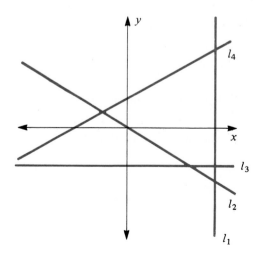

FIGURE 1.27

43. A manufacturing company finds that its cost C (in dollars) is given by

$$C = 5.80x + 1500.$$

where x is the number of units produced.

a. Draw the graph of the cost equation.
b. What is the slope of the graph?
c. What is the cost C when x is 1000?

44. The cost C (in dollars) of making x clocks is given by

$$C = 22.75x + 250.$$

a. Draw the graph of the cost equation.
b. What is the slope of the graph?
c. What is the cost C when 50 clocks are made?

1.3 Writing the Equation of a Line

In the previous section we considered graphing a given linear equation; here we discuss the problem of writing the equation of a line. Understanding how to write the equation of a line helps us with applications requiring an equation that expresses the dependency of one quantity on another. For example, suppose the value of a stamp increases with age. Given sufficient information, we can write an equation that shows how the value of the stamp depends on its age. We will return to this type of problem later.

Now consider the problem of writing the equation of a line ℓ in the coordinate plane. We consider the two cases, a vertical line and a nonvertical line. If the line is vertical, all points on it have the same abscissa a with no restriction on the ordinate (see figure 1.28). Thus any point on this vertical line must satisfy the condition

$$x = a$$

which is the general form of the equation of a vertical line. For example, the equation of the vertical line through the point $(4, -3)$ is

$$x = 4.$$

Now consider a nonvertical line ℓ, for example, the one given in figure 1.29. Let us write an equation that a point $P(x, y)$ must satisfy in order to be on ℓ. The condition we will use is that the slope will be the same for any two distinct points chosen on ℓ. Consider the points $P_1(-4, 5)$ and $Q_1(4, 1)$, which are two points on ℓ. The slope of the line determined by P_1 and Q_1 is

$$m = \frac{1 - 5}{4 - (-4)} = -\frac{1}{2}.$$

For a point $P(x, y)$ to be on ℓ, the slope determined by it and, for example, $Q_1(4, 1)$ must also be $-\frac{1}{2}$. Therefore, we have

$$\frac{y - 1}{x - 4} = -\frac{1}{2}$$

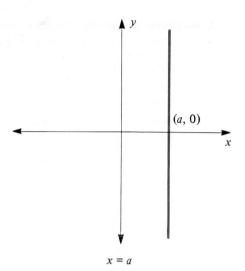

FIGURE 1.28

$x = a$

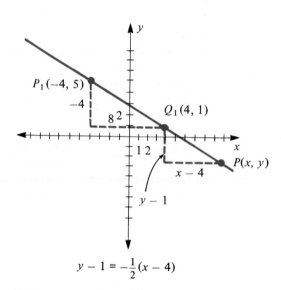

FIGURE 1.29

$$y - 1 = -\frac{1}{2}(x - 4)$$

and therefore

$$y - 1 = -\frac{1}{2}(x - 4).$$

This idea can be used in general to obtain the following form of the equation of a non-vertical line:

$$y - y_1 = m(x - x_1). \tag{1}$$

This equation is called the **point-slope** form.

As the name indicates, a line is uniquely determined when its slope m and one point (x_1, y_1) are given.

EXAMPLE 1

Write the equation of a line that has slope -2 and goes through the point $(4, -6)$.

We can write the equation in the point-slope form. Using $m = -2, x_1 = 4$, and $y_1 = -6$ in equation (1), we have

$$y - (-6) = -2(x - 4).$$

This equation can also be written in the equivalent form

$$y = -2x + 2.$$

Equation (1) can be used to find the equation of a line when two points are given.

EXAMPLE 2

Find the equation of the line that passes through the two points $(-1, 5)$ and $(-4, -2)$.

The slope of the line through the given points is

$$m = \frac{-2 - 5}{-4 - (-1)} = \frac{7}{3}.$$

Therefore, setting $m = \frac{7}{3}$ in the point-slope form (1), we have

$$y - y_1 = \frac{7}{3}(x - x_1).$$

The next step is to substitute one of the two given points for (x_1, y_1). Either point may be used because they both lie on the line. Therefore, selecting $x_1 = -1$ and $y_1 = 5$, we have

$$y - 5 = \frac{7}{3}(x + 1).$$

This equation can be written in the equivalent form

$$y = \frac{7}{3}x + \frac{22}{3}.$$

Another basic form of the equation of a line may be obtained from the point-slope form. A non-vertical line intersects the y-axis in exactly one point. The coordinates of this point are given by an ordered pair of the form $(0, b)$, where b is called the **y-intercept** of the line.

Suppose that the equation of a line is given in the point-slope form $y - y_1 = m(x - x_1)$. Using $(0, b)$ for the point (x_1, y_1), we have

$$y - b = m(x - 0)$$

which can be written as

$$y = mx + b. \tag{2}$$

Equation (2) is called the **slope-intercept** form.

EXAMPLES

3. Find the equation of the line that has slope -5 and y-intercept $-\frac{1}{3}$. Setting $m = -5$ and $b = -\frac{1}{3}$ in equation (2), the desired equation is

$$y = -5x - \frac{1}{3}.$$

4. Use the slope-intercept form to find the equation of the line passing through $(1, -4)$ and $(-3, 2)$.

We first find the slope m.

$$m = \frac{2 - (-4)}{-3 - 1} = -\frac{3}{2}.$$

Substituting this value of m into the slope-intercept form (2), we have

$$y = -\frac{3}{2}x + b.$$

Our next step is to find b. Since $(1, -4)$ and $(-3, 2)$ both lie on the line, the coordinates of both points must satisfy the equation. Taking $x = 1$ and $y = -4$, we have

$$-4 = -\frac{3}{2}(1) + b.$$

Therefore

$$b = -\frac{5}{2},$$

and the desired equation is

$$y = -\frac{3}{2}x - \frac{5}{2}.$$

EXERCISES 1.3

For exercises 1–4 use the point-slope form to find the equation of the line with slope m and passing through the given point P. Simplify your answer to the form $y = mx + b$.

1. $m = 7$ and $P(5, 2)$

2. $m = \frac{2}{5}$ and $P(10, -4)$

3. $m = -2$ and $P(11, 13)$

4. $m = -0.75$ and $P(0, 18.5)$

In each of the exercises 5–10, use the point-slope form to find the equation of the line passing through the two given points P and Q. Simplify your answer to the form $y = mx + b$.

5. $P(0, 4)$ and $Q(5, 9)$

6. $P(6, 10)$ and $Q(7, 4)$

7. $P(7, -6)$ and $Q(-2, -6)$

8. $P(2, 3)$ and $Q(5, 8)$

9. $P(0, 0)$ and $Q\left(\frac{1}{16}, -\frac{5}{16}\right)$

10. $P\left(4, \frac{3}{2}\right)$ and $Q\left(-2, \frac{1}{3}\right)$

Use the slope-intercept form to find the equation of the line for the given slope m and the given y-intercept b in each of the exercises 11–14.

11. $m = -3$ and $b = -12$ 12. $m = \dfrac{9}{2}$ and $b = \dfrac{1}{5}$

13. $m = 3.7$ and $b = 7.5$ 14. $m = 0$ and $b = 0$

For exercises 15–20 use the slope-intercept form to find the equation of the line passing through the two given points.

15. $P(1,1)$ and $Q(0,2)$ 16. $P(2,4)$ and $Q(-5,5)$

17. $P(20,-2)$ and $Q(15,8)$ 18. $P(8.2,3.0)$ and $Q(15.2,5.1)$

19. $P\left(\dfrac{1}{2},8\right)$ and $Q\left(\dfrac{7}{4},6\right)$ 20. $P\left(\dfrac{19}{7},\dfrac{32}{5}\right)$ and $Q\left(\dfrac{32}{5},\dfrac{19}{7}\right)$

Write the equation of the line graphed in each of the exercises 21–26. Whenever possible, write your answer in the form $y = mx + b$.

21.

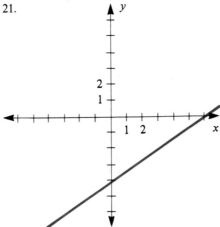

22.

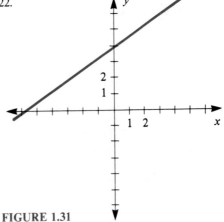

FIGURE 1.30

FIGURE 1.31

23.

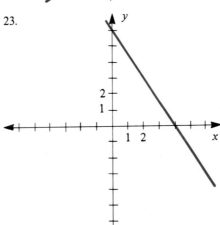

24.

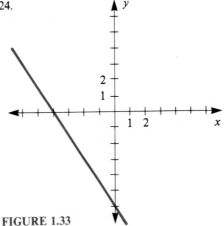

FIGURE 1.32

FIGURE 1.33

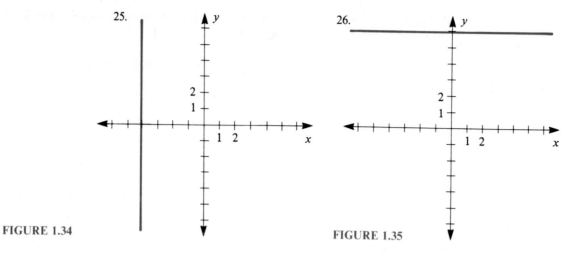

FIGURE 1.34   FIGURE 1.35

1.4 Logic

Language is used to convey meaning—a way to communicate. It is also used as the framework to solve problems. Problem solving involves reaching valid conclusions through logical reasoning. Logical methods are central to problem solving in any area of inquiry. We use logic to prove theorems in mathematics and to write programs in computer science. In the natural and social sciences, experiments are performed and conclusions drawn using valid arguments.

The building blocks of logic are sentences that are called statements. A **statement** is a sentence that is either true or false. Examples of statements are the following.

> I am taking mathematics this semester.
> $5 + 7 = 12$.
> The moon is made out of cheese.

Sentences that are questions or commands or are vague are not statements. For example, consider the following three sentences.

> Are you happy?
> Hit the ball.
> He is tall.

The first sentence is a question and the second sentence is a command, so neither one is a statement. The third sentence is vague as it is not clear to whom the "he" refers. Therefore the third sentence is not a statement.

It is convenient to introduce symbolism in our study. The symbolism of mathematics involves letters such as $x, y, z, \ldots$ to represent variables. A variable can be replaced by any number from a designated set. In symbolic logic letters such as *p, q, r,* ... are used to represent **propositional variables** that can be replaced by statements. When propositional variables are replaced by statements, we write, for example

p:	The light in Rita's room is on.	
q:	Rita is in the room.	**(1)**
r:	The door to Rita's room is closed.	
s:	Rita is studying mathematics.	

In mathematics, we combine or connect variables by operations such as addition, subtraction, multiplication, and division. In symbolic logic, we combine propositional variables by means of **logical connectives** to obtain **compound statements**. Logical connectives are operations on propositional variables. The most common connectives are listed in table 1.2.

TABLE 1.2

Logical connective	Meaning	Symbol
negation	not	$\sim$
conjunction	and	$\wedge$
disjunction	or	$\vee$
implication	if ... then ...	$\rightarrow$
equivalence	if and only if	$\leftrightarrow$

Strictly speaking, the operation of negation is not a connective, since $\sim$ simply negates a statement and does not connect one statement to another statement.

EXAMPLE 1

Consider the statements in (1). Then

$\sim p$: The light in Rita's room is not on.

$r \wedge s$: The door to Rita's room is closed and she is studying mathematics.

$q \vee r$: Either Rita is in the room or the door to her room is closed.

$s \rightarrow p$: If Rita is studying mathematics, then the light is on.

The next example illustrates how an English statement can be symbolized by breaking it down into simpler or component statements that are combined by connectives.

EXAMPLE 2

Let

p: It is raining.
q: It is hot.

Symbolize the following statements in terms of p and q.

a. It is hot and raining.

b. It is either hot or it is not raining.

c. It is neither hot nor raining.

d. It is not true that it is either raining or hot.

e. If it is raining, then it is not hot.

The answers are:

a. $q \wedge p$ b. $q \vee \sim p$ c. $\sim q \wedge \sim p$ d. $\sim(p \vee q)$ e. $p \rightarrow \sim q$.

The next two examples illustrate compound statements involving three basic statements.

EXAMPLES

3. Consider the three statements

p: My car is new.
q: My car is defective.
r: My car was built on a Friday.

Then

$p \wedge q \wedge r$:	My new defective car was built on a Friday.
$(p \wedge q) \rightarrow r$:	If my new car is defective, then it was built on a Friday.
$(p \wedge r) \rightarrow q$:	If my new car was built on a Friday, then it is defective.
$[p \wedge (\sim q)] \rightarrow \sim r$:	If my new car is not defective, then it was not built on a Friday.
$(p \wedge q) \leftrightarrow r$:	My new car is defective if and only if it was built on a Friday.

4. Consider the three statements

p: Sharon is attending business school.
q: Sharon is taking calculus.
r: Sharon is a good student.

Symbolize the following statements in terms of p, q, and r.

a. Sharon is either a good student attending business school or she is not taking calculus.

b. If Sharon is not a good student, then she is attending business school and not taking calculus.

c. Sharon is taking calculus in business school if and only if she is a good student.

d. If Sharon is not taking calculus and is a good student, then she is not in business school.

The answers are:

a. $(r \wedge p) \vee \sim q$ b. $\sim r \rightarrow (p \wedge \sim q)$ c. $(q \wedge p) \leftrightarrow r$ d. $(\sim q \wedge r) \rightarrow \sim p$

Observe that when statements p, q, r, ... are combined using logical connectives, the resulting compound statement is itself a statement. Therefore it is either true or false. Consider, for example, the negation of a statement p. If p is a true statement, then $\sim p$ is a false statement. If p is a false statement, then $\sim p$ is a true statement. We can summarize this discussion in the **truth table** shown as table 1.3.

TABLE 1.3   *Negation*

p	$\sim p$
T	F
F	T

EXAMPLE 5

a. Let
$$p: \quad 2^3 = 8.$$
The statement p is true, so the statement $\sim p$ is false. Namely
$$\sim p: \quad 2^3 \neq 8$$
is false.

b. Let
$$p: \quad \text{New York is the world's largest city.}$$
In this case, the statement p is false, so the statement $\sim p$ is true. Namely
$$\sim p: \quad \text{New York is not the world's largest city}$$
is true.

We next construct the truth table for the conjunction of two statements p and q. Note that there are four cases to be considered when forming this table. The compound statement $p \wedge q$ is true only when both p and q are true. (See table 1.4.)

TABLE 1.4   *Conjunction*

p	q	$p \wedge q$
T	T	T
T	F	F
F	T	F
F	F	F

EXAMPLE 6

a. Let
$$p: \quad 2 + 2 = 4.$$
$$q: \quad -3 \text{ is a negative number.}$$
Since both statements are true, the conjunction $p \wedge q$ is also true. That is,
$$p \wedge q: \quad 2 + 2 = 4 \text{ and } -3 \text{ is a negative number}$$
is a true statement.

b. Let

$$p: \quad 2 + 5 = 7.$$
$$q: \quad 0 \text{ is a positive number.}$$

Since p is true and q is false, the conjunction $p \wedge q$ is false. That is,

$$p \wedge q: \quad 2 + 5 = 7 \text{ and } 0 \text{ is a positive number}$$

is a false statement.

We now consider the truth table for the disjunction of two statements p and q. The compound statement $p \vee q$ is true if at least one of the statements p or q is true; $p \vee q$ is false when both p and q are false. (See table 1.5.)

TABLE 1.5 *Disjunction*

p	q	$p \vee q$
T	T	T
T	F	T
F	T	T
F	F	F

EXAMPLE 7

Let

$$p: \quad 7 > 5.$$
$$q: \quad -2 > 0.$$

Now the disjunction of p and q,

$$p \vee q: \quad \text{Either } 7 > 5 \text{ or } -2 > 0$$

is a true statement, since p is true and q is false. However, the disjunction of $\sim p$ and q,

$$\sim p \vee q: \quad \text{Either } 7 \not> 5 \text{ or } -2 > 0$$

is a false statement, since $\sim p$ and q are both false.

In everyday language, the connective "or" can be used in two different ways. For example, let

p: I will take a plane to the conference.
q: I will take a train to the conference.

Then the statement

p or q: I will either take a plane or a train to the conference.

is a use of the connective "or" in the **exclusive** sense, since p and q will not both occur.

The second meaning of the connective "or" is the **inclusive** sense. For example, let

> p: Scott passed mathematics.
> q: Scott passed accounting.

Then the statement

> p or q: Either Scott passed mathematics or accounting.

could mean that Scott passed both mathematics and accounting. In symbolic logic, the connective " $\vee$ " is always used in the inclusive sense.

We now consider the truth table for the implication. If p is a true statement, then to say that $p \to q$ is false, it must follow that q is also false. In fact, the implication $p \to q$ is false only when p is true and q is false. (See table 1.6.)

TABLE 1.6   *Implication*

p	q	$p \to q$
T	T	T
T	F	F
F	T	T
F	F	T

EXAMPLE 8

Let

> p: This computer has 48K memory.
> q: $2 + 7 = 9$.

a. Suppose p is a true statement. Then the implication

> $p \to q$: If this computer has 48K memory, then $2 + 7 = 9$.

is a true statement, since q is true. However, the implication

> $p \to \sim q$: If this computer has 48K memory, then $2 + 7 \neq 9$.

is a false statement, since $\sim q$ is false.

b. Suppose p is a false statement. Then the implications $p \to q$ and $p \to \sim q$ are both true statements.

In everyday language when we make a statement of the form "if p, then q," we are assuming that p causes q. However, in symbolic logic, the cause and effect relationship is not assumed. For example, the statement

> If $2 + 7 \neq 9$, then this computer has 48K memory.

is true, even though the statement $2 + 7 \neq 9$ does not cause the computer to have 48K memory.

Suppose we are given the implication $p \to q$. The **converse** of $p \to q$ is the statement $q \to p$. The **contrapositive** of $p \to q$ is the statement $\sim q \to \sim p$.

EXAMPLE 9

Let

p: It is snowing.
q: The streets are slippery.

Let us form the converse and contrapositive of the implication $p \rightarrow q$. The converse of $p \rightarrow q$ is the statement

$q \rightarrow p$: If the streets are slippery, then it is snowing.

The contrapositive of $p \rightarrow q$ is the statement

$\sim q \rightarrow \sim p$: If the streets are not slippery, then it is not snowing.

Given the implication $p \rightarrow q$, we sometimes say q, *only if p*. The compound statement *p if and only if q*, denoted by $p \leftrightarrow q$, means that p and q are **equivalent**. The statement $p \leftrightarrow q$ is true when either p and q are both true or both false. The truth table for the "if and only if" connective $\leftrightarrow$ is given in table 1.7.

TABLE 1.7 *Equivalence*

p	q	$p \leftrightarrow q$
T	T	T
T	F	F
F	T	F
F	F	T

EXAMPLE 10

Let

p: $5 < 7$.
q: $5/2 < 7/2$.

Then $p \leftrightarrow q$ is a true statement, since both p and q are true statements.

We can now "break down" compound statements by identifying its component statements and by using the logical connectives. By means of truth tables, we can determine when such statements are true or false.

EXAMPLE 11

Determine the truth table for $p \wedge \sim q$.

Since two letters are involved, we need a column for each of them. Note that in the expression, we have the negation of q. Therefore, we need a column for $\sim q$. The final column is labeled $p \wedge \sim q$, as shown in table 1.8(a).

TABLE 1.8(a)

p	q	$\sim q$	$p \wedge \sim q$

Now, in the first two columns, we label the four possible combinations of the truth or falsity of p and q. We then fill in the other columns using the basic truth tables defining the appropriate connectives. (See table 1.8(b).)

TABLE 1.8(b)

p	q	$\sim q$	$p \wedge \sim q$
T	T	F	F
T	F	T	T
F	T	F	F
F	F	T	F

The encircled column in table 1.8(b) indicates the truth or falsity of $p \wedge \sim q$.

As we saw in example 11, the truth value of the entire statement $p \wedge \sim q$ is true or false depending on the truth values of p and q. The next example involves three component statements p, q, and r.

EXAMPLE 12

Determine the truth table for $(q \rightarrow r) \vee p$.

With three component statements, we have eight possible combinations of T's and F's. We fill in the columns with the results shown in table 1.9.

TABLE 1.9

p	q	r	$(q \rightarrow r) \vee p$	
T	T	T	T	T
T	T	F	F	T
T	F	T	T	T
T	F	F	F	T
F	T	T	T	T
F	T	F	F	F
F	F	T	T	T
F	F	F	T	T

Note that the final (encircled) column in table 1.9 contains all T's, except for one F. A compound statement whose truth table has a final column consisting of all T's is called a **tautology**. A tautology is true regardless of the truth values of the component statements.

EXAMPLE 13

Show that $\sim(p \wedge \sim p)$ is a tautology.

The truth table of this statement is given in table 1.10.

TABLE 1.10

p	$\sim(p \wedge \sim p)$		
T	T	F	F
F	T	F	T

Note that the encircled column has all T's. Therefore $\sim(p \wedge \sim p)$ is a tautology.

It can be shown that $(p \rightarrow q) \leftrightarrow (\sim q \rightarrow \sim p)$ is a tautology (see exercise 21c of this section). Note that this statement involves the equivalence connective. Since the equivalence is a tautology, the two sides are said to be **logically equivalent**. The two sides are just alternate ways to make the same statement. Note that the left-hand side of this equivalence is an implication and the right-hand side is its contrapositive. Therefore an implication and its contrapositive are logically equivalent. This equivalency is useful in mathematics when it is easier to prove the contrapositive of a statement rather than the original statement itself.

We conclude this section with a discussion about arguments. An **argument** consists of statements $p_1, p_2, \ldots, p_n$ called the **hypotheses** and a statement q called the **conclusion**. An example of an argument with two hypotheses is the following:

$$\begin{aligned} \textit{hypotheses:} \quad &p \rightarrow q \\ &\underline{p} \\ \textit{conclusion:} \quad &q \end{aligned} \tag{2}$$

We say that the argument is **valid** if

$$p_1 \wedge p_2 \wedge \cdots \wedge p_n \rightarrow q$$

is a tautology; otherwise, the argument is **invalid**.

EXAMPLE 14

Let us test argument (2) for validity

With $p_1 = p \rightarrow q$ and $p_2 = p$, we form the implication $p_1 \wedge p_2 \rightarrow q$,

$$[(p \rightarrow q) \wedge p] \rightarrow q$$

and construct the truth table for it. (See table 1.11.)

TABLE 1.11

p	q	$[(p \rightarrow q)$	$\wedge$	$p]$	$\rightarrow q$
T	T	T	T	T	T
T	F	F	F	F	T
F	T	T	T	F	T
F	F	T	T	F	T

Since the encircled column has all T's, $[(p \rightarrow q) \wedge p] \rightarrow q$ is a tautology. Therefore the argument is valid.

The argument in example 14 is valid because of the form of the argument. It is valid regardless of whether the statements p and q are true or false. For example, let

p: The sun is shining.
q: $2 + 3 = 6$.

With these particular statements, argument (2) is the following.

hypotheses:　If the sun is shining, then $2 + 3 = 6$.
　　　　　　　The sun is shining.

conclusion:　$2 + 3 = 6$.

This argument is valid, although the conclusion is false.

EXAMPLE 15　　Determine whether the following argument is valid or not.

hypotheses:　If George is not athletic, then he is not a big strong boy.
　　　　　　　George is a strong boy.

conclusion:　George is a big athletic boy.

First, we identify the component statements involved in the argument and label them with letters. Let

p:　George is an athletic boy.
q:　George is a big boy.
r:　George is a strong boy.

Now, in terms of p, q, and r, the argument has the symbolic form

hypotheses:　$\sim p \to \sim (q \wedge r)$
　　　　　　　　　r

conclusion:　$q \wedge p$

We next form the truth table for $([\sim p \to \sim (q \wedge r)] \wedge r) \to (q \wedge p)$. (See table 1.12.)

TABLE 1.12

p	q	r	$([\sim p$	$\to$	$\sim (q \wedge r)]$	$\wedge r$	$\to$	$(q \wedge p)$
T	T	T	F	T	F	T	T	T
T	T	F	F	T	T	F	F	T
T	F	T	F	T	T	F	T	F
T	F	F	F	T	T	F	T	F
F	T	T	T	F	F	T	F	F
F	T	F	T	T	T	F	F	F
F	F	T	T	T	T	F	T	F
F	F	F	T	T	T	F	F	F

Since the encircled column in table 1.12 does not contain all T's, the argument is not valid.

EXERCISES 1.4

---

1. Determine which of the following are statements.

a. It is cold.

b. She is a good student.

   c. Zero is a positive number.

   d. Sit down.

   e. It is neither cold nor snowing.

   f. Are you ready?

2. Determine which of the following are statements.

   a. $3 < 1$.

   b. If $3 < 1$, then $0 < 2$.

   c. What does it cost?

   d. Penny is tall.

   e. Go fish.

   f. Joe has a compact disc player.

3. Let

   > p: Jesse is smart.
   > q: Jesse is a lawyer.

   Express in words each of the following symbolic statements.

a. $p \wedge q$	b. $p \vee q$	c. $p \wedge \sim q$	d. $\sim q \wedge \sim p$
e. $\sim(p \wedge q)$	f. $\sim(\sim p \vee q)$	g. $p \rightarrow q$	h. $q \rightarrow q$
i. $\sim q \rightarrow \sim p$	j. $p \rightarrow (p \wedge q)$	k. $p \leftrightarrow \sim q$	l. $\sim(\sim q \rightarrow p)$

4. Let

   > p: Beth is a business major.
   > q: Beth is a senior.

   Express in words each of the following symbolic statements.

a. $\sim p$	b. $q \vee p$	c. $q \wedge \sim p$	d. $p \rightarrow q$
e. $(q \wedge p) \rightarrow \sim p$	f. $p \leftrightarrow q$	g. $\sim p \wedge \sim q$	h. $\sim p \vee \sim q$
i. $\sim p \rightarrow \sim q$	j. $\sim q \rightarrow \sim p$	k. $\sim(\sim p \rightarrow q)$	l. $\sim(p \leftrightarrow \sim q)$

5. Let

   > p: Scott attends class.
   > q: Scott does the homework.

   Symbolize the following statements in terms of p and q.

   a. Scott attends class and does the homework.

   b. Scott does not do the homework.

   c. Scott either attends class or does the homework.

   d. If Scott attends class, then he does the homework.

   e. Scott neither attends class nor does he do the homework.

   f. It is not the case that Scott either does the homework or attends class.

   g. If Scott does not do the homework, then he does not attend class.

   h. Scott attends class if and only if he does the homework.

6. Let

   > p: Sherry is hot.
   > q: Sherry is swimming.

Symbolize the following statements in terms of p and q.

a. Sherry is not swimming.

b. Sherry is hot and not swimming.

c. Sherry is neither hot nor swimming.

d. If Sherry is swimming, then she is not hot.

e. It is not the case that Sherry is either swimming or hot.

f. Sherry is swimming if and only if she is hot.

g. If Sherry is not hot, then she is not swimming.

h. Sherry is either not swimming or she is hot.

7. Let

    p: Ted is angry.
    q: Ted is tired.
    r: Ted is playing tennis.

Express in words the following compound statements.

a. $p \wedge q \wedge \sim r$ b. $p \vee q \vee r$ c. $(\sim p \wedge \sim q) \rightarrow \sim r$

d. $r \rightarrow (p \vee q)$ e. $(p \wedge r) \leftrightarrow q$

8. Let

    p: $2 + 2 = 5$.
    q: Tom is a mathematics major.
    r: It is hot today.

Express in words the following compound statements.

a. $(r \vee q) \wedge \sim p$ b. $(p \wedge \sim q) \rightarrow r$ c. $(p \wedge q) \rightarrow r$

d. $(r \wedge (\sim q)) \rightarrow \sim p$ e. $(r \vee q) \leftrightarrow p$

9. Let

    p: Roy drives to school.
    q: Roy parks in the faculty parking lot.
    r: Roy gets a parking ticket.

Symbolize the following statements in terms of p, q, and r.

a. Roy drives to school and parks in the faculty parking lot, but does not get a parking ticket.

b. Roy drives to school but does not park in the faculty parking lot and does not get a parking ticket.

c. If Roy drives to school and parks in the faculty parking lot, then he gets a parking ticket.

d. If Roy drives to school and does not park in the faculty parking lot, then he does not get a parking ticket.

10. Let

    p: Nancy jogs each morning.
    q: Nancy has sore feet.
    r: Nancy is happy.

Symbolize the following statements in terms of p, q, and r.

a. If Nancy jogs each morning, then she either is happy or has sore feet.

    b. Nancy jogs each morning and has sore feet, but she is happy.

    c. Nancy is happy if and only if she jogs each morning and does not have sore feet.

    d. If Nancy jogs each morning and has sore feet, then she is not happy.

11. Consider the following statement.

If John is a hard worker, then he is tired.

Form the converse and contrapositive of this statement.

12. Consider the following statement.

If Paula is a computer science major, then she takes finite mathematics.

Form the converse and contrapositive of this statement.

13. Consider the following statement.

If I work hard, then I am either rich or famous.

Form the converse and contrapositive of this statement.

14. Consider the following statement.

If it does not rain or snow, then I am going for a walk.

Form the converse and contrapositive of this statement.

15. Determine if the given statement is true or false.

    a. $2 + 2 = 5$ and 7 is a positive number.

    b. Either $2 + 2 = 5$ or 7 is a positive number.

    c. $2 + 2 \neq 5$ and 7 is a positive number.

    d. If $2 + 2 = 5$, then 7 is a positive number.

    e. If $2 + 2 = 5$, then 7 is a negative number.

    f. $2 + 2 \neq 5$ if and only if 7 is not a positive number.

16. Determine if the given statement is true or false.

    a. Either 3 is a rational number or $1 > 0$.

    b. $5 + 2 = 7$ and $\sqrt{2}$ is a rational number.

    c. If 3 is a rational number, then $1 > 0$.

    d. If 3 is a rational number, then 1 is a negative number.

    e. If 3 is not a rational number, then $1 > 0$.

    f. 3 is a rational number if and only if $1 > 0$.

17. Determine the truth table for each of the following.

    a. $\sim p \wedge \sim q$ b. $\sim p \vee q$ c. $\sim(p \vee q)$

    d. $\sim(p \rightarrow q)$ e. $p \leftrightarrow \sim q$ f. $(q \wedge p) \rightarrow \sim p$

18. Determine the truth table for each of the following.

    a. $(p \wedge q) \rightarrow q$ b. $\sim p \rightarrow q$ c. $\sim(p \rightarrow \sim q)$

    d. $(p \wedge q) \vee p$ e. $\sim(\sim p \vee q)$ f. $\sim(p \wedge q) \rightarrow \sim q$

19. Determine the truth table for each of the following.

    a. $(p \wedge q) \rightarrow r$ b. $(p \wedge q) \rightarrow (r \vee p)$ c. $(r \wedge \sim q) \leftrightarrow \sim p$

20. Determine the truth table for each of the following.

    a. $p \wedge (q \vee r)$ b. $(p \wedge q) \wedge \sim r$ c. $\sim p \rightarrow (q \vee r)$

21. Show that the following are tautologies.

    a. $(p \vee q) \leftrightarrow (q \vee p)$ b. $\sim \sim p \leftrightarrow p$ c. $(p \rightarrow q) \leftrightarrow (\sim q \rightarrow \sim p)$

    d. $\sim (p \rightarrow q) \rightarrow p$ e. $[(p \rightarrow q) \wedge (q \rightarrow r)] \rightarrow (p \rightarrow r)$

22. Show that the following are tautologies.

    a. $(p \wedge q) \leftrightarrow (q \wedge p)$ b. $(p \vee p) \leftrightarrow p$ c. $p \rightarrow (p \vee q)$

    d. $[p \wedge (p \rightarrow q)] \rightarrow q$ e. $[p \vee (q \wedge r)] \leftrightarrow [(p \vee q) \wedge (p \vee r)]$

23. Show that $\sim (p \rightarrow q)$ and $p \wedge \sim q$ are logically equivalent.

24. Show that $p \rightarrow q$ and $\sim p \vee q$ are logically equivalent.

For exercises 25–32 determine if the given argument is valid or not valid.

25. *hypotheses:* $p \rightarrow q$ 26. *hypotheses:* $q \rightarrow p$

               $\sim q \rightarrow p$ $p \rightarrow q$

    conclusion: $\sim p$ *conclusion:* q

27. *hypotheses:* $p \vee q$ 28. *hypotheses:* $\sim p$

               p $p \vee q$

    conclusion: q *conclusion:* $\sim q$

29. *hypotheses:* $\sim q$ 30. *hypotheses:* $p \vee \sim r$

               $p \rightarrow q$ r

    conclusion: $\sim p$ *conclusion:* p

31. *hypotheses:* $p \rightarrow q$ 32. *hypotheses:* $\sim p \rightarrow \sim q$

               $q \rightarrow r$ $r \rightarrow \sim p$

    conclusion: $p \rightarrow r$ *conclusion:* $r \rightarrow \sim q$

For exercises 33–46 determine if the given argument is valid or not valid.

33. *hypotheses:*   If Ron sleeps on a waterbed, then he gets seasick.

                    Ron gets seasick.

    conclusion: Ron slept on a waterbed.

34. *hypotheses:*   If Diane does the homework every day, then she will pass the course.

                    Diane did not pass the course.

    conclusion: Diane did not do the homework every day.

35. *hypotheses:*   If $2 + 2 = 4$, then Tom is a mathematician.

                    Tom is not a mathematician.

    conclusion: $2 + 2 \neq 4$.

36. *hypotheses:*   If $2 + 2 = 5$, then one is less than zero.

                    One is greater than zero.

    conclusion: $2 + 2 = 5$

37. *hypotheses*:   If Bill does not play tennis, then he is not tired.
    Bill played tennis.

    conclusion: Bill is tired.

38. *hypotheses*:   If Jenny is taking finite mathematics, then she is a business major.
    Jenny is a business major.

    conclusion: Jenny is taking finite mathematics.

39. *hypotheses*:   Either Sam drives a big car or a black car.
    Sam does not drive a big car.

    conclusion: Sam drives a black car.

40. *hypotheses*:   If Maureen does aerobics every day, then she is in good physical condition.
    She is not in good physical condition.

    conclusion: She does not do aerobics every day.

41. *hypotheses*:   If Tammy jogs each morning, then she is tired.
    If Tammy is tired, then she is overweight.

    conclusion: If Tammy jogs each morning, then she is overweight.

42. *hypotheses*:   If Tammy jogs each morning, then she is tired.
    If Tammy is overweight, then she is tired.

    conclusion: If Tammy is overweight, then she jogs each morning.

43. *hypotheses*:   If Jodi does not have a sinus infection, then she neither smokes cigarettes nor goes to the doctor.
    Jodi either does not smoke cigarettes or does not have a sinus infection.

    conclusion: Jodi does not go to the doctor.

44. *hypotheses*:   If either Isaac writes science fiction novels or does not work hard, then he is not famous.
    Isaac is famous.

    conclusion: Isaac does not write science fiction novels and works hard.

45. *hypotheses*:   If Jason is a butcher and has a sharp knife, then he sells prime beef.
    Jason has a sharp knife.

    conclusion: Jason sells prime beef.

46. *hypotheses*:   If Jack is rich, then he drives a big black car.
    Jack drives neither a big car nor a black car.

    conclusion: Jack is not rich.

47. The connective $\veebar$ means the word "or" used in the exclusive sense. Construct the truth table for $\veebar$.

48. Refer to exercise 47. Determine the truth tables for the following compound statements.

    a. $(p \rightarrow \sim q) \veebar q$ b. $(r \wedge p) \vee (\sim q)$

SUMMARY OF TERMS

Equations and Inequalities

    Solution of an Equation
    Equivalent Equations
    Solution Set of an Inequality
    Equivalent Inequalities

Cartesian Coordinate System

    Ordered Pair
    Coordinate Axes
    x-axis, y-axis
    Coordinate Plane
    Cartesian System
    Origin
    Scale
    Coordinates of a Point P
    Abscissa, Ordinate
    Quadrants
    Graphing an Equation
    Linear Equation
    Slope
    Run, Rise

Linear Functions

    Point-Slope Form
      $y - y_1 = m(x - x_1)$
    y-Intercept
    Slope-Intercept Form $y = mx + b$

Logic

    Statement
    Propositional Variables
    Compound Statements
    Logical Connectives
    Truth Table
    Negation
    Conjunction
    Disjunction
    Inclusive or
    Exclusive or
    Implication
    Converse
    Contrapositive
    Equivalence
    Tautology
    Logically Equivalent
    Argument
    Hypotheses
    Conclusion
    Valid Argument

REVIEW EXERCISES (CH. 1)

1.1 Equations and Inequalities

For exercises 1–4 solve each of the equations for x.

1. $2x - 3 = 9$

2. $4(x - 1) + 5 = 17$

3. $0.3x - 0.7 = -0.6x + 1.1$

4. $\dfrac{3}{2}x - \dfrac{1}{2} = 4\left(x + \dfrac{3}{2}\right)$

For exercises 5–8 solve the inequalities for x and graph the solution set on the real number line.

5. $3x + 4 < 10$

6. $-2x + 5 \geq 4x - 3$

7. $-5(2x - 5) \leq 4(5 - x) + 23$

8. $4(2 - x) + 7 > 3(2x - 5)$

For exercises 9–12 let x represent the number of acres of corn a farmer plans to grow. Symbolize the given statement in terms of an inequality.

9. There are at most 150 acres available for corn.

10. There are at least 150 acres available for corn.

11. There are no fewer than 150 acres available for corn.

12. There are no more than 150 acres available for corn.

1.2 Coordinate Systems: Graphing a Linear Equation

For exercises 1–4 plot the points in a coordinate plane.

1. $(2, 5)$ 2. $(2, -5)$ 3. $(-5, 2)$ 4. $\left(-\dfrac{1}{2}, \dfrac{3}{2}\right)$

For exercises 5–8 find the slope of the line passing through the two given points.

5. $P(5, 9)$ and $Q(2, 3)$

6. $P(5, -4)$ and $Q(-2, -4)$

7. $P\left(-2, \dfrac{1}{2}\right)$ and $Q\left(\dfrac{3}{4}, -1\right)$

8. $P(0.5, 0.25)$ and $Q(1.25, 0.75)$

9. What is true about the abscissa of a point if the point is to the left of the y-axis?

10. What is true about the ordinate of a point if the point is above the x-axis?

11. What is true about the coordinates of a point if the point is in quadrant III?

12. What is true about the coordinates of a point if the point is on the negative y-axis?

13. If the slope of a line is negative, what must be true about the rise if the run is positive?

14. If the slope of a line is positive and if we move to the right from a point on the line, then which way must we move vertically to get back to the line?

15. Consider the equation $2x + 7y = 28$. Determine the pair (x, y) that satisfies the equation when

a. $x = 0$ b. $y = 1$ c. $x = -1$ d. $y = -3$

16. Consider the equation $\frac{2}{5}x + \frac{7}{4}y = 1$. Determine the pair (x, y) that satisfies the equation when

a. $x = 15$ b. $y = 16$ c. $x = \dfrac{1}{4}$ d. $y = \dfrac{1}{5}$

For exercises 17–20 find the slope m and the y-intercept b for each of the equations and draw the graph of the equation.

17. $y = \dfrac{3}{2}x$

18. $y = 3x - 7$

19. $6x - 2y = 1$

20. $\dfrac{2}{3}x + \dfrac{5}{4}y = 6$

1.3   Writing the Equation of a Line

For exercises 1–4 write the equation of the line for the slope m and the y-intercept b.

1. $m = -7$ and $b = -3$

2. $m = \dfrac{2}{3}$ and $b = \dfrac{7}{6}$

3. $m = 0$ and $b = 15$

4. $m = -1.45$ and $b = 275$

For exercises 5–8 write the equation of the line for the given slope m and the given point P.

5. $m = -4$ and $P(2, -5)$

6. $m = 0$ and $P(7.9, 3.5)$

7. $m = 5$ and $P\left(-\dfrac{7}{5}, 2\right)$

8. $m = \dfrac{1}{4}$ and $P(5, 3)$

For exercises 9–12 write the equation of the line that passes through the two given points P and Q.

9. $P(2, 4)$ and $Q(-1, -2)$

10. $P(2, -7)$ and $Q(2, 8)$

11. $P(4, 0)$ and $Q(0, 5)$

12. $P\left(-\dfrac{1}{2}, 2\right)$ and $Q\left(-\dfrac{1}{4}, -6\right)$

1.4   Logic

Exercises 1 and 2 refer to the following statements.

p: Bob does the homework regularly.
q: Bob gets an A on the next test.

1. Express in words the following compound statements.

   a. $p \rightarrow q$ b. $\sim q \wedge p$ c. $q \rightarrow \sim p$

2. Symbolize the following compound statements in terms of p and q.

   a. If Bob gets an A on the next test, then he had done the homework regularly.

   b. Bob does the homework regularly if and only if he gets an A on the next test.

   c. Bob neither does the homework regularly nor will he get an A on the next test.

3. Determine if the following statements are true or false.

   a. $2 + 5 = 7$ and $4 + 5 = 10$.

   b. It is not the case that either $2 + 5 = 7$ or $4 + 5 = 10$.

   c. If $2 + 5 = 7$, then $4 + 5 = 10$.

4. Determine truth tables for the following statements.

   a. $\sim p \wedge q$ b. $(q \wedge r) \rightarrow (p \rightarrow q)$

5. Show that $[p \wedge (q \vee r)] \leftrightarrow [(p \wedge q) \vee (p \wedge r)]$ is a tautology.

For exercises 6–9 determine if the given argument is valid or not valid.

6. *hypotheses:* $p \wedge q$
                    $\dfrac{\sim p}{}$

   conclusion: q

7. *hypotheses:* $\sim q$
                    $\dfrac{p \rightarrow q}{}$

   conclusion: $\sim p$

8. *hypotheses*: $r \to \sim p$
                     $\sim p \to \sim q$

   conclusion: $\sim q \to r$

9. *hypotheses*: $(r \wedge q) \to \sim p$
                     p

   conclusion: $\sim r \vee \sim q$

10. *hypotheses*: Either $2 + 3 = 5$ or $\sqrt{2}$ is a rational number.
                   $2 + 3 \neq 5$.

    conclusion: $\sqrt{2}$ is a rational number.

11. *hypotheses*: If Joy works at the computer center, then she is a student.
                   Joy is not a student.

    conclusion: Joy does not work at the computer center.

12. *hypotheses*: If I bought a new car that is defective, then it was built on a Friday.
                   The car was not built on a Friday.

    conclusion: My car is either not new or not defective.

CHAPTER 2

Applications of Linear Equations

Introduction

We are familiar with models such as prototypes of cars, airplanes, and spacecraft used for testing design concepts. In chemistry, molecules can be modeled with balls representing the atoms and connecting rods representing the chemical bonds between them. For example, the methane molecule, the simplest hydrocarbon, can be modeled as shown in figure 2.1.

Not all models have to be of a physical nature. A model is simply a replica, a symbolic representation, of the significant aspects of a situation. When mathematics is applied to a situation, a mathematical description is obtained. In keeping with the general meaning of the term, we also call a symbolic representation that uses mathematical symbols and language a model, specifically a **mathematical model**.

Applications of mathematics yield models belonging to one of two general categories —**deterministic** and **probabilistic**. Deterministic models give, based upon the assumptions made, exact information about the situation, whereas probability models give information with some degree of uncertainty. Along with probability theory and statistics, models of the latter type are included in our study of game theory and Markov chains. Our study of deterministic models includes linear forms, systems of linear equations, matrices, and linear programming. Our study in these areas of mathematics includes tools and techniques to help us derive solutions and conclusions in our applications.

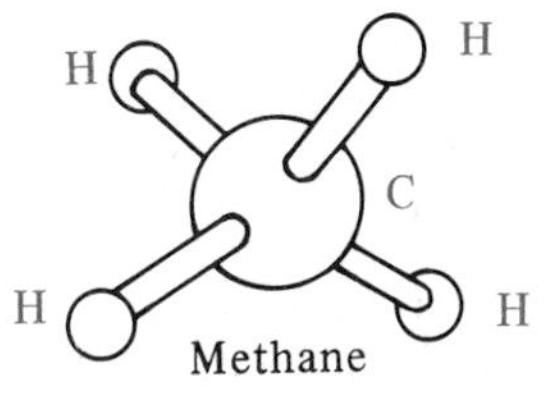

FIGURE 2.1

Methane

51

2.1 Linear Functions

You should be familiar with writing the equation of a line when given certain information. For example, when given two points, we can write the equation of the line that goes through the points. Similarly we can write the equation when given the slope of the line and one point on the line. These concepts are reviewed in chapter 1. We consider now how equations of a line can be used to model certain real-world situations.

Many applications involve the relationship between two quantities, but further, where one of the quantities can be uniquely determined from the other. For example, the distance traveled in a jet flying at 450 miles per hour is uniquely determined from the time in flight. The area of a square depends on the length of an edge, and a unique value for the area can be found from a given length of an edge.

A relationship between two quantities in which one of the quantities can be uniquely determined from the other is given a special name. In particular, we call such a dependency a **function** and say that the one quantity "is a function of" the other quantity.

As noted, the area of a square A is a function of the length of an edge e. Furthermore, we should be familiar with the formula

$$A = e^2$$

that describes this dependency. If D denotes the distance a jet flying at 450 miles per hour has traveled in t hours, then D is a function of t, and this function can be described by the equation

$$D = 450t.$$

We should recognize that this latter equation is that of a line. Note also that the area equation is not linear. Our interest here is with functions that can be described by linear equations (see figure 2.2).

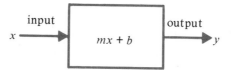

FIGURE 2.2

Definition

■ If y is a function of x that can be described by a linear equation

$$y = mx + b,$$

we call it a **linear function**. Further, y is called the **dependent variable** and x is called the **independent variable**. ■

As with our distance equation, linear functions arise quite naturally in many applications.

EXAMPLE 1

An operator of a winding machine in a textile mill is paid $28 a day plus a bonus of $0.05 for each pound she winds. Thus the operator's pay depends on the amount of yarn she winds. Let us express this function with an equation.

We first introduce symbols for the independent and dependent variables. Let

$$x = \text{the pounds of yarn wound during a day}$$

and

$$y = \text{the operator's income for a day.}$$

Next, since y is dependent on x, we look for an equation that expresses y in terms of x.

Since the operator receives $0.05 for each pound of yarn that she winds, we multiply this rate by the number of pounds to obtain her bonus. This particular operator's bonus is $0.05x$. Next the operator's income for a day is stated as $28 plus this bonus. Therefore her income is described by the equation

$$y = 0.05x + 28.$$

We can use this equation to obtain, for example, the operator's income on a day that she winds 400 pounds of yarn. In this case $x = 400$ and $y = 0.05(400) + 28 = \$48$.

The graph of the equation $y = 0.05x + 28$ from example 1 is a line with slope $m = 0.05$ (see figure 2.3a). However, since the dependent variable x represents the number of pounds of yarn wound, there is the natural restriction on x that it must be greater than or equal to zero ($x \geq 0$). Therefore the **graph of the function** is that part of the line indicated in figure 2.3b.

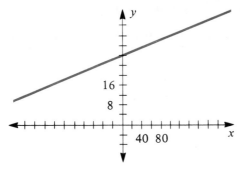

(a) Graph of the equation
$y = 0.05x + 28$

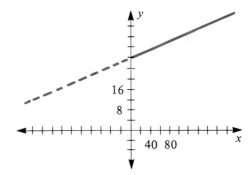

(b) Graph of the linear function
$y = 0.05x + 28 \quad (x \geqslant 0)$

FIGURE 2.3

Recall that the slope of a line is the ratio of the change in the y-coordinate to the change in the x-coordinate that takes place as a point moves along the line. Thus, if the x-coordinate increases by 1, the y-coordinate must change by the slope m. The manner in which the operator's income varies with the number of pounds wound agrees with this concept. Specifically her income increases by the constant rate $0.05 with each additional pound that she winds. Therefore it is natural for the operator's income to be described by a linear equation and for her bonus of $0.05 per pound to be the slope of this line. The presence of a constant rate of change is what distinguishes a linear function from others.

Theorem 1 ■ 1. If y is a function of x such that y changes with x at a constant rate m, then this function is linear and can be expressed by an equation of the form $y = mx + b$.

2. Also, if y is a linear function of x expressed by the equation $y = mx + b$, then y changes with x at the constant rate m. ■

EXAMPLES

2. At 8:00 A.M. Jean begins driving from Alburtis to Westline, which is 276 miles away. From past trips, she can expect to drive an average of 46 miles per hour. Let D denote her distance from Westline and let t denote the time in hours. Then D is a function of t. Write an equation that describes this function.

We recognize that D is changing with t at a constant rate. In particular, for each hour of travel the distance decreases by 46 miles. Therefore, by part 1 of the theorem, D is a linear function of t and can be expressed by an equation of the form

$$D = mt + b$$

where m is the constant rate.

We now determine m and b. Since D decreases as time increases, the rate (slope) is $m = -46$. Furthermore, at time $t = 0$(8:00 A.M.), Jean is 276 miles from Westline, so the data point $(0, 276)$ must satisfy the linear equation $D = -46t + b$. But this means that the D-intercept is 276, so $b = 276$ and we have

$$D = -46t + 276.$$

3. Temperature measured in Celsius degrees C can be determined from temperature measured in Fahrenheit degrees F. Furthermore, C is a linear function of F. Write the linear equation that expresses the dependency of C on F.

The desired linear equation is of the form

$$C = mF + b.$$

We determine the two constants m and b by finding two data points. These two data points, in turn, are determined by the freezing point and the boiling point of water. When $F = 32$, $C = 0$, and when $F = 212$, $C = 100$. Therefore, two data points are $(32, 0)$ and $(212, 100)$. We can now think of our problem as writing the equation of the line that goes through the two points.

The slope m of this line is given by

$$m = \frac{100 - 0}{212 - 32} = \frac{5}{9}$$

and

$$C = \frac{5}{9}F + b.$$

Using the data point $(32, 0)$ to find b, we have

$$0 = \frac{5}{9}(32) + b \qquad \text{or} \qquad b = -\frac{5}{9}(32).$$

Therefore, the linear equation is

$$C = \frac{5}{9}F - \frac{5}{9}(32)$$

or equivalently, $$C = \frac{5}{9}(F - 32).$$

4. A painting was appraised at $3000 6 years ago, and 1 year ago it was appraised at $4500. Assuming the appraised value A of this painting increases linearly with time t, determine the equation that describes this relation.

Since A is a function of t and we are assuming that this dependency is linear, our equation takes the form

$$A = mt + b.$$

We can use the two data points $(-6, 3000)$ and $(-1, 4500)$ to determine m and one of these points to determine b. In particular,

$$m = \frac{4500 - 3000}{(-1) - (-6)} = 300;$$

that is, we are assuming that the painting's value is increasing at the constant rate of $300 per year.

Using the point $(-6, 3000)$, we have

$$3000 = 300(-6) + b$$

and so, $$b = 4800.$$

Therefore the desired equation is

$$A = 300t + 4800.$$

As we have seen, sometimes a relationship between two quantities is naturally linear; at other times, linearity is assumed. In either case, the dependent variable varies with the independent variable at a constant rate. Furthermore we can use our knowledge of writing the equation of a line to give the model for a linear function. For example, we used two data points in examples 3 and 4 and one data point and the slope in example 2 to write the desired equations. If, on the other hand, the dependent variable does not vary with the independent variable at a constant rate, then the function cannot be modeled with a linear equation, as we see in our next example.

EXAMPLE 5

An initial deposit of $100 in a savings account, compounded quarterly at 6 percent per year, results in the amounts given in table 2.1.

TABLE 2.1

Months (t)	Amount on deposit at the end of t months (A)
3	$101.50
6	$103.02
9	$104.57
12	$106.14

The method for computing the values in the table is developed in chapter 7. If A depends on t, determine if this relationship is linear.

We see from table 2.1 that over the first quarter ($t = 0$ to $t = 3$) the amount A grows by \$1.50 ($A = 100$ to $A = 101.50$). However, over the next quarter ($t = 3$ to $t = 6$), the amount increases by \$1.52 ($A = 101.50$ to $A = 103.02$). Therefore the rate is not constant and we cannot use a linear equation to describe the relationship between A and t.

EXERCISES 2.1

1. Each ounce of potatoes contains 17 grams of carbohydrates. Express grams of carbohydrates as a function of ounces of potatoes.

2. The circumference of a circle is a linear function of its radius. Write this linear function.

3. In a particular state the income tax is 2.4% of gross income. Express the income tax as a function of gross income.

4. In a certain locality real estate is taxed at the rate of \$81 per \$1000 of assessed value. Write the linear equation that expresses the real estate tax as a function of the assessed value.

5. The total length of a certain species of snake is approximately $7\frac{1}{4}$ times the length of its tail, provided that the tail length is at least 30 millimeters. Write the function that relates the total length to the tail length.

6. The value of a new automobile at the end of one year is 78% of its original price, if the original price is between \$6500 and \$9000. Express the value at the end of one year of such automobiles as a function of their original price.

7. A men's clothing store is having a winter sale on all items. The sale price on each item is the list price discounted by 20%. Write the function that relates the sale price to the list price. Also find the sale price of a sport coat if its list price is \$150.

8. A & E Electric is having a sale on all its light fixtures. Specifically it is marking prices down by 25%. Express the marked-down price as a function of the original price. Also find the sale price of a ceiling light if it originally sold for \$140.

9. Each month a salesman makes \$300 plus \$2 on every dictionary he sells. Express by an equation the salesman's monthly income as a function of the number of dictionaries sold during the month. Also find the salesman's income if he sells 125 dictionaries during one month.

10. A technical representative of Standard Equipment has a weekly expense account of \$75 plus 20¢ for each mile she drives when visiting customers. Express the representative's total weekly allowance in terms of the number of miles driven in a week. If the representative drives 275 miles in one week, what will her expense allowance be for that week?

11. A bus travels regularly from Milwaukee to Chicago, a distance of 90 miles. If the bus averages 50 miles per hour, express the distance that the bus is from Chicago in terms of the time it left Milwaukee. Next find the distance the bus is from Chicago after it has been traveling for 30 minutes.

12. Marilyn commutes from home to college in Oshkosh, which is a distance of 25 miles. Based on previous trips, she can expect to drive an average of 42 miles per hour. For a

given trip, express the distance that she is from Oshkosh as a function of time. Also find how far she is from Oshkosh 15 minutes after leaving home.

13. In 1981 a comic book was valued at $675, and in 1984 it was worth $1260. Assuming that the value of this comic book increases linearly with time, find the equation that describes this relation. Also what would be the value of the comic book in 1987, assuming there is no change in the rate of increase?

14. In 1980 a stamp was worth $200, and in 1985 it was valued at $320. Assuming that the value of this stamp increases linearly with time, find an equation that describes this function. Also find the value of the stamp in 1988, assuming the value changes at the same rate.

15. Mosquitoes make up part of a Purple Martin's diet. Suppose a Purple Martin eats mosquitoes at the constant rate of 10 per minute, and suppose that the mosquito population in a small area is 100,000. Express the mosquito population as a function of time after a single Martin is introduced into the area. Assume that the change in the population is due only to the Martin's feeding.

16. Water samples taken from Lake Erie in 1982 showed the presence of mercury compounds in the amount of 12 parts per 1000. When samples were taken in 1984, the presence of mercury compounds had decreased to 8 parts per 1000. Assuming that the decrease in mercury compounds will continue at a constant rate, express the parts per 1000 of mercury compounds in Lake Erie as a function of time. Also find the year in which there will be 2 parts per 1000 of mercury compounds in Lake Erie.

17. On a visit to a large city, a traveler took a taxi from the train station to her hotel, a trip of 15 minutes duration. The fare for this trip was $3.25. When the visitor took a taxi from her hotel to the theater, the trip took 5 minutes and cost $1.75. We see that the taxi fare is a function of the number of minutes in the taxi. Assume that this dependency is linear, which is not an unreasonable assumption.

    a. Write the equation that models this function.

    b. What does the taxi company charge for each minute of travel?

18. A homeowner received a gas bill one month for $45.26 in which 73 thousand cubic feet were used. Her bill the next month was $42.16 for 68 thousand cubic feet. Assume that the rate was the same for both months.

    a. What was this rate?

    b. Write the equation that expresses the dependency of the amount owed on the number of thousand cubic feet of gas used.

19. The current average American diet consists of 42% fat and 12% protein. The dietary goals for the average diet 10 years from now are 30% fat and 12% protein. Assume that these goals are to be reached in a linear fashion.

    a. Express the percentage of fat in the diet as a function of time in years.

    b. Express the percentage of protein in the diet as a function of time in years.

20. A person has $15,000 deposited between two types of accounts, a regular savings account and a money management account. The regular savings account pays 5% interest, and the money management account pays 9.5% interest. Let I represent the total interest received for one month from both accounts. Write an equation that expresses I as a function of the money deposited in the regular savings account.

21. Olean and Binghamton are 150 miles apart. Steve leaves Olean at 8:00 A.M. and travels at 48 miles per hour toward Binghamton. At 8:15 A.M. Gary leaves Olean for Binghamton, driving on the same road as Steve but at 54 miles per hour.

a. Let D denote the distance Steve is from Olean and t denote time with $t = 0$ meaning 8:15 A.M. Write the equation that expresses D as a function of t.

b. Let d denote the distance Gary is from Olean and t denote time with $t = 0$ meaning 8:15 A.M. Write the equation that expresses d as a function of t.

c. Write the equation that expresses the distance between Steve and Gary as a function of time t.

d. At what time will Gary catch Steve?

22. Two joggers who are 6 miles apart start jogging toward each other at the same time. The joggers rates are 7 miles per hour and 8 miles per hour, respectively.

a. Express the distance between them as a function of time.

b. At what time will the two joggers meet?

23. An assembly line worker makes $80 per day plus a bonus of 50¢ for each item assembled over 90 items.

a. Express the worker's daily pay as a function of the total number of items assembled.

b. What would be the worker's pay on a day when he assembles only 75 items?

c. What would be the worker's pay on a day when he assembles 120 items?

24. New York State provides a reduction of income taxes due when the income exceeds $17,000. For example, the amount of reduction for incomes between $19,000 and $21,000 is $20 plus 2% of the excess over $19,000.

a. Express the amount of reduction as a function of total income for incomes in the range of 19,000 and 21,000 dollars.

b. What would be the amount of reduction for an income of $19,000?

c. What would be the amount of reduction for an income of $20,500?

25. For a given temperature, the apparent temperature, commonly called windchill factor, is a function of wind velocity. For example, table 2.2 shows the apparent temperatures when the actual temperature is 20°F.

TABLE 2.2

Wind velocity (mph)	Apparent temperature (°F)
0	20
5	16
10	−4
20	−18

Is the apparent temperature a linear function of wind velocity?

26. For each of the following parts, two quantities are described. State whether or not it is reasonable to use a linear equation to model the relation between the quantities.

a. Grams of protein and ounces of rice.

b. The average resale value of a particular model of an American automobile and its age.

c. The perimeter of a square and the length of an edge.

d. The total consumer demand for beef in pounds and the average price per pound.

e. The height of mercury in a particular thermometer and the temperature.

2.2 Applications of Linear Functions to Business

Many elementary situations in business and economics use linear functions. The next four examples illustrate how linear functions can be applied in significant ways.

EXAMPLE 1

Straight-Line Depreciation.   A new winding machine in a textile mill costs $2500. The machine has a **useful life** of 8 years. At the end of 8 years, the machine has a **salvage value** of $200. The salvage value is the amount the mill can obtain by selling the machine as a used piece of equipment or as scrap metal. Assume that the **depreciation** (loss in value through use) is linearly dependent on time and write the linear equation that expresses this linear function.

Let t denote the number of years since the machine was purchased by the mill and let V denote the value (in dollars). Assuming that V is a linear function of t means that V can be expressed as

$$V = mt + b.$$

When $t = 0$, $V = 2500$ and when $t = 8$, $V = 200$. We therefore have the two data points $(0, 2500)$ and $(8, 200)$. The slope (rate of depreciation) is given by

$$m = \frac{200 - 2500}{8 - 0} = -287.50.$$

Thus, the value of the machine is decreasing at the rate of $287.50 per year. Our equation thus far is

$$V = -287.50t + b.$$

Since one of our data points is $(0, 2500)$, we see that the V-intercept is 2500 and our linear equation is thus

$$V = -287.50t + 2500.$$

The graph of $V = -287.50t + 2500$ is given in figure 2.4. The horizontal axis corresponds to the independent variable t and the vertical axis corresponds to the dependent variable V.

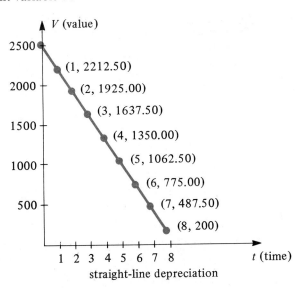

FIGURE 2.4   straight-line depreciation

Straight-line depreciation is used when the amount deducted for use is the same for each time period. Other depreciation methods are also acceptable to the Internal Revenue Service. However, they cannot be expressed by a linear equation since the amount deducted does not remain constant over the useful life. For example, in the **sum-of-the-years-digits** method, the amount deducted changes each year over the useful life. This method as applied to the winding machine in example 1 is shown graphically in figure 2.5. Note that the graph is not a straight line, indicating that this depreciation method is not linear.

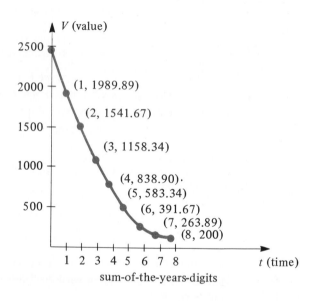

FIGURE 2.5

sum-of-the-years-digits

EXAMPLE 2

Cost–Revenue–Profit. Consider a company that makes electric coils. The **total cost** of making the coils is the sum of the **fixed cost** and the **variable cost**. The fixed cost (for example, the cost of machinery) is independent of the number of coils produced. On the other hand, the variable cost depends on the cost per coil as well as the number of coils made.

Suppose that the fixed cost for the company is $300 a day. Further, if it costs the company $1.50 per coil and if the company makes x coils in a day, then the variable cost is $1.50x$. Letting C denote the total cost, by the formula

$$\text{total cost} = \text{variable cost} + \text{fixed cost}$$

we have
$$C = 1.50x + 300.$$

If the company sells their coils at $4 each, the **revenue** obtained by selling x coils is given by

$$R = 4x.$$

The **profit** for the company from selling x coils is given by

$$\text{profit} = \text{revenue} - \text{cost}.$$

Therefore, if P denotes the profit, we have

$$P = 4x - (1.50x + 300)$$

$$= 2.50x - 300.$$

In the competitive marketplace, the price of an item affects the demand for it. For example, as the price for video cassette recorders decreases, the demand for them increases. We assume that the demand for an item and the price per item are related by an equation. The equation that expresses this relation is called a **demand equation**. In the next example we assume that the demand equation is linear. However, we should be aware that economists usually graph a demand equation with price p on the vertical axis and quantity q on the horizontal axis. Since in mathematics the vertical axis is used for the dependent variable, we will need to reverse our thinking. Thus we will consider price as a function of quantity in order to obtain an equation whose graph is consistent with the graph an economist would draw.

EXAMPLE 3

Linear Demand Equations. Whan a company sold radios at $45 each, the demand was 2000 radios per month. After the price rose to $65 each, the demand decreased to 1500 radios per month. Assuming that the price p per radio is linearly dependent on the demand q (quantity of radios), find the linear equation that expresses this linear function.

The demand equation is of the form

$$p = mq + b.$$

From the information given, we obtain the two data points (2000, 45) and (1500, 65). Therefore the slope is

$$m = \frac{65 - 45}{1500 - 2000} = -\frac{1}{25}$$

and our equation becomes

$$p = -\frac{1}{25}q + b.$$

Using either data point to find b, we obtain $b = 125$; thus the demand equation is

$$p = -\frac{1}{25}q + 125.$$

Let us graph the linear demand function obtained in example 3. The graph of the linear *equation* $p = -\frac{1}{25}q + 125$ is the straight line as shown in figure 2.6a. However, since q represents the number of radios, then q is non-negative. Similarly, since p represents the unit price, p is non-negative. Therefore the graph of the linear *function* is restricted to quadrant I as shown in figure 2.6b.

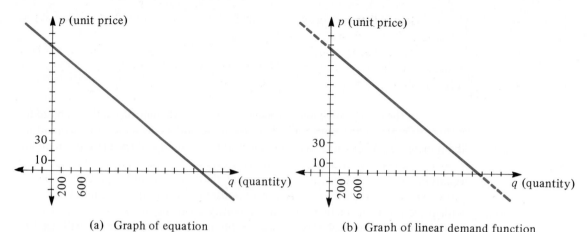

(a) Graph of equation

(b) Graph of linear demand function

FIGURE 2.6

$$p = -\frac{1}{25}q + 125$$

$$p = -\frac{1}{25}q + 125$$

Note that the slope of the line in figure 2.6b is negative; that is, the price decreases as the demand increases. This reflects the fact that high prices correspond to low demand and low prices correspond to high demand.

There is another phenomenon occurring in the marketplace: the relationship between the supply of an item and the unit price. For example, when the price of soybeans is low, many farmers elect not to sell their harvest immediately but instead wait for the price to increase. This results in a decrease in the supply of soybeans. The equation relating the supply of an item and the unit price is called a **supply equation**. Again, as done with demand equations, we reverse our thinking and write our equations with price dependent on supply. This is done so that when graphing a supply equation price will be on the vertical axis, which is how an economist would graph the equation.

EXAMPLE 4

Linear Supply Equation. When the price of gasoline in Iowa City was $1.00 a gallon, the supply was 150,000 gallons. When the price increased to $1.20 a gallon, the supply was 175,000 gallons. Find the supply equation, assuming that it is linear.

Let p denote the price per gallon (in cents) and let q denote the supply of gasoline (in thousands of gallons). Then the supply equation is of the form

$$p = mq + b.$$

From the information given, we have the two data points $(150, 100)$ and $(175, 120)$. Therefore the slope is

$$m = \frac{120 - 100}{175 - 150} = 0.8.$$

Using either data point to find b, we have $b = -20$, and therefore

$$p = 0.8q - 20.$$

Remembering that p and q are both non-negative, we graph the supply equation of example 4 as in figure 2.7.

Note that the slope of the line in figure 2.7 is positive; that is, the price increases as the supply increases. This reflects the fact that low prices correspond to low supply and high prices correspond to high supply.

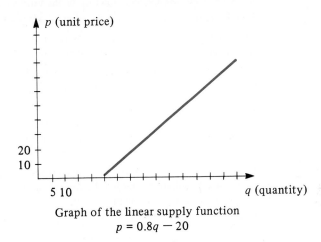

Graph of the linear supply function
$$p = 0.8q - 20$$

FIGURE 2.7

EXERCISES 2.2

---

1. A bottling machine in a brewery originally cost $26,500. If the salvage value is $4000 at the end of 15 years, what equation expresses the straight-line depreciation for this machine?

2. The tennis pro at the Lehigh Country Club buys a ball machine for $895. If the estimated salvage value is $100 at the end of 5 years, what equation expresses the straight-line depreciation for this machine?

3. A small business uses the method of straight-line depreciation for one of its sewing machines that cost originally $778.50. A useful life of 7 years and a salvage value of $110 is used. Write the equation that expresses the value of the sewing machine as a function of the number of years used. Also, determine the value of the machine after it has been used for 3 years by the business.

4. A real estate investor depreciates the cost of an apartment house over 25 years using the straight-line method. If she uses $92,000 as the original cost and no salvage value, then

   a. write the equation that relates the remaining book value to time in years;

   b. write the equation that describes the accumulated depreciation as a function of the years in use.

5. A company that produces belts has a fixed cost of $660 and a variable cost given by $0.75 per belt. The company sells its belts for $1.85 each.

   a. Express by a linear equation the total cost of producing x belts.

    b. Express by a linear equation the revenue from selling x belts.

    c. Express the profit from selling x belts.

6. A company that makes frozen TV dinners has a fixed cost of $25,800 and a variable cost of $1.30 per dinner. The company sells its TV dinners for $2.50 each.

    a. Express the total cost as a function of the number of TV dinners made.

    b. Express the revenue as a function of the number of TV dinners sold.

    c. Express the profit as a function of the number of TV dinners made and sold.

7. One department of a mill makes wood interior doors. The fixed cost charged to this department is $15,000. The department's variable cost comes from materials and labor.

    a. If each door costs $12 in materials and $15 in labor, what is the total cost function?

    b. If each door is sold for $87, what is the profit function?

8. Trico sells well pumps which it buys for $3500 each and resells for $5000. Trico has no other variable cost, but it has a fixed cost of $25,000 in building and truck depreciation. Write the equation that expresses Trico's profit as a function of the number of pumps sold.

9. A small business makes and sells wooden toys. If it makes up to 20 of these toys, their total cost is $296 plus $2 per toy. If, however, it makes 20 or more of these toys, the total cost is $336, the cost for the first 20 plus $1.50 for each additional toy over 20. Write the equation that expresses the business's total cost as a function of the number of toys made and draw the graph of this function.

10. Consider the business described in exercise 9. Suppose the toys are sold for $10 a piece. Give the profit function for this business and draw its graph.

11. When a record company sold albums at $5 each, the demand in Oakland was 25,000 albums. After the price rose to $6 each, the demand dropped to 17,000 albums. Assuming that the price p per album is linearly dependent on the demand q (in thousands of albums), find the demand equation and draw its graph.

12. When an orchard sold apples at 50 cents per pound, the demand was 15,000 pounds. When the price was increased to 60 cents per pound, the demand was 12,000 pounds. Assuming that the price p per pound is linearly dependent on the demand q (in thousands of pounds), find the demand equation and draw its graph.

13. The price of halibut was $4.50 per pound in Portland and 18,000 pounds was sold at this price. However, when the fishery increased its price to $4.95 per pound, the demand dropped to 14,000 pounds. Write the demand equation, assuming linearity.

14. When coffee beans were priced at $5 per pound, the demand was 14,000 pounds, but when the price dropped to $4.30 per pound, the demand rose to 17,500 pounds. Write the demand equation, assuming linearity.

15. When the price of gasoline in a Midwest region was $1.25 a gallon, the supply was 200,000 gallons. When the price increased to $1.30 a gallon, the supply increased to 250,000 gallons. Assuming that the price p per gallon is linearly dependent on the supply q (in thousands of gallons), find the supply equation and draw its graph.

16. When the price of soft-shell crabs was $4.35 per pound in Tacoma, the supply was 1200 pounds. But when the price dropped to $3.95 per pound, the supply was 800 pounds. Find the supply equation, assuming it is linear, and draw its graph.

17. When halibut was priced at $4.50 per pound in Portland, the supply was 15,000 pounds. After the price increased to $4.95 per pound, the supply was 20,000 pounds. Write the supply equation, assuming linearity.

18. The following two points of information on the supply of coffee beans is observed. When priced at \$4.30 per pound, the supply was 15,000 pounds. When the price rose to \$5 per pound, the supply rose to 20,000 pounds. If price varies linearly with supply, give the equation that expresses this relation.

2.3 Solving Systems of Two Equations in Two Unknowns

Recall that a linear equation in x and y can be written in the form

$$Ax + By = C.$$

Recall also that the graph of a linear equation in two variables in a two-dimensional coordinate system is a line. The graph is a visual display of the solutions to the equation; in particular, the coordinates of each point on the line satisfy the equation. For example, the point $(3, 2)$ is a solution to the equation $2x + 3y = 12$ since $2(3) + 3(2) = 12$, whereas the point $(0, 0)$ is not a solution since $2(0) + 3(0) \neq 12$. Therefore a linear equation in two variables has infinitely many solutions (see figure 2.8).

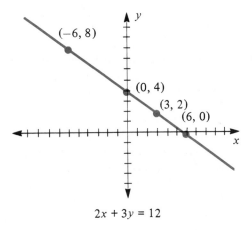

FIGURE 2.8 $2x + 3y = 12$

We now direct our attention in this section to systems of two linear equations in two unknowns and their solutions.

An example of a **system** of two linear equations in x and y is

$$2x + 3y = \ \ 21$$
$$-5x + 2y = -5.$$

A **solution** of the system is any ordered pair (x, y) that satisfies both equations and, therefore, represents a point on the graph of both lines. Observe that $(3, 5)$ is a solution of the given system since

$$2(3) + 3(5) = \ \ 21$$

and $$-5(3) + 2(5) = -5.$$

Therefore the solution $(3, 5)$ gives the coordinates of a point common to both lines.

In general, the graph of each equation in a system of two equations in two unknowns

$$Ax + By = C$$

$$Dx + Ey = F$$

is a line. Since two lines in a plane can either be parallel, coincide, or intersect in a single point, the system of linear equations must have either no solutions, infinitely many solutions, or one solution (see figure 2.9). In the case when a system has at least one solution, it is said to be **consistent**; otherwise, the system is said to be **inconsistent**.

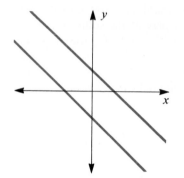

(a) Parallel lines inconsistent
no solutions

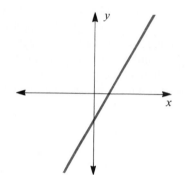

(b) Coincident lines consistent
many solutions

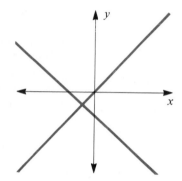

(c) Intersecting lines consistent
one solution

FIGURE 2.9

We solve a given system by finding the set of all solutions, called the **solution set**, of the system. From our discussion we see that the solution set of a system can be the empty set, can contain infinitely many points, or can contain a single point. We present two methods for finding the solution set of a system of two equations in two unknowns. The two methods are called the **elimination method** and the **substitution method**. In chapter 4 we present other methods for solving larger systems of equations. Let us begin with the elimination method, the steps of which follow.

Elimination Method

Step 1: Multiply one or both equations in the system by appropriate non-zero constants so that when the equations are added, one of the unknowns is eliminated and we obtain one equation in at most one unknown.

Step 2: Add the two equations obtained with the multiplication in step 1. Solve the resulting equation in one unknown for that unknown and proceed to step 3. If both unknowns are eliminated when the equations are added, the system is either inconsistent or has infinitely many solutions, as we shall see in the examples.

Step 3: Substitute the value obtained from step 2 into either of the original equations and solve for the other unknown. In this case, the system has one solution and the values obtained here and from step 2 make up that solution.

The system resulting from an application of step 1 of the elimination method is said to be **equivalent** to the original system of equations. Observe that the graph of a linear equation will not change when the equation is multiplied by a non-zero constant. Therefore the equivalent system of two equations obtained in step 1 has the same solution set as the original system.

EXAMPLE 1

Use the elimination method to find the solution set of the system

$$-3x + 4y = 11$$
$$2x - 3y = -8.$$

Step 1: We can eliminate either the unknown x or the unknown y. We choose to eliminate x. Therefore, we multiply the first equation by 2 and the second equation by 3 to obtain the equivalent system

$$-6x + 8y = 22$$
$$6x - 9y = -24.$$

Step 2: By adding these equations, we eliminate x to obtain one equation in y, namely,

$$-6x + 8y = 22$$
$$\underline{6x - 9y = -24}$$
$$-y = -2$$

and

$$y = 2.$$

Step 3: Next, replace y by 2 in either of the original two equations. Picking the first equation, we have

$$-3x + 4(2) = 11.$$

Solving this equation for x, we have $x = -1$. Thus the (unique) solution to the system is $(-1, 2)$.

Substitution Method

Step 1: Select one of the two equations in the system and solve for one of the unknowns in terms of the other unknown.

Step 2: Substitute the expression obtained in step 1 for the same unknown in the other equation, solve the resulting equation for the other unknown, and proceed to step 3. If, however, upon substituting the expression we obtain an equation without any unknowns, then the system is inconsistent or has infinitely many solutions.

Step 3: Substitute the number found in step 2 into the equation derived in step 1 and compute the value of the second unknown. In this case, the system has one solution and the values obtained here and from step 2 make up the solution.

EXAMPLE 2

Solve the system

$$2x - y = -3$$
$$6x + 4y = 5$$

by the substitution method.

Step 1: We select the first equation to solve for y, giving

$$y = 2x + 3.$$

Step 2: Substituting this expression for y into the second equation, we obtain the equation in x

$$6x + 4(2x + 3) = 5.$$

Solving this equation for x, we have $x = -\frac{1}{2}$.

Step 3: Replacing the value for x found in step 2, namely, $x = -\frac{1}{2}$, into the expression for y found in step 1, we have

$$y = 2\left(-\frac{1}{2}\right) + 3 = 2.$$

Therefore the solution set of the system contains the single point $(-\frac{1}{2}, 2)$.

Examples 1 and 2 illustrate systems having unique solutions. We consider in the next two examples systems having, respectively, no solutions and infinitely many solutions.

EXAMPLES

3. Consider the system

$$-2x + 3y = -\frac{1}{3}$$

$$\frac{1}{2}x - \frac{3}{4}y = 1.$$

Let us use the elimination method to find the solution set for this system.

Step 1: We multiply the second equation by 4 to obtain the equivalent system

$$-2x + 3y = -\frac{1}{3}$$

$$2x - 3y = 4.$$

Step 2: Adding the two equations of step 1, we obtain $0 = 11/3$. The solution process can stop at this step since we obtained an equation without any unknowns. Furthermore the inconsistency we derived tells us that the original system has no solutions; that is, the solution set is empty.

4. Consider the system

$$x - 2y = -5$$
$$-3x + 6y = 15.$$

We use the elimination method on this system to find the solution set.

Step 1: Multiplying the first equation by 3 yields the equivalent system.

$$3x - 6y = -15$$
$$-3x + 6y = 15.$$

Step 2: Adding the two equations gives the equation $0 = 0$. We need not go any further in the solution process since we arrived at an equation without any unknowns. Furthermore the statement $0 = 0$, which can be true for any numbers, tells us that the original system has infinitely many solutions; any point in the graph of one of the equations is in the graph of the other and, therefore, is a solution to the system.

We now have seen examples of the three possibilities for the solution set of a system of 2 equations in 2 unknowns. The solution set can easily be determined by either of our two methods, substitution or elimination. Furthermore, as seen in example 3, if we derive an equation of the form $0 = b$, where $b \neq 0$, then we can conclude that the solution set is empty. As seen in example 4, if we derive the equation $0 = 0$, then we can conclude that the two equations represent the same line and the solution set has infinitely many solutions. Note, however, that when we consider larger systems of linear equations in chapter 4, we cannot necessarily conclude that the system has infinitely many solutions if we derive the equation $0 = 0$.

EXERCISES 2.3

For exercises 1 and 2, determine whether or not the stated solution set is correct for the given system.

1. a. $6x + y = 18$

   $5x + \dfrac{1}{3}y = 13$

   Solution set contains only the point $(\frac{7}{3}, 4)$.

   b. $2x - 3y = 0$

   $-8x + 12y = 0$

   Solution set contains infinitely many solutions.

   c. $11x + 4y = 56$

   $8x + 3y = 45$

   Solution set is empty.

   d. $7x - 3y = 40$

   $3x + 7y = 21$

   Solution set contains only the point $(7, 3)$.

2. a. $\dfrac{1}{2}x + \dfrac{1}{3}y = 2$

     $\dfrac{1}{3}x + \dfrac{1}{2}y = 6$

     Solution set contains
infinitely many solutions.

  c. $\dfrac{7}{3}x + \dfrac{5}{4}y = 12$

     $4x + 3y = 0$

     Solution set contains
only the point $(-3, 4)$.

b. $0.1x + 0.3y = 13$

   $0.2x + 0.5y = 24$

   Solution set contains
only the point $(70, 20)$.

d. $13x + 39y = 26$

   $2x + 6y = 5$

   Solution set is empty.

For exercises 3–24 find the solution set for each of the systems of linear equations.

3. $3x + 6y = 21$
   $x - 6y = -5$

4. $3x + y = 50$
   $y = 5x + 2$

5. $2x + y = 5$
   $x = y - 2$

6. $x - y = \dfrac{1}{2}$
   $4x + 2y = 14$

7. $x - 6y = 5$
   $-\dfrac{2}{3}x + 4y = -7$

8. $x = -7y - 16$
   $y = 7x - 38$

9. $3x + 4y = 20$
   $y = -\dfrac{3}{4}x + 5$

10. $2x + 5y = 13$
   $5x + 2y = -41$

11. $\dfrac{7}{5}x + \dfrac{4}{9}y = 15$
   $3x + y = 33$

12. $x + \dfrac{1}{2}y = 9$
   $\dfrac{1}{3}x + \dfrac{1}{6}y = -3$

13. $5x + 11y = 59.5$
   $7x + 3y = 58.5$

14. $9x + 16y = 96$
   $13x + 12y = 97$

15. $\dfrac{1}{3}x + \dfrac{1}{4}y = 1$
   $\dfrac{1}{4}x + \dfrac{1}{3}y = 1$

16. $\dfrac{2}{5}x + \dfrac{1}{7}y = 0$
   $\dfrac{1}{3}x + \dfrac{3}{5}y = 0$

17. $27x + 45y = 36$
   $12x + 20y = 12$

18. $27x + 45y = 18$
   $12x + 20y = 8$

19. $30x + 5y = 381$
   $10x + 15y = 239$

20. $5x + 2y = 49$
   $15x + 8y = 170$

21. $0.1x + 0.3y = 7.9$
   $0.4x + 0.5y = 16.9$

22. $0.5x + 0.3y = 17.0$
   $0.2x + 0.7y = 15.5$

23. $0.36x + 0.02y = 57.4$
   $0.09x + 0.15y = 39.0$

24. $0.7x + 1.1y = 7.95$
   $2.1x + 1.5y = 17.37$

2.4 Applications of Systems of Two Equations

With techniques for solving systems of two equations in two unknown at hand, we now turn our attention to applications.

In section 2.2 we considered cost and revenue functions. In particular, in example 2 of section 2.2, the total cost for producing x electric coils was determined as

$$C = 1.50x + 300$$

and the total revenue as

$$R = 4x.$$

Figure 2.10a shows the graphs of these two functions in the same coordinate system. The point of intersection of the two graphs is where $R = C$. The level of production x where cost equals revenue (i.e., the x-coordinate of the point of intersection) is called the **break-even point**.

For production levels less than the break-even point, cost exceeds revenue and the business is operating at a loss. On the other hand, for production levels greater than the break-even point, revenue exceeds cost and the business is operating at a profit. Furthermore the amount of profit (or loss) is the difference

$$P = R - C.$$

For any production x, the size of profit (or loss) is, therefore, the difference in the ordinates of the points on the graphs. Thus, as we see from figure 2.10b, the loss becomes greater as the production level moves to the left of the break-even point, and the profit becomes larger as the production level moves to the right of the break-even point. We see also that an alternate definition of break-even point is the production level x for which the profit is zero.

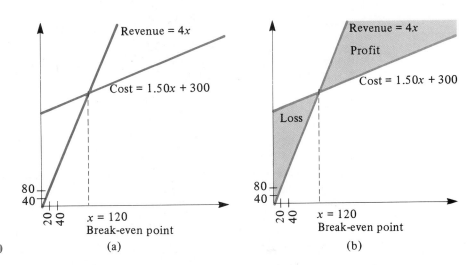

FIGURE 2.10   (a)   (b)

It is not difficult to find the break-even point once we have the cost and revenue functions. We simply set the cost expression equal to the revenue expression and solve the resulting equation for x. In doing so for our example, we have

$$4x = 1.50x + 300.$$

Solving, we have for the break-even point

$$2.50x = 300$$

$$x = 120.$$

EXAMPLE 1

Blackmore Tool Company makes tool kits with cost and revenue functions given by

$$C = 5x + 5250$$

$$R = 20x$$

where x is the number of kits produced and sold in a week. Let us find the break-even point for the company.

Setting $R = C$, we have

$$20x = 5x + 5250.$$

Solving this equation for x, we see that the break-even point is 350 kits. Therefore, if the company produces and sells less than 350 kits in a week, it will be operating at a loss; whereas, if more than 350 kits are produced and sold, a profit will be realized.

Supply and demand functions were also discussed in section 2.2. Our next application deals with the supply and demand of a single commodity in a competitive marketplace.

Suppose that in the same market a given commodity obeys both a demand equation and a supply equation. Suppose, for example, the demand equation for scallops in Baltimore is given by

$$p = -\frac{1}{5}q + 4 \qquad \text{(demand)}$$

where p is the price per pound (in dollars) and q is the number of pounds (in thousands of pounds) of scallops.

On the other hand, the supply equation for scallops is given by

$$p = 3q - 6. \qquad \text{(supply)}$$

If the demand and supply equations are graphed in the same pq-coordinate plane, we see that these lines intersect (see figure 2.11). The **market equilibrium** corresponds to that point of intersection; namely, where the demand is equal to the supply. To find the coordinates of this equilibrium point, we set demand equal to supply, or

$$-\frac{1}{5}q + 4 = 3q - 6.$$

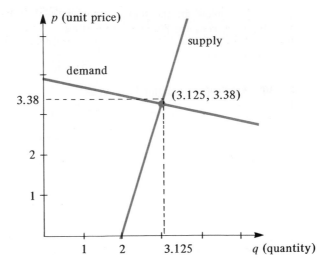

FIGURE 2.11

Solving for q, we have

$$q = 3.125.$$

To find the p-coordinate of the equilibrium point, we substitute $q = 3.125$ into either the demand or the supply equation. For example, using the demand equation, we have

$$p = -\frac{1}{5}(3.125) + 4$$

$$= 3.38.$$

We call 3125 pounds the **equilibrium quantity** and \$3.38 the **equilibrium price** (see figure 2.11).

EXAMPLE 2

In Little Rock, when wheat germ sold for \$1.25 per jar, the demand was 12,000 jars, but when the price increased to \$1.75 per jar, the demand dropped to 8000 jars. Furthermore, at the price of \$1.25 per jar, the supply was 10,000 jars, and at the price of \$1.75 per jar, the supply was 12,000 jars. Assuming the price p per jar is linearly dependent on the demand q (in thousands of jars) and p is also linearly dependent on supply q, find the equilibrium quantity and equilibrium price.

First we must write the demand and supply equations. For the demand equation, we use the data points $(12, 1.25)$ and $(8, 1.75)$. Thus the slope of the demand equation is

$$m = \frac{1.75 - 1.25}{8 - 12} = -0.125.$$

Therefore we have for the demand equation

$$p - 1.25 = -0.125(q - 12)$$

and

$$p = -0.125q + 2.75. \qquad \text{(demand)}$$

Now for the supply equation, we use the data points $(10, 1.25)$ and $(12, 1.75)$, from which we obtain the slope

$$m = \frac{1.75 - 1.25}{12 - 10} = 0.25.$$

Thus the supply equation is found as

$$p - 1.25 = 0.25(q - 10)$$

and
$$p = 0.25q - 1.25. \qquad \text{(supply)}$$

Setting the demand price equal to the supply price, we obtain the equation in q

$$-0.125q + 2.75 = 0.25q - 1.25$$

and
$$q = 10.667.$$

Substituting this value for q in the demand equation, we find the price p to be

$$p = 1.42.$$

Therefore the equilibrium quantity is 10,667 jars and the equilibrium price is $1.42 per jar.

In each of the examples, break-even analysis and market equilibrium, we have two functions of *one* variable, and we wish to find the point common to both functions. There are many situations in which one quantity depends on several other quantities. As in the case of a linear function of one variable, equations may be found that describe a function of two variables, as we see next.

A manufacturer of glass laboratory apparatus makes special retorts and titration pipets. Natural gas and labor are used in the production of each type of apparatus. The particular production requirements are as follows: one retort requires 1 man-hour and 10 cubic feet of gas, whereas one pipet requires 2 man-hours and 15 cubic feet of gas. Therefore the amount of labor L and the amount of natural gas NG used in production are each dependent on the number of retorts and the number of pipets produced. To obtain a mathematical model of these dependencies, first let

$$x = \text{the number of special retorts produced}$$

and
$$y = \text{the number of titration pipets produced.}$$

Since each retort requires 1 man-hour, the number of man-hours needed to produce x of them is x. Furthermore, since each pipet requires 2 man-hours, the labor needed to produce y pipets is two times y. Finally, since

$$L = (\text{man-hours used for retorts}) + (\text{man-hours used for pipets}),$$

we have
$$L = x + 2y.$$

Using a similar discussion, we find that

$$NG = 10x + 15y.$$

We see from our equations that L and NG are each a *function of two variables x and y*; that is, unique values for L and NG can be determined for each given pair of values (x, y). Furthermore these two functions are of a special form.

Definition

■ A function of two variables x and y is called a **linear form** if the relationship between the **dependent variable** T and the **independent variables** x and y can be expressed in the form

$$T = ax + by + c$$

where a, b, and c are constants. (See figure 2.12.) ■

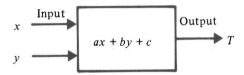

FIGURE 2.12

EXAMPLE 3

Woodhaven Farms grows soybeans and corn. It costs $120 to plant one acre of soybeans and $100 to plant one acre of corn.

a. Express the total planting cost as a function of the number of acres of soybeans and the number of acres of corn planted.

b. If Woodhaven Farms plants 125 acres of soybeans and 175 acres of corn, what is the total planting cost?

To obtain an expression for

$$T = \text{the total planting cost,}$$

we let

$$x = \text{the number of acres of soybeans planted}$$

$$y = \text{the number of acres of corn planted.}$$

Now, since one acre of soybeans costs $120 to plant, x acres would cost

$$(\text{cost per acre}) \times (\text{number of acres}) = 120x.$$

Similarly, since one acre of corn costs $100 to plant, y acres would cost $100y$ dollars to plant. The total planting cost would be the sum of the individual costs; that is,

$$T = 120x + 100y.$$

For part (b), we evaluate the linear form T at $(x, y) = (125, 175)$ giving

$$T = 120(125) + 100(175)$$

$$= \$32,500.$$

We saw that a break-even point is found by requiring cost to be equal to revenue. This requirement is obtained by setting the cost function equal to the revenue function. In situations where we have two linear forms, it may be of interest to find a pair (x, y) that meets certain requirements on the two functions, as we see next.

We saw that the manufacturer of glass laboratory apparatus can represent the amounts of labor and natural gas used in production by the linear forms

$$L = \quad x + 2y$$

$$NG = 10x + 15y.$$

Suppose, of these two resources, 575 man-hours and 5000 cubic feet of natural gas are available for production. Now one question the manufacturer may wish to answer is: "How many of each type of apparatus can be produced in order to use *completely* the man-hours and the gas it has available?"

The answer to this question can be found by requiring the function L be equal to 575 and the function NG be equal to 5000; that is, we require that $L = 575$ and $NG = 5000$. Therefore the mathematical formulation of the question is to determine, if possible, x and y that satisfy the system of equations

$$x + 2y = \quad 575 \quad \text{(labor)}$$

$$10x + 15y = 5000 \quad \text{(natural gas).}$$

Using techniques for solving a system of two linear equations in two unknowns we can obtain an answer to the manufacturer's question. In particular, solving the system by elimination, we find $x = 275$ and $y = 150$. Therefore, by producing 275 retorts and 150 pipets, the manufacturer will use all the labor and natural gas it has available.

EXAMPLES

4. Woodhaven Farms has available 150 acres for planting soybeans or corn or a combination of both. It costs \$120 to plant one acre of soybeans and \$100 to plant one acre of corn. Woodhaven has available \$16,300 to cover planting costs. How many acres of each crop should Woodhaven Farms plant so that the two resources, acreage and money for planting, are used completely?

First, let the two unknowns be represented as follows:

$$x = \text{the number of acres of soybeans to be planted}$$

and $\quad y = \text{the number of acres of corn to be planted.}$

One condition of the problem comes from the planting cost. The linear form for planting cost is given in example 3 as

$$T = 120x + 100y.$$

The second condition is imposed on the number of acres to be planted. It is easily seen that the linear form for acreage is given by

$$A = x + y.$$

Now imposing the conditions that \$16,300 be spent planting the 150 available acres (i.e., $T = 16,300$ and $A = 150$), we have the mathematical model of our problem, namely, find x and y such that

$$120x + 100y = 16,300$$
$$x + \quad y = 150.$$

Solving this system, for example, by substitution, we find that $x = 65$ and $y = 85$. Therefore Woodhaven Farms should plant 65 acres in soybeans and 85 acres in corn, in which case it will plant a total of 150 acres and spend $16,300.

5. The Heavy Metal Company makes two types of alloys. Each of these contains, among other metals, copper and zinc. Each ton of the first alloy contains $\frac{1}{2}$ ton of copper and $\frac{1}{3}$ ton of zinc, whereas each ton of the second type of alloy contains $\frac{3}{4}$ ton of copper and $\frac{1}{6}$ ton of zinc. Suppose on a particular day, the company has available 65.5 tons of copper and 30.5 tons of zinc. How many tons of each type of alloy can be made so that the available copper and zinc are used completely?

We obtain a mathematical formulation of our problem by first letting

$$x = \text{the number of tons of the first alloy made each day}$$

and $y = \text{the number of tons of the second alloy made each day.}$

Next we want to obtain expressions for the total tons of copper and the total tons of zinc used in production each day. If we denote the total tons of copper used each day as C, then

$$C = (\text{copper per ton of first alloy}) \cdot (\text{tons of first alloy})$$

$$+ (\text{copper per ton of second alloy}) \cdot (\text{tons of second alloy})$$

$$= \frac{1}{2}x + \frac{3}{4}y.$$

Also, if we denote the total tons of zinc used each day as Z, then we have

$$Z = (\text{zinc per ton of first alloy}) \cdot (\text{tons of first alloy})$$

$$+ (\text{zinc per ton of second alloy}) \cdot (\text{tons of second alloy})$$

$$= \frac{1}{3}x + \frac{1}{6}y.$$

The requirements of the problem are expressed by setting the functions for C and Z equal to 65.5 and 30.5, respectively. In doing so we have the system of linear equations

$$\frac{1}{2}x + \frac{3}{4}y = 65.5$$

$$\frac{1}{3}x + \frac{1}{6}y = 30.5.$$

Before solving this system by elimination, let us clear fractions to obtain an easier system to work with. Thus, multiplying the first equation by 4 and the second equation by 6, we obtain the equivalent system

$$2x + 3y = 262$$

$$2x + y = 183.$$

Now solving this equivalent system by elimination, first eliminating x and solving for y, we obtain the solution $y = 39.5$ and $x = 71.75$. Therefore, if Heavy Metals produces 71.75 tons of the first alloy and 39.5 tons of the second alloy, it will use completely its available copper and zinc.

We see from our production problems that conditions arise naturally that in turn require setting a linear form equal to a constant. In the next chapter we will see how conditions can arise in which we require that a linear form have values no larger than some constant. Conditions of a problem can arise in other ways, as demonstrated with our last example in this section.

EXAMPLE 6

Sherry has $9000 available to invest between two stocks, Calico which sells for $31 per share and Taco Times which sells for $8 per share. Sherry's investment strategy is to buy 150 more shares of Taco Times than Calico. How many shares of each stock should Sherry buy?

We begin first by representing the unknowns of the problem as

$$x = \text{the number of shares of Calico to buy}$$

and $\qquad y = \text{the number of shares of Taco Times to buy.}$

Now, from the information given, we see that two requirements must be met. The first requirement is on the total cost of the shares and the second requirement is on the number of shares to purchase.

We can represent the total cost (T) of the shares by the linear form

$$T = 31x + 8y.$$

Imposing the condition that $9000 is invested, we have the equation

$$31x + 8y = 9000.$$

The second condition, the number of shares of Taco Times exceeds the number of shares of Calico by 150, is expressed by the equation

$$y = x + 150.$$

Therefore the solution is found by solving the system of equations

$$31x + 8y = 9000$$

$$y = x + 150.$$

Using the substitution method, we find that $x = 200$ and $y = 350$. Therefore Sherry should buy 200 shares of Calico and 350 shares of Taco Times, in which case she will invest $9000 and meet her investment strategy.

EXERCISES 2.4

1. Find the break-even point for the company described in exercise 5 of section 2.2. Also draw the graphs of the cost and revenue functions in the same coordinate system and show the regions of profit and loss.

2. Find the break-even point for the company described in exercise 6 of section 2.2. Also draw the graphs of the cost and revenue functions in the same coordinate system and show the regions of profit and loss.

3. A tool kit manufacturer has a fixed cost of $5250 and a variable cost of $5 per tool kit. The manufacturer sells the kits for $20 each. Find the manufacturer's break-even point.

4. Hanley Brick has a fixed cost of $22,500, which is the depreciation cost on its building and machinery. Hanley's variable cost is 14¢ per brick. Hanley sells its bricks for 32¢ each. What is Hanley's break-even point?

5. Find the break-even point for the wooden-toy maker described in exercises 9 and 10 of section 2.2.

6. Dauber is an outfitter who operates wilderness canoe trips. For each trip he outfits and guides, the costs to him are $1250 in fixed costs plus $100 for each person up to and including 10 people going on the trip. If, however, more than 10 people go on the trip, Dauber's total cost is $2250, the cost for the first 10 people, plus $90 for each additional person over 10. Dauber charges $180 per person independent of how many people are on the trip. Find Dauber's break-even point for a trip.

7. In Topeka the demand equation for coffee beans is

$$p = -\frac{1}{4}q + 11$$

and the supply equation is given by

$$p = 2q - 7,$$

where p is the price per pound (in dollars) and q is the quantity (in thousands of pounds). Find the equilibrium quantity and the equilibrium price. Also draw the graphs of the demand and supply equations in the same coordinate system.

8. In the Greater Houston area, the demand equation for liquid fructose is

$$p = -0.125q + 5$$

and the supply equation is given by

$$p = 0.35q - 0.7$$

where p is the price per bottle (in dollars) and q is the number of bottles (in thousands) of fructose. Find the equilibrium quantity and the equilibrium price. Also draw the graphs of the demand and supply equations in the same coordinate system.

9. In the Bradford area when lobsters were selling for $6.60 per pound, the demand was for 3800 pounds. When the price dropped to $5.10 per pound, the demand went up to 4300 pounds. On the other hand, the supply at $6.60 per pound was 4240 pounds, and the supply at $5.10 per pound was 3640 pounds. Find the equilibrium quantity and the equilibrium price.

10. A rancher and a meat packing house are negotiating a beef sale. The rancher states that he would only sell 100 head of cattle at 30¢ per pound but would sell 200 head at 60¢ per pound. On the other hand, the packing house offers to buy 200 head at 30¢ per pound, but only 50 head at 60¢ per pound. Assuming that the demand and supply equations are linear, what price should they agree on that is satisfactory to both?

11. A company owns two bicycle factories, A and B. The weekly cost (in dollars) to run factory A is $110x + 1200$, where x is the number of bicycles produced in a week. The weekly cost (in dollars) to run factory B is $90y + 1350$, where y is the number of bicycles produced in a week.

    a. Write an equation that represents the company's total cost of running both factories for a week.

    b. If factory A produces 225 bicycles and factory B produces 155 bicycles during a week, what is the company's total cost for this week?

12. A wood stove company makes two types of stoves, basic and deluxe. The company's fixed cost per day is $525 and it costs the company $250 to make a basic stove and $450 to make a deluxe stove.

    a. Write an equation that describes the dependency of the total cost for a day on the number of stoves of each type made in a day.

    b. If the company makes 12 basic stoves and 7 deluxe stoves during one day, what is their total cost for this day?

13. A person has a regular savings account that pays interest at the rate of $\frac{1}{2}\%$ per month and a certificate of deposit that pays $\frac{3}{4}\%$ per month.

    a. Express the total interest obtained during a month as a function of the amounts in the savings account and the certificate of deposit.

    b. If the person has $1500 in the savings account and $10,000 in a certificate of deposit during one month, how much interest will this person receive for this month?

14. A diet consists of rice and chicken breasts. Each unit of rice has 350 calories and 3 grams of protein, whereas each unit of chicken breasts has 86 calories and 19 grams of protein.

    a. Write the linear form that expresses the dependency of the number of calories on the units of rice and units of chicken breasts.

    b. Write the linear form that expresses the dependency of the grams of protein on the units of rice and units of chicken breasts.

15. Allegheny Cabinets makes unfinished tables and bookcases. It takes 12 board-feet of lumber for each table and 28 board-feet for each bookcase. Also, it takes 2 man-hours to cut and assemble each table, whereas it takes 3 man-hours for each bookcase. If Allegheny has available in one week 840 board-feet of lumber and 115 man-hours of labor, how many tables and bookcases can Allegheny make in this week so that all the lumber and available man-hours are used?

16. Paul's Pizza makes frozen pizzas in two sizes, 9″ and 12″. Each 9″ pizza uses 18 ounces of dough and one cup of sauce, whereas each 12″ pizza uses 24 ounces of dough and $1\frac{1}{2}$ cups of sauce. Paul has on hand 102 pounds of dough and 96 cups of sauce (and an unlimited amount of cheese). How many pizzas of each size can Paul make so that he uses all the dough and sauce that is available?

17. Crestfield Farms has 375 acres in which it plans to plant barley or wheat or a combination of both. It costs the farm $110 to plant one acre of barley and $80 to plant one acre of wheat; the farm has $35,100 available to cover these costs. How many acres of each crop should be planted so as to use completely the available money for planting the 375 acres?

18. Leisure Village Development Corporation plans to develop 120 acres into 2-acre and 3-acre building lots. It costs the corporation $2500 and $3500, respectively, to develop each size lot. If Leisure Village has available $144,500 to develop the 120 acres, how many lots of each size can be developed?

19. Beth Metals makes two types of stainless steel. Each grade is composed of nickel and steel. Each ton of the lesser grade is 5% nickel and 95% steel, whereas each ton of the better grade is 15% nickel and 85% steel. If Beth Metals has available 4.7 tons of nickel and 48.3 tons of steel, how many tons of each grade of stainless steel can be made?

20. Computer Forms Company makes standard form punch cards and standard form paper for computers. Both the punch cards and the paper must be cut and then have lines printed on them. One box of punch cards requires 0.25 man-hours of cutting time and 0.30 man-hours of printing time, whereas each box of paper requires 0.15 man-hours of cutting time and 0.20 man-hours of printing time. Suppose the company has available during a particular week 368 man-hours in the cutting department and 462 man-hours in the print shop. How many boxes of punch cards and boxes of paper can be made to utilize completely the man-hours available?

21. Consider the diet of rice and chicken breasts described in exercise 14 of this section. If it is desired to obtain 1070 calories and 32 grams of protein, how many units of each type food should be used to obtain these amounts exactly?

22. Erica is planning a meal of steak and salad. Each unit of steak contains 200 calories and 3 grams of vitamins, whereas each unit of salad contains 20 calories and 4 grams of vitamins. If the meal is to supply exactly 500 calories and 26 grams of vitamins, how many units of steak and units of salad should be eaten?

23. An individual decides to divide $15,000 between two bank accounts, a regular savings account that pays 6% annually and a money management account that pays 9% annually. If the individual wants to obtain $1200 in interest, how much money should be deposited in each account?

24. Jean currently has $50,000 invested between two medium-risk investments and is realizing a total return of $450 per month on her money. However Jean wishes to change her investment strategy and divide the $50,000 between a certificate of deposit that returns $\frac{3}{4}$% per month and an oil income partnership that returns $1\frac{1}{4}$% per month. How should she divide the $50,000 between these two new investments so as she continues to receive $450 per month?

25. The Deerfield Company manufactures a vegetable fertilizer that it guarantees to have 18 pounds of nitrogen and 24 pounds of phosphoric acid in each bag. Two ingredients, A and B, are used to make the fertilizer. Ingredient A contains 40% nitrogen and 30% phosphoric acid, whereas ingredient B contains 20% nitrogen and 50% phosphoric acid. How many pounds of each ingredient should the company use per bag in order to meet exactly the nitrogen and phosphoric acid requirements?

26. At a copper smelter, water drawn from the river is used in a certain stage of the production process. As it passes through the smelter it becomes 3% acid. An environmental agency requires that the acid concentration of any water returned to the river be no more than 0.1% acid. Therefore the water coming from the smelter is put into a holding pond and mixed with fresh water before being released into the river. The capacity of the holding pond is 150,000 gallons. How much fresh water and acidic water must be mixed together in this pond so that when the water is released into the river it is no more than 0.1% acid?

27. A small state university wishes to maintain a student-faculty ratio of 1 teaching faculty member to every 20 undergraduate students and 1 teaching faculty member to every 10 graduate students. The university receives state appropriations to cover costs over and above those covered by tuition. Specifically the university needs state funding of $700 for every undergraduate student and $900 for every graduate student. If there are 320 teaching faculty at the university and the state appropriation will be $4,255,000, how many undergraduate and graduate students can be at the university?

28. A recommended daily dietary requirement for an active male weighing about 160 pounds includes 3000 calories and 70 grams of protein. Suppose a male who fits into this category plans to eat only a combination of rice and almonds while on a backpacking trip. It is

known that 100 grams of rice supplies 360 calories and 7.5 grams of protein, whereas 100 grams of almonds supplies 598 calories and 18.6 grams of protein. How many grams of each food should he eat daily to satisfy exactly his requirement in calories and protein?

29. An individual who wishes to invest $10,000 has decided to divide her money between two types of investments, one that returns 13% annually and the other, a lower risk investment, that yields 8% annually. The investor wishes to diversify the investment by not putting all her money into just one investment. This individual would like to realize an annual income of at least $1000 on the $10,000. What is the maximum amount that she can invest at 8% to achieve this goal?

30. A craftsman working in leather can produce three pocketbooks and two wallets from a hide costing $80. From a smaller hide that costs $55, he can make two pocketbooks and three wallets. Suppose he needs to fill orders for 9 pocketbooks and 11 wallets. If he is operating his business on a limited budget, what is the least amount he can spend on hides and yet still have enough leather to fill his orders?

31. Suppose there are two countries, "Us" and "Them," engaged in trade with each other. Each country's wealth is measured by the amount of gold it holds. However, during a trade period, "Us" retains $\frac{3}{4}$ of the gold it holds at the beginning of the period and spends the other $\frac{1}{4}$ on imports from "Them." "Them" retains $\frac{2}{5}$ of the gold it holds at the beginning of the period and gives the other $\frac{3}{5}$ to "Us" for imports. Each country also adds to its wealth gold that is mined within its own country. If, during a trade period, "Us" mines 1700 bars of gold and "Them" mines 400 bars, how many bars of gold must each country hold at the beginning of a trade period so that at the end of the period each is twice as wealthy?

32. The population in a certain area is divided into two types of sectors, urban and non-urban. Suppose that, during a single time period, $\frac{3}{5}$ of the urban population moves into non-urban sectors and $\frac{1}{7}$ of the non-urban population moves into urban sectors. Assume that changes in the population distribution are due only to moving. What must be the ratio of urban population to non-urban population so that the number of people living in each sector is the same at the end of a time period as in the beginning of the time period?

SUMMARY OF TERMS

Linear Function
  Independent Variable
  Dependent Variable
  Graph
  Constant Rate of Change

Systems of Linear Equations
  Solution Set
  Inconsistent
  Consistent
  Equivalent Systems
  Elimination Method
  Substitution Method
  Linear Form
  Function of Two Variables

Applications in Business and Economics
  Straight-Line Depreciation
  Useful Life
  Salvage Value
  Total Cost = Fixed Cost + Variable Cost
  Profit = Revenue − Cost
  Break-Even Point
  Demand Equation
  Supply Equation
  Market Equilibrium
  Equilibrium Quantity
  Equilibrium Price

REVIEW EXERCISES (CH. 2)

---

1. A major applicance store is having a spring sale on all items in the store. The sale price is 15% off of the list price. Write a function that relates the sale price to the list price.

2. A salesperson's weekly salary is $375 plus a commission of 10% on gross sales. Write an equation that expresses the salesperson's total weekly salary as a function of gross sales for the week.

3. Joe drives from home to his office, which is 32 miles away. He knows that he averages 52 miles per hour on a given trip. Express the distance that Joe is from his office in terms of time.

4. In 1980 an antique oak desk was appraised at $1275, and in 1984 it was worth $1595. Assuming that the value of the desk increases linearly with time, find an equation that describes this function.

5. A wheat combine originally cost $35,800. If the salvage value is $10,000 at the end of 12 years, what is the equation that expresses the straight-line depreciation?

6. A company that manufactures piano rolls has a fixed cost of $1750 and a variable cost of $1.25 per roll. The company sells its product at $4.45 per roll.

   a. Express the total cost of producing x piano rolls by a linear equation.

   b. Express the revenue of selling x piano rolls by a linear equation.

   c. Express the profit of selling x piano rolls by a linear equation.

7. When the price of halibut in Portland was $4.75 per pound, the demand was 18,000 pounds. But when the price rose to $5.45 per pound, the demand was 14,000 pounds. Write the demand equation assuming linearity.

8. In Pittsburgh the demand equation for wild rice is

$$p = -\frac{1}{3}q + 6$$

and the supply equation is given by

$$p = q - 4$$

where p is the price per pound and q is the quantity (in thousands of pounds).

   a. Find the equilibrium quantity and the equilibrium price.

   b. Draw the graphs of the demand and supply equations in the same coordinate system.

9. Solve the system of linear equations

$$5x + 7y = 265$$
$$3x + 2y = 115.$$

10. Solve the system of linear equations

$$\frac{3}{7}x + \frac{2}{5}y = 25$$

$$\frac{1}{3}x + \frac{3}{4}y = 37.$$

11. Derby Feed Mills makes two types of dog food, High-Protein and Regular. Each type of food is made from beef by-products and grain. It takes $\frac{5}{8}$ ton of beef by-products and $\frac{3}{8}$ ton of grain to make 1 ton of High-Protein and $\frac{1}{3}$ ton of beef by-products and $\frac{2}{3}$ ton of grain to make 1 ton of Regular.

   a. Express by a linear form the tons of beef by-products used as a function of the tons of High-Protein and tons of Regular that are made.

   b. Express by a linear form the tons of grain used as a function of the tons of High-Protein and tons of Regular that are made.

12. Refer to exercise 11. How many tons of each type of food can be made from 94.8 tons of beef by-products and 88.8 tons of grain?

13. An individual plans to divide $10,000 between two investments, one that pays 14% annually and another that pays 9% annually. How much must be invested in each type of investment in order to obtain exactly $1250 per year in income?

CHAPTER 3

Linear Programming: Graphical Method

Introduction

Linear programming is an important tool of applied mathematics. During World War II an Air Force research group, working on military logistics problems, developed planning techniques that later evolved into linear programming. Modern applications of linear programming can be found extensively in operations research, for example, in determining how a manufacturer could best use a supply of raw materials to achieve the largest possible profit.

In section 3.1, we begin our study with the graph of a system of linear inequalities. Section 3.2 deals with the graphical method for solving a linear programming problem and applications. In addition to being used to solve linear programming problems in two variables, the graphical method is helpful in understanding the algebraic method given in chapter 6.

3.1 Graphing Linear Inequalities

We begin by considering again the production problem given in chapter 2. A manufacturer of glass laboratory apparatus makes special retorts and titration pipets. Natural gas and labor are used in the production of each item. The manufacturer has available each week 575 man-hours of labor and 5000 cubic feet of natural gas. Also the production requirements are as follows: one retort requires 1 man-hour of labor and 10 cubic feet of gas, whereas one pipet requires 2 man-hours of labor and 15 cubic feet of gas. As considered in chapter 2, the manufacturer may wish to determine how many units of each apparatus it could produce so as to use completely the available labor and natural gas.

By letting

x = the number of special retorts to be produced

and $\qquad$ $y =$ the number of titration pipets to be produced,

the amount L of labor and the amount NG of natural gas used in production are represented by the linear forms $L = x + 2y$ and $NG = 10x + 15y$. To determine the production program that uses the available labor and natural gas completely, we solve the system of linear equations

$$
\begin{aligned}
x + \ 2y &= \ 575 \qquad \text{(labor)} \\
10x + 15y &= 5000 \qquad \text{(gas)}
\end{aligned}
$$

$\qquad$ (1)

obtaining the unique solution $x = 275$ and $y = 150$.

$\qquad$ Now consider an alternate question. Suppose that labor and natural gas are in limited supply; the manufacturer may then ask: "How many units of each apparatus should be produced so as not to exceed the available labor and natural gas?"

$\qquad$ Instead of requiring that the labor used in production be exactly 575 man-hours, this alternate question requires that it be no greater than 575. This requirement can be written as:

$$
x + 2y \leq 575 \qquad \text{(labor)}.
$$

Also the alternate requirement on the natural gas used in production is

$$
10x + 15y \leq 5000 \qquad \text{(gas)}.
$$

$\qquad$ Each of the alternate requirements on labor and natural gas is expressed by a linear inequality. By a **linear inequality in two variables** we mean a statement that can be expressed in one of the following forms:

$$
Ax + By \leq C
$$
$$
Ax + By < C
$$
$$
Ax + By \geq C
$$
$$
Ax + By > C.
$$

The coefficients A and B are real numbers that are not both zero, and the right-hand side C is a real number.

$\qquad$ Since, in our production situation, x and y represent units of apparatus to be produced, we also have the **natural conditions** that $x \geq 0$ and $y \geq 0$. Thus a production program (x, y) must satisfy each of the linear inequalities

$$
\begin{aligned}
x + \ 2y &\leq \ 575 \\
10x + 15y &\leq 5000 \\
x \geq 0 \qquad \text{and} &\qquad y \geq 0
\end{aligned}
$$

$\qquad$ (2)

called a **system of linear inequalities in two variables**.

$\qquad$ Any ordered pair (t_1, t_2) of real numbers is a **solution** to a system of linear inequalities in two variables if we obtain a true statement when we set $x = t_1$ and $y = t_2$ in each of the inequalities.

We see that $x = 275$ and $y = 100$, the solution to the system of equations (1) is also a solution to the system of inequalities (2). However the system of inequalities has many more solutions. For example, $(175, 200)$ is also a solution to (2), since it is true that

$$175 + 2(200) = 575 \leq 575$$

$$10(75) + 15(200) = 4750 \leq 5000$$

$$175 \geq 0 \quad \text{and} \quad 200 \geq 0.$$

Other examples of solutions to (2) are $(0, 0)$, $(500, 0)$, and $(100, 100)$. When considering a system of linear inequalities as a model for a described situation, we say that any one of the solutions to the system is a **feasible solution** (or **feasible program**).

We want to display in a Cartesian plane the solutions to a system of linear inequalities, that is, **graph** a system of linear inequalities.

EXAMPLE 1

Graph the linear inequality $2x + y \leq 10$.

The meaning of $\leq$ tells us that a pair (x, y) satisfies this inequality if either the number $2x + y$ is the same as 10 or is less than 10. The points for which the first condition is true are those on the line $2x + y = 10$.

To locate the points for which $2x + y < 10$, first consider a point on the line $2x + y = 10$ (see figure 3.1). By starting at a point on the line, we realize as we move below this point that x remains the same and y becomes smaller; therefore the number $2x + y$ becomes smaller. Since $2x + y$ is 10 at a point on the line, $2x + y$ is less than 10 at points below the line. Thus the graph of the linear inequality $2x + y \leq 10$ is the region of points on the line $2x + y = 10$ and those below it. We indicate the graph by shading this region as shown in figure 3.1.

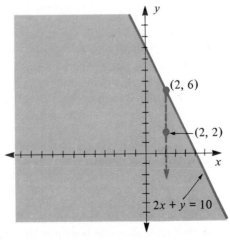

FIGURE 3.1

Graph of $2x + y \leqslant 10$

A line $Ax + By = C$ divides the coordinate plane into two regions called **half-planes**. The line $Ax + By = C$ is called the **boundary line**. The points in one of the half-planes satisfy the linear inequality $Ax + By < C$ and the points in the other half-plane satisfy $Ax + By > C$ (see figure 3.2). Therefore the graph of a linear inequality $Ax + By \leq C$ or $Ax + By \geq C$ is one of the two half-planes determined by the boundary line $Ax + By = C$ and includes the line. For an inequality $Ax + By < C$ or $Ax + By > C$, we do not include the boundary line $Ax + By = C$ in the graph.

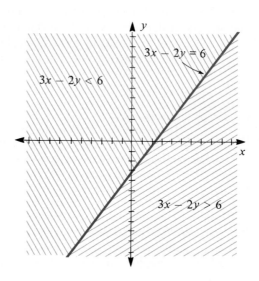

FIGURE 3.2

Graphing a Linear Inequality

1. Graph the boundary line $Ax + By = C$. Draw a solid line if the inequality is $\leq$ or $\geq$; otherwise draw a dotted line.

2. Pick a point in one of the half-planes determined by the boundary line. If the coordinates of this point satisfy the inequality, then all points in this half-plane are also included in the graph. Otherwise the half-plane on the other side of the boundary line belongs to the graph.

EXAMPLE 2

Graph the linear inequality $5x - 3y \leq 15$.
　　We first graph the boundary line $5x - 3y = 15$ as a solid line. Next we test the inequality at a point in one of the half-planes, such as at the point $(0, 0)$, which is in the half-plane above the line. Since $5(0) - 3(0) = 0 < 15$, all points in the same half-plane as $(0, 0)$ will also satisfy the inequality. Thus the graph of the inequality is the half-plane above the boundary line and includes the line (see figure 3.3).

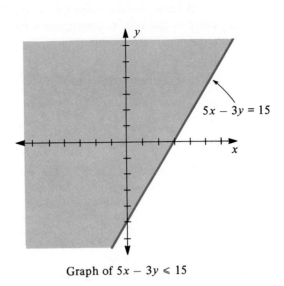

FIGURE 3.3

Graph of $5x - 3y \leqslant 15$

The point $(0, 0)$ is convenient for testing the inequality if it is not on the boundary line. If $(0, 0)$ is on the boundary line, we must select some other point.

EXAMPLE 3

Graph the linear inequality $2x - 3y > 6$.

We graph the boundary line $2x - 3y = 6$ but draw it as a dotted line. Next we test the inequality at the point $(0, 0)$ and find that the inequality is not satisfied at this point. Thus the graph is the half-plane that does not contain $(0, 0)$, and we shade this region (see figure 3.4).

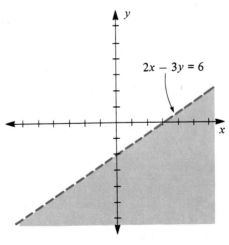

FIGURE 3.4

Graph of $2x - 3y > 6$

We now turn our attention to graphing a system of linear inequalities. Each inequality in the system determines a half-plane; the graph of the system is the intersection of these half-planes.

EXAMPLE 4

Graph the system of linear inequalities

$$x + \ y \le 12$$
$$2x + 5y \le 30$$
$$x \ge 0 \quad \text{and} \quad y \ge 0.$$

Since $x \ge 0$ and $y \ge 0$, the graph of the system is in the first quadrant including the non-negative x and y axes.

Next we graph the region determined by the inequality $x + y \le 12$. Instead of shading the region, we draw an arrow to indicate the side of the boundary line $x + y = 12$ on which the half-plane lies. Also we graph on the same plane the region corresponding to $2x + 5y \le 30$; again we only draw an arrow in place of the shading (see figure 3.5a).

Finally we look for the intersection of these half-planes, which is above the x-axis, to the right of the y-axis, below the lines $x + y = 12$ and $2x + 5y = 30$. We complete the graph by shading this area as shown in figure 3.5b. The boundary lines of this shaded area are included in the graph.

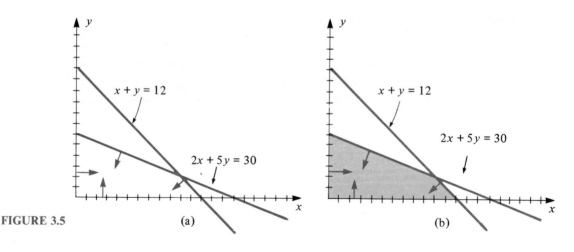

FIGURE 3.5   (a)   (b)

The graph of the system of inequalities in figure 3.5 is called a **convex polygonal set**. A region of the cartesian plane is a **convex set** if for any two points in the region, the line segment joining the points is completely contained in the region. The set shown in figure 3.6 is not convex since the line segment joining the points P_1 and P_2 does not lie completely in the set.

Furthermore a convex set in a plane is **polygonal** if the boundaries of the region are straight-line segments. The points of intersection of the boundary segments are called **vertices** of the polygonal region. The convex polygonal set shown in figure 3.5 has four vertices, the points $(0, 0)$, $(0, 6)$, $(10, 2)$, and $(12, 0)$.

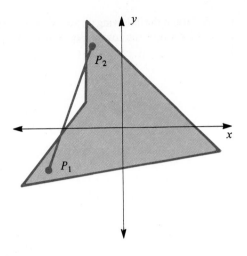

FIGURE 3.6

Always, if the graph of a system of linear inequalities is not empty, then it is a convex polygonal set.

EXAMPLES

5. Graph the system of linear inequalities

$$2x - y \geq \quad 2$$
$$x + y \leq -2$$
$$x \geq 0 \quad \text{and} \quad y \geq 0.$$

The graph of the region determined by $2x - y \geq 2$ and $x + y \leq -2$ is shown in figure 3.7. We see that there does not exist a pair (x, y) which can satisfy the non-negative conditions $x \geq 0$ and $y \geq 0$ and also satisfy the first two inequalities in the system. Thus the graph of this system is empty.

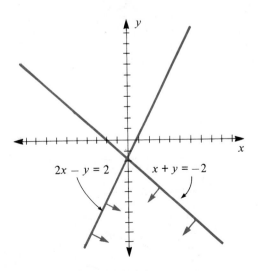

FIGURE 3.7

6. Graph the following system of linear inequalities; if the set is not empty, give the vertices of the convex polygonal set.

$$8x + 5y \leq 120$$

$$x + y \leq 18$$

$$2x + 4y \leq 59$$

$$x \geq 0 \quad \text{and} \quad y \geq 0.$$

The region determined by the system of inequalities is the intersection of the half-planes below the boundary lines $8x + 5y = 120$, $x + y = 18$, and $2x + 4y = 59$, above the x-axis and to the right of the y-axis; the graph also includes the boundary line segments (see figure 3.8).

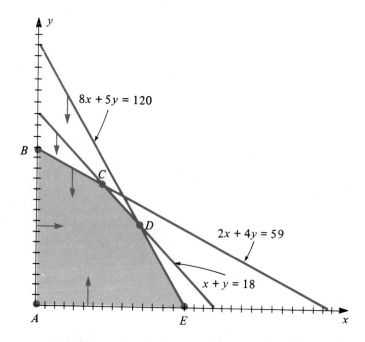

FIGURE 3.8

The graph of the system is a convex polygonal set. The vertices of this set are at the points of intersection of pairs of boundary lines. We see from the graph which pairs of boundary lines intersect at a vertex; the coordinates of the vertex are obtained by solving the system of linear equations corresponding to the pair of boundary lines, as shown in table 3.1

TABLE 3.1

Vertex	At intersection of boundary lines (see figure 3.8)	Coordinates of vertex (solution of system of equations)
A	$x = 0$ and $y = 0$	$(0, 0)$
B	$x = 0$ and $2x + 4y = 59$	$(0, 14.75)$
C	$x + y = 18$ and $2x + 4y = 59$	$(6.5, 11.5)$
D	$x + y = 18$ and $8x + 5y = 120$	$(10, 8)$
E	$y = 0$ and $8x + 5y = 120$	$(15, 0)$

We can now display the set of feasible solutions to our production situation described earlier. The graph of the system

$$x + 2y \le 575$$

$$10x + 15y \le 5000$$

$$x \ge 0 \quad \text{and} \quad y \ge 0$$

is shown in figure 3.9.

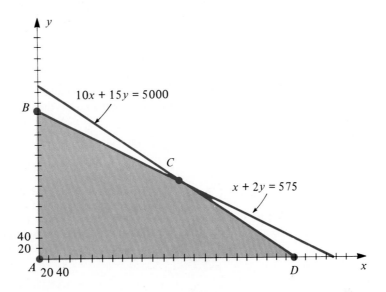

FIGURE 3.9

The region of feasible production programs is a convex polygonal set, and we can see in figure 3.9 which pair of boundary lines intersect at a vertex. Solving the system of equations corresponding to these pairs of boundary lines, we obtain the coordinates of the vertices, as shown in table 3.2.

TABLE 3.2

Vertex	At intersection of boundary lines (system of linear equations)	Coordinates of vertex (solution of system)
A	$x = 0$ and $y = 0$	$(0, 0)$
B	$x = 0$ and $x + 2y = 575$	$(0, 287.5)$
C	$10x + 15y = 5000$ $x + 2y = 575$	$(275, 150)$
D	$10x + 15y = 5000$ and $y = 0$	$(500, 0)$

EXAMPLE 7

A backpacker is planning a wilderness trip on which he plans to eat only a combination of cereal and nuts. The recommended minimum daily requirement of calories and protein is 3000 calories and 2.36 ounces of protein. Each ounce of cereal supplies 120 calories and 0.04 ounce of protein, whereas each ounce of nuts supplies 60 calories and 0.12 ounce of protein. Display in a Cartesian plane the set of feasible diets (combination of cereal and nuts) that meet or exceed the daily requirement of calories and protein.

Let

$$C = \text{the ounces of cereal in the daily diet}$$

and

$$N = \text{the ounces of nuts in the daily diet.}$$

In terms of C and N we can represent the calories and the ounces of protein in the diet. Since cereal is contributing calories at the rate of 120 per ounce, and nuts are contributing calories at the rate of 60 per ounce, we have that the calories in the diet can be represented by

$$\text{calories} = 120C + 60N.$$

Also the ounces of protein in the diet can be represented by a linear form,

$$\text{protein} = 0.04C + 0.12N.$$

Since a diet is feasible when the number of calories is no less than 3000, and the number of ounces of protein is no less than 2.36, we must pick C and N so that

$$
\begin{aligned}
120C + \quad 60N &\geq 3000 \\
0.04C + 0.12N &\geq \quad 2.36 \\
C \geq 0 \quad \text{and} \quad N &\geq 0.
\end{aligned}
\qquad (3)
$$

We must include the non-negative restrictions on C and N since they represent ounces of food. Thus, a feasible diet (C, N) is any solution to the system of linear inequalities (3). The graph of the set of feasible solutions is given in figure 3.10.

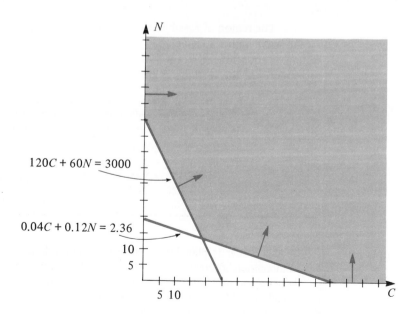

FIGURE 3.10

The region of feasible solutions in figure 3.10 is **unbounded**, whereas the region in figure 3.9 is **bounded**. A convex polygonal set in the plane is **bounded** if it can be enclosed within a circle with center at $(0,0)$; if a region cannot be enclosed within such a circle, then it is **unbounded** (see figure 3.11).

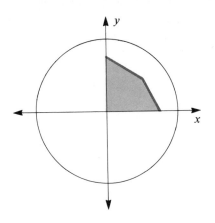

FIGURE 3.11   (a) Bounded region   (b) Unbounded region

EXERCISES 3.1

For exercises 1–10 graph the linear inequality.

1. $3x + 2y \leq 12$

2. $3x + 2y > 12$

3. $x \geq y$

4. $y \leq 2x + 1$

5. $2x + 5y < 10$

6. $4x - 2y \geq 16$

7. $-8x + 3y \leq 24$

8. $2x - 9y \geq 18$

9. $\dfrac{6}{5}x - \dfrac{4}{5}y \geq 6$

10. $\dfrac{5}{6}x + \dfrac{3}{7}y < 15$

For exercises 11–28 graph the given system of linear inequalities. Also find the coordinates of the vertices of the region, if the region is non-empty.

11. $y \leq 5$
    $x \leq 4$
    $x \geq 0$ and $y \geq 0$

12. $y \leq 6$
    $x \geq 2$
    $x \geq 0$ and $y \geq 0$

13. $2x + 3y \leq 15$
    $y \geq x$
    $x \geq 0$ and $y \geq 0$

14. $x + y \geq 3$
    $y \leq 2x$
    $x \geq 0$ and $y \geq 0$

15. $5x + 9y \leq 90$
    $15x + 9y \leq 180$
    $x \geq 0$ and $y \geq 0$

16. $3x + 2y \geq 24$
    $-7x + 15y \leq 270$
    $x \geq 0$ and $y \geq 0$

17. $4x + 7y \geq 63$
$5x + 3y \geq 50$
$x \geq 0$ and $y \geq 0$

18. $-x + 2y \leq 8$
$7x + 4y \leq 70$
$x \geq 0$ and $y \geq 0$

19. $-3x + y \geq -5$
$x + y \geq 7$
$x \geq 0$ and $y \geq 0$

20. $-2x + y \leq 5$
$x - 2y \geq -16$
$x \geq 0$ and $y \geq 0$

21. $2x + y \leq 16$
$x + y \leq 14$
$x \leq 6$
$x \geq 0$ and $y \geq 0$

22. $x + y \leq 13$
$-4x + 3y \geq -24$
$y \leq 8$
$x \geq 0$ and $y \geq 0$

23. $5x + 4y \geq 60$
$5x + 8y \geq 100$
$5x + 14y \geq 130$
$x \geq 0$ and $y \geq 0$

24. $-6x + 3y \leq 1$
$4x + 6y \leq 7$
$22x + 39y \leq 28$
$x \geq 0$ and $y \geq 0$

25. $x + 3y \leq 26$
$4x + 3y \leq 44$
$2x + 3y \leq 28$
$x \geq 0$ and $y \geq 0$

26. $5x + 4y \geq 60$
$5x + 8y \geq 100$
$5x + 14y \geq 130$
$x \geq 0$ and $y \geq 0$

27. $x + y \leq 19$
$5x + 8y \geq 80$
$x + 4y \leq 32$
$x \geq 0$ and $y \geq 0$

28. $3x + 2y \geq 34$
$x + y \geq 15$
$3x + 8y \geq 60$
$x \geq 0$ and $y \geq 0$

29. In the odd-numbered exercises 11–27 state which of the graphs are bounded and which are not.

30. In the even-numbered exercises 12–28 state which of the graphs are bounded and which are not.

31. Linda, who owns the Apple Hill Shoppe, wants her weekly payroll to be no more than $1026. She pays $4.00 per regular hour and $5.40 per overtime hour. Also she wants the number of overtime hours to be at most $\frac{2}{3}$ times the number of regular hours. Let

$$x = \text{the number of regular hours scheduled in a week}$$

and $\qquad y = \text{the number of overtime hours scheduled in a week.}$

a. Give a system of linear inequalities in terms of x and y that models the given information.

b. Graph the system written in part (a), and find the coordinates of the vertices of the region.

32. Jodi wants to invest up to $10,000 in a combination of two types of mutual funds, performance and growth. She wants to invest at least $2000 in the growth fund, but the amount should not exceed the amount invested in the performance fund. Let

$$x = \text{the amount invested in the performance fund}$$

and $\qquad y = \text{the amount invested in the growth fund.}$

a. Model the given information with a system of linear inequalities involving x and y.

b. Display by a graph the feasible amounts of money Jodi can invest in the two types of mutual funds as determined by the system of part (a), and find the coordinates of the region of feasible amounts.

33. The Tennessee Chainsaw Company makes two types of saws, a utility model and a deluxe model. Two machines, A and B, are used in the production of each model. The number of hours needed on each machine to produce one dozen of each model of chainsaw and the maximum number of hours available each day for each machine are given in table 3.3.

TABLE 3.3

	Utility model	Deluxe model	Hours available
Machine A	2	3	9
Machine B	3	1	10

  a. Model the given information with a system of linear inequalities.

  b. Display by a graph the feasible number (in dozens) of chainsaws of each type that can be made each day.

34. Joe must follow a diet that contains at least 125 milligrams of vitamin B_1 and at least 150 milligrams of vitamin B_2 each day. Two food supplements, I and II, are used to obtain these requirements. The milligrams of B_1 and B_2 contained in each packet of supplements are given in table 3.4.

TABLE 3.4

	B_1 (milligrams)	B_2 (milligrams)
Supplement I	15	15
Supplement II	10	20

  a. Model the number of packets of supplements needed to meet the stated requirements of vitamins by a system of linear inequalities.

  b. Display by a graph the feasible units of each supplement that Joe should take each day so as to meet or exceed the stated requirements of vitamins B_1 and B_2.

35. Petro Chemicals makes two types of plastic, medium-density and high-density. Each type of plastic is made from two chemicals, A and B. Each pound of medium-density plastic requires 0.5 pound of chemical A and 0.5 pound of chemical B, whereas each pound of high-density plastic requires 0.7 pound of chemical A and 0.3 pound of chemical B. Petro Chemicals has available at most 2100 pounds of chemical A and 1500 pounds of chemical B. Display with a graph the feasible number of pounds of each type of plastic that can be made from the available chemicals.

36. A dietician is planning a dinner that is to consist of a cheese dish and a green vegetable. The dietician wants the dinner to supply at least 1500 milligrams of calcium and 6 milligrams of iron. Each ounce of the cheese dish supplies 100 milligrams of calcium and 0.25 milligram of iron, whereas each ounce of the green vegetable supplies 20 milligrams of calcium and 0.5 milligram of iron. Display with a graph the feasible ounces of each food; that is, the ounces of the cheese dish and the ounces of the green vegetable that should be eaten so as to meet or exceed the stated requirements of calcium and iron.

37. Refer to exercise 35. Suppose in addition to the limitations of the amounts of chemicals available, the amount of labor available to make the plastic is also limited. Petro Chemicals has up to 340 man-hours available, and it takes 0.1 man-hour to make one pound of either type of plastic. Display with a graph the feasible number of pounds of each type plastic that can be made from the available chemicals and labor.

38. Refer to exercise 36. Suppose in addition to the calcium and iron requirements, the dietician wants the dinner to supply at least 72 grams of protein. Each ounce of the cheese dish supplies 4 grams of protein, and each ounce of the green vegetable supplies 2 grams of protein. Display with a graph the feasible diets including this added condition.

3.2 Graphical Solution of Linear Programming Problems

Recall our discussion in section 3.1 on the number of retorts and pipets the manufacturer could make with the available quantities of natural gas and labor. The feasible production programs are the solutions to the system of linear inequalities

$$x + 2y \leq 575$$

$$10x + 15y \leq 5000$$

$$x \geq 0 \quad \text{and} \quad y \geq 0$$

where x = the number of retorts to be produced

and y = the number of pipets to be produced.

Since there are an infinite number of feasible solutions to the system, we may now ask whether one is better than all others. Therefore let us consider a further requirement on the choice of a production program.

Profit is one concern to a manufacturer. Therefore the manufacturer may wish to obtain a feasible solution that also yields the largest profit. Suppose the profit on each retort is $20 and the profit on each pipet is $18. With this added requirement, we see of the two solutions (275, 150) and (500, 0), the second solution gives a larger profit. The profit when $x = 500$ and $y = 0$ is $20(500) + 18(0) = $10,000$, whereas the profit when $x = 275$ and $y = 150$ is $20(275) + 18(150) = 8200.

The profit for x retorts and y pipets is represented by the linear form $P = 20x + 18y$. Thus we can model the production problem of picking a program for which the available quantities of natural gas and labor are not exceeded and the profit is largest by the following mathematical problem: Determine among the solutions to the system of linear inequalities

$$x + 2y \leq 575$$

$$10x + 15y \leq 5000$$

one that satisfies the non-negative conditions

$$x \geq 0 \quad \text{and} \quad y \geq 0$$

and maximizes the quantity

$$P = 20x + 18y.$$

The mathematical problem just stated is an example of a linear programming problem in two variables.

Definition   ■   A **linear programming problem in two variables** is to determine, if possible, from among the solutions to a system of linear inequalities in two variables, one that satisfies the non-negative conditions

$$x \geq 0 \quad \text{and} \quad y \geq 0$$

and for which a number of the form

$$z = ax + by$$

is optimal (either maximum or minimum, depending on the problem).

The inequalities to be satisfied are called **constraints**, and the linear form $z = ax + by$ is called the **objective function** of the problem. A solution to the system of linear inequalities that also satisfies the non-negative conditions is called **feasible**. A feasible solution, if any, for which the objective function is optimal, is called an **optimal solution** and the value of the objective function at an optimal solution is called the **optimal value**.   ■

The glass manufacturer's production problem çan be modeled as a linear programming problem. We shall consider other applications after studying a method for solving a linear programming problem in two variables. Let us continue with our production problem and discuss how a solution can be obtained.

It is clear that the program represented by the point $(0,0)$, the program of making nothing, yields no profit. Therefore we begin with a program represented by a point (x, y) in the region of feasible solutions that yields a positive profit. For example, we see from figure 3.12a that $(300, 0)$ is such a point, and the value of P at this point is $20(300) + 18(0) = 6000$. This value for P can be attained at other points in the region; in fact, $P = 6000$ at all points in the region and on the line $20x + 18y = 6000$ (see figure 3.12a).

From graphing linear inequalities, we see that P is greater than 6000 at points in the half-plane above the line $20x + 18y = 6000$. Since our objective is for P to be as large as possible, we can eliminate those points in the half-plane $20x + 18y < 6000$ as candidates for an optimal solution, as we see in figure 3.12b.

We now move to a point in the region of feasible solutions and in the half-plane $20x + 18y > 6000$. Such a point, for example, is $(400, 0)$, and the value of P at this point is 8000. The value of P is also 8000 at feasible solutions on the line $20x + 18y = 8000$. As before, we see from figure 3.12c that those points in the region of feasible solutions and below the line $20x + 18y = 8000$ can be eliminated as candidates for an optimal solution.

Let us now move to the vertex $D(500, 0)$ which is in the half-plane $20x + 18y > 8000$. The value of P at D is 10,000. Furthermore all points in the region of feasible solutions, except D, are in the half-plane $20x + 18y < 10,000$ (see figure 3.12d). Therefore an optimal solution for our problem exists at the vertex $D(500, 0)$, and the optimal value for P is 10,000.

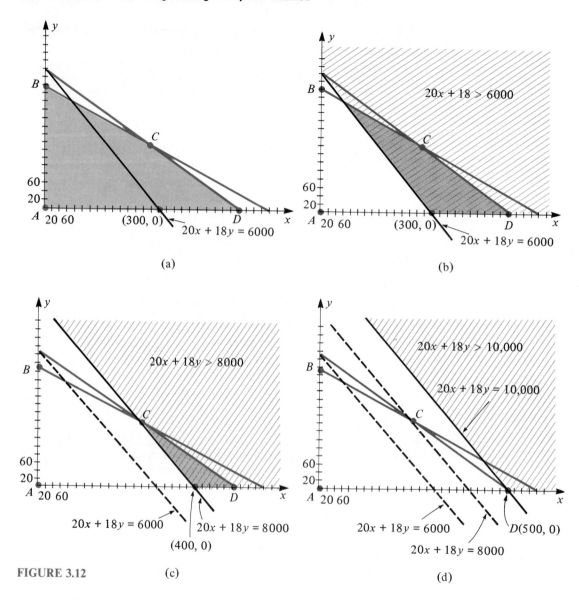

FIGURE 3.12

(a)

(b)

(c)

(d)

Our production problem is an example of what is called a **standard maximization problem**. A standard maximization problem is one in which we are to maximize the objective function over a region of feasible solutions determined from the natural conditions $x \geq 0$ and $y \geq 0$ and constraints of the form

$$a_{1i}x + a_{2i}y \leq b_i, \quad \text{where } b_i \geq 0.$$

As seen with our production problem, if the region of feasible solutions is non-empty and bounded, then the problem will have an optimal solution, and it will occur at one of the vertices of the region of feasible solutions.

Another type of problem that is of interest to us is a **standard minimization problem**. Such a problem is one in which we are to minimize an objective function where the coefficients are non-negative. Furthermore the region of feasible solutions is determined from the natural conditions $x \geq 0$ and $y \geq 0$ and constraints of the form

$$a_{1i}x + a_{2i}y \geq b_i.$$

Note that there is no restriction on the right-hand side b_i. It happens that if the region of feasible solutions of a standard minimization problem is non-empty, the problem has an optimal value and it will also occur at one of the vertices of the region of feasible solutions. Let us see why this is the case using the following problem:

$$\text{Minimize} \quad z = 3x + 2y$$

$$\text{subject to} \quad 4x + \ y \geq 31$$

$$x + 5y \geq 41$$

$$x \geq 0 \quad \text{and} \quad y \geq 0.$$

The region of feasible solutions and its vertices are given in figure 3.13a. We evaluate the objective function at each of the vertices and summarize our results in table 3.5. We can see in this table that the smallest value of the objective function at the vertices of the region is attained at B. However, we need to look at the situation a little closer to see that the problem does indeed have an optimal solution and it occurs at the vertex B.

Let us draw the line where the objective function has the value 32, that is, the line $3x + 2y = 32$. In doing so, we see that all points in the region of feasible solutions, except the vertex B, lie in the half-plane for which $3x + 2y > 32$ (see figure 3.13b). Therefore we can conclude that 32 is the smallest value the objective function can attain over the region of feasible solutions. Thus the problem has an optimal solution, namely, $z = 32$ occurring at $B(6, 7)$.

TABLE 3.5

Vertex	At intersection of boundary lines (system of equations)	Coordinates of vertex (solution of system)	Value of objective function
A	$x = 0$ $4x + \ y = 31$	$(0, 31)$	$z = 3(0) + 2(31) = 62$
B	$4x + \ y = 31$ $x + 5y = 41$	$(6, 7)$	$z = 3(6) + 2(7) = 32$
C	$y = 0$ $x + 5y = 41$	$(41, 0)$	$z = 3(41) + 2(0) = 123$

The discussion of the two preceding examples helps in understanding the following theorem, which gives the basis for the method used in this chapter for solving a linear programming problem.

Theorem   ■ (Standard Maximization Problem)   If the region of feasible solutions is non-empty and bounded, the problem has an optimal solution and it occurs at a vertex of the region.

(Standard Minimization Problem)   If the region of feasible solutions is non-empty, the problem has an optimal solution and it occurs at a vertex of the region.

(Other Problems)   If the region of feasible solutions is non-empty and bounded, the problem will have a solution and it will occur at a vertex of the region. On the other hand, if the region is non-empty but unbounded, the problem may not have an optimal solution; however, if it does, it must also occur at a vertex of the region. ■

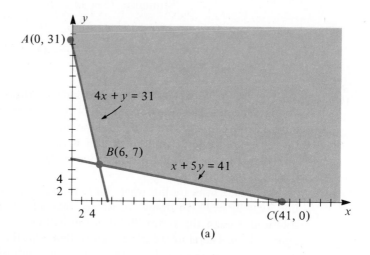

(a)

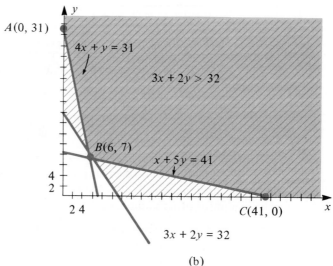

FIGURE 3.13

(b)

We apply the theorem in the following way.

Steps of the Graphical Method of Solution

1. Graph the region of feasible solutions.
2. If the graph is non-empty, then determine, by solving systems of two linear equations, the vertices of the convex polygonal set obtained in step 1 and go to step 3. Otherwise, the problem has no solution.
3. Evaluate the objective function at each of the vertices determined in step 2.
4. Select the largest (smallest) value from those determined in step 3. If the region of feasible solutions is bounded, then this value is the optimal value for the objective function and a vertex at which this value is attained is an optimal solution. If the region is unbounded but the problem is a standard minimization problem, then also this value is the optimal value for the objective function and a vertex at which this value occurs is an optimal solution. In any other case, a further analysis must be done.

EXAMPLES

1. Solve, using the graphical method, the linear programming problem: Determine the maximal value for $z = 3x + 7y$, where (x, y) must satisfy

$$3x + y \leq 27$$

$$x + 2y \leq 24$$

$$x \geq 0 \quad \text{and} \quad y \geq 0.$$

Step 1: The graph of the region of feasible solutions is given in figure 3.14.

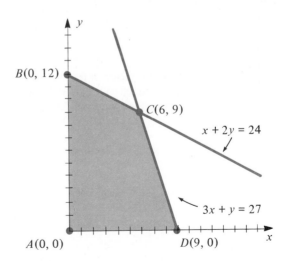

FIGURE 3.14

Step 2: We see in figure 3.14 that the region of feasible solutions is non-empty and bounded. Furthermore the pairs of boundary lines that intersect at vertices can also be seen in figure 3.14. Next we solve these systems of linear equations for the coordinates of the vertices. These results are summarized in table 3.6.

TABLE 3.6

Vertex	At intersection of boundary lines (system of equations)	Coordinates of vertex (solution of system)
A	$x = 0$ $y = 0$	$(0, 0)$
B	$x = 0$ $x + 2y = 24$	$(0, 12)$
C	$x + 2y = 24$ $3x + y = 27$	$(6, 9)$
D	$y = 0$ $3x + y = 27$	$(9, 0)$

Step 3: We next evaluate the objective function at each of the vertices of the region of feasible solutions. By adding another column to our table of information, we obtain table 3.7.

TABLE 3.7

Vertex	At intersection of boundary lines (system of equations)	Coordinates of vertex (solution of system)	Value of objective function
A	$x = 0$ $y = 0$	$(0, 0)$	$z = 3(0) + 7(0) = 0$
B	$x = 0$ $x + 2y = 24$	$(0, 12)$	$z = 3(0) + 7(12) = 84$
C	$x + 2y = 24$ $3x + y = 27$	$(6, 9)$	$z = 3(6) + 7(9) = 81$
D	$y = 0$ $3x + y = 27$	$(9, 0)$	$z = 3(9) + 7(0) = 27$

Step 4: Since the region of feasible solutions is bounded, the objective function has an optimal value over this region. Furthermore, since we wish to maximize z, we see from table 3.7 that the optimal value is 84 and it is attained at the vertex $x = 0$ and $y = 12$.

2. Solve by the graphical method the linear programming problem:

$$\text{Maximize} \quad z = x + 2y,$$
$$\text{where} \quad 3x + y \leq 27$$
$$x + 2y \leq 24$$
$$x \geq 0 \quad \text{and} \quad y \geq 0.$$

Steps 1, 2: Since the region of feasible solutions is the same as that in example 1, the necessary information on vertices of the region is given in table 3.6.
Step 3: We evaluate the objective function at each of the vertices, the results of which are given in table 3.8.

TABLE 3.8

Vertex	At intersection of boundary lines (system of equations)	Coordinates of vertex (solution of system)	Value of objective function
A	$x = 0$ $y = 0$	$(0, 0)$	$z = 0 + 2(0) = 0$
B	$x = 0$ $x + 2y = 24$	$(0, 12)$	$z = 0 + 2(12) = 24$
C	$x + 2y = 24$ $3x + y = 27$	$(6, 9)$	$z = 6 + 2(9) = 24$
D	$y = 0$ $3x + y = 27$	$(9, 0)$	$z = 9 + 2(0) = 9$

Step 4: Since the region is bounded and since, as we see from table 3.8, the largest value of z at the vertices of the region is 24, the optimal value exists and it is 24. Furthermore this optimal value is attained at two vertices, namely, *B* and *C*. In fact, all points in the region on the boundary line $x + 2y = 24$ are optimal solutions.

3. Solve by the graphical method the linear programming problem:

$$\text{Minimize} \quad z = 3x + 2y$$

$$\text{subject to} \quad 9x + 7y \geq 157$$

$$9x + 5y \geq 125$$

$$7x + 8y \geq 140$$

$$x \geq 0 \quad \text{and} \quad y \geq 0.$$

Steps 1, 2: The region of feasible solutions and its vertices are given in figure 3.15. Since the region is non-empty, even though it is unbounded, we proceed to step 3.

Step 3: We evaluate the objective function at each of the vertices and summarize our results in table 3.9.

TABLE 3.9

Vertex	At intersection of boundary lines (system of equations)	Coordinates of vertex (solution of system)	Value of objective function
A	$y = 0$ $7x + 8y = 140$	$(20, 0)$	$z = 3(20) + 2(0) = 60$
B	$7x + 8y = 140$ $9x + 7y = 157$	$(12, 7)$	$z = 3(12) + 2(7) = 50$
C	$9x + 7y = 157$ $9x + 5y = 125$	$(5, 16)$	$z = 3(5) + 2(16) = 47$
D	$x = 0$ $9x + 5y = 125$	$(0, 25)$	$z = 3(0) + 2(25) = 50$

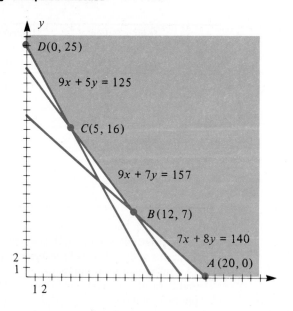

FIGURE 3.15

Step 4: From table 3.9 we see that the smallest of the values of the objective function at vertices is $z = 47$ and it occurs at the vertex $C(5, 16)$. Since the problem is a standard minimization problem, this value is the optimal value of the objective function. Thus the problem has a solution and it is $z = 47$ occurring at $C(5, 16)$.

Not all linear programming problems have a solution. The problem may fail to have a solution when the region of feasible solutions is empty; example 5 of section 3.1 has such a region. Another way in which a problem may fail to have an optimal solution is when the region of feasible solutions is non-empty but unbounded. In this case, if the objective function is to be maximized, it may be possible to make its value arbitrarily large, as illustrated in the next example.

EXAMPLE 4

Consider the linear programming problem:

$$\text{Maximize} \quad z = 2x + y,$$
$$\text{subject to} \quad -2x + y \leq 1$$
$$-x + y \leq 2$$
$$x \geq 0 \quad \text{and} \quad y \geq 0.$$

The region of feasible solutions is given in figure 3.16, which is an unbounded region.

The values of the objective function at the vertices $A(0, 0)$, $B(0, 1)$ and $C(3, 1)$ are 0, 1, and 7, respectively. By drawing in the line where the objective function attains the largest of these values, namely the line $2x + y = 7$, we see in figure 3.16 that there are many feasible solutions in the half-plane for which

$2x + y > 7$. Therefore the objective function can attain any large value and the problem does not have an optimal value.

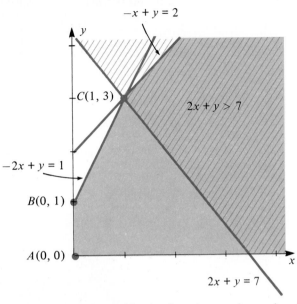

FIGURE 3.16

Whenever the value of the objective function can be made arbitrarily large, as in example 4, we say that the problem has an **unbounded solution**.

In linear programming problems, one of the following situations may occur, as illustrated by the previous examples.

1. The problem has no feasible solutions (see example 5 in section 3.1).
2. The problem has an unbounded solution (example 4).
3. The problem has a unique optimal solution (examples 1 and 3).
4. The problem has an optimal solution but it is not unique (example 2).

Linear programming problems that are models of real-world situations generally have solutions. For example, it is highly unlikely that an unbounded solution would arise when we wish to maximize profit since the set of feasible programs would be limited by the constraints of available resources. However, errors could arise in the formulation of the model thus creating a linear programming problem with no optimal solution. Therefore the method we use should handle any type of problem—even those with no optimal solution. As seen in the examples, when solving a problem in two variables, the graphical method will either give an optimal solution, or indicate that there are no feasible solutions or that the problem has an unbounded solution. In chapter 6 we give the simplex method for solving a linear programming problem. This algebraic procedure can also be applied to any one of these type problems.

We close this section with several applications.

5. *A Diet Problem.*   Consider again example 7 in section 3.1. In that example we determined the feasible diets for a backpacker who was planning to eat only a combination of cereal and nuts on a wilderness trip. Recall that a feasible diet is one that satisfies the system of linear inequalities

$$120C + 12N \geq 3000$$

$$0.04C + 0.12N \geq 2.36$$

$$C \geq 0 \quad \text{and} \quad N \geq 0,$$

where C = the ounces of cereal in the daily diet

and N = the ounces of nuts in the daily diet.

The graph of this system is given again in figure 3.17.

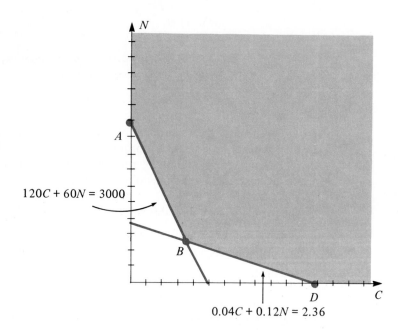

FIGURE 3.17

A concern of a backpacker is the weight that must be carried. Therefore solve the problem by finding a feasible diet that minimizes the weight

$$w = C + N.$$

Even though the region of feasible solutions is unbounded, as seen in figure 3.17, the problem has an optimal solution since it is a standard minimization problem. Thus let us find the coordinates of the vertices and evaluate the objective function at each.

To make calculations easier, we observe that the boundary lines $120C + 60N = 3000$ and $0.04C + 0.12N = 2.36$ are equivalent to $2C + N = 50$

and $C + 3N = 59$, respectively. Solving the systems of equations for the coordinates of the vertices, we obtain $A(0, 50)$, $B(18.2, 13.6)$, and $D(59, 0)$.

Evaluating the objective function $w = C + N$ at each of these vertices, we see that the smallest weight is $18.2 + 13.6 = 31.8$. Therefore the backpacker should plan to carry 18.2 ounces of cereal and 13.6 ounces of nuts for each day of the trip.

6. *An Investment Problem.* An investor has up to $10,000 available to invest in two different types of securities, one that yields 8% per year and another that yields 13% per year. Although the investor wants the largest possible income per year from the investments, she has restrictions on how the money is to be invested. She does not want to invest more than $2000 in the security paying 13% and the amount invested in the security paying 8% has to be at least twice that invested in the 13% security. Find an investment program for this investor.

We begin our model with defining the unknowns. Let

$$x = \text{the amount invested at } 8\% \text{ per year}$$

and

$$y = \text{the amount invested at } 13\% \text{ per year.}$$

Two of the restrictions on the choice of investment plans are seen to be $x + y \le 10,000$ and $y \le 2000$. The condition that x be at least twice y translates into $x \ge 2y$. Including the natural conditions on x and y, we see that a feasible investment program is one satisfying the system

$$x + y \le 10,000$$

$$y \le 2000$$

$$x \ge 2y$$

(4)

$$x \ge 0 \quad \text{and} \quad y \ge 0.$$

Now consider the objective function. Since the income on each dollar invested at 8% is 0.08 and since the income on each dollar invested at 13% is 0.13, the total income can be expressed by the linear form $I = 0.08x + 0.13y$. Therefore our problem is to find a feasible investment program (x, y) that satisfies the system of linear inequalities (4), for which the income

$$I = 0.08x + 0.13y$$

is maximum.

The graph of feasible solutions, along with the vertices of the region, is shown in figure 3.18.

The values of the objective function at the vertices are:

at $A(0, 0)$	$I = 0.08(0)$	$+ 0.13(0)$	$= 0$
at $B(4000, 2000)$	$I = 0.08(4000)$	$+ 0.13(2000)$	$= 580$
at $C(8000, 2000)$	$I = 0.08(8000)$	$+ 0.13(2000)$	$= 900$
at $D(10,000, 0)$	$I = 0.08(10,000)$	$+ 0.13(0)$	$= 800.$

Thus the investor should invest $8000 at 8% and $2000 at 13%, optimizing income at $900.

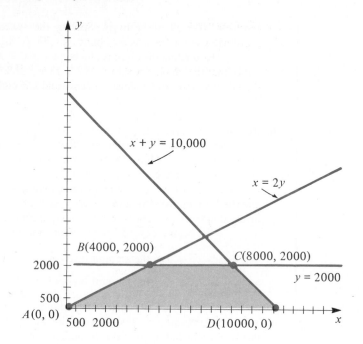

FIGURE 3.18

Certain applications of linear programming require that the values of some or all of the unknowns be integers. For example, in our production problem we are seeking a number of retorts and pipets, and, therefore, we have the implicit constraint that both the x and y values of the solution be integers. This sort of problem is called an **integer programming problem**. We can still try to find a solution to an integer problem using the technique in this section where we solve the problem without the integer requirement. If we find that the optimal solution to the linear programming problem satisfies the integer condition, then we have the answer to the integer programming problem, as in the case of our production problem. On the other hand, if the optimal solution to the linear programming problem does not satisfy the integer condition, then we cannot simply round off the answer to the nearest integer; there are examples where rounding will not yield the correct result. There are methods beyond the scope of this book that are designed for solving integer programming problems.

EXERCISES 3.2

1. Determine from among the feasible solutions to the system

$$4x + 7y \leq 63$$

$$5x + 3y \leq 50$$

$$x \geq 0 \quad \text{and} \quad y \geq 0$$

those that maximize the objective function

a. $z = x + 3y$ b. $z = 2x + 3y$ c. $z = 5x + 3y$

2. Determine from among the feasible solutions to the system

$$-7x + 10y \leq 110$$

$$9x - 2y \geq 54$$

$$x \geq 0 \quad \text{and} \quad y \geq 0$$

those that minimize the objective function

a. $z = 4x + 9y$ b. $z = 4x + 5y$ c. $z = 4x + 7y$

3. Determine from among the feasible solutions to the system

$$x + 2y \leq 24$$

$$4x + 5y \leq 72$$

$$2x + y \leq 30$$

$$x \geq 0 \quad \text{and} \quad y \geq 0$$

those that maximize the objective function

a. $z = 3x + 5y$ b. $z = 4x + 5y$ c. $z = 12x + 5y$

4. Determine from among the solutions to the system

$$4x + 9y \geq 72$$

$$-x + 2y \leq 16$$

$$3y \leq 4x$$

$$x \geq 0 \quad \text{and} \quad y \geq 0$$

those, if any, that

a. maximizes $z = 5x + 3y$ b. minimizes $z = 5x + 3y$

For exercises 5–20 determine the solution to the linear programming problem using the graphical method.

5. Maximize $z = x + y$
   subject to $7x - 3y \geq 0$
   $$x \leq 3$$
   $$x \geq 0 \quad \text{and} \quad y \geq 0$$

6. Maximize $z = x + 3y$
   subject to $3x - 5y \leq 0$
   $$y \leq 6$$
   $$x \geq 0 \quad \text{and} \quad y \geq 0$$

7. Maximize $z = 3x + y$
   subject to $x + 3y \leq 24$
   $$7x + 2y \leq 35$$
   $$x \geq 0 \quad \text{and} \quad y \geq 0$$

8. Maximize $z = x - 2y$
   subject to $-3x + 5y \leq 20$
   $$7x + 4y \leq 63$$
   $$x \geq 0 \quad \text{and} \quad y \geq 0$$

9. Minimize $z = 7x + 5y$
   subject to $x + y \geq 8$
   $$5x + 3y \geq 30$$
   $$x \geq 0 \quad \text{and} \quad y \geq 0$$

10. Minimize $z = 5x + 2y$
    subject to $11x + 26y \geq 183$
    $$17x + 10y \geq 147$$
    $$x \geq 0 \quad \text{and} \quad y \geq 0$$

11. Maximize $z = 5x + 7y$
    subject to $3x + 4y \leq 40$
    $x \leq 2y$
    $y \leq 8$
    $x \geq 0$ and $y \geq 0$

12. Maximize $z = \frac{1}{4}x + \frac{1}{5}y$
    subject to $-5x + 4y \leq 40$
    $5x + 4y \leq 80$
    $x \leq 8$
    $x \geq 0$ and $y \geq 0$

13. Minimize $z = 7x + 12y$
    subject to $2x + 3y \geq 16$
    $y \leq 10$
    $x \leq 5$
    $x \geq 0$ and $y \geq 0$

14. Minimize $z = 6x + 3y$
    subject to $2x + y \geq 12$
    $x + y \geq 8$
    $y \leq 9$
    $x \geq 0$ and $y \geq 0$

15. Maximize $z = \frac{1}{3}x + y$

    subject to $2x + 9y \leq 99$
    $4x - 3y \leq 9$
    $-x + 12y \geq 24$
    $x \geq 0$ and $y \geq 0$

16. Maximize $z = 4x + 3y$

    subject to $\frac{3}{2}x + y \leq 15$

    $x + \frac{3}{7}y \leq \frac{65}{7}$

    $x + y \leq 13$
    $x \geq 0$ and $y \geq 0$

17. Minimize $z = 3x + y$
    subject to $5x + 4y \geq 52$
    $7x + 3y \geq 52$
    $3x + 5y \geq 39$
    $x \geq 0$ and $y \geq 0$

18. Minimize $z = \frac{1}{2}x + \frac{1}{6}y$

    subject to $3x + y \geq 7$
    $x + 2y \geq 4$
    $x \leq 6$
    $x \geq 0$ and $y \geq 0$

19. Minimize $z = 3x - 5y$
    subject to $3x + 4y \leq 61$
    $-5x + 3y \leq 24$
    $5x + 7y \geq 35$
    $x \leq 7$
    $x \geq 0$ and $y \geq 0$

20. Maximize $z = x + 3y$
    subject to $4x + 7y \geq 90$
    $3x - y \geq 5$
    $y \leq 6$
    $x \geq 0$ and $y \geq 0$

21. Find the maximum and minimum values of

$$z = \frac{1}{2}x + 5y$$

over the region defined by

$$x + y \geq 1$$
$$x + 3y \leq 9$$
$$x + y \leq 5$$
$$x \geq 0 \quad \text{and} \quad y \geq 0.$$

22. Find the maximum and minimum values of

$$z = -x + \frac{1}{2}x$$

over the region defined by

$$-3x + 7y \leq 77$$

$$14x + 3y \leq 140$$

$$6x + 5y \geq 30$$

$$x \geq 0 \quad \text{and} \quad y \geq 0.$$

For exercises 23–36, model each as a linear programming problem, making sure to state what each variable in the problem represents. Then solve the problem using the graphical method. Note that the regions of feasible solutions for exercises 23–28 were given as exercises in section 3.1; the details are repeated here.

23. Linda, who owns the Apple Hill Shoppe, wants her weekly payroll to be no more than $1026. She pays $4.00 per regular hour and $5.40 per overtime hour. Also she wants the number of overtime hours to be at most $\frac{2}{3}$ times the number of regular hours. If the store makes $19 on each regular hour and $27 on each overtime hour, how many hours of each type should Linda schedule each week in order to maximize what she makes? What is this maximum amount?

24. Jodi wants to invest up to $10,000 in a combination of two types of mutual funds, performance and growth. She wants to invest at least $2000 in the growth fund, but the amount should not exceed the amount invested in the performance fund. If the performance fund returns 18% annually on the amount invested and the growth fund pays 12% annually, how much should Jodi invest in each fund to maximize her anticipated annual income? What is the maximum income?

25. The Tennessee Chainsaw Company makes two types of saws, a utility model and a deluxe model. Two machines, A and B, are used in the production of each model. It takes 2 hours on machine A and 3 hours on machine B to produce one dozen of the utility model, whereas it takes 3 hours on machine A and 1 hour on machine B to produce one dozen of the deluxe model. In a day there are 9 hours available on machine A and 10 hours available on machine B. If the company makes a profit of $666 per dozen on the utility model and $855 per dozen on the deluxe model, how many dozen of each model should the company make each day so that its profit is maximum? What is its maximum daily profit?

26. Joe must follow a diet that contains at least 125 milligrams of vitamin B_1 and at least 150 milligrams of vitamin B_2 each day. Two food supplements, I and II, are used to obtain these requirements. Each packet of supplement I contains 15 milligrams of B_1 and 15 milligrams of B_2, and each packet of supplement II contains 10 milligrams of B_1 and 20 milligrams of B_2. If each packet of supplement I costs $.60 and each packet of supplement II costs $.40, then how many packets of each supplement should Joe use a day to meet or exceed the vitamin requirements while keeping the daily cost minimal? What is the minimum cost?

27. Petro Chemicals makes two types of plastic, medium-density and high-density. Each type of plastic is made from two chemicals, A and B. Each pound of medium-density plastic requires 0.5 pound of chemical A and 0.5 pound of chemical B, whereas each pound of high-density plastic requires 0.7 pound of chemical A and 0.3 pound of chemical B. Furthermore, each pound of medium-density and each pound of high-density plastic requires 0.1 man-hour of labor. Petro Chemicals has available at most 2100 pounds of

chemical A, 1500 pounds of chemical B, and 340 man-hours of labor. If Petro Chemicals makes a profit of $52 on each pound of medium-density plastic and $70 on each pound of high-density plastic, how many pounds of each type of plastic should the company produce in order to achieve the largest profit but not exceed its available resources of chemicals and labor? What would be the maximum profit?

28. A dietician is planning a dinner that is to consist of a cheese dish and a green vegetable. The dietician wants the dinner to supply at least 1500 milligrams of calcium, 6 milligrams of iron, and 72 grams of protein. Each ounce of the cheese dish supplies 100 milligrams of calcium, 0.25 milligram of iron, and 4 grams of protein. Each ounce of the green vegetable supplies 20 milligrams of calcium, 0.5 milligram of iron, and 2 grams of protein. Furthermore, each ounce of the cheese dish contains 100 calories, and each ounce of the green vegetable contains 40 calories. How many ounces of each type of food should be eaten so that the stated requirements in calcium, iron, and protein are met but the number of calories is minimal?

29. Pottery Makers produces ceramic bowls and plates. Each item goes through two phases of production, molding and firing. The molding times are $\frac{1}{5}$ man-hour for each bowl and $\frac{1}{10}$ man-hour for each plate. Each bowl and plate requires $\frac{5}{100}$ man-hour for firing. Pottery Makers has at most 72 man-hours available in the molding shop and at most 26 man-hours available in the firing shop. If Pottery Makers realizes a profit of $1.75 on each bowl and $1.00 on each plate, how many of each item should be made so that the largest profit is achieved but the limits on man-hours are not exceeded? What is the maximum profit?

30. King's Manufacturing Company makes plastic food trays and plastic defroster ducts. Each product requires manufacturing time on both the molding machine and the trimming machine. Each case of food trays requires $\frac{1}{2}$ hour on the molding machine and $\frac{1}{3}$ hour on the trimming machine, whereas each case of the defroster ducts requires $\frac{1}{8}$ hour on the molding machine and $\frac{1}{6}$ hour on the trimming machine. King's Manufacturing has available each day up to 11 hours on the molding machine and 8 hours on the trimming machine. If the company makes a profit of $32 per case of food trays and $15 per case of defroster ducts, how many cases of each should be made to achieve a maximum profit? What is the maximum profit?

31. Pendrake Lawn Fertilizer is guaranteed to have at least 60 pounds of phosphoric acid and at least 30 pounds of potash in each bag of its fertilizer. Two ingredients, A and B, are used to make the fertilizer. Ingredient A is 70% phosphoric acid and 30% potash, whereas ingredient B is 50% phosphoric acid and 50% potash. If ingredient A costs 30 cents per pound and ingredient B costs 40 cents per pound, how many pounds of each ingredient should a bag of fertilizer contain to satisfy the requirements of phosphoric acid and potash and minimize the total cost? What is the minimum total cost?

32. Whitehall Feed Company guarantees at least 10 pounds of protein, 18 pounds of fat, and 9.75 pounds of carbohydrates in each bag of its feed. Two ingredients, A and B, are used to make the feed. Ingredient A is 40% protein, 45% fat, and 15% carbohydrates. Ingredient B is 10% protein, 30% fat, and 60% carbohydrates. If ingredient A costs 48 cents per pound and ingredient B costs 42 cents per pound, how many pounds of each ingredient should a bag of feed contain to satisfy the requirements of protein, fat, and carbohydrates and minimize the total cost? What would be the minimum total cost?

33. Erica is planning a meal of steak and salad. Each unit of steak contains 200 calories, 3 grams of vitamins, and 2 grams of protein, whereas each unit of salad contains 20

calories, 4 grams of vitamins, and 1 gram of protein. The meal is to supply at least 500 calories, 24 grams of vitamins, and 11 grams of protein. If each unit of steak costs $1.20 and each unit of salad costs $0.20, how many units of each food should Erica plan for the meal so that the requirements in calories, vitamins, and protein are met or exceeded but the cost is minimal? What is the minimum cost?

34. State University is having difficulty servicing its students with its present computer system. The university's administration has decided to purchase additional computer terminals. At most, 20 new terminals can be added to the system. In addition, either slower terminals can be purchased at $1000 apiece or faster terminals can be purchased at $1600 apiece. The administration has budgeted $27,200 for the purchase of the new terminals. A faster terminal operates $1\frac{1}{2}$ times faster than a slower terminal; that is, for each slower terminal a minute of use is equivalent to 1 student-minute of service, whereas for each faster terminal a minute of use is equivalent to $1\frac{1}{2}$ student-minutes of service. How many of each type of terminal should be purchased so that the student-minutes of service available each hour is maximum?

35. Satellite Paper Company makes two grades of paper, newsprint and mando stock. To satisfy customer orders, the company must produce at least 8 tons of newsprint and at least 5 tons of mando stock paper per day. The total amount of paper that can be produced in a day is 30 tons; however, the amount of mando stock the company can produce is no more than twice the amount of newsprint it can produce. If the company makes $1500 on each ton of newsprint and $1800 on each ton of mando stock, how many tons of each grade paper should it produce in a day to maximize its profit?

36. Ms. Smith has available up to $20,000 to invest in a combination of different types of investments, a low-risk investment that pays 12% annually and a high-risk investment that pays 20% annually. Ms. Smith does not want to invest more than $6000 in the high-risk investment, and the amount invested in the low-risk investment can be no less than the amount invested in the high-risk investment. How much should Ms. Smith invest in each type of investment to maximize the anticipated income?

SUMMARY OF TERMS

Linear Inequalities and Systems of Linear Inequalities
   Graph of a Linear Inequality
   Graph of a System of
      Linear Inequalities
   Boundary Lines
   Convex Polygonal Set
   Vertices
   Solution to a System of Linear
      Inequalities
   Half-Planes
   Bounded Region
   Unbounded Region

Linear Programming Problems in Two Variables
   Constraints
   Natural Conditions
      (Non-Negative Conditions)
   Feasible Solutions
   Objective Function
   Integer Programming Problem
   Optimal Solution
   Optimal Value
   Unbounded Solution
   Graphical Method of Solution
   Standard Maximization Problem
   Standard Minimization Problem

REVIEW EXERCISES (CH. 3)

For exercises 1–4 graph the given system of linear inequalities. Also find the coordinates of the vertices of the region.

1. $-2x + 5y \leq 20$
   $6x + 2y \leq 42$
   $x \geq 0$ and $y \geq 0$

2. $17x + 8y \leq 104$
   $8y \geq 9x$
   $x \geq 0$ and $y \geq 0$

3. $3x + 2y \geq 80$
   $3x + 4y \geq 100$
   $x + 4y \geq 40$
   $x \geq 0$ and $y \geq 0$

4. $3x + 4y \geq 28$
   $x + 2y \geq 12$
   $-x + 2y \leq 22$
   $x \geq 0$ and $y \geq 0$

For exercises 5–12 solve the linear programming problem using the graphical method

5. Maximize $z = \frac{1}{4}x + \frac{1}{3}y$
   subject to $x + 2y \leq 70$
   $4y \leq 5x$
   $x \geq 0$ and $y \geq 0$

6. Maximize $z = 6x + 3y$
   subject to $2x + 3y \leq 3$
   $4x \geq 3y$
   $x \geq 0$ and $y \geq 0$

7. Maximize $z = 2x + 4y$
   subject to $10x - 8y \leq 15$
   $5x + 7y \leq 35$
   $x \geq 0$ and $y \geq 0$

8. Maximize $z = 10x + 20y$
   subject to $-x + 5y \leq 175$
   $9x + 5y \leq 675$
   $x \geq 0$ and $y \geq 0$

9. Maximize $z = 3x + 2y$
   subject to $x + 5y \leq 135$
   $11x + 10y \leq 405$
   $13x - 10y \leq 195$
   $x \geq 0$ and $y \geq 0$

10. Maximize $z = 4x + 5y$
    subject to $-x + y \leq 8$
    $7x + 6y \leq 113$
    $3x - 2y \leq 21$
    $x \geq 0$ and $y \geq 0$

11. Minimize $z = \frac{3}{2}x + \frac{3}{2}y$
    subject to $x + y \geq 3$
    $-3x + 4y \leq 32$
    $y \leq 11$
    $x \geq 0$ and $y \geq 0$

12. Minimize $z = x + \frac{1}{3}y$
    subject to $3x + y \geq 20$
    $x + y \geq 10$
    $3x + 7y \geq 42$
    $x \geq 0$ and $y \geq 0$

13. Juicy Orchards makes apple cider and vinegar. Two machines, the masher and the bottler, are used in the manufacture of each product. The number of hours on each machine to produce a crate of each product is given in table 3.10.

TABLE 3.10

	Apple cider	Vinegar
masher	2	5
bottler	3	4

The masher can be used at most 9 hours a day, and the bottler can be used at most 10 hours a day. Furthermore Juicy Orchards must make at least $\frac{1}{2}$ crate of vinegar each day. There is a profit of $300 and $250 on each crate of apple cider and vinegar, respectively. How many crates of each product should be made each day to maximize the total daily profit? What is the maximum profit?

14. Jesse wants to invest at the most $20,000 in municipal bonds and certificates of deposit. The amount invested in municipal bonds must not exceed the amount invested in certificates of deposit. Also Jesse wants to invest at the most $12,000 in certificates of deposit. If the municipal bonds yield 12% and the certificates of deposit yield 9%, how much should Jesse invest in each one to maximize the annual interest? What is the maximum annual interest?

15. Feline Pet Food Company guarantees that each bottle of its pet care supplement for cats contains at least 12 grams of vitamin B_1 and at least 18 grams of vitamin B_2. Two ingredients, I and II, are used to make this supplement. Ingredient I contains 10% vitamin B_1 and 30% vitamin B_2, whereas ingredient II contains 20% vitamin B_1 and 10% vitamin B_2. The company wants each bottle to contain at most 60 grams of ingredient II. If each gram of ingredient I costs 2 cents and each gram of ingredient II costs 2.5 cents, how many grams of each ingredient should be used per bottle to minimize the total cost? What is the minimum total cost?

16. Joan is planning a meal that consists of a casserole and a vegetable. She wants the meal to supply at least 400 milligrams of calcium, 40 milligrams of protein, and 5 milligrams of iron. The amount, in milligrams, of calcium, protein, and iron per ounce of the casserole and vegetable are given in table 3.11.

TABLE 3.11

	Casserole	Vegetable
calcium	160	20
protein	8	4
iron	$\frac{1}{2}$	$\frac{3}{4}$

Furthermore each ounce of the casserole contains 90 calories and each ounce of the vegetable contains 60 calories. How many ounces of each type of food must be eaten so that the stated requirements in calcium, protein, and iron are met but the number of calories eaten is minimal? What is the minimum number of calories?

CHAPTER 4

Systems of Linear Equations

Introduction

We continue our study of linear models with extensions to situations that give rise to n unknowns. In particular, this chapter is devoted to the consideration of linear forms in n variables and the related systems of linear equations in n unknowns. In this chapter we also develop an elimination method for solving systems of linear equations in n unknowns.

4.1 Linear Forms and Systems of Linear Equations

Linear forms in two variables are discussed in chapter 2. A linear form describing a function of three variables is, in general,

$$T = ax + by + cz + d.$$

For example, a sporting goods company manufactures three types of tennis rackets, wooden, metal, and graphite. The number of man-hours needed to make wooden rackets is constant; this is also true for metal and graphite rackets. In particular, it takes 8 man-hours to make one wooden racket, 9 man-hours to make one metal racket, and 11 man-hours to make one graphite racket. The total man-hours T is a function of the rackets made and is the sum of the hours spent in making the three types of rackets. Let us write an equation that describes this function.

Let x, y, and z represent, respectively, the number of wooden, metal, and graphite rackets made. Then the number of man-hours needed to make x wooden rackets is $8x$. Likewise it would take $9y$ and $11z$ man-hours to produce y metal rackets and z graphite rackets, respectively. Thus T can be expressed as a function of

x, y, and z by the linear form

$$T = 8x + 9y + 11z.$$

We can find the total man-hours needed to produce, for example, 10 wooden rackets, 12 metal rackets, and 7 graphite rackets by substituting these numbers into the linear form. In particular, we have

$$T = 8(10) + 9(12) + 11(7) = 265.$$

Linear forms can easily be extended to n variables by introducing **subscripted variables** x_1, x_2, x_3, x_4, etc. in place of simple variables x, y, z, w, etc. An example of a linear form in four variables is

$$T = 4x_1 + 11x_2 + 7x_3 + x_4.$$

In general, we have the following definition.

Definition ■ A **linear form** in n variables is

$$T = a_1 x_1 + a_2 x_2 + \cdots + a_n x_n + b,$$

where $a_1, a_2, \ldots, a_n$ and b are constants. The linear form expresses the **dependent variable** T as a function of the **independent variables** $x_1, x_2, \ldots, x_n$. (See figure 4.1.) ■

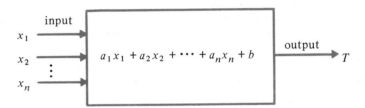

FIGURE 4.1

As seen in the case $n = 2$, we usually use x and y for the independent variables in place of x_1 and x_2. Similarly, when $n = 3$, we usually use x, y, and z as independent variables.

EXAMPLE 1

A person has money invested between a savings account, a certificate of deposit, a money management fund, and an oil income partnership. The four investments return, respectively, 6%, 9%, 12%, and 15% per year. Let us write an equation that describes the total monthly income from these four investments.

First, we recognize that the total monthly income is a function of the amounts invested in the four vehicles. Thus we choose four independent variables x_1, x_2, x_3, and x_4 to represent, respectively, the monies invested in the savings account, the certificate of deposit, the money management fund, and the oil income partnership.

Next the monthly return on one of the investments is the monthly rate times the amount invested. Therefore we must convert each of the annual rates to a monthly rate. The monthly rate of return from the savings account is obtained by dividing the annual rate by 12; that is, the monthly rate for the savings account is $0.06/12 = 0.005$. The monthly rates for the other three investments are obtained in a like manner and are $0.09/12 = 0.0075$, $0.12/12 = 0.01$, and $0.15/12 = 0.125$, respectively. Therefore the total monthly income I, being the sum of the individual incomes, is described by the linear form

$$I = 0.005x_1 + 0.0075x_2 + 0.01x_3 + 0.0125x_4.$$

As seen in chapter 2, linear forms give rise in a natural way to linear equations. Generally, the problem posed is to find values for the independent variables that cause the function to achieve a given value. For example, suppose the sporting goods company that makes tennis rackets has available 265 man-hours and asks: "How many wooden, metal and graphite rackets it can produce so as to use completely the available man-hours?" This question can be stated mathematically as follows: find x, y, and z so that

$$8x + 9y + 11z = 265;$$

in other words, the company needs to find solutions to a linear equation in three unknowns.

Definition   ■   An equation of the form

$$a_1x_1 + a_2x_2 + \cdots + a_nx_n = b$$

is called a **linear equation** in the n **variables (unknowns)** $x_1, x_2, \ldots, x_n$. The **coefficients** $a_1, a_2, \ldots, a_n$ of the variables and the **right-hand side** b of the equation are real numbers.   ■

A solution to the linear equation $8x + 9y + 11z = 265$ is any ordered triple of numbers that satisfies the equation. We see that $(10, 12, 7)$ is a solution, whereas $(5, 15, 10)$ is not a solution since $8(5) + 9(15) + 11(10) \neq 265$.

In general, we say that an ordered set of n real numbers $(t_1, t_2, \ldots, t_n)$ is a **solution** to the linear equation $a_1x_1 + a_2x_2 + \cdots + a_nx_n = b$ if we obtain a true equality when we set $x_1 = t_1, x_2 = t_2, \ldots, x_n = t_n$ in the equation.

EXAMPLE 2
a. We see that $(-1, 2, 3, 1)$ is a solution to the equation $2x_1 + x_2 - x_3 + 3x_4 = 0$ since $2(-1) + (2) - (3) + 3(1) = 0$.

b. The triple $(\frac{3}{2}, 1, -2)$ is not a solution to $4x + 3y + z = 4$ since $4(\frac{3}{2}) + 3(1) + (-2) \neq 4$.

Recall that the graph of a linear equation in two variables in a two-dimensional coordinate system is a line. Although we will not do any graphing in three-dimensional space, it can be shown that the graph of a linear equation in three variables is a plane; each point on this plane represents a solution to the equation. Thus, as in the case of two variables, a linear equation in three variables has infinitely many solutions. In general, a linear equation in two or more variables has an infinite number of solutions.

We have seen in chapter 2 how a system of linear equations in two unknowns could arise by placing requirements on linear forms in two variables. We now direct our attention to systems of linear equations in n variables and their solutions.

Definition ■ A system of m linear equations in n unknowns takes the form

$$a_{11}x_1 + a_{12}x_2 + \cdots + a_{1n}x_n = b_1$$
$$a_{21}x_1 + a_{22}x_2 + \cdots + a_{2n}x_n = b_2$$
$$\vdots \qquad\qquad \vdots$$
$$a_{m1}x_1 + a_{m2}x_2 + \cdots + a_{mn}x_n = b_m. \quad ■$$

A specific example of a system of four equations in four unknowns is

$$x_1 + 15x_2 - 7x_3 + 3x_4 = 0$$
$$8x_1 \qquad\qquad + 35x_3 - x_4 = 9$$
$$-11x_2 \qquad\qquad + 18x_4 = -6$$
$$5x_3 - 13x_4 = 10.$$

Definition ■ A system of m linear equations in n unknowns has a **solution** $x_1 = t_1$, $x_2 = t_2, \ldots, x_n = t_n$ if each of the m equations in the system is satisfied by this set of numbers. A system that has at least one solution is said to be **consistent**; otherwise, the system is said to be **inconsistent**. ■

EXAMPLE 3

We see that $x = 2$, $y = 2$, and $z = 1$ is a solution to the system of two equations in three unknowns

$$x + 2y - z = 5$$
$$6x - 3y - 8z = -2$$

since *both* equations are satisfied by this set of numbers. On the other hand, since $6(3) - 3(1) - 8(0) \neq -2$, we have that $x = 3$, $y = 1$, and $z = 0$ is not a solution to the system even though the first equation is satisfied by this set of numbers.

The system in example 3 is consistent since it has at least one solution, namely $(2, 2, 1)$. In chapter 2 we saw systems that were inconsistent and systems that had infinitely many solutions. As in the case of a system with two unknowns, the solution set of a system with n unknowns may be empty, contain infinitely many solutions, or contain a single solution.

We return in the next section to applications that can be modeled by a system of linear equations. For the remainder of this section we extend the elimination method of chapter 2 to systems of linear equations in three unknowns. Our discussion provides the basis for the matrix method developed in the next section.

Let us begin by considering again the system

$$x + 2y = 575$$
$$10x + 15y = 5000. \tag{1}$$

Applying the elimination method we derive an equation in y alone by adding -10 times the first equation to the second equation. In particular, we have

$$-10x - 20y = -5750$$
$$\underline{10x + 15y = 5000}$$
$$-5y = -750.$$

Solving for y, we have $y = 150$. Next, substituting this value for y back in the first equation

$$x + 2y = 575$$

we obtain the value for x, namely $x = 275$.

We should keep in mind that the system (1) is solved by using an equivalent system, namely

$$x + 2y = 575$$
$$-5y = -750. \tag{2}$$

The equivalent system (2) is derived from (1) by replacing the second equation in (1) by an equation in which x has been eliminated.

In general, our procedure for solving a system of n linear equations in n unknowns is to derive an **equivalent** system, one that has the same solutions, and that is in a special form. We derive an equivalent system by performing one or more of the following types of **operations on equations**.

Operations on Equations

1. Interchange the order in which two equations appear in the system.
2. Replace an equation in the system by the equation multiplied by a non-zero constant.
3. Replace an equation in the system by the sum of it and a non-zero constant times another equation in the system.

Using these operations, we wish to derive an equivalent system in which the first unknown has been eliminated from all but the first equation, the second unknown has been eliminated from all but the second and possibly first equations, and so forth. The resulting equivalent system is called a system in **triangular form**.

For a system in triangular form, if none of the coefficients of the variables x_1 through x_n in equations 1 through n respectively is zero, then the system can be easily solved using **back substitution**. For example, the system

$$2x_1 + 3x_2 + 4x_3 = 20$$
$$x_2 + 5x_3 = 7$$
$$3x_3 = 3$$

is in triangular form. We obtain the solution to the system by solving the third equation for x_3, yielding $x_3 = 1$. Substituting this value back into the second equation and solving for x_2, we have $x_2 = 7 - 5(1) = 2$. Finally, substituting these two values back into the first equation, we find $x_1 = \frac{1}{2}[20 - 3(2) - 4(1)] = 5$.

Let us now illustrate the entire procedure of deriving an equivalent system in triangular form and solving, if possible, by back substitution.

EXAMPLE 4

Derive from the system

$$x + 2y + 4z = -7$$
$$-x - 2y + z = 2 \tag{3}$$
$$2x + 6y + 2z = -12$$

an equivalent system in triangular form and then solve the system, if possible, by back substitution.

First we eliminate x in the second equation of (3); in particular, we replace $-x - 2y + z = 2$ by the sum of it and the first equation to obtain the equivalent system

$$x + 2y + 4z = -7$$
$$5z = -5 \tag{4}$$
$$2x + 6y + 2z = -12.$$

We now wish to replace the third equation of (4) by one in which the x term is eliminated. We add -2 times the first equation to it, yielding the system

$$x + 2y + 4z = -7$$
$$5z = -5 \tag{5}$$
$$2y - 6z = 2.$$

Since the second equation of (5) contains only the third unknown, we make that the third equation of our system. We now obtain the equivalent system in triangular form

$$x + 2y + 4z = -7$$
$$2y - 6z = 2$$
$$5z = -5.$$

By back substitution, we have the solution $z = -1$, $y = \frac{1}{2}[2 + 6(-1)] = -2$, and $x = -7 - 2(-2) - 4(-1) = 1$.

EXERCISES 4.1

---

1. Crossfire Enterprises makes pivotal gear housings, air cleaners, and balance assemblies. It costs the company $4, $7, and $12 to make one dozen pivotal gear housings, one dozen air cleaners, and one dozen balance assemblies, respectively. The company sells its products for $6, $11, and $15 per dozen of pivotal gear housings, air cleaners, and balance

assemblies, respectively. Let

x = the number (in dozens) of pivotal gear housings made and sold each day

y = the number (in dozens) of air cleaners made and sold each day

z = the number (in dozens) of balance assemblies made and sold each day.

a. Express the total daily cost C as a function of x, y, and z.

b. Express the total daily revenue R as a function of x, y, and z.

c. Express the total daily profit P as a function of x, y, and z.

d. During a particular day, the company makes 50 dozen pivotal gear housings, 65 dozen air cleaners, and 78 dozen balance assemblies. Find the cost, revenue, and profit for this day's production.

2. The DBS Company makes hydraulic tanks, wind-doors, and ram controls. It costs the company $23, $12, and $45 for each hydraulic tank, wind-door, and ram control, respectively. The company sells each hydraulic tank for $35, each wind-door for $21, and each ram control for $69. Let

x_1 = the number of hydraulic tanks made and sold each day

x_2 = the number of wind-doors made and sold each day

x_3 = the number of ram controls made and sold each day.

a. Express the total daily cost C as a function of x_1, x_2, and x_3.

b. Express the total daily revenue R as a function of x_1, x_2, and x_3.

c. Express the total daily profit P as a function of x_1, x_2, and x_3.

d. During a particular day, the company makes 35 hydraulic tanks, 41 wind-doors, and 53 ram controls. Find the cost, revenue, and profit for this day's production.

3. A diet consists of rice, chicken breasts, cottage cheese, and yogurt. The calories per unit serving of each food are given in table 4.1.

TABLE 4.1

Food	Calories per unit serving
rice	350
chicken breasts	86
cottage cheese	85
yogurt	123

Let

x_1 = the number of servings of rice eaten each day

x_2 = the number of servings of chicken breasts eaten each day

x_3 = the number of servings of cottage cheese eaten each day

x_4 = the number of servings of yogurt eaten each day.

a. Express the total daily calories TC obtained from this diet as a function of x_1, x_2, x_3, and x_4.

   b. How many total daily calories are obtained from a diet that consists of 6 servings of rice, 8 servings of chicken breasts, 10 servings of cottage cheese, and 9 servings of yogurt per day?

4. Scorpio Ventures makes a food supplement from four ingredients: A, B, C, and D. The amount of vitamin B_1 (in milligrams) per cup of each ingredient is given in table 4.2.

TABLE 4.2

Ingredient	Vitamin B_1
A	50
B	27
C	30
D	45

In a batch of this food supplement, let

$$x_1 = \text{the number of cups of ingredient A}$$

$$x_2 = \text{the number of cups of ingredient B}$$

$$x_3 = \text{the number of cups of ingredient C}$$

$$x_4 = \text{the number of cups of ingredient D.}$$

   a. Express the total amount TA of vitamin B_1 in a batch of this food supplement.

   b. How many milligrams of vitamin B_1 are there in a batch of the food supplement, if 13 cups of ingredient A, 7 cups of ingredient B, 9 cups of ingredient C, and 10 cups of ingredient D are used?

5. The Spellstone Company owns four bicycle factories. The weekly fixed and variable costs (in dollars) for each factory are given in table 4.3.

TABLE 4.3

Factory	Fixed cost	Variable cost
1	1250	105
2	1500	90
3	1335	102
4	980	114

   a. Express the total weekly cost TC as a function of the number of bicycles made at each factory in a week.

   b. During a particular week, 210, 182, 165, and 93 bicycles were made at factories 1, 2, 3, and 4, respectively. What is the company's total cost for this week?

6. Aztec Supplies makes vacuum pumps at three factories. The weekly fixed and variable costs (in dollars) for each factory are given in table 4.4.

TABLE 4.4

Factory	Fixed cost	Variable cost
1	9,284	36
2	10,535	31
3	10,659	29

a. Express the total weekly cost TC as a function of the number of vacuum pumps made at each factory in a week.

b. During a particular week, 200 pumps were made at factory 1, 250 pumps made at factory 2, and 195 pumps made at factory 3. What is the company's total cost for the week?

7. Al owns a restaurant called The Pizza Place. He sells pizza in the following three sizes: 10-inch, 13-inch, and 15-inch. The cost for ingredients to make a pizza and the price charged for the pizza are given in table 4.5.

TABLE 4.5

Pizza	Al's cost	Price charged
10″	$0.60	$2.50
13″	$1.20	$4.00
15″	$1.85	$5.25

a. Express the weekly cost for ingredients as a function of the number of pizzas of each size made in a week.

b. Express the weekly revenue from sales as a function of the number of pizzas of each size made in a week.

c. Express the weekly profit as a function of the number of pizzas of each size made in a week.

8. Sunrise Development Corporation sells 1-acre, 2-acre, and 3-acre building lots. It costs $1500, $2500, and $3000, respectively, to develop each size lot. The corporation sells the lots at the following prices: $2300 for each 1-acre lot, $3500 for each 2-acre lot, and $4500 for each 3-acre lot.

a. Express the total cost of developing the lots as a function of the number of lots developed in each size.

b. Express the total revenue that is possible from sales as a function of the number of lots developed in each size.

c. Express the total profit that is possible as a function of the number of lots developed in each size.

9. Consider the system of linear equations

$$2x - 3y + 4z = 1$$
$$x - 2y + z = 2$$
$$4x - 7y + 6z = 5.$$

Determine whether or not each of the following ordered sets is a solution to the system.

a. $(-4, -3, 0)$
b. $(1, -1, 1)$
c. $(6, 1, -2)$
d. $(-9, -5, 1)$

10. Consider the system of linear equations

$$x_1 + 4x_2 - 2x_3 - x_4 = 0$$
$$3x_2 - x_3 + 2x_4 = 0$$
$$2x_1 - x_2 - 2x_3 - 10x_4 = 0.$$

Determine whether or not each of the following ordered sets is a solution to the system.

a. $(0,0,0,0)$
b. $(-37, 4, -6, -9)$
c. $(-21, 12, 18, -9)$

For exercises 11 and 12 verify that the given pair of linear systems are equivalent by determining a sequence of operations on the equations.

11. a. $\begin{aligned} -2x + 4y &= 9 \\ x - 3y &= 7 \end{aligned}$ and $\begin{aligned} x - 3y &= 7 \\ -2y &= 23 \end{aligned}$

    b. $\begin{aligned} x + 2y - z &= 5 \\ 2x + 5y + z &= 11 \\ -3x - 8y - 2z &= -9 \end{aligned}$ and $\begin{aligned} x + 2y - z &= 5 \\ y + 3z &= 1 \\ -2y - 5z &= 6 \end{aligned}$

12. a. $\begin{aligned} 2x - y + 3z &= 1 \\ -2x + 4y - 6x &= 5 \\ 4x - 5y + 10z &= 0 \end{aligned}$ and $\begin{aligned} 2x - y + 3z &= 1 \\ y - z &= 2 \\ z &= 4 \end{aligned}$

    b. $\begin{aligned} 3x_2 - 6x_3 + 9x_4 &= 12 \\ x_1 - 2x_2 + 5x_3 - x_4 &= 0 \\ -x_2 + 4x_3 - 2x_4 &= -1 \\ -3x_1 + 8x_2 - 17x_3 + 11x_4 &= 10 \end{aligned}$ and $\begin{aligned} x_1 - 2x_2 + 5x_3 - x_4 &= 0 \\ x_2 - 2x_3 + 3x_4 &= 4 \\ x_3 + \frac{1}{2}x_4 &= \frac{3}{2} \\ x_4 &= -1 \end{aligned}$

For exercises 13–16 determine the solution to the system in triangular form using back substitution.

13. $\begin{aligned} 2x + 3y &= 17 \\ 4y &= 12 \end{aligned}$

14. $\begin{aligned} 5x - 6y &= 20 \\ -3y &= -10 \end{aligned}$

15. $\begin{aligned} 2x - y + 3z &= 1 \\ y - z &= 2 \\ z &= 4 \end{aligned}$

16. $\begin{aligned} x - 4y + 3z &= 11 \\ 2y - z &= -4 \\ 4z &= 8 \end{aligned}$

For exercises 17–22 obtain from the system an equivalent system in triangular form and then solve the system by back substitution.

17. $\begin{aligned} x + 3y - 2z &= -11 \\ 2x - y + z &= 0 \\ -x + 4y + 5z &= 13 \end{aligned}$

18. $\begin{aligned} x + 3y - 2z &= 9 \\ 2x - y + 10z &= 18 \\ 3x + 5y - 3z &= 22 \end{aligned}$

19. $\begin{aligned} -2x + y + z &= -9 \\ 6x + 4y + 5z &= 5 \\ x + 2y - 3z &= 2 \end{aligned}$

20. $\begin{aligned} x - 2y + z &= 0 \\ 3x + 5y - z &= 0 \\ -2x + y + z &= 0 \end{aligned}$

21. $\begin{aligned} x - 2y + 4z &= -2 \\ 3x + 4y + 3z &= 1 \\ -x + 8y + 5z &= -1 \end{aligned}$

22. $\begin{aligned} -x + 2y + 2z &= 2 \\ 3x + 6y - 3z &= -1 \\ 2x + 2y + 4z &= 5 \end{aligned}$

4.2 Augmented Matrices and Gaussian Elimination

It is not difficult to imagine how tedious the solution process would be if we had to keep writing equations when solving even a system of six equations in six unknowns. Even though we will encounter systems no larger than these, systems involving, for example, 100 equations in 100 unknowns do occur in real-world situations.

Therefore we want to look at a way of organizing the data so that the elimination method of section 4.1 can be easily performed. From our discussion in section 4.1, we see that the important information when solving a system of linear equations is the coefficients of the variables and the numbers on the right-hand sides of the equations. Therefore the amount of writing in the elimination process can be reduced greatly by not carrying the variables along in the equations. However, to do this we need a way of keeping the coefficients and the right-hand sides in their appropriate places. The method of organizing the data is to use a **matrix** representation of the system of linear equations. Matrices not only provide a savings on writing, but also they provide the necessary data organization so that the elimination process can be "automated." An automated procedure can be carried out on a computer, which is the tool we would want to use when solving large systems. Let us begin then with a definition of a matrix.

Definition ■ An m-by-n **matrix** is a rectangular array of numbers with m rows and n columns. The **size** of a matrix is described by the number of rows followed by the number of columns and is indicated by writing $m \times n$. An **entry** is located in a matrix by stating the row followed by the column in which it appears; the (i, j) entry appears in the ith row and the jth column. ■

Throughout the text uppercase letters are used for the names of matrices and lowercase letters, possibly subscripted, are used for the names of entries.

EXAMPLE 1 Consider the 2-by-3 matrix

$$\begin{bmatrix} -1 & 2 & 4 \\ 0 & 5 & -3 \end{bmatrix}.$$

The $(2, 1)$ entry is 0, whereas the $(1, 2)$ entry is 2.

Since any linear equation can be written in a standard form, namely

$$a_1 x_1 + a_2 x_2 + \cdots + a_n x_n = b,$$

it can be conveniently given by displaying only the coefficients of the variables and the right-hand side of the equation. We do this by using the row of numbers

$$[a_1 \quad a_2 \quad \cdots \quad a_n | b],$$

called an **augmented matrix**. For example, $[2 \quad 3 | 6]$ could be considered as a representation of the linear equation $2x + 3y = 6$. To display the linear equation $x_1 - 2x_2 + 3x_3 - 4x_4 = 7$, we use the augmented matrix $[1 \quad -2 \quad 3 \quad -4 | 7]$.

When we use an augmented matrix to represent a linear equation, we understand that the last entry of the row is the right-hand side of the equation and the other entries are the coefficients of the unknowns; the first entry is the coefficient of the first unknown, and so forth. To give an augmented matrix for $3y = -5x + 6$, we must first rewrite the equation in the standard form $5x + 3y = 6$; the augmented matrix for the equation is $[5 \quad 3 | 6]$.

As done with a single equation, we can represent a system

$$a_{11}x_1 + a_{12}x_2 + \cdots + a_{1n}x_n = b_1$$
$$a_{21}x_1 + a_{22}x_2 + \cdots + a_{2n}x_n = b_2$$
$$\vdots \qquad\qquad\qquad\qquad \vdots$$
$$a_{m1}x_1 + a_{m2}x_2 + \cdots + a_{mn}x_n = b_m$$

by an $m \times (n + 1)$ matrix. The array of coefficients and right-hand sides is written as

$$\begin{bmatrix} a_{11} & a_{12} & \cdots & a_{1n} & b_1 \\ a_{21} & a_{22} & \cdots & a_{2n} & b_2 \\ & \vdots & & & \vdots \\ a_{m1} & a_{m2} & \cdots & a_{mn} & b_m \end{bmatrix}.$$

The preceding array is called the **augmented matrix** for the system of linear equations. Each row of the augmented matrix corresponds to an equation within the system; the first row represents the first equation of the system and, in general, the ith row of the matrix represents the ith equation of the system.

EXAMPLE 2

The augmented matrix for the system

$$x + 2y - z = 5$$
$$6x - 3y - 8z = -2$$

is

$$\begin{bmatrix} 1 & 2 & -1 & 5 \\ 6 & -3 & -8 & -2 \end{bmatrix}.$$

The system of linear equations represented by the augmented matrix

$$\begin{bmatrix} 5 & 4 & -1 & 7 & 1 \\ 2 & -3 & 0 & -5 & 8 \\ 6 & -2 & 9 & 3 & -10 \\ 0 & 0 & -4 & -7 & -6 \end{bmatrix}$$

is

$$5x_1 + 4x_2 - x_3 + 7x_4 = 1$$
$$2x_1 - 3x_2 \qquad\quad - 5x_4 = 8$$
$$6x_1 - 2x_2 + 9x_3 + 3x_4 = -10$$
$$-4x_3 - 7x_4 = -6.$$

Performing an operation on equations in a system is simply doing arithmetic operations on the coefficients of the unknowns and the right-hand sides of the equations. Since a system of linear equations can be conveniently displayed by its augmented matrix, we can abbreviate the writing when deriving an equivalent system by working with the rows of the augmented matrix. The manner in which we operate on the rows corresponds directly to the operations we can perform on the equations within the system. We state these operations in terms of the rows of a matrix.

Definition
■ Each of the following is called a **row operation** on a matrix.

1. Interchange the order in which two rows appear in the matrix.
2. Replace a row in the matrix by non-zero constant times the row.
3. Replace a row in the matrix by the sum of it and a non-zero constant times another row in the matrix. ■

Definition
■ Two $m \times n$ matrices are **row equivalent** if one can obtained from the other by performing a finite number of row operations. ■

When solving a system of n linear equations in n unknowns, we perform row operations on the augmented matrix in order to obtain a row equivalent augmented matrix that represents a system in triangular form. Since row operations on an augmented matrix agree with operations on the equations in the system, the system represented by the derived row equivalent augmented matrix has the same solutions as the original system. We can illustrate these remarks by looking again at example 4 in section 4.1.

EXAMPLE 3

The augmented matrix for the system

$$x + 2y + 4z = -7$$
$$-x - 2y + z = 2$$
$$2x + 6y + 2z = -12$$

is

$$\begin{bmatrix} 1 & 2 & 4 & -7 \\ -1 & -2 & 1 & 2 \\ 2 & 6 & 2 & -12 \end{bmatrix}$$

We now do the row operations on the augmented matrix that mirror the equation operations done in example 4 in section 4.1 and are shown in table 4.5.

The final augmented matrix is in the desired form, one that represents a system in triangular form. By the interpretation of the rows, we can now solve the system by back substitution, yielding the solution $z = -1$, $y = -2$, and $x = 1$. Furthermore, since the two augmented matrices

$$\begin{bmatrix} 1 & 2 & 4 & -7 \\ -1 & -2 & 1 & 2 \\ 2 & 6 & 2 & -12 \end{bmatrix} \quad \text{and} \quad \begin{bmatrix} 1 & 2 & 4 & -7 \\ 0 & 2 & -6 & 2 \\ 0 & 0 & 5 & -5 \end{bmatrix}$$

TABLE 4.5

Equivalent systems	Steps	Equivalent matrices
$x + 2y + 4z = -7$ $-x - 2y + z = 2$ $2x + 6y + 2z = -12$		$\begin{bmatrix} 1 & 2 & 4 & -7 \\ -1 & -2 & 1 & 2 \\ 2 & 6 & 2 & -12 \end{bmatrix}$
$x + 2y + 4z = -7$ $5z = -5$ $2x + 6y + 2z = -12$	add the first row to the second row	$\begin{bmatrix} 1 & 2 & 4 & -7 \\ 0 & 0 & 5 & -5 \\ 2 & 6 & 2 & -12 \end{bmatrix}$
$x + 2y + 4z = -7$ $5z = -5$ $2y - 6z = 2$	add -2 times the first row to the third row	$\begin{bmatrix} 1 & 2 & 4 & -7 \\ 0 & 0 & 5 & -5 \\ 0 & 2 & -6 & 2 \end{bmatrix}$
$x + 2y + 4z = -7$ $2y - 6z = 2$ $5z = -5$	interchange the second and third rows	$\begin{bmatrix} 1 & 2 & 4 & -7 \\ 0 & 2 & -6 & 2 \\ 0 & 0 & 5 & -5 \end{bmatrix}$

are row equivalent, they represent two systems of linear equations with the same solutions. Therefore $x = 1$, $y = -2$, and $z = -1$ is also the only solution to the original system.

If an augmented matrix represents a system in triangular form, the solution, if any, can be obtained through back substitution. If the matrix does not represent a system in triangular form, we wish to choose row operations so as to obtain a row equivalent matrix in the desired form, as done in example 3. To this end, we employ a procedure called **Gaussian elimination**. In the next example, we introduce convenient notation for describing the row operations being performed in the elimination processes.

EXAMPLES

4. Solve, if possible, the system

$$2x_1 - 2x_2 + 4x_3 - 6x_4 = 10$$
$$2x_1 - 2x_2 + 5x_3 - 5x_4 = 9$$
$$x_2 - x_3 = 5$$
$$-3x_1 + 2x_2 + x_3 + 16x_4 = -18$$

using Gaussian elimination and back substitution.

The augmented matrix for the system is

$$A = \begin{bmatrix} 2 & -2 & 4 & -6 & 10 \\ 2 & -2 & 5 & -5 & 9 \\ 0 & 1 & -1 & 0 & 5 \\ -3 & 2 & 1 & 16 & -18 \end{bmatrix}.$$

We first perform row operations on A to obtain a row equivalent matrix A' that represents a system in triangular form.

Steps

$$\begin{bmatrix} 2 & -2 & 4 & -6 & | & 10 \\ 2 & -2 & 5 & -5 & | & 9 \\ 0 & 1 & -1 & 0 & | & 5 \\ -3 & 2 & 1 & 16 & | & -18 \end{bmatrix}$$

$$\xrightarrow[\frac{3}{2}R_1 + R_4]{-R_1 + R_2} \begin{bmatrix} 2 & -2 & 4 & -6 & | & 10 \\ 0 & 0 & 1 & 1 & | & -1 \\ 0 & 1 & -1 & 0 & | & 5 \\ 0 & -1 & 7 & 7 & | & -3 \end{bmatrix}$$

Explanation of Steps

add -1 times the first row (R_1) to the second row (R_2) and add $\frac{3}{2}$ times the first row (R_1) to the fourth row (R_4)

$$\xrightarrow{R_2 \leftrightarrow R_3} \begin{bmatrix} 2 & -2 & 4 & -6 & | & 10 \\ 0 & 1 & -1 & 0 & | & 5 \\ 0 & 0 & 1 & 1 & | & -1 \\ 0 & -1 & 7 & 7 & | & -3 \end{bmatrix}$$

interchange the second row (R_2) with the third row (R_3)

$$\xrightarrow{R_2 + R_4} \begin{bmatrix} 2 & -2 & 4 & -6 & | & 10 \\ 0 & 1 & -1 & 0 & | & 5 \\ 0 & 0 & 1 & 1 & | & -1 \\ 0 & 0 & 6 & 7 & | & 2 \end{bmatrix}$$

add the second row (R_2) to the fourth row (R_4)

$$\xrightarrow{-6R_3 + R_4} \begin{bmatrix} 2 & -2 & 4 & -6 & | & 10 \\ 0 & 1 & -1 & 0 & | & 5 \\ 0 & 0 & 1 & 1 & | & -1 \\ 0 & 0 & 0 & 1 & | & 8 \end{bmatrix}$$

add -6 times the third row (R_3) to the fourth row (R_4)

A'

We now obtain the solution to our original system using A' with back substitution.

The last row of A' represents the equation $x_4 = 8$. Placing this value into the equation $x_3 + x_4 = -1$, represented by the third row of A', we obtain $x_3 = -9$.

Since the second row of A' represents the equation $x_2 - x_3 = 5$, we obtain $x_2 = -4$. Finally, from the first equation

$$2x_1 - 2x_2 + 4x_3 - 6x_4 = 10$$

and the values we obtained thus far, we have

$$x_1 = \frac{1}{2}[10 + 2(-4) - 4(-9) + 6(8)] = 43.$$

5. Let us solve, if possible, the system

$$x + 2y + 3z = 4$$
$$3x + 7y + 8z = 14$$
$$-x \quad\quad - 5z = 1$$

using Gaussian elimination with back substitution.

The following gives the sequence of equivalent augmented matrices and the steps used to obtain them.

Steps	Explanation of Steps

$$\begin{bmatrix} 1 & 2 & 3 & | & 4 \\ 3 & 7 & 8 & | & 14 \\ -1 & 0 & -5 & | & 1 \end{bmatrix}$$

$$\xrightarrow[R_1+R_3]{-3R_1+R_2} \begin{bmatrix} 1 & 2 & 3 & | & 4 \\ 0 & 1 & -1 & | & 2 \\ 0 & 2 & -2 & | & 5 \end{bmatrix}$$ reduce the entries in column 1 below the $(1,1)$ entry to zero

$$\xrightarrow{-2R_1+R_3} \begin{bmatrix} 1 & 2 & 3 & | & 4 \\ 0 & 1 & -1 & | & 2 \\ 0 & 0 & 0 & | & 1 \end{bmatrix}$$ reduce the entries in column 2 below the $(2,2)$ entry to zero

The third row of the final augmented matrix represents the equation $0z = 1$. It is clear from this equation that the system has no solutions, and therefore the original system has no solutions.

As in example 5, it is not difficult to recognize when a system is inconsistent, for example, if we derive an augmented matrix in which one row has all zeros except for the final entry, then the system has no solutions. The next section gives an example of a system that has infinitely many solutions. Let us now look at some applications of systems of linear equations.

EXAMPLES

6. *A Mixture Problem.* The Lobo Preserve wishes to blend three types of food, grade A, grade B, and grade C, to obtain a mixture for feeding to its wolves. The mixture must be 24.5% protein and 10.8% fat. Grade A food is 26% protein and 12% fat; grade B food is 22% protein and 8% fat, and grade C food is 20% protein and 9% fat. How many pounds of each type food should be used to obtain 50 pounds of the desired mixture?

Let us begin our model with the unknowns

$$A = \text{pounds of grade A food in the mixture,}$$
$$B = \text{pounds of grade B food in the mixture,}$$
and $$C = \text{pounds of grade C food in the mixture.}$$

One requirement is that the total pounds must be equal to 50, which can be written as
$$A + B + C = 50.$$

The other requirements on the choice of A, B, and C pertain to the protein and fat they contribute to the mixture. Using our knowledge on writing linear forms, we have

$$\text{protein in mixture} = 0.26A + 0.22B + 0.20C$$
and $$\text{fat in mixture} = 0.12A + 0.08B + 0.09C.$$

Since the 50 pounds of mixture must contain $(0.245)(50) = 12.25$ units of protein and $(0.108)(50) = 5.4$ units of fat, we set our linear forms to these numbers to obtain the equations

$$0.26A + 0.22B + 0.20C = 12.25$$

$$0.12A + 0.08B + 0.09C = 5.4.$$

Thus our problem is to solve a system of three linear equations in three unknowns with the augmented matrix being as follows:

$$\begin{bmatrix} 1 & 1 & 1 & 50 \\ 0.26 & 0.22 & 0.20 & 12.25 \\ 0.12 & 0.08 & 0.09 & 5.40 \end{bmatrix}$$

Performing Gaussian elimination, we obtain the following sequence of matrices.

$$\begin{bmatrix} 1 & 1 & 1 & 50 \\ 0.26 & 0.22 & 0.20 & 12.25 \\ 0.12 & 0.08 & 0.09 & 5.40 \end{bmatrix}$$

$$\xrightarrow[\substack{100R_2 \\ 100R_3}]{} \begin{bmatrix} 1 & 1 & 1 & 50 \\ 26 & 22 & 20 & 1225 \\ 12 & 8 & 9 & 540 \end{bmatrix}$$

$$\xrightarrow[\substack{-26R_1 + R_2 \\ -12R_1 + R_3}]{} \begin{bmatrix} 1 & 1 & 1 & 50 \\ 0 & -4 & -6 & -75 \\ 0 & -4 & -3 & -60 \end{bmatrix}$$

$$\xrightarrow[\substack{-R_2 + R_3}]{} \begin{bmatrix} 1 & 1 & 1 & 50 \\ 0 & -4 & -6 & -75 \\ 0 & 0 & 3 & 15 \end{bmatrix}.$$

Observe that we made the computation easier by first clearing the decimals with multiplying rows 2 and 3 by 100. Now, with the last matrix and back substitution, we obtain the solution

$$C = 15/3 \qquad\qquad = 5$$

$$B = [-75 + 6(5)]/(-4) = 11.25$$

$$A = 50 - 11.25 - 5 \qquad = 33.75.$$

Therefore, by mixing 33.75 pounds of grade A, 11.25 pounds of grade B, and 5 pounds of grade C food, 50 pounds of mixture will be obtained that is 24.5% protein and 10.8% fat.

7. *A Production Problem.* A leather goods company manufactures wallets, belts, purses, and key fobs. The items go through four phases of production: cutting, design work, staining, and assembling. The amount of time needed in each phase to produce one unit of each item and the total amount of time available for each phase are given in table 4.6. How many items should the company produce so that the available time is completely used?

TABLE 4.6

Phase	Time required per unit (man-hours)				Time available
	Wallet	Belt	Purse	Key fob	
cutting	$\frac{1}{8}$	$\frac{1}{10}$	$\frac{1}{4}$	$\frac{1}{20}$	39
design work	$\frac{1}{2}$	$\frac{1}{2}$	$\frac{3}{4}$	$\frac{1}{4}$	161
staining	$\frac{1}{8}$	$\frac{1}{12}$	$\frac{1}{4}$	$\frac{1}{20}$	37
assembling	$\frac{1}{4}$	$\frac{1}{15}$	$\frac{1}{2}$	$\frac{1}{25}$	56

Let

$$w = \text{the number of wallets to be made,}$$

$$b = \text{the number of belts to be made,}$$

$$p = \text{the number of purses to be made,}$$

and

$$k = \text{the number of key fobs to be made.}$$

From our work with linear forms, we see that the dependencies of time used on these four variables are

$$\text{cutting} = \frac{1}{8}w + \frac{1}{10}b + \frac{1}{4}p + \frac{1}{20}k$$

$$\text{design work} = \frac{1}{2}w + \frac{1}{2}b + \frac{3}{4}p + \frac{1}{4}k$$

$$\text{staining} = \frac{1}{8}w + \frac{1}{12}b + \frac{1}{4}p + \frac{1}{20}k$$

$$\text{assembling} = \frac{1}{4}w + \frac{1}{15}b + \frac{1}{2}p + \frac{1}{25}k.$$

Now we impose the requirement that these linear forms be equal to 39, 161, 37, and 56, respectively, yielding the system of linear equations with the augmented matrix

$$\begin{bmatrix} \frac{1}{8} & \frac{1}{10} & \frac{1}{4} & \frac{1}{20} & | & 39 \\ \frac{1}{2} & \frac{1}{2} & \frac{3}{4} & \frac{1}{4} & | & 161 \\ \frac{1}{8} & \frac{1}{12} & \frac{1}{4} & \frac{1}{20} & | & 37 \\ \frac{1}{4} & \frac{1}{15} & \frac{1}{2} & \frac{1}{25} & | & 56 \end{bmatrix}.$$

We can eliminate the fractions by multiplying the first, second, third, and fourth rows by 40, 4, 120, and 300, respectively, and we have the augmented matrix

$$\begin{bmatrix} 5 & 4 & 10 & 2 & | & 1560 \\ 2 & 2 & 3 & 1 & | & 644 \\ 15 & 10 & 30 & 6 & | & 4440 \\ 75 & 20 & 150 & 12 & | & 16800 \end{bmatrix}.$$

A computer program was used to solve the system with the latter augmented matrix. The computer printout is given in figure 4.2. We see from these results that the company should make $w = 80$ wallets, $b = 120$ belts, $p = 48$ purses, and $k = 100$ key fobs.

```
THIS PROGRAM SOLVES BY GAUSSIAN ELIMINATION WITH
BACK SUBSTITUTION A SQUARE SYSTEM OF LINEAR EQUATIONS.
THE NUMBER OF EQUATIONS (AND THEREFORE THE NUMBER OF
VARIABLES ALSO) CAN BE UP TO 10.

How many equations does the system have?
!4
Do you want the matrix after each elimination - Y or N?
!Y
Enter a row of the augmented matrix at each prompt.
!5, 4, 10, 2, 1560
!2, 2, 3, 1, 644
!15, 10, 30, 6, 4440
!75, 20, 150, 12, 16800
```

```
                    THE AUGMENTED MATRIX IS
                    === ========= ====== ==

X(1)        X(2)        X(3)        X(4)        RIGHT SIDE
5           4           10          2           1560
2           2           3           1           644
15          10          30          6           4440
75          20          150         12          16800

               AUGMENTED MATRIX AFTER ELIMINATION OF
                       VARIABLES IN COLUMN 1
               =======================================

X(1)        X(2)        X(3)        X(4)        RIGHT SIDE
5           4           10          2           1560
0           .4          -1          .2          20
0           -2          0           0           -240
0           -40         0           -18         -6600

               AUGMENTED MATRIX AFTER ELIMINATION OF
                       VARIABLES IN COLUMN 2
               =======================================

X(1)        X(2)        X(3)        X(4)        RIGHT SIDE
5           4           10          2           1560
0           .4          -1          .2          20
0           0           -5          1           -140
0           0           -100        2           -4600

               AUGMENTED MATRIX AFTER ELIMINATION OF
                       VARIABLES IN COLUMN 3
               =======================================

X(1)        X(2)        X(3)        X(4)        RIGHT SIDE
5           4           10          2           1560
0           .4          -1          .2          20
0           0           -5          1           -140
0           0           0           -18         -1800

THE SOLUTION TO THE SYSTEM IS
=============================
        X(1) = 80
        X(2) = 120
        X(3) = 48
        X(4) = 100
```

FIGURE 4.2

In example 7 we illustrated the value of using a computer program, especially when the calculations become tedious. Using a computer in this way allows us to devote more energy to what is truly the human role in problem solving—that is, describing the problem, gathering the necessary information, and putting it into a form suitable for input to the computer program.

EXERCISES 4.2

In exercises 1 and 2 give the augmented matrix for the system of linear equations.

1. a. $\begin{aligned} -x \quad\quad + 6z &= \quad 0 \\ 4x + 5y - 2z &= \quad 1 \\ 2x - 6y \quad\quad &= -3 \end{aligned}$

   b. $\begin{aligned} 8x_1 - \quad x_2 + 3x_3 + \quad x_4 &= \quad 2 \\ 3x_2 - \quad x_3 + 2x_4 &= -3 \\ x_1 + 4x_2 - 2x_3 - \quad x_4 &= \quad 1 \end{aligned}$

2. a. $\begin{aligned} 2x + 7y - \quad z &= 0 \\ 5y + 6z &= 0 \\ 6x \quad\quad - 4z &= 0 \end{aligned}$

   b. $\begin{aligned} x_1 + 2x_2 \quad\quad + 4x_4 &= 8 \\ 3x_2 + 5x_3 - \quad x_4 &= 7 \\ 2x_3 + \quad x_4 &= 6 \end{aligned}$

For exercises 3 and 4 write the system of linear equations that has the given augmented matrix.

3. a. $\left[\begin{array}{cc|c} 1 & -1 & 4 \\ 0 & 1 & 2 \end{array}\right]$

   b. $\left[\begin{array}{ccc|c} 1 & 2 & -1 & 3 \\ 2 & 0 & 5 & -7 \\ 0 & -3 & 4 & 2 \end{array}\right]$

4. a. $\left[\begin{array}{ccc|c} 1 & -\frac{1}{2} & \frac{3}{2} & \frac{1}{2} \\ 0 & 1 & -1 & 2 \\ 0 & 0 & 1 & 4 \end{array}\right]$

   b. $\left[\begin{array}{ccc|c} 0 & 3 & 4 & 5 & 0 \\ 2 & 1 & -2 & 6 & 0 \\ 3 & -1 & 1 & 0 & 0 \end{array}\right]$

For exercises 5 and 6 verify that the given pair of matrices are row equivalent by determining a sequence of row operations.

5. a. $\left[\begin{array}{cc|c} 2 & -2 & 8 \\ 3 & 0 & 18 \end{array}\right]$ and $\left[\begin{array}{cc|c} 1 & -1 & 4 \\ 0 & 1 & 2 \end{array}\right]$

   b. $\left[\begin{array}{ccc|c} 2 & -1 & 3 & 1 \\ -2 & 4 & -6 & 5 \\ 4 & -5 & 10 & 0 \end{array}\right]$ and $\left[\begin{array}{ccc|c} 1 & -\frac{1}{2} & \frac{3}{2} & \frac{1}{2} \\ 0 & 1 & -1 & 2 \\ 0 & 0 & 1 & 4 \end{array}\right]$

6. a. $\left[\begin{array}{ccc|c} 0 & \frac{1}{3} & 1 & 0 \\ 2 & -2 & 8 & 0 \end{array}\right]$ and $\left[\begin{array}{ccc|c} 1 & -1 & 4 & 0 \\ 0 & 1 & 3 & 0 \end{array}\right]$

   b. $\left[\begin{array}{cccc|c} 2 & 0 & 3 & 4 & -4 \\ 2 & 0 & 2 & 4 & -6 \\ 1 & 0 & 3 & 0 & 5 \end{array}\right]$ and $\left[\begin{array}{cccc|c} 1 & 0 & 1 & 2 & -3 \\ 0 & 0 & 1 & -1 & 4 \\ 0 & 0 & 0 & 1 & -2 \end{array}\right]$

For exercises 7–12 each matrix is the augmented matrix for a sytem of linear equations. Solve the system, if possible, by back substitution.

7. $\begin{bmatrix} 1 & -1 & | & 4 \\ 0 & 1 & | & 2 \end{bmatrix}$

8. $\begin{bmatrix} 1 & 3 & | & 0 \\ 0 & 0 & | & 4 \end{bmatrix}$

9. $\begin{bmatrix} 1 & -\frac{1}{2} & \frac{3}{2} & | & \frac{1}{2} \\ 0 & 1 & -1 & | & 2 \\ 0 & 0 & 1 & | & 4 \end{bmatrix}$

10. $\begin{bmatrix} 1 & 0 & -2 & | & 1 \\ 0 & 2 & 8 & | & 6 \\ 0 & 0 & 1 & | & 2 \end{bmatrix}$

11. $\begin{bmatrix} -1 & -1 & -2 & | & 3 \\ 0 & 4 & -4 & | & 16 \\ 0 & 0 & 0 & | & 5 \end{bmatrix}$

12. $\begin{bmatrix} 1 & 1 & 2 & | & -3 \\ 0 & 1 & -1 & | & 4 \\ 0 & 0 & -3 & | & 6 \end{bmatrix}$

For exercises 13–18 solve the system, if possible, by using Gaussian elimination and back substitution.

13. $\begin{aligned} -2x + 3y + 4z &= -8 \\ 3x + 4y + 7z &= -5 \\ 2x - 5y + 6z &= 12 \end{aligned}$

14. $\begin{aligned} 2x - y + 3z &= 1.4 \\ 3x + 2z &= 1.2 \\ 4y - 7z &= -10 \end{aligned}$

15. $\begin{aligned} x_1 - 2x_2 + 3x_3 + 4x_4 &= 8 \\ 2x_2 - 10x_4 &= 20 \\ -2x_3 + 2x_4 &= -6 \\ 3x_1 - 4x_2 + 9x_3 + 6x_4 &= 0 \end{aligned}$

16. $\begin{aligned} x_1 + 2x_2 + 3x_3 + 4x_4 &= -3 \\ -0.4x_1 + 0.6x_3 + 0.3x_4 &= -3 \\ 0.2x_3 - 0.2x_4 &= 0 \\ 0.2x_1 - 0.4x_2 + 0.9x_3 + 0.1x_4 &= -3 \end{aligned}$

17. $\begin{aligned} 2x_1 + x_2 + 2x_3 + 3x_4 + x_5 &= 1 \\ x_2 + x_3 + 2x_4 + 2x_5 &= -1 \\ 2x_1 + x_2 + 3x_3 + x_4 &= 1 \\ 4x_1 + 3x_2 + 6x_3 + 6x_4 + 4x_5 &= -1 \\ 3x_4 + x_1 &= 1 \end{aligned}$

18. $\begin{aligned} x_1 + 9x_2 - 7x_3 + 2x_4 + 4x_5 &= 4 \\ 2x_1 - 3x_2 + 5x_3 + x_4 - x_5 &= 6 \\ 5x_1 - x_2 + 3x_3 - 2x_4 &= -2 \\ x_1 - 12x_2 + 12x_3 - x_4 - 5x_5 &= -10 \\ 3x_1 + 4x_2 - 2x_3 + 2x_5 &= -3 \end{aligned}$

19. Orion Manufacturing Company makes lift rods, slip nuts, and trap plugs. Each item requires three machines in its manufacture. The number of man-hours required of each machine for a box of each item is given in table 4.7.

TABLE 4.7

	Lift rods	Slip nuts	Trap plugs
machine 1	$\frac{1}{2}$	$\frac{3}{4}$	1
machine 2	$\frac{1}{2}$	$\frac{1}{2}$	$\frac{1}{2}$
machine 3	$\frac{1}{4}$	$\frac{1}{2}$	1

Let

$$x = \text{the number of boxes of lift rods made in a day}$$

$$y = \text{the number of boxes of slip nuts made in a day}$$

$$z = \text{the number of boxes of trap plugs made in a day.}$$

a. Let T_i be the time needed on machine i, $i = 1, 2, 3$, for the daily production of the three items. Express each T_i as a function of x, y, and z.

b. Suppose machine 1 can be used 14 hours a day, machine 2 can be used 10 hours a day, and machine 3 can be used 10 hours a day. How many boxes of each item can be made in a day under these restrictions on the three machines?

20. The Avalone Production Company makes flash heads, battery compartments, and calculator dials. Each item requires three machines in its manufacture. The number of man-hours required on each machine for a carton of each item is given in table 4.8.

TABLE 4.8

	Flash heads	Battery compartments	Calculator dials
machine 1	$\frac{1}{6}$	$\frac{1}{3}$	$\frac{1}{2}$
machine 2	$\frac{1}{6}$	$\frac{2}{3}$	$\frac{1}{2}$
machine 3	$\frac{1}{2}$	$\frac{1}{3}$	$\frac{1}{2}$

Let

x = the number of cartons of flash heads made in a day

y = the number of cartons of battery compartments made in a day

z = the number of cartons of calculator dials made in a day.

a. Let T_i be the time needed on machine i, $i = 1, 2, 3$, for the daily production of the three items. Express each T_i as a function of x, y, and z.

b. Suppose machine 1 can be used 10 hours a day, machine 2 can be used 13 hours a day, and machine 3 can be used 14 hours a day. How many cartons of each item can be made in a day under these restrictions on the three machines?

21. Hillcrest Farms has 375 acres, which it plans to plant in soybeans, corn, and wheat. It costs the farm $120 to plant one acre of soybeans, $100 to plant one acre of corn, and $90 to plant one acre of wheat. The farm has $41,515 available to cover these costs. Furthermore, Hillcrest Farms anticipates an income of $400 on each acre of wheat, $480 on each acre of soybeans, and $400 on each acre of corn. How many acres of each crop should be planted so that the entire acreage is used, all the money available for planting is used, and the anticipated income is $167,260?

22. Parkland Kennels wishes to blend three types of dog food, High-Protein, Regular, and Low-Protein, to obtain a food that is 24.5% protein and 11% fat. The percentage of protein and fat for each type food is given in table 4.9.

TABLE 4.9

	High-Protein	Regular	Low-Protein
protein (%)	26	22	20
fat (%)	12	9	8

How many pounds of each type food should be blended to obtain a 50-pound mixture that is 24.5% protein and 11% fat?

23. Consider the system

$$x \qquad + \ z = -3$$
$$-2x + \ y + 2z = -7$$
$$-x - 2y + 4z = \quad a.$$

Determine a so that the system has $x = -1$, $y = -5$, and $z = -2$ as its solution.

4.3 Gauss–Jordan Reduction

Gaussian elimination with back substitution is described in the previous section. In this section we give a second method for solving a system of linear equations, called **Gauss–Jordan reduction**, convenient for solving a linear system by hand. We also use Gauss–Jordan reduction when finding the inverse of a matrix.

Consider again the augmented matrix derived in example 4 in section 4.2

$$A = \begin{bmatrix} 2 & -2 & 4 & -6 & | & 10 \\ 0 & 1 & -1 & 0 & | & 5 \\ 0 & 0 & 1 & 1 & | & -1 \\ 0 & 0 & 0 & 1 & | & 8 \end{bmatrix}.$$

The last two rows of A represent the equations $x_3 + x_4 = -1$ and $x_4 = 8$, respectively. Using back substitution we found that $x_3 = -1 - 8 = -9$. As an alternate method for finding the value for x_3, let us eliminate x_4 from the equation $x_3 + x_4 = -1$ by subtracting $x_4 = 8$ from the equation. In doing so we have $x_3 = -9$. In terms of the rows of A this is comparable to subtracting the fourth row from the third row and obtaining the row equivalent matrix

$$A_1 = \begin{bmatrix} 2 & -2 & 4 & -6 & | & 10 \\ 0 & 1 & -1 & 0 & | & 5 \\ 0 & 0 & 1 & 0 & | & -9 \\ 0 & 0 & 0 & 1 & | & 8 \end{bmatrix}.$$

We see immediately from the third and fourth rows of A_1 that $x_3 = -9$ and $x_4 = 8$.

Next, to find x_2, eliminate all variables except x_2 from the second equation, represented by the second row of A_1. The row operation that accomplishes this is to add the third row of A_1 to the second row, which gives the row equivalent matrix

$$A_2 = \begin{bmatrix} 2 & -2 & 4 & -6 & | & 10 \\ 0 & 1 & 0 & 0 & | & -4 \\ 0 & 0 & 1 & 0 & | & -9 \\ 0 & 0 & 0 & 1 & | & 8 \end{bmatrix}.$$

We see from the second row of A_2 that $x_2 = -4$.

The value for x_1 can be found by eliminating from the first equation (the first row of A_2), all variables except x_1. This is accomplished by the following three row operations: add to the first row 6 times the fourth row, -4 times the third row, and 2 times the second row. Performing these row operations on A_2 and multiplying the first row by $\frac{1}{2}$ gives the row equivalent matrix

$$A_3 = \begin{bmatrix} 1 & 0 & 0 & 0 & | & 43 \\ 0 & 1 & 0 & 0 & | & -4 \\ 0 & 0 & 1 & 0 & | & -9 \\ 0 & 0 & 0 & 1 & | & 8 \end{bmatrix}.$$

Since matrix A_3 is row equivalent to matrix A, the solution to the system represented by A can be obtained from A_3. However, A_3 is in a more simplified form and the solution can be read directly from it. The equations represented by A_3 are $x_1 = 43$, $x_2 = -4$, $x_3 = -9$, and $x_4 = 8$.

From the preceding example we see that an alternative to back substitution is to perform further row operations on the matrix to obtain a matrix in a more simplified form.

Definition

■ A matrix is in **row-echelon form** if all three of the following conditions are satisfied.

1. any row in which all entries are zero appears at the bottom of the matrix; that is, below any other row that has at least one non-zero entry

2. if a row has a non-zero entry, then the first non-zero entry in that row is 1, called the **leading** 1 of the row

3. for any two rows which have non-zero entries, the leading 1 of the lower row is to the right of the leading 1 of the upper row. ■

A matrix is in **reduced row-echelon form** if it is in row-echelon form and

4. if a row has a non-zero entry and the leading 1 of this row is in column j, then every entry in the jth column, above and below the leading 1, is zero.

EXAMPLES

1. The matrix

$$\begin{bmatrix} 1 & 2 & 4 & -7 \\ 0 & 1 & -3 & 1 \\ 0 & 0 & 1 & -1 \end{bmatrix}$$

is in row-echelon form. Other examples of matrices in row-echelon form are

a. $\begin{bmatrix} 1 & 0 & 1 & 7 \\ 0 & 0 & 1 & 2 \\ 0 & 0 & 0 & 0 \end{bmatrix}$ 　 b. $\begin{bmatrix} 1 & 0 \\ 0 & 1 \end{bmatrix}$ 　 c. $\begin{bmatrix} 0 & 0 & 0 \\ 0 & 0 & 0 \end{bmatrix}$.

A matrix that does not satisfy the first condition and, therefore, is not in row-echelon form is

$$\begin{bmatrix} 1 & 0 & 1 \\ 0 & 0 & 0 \\ 0 & 1 & 0 \end{bmatrix}.$$

Examples of matrices not in row-echelon form because they do not satisfy the second and third conditions, respectively, are

$$\begin{bmatrix} 0 & 1 & 2 & 3 \\ 0 & 0 & 4 & 1 \end{bmatrix} \quad \text{and} \quad \begin{bmatrix} 1 & 0 & 1 \\ 0 & 1 & 2 \\ 0 & 1 & 3 \end{bmatrix}.$$

2. The following matrices are in reduced row-echelon form.

a. $\begin{bmatrix} 1 & 0 & 0 \\ 0 & 1 & 0 \\ 0 & 0 & 1 \end{bmatrix}$
b. $\begin{bmatrix} 0 & 1 & 2 & 0 & 3 \\ 0 & 0 & 0 & 1 & 4 \\ 0 & 0 & 0 & 0 & 0 \end{bmatrix}$

c. $\begin{bmatrix} 1 & 0 & -1 \\ 0 & 1 & 2 \end{bmatrix}$
d. $\begin{bmatrix} 0 & 0 \\ 0 & 0 \end{bmatrix}$

A matrix in row-echelon form but not in *reduced* row-echelon form is

$$\begin{bmatrix} 1 & 0 & 1 & 7 \\ 0 & 0 & 1 & 2 \\ 0 & 0 & 0 & 0 \end{bmatrix}.$$

As we have seen, once an augmented matrix is derived in reduced row-echelon form we can easily obtain the solutions to the system directly from the matrix. The elimination procedure producing a row equivalent matrix in reduced row-echelon form is called **Gauss–Jordan reduction**. This procedure is now discussed using the following system.

Consider the system of linear equations

$$2x_1 + 4x_2 + 8x_3 = -14$$
$$-x_1 - 2x_2 + x_3 = 2$$
$$2x_1 + 6x_2 + 2x_3 = -12$$

for which the augmented matrix is

$$\left[\begin{array}{ccc|c} ② & 4 & 8 & -14 \\ -1 & -2 & 1 & 2 \\ 2 & 6 & 2 & -12 \end{array}\right].$$

We perform row operations on the augmented matrix to obtain an equivalent matrix in reduced row-echelon form. We first work with column 1 to change the (1, 1) entry, called a **pivot**, to 1 and then change the other entries in the column of the pivot to zeros. Therefore, let us multiply row 1 by $\frac{1}{2}$, which gives the equivalent augmented matrix

$$\left[\begin{array}{ccc|c} ① & 2 & 4 & -7 \\ -1 & -2 & 1 & 2 \\ 2 & 6 & 2 & -12 \end{array}\right].$$

Next we reduce the entries below the pivot to zero by adding the first row to the second and adding -2 times the first row to the third. In doing so we derive the following augmented matrix.

$$\left[\begin{array}{ccc|c} 1 & 2 & 4 & -7 \\ 0 & 0 & 5 & -5 \\ 0 & 2 & -6 & 2 \end{array}\right].$$

Having changed column 1 by row operations to the column

$$\begin{bmatrix} 1 \\ 0 \\ 0 \end{bmatrix},$$

we now focus on column 2 and the $(2, 2)$ entry, which is our new pivot. We want to perform row operations on the augmented matrix to obtain 1 in the pivot and zeros above and below it in column 2; that is, we want to perform row operations in order to change

$$\begin{bmatrix} 2 \\ 0 \\ 2 \end{bmatrix} \quad \text{to} \quad \begin{bmatrix} 0 \\ 1 \\ 0 \end{bmatrix}.$$

However, since the pivot is now 0, this change can only be accomplished with first doing a row interchange. But since we want the $(1, 1)$ entry left as is, the only row interchange available is interchanging row 2 with the one below it. Therefore, by interchanging rows 2 and 3, we derive the equivalent augmented matrix

$$\begin{bmatrix} 1 & 2 & 4 & | & -7 \\ 0 & 2 & -6 & | & 2 \\ 0 & 0 & 5 & | & -5 \end{bmatrix}.$$

From here Gauss–Jordan reduction follows.

Steps	*Explanation of Steps*

$$\begin{bmatrix} 1 & 2 & 4 & | & -7 \\ 0 & ② & -6 & | & 2 \\ 0 & 0 & 5 & | & -5 \end{bmatrix}$$

$\xrightarrow{\frac{1}{2}R_2}$
$$\begin{bmatrix} 1 & 2 & 4 & | & -7 \\ 0 & ① & -3 & | & 1 \\ 0 & 0 & 5 & | & -5 \end{bmatrix}$$
multiply row 2 by $\frac{1}{2}$ to change the pivot from 2 to 1

$\xrightarrow{-2R_2 + R_1}$
$$\begin{bmatrix} 1 & 0 & 10 & | & -9 \\ 0 & 1 & -3 & | & 1 \\ 0 & 0 & ⑤ & | & -5 \end{bmatrix}$$
reduce the entries in the column of the pivot, other than the pivot, to zeros

$\xrightarrow{\frac{1}{5}R_3}$
$$\begin{bmatrix} 1 & 0 & 10 & | & -9 \\ 0 & 1 & -3 & | & 1 \\ 0 & 0 & ① & | & -1 \end{bmatrix}$$
multiply row 3 by $\frac{1}{5}$ to change the new pivot from 5 to 1

$\xrightarrow[-10R_3 + R_1]{3R_3 + R_2}$
$$\begin{bmatrix} 1 & 0 & 0 & | & 1 \\ 0 & 1 & 0 & | & -2 \\ 0 & 0 & 1 & | & -1 \end{bmatrix}$$
reduce the entries in the column of the pivot, other than the pivot, to zeros

The last matrix is in reduced row-echelon form, from which we can easily read the solution. In particular, we have that

$$x_1 = 1, \; x_2 = -2, \quad \text{and} \quad x_3 = -1.$$

With this last example in mind, we can state the steps of Gauss–Jordan reduction.

Steps of Gauss–Jordan Reduction

Step 1: Consider each of the entries $(1, 1), (2, 2)$, etc. in succession as the **pivot**. If the pivot is non-zero, proceed to step 2. Otherwise, interchange the row of the pivot with the first row *below* it having a non-zero entry in the column of the pivot and proceed to step 2.

Step 2: Change the pivot to 1 by multiplying the row of the pivot by the reciprocal of the pivot. Next, using row operations, reduce the entries above and below the pivot to zeros and proceed to step 3.

Step 3: Repeat steps 1 and 2 with the next pivot until a matrix in reduced row-echelon form is derived or we can determine that the system has no solutions.

EXAMPLES

3. Solve the system of linear equations

$$3x_1 - 2x_2 + 6x_3 + x_4 = 9$$

$$x_1 + \frac{1}{3}x_2 + 2x_3 + \frac{1}{3}x_4 = 5$$

$$3x_1 - 4x_2 + 6x_3 + 3x_4 = 9$$

$$2x_1 - \frac{13}{3}x_2 + 10x_3 + \frac{5}{3}x_4 = 11$$

using Gauss–Jordan reduction.

We give the augmented matrix and the steps of Gauss–Jordan reduction as follows:

Steps			*Explanation of Steps*

$$\begin{bmatrix} ③ & -2 & 6 & 1 & | & 9 \\ 1 & \frac{1}{3} & 2 & \frac{1}{3} & | & 5 \\ 3 & -4 & 6 & 3 & | & 9 \\ 2 & -\frac{13}{3} & 10 & \frac{5}{3} & | & 11 \end{bmatrix}$$

$$\xrightarrow{\frac{1}{3}R_1} \begin{bmatrix} ① & -\frac{2}{3} & 2 & \frac{1}{3} & | & 3 \\ 1 & \frac{1}{3} & 2 & \frac{1}{3} & | & 5 \\ 3 & -4 & 6 & 3 & | & 9 \\ 2 & -\frac{13}{3} & 10 & \frac{5}{3} & | & 11 \end{bmatrix}$$

multiply row 1 by $\frac{1}{3}$ to change the pivot to 1

$$\xrightarrow[\substack{-R_1+R_2 \\ -3R_1+R_3 \\ -2R_1+R_4}]{} \begin{bmatrix} 1 & -\frac{2}{3} & 2 & \frac{1}{3} & | & 3 \\ 0 & ① & 0 & 0 & | & 2 \\ 0 & -2 & 0 & 2 & | & 0 \\ 0 & -3 & 6 & 1 & | & 5 \end{bmatrix}$$

reduce the entries below the pivot in column 1 to zeros

$$\xrightarrow[\substack{\frac{1}{4}R_2+R_1 \\ 2R_2+R_3 \\ 3R_2+R_4}]{}
\begin{bmatrix}
1 & 0 & 2 & \frac{1}{3} & \frac{13}{3} \\
0 & 1 & 0 & 0 & 2 \\
0 & 0 & ⓪ & 2 & 4 \\
0 & 0 & 6 & 1 & 11
\end{bmatrix}$$

reduce the entries above and below the pivot in column 2 to zeros

$$\xrightarrow[\substack{R_3 \leftrightarrow R_4 \\ \frac{1}{6}R_3}]{}
\begin{bmatrix}
1 & 0 & 2 & \frac{1}{3} & \frac{13}{3} \\
0 & 1 & 0 & 0 & 2 \\
0 & 0 & ① & \frac{1}{6} & \frac{11}{6} \\
0 & 0 & 0 & 2 & 4
\end{bmatrix}$$

interchange rows 3 and 4 and multiply the new row 3 by $\frac{1}{6}$ to obtain a pivot of 1 in the $(3,3)$ entry

$$\xrightarrow[-2R_3+R_1]{}
\begin{bmatrix}
1 & 0 & 0 & 0 & \frac{2}{3} \\
0 & 1 & 0 & 0 & 2 \\
0 & 0 & 1 & \frac{1}{6} & \frac{11}{6} \\
0 & 0 & 0 & ② & 4
\end{bmatrix}$$

reduce the entries above and below the pivot in column 3 to zeros

$$\xrightarrow[\frac{1}{2}R_4]{}
\begin{bmatrix}
1 & 0 & 0 & 0 & \frac{2}{3} \\
0 & 1 & 0 & 0 & 2 \\
0 & 0 & 1 & \frac{1}{6} & \frac{11}{6} \\
0 & 0 & 0 & ① & 2
\end{bmatrix}$$

multiply row 4 by $\frac{1}{2}$ to change the pivot to 1

$$\xrightarrow[-\frac{1}{6}R_4+R_3]{}
\begin{bmatrix}
1 & 0 & 0 & 0 & \frac{2}{3} \\
0 & 1 & 0 & 0 & 2 \\
0 & 0 & 1 & 0 & \frac{3}{2} \\
0 & 0 & 0 & 1 & 2
\end{bmatrix}$$

reduce the entries above the pivot to zeros

The last matrix is in reduced row-echelon form and the reduction process stops. From this matrix we read the solution

$$x_1 = \frac{2}{3}, \quad x_2 = 2, \quad x_3 = \frac{3}{2}, \quad \text{and} \quad x_4 = 2.$$

4. Solve the system with augmented matrix

$$\begin{bmatrix}
2 & 1 & 4 & -8 \\
-3 & 1 & -11 & 22 \\
2 & -3 & 12 & -19
\end{bmatrix}$$

using Gauss–Jordan reduction.

Steps	*Explanation of Steps*

$$\begin{bmatrix}
② & 1 & 4 & -8 \\
-3 & 1 & -11 & 22 \\
2 & -3 & 12 & -19
\end{bmatrix}$$

$$\xrightarrow[\substack{\frac{1}{2}R_1 \\ 3R_1+R_2 \\ -2R_1+R_3}]{}
\begin{bmatrix}
1 & \frac{1}{2} & 2 & -4 \\
0 & \frac{5}{2} & -5 & 10 \\
0 & -4 & 8 & -11
\end{bmatrix}$$

multiply row 1 by $\frac{1}{2}$ to change the pivot to 1 and reduce the entries below it to zeros

$$\xrightarrow[\substack{\frac{2}{5}R_2 \\ -\frac{1}{2}R_2+R_1 \\ 4R_2+R_3}]{}
\begin{bmatrix}
1 & 0 & 3 & -6 \\
0 & 1 & -2 & 4 \\
0 & 0 & 0 & 5
\end{bmatrix}$$

multiply row 2 by $\frac{2}{5}$ to change the pivot to 1 and reduce the entries above and below it to zeros

The reduction process can stop with the last matrix since the $(3, 3)$ entry is zero and there is no row *below* it to interchange with. Furthermore, the third row of this final matrix represents the equation

$$0x + 0x + 0x = 5$$

which cannot be satisfied by any real numbers. Therefore the system has no solutions.

We have thus far been considering **square systems**—systems in which the number of unknowns and the number of equations are the same. In our next example we apply Gauss–Jordan reduction to solve a system that is not square.

EXAMPLE 5

Solve the system of linear equations

$$
\begin{aligned}
x_1 - x_2 - x_3 \quad\;\;\;\; &= 3 \\
2x_1 + 3x_2 \quad\quad\;\; - x_4 &= 5 \\
5x_2 + 3x_3 + x_4 &= 1
\end{aligned}
$$

using Gauss–Jordan reduction.

$$
\overset{A}{
\begin{bmatrix}
① & -1 & -1 & 0 & 3 \\
2 & 3 & 0 & -1 & 5 \\
0 & 5 & 3 & 1 & 1
\end{bmatrix}
}
$$

$$
\xrightarrow{-2R_1 + R_2}
\begin{bmatrix}
1 & -1 & -1 & 0 & 3 \\
0 & ⑤ & 2 & -1 & -1 \\
0 & 5 & 3 & 1 & 1
\end{bmatrix}
$$

$$
\xrightarrow[\substack{R_2 + R_1 \\ -5R_2 + R_3}]{\tfrac{1}{5}R_2}
\begin{bmatrix}
1 & 0 & -\tfrac{3}{5} & -\tfrac{1}{5} & \tfrac{14}{5} \\
0 & 1 & \tfrac{2}{5} & -\tfrac{1}{5} & -\tfrac{1}{5} \\
0 & 0 & ① & 2 & 2
\end{bmatrix}
\xrightarrow[\tfrac{3}{5}R_3 + R_1]{-\tfrac{2}{5}R_3 + R_2}
\begin{bmatrix}
1 & 0 & 0 & 1 & 4 \\
0 & 1 & 0 & -1 & -1 \\
0 & 0 & 1 & 2 & 2
\end{bmatrix}.
$$

$$A'$$

The final matrix A' is in reduced row-echelon form. We see in matrix A' that not all of the variables appear in one and only one equation with coefficient 1; that is, not all of the variables correspond to a **leading one** in A'. The variables that correspond to a leading one in our example are x_1, x_2, and x_3. Furthermore, the system represented by the augmented matrix A', which is equivalent to the original system, is

$$
\begin{aligned}
x_1 \quad\quad\quad + x_4 &= 4 \\
x_2 \quad\; - x_4 &= -1 \\
x_3 + 2x_4 &= 2.
\end{aligned}
$$

We observe that the system has infinitely many solutions. In particular, no matter what value is assigned to x_4, we can find values for the other variables. For example, solving the first equation for x_1, we have $x_1 = 4 - x_4$. Thus, solving the equations for the variables that correspond to a leading one and setting $x_4 = t$, as an arbitrary real number, we have

$$x_1 = 4 - t$$

$$x_2 = -1 + t$$

$$x_3 = 2 - 2t$$

$$x_4 = t, \quad \text{any real number.}$$

These last equations describe all solutions to the system. Since there are an infinite number of values that can be assigned to the **parameter** t, and therefore to x_4, the system has infinitely many solutions of the preceding form. Examples of two such solutions are

$$x_1 = 4, \quad x_2 = -1, \quad x_3 = 2, \quad x_4 = 0,$$

obtained from setting $t = 0$, and

$$x_1 = 0, \quad x_2 = 3, \quad x_3 = -6, \quad x_4 = 4,$$

obtained from setting $t = 4$.

Example 5 illustrates a system that has infinitely many solutions. As seen in the final matrix A', obtained through Gauss–Jordan reduction, the number of leading ones is less than the number of unknowns in the system. The unknowns corresponding to the leading ones are called **basic variables**. As we saw in the example, if not all the unknowns of the system are basic variables and if there is no row of the reduced row-echelon matrix of the form $[0 \quad 0 \cdots 0 \,|\, b]$, where $b \neq 0$, then the system has an infinite number of solutions. In this case, the general solution to the system can be obtained as follows:

General Solution to a System with Infinitely Many Solutions

Step 1: Using the reduced row-echelon augmented matrix obtained through Gauss–Jordan reduction, solve the equations with leading ones for the basic variables.

Step 2: Assign a parameter, an arbitary real number, to each of the variables that is not basic.

Step 3: The **parametric equations** obtained in step 2 give the general solution to the system. A particular solution to the system can be obtained by giving specific values to the parameters.

EXAMPLES

6. Consider the system

$$\begin{aligned} x_1 - 2x_2 + 3x_4 &= 4 \\ 2x_1 - 4x_2 + x_3 + 2x_4 &= 3 \\ -5x_1 + 10x_2 - 3x_3 - 3x_4 &= -5 \\ x_1 - 2x_2 + x_3 - x_4 &= -1. \end{aligned}$$

Let us solve the system using Gauss–Jordan reduction.

First, the augmented matrix for the system is

$$\left[\begin{array}{cccc|c} 1 & -2 & 0 & 3 & 4 \\ 2 & -4 & 1 & 2 & 3 \\ -5 & 10 & -3 & -3 & -5 \\ 1 & -2 & 1 & -1 & -1 \end{array}\right].$$

Using row operations, it is not difficult to derive the following augmented matrix, which is in reduced row-echelon form.

$$\left[\begin{array}{cccc|c} 1 & -2 & 0 & 3 & 4 \\ 0 & 0 & 1 & -4 & -5 \\ 0 & 0 & 0 & 0 & 0 \\ 0 & 0 & 0 & 0 & 0 \end{array}\right]$$

Since the number of leading ones, and therefore the number of basic variables, is less than the number of unknowns, and since there is no equation that gives an inconsistency, the system has infinitely many solutions. Let us now obtain the general solution for the system by deriving the parametric equations.

For convenience we first label each column of the augmented matrix with its corresponding variable and circle the leading ones, as shown.

$$\begin{array}{cccc} x_1 & x_2 & x_3 & x_4 \end{array}$$
$$\left[\begin{array}{cccc|c} ① & -2 & 0 & 3 & 4 \\ 0 & 0 & ① & -4 & -5 \\ 0 & 0 & 0 & 0 & 0 \\ 0 & 0 & 0 & 0 & 0 \end{array}\right].$$

Now solving the equations represented by the first two rows for the basic variables x_1 and x_3, we have

$$x_1 = \quad 4 + 2x_2 - 3x_4$$
$$x_3 = -5 \qquad\quad + 4x_4.$$

Next, setting the variables x_2 and x_4 equal to parameters s and t, respectively, we obtain the parametric equations

$$x_1 = 4 + 2s - 3t$$
$$x_2 = s, \quad \text{any real number}$$
$$x_3 = -5 + 4t$$
$$x_4 = t, \quad \text{any real number.}$$

These equations give the general solution for the system. Again, any particular solution is obtained by assigning specific values to the parameters. For example, setting $s = 1$ and $t = 2$, we have the particular solution

$$x_1 = 0, \quad x_2 = 1, \quad x_3 = 3, \quad x_4 = 2.$$

7. Solve the 3-by-4 system

$$x_1 - x_2 - x_3 \qquad = 0$$
$$2x_1 + 3x_2 \qquad -x_4 = 0$$
$$5x_2 + 3x_3 + x_4 = 0.$$

The augmented matrix for the system is

$$B = \begin{bmatrix} 1 & -1 & -1 & 0 & \bigm| & 0 \\ 2 & 3 & 0 & -1 & \bigm| & 0 \\ 0 & 5 & 3 & 1 & \bigm| & 0 \end{bmatrix}.$$

The system differs from the one given in example 5 only in the right-hand sides of the equations. Therefore, the row operations we perform on the augmented matrix A of example 5 will be the same as those we use on the matrix B. Furthermore, since each entry of the last column of B is zero, these row operations do not affect this column of zeros and, therefore, the last column of the derived matrix will also contain all zeros. Thus, considering matrix A' of example 5, we realize that the matrix in reduced row-echelon form that is row equivalent to B is

$$
\begin{array}{cccc}
x_1 & x_2 & x_3 & x_4
\end{array}
$$
$$B' = \begin{bmatrix} ① & 0 & 0 & 1 & \bigm| & 0 \\ 0 & ① & 0 & -1 & \bigm| & 0 \\ 0 & 0 & ① & 2 & \bigm| & 0 \end{bmatrix}.$$

The leading ones in B' have been circled and the columns corresponding to variables have been labeled for easy identification. Solving for the basic variables, we have

$$x_1 = -x_4, \quad x_2 = x_4, \quad \text{and} \quad x_3 = -2x_4.$$

Assigning the parameter t to x_4 gives the general solution for the system, namely,

$$x_1 = -t$$
$$x_2 = t$$
$$x_3 = -2t$$
$$x_4 = t, \quad \text{any real number.}$$

The system of linear equations in example 7 is a special type—one in which each of the right-hand sides is zero. Such a system is called **homogeneous**. Every homogeneous system has at least one solution, namely, the solution in which each unknown equals zero. The solution in which $x_1 = x_2 = \cdots = x_n = 0$ is called the **trivial solution**. The trivial solution is obtained from the general solution in example 7 by setting the parameter t equal to zero. Example 7 exemplifies a homogeneous system that also has **non-trivial** solutions. For example, letting $t = 1$ in the general solution of example 7, we obtain the particular non-trivial solution $x_1 = -1, x_2 = 1, x_3 = -2,$ and $x_4 = 1$.

Not every homogeneous system will have a non-trivial solution. For example, the system represented by the augmented matrix

$$\left[\begin{array}{ccc|c} 1 & 0 & 0 & 0 \\ 0 & 1 & 0 & 0 \\ 0 & 0 & 1 & 0 \end{array}\right]$$

has only the trivial solution. In fact, for an n-by-n homogeneous system to have only the trivial solution, the augmented matrix for the system, when it is in reduced row-echelon form, must be similiar to the preceding matrix.

We return to our discussion on the number of solutions to a system of linear equations in chapter 5. We conclude this section and chapter with two more applications that lead to solving systems of linear equations.

EXAMPLES

8. *An Investment Problem.* An investor has \$50,000 available to invest in three types of securities, one that yields 8% annually, another that yields 10% annually, and the third that yields 13% annually. How much can be invested in each type of security in order to obtain an annual income of \$5500?

Let
$$x = \text{the amount invested at } 8\%$$
$$y = \text{the amount invested at } 10\%$$
$$z = \text{the amount invested at } 13\%.$$

The income is a function of x, y, and z; considering the rates of return, it can be written as $0.08x + 0.10y + 0.13z$. Imposing the conditions that the income be \$5500 and the total investment be \$50,000, we have the system of linear equations

$$x + \quad y + \quad z = 50{,}000$$
$$0.08x + 0.10y + 0.13z = \quad 5500.$$

Applying Gauss–Jordan reduction to the augmented matrix, we obtain the equivalent matrix in reduced row-echelon form

$$\begin{array}{ccc} x & y & z \end{array}$$
$$A = \left[\begin{array}{ccc|c} 1 & 0 & -1.5 & -25{,}000 \\ 0 & 1 & 2.5 & 75{,}000 \end{array}\right].$$

Solving for the basic variables x and y and setting $z = t$, we have the parametric equations

$$x = 1.5t - 25{,}000$$
$$y = -2.5t + 75{,}000$$
$$z = t, \quad \text{any real number.}$$

However, not every choice of a value for t yields a meaningful solution to the problem. Since x, y, and z represent amounts of money to be invested, we must choose t, when obtaining a particular solution, so that x, y, and z are non-negative. In particular, we must first restrict $t \geq 0$. Second, for $y = 75{,}000 -$

$2.5t \geq 0$, we must have $2.5t \leq 75{,}000$ and, therefore, $t \leq 30{,}000$. Finally, for $x = 1.5t - 25{,}000 \geq 0$, we must have $1.5t \geq 25{,}000$ and, therefore, $t \geq 16{,}667$. Thus we must restrict t such that $16{,}667 \leq t \leq 30{,}000$. With this restriction on the values for the parameter t, the solutions to our investment problem are described by

$$x = 1.5t - 25{,}000$$

$$y = -2.5t + 75{,}000$$

$$z = t, \quad \text{where } 16{,}667 \leq t \leq 30{,}000.$$

9. *A Shipment Problem.* The Forest Paper Company has two pulp mills, A and B, and two finishing plants, 1 and 2. Mill A can produce and supply to the finishing plants 100 tons of unfinished paper in a week, whereas mill B can produce and supply 150 tons of unfinished paper in a week. Plant 1 can finish 120 tons of paper in a week, whereas plant 2 can finish 130 tons of paper in a week. A concern of the company is to determine the amount of unfinished paper to ship from each pulp mill to each finishing plant. Let us determine a shipping plan.

The unknowns are the amounts of paper to ship from each mill to each plant. Thus, to construct our model, we let

A_i = the amount of unfinished paper to ship to plant i from mill A

B_i = the amount of unfinished paper to ship to plant i from mill B

where $i = 1$ and 2. There are four unknowns to be determined, A_1, A_2, B_1, and B_2.

We can picture the amount of unfinished paper to be shipped as shown in figure 4.3. The conditions that must be satisfied are obtained from requiring that each mill ship what it can supply and each plant receive what it can handle, which gives us the system of equations

$$A_1 + A_2 \qquad\qquad = 100$$

$$B_1 + B_2 = 150$$

$$A_1 \qquad + B_1 \qquad = 120$$

$$A_2 \qquad + B_2 = 130.$$

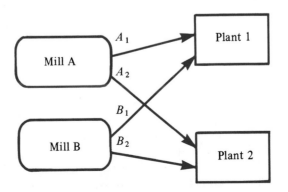

FIGURE 4.3

The augmented matrix for this system is

$$\left[\begin{array}{cccc|c} 1 & 1 & 0 & 0 & 100 \\ 0 & 0 & 1 & 1 & 150 \\ 1 & 0 & 1 & 0 & 120 \\ 0 & 1 & 0 & 1 & 130 \end{array}\right].$$

Performing Gauss–Jordan reduction on this matrix, we obtain the equivalent augmented matrix in reduced row-echelon form

$$\begin{array}{cccc} A_1 & A_2 & B_1 & B_2 \end{array}$$
$$\left[\begin{array}{cccc|c} ① & 0 & 0 & -1 & -30 \\ 0 & ① & 0 & 1 & 130 \\ 0 & 0 & ① & 1 & 150 \\ 0 & 0 & 0 & 0 & 0 \end{array}\right].$$

Solving for the basic variables A_1, A_2, and B_1 in terms of B_2 and setting $B_2 = t$, we obtain the parametric equations

$$A_1 = t - 30$$
$$A_2 = -t + 130$$
$$B_1 = -t + 150$$
$$B_2 = t.$$

Since a solution represents an amount of paper to be shipped, we must choose t such that each unknown is non-negative. From $A_1 = t - 30$, we must have $t \geq 30$. To ensure that $A_2 = -t + 130$ and $B_1 = -t + 150$ are non-negative, we must also select t such that $t \leq 130$. Therefore, for every choice of t where $30 \leq t \leq 130$ we obtain a shipping plan. For example, if we select $t = 30$, we have the plan shown in table 4.10.

TABLE 4.10

| | Ship to | |
From mill	Plant 1	Plant 2
A	0	100
B	120	30

In general, the possible shipping plans are indicated in table 4.11.

TABLE 4.11

| | Ship to | |
From mill	Plant 1	Plant 2
A	$t - 30$	$-t + 130$
B	$-t + 150$	t
	where $30 \leq t \leq 130$	

EXERCISES 4.3

For exercises 1 and 2 state whether or not the given matrix is in row-echelon form. If the matrix is not, give the reason why.

1. a. $\begin{bmatrix} 1 & 0 \\ 0 & 1 \end{bmatrix}$

   b. $\begin{bmatrix} 1 & 0 & 0 \\ 0 & 1 & 0 \\ 0 & 1 & 0 \end{bmatrix}$

   c. $\begin{bmatrix} 1 & 0 & 2 & 3 \\ 0 & 1 & 0 & 2 \\ 0 & 0 & 0 & 1 \end{bmatrix}$

   d. $\begin{bmatrix} 1 & 3 & 4 \\ 0 & 2 & 1 \\ 0 & 0 & 0 \end{bmatrix}$

2. a. $\begin{bmatrix} 0 & 1 & -2 & 0 \\ 0 & 1 & 0 & 2 \\ 0 & 0 & 1 & 3 \end{bmatrix}$

   b. $\begin{bmatrix} 1 & 0 & 0 & 2 & 3 \\ 0 & 0 & 1 & 5 & 7 \\ 0 & 0 & 0 & 0 & 0 \end{bmatrix}$

For exercises 3 and 4 state whether or not the given matrix is in reduced row-echelon form. If it is not, give the reason why.

3. a. $\begin{bmatrix} 1 & 2 \\ 0 & 1 \end{bmatrix}$

   b. $\begin{bmatrix} 1 & 0 \\ 0 & 1 \end{bmatrix}$

   c. $\begin{bmatrix} 1 & 0 & 2 & 0 \\ 0 & 1 & 0 & 0 \\ 0 & 0 & 0 & 1 \end{bmatrix}$

   d. $\begin{bmatrix} 1 & 0 & 2 \\ 0 & 2 & 1 \\ 0 & 0 & 0 \end{bmatrix}$

4. a. $\begin{bmatrix} 1 & 0 & 2 \\ 0 & 1 & 0 \\ 0 & 0 & 0 \end{bmatrix}$

   b. $\begin{bmatrix} 1 & 3 & 0 & 5 \\ 0 & 0 & 1 & 0 \\ 0 & 0 & 0 & 1 \end{bmatrix}$

For exercises 5 and 6, the given matrix, which is in reduced row-echelon form, is the augmented matrix for a system of linear equations. Solve the system.

5. a. $\left[\begin{array}{cccc|c} 1 & 0 & 0 & 0 & 0 \\ 0 & 1 & 0 & 0 & 6 \\ 0 & 0 & 1 & 0 & -8 \\ 0 & 0 & 0 & 1 & 0 \end{array}\right]$

   b. $\left[\begin{array}{ccccc|c} 1 & 0 & 0 & 0 & 3 & 6 \\ 0 & 0 & 1 & 0 & 0 & -1 \\ 0 & 0 & 0 & 1 & -2 & -7 \\ 0 & 0 & 0 & 0 & 0 & 0 \end{array}\right]$

6. a. $\left[\begin{array}{ccc|c} 1 & 0 & 0 & 2 \\ 0 & 1 & 0 & -4 \\ 0 & 0 & 1 & 3 \end{array}\right]$

   b. $\left[\begin{array}{cccc|c} 1 & 0 & 2 & 0 & 3 \\ 0 & 1 & -1 & 0 & -2 \\ 0 & 0 & 0 & 1 & 5 \end{array}\right]$

For exercises 7–14 the given matrix is the augmented matrix for a system of linear equations. Solve the system using Gauss–Jordan reduction.

7. $\left[\begin{array}{ccc|c} 1 & -1 & 2 & 1 \\ 0 & 2 & \frac{1}{2} & -3 \end{array}\right]$

8. $\left[\begin{array}{ccc|c} 1 & 2 & -1 & 3 \\ -1 & 3 & -2 & 1 \end{array}\right]$

9. $\begin{bmatrix} 1 & -1 & -2 & \bigm| & 9 \\ 2 & -3 & 1 & \bigm| & 5 \\ 3 & -1 & 3 & \bigm| & -4 \end{bmatrix}$

10. $\begin{bmatrix} 3 & 2 & -1 & \bigm| & 3 \\ 1 & 2 & -1 & \bigm| & 3 \\ 0 & 1 & 0 & \bigm| & 1 \end{bmatrix}$

11. $\begin{bmatrix} 1 & 2 & -1 & \bigm| & 4 \\ -3 & 1 & 0 & \bigm| & 0 \\ 2 & -4 & 6 & \bigm| & -8 \end{bmatrix}$

12. $\begin{bmatrix} 1 & -6 & -5 & \bigm| & -3 \\ 3 & 2 & 1 & \bigm| & -5 \\ -2 & 3 & 5 & \bigm| & 2 \end{bmatrix}$

13. $\begin{bmatrix} 0 & 2 & 0 & -10 & \bigm| & -5 \\ 1 & -2 & 3 & 4 & \bigm| & -5 \\ 0 & 0 & -2 & 2 & \bigm| & 7 \\ 3 & -4 & 9 & 6 & \bigm| & -18 \end{bmatrix}$

14. $\begin{bmatrix} 2 & -1 & 3 & 4 & \bigm| & -2 \\ 4 & 0 & 2 & 4 & \bigm| & -1 \\ 0 & 4 & -7 & 3 & \bigm| & -5 \end{bmatrix}$

In exercises 15–34 find the solution, if it exists, using Gauss–Jordan reduction.

15.
$$\begin{aligned} x + y - 7z &= -1 \\ 3x - 6y + 21z &= 0 \\ -9x - 3y + 14z &= 4 \end{aligned}$$

16.
$$\begin{aligned} x + 2y - 3z &= -15 \\ 2x + 4y + 5z &= 14 \\ -3x - 5y - 2z &= 0 \end{aligned}$$

17.
$$\begin{aligned} -x + 3y - 4z &= -5 \\ 4x - y + 5z &= 7 \\ x + 8y - 7z &= 6 \end{aligned}$$

18.
$$\begin{aligned} x - 2y + z &= 1 \\ 3x + y - 3z &= -2 \\ -7x - 7y + 11z &= 7 \end{aligned}$$

19.
$$\begin{aligned} x - 4y + z &= -6 \\ 2x + y - 5z &= 5 \\ -x + 2y + 7z &= -1 \end{aligned}$$

20.
$$\begin{aligned} 2x + y + 4z &= -3 \\ -x - 3y + 5z &= 16 \\ 5x - 6y - 6z &= 2 \end{aligned}$$

21.
$$\begin{aligned} x - 3y + z &= -1 \\ 3x - y + 2z &= 5 \\ 5x + y + 3z &= 11 \end{aligned}$$

22.
$$\begin{aligned} 2x - y + z &= -3 \\ 4x - 3y - 2z &= -1 \\ -2x + 2y + 3z &= -2 \end{aligned}$$

23.
$$\begin{aligned} x - 2y + z &= -4 \\ -3x + 6y - z &= -1 \end{aligned}$$

24.
$$\begin{aligned} 2x - 3y + z &= 5 \\ 2y - 3z &= -1 \end{aligned}$$

25.
$$\begin{aligned} x_1 - 2x_2 + x_3 &= 1 \\ -2x_1 + x_2 + 4x_3 &= -5 \end{aligned}$$

26.
$$\begin{aligned} x - 2y + z &= 1 \\ -2x + y + 4z &= -5 \end{aligned}$$

27.
$$\begin{aligned} x_1 - 2x_2 - x_3 &= 0 \\ -2x_1 + 4x_2 + x_4 &= 2 \end{aligned}$$

28.
$$\begin{aligned} x_1 + x_2 - 2x_3 + x_4 &= -1 \\ 3x_1 + 3x_2 - x_3 &= -2 \end{aligned}$$

29.
$$\begin{aligned} x_1 \qquad\quad - 3x_4 + x_5 &= 0 \\ 3x_1 + 9x_2 \qquad - x_4 &= 6 \\ x_2 - x_3 \qquad + 2x_5 &= -1 \end{aligned}$$

30.
$$\begin{aligned} 4x_2 - x_3 + x_4 &= -3 \\ x_1 \qquad + 3x_3 \qquad - 2x_5 &= -1 \end{aligned}$$

31.
$$\begin{aligned} x_1 - 2x_2 + x_3 - x_4 &= 0 \\ 3x_1 + 6x_2 - 2x_3 + x_4 &= 1 \\ x_2 + 5x_3 + 2x_4 &= -1 \end{aligned}$$

32.
$$\begin{aligned} x_1 + 2x_2 - x_3 + x_4 &= -1 \\ x_2 + x_3 - x_4 &= 0 \\ -2x_1 - x_2 \qquad + 4x_4 &= 3 \end{aligned}$$

33.
$$\begin{aligned} x_1 - 2x_2 \qquad + x_4 &= -1 \\ x_3 \qquad - 5x_5 &= 3 \\ 3x_1 - 6x_2 - x_3 + x_4 - x_5 &= -2 \\ - 2x_3 - x_4 + 2x_5 &= 1 \end{aligned}$$

34.
$$\begin{aligned} x_1 + 2x_2 - x_3 + x_4 - 2x_5 &= -1 \\ x_2 + x_3 - 2x_4 &= 0 \\ -x_1 - x_2 \qquad - x_4 + x_5 &= 2 \\ x_3 + 2x_4 - x_5 &= -3 \end{aligned}$$

For exercises 35–44 solve the problem using Gauss–Jordan reduction. Make sure to state what each variable represents.

35. White Birch Enterprises makes plaques, trophy bases, and picture frames. Each item

requires machining, polishing, and finishing. The number of hours needed for each crate of each item is given in table 4.12.

TABLE 4.12

	Plaques	Trophy bases	Picture frames
machining	$\frac{1}{4}$	$\frac{1}{2}$	$\frac{1}{4}$
polishing	$\frac{1}{4}$	$\frac{1}{2}$	$\frac{1}{2}$
finishing	$\frac{1}{8}$	$\frac{1}{2}$	$\frac{1}{4}$

On a given shift, the amount of time available for machining, polishing, and finishing are, respectively, 12, 15, and 10 hours. How many plaques, trophy bases, and picture frames can be made during this shift?

36. Starshine Optics makes running lights, tilting knobs, and dubbing controls. Each item requires three machines in its manufacture. The number of hours needed on each machine to make a case of each item is shown in table 4.13.

TABLE 4.13

	Running lights	Tilting knobs	Dubbing controls
machine A	$\frac{1}{3}$	$\frac{1}{9}$	1
machine B	$\frac{1}{3}$	$\frac{2}{9}$	$\frac{2}{3}$
machine C	$\frac{2}{3}$	$\frac{1}{3}$	1

Each day, the amount of time available on each machine is as follows: 10 hours on machine A, 9 hours on machine B, and 15 hours on machine C. How many cases of each item can be made per day?

37. The Imperial Cigar Company makes three brands of cigars: the Lady Jane, the Slim, and the Concord. Three machines, the blender·, the chopper, and the binder, are used in the manufacture of each brand. The number of hours needed on each machine to make a case of each brand is shown in table 4.14.

TABLE 4.14

	Lady Jane	Slim	Concord
blender	1	0.2	0.4
chopper	0.4	0.2	0.4
binder	1	0.8	0.2

Each day, the amount of time available for blending, chopping, and binding are, respectively, 14, 11.6, and 16 hours. How many cases of each brand can be made per day?

38. Triax Medical Supplies makes thrust tubes, cap mountings, and air scoops. Each item requires three instruments in its manufacture. The number of hours needed in each department to make a batch of each item is shown in table 4.15.

TABLE 4.15

	Thrust tubes	Cap mountings	Air scoops
instrument A	1	0.3	0.6
instrument B	0.6	0.3	1
instrument C	1	0.6	0.6

Each shift, the amount of time available on each instrument is as follows: 6.1 hours on instrument A, 5.3 hours on instrument B, and 7 hours on instrument C. How many batches of each item can be made during each shift?

39. A 100-pound mixture is to be made from three types of foods, A, B, and C. Food A has 25 micrograms of selenium and 200 calories per pound; food B has 30 micrograms of selenium and 100 calories per pound; food C has 15 micrograms of selenium and 150 calories per pound. If the mixture is to have 2475 micrograms of selenium and 14,250 calories, how many pounds of each type of food should be used?

40. A 50-pound mixture is to be made from three types of food supplements, 1, 2, and 3. Supplement 1 has 150 micrograms of GTF-chromium and 50 units of vitamin K; supplement 2 has 200 micrograms of GTF-chromium and 75 units of vitamin K; supplement 3 has 175 micrograms of GTF-chromium and 100 units of vitamin K. If the mixture is to have 8375 micrograms of GTF-chromium and 3500 units of vitamin K, how many pounds of each type of food supplement should be used?

41. Styx Boating Company ships pleasure boats from two factories, A and B, to two warehouses, 1 and 2. Each week factory A makes and ships 400 boats, and factory B makes and ships 200 boats. Each week warehouse 1 has room for 500 additional boats, and warehouse 2 has room for 100 additional boats. Determine a shipping plan.

42. Salisbury Meat Company ships beef from two packing plants, A and B, to two supermarket chains, Continental and Fairway. Plant A can supply 8 tons of beef per week, whereas plant B can supply 14 tons per week. Continental requires 10 tons of beef per week and Fairway requires 12 tons of beef per week. How should Salisbury plan its shipments so as to satisfy the supermarket chains' requirements?

43. Luther has $30,000 to invest and wishes to distribute this amount among four types of investments paying 7%, 8%, 9.5%, and 10.5% annually. He wants to balance the amount invested in the higher yield securities against the amount invested in the lower yield securities; that is, the combined amount invested at 9.5% and 10.5% must be equal to the combined amount invested at 7% and 8%. If his objective is to realize an annual income of $2500, then how much should be invested in each type of security?

44. A dietician is planning the daily menu for a hospital patient and plans to use a combination of four types of food. The patient's daily requirement of calories and protein and the amount of calories and grams of protein supplied by one ounce of each type food is shown in table 4.16.

TABLE 4.16

	Per ounce of				
	Type 1	Type 2	Type 3	Type 4	Required
grams of protein	4	3	3	2	72
calories	120	120	150	60	3000

How many ounces of each type of food should be planned in order to obtain the exact amount of calories and grams of protein required?

45. Suppose the demand for a certain product of a firm is a function of the variables x, y and z and the dependency can be expressed in the form $D = ax + by + cz + d$. Also observations were made on the value of D for certain values of the variables x, y, and z, as given in table 4.17.

TABLE 4.17

x	y	z	D
1	1	1	8.0
0	3	4	13.8
3	0	2	17.1
1	2	0	5.9

Determine a, b, c, and d so that the function fits the observations.

46. Repeat exercise 45 for the data points in table 4.18.

TABLE 4.18

x	y	z	D
1	1	1	14.4
0	3	4	14.5
3	0	2	14.6
1	2	0	22.2

47. Eclipse Manufacturers ships truck cabs from two factories, A and B, to three finishing plants, 1, 2, and 3. Each day factory A makes and ships 120 truck cabs and factory B makes and ships 400 truck cabs. Each day, finishing plant 1 needs 200 additional truck cabs, finishing plant 2 needs 150 additional truck cabs, and finishing plant 3 needs 170 additional truck cabs.

   a. Draw a diagram similar to the one in figure 4.3 to illustrate this situation.

   b. Determine a shipping schedule.

48. The Ebony Company ships pianos from two factories, A and B, to three warehouses, 1, 2, and 3. Each week factory A makes and ships 15 pianos and factory B makes and ships 30 pianos. Each week, warehouse 1 has room for 15 additional pianos, warehouse 2 has room for 12 additional pianos, and warehouse 3 has room for 18 additional pianos.

   a. Draw a diagram similar to the one in figure 4.3 to illustrate this situation.

   b. Determine a shipping schedule.

49. Consider the following system.

$$x - y + z = 1$$
$$3x - 2y + 3z = -2$$
$$x \quad + z = a$$

Determine a so that the system has infinitely many solutions.

SUMMARY OF TERMS

Linear Forms in n Variables
 Independent Variable
 Dependent Variable

Systems of Linear Equations
 Homogeneous System
 Triangular Form
 Solution to a System
 Consistent System
 Inconsistent System
 Trivial Solution
 Non-Trivial Solution
 Operations on Equations
 Equivalent Systems

Gaussian Elimination
 Back Substitution
 Square Systems

Linear Equations
 Variables (Unknowns)
 Subscripted Variables
 Coefficients of Variables
 Right-Hand Side of Equation
 Solution to an Equation

Matrix Representation
 Augmented Matrix
 Size of a Matrix
 Entry of a Matrix
 Row-Echelon Form
 Reduced Row-Echelon Form
 Row Operations
 Row Equivalent Matrices

Gauss–Jordan Reduction
 Pivot
 Leading Ones
 Basic Variables
 Parameter
 Parametric Equations

REVIEW EXERCISES (CH. 4)

---

1. Salisbury Meat Company makes three types of prepared meat products, A, B, and C. The cost of ingredients for each pound of A, B, and C is $0.22, $0.43, and $0.31, respectively.

   a. Express the total cost of ingredients as a function of the number of pounds of each type of meat product made.

   b. If the company makes 150 pounds of A, 200 pounds of B, and 120 pounds of C, what is the total cost of its ingredients?

2. Sue has money invested in four types of investments, a regular savings account, a certificate of deposit, a money market fund, and an oil income partnership. The rates of income per month these four investments pay are as follows: $\frac{1}{2}\%$ for the regular savings account, $\frac{5}{6}\%$ for the certificate of deposit, $1\frac{1}{4}\%$ for the money market fund, and $1\frac{1}{2}\%$ for the oil income partnership.

   a. Express the monthly income as a function of the amount invested in each type of investment.

   b. If Sue has $3500 in the regular savings account, $10,000 in the certificate of deposit, $5600 in the money market fund, and $15,000 in the oil income partnership, how much is the monthly income?

3. Consider the system of linear equations

$$0.5x_1 + 0.2x_2 + 0.1x_3 + 0.7x_4 = 6$$
$$0.1x_1 \qquad + 0.3x_3 + 0.8x_4 = 7.5$$
$$0.4x_2 + 0.2x_3 \qquad = 3.$$

Determine whether or not each of the following ordered sets is a solution to the system.

a. $(-23, 27, -39, 25)$ b. $(-5, 7.5, 0, 10)$ c. $(3.6, -3.9, 7.8, 6)$

4. Use Gaussian elimination with back substitution to solve the following system.

$$x - 2y + 3z = -6$$
$$2x + y - z = 5$$
$$-x + 3y + z = 4$$

5. Use Gauss–Jordan reduction to find the solution to the following system.

$$x_1 - 2x_2 + x_3 + 11x_4 = 5$$
$$x_2 + x_3 - 5x_4 = -2$$
$$x_3 + x_4 = 1$$

6. Use Gauss–Jordan reduction to find the solution to the following system.

$$x_1 - x_2 \qquad - 2x_4 \qquad = 2$$
$$-5x_1 + 5x_2 - x_3 + 9x_4 \qquad = -9$$
$$3x_1 - 3x_2 + 2x_3 - 4x_4 + x_5 = 7$$
$$-x_1 + x_2 - x_3 + x_4 - x_5 = -4$$

7. Use Gauss–Jordan reduction to find the solution to the following system.

$$x_1 - 2x_2 + x_3 + 3x_4 + 2x_5 = -1$$
$$-2x_1 + 4x_2 - x_3 - 3x_4 - x_5 = -3$$
$$2x_2 - x_3 - 3x_4 + x_5 = 0$$
$$-x_1 + 2x_2 \qquad + 4x_4 \qquad = 3$$

8. Use Gauss–Jordan reduction to find the solution to the following system.

$$2x - 5y + 10z = -4$$
$$-x + 2y - 8z = 5$$

9. Use Gauss–Jordan reduction to find the solution if it exists, of the following system.

$$x_1 - 3x_2 + x_3 \qquad - x_5 = 1$$
$$3x_1 - 9x_2 - 2x_3 + x_4 - 2x_5 = 3$$
$$-2x_1 + 6x_2 - x_3 + 3x_4 + 5x_5 = -5$$
$$-5x_1 + 15x_2 + 6x_3 - 2x_4 + 3x_5 = 7$$

10. Timestop Photo makes pinch rollers, pressure pads, and eyepiece guards. Each item requires three departments in its manufacture. The number of hours needed in each department to make a crate of each item is shown in table 4.19.

TABLE 4.19

	Pinch rollers	Pressure pads	Eyepiece guards
department A	1	$\frac{1}{5}$	$\frac{2}{5}$
department B	$\frac{3}{5}$	$\frac{2}{5}$	$\frac{1}{5}$
department C	1	$\frac{3}{10}$	$\frac{3}{5}$

Each week, the amount of time available in each department is as follows: 25 hours in department A, 17 hours in department B, and 30 hours in department C. How many crates of each item can be made each week?

11. Sunrise Development Corporation has a 240-acre tract of land that it plans to subdivide into 1-acre, 2-acre, and 3-acre building lots. It costs $1500, $2500, and $3000 to develop a 1-acre lot, a 2-acre lot, and a 3-acre lot, respectively. The corporation has available $255,000 to cover these development costs. How many lots of each size should be planned if the entire 240 acres is developed and the entire $255,000 is used?

12. Refer to exercise 11. Suppose the company wants, in addition to the conditions given in exercise 11, the number of 1-acre lots to be equal to the number of 2-acre lots. Now how many lots of each size should be planned?

13. A 50-pound bag of lawn fertilizer should contain 22.2 pounds of phosphoric acid and 18.6 pounds of potash. Three ingredients, A, B, and C, are used to make the fertilizer. Ingredient A contains 30% phosphoric acid and 60% potash; ingredient B contains 60% phosphoric acid and 30% potash; ingredient C contains 30% phosphoric acid and 30% potash. How many pounds of each ingredient should be used in order to satisfy the requirements of phosphoric acid and potash?

14. Applewood Computers ships microcomputers from two factories, A and B, to two warehouses, 1 and 2. Each day, factory A makes and ships 1450 microcomputers and factory B makes and ships 1200 microcomputers. Each day, warehouse 1 has room for 850 additional microcomputers and warehouse 2 has room for 1800 additional microcomputers. Determine a shipping plan.

CHAPTER 5

Matrix Algebra with Applications

Introduction

Recall from chapter 4 that an $m \times n$ matrix is defined as a rectangular array of numbers. In particular, an $m \times n$ matrix is an array with m rows and n columns in the form

$$A = \begin{bmatrix} a_{11} & a_{12} & \cdots & a_{1n} \\ a_{21} & a_{22} & \cdots & a_{2n} \\ \vdots & \vdots & & \vdots \\ a_{m1} & a_{m2} & \cdots & a_{mn} \end{bmatrix}.$$

A $1 \times n$ matrix, a matrix with one row, is called a **row matrix** and an $m \times 1$ matrix, a matrix with one column, is called a **column matrix**. To indicate an arbitrary matrix, we will at times simply write $A = (a_{ij})$.

So far matrices have been considered in the context of systems of linear equations. In this chapter further matrix notation and arithmetic operations on matrices are developed. These new concepts enable us both to obtain additional results on systems of linear equations and to consider other matrix applications.

5.1 Matrix Operations

As we saw in chapter 4, an augmented matrix is a useful way of representing a system of linear equations. A matrix is also a convenient way of displaying other types of information.

EXAMPLE 1

Acme Feed Mills makes and packages four types of dog food: high protein (H), medium protein (M), low protein (L), and puppy food (P). During June the mill shipped to distributor X 160 bags of high protein, 100 bags of medium protein, 60 bags of low protein, and 80 bags of puppy food. Also in June, the mill shipped dog food to distributors Y and Z. The number of bags shipped to the three distributors is given by the matrix

Number of bags

$$
\begin{array}{cc}
& \begin{array}{cccc} \text{H} & \text{M} & \text{L} & \text{P} \end{array} \\
\begin{array}{c} \text{shipped to X} \\ \text{shipped to Y} \\ \text{shipped to Z} \end{array} &
\begin{bmatrix}
160 & 100 & 60 & 80 \\
400 & 80 & 200 & 120 \\
240 & 160 & 120 & 80
\end{bmatrix}.
\end{array}
$$

With the definition of matrix equality and addition, we can see the convenience of displaying the information in example 1 by a matrix.

Definition

■ Two matrices are **equal** if they are the same size and if corresponding entries are equal. In particular, if $A = [a_{ij}]$ and $B = [b_{ij}]$ are two $m \times n$ matrices, then $A = B$ provided that $a_{ij} = b_{ij}$ for every i and j. ■

EXAMPLE 2

a. In order for

$$\begin{bmatrix} 2 & x \\ y & 3 \end{bmatrix} \quad \text{to equal} \quad \begin{bmatrix} 2 & 4 \\ -1 & 3 \end{bmatrix}$$

we must have $x = 4$ and $y = -1$.

b. The two matrices

$$\begin{bmatrix} 5 & -2 \\ 0 & 3 \\ 1 & 4 \end{bmatrix} \quad \text{and} \quad \begin{bmatrix} 5 & -2 \\ 0 & 3 \end{bmatrix}$$

cannot be equal since they are of different sizes.

Definition

■ Let A and B be two $m \times n$ matrices. The **sum** $A + B$ is the $m \times n$ matrix in which its (i, j) entry is obtained by adding the (i, j) entries of A and B. ■

Note that the definition of matrix sum allows only for adding matrices of the same size.

EXAMPLES

3.

a. $\begin{bmatrix} 2 & -1 & 0 \\ 4 & 5 & 3 \end{bmatrix} + \begin{bmatrix} -2 & 2 & 3 \\ 1 & -3 & 4 \end{bmatrix} = \begin{bmatrix} 2-2 & -1+2 & 0+3 \\ 4+1 & 5-3 & 3+4 \end{bmatrix}$

$= \begin{bmatrix} 0 & 1 & 3 \\ 5 & 2 & 7 \end{bmatrix}.$

b. $\begin{bmatrix} 2 & -1 \\ -3 & 2 \end{bmatrix} + \begin{bmatrix} 1 \\ 0 \end{bmatrix}$ cannot be done since the matrices are not the same size.

c. $\begin{bmatrix} 3 \\ -2 \\ 1 \end{bmatrix} + \begin{bmatrix} 5 \\ 3 \\ -1 \end{bmatrix} = \begin{bmatrix} 8 \\ 1 \\ 0 \end{bmatrix}$.

4. Consider again the matrix of example 1. In addition to this matrix of shipments, we also have the following matrices of shipments made by Acme Feed Mills in July and August.

July

$$
\begin{array}{c c c c c}
 & H & M & L & P \\
\text{shipped to X} & 120 & 60 & 20 & 40 \\
\text{shipped to Y} & 340 & 180 & 120 & 80 \\
\text{shipped to Z} & 160 & 140 & 80 & 60
\end{array}
$$

August

$$
\begin{array}{c c c c c}
 & H & M & L & P \\
\text{shipped to X} & 200 & 120 & 80 & 100 \\
\text{shipped to Y} & 480 & 100 & 180 & 80 \\
\text{shipped to Z} & 220 & 160 & 140 & 100
\end{array}.
$$

We can obtain the total number of bags of each type of food shipped to each distributor during the summer quarter (June, July, and August) by finding the sum of the three matrices. We have

$$
\underset{\text{July}}{\begin{bmatrix} 120 & 60 & 20 & 40 \\ 340 & 180 & 120 & 80 \\ 160 & 140 & 80 & 60 \end{bmatrix}} + \underset{\text{August}}{\begin{bmatrix} 200 & 120 & 80 & 100 \\ 480 & 100 & 180 & 80 \\ 220 & 160 & 140 & 100 \end{bmatrix}} = \underset{\text{July and August}}{\begin{bmatrix} 320 & 180 & 100 & 140 \\ 820 & 280 & 300 & 160 \\ 380 & 300 & 220 & 160 \end{bmatrix}}
$$

and

$$
\underset{\text{June}}{\begin{bmatrix} 160 & 100 & 60 & 80 \\ 400 & 80 & 200 & 120 \\ 240 & 160 & 120 & 80 \end{bmatrix}} + \underset{\text{July and August}}{\begin{bmatrix} 320 & 180 & 100 & 140 \\ 820 & 280 & 300 & 160 \\ 380 & 300 & 220 & 160 \end{bmatrix}}
$$

Number of bags
June, July, and August

$$
= \begin{array}{c c c c c}
 & H & M & L & P \\
\text{shipped to X} & 480 & 280 & 160 & 220 \\
\text{shipped to Y} & 1220 & 360 & 500 & 280 \\
\text{shipped to Z} & 620 & 460 & 340 & 240
\end{array}.
$$

In example 4, we found the sum of the three matrices by first adding the July and August shipment matrices and then adding to this result the June shipment matrix. However, it does not matter which two matrices we add first, as you can see from the following.

June and July

$$\begin{bmatrix} 280 & 160 & 80 & 120 \\ 740 & 260 & 320 & 200 \\ 400 & 300 & 200 & 140 \end{bmatrix}$$

August

$$+ \begin{bmatrix} 200 & 120 & 80 & 100 \\ 480 & 100 & 180 & 80 \\ 220 & 160 & 140 & 100 \end{bmatrix} =$$

June, July, and August

$$\begin{bmatrix} 480 & 280 & 160 & 220 \\ 1220 & 360 & 500 & 280 \\ 620 & 460 & 340 & 240 \end{bmatrix}.$$

Other properties of addition of real numbers carry over to matrix addition.

EXAMPLE 5

Let

$$A = \begin{bmatrix} 4 & 5 \\ 3 & -2 \\ -1 & 0 \end{bmatrix}, \quad B = \begin{bmatrix} -3 & 6 \\ 0 & 7 \\ 1 & 2 \end{bmatrix}, \quad \text{and} \quad \mathbf{0} = \begin{bmatrix} 0 & 0 \\ 0 & 0 \\ 0 & 0 \end{bmatrix}.$$

We have that

$$A + B = \begin{bmatrix} 4 + (-3) & 5 + 6 \\ 3 + 0 & -2 + 7 \\ -1 + 1 & 0 + 2 \end{bmatrix} = \begin{bmatrix} 1 & 11 \\ 3 & 5 \\ 0 & 2 \end{bmatrix}$$

and

$$B + A = \begin{bmatrix} -3 + 4 & 6 + 5 \\ 0 + 3 & 7 + (-2) \\ 1 + (-1) & 2 + 0 \end{bmatrix} = \begin{bmatrix} 1 & 11 \\ 3 & 5 \\ 0 & 2 \end{bmatrix}.$$

Also,

$$A + \mathbf{0} = \begin{bmatrix} 4 & 5 \\ 3 & -2 \\ -1 & 0 \end{bmatrix} + \begin{bmatrix} 0 & 0 \\ 0 & 0 \\ 0 & 0 \end{bmatrix} = \begin{bmatrix} 4 & 5 \\ 3 & -2 \\ -1 & 0 \end{bmatrix} = A.$$

Finally,

$$A + (-A) = \begin{bmatrix} 4 & 5 \\ 3 & -2 \\ -1 & 0 \end{bmatrix} + \begin{bmatrix} -4 & -5 \\ -3 & 2 \\ 1 & 0 \end{bmatrix}$$

$$= \begin{bmatrix} 4 + (-4) & 5 + (-5) \\ 3 + (-3) & -2 + 2 \\ -1 + 1 & 0 + 0 \end{bmatrix} = \begin{bmatrix} 0 & 0 \\ 0 & 0 \\ 0 & 0 \end{bmatrix}.$$

The matrix **0** of example 5 is called a zero matrix. Any $m \times n$ matrix in which each entry is the number zero is called a **zero matrix** and is denoted by **0**.

We have shown by example that certain properties are true for matrix addition and we can accept that they are true for other matrices.

Theorem 1   ■   Let A, B, and C be matrices of the same size. Then

   a. $A + B = B + A$ (commutative property)
   b. $A + (B + C) = (A + B) + C$ (associative property)
   c. $A + \mathbf{0} = \mathbf{0} + A = A$ (additive identity)
   d. there is a unique matrix $-A$, where the   (additive inverse)
      entries are the negatives of the entries of
      A, such that $A + (-A) = \mathbf{0}$.   ■

   Matrix **subtraction** is defined by using the additive inverse. For two matrices of the same size, $A - B = A + (-B)$.

EXAMPLE 6

$$\begin{bmatrix} 3 & 2 \\ -1 & 4 \\ 5 & -2 \end{bmatrix} - \begin{bmatrix} 1 & -6 \\ 0 & 2 \\ -3 & 4 \end{bmatrix} = \begin{bmatrix} 3 & 2 \\ -1 & 4 \\ 5 & -2 \end{bmatrix} + \begin{bmatrix} -1 & 6 \\ 0 & -2 \\ 3 & -4 \end{bmatrix} = \begin{bmatrix} 2 & 8 \\ -1 & 2 \\ 8 & -6 \end{bmatrix}.$$

   As seen in example 6, adding to A the additive inverse of B is the same as subtracting the entries of B from the corresponding entries of A.

   Sometimes we need to multiply each entry of a matrix by the same number, as we see in the next example.

EXAMPLE 7   Consider the shipment matrix of example 4. The entries of the matrix represent the number of 50-pound bags shipped. Since 50 pounds is $\frac{50}{2000} = \frac{1}{40}$ of a ton, the number of tons of each type of food shipped to each distributor is obtained by multiplying each entry of the shipment matrix by $\frac{1}{40}$.

Number of tons

$$\frac{1}{40}\begin{bmatrix} 480 & 280 & 160 & 220 \\ 1220 & 360 & 500 & 280 \\ 620 & 460 & 340 & 240 \end{bmatrix} = \begin{matrix} X \\ Y \\ Z \end{matrix} \begin{bmatrix} H & M & L & P \\ 12.0 & 7.0 & 4.0 & 5.5 \\ 30.5 & 9.0 & 12.5 & 7.0 \\ 15.5 & 11.5 & 8.5 & 6.0 \end{bmatrix}.$$

Definition   ■   Let A be an $m \times n$ matrix and let b be a real number, called a **scalar**. The product bA is the $m \times n$ matrix obtained by multiplying *each* entry of A by the scalar b.   ■

EXAMPLE 8

   a. $2\begin{bmatrix} 1 & 3 \\ 0 & -1 \end{bmatrix} = \begin{bmatrix} 2 & 6 \\ 0 & -2 \end{bmatrix}.$

   b. $\begin{bmatrix} 3 \\ 6 \\ -12 \end{bmatrix} = 3\begin{bmatrix} 1 \\ 2 \\ -4 \end{bmatrix}.$

   The last operation we consider in this section is applied to a single matrix. This operation will be useful in section 5.4 and chapter 6.

Definition ■ Consider an $m \times n$ matrix $A = [a_{ij}]$. The **transpose** of A, denoted by A^t, is the $n \times m$ matrix obtained by interchanging the rows and columns of A. In particular, the jth row of A^t is $[a_{1j}a_{2j}, \ldots, a_{mj}]$ where

$$\begin{bmatrix} a_{1j} \\ a_{2j} \\ \vdots \\ a_{mj} \end{bmatrix}$$

is the jth column of A. ■

EXAMPLE 9 | Consider the matrices

$$A = \begin{bmatrix} 2 & 3 & 1 \\ 4 & 5 & 6 \end{bmatrix}, \quad B = \begin{bmatrix} 1 & 2 \\ 0 & 4 \end{bmatrix}, \quad C = \begin{bmatrix} 1 \\ 2 \\ 3 \end{bmatrix}, \quad \text{and} \quad D = \begin{bmatrix} 1 & 0 & 3 \\ 0 & 2 & 5 \\ 3 & 5 & 0 \end{bmatrix}.$$

The transpose of each matrix follows.

$$A^t = \begin{bmatrix} 2 & 4 \\ 3 & 5 \\ 1 & 6 \end{bmatrix} \quad B^t = \begin{bmatrix} 1 & 0 \\ 2 & 4 \end{bmatrix} \quad C^t = [1 \quad 2 \quad 3] \quad D^t = \begin{bmatrix} 1 & 0 & 3 \\ 0 & 2 & 5 \\ 3 & 5 & 0 \end{bmatrix}.$$

EXERCISES 5.1

For exercises 1–6 let

$$A = \begin{bmatrix} 1 & 2 & 0 \\ -1 & 3 & 4 \end{bmatrix}, \quad B = \begin{bmatrix} 2 & -1 & -2 \\ 3 & 5 & 0 \end{bmatrix},$$

$$C = \begin{bmatrix} 1 & 2 \\ -2 & -1 \\ 5 & 4 \end{bmatrix}, \quad \text{and} \quad D = \begin{bmatrix} 1 & 0 \\ 0 & 1 \\ 1 & 1 \end{bmatrix}$$

and compute the indicated matrix, if it is defined.

1. $A + B$ 2. $C - D$ 3. $A + C$
4. $3D$ 5. $A + 2B$ 6. $3C - 2D$

For exercises 7–12 let

$$A = \begin{bmatrix} 3 & 0 \\ 0 & 4 \end{bmatrix}, \quad B = \begin{bmatrix} 1 & -2 & 4 \\ -5 & 0 & 3 \\ 0 & 4 & -1 \end{bmatrix}, \quad \text{and} \quad C = \begin{bmatrix} 2 & 4 & 5 \\ 6 & 2 & 1 \\ 3 & 2 & 1 \end{bmatrix}$$

and compute the indicated matrix, if it is defined.

7. $B + C$ 8. $A - B$ 9. $\frac{1}{2}C$

10. $-3A$ 11. $B - C$ 12. $2B + 4C$

13. Let

$$A = \begin{bmatrix} -1 & -2 & 0 & 5 \\ 2 & 0 & 7 & -9 \\ 3 & -1 & 4 & 8 \end{bmatrix}.$$

Determine $-A$ and verify that $A + (-A)$ is the 3×4 zero matrix.

14. Let

$$A = \begin{bmatrix} 1 & 0 & 3 \\ -2 & 3 & -5 \end{bmatrix}.$$

Determine $-A$ and verify that $A + (-A)$ is the 2×3 zero matrix.

15. Let

$$A = \begin{bmatrix} 1 & -2 \\ -3 & 4 \end{bmatrix} \quad \text{and} \quad B = \begin{bmatrix} 0 & 5 \\ 6 & -2 \end{bmatrix}.$$

Verify that $A + B = B + A$.

16. Let

$$A = \begin{bmatrix} 1 & -1 \\ 2 & 3 \\ 0 & 4 \end{bmatrix} \quad \text{and} \quad B = \begin{bmatrix} 2 & 3 \\ -1 & 5 \\ -2 & 0 \end{bmatrix}.$$

Verify that $A + B = B + A$.

17. Let

$$A = \begin{bmatrix} 1 & -2 \\ -3 & 4 \end{bmatrix}, \quad B = \begin{bmatrix} 0 & 5 \\ 6 & -2 \end{bmatrix}, \quad \text{and} \quad C = \begin{bmatrix} 7 & -3 \\ 8 & 6 \end{bmatrix}.$$

Verify that $A + (B + C) = (A + B) + C$.

18. Let

$$A = \begin{bmatrix} 1 & -1 \\ 2 & 3 \\ 0 & 4 \end{bmatrix}, \quad B = \begin{bmatrix} 2 & 3 \\ -1 & 5 \\ -2 & 0 \end{bmatrix}, \quad \text{and} \quad C = \begin{bmatrix} 1 & 0 \\ -2 & -3 \\ 4 & 5 \end{bmatrix}.$$

Verify that $A + (B + C) = (A + B) + C$.

For exercises 19–24 write the transpose of the given matrix.

19. $[2 \quad -3 \quad 4 \quad 9]$

20. $[0 \quad 1 \quad -1]$

21. $\begin{bmatrix} 5 \\ -6 \\ 7 \end{bmatrix}$

22. $\begin{bmatrix} 4 \\ -3 \\ 0 \\ 1 \end{bmatrix}$

23. $\begin{bmatrix} 1 & 0 & -2 \\ 3 & 0 & 4 \\ 2 & 0 & -5 \end{bmatrix}$

24. $\begin{bmatrix} 1 & 0 & 0 \\ 0 & 1 & 0 \\ 0 & 0 & 1 \end{bmatrix}$

25. The Regent Shoe Company has three plants located at Kutztown, Reading, and Allentown. Each plant produces men's, women's, and children's shoes. During one hour of operation the plant at Kutztown makes 15 pairs of men's shoes, 15 pairs of women's shoes, and 20 pairs of children's shoes. The Reading plant produces 20 pairs of men's shoes, 40 pairs of women's shoes, and 16 pairs of children's shoes in one hour, whereas in one hour the Allentown plant produces 30 pairs of men's shoes, 25 pairs of women's shoes, and 17 pairs of children's shoes. Express this information in a matrix, labeling the rows and columns appropriately.

26. Suppose a meal consists of three foods, a meat, a green vegetable, and a potato. One unit of the meat contains 7 units of protein and 35 calories, one unit of the green vegetable contains 1.5 units of protein and 10 calories, and one unit of the potato contains 0.5 unit of protein and 65 calories. Express this information in a matrix, labeling the rows and columns appropriately.

27. A salesperson for a microcomputer company which makes four models has the Fischer and the Computer World accounts. During one week, the salesperson sold Fischer 2 of the first model, none of the second model, 4 of the third model, and 5 of the fourth model. The sales during this same week to Computer World were none of the first or fourth models, but 3 of the second model and 7 of the third model. Express this information in a matrix, labeling the rows and columns appropriately.

28. One-Way Freight, a local trucking company, has two trucks. One of these trucks was driven 150 miles on Monday, whereas on the other four days of this week it was driven, respectively, 375 miles, 210 miles, 90 miles, and 175 miles. The second truck was not driven on either Tuesday or Friday, but it was driven 80 miles on Monday, 200 miles on Wednesday, and 330 miles on Thursday. Display this information in a matrix, labeling the rows and columns appropriately.

29. Refer to exercise 25. During the first day in September each of Regent's three plants operated 8 hours. Express by scalar multiplication the total number of pairs of each type of shoe produced by each of the three plants during the 8 hours of operation.

30. Refer to exercise 28. Suppose it costs One-Way Freight 50¢ per mile to operate either of its two trucks. Express by scalar multiplication the operating cost for each truck on each of the five days.

5.2 Row-Column Product

The first matrix multiplication operation we define is a row matrix times a column matrix.

Definition   ■ Consider a $1 \times n$ row matrix $[a_1 a_2, \ldots, a_n]$ and an $n \times 1$ column matrix

$$\begin{bmatrix} b_1 \\ b_2 \\ \vdots \\ b_n \end{bmatrix}.$$

The **row-column product** is given by

$$[a_1 a_2, \ldots, a_n] \begin{bmatrix} b_1 \\ b_2 \\ \vdots \\ b_n \end{bmatrix} = a_1 b_1 + a_2 b_2 + \cdots + a_n b_n. \quad \blacksquare$$

A row-column product is a number. To be able to compute this product, the number of entries in both the row and column matrices must be the same; otherwise, the multiplication cannot be performed.

EXAMPLE 1

a.

$$[4 \quad -1 \quad 3 \quad -2] \begin{bmatrix} 1 \\ -2 \\ 5 \\ 4 \end{bmatrix} = (4)(1) + (-1)(-2) + (3)(5) + (-2)(4) = 13.$$

b.

$$[2 \quad -1 \quad 5] \begin{bmatrix} -2 \\ 1 \\ 1 \end{bmatrix} = (2)(-2) + (-1)(1) + (5)(1) = 0.$$

c.

$$[3 \quad 4 \quad 1 \quad 2] \begin{bmatrix} -1 \\ 5 \\ -2 \end{bmatrix}$$ cannot be done since the number of entries in the row matrix is not the same as the number of entries in the column matrix.

In many applications we take the products of corresponding entries in two lists and sum the products; that is, we perform a row-column multiplication.

EXAMPLE 2

Recall from example 1 in section 5.1 the number of bags of dog food shipped in June from Acme Feed Mills to distributor X. This information is displayed in the row matrix

Number of bags shipped

$$\begin{matrix} H & M & L & P \\ [160 & 100 & 60 & 80]. \end{matrix}$$

The price per bag charged to the distributor by the mill is $7.60 for high protein, $7.25 for medium protein, $6.80 for low protein, and $10.60 for puppy food. We can display the price per bag by the column matrix

Prices

$$\begin{matrix} H \\ M \\ L \\ P \end{matrix} \begin{bmatrix} 7.60 \\ 7.25 \\ 6.80 \\ 10.60 \end{bmatrix}.$$

This column of prices is called a **price vector**.

The money received from distributor X for the June shipment is obtained from the row matrix of shipments times the price vector

$$[160 \quad 100 \quad 60 \quad 80] \begin{bmatrix} 7.60 \\ 7.25 \\ 6.80 \\ 10.60 \end{bmatrix} = 160(7.60) + 100(7.25) + 60(6.80) + 80(10.60)$$

$$= \$3197.$$

In some applications we must perform several row-column multiplications, as we see in the next examples.

EXAMPLE 3

Consider again example 1 from section 5.1 and the previous example. In addition to distributor X, the Acme Feed Mills shipped dog food to two other distributors in June, Y and Z. The number of bags shipped to the three distributors is given by the matrix

Number of bags

	H	M	L	P
shipped to X	160	100	60	80
shipped to Y	400	80	200	120
shipped to Z	240	160	120	80

As in example 2, we obtain the amount paid by each distributor by taking each row of the shipment matrix times the price vector. However, instead of displaying the results as three separate matrix products, we give it more concisely as

$$\begin{bmatrix} 160 & 100 & 60 & 80 \\ 400 & 80 & 200 & 120 \\ 240 & 160 & 120 & 80 \end{bmatrix} \begin{bmatrix} 7.60 \\ 7.25 \\ 6.80 \\ 10.60 \end{bmatrix} = \begin{bmatrix} 3197 \\ 6252 \\ 4648 \end{bmatrix}.$$

It is understood that the first entry, 3197, is the amount paid by distributor X, obtained from the first row of the shipment matrix times the price vector, and so forth.

The row-column product provides an alternate way of writing a linear equation. For example, consider the linear equation $2x + 3y - 4z = 8$. Using the 1×3 row matrix of coefficients $[2 \quad 3 \quad -4]$ and the 3×1 column matrix of variables

$$\begin{bmatrix} x \\ y \\ z \end{bmatrix},$$

we have that

$$[2 \quad 3 \quad -4] \begin{bmatrix} x \\ y \\ z \end{bmatrix} = 2x + 3y - 4z.$$

Therefore we can represent the linear equation by the matrix equation

$$[2 \quad 3 \quad -4] \begin{bmatrix} x \\ y \\ z \end{bmatrix} = 8.$$

Consider now an m by n system of linear equations

$$a_{11}x_1 + a_{12}x_2 + \cdots + a_{1n}x_n = b_1$$
$$a_{21}x_1 + a_{22}x_2 + \cdots + a_{2n}x_n = b_2$$
$$\vdots \qquad \vdots \qquad \qquad \vdots \qquad \vdots$$
$$a_{m1}x_1 + a_{m2}x_2 + \cdots + a_{mn}x_n = b_m.$$

Let

$$A = \begin{bmatrix} a_{11} & a_{12} & \cdots & a_{1n} \\ a_{21} & a_{22} & \cdots & a_{2n} \\ \vdots & \vdots & & \vdots \\ a_{m1} & a_{m2} & \cdots & a_{mn} \end{bmatrix}, \quad X = \begin{bmatrix} x_1 \\ x_2 \\ \vdots \\ x_n \end{bmatrix}, \quad \text{and} \quad B = \begin{bmatrix} b_1 \\ b_2 \\ \vdots \\ b_m \end{bmatrix}.$$

The system can be given by the matrix equation

$$AX = B.$$

A is called the **coefficient matrix** for the system.

EXAMPLE 4

Write the 2×3 system

$$2x + 3y - 4z = \quad 5$$
$$- \quad y + 2z = -3$$

as a matrix equation.

The coefficient matrix for the system is the 2×3 matrix

$$\begin{bmatrix} 2 & 3 & -4 \\ 0 & -1 & 2 \end{bmatrix}.$$

Furthermore, the matrix equation

$$\begin{bmatrix} 2 & 3 & -4 \\ 0 & -1 & 2 \end{bmatrix} \begin{bmatrix} x \\ y \\ z \end{bmatrix} = \begin{bmatrix} 5 \\ -3 \end{bmatrix}$$

represents the system.

Recall that a solution to an $m \times n$ system of linear equations is any ordered set of n numbers that satisfies each equation in the system. In terms of a matrix equation we have the following definition.

Definition ∎ Let A be an $m \times n$ matrix. A **solution** to the matrix equation $AX = B$ is any $n \times 1$ column matrix that satisfies the equation. ∎

EXAMPLE 5

The 3×1 column matrix

$$X = \begin{bmatrix} -1 \\ 1 \\ -1 \end{bmatrix}$$

is a solution to the matrix equation of example 1 since

$$\begin{bmatrix} 2 & 3 & -4 \\ 0 & -1 & 2 \end{bmatrix} \begin{bmatrix} -1 \\ 1 \\ -1 \end{bmatrix} = \begin{bmatrix} -2 + 3 + 4 \\ 0 - 1 - 2 \end{bmatrix} = \begin{bmatrix} 5 \\ -3 \end{bmatrix}.$$

EXERCISES 5.2

For exercises 1–8 compute the indicated row-column product if it is defined.

1. $\begin{bmatrix} 2 & -3 & 1 \end{bmatrix} \begin{bmatrix} -2 \\ 4 \\ 6 \end{bmatrix}$

2. $\begin{bmatrix} 1 & 0 & 4 & -5 \end{bmatrix} \begin{bmatrix} 2 \\ 3 \\ 0 \\ -2 \end{bmatrix}$

3. $\begin{bmatrix} 3 & -1 & 5 \end{bmatrix} \begin{bmatrix} 6 \\ -3 \\ -4 \\ 2 \end{bmatrix}$

4. $\begin{bmatrix} 0 & 1 & 0 \end{bmatrix} \begin{bmatrix} 6 \\ -5 \\ 7 \end{bmatrix}$

5. $\begin{bmatrix} 0 & 1 & 0 & 1 \end{bmatrix} \begin{bmatrix} 4 \\ -2 \\ 5 \\ -8 \end{bmatrix}$

6. $\begin{bmatrix} 2 & -4 & 3 & -5 \end{bmatrix} \begin{bmatrix} -8 \\ 4 \\ 3 \end{bmatrix}$

7. $\begin{bmatrix} 2 & -4 & 3 & -5 \end{bmatrix} \begin{bmatrix} -6 \\ 3 \\ -4 \\ 2 \end{bmatrix}$

8. $\begin{bmatrix} 2 & -4 & 3 & -5 \end{bmatrix} \begin{bmatrix} 7 \\ 5 \\ -2 \\ 1 \end{bmatrix}$

9. Refer to example 2. Suppose distributor X sells its entire June shipment of dog food to dealers. The price per bag charged to a dealer by the distributor is $9.75 for high protein, $9.20 for medium protein, $8.75 for low protein, and $12.20 for puppy food. Express the amount received from dealers as a row-column product.

10. The Banner Tool Company makes angle grinders, circular saws, and hammer drills. The rates at which workers are needed to produce these items are: 2 man-hours for one angle grinder, 3 man-hours for one circular saw, and 1 man-hour for one hammer drill. In one day the company produces 40 angle grinders, 35 circular saws, and 60 hammer drills. Express as a row-column product the total amount of labor needed in a single day's production.

11. A seamstress makes girl's blouses and boy's shirts. On one day she made 2 blouses and 3 shirts, whereas on two other days she made, respectively, 4 blouses and 2 shirts and 3 blouses and 5 shirts. If each blouse requires 2.5 yards of material and each shirt requires 3 yards of material, express by a row-column product the total yards used each day in making the blouses and shirts.

12. Refer to exercise 27 of section 5.1. If the salesperson receives a commission of $75 on each of the first model microcomputer she sells, $125 on each of the second model she sells, $80 on each of the third model, and $130 on each of the fourth model, express by a row-column product the total commission from sales to Fischer and the total commission from sales to Computer World she would receive.

For exercises 13–18 write the equation or system of equations as a matrix equation.

13. $2x + 3y - 4z = 7$

14. $4x_1 - 2x_2 + 5x_3 + 8x_4 = 10$

15. $\begin{aligned} 2x + 3y - 4z &= 7 \\ x \quad\quad + 2z &= 5 \\ 4y - \quad z &= 8 \end{aligned}$

16. $\begin{aligned} x_2 + 4x_3 - 2x_4 &= -1 \\ 2x_1 - 4x_2 \quad\quad + x_4 &= 2 \\ 3x_1 + 5x_2 - \quad x_3 + 7x_4 &= 3 \end{aligned}$

17. $\begin{aligned} 2x_1 + 3x_2 \quad\quad - x_4 &= 7 \\ x_2 + 5x_3 \quad\quad &= 8 \end{aligned}$

18. $\begin{aligned} x_1 + 5x_2 - 6x_3 + 7x_4 &= 10 \\ 2x_2 + 9x_3 - 4x_4 &= 11 \\ x_3 + 2x_4 &= 8 \\ 5x_4 &= 10 \end{aligned}$

19. Let

$$A = \begin{bmatrix} 1 & -3 & 5 & 4 \\ 0 & 1 & 0 & 2 \\ 0 & 0 & 1 & -6 \end{bmatrix} \quad \text{and} \quad B = \begin{bmatrix} 6 \\ 3 \\ 4 \end{bmatrix}.$$

a. Verify that

$$X_1 = \begin{bmatrix} -40 \\ -2 \\ 6 \\ 1 \end{bmatrix}$$

is a solution to $AX = \mathbf{0}$.

b. Verify that

$$X_2 = \begin{bmatrix} -5 \\ 3 \\ 4 \\ 0 \end{bmatrix}$$

is a solution to $AX = B$.

20. Let

$$A = \begin{bmatrix} 1 & -2 & 3 \\ 0 & 3 & -1 \\ 3 & 0 & 7 \end{bmatrix}, \quad B = \begin{bmatrix} 0 \\ 5 \\ 10 \end{bmatrix}, \quad X_1 = \begin{bmatrix} 7 \\ -1 \\ -3 \end{bmatrix}, \quad \text{and} \quad X_2 = \begin{bmatrix} 1 \\ 2 \\ 1 \end{bmatrix}.$$

a. Verify that X_1 is a solution to $AX = 0$.

b. Verify that X_2 is a solution to $AX = B$.

21. Let

$$A = \begin{bmatrix} 1 & 0 & 3 & -4 \\ 0 & 1 & -2 & 5 \end{bmatrix} \quad \text{and} \quad X_1 = \begin{bmatrix} -2 \\ -1 \\ 2 \\ 1 \end{bmatrix}.$$

a. Verify that $AX_1 = 0$.

b. Verify that $A(2X_1) = 0$.

22. Let

$$A = \begin{bmatrix} 1 & 0 & 3 & -4 \\ 0 & 1 & -2 & 5 \end{bmatrix} \quad \text{and} \quad B = \begin{bmatrix} 19 \\ -15 \end{bmatrix}$$

and let

$$X_1 = \begin{bmatrix} -2 \\ -1 \\ 2 \\ 1 \end{bmatrix} \quad \text{and} \quad X_2 = \begin{bmatrix} 3 \\ -2 \\ 4 \\ -1 \end{bmatrix}.$$

a. Verify that $AX_2 = B$.

b. In exercise 21 we verified that $AX_1 = 0$. Now verify that $A(X_1 + X_2) = B$.

5.3 Matrix Multiplication

With row-column products, as given in section 5.2, firmly in mind, we now define matrix multiplication in general.

Definition

■ Let $A = [a_{ij}]$ be an $m \times k$ matrix and let $B = [b_{ij}]$ be a $k \times n$ matrix. The **matrix product** of A and B, denoted by AB, is the $m \times n$ matrix where the (i, j) entry, obtained by taking the row-column product of the ith row of A with the jth column of B, is

$$[a_{i1} a_{i2}, \ldots, a_{ik}] \begin{bmatrix} b_{1j} \\ b_{2j} \\ \vdots \\ b_{kj} \end{bmatrix} = a_{i1} b_{1j} + a_{i2} b_{2j} + \cdots + a_{ik} b_{kj}. \quad ■$$

For the product AB to be defined, the number of columns of A must be the same as the number of rows of B. Furthermore, the number of rows of AB is the same as the number of rows of A, and the number of columns of AB is the same as the number of columns of B.

EXAMPLE 1

Compute the matrix product

$$AB = \begin{bmatrix} 2 & -3 & 4 \\ 1 & 5 & 3 \end{bmatrix} \begin{bmatrix} -1 & 2 \\ 2 & -4 \\ 3 & 1 \end{bmatrix}.$$

Since B has two columns, the product AB also has two columns. Also, since A has two rows, each of these columns in the product contains two rows.

The first column of AB is obtained from A times the first column of B. In particular, the first column of AB is

$$\begin{bmatrix} 2 & -3 & 4 \\ 1 & 5 & 3 \end{bmatrix} \begin{bmatrix} -1 \\ 2 \\ 3 \end{bmatrix} = \begin{bmatrix} 2(-1) - 3(2) + 4(3) \\ 1(-1) + 5(2) + 3(3) \end{bmatrix} = \begin{bmatrix} 4 \\ 18 \end{bmatrix}.$$

Similarly, we obtain the second column of AB as follows:

$$\begin{bmatrix} 2 & -3 & 4 \\ 1 & 5 & 3 \end{bmatrix} \begin{bmatrix} 2 \\ -4 \\ 1 \end{bmatrix} = \begin{bmatrix} 2(2) - 3(-4) + 4(1) \\ 1(2) + 5(-4) + 3(1) \end{bmatrix} = \begin{bmatrix} 20 \\ -15 \end{bmatrix}.$$

Thus,

$$AB = \begin{bmatrix} 4 & 20 \\ 18 & -15 \end{bmatrix}.$$

It is important to keep in mind when a matrix product is defined and the size of the product matrix when it is defined. This point is reiterated in the following diagram.

$$
\begin{array}{ccccc}
A & \times & B & = & AB \\
m \times k & & k \times n & & m \times n
\end{array}
$$

must be equal

size of product matrix $m \times n$

EXAMPLES

2.

a.

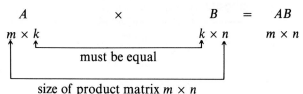

$$
\underset{2 \times 3}{A} \qquad \underset{3 \times 4}{B} \qquad \underset{2 \times 4}{AB}
$$

For example, the $(1, 3)$ entry of AB is the product of the first row of A with the third column of B and is

$$5(2) - 1(0) + 2(4) = 18.$$

b.
$$\begin{bmatrix} 1 & 3 & 2 \\ 3 & 4 & 0 \\ 0 & 1 & 5 \end{bmatrix} \begin{bmatrix} 1 & 0 \\ 1 & 1 \\ 0 & 1 \end{bmatrix} = \begin{bmatrix} 4 & 5 \\ 7 & 4 \\ 1 & 6 \end{bmatrix}$$

$ 3 \times 3 3 \times 2 3 \times 2$

c.
$$\begin{bmatrix} 1 & 0 & 1 \\ 0 & 1 & 1 \end{bmatrix} \begin{bmatrix} 2 & -1 \\ 1 & 0 \end{bmatrix}$$
cannot be done since the first matrix has three columns and the second matrix has two rows.

$ 2 \times 3 2 \times 2$

3. A survey is taken of shoppers to determine their buying habits with respect to three detergents, brands X, Y, and Z. The results of the survey are indicated in the matrix P.

$$\begin{array}{c} \text{Changes to} \\ \begin{array}{ccc} & \text{X} & \text{Y} & \text{Z} \end{array} \\ P = \begin{array}{c} \text{X} \\ \text{Y} \\ \text{Z} \end{array} \begin{bmatrix} 0.60 & 0.20 & 0.20 \\ 0.10 & 0.70 & 0.20 \\ 0.40 & 0.20 & 0.40 \end{bmatrix}. \end{array}$$

We read from P, for example, that of the users of brand X, 60% stayed with it the next month, whereas 20% changed to brand Y, and 20% changed to brand Z the next month. The matrix P is called a **transition matrix**.

Suppose in a group of 1000 shoppers, 400 presently use brand X, 200 use brand Y, and 400 use brand Z. This information is displayed in the row matrix

$$\begin{array}{c} \text{Distribution of shoppers} \\ \begin{array}{ccc} \text{X} & \text{Y} & \text{Z} \end{array} \\ [400 \quad 200 \quad 400], \end{array}$$

called a **distribution vector**.

Let us now find the number of shoppers who will be using the different brands the next month.

The desired results are obtained by taking the distribution vector times each column of the transition matrix. Therefore the desired results are found with the matrix product

$$[400 \quad 200 \quad 400] \begin{bmatrix} 0.60 & 0.20 & 0.20 \\ 0.10 & 0.70 & 0.20 \\ 0.40 & 0.20 & 0.40 \end{bmatrix} = [420 \quad 300 \quad 280].$$

Thus we see that after the first month of transition 420 of the 1000 shoppers will be using brand X, 300 will be using brand Y, and 280 will be using brand Z.

4. Three of the several requirements of a diet are protein, iron, and calories.

Suppose these dietary requirements are obtained from two foods, T_1 and T_2. One ounce of T_1 contains 3 units of protein, 0.5 unit of iron, and 40 calories; whereas, one ounce of T_2 contains 1 unit of protein, 0.1 unit of iron, and 60 calories. We display this information in the following matrix.

$$S = \begin{array}{c} \text{units of protein} \\ \text{units of iron} \\ \text{calories} \end{array} \overset{\begin{array}{cc} T_1 & T_2 \end{array}}{\begin{bmatrix} 3.0 & 1.0 \\ 0.5 & 0.1 \\ 40.0 & 60.0 \end{bmatrix}}.$$

Per ounce of

Suppose that the number of ounces of each type of food eaten by a person for breakfast (B), lunch (L), and dinner (D) is given in the following matrix.

Eaten for

$$E = \begin{array}{c} \text{ounces of } T_1 \\ \text{ounces of } T_2 \end{array} \overset{\begin{array}{ccc} B & L & D \end{array}}{\begin{bmatrix} 11 & 7 & 10 \\ 9 & 8 & 12 \end{bmatrix}}.$$

When we take the units of protein per ounce times the number of ounces eaten, we obtain the number of units of protein acquired. Thus, the number of units of protein obtained from breakfast is $(3.0)(11) + (1.0)(9) = 42$, which is the product of the first row of S and the first column of E. To compute the number of units of protein obtained from lunch we take the product of the first row of S and the second column of E. In general, the number of units of protein acquired from the three meals is found in the first row of the product matrix SE. The number of units of iron and the number of calories are found in the second and third rows of SE, respectively.

$$SE = \begin{bmatrix} 3.0 & 1.0 \\ 0.5 & 0.1 \\ 40.0 & 60.0 \end{bmatrix}\begin{bmatrix} 11 & 7 & 10 \\ 9 & 8 & 12 \end{bmatrix}$$

$$= \begin{array}{c} \text{units of protein} \\ \text{units of iron} \\ \text{calories} \end{array} \overset{\begin{array}{ccc} B & L & D \end{array}}{\begin{bmatrix} 42.0 & 29.0 & 42.0 \\ 6.4 & 4.3 & 6.2 \\ 980.0 & 760.0 & 1120.0 \end{bmatrix}}.$$

Note that the labels of the rows of the product SE are the same as those of S and the labels of the columns of SE are the same as those of E.

We should not expect properties of multiplication of real numbers to carry over to matrix multiplication. For example, from properties of real numbers, if $ab = 0$ then $a = 0$ or $b = 0$. However, a similar result does not exist for matrix multiplication as indicated by the product

$$\begin{bmatrix} 2 & 3 \\ 4 & 6 \end{bmatrix}\begin{bmatrix} 3 & -1 \\ -2 & \frac{2}{3} \end{bmatrix} = \begin{bmatrix} 0 & 0 \\ 0 & 0 \end{bmatrix}.$$

Even though the product is the zero matrix neither of the two matrices multiplied is the zero matrix.

Matrix multiplication is also not, in general, commutative. For two matrices A and B, AB may not be equal to BA even when these two products are defined. For example,

$$\begin{bmatrix} 1 & 2 \\ -3 & 4 \end{bmatrix} \begin{bmatrix} 2 & -1 \\ -4 & 3 \end{bmatrix} = \begin{bmatrix} -6 & 5 \\ -22 & 15 \end{bmatrix}$$

whereas

$$\begin{bmatrix} 2 & -1 \\ -4 & 3 \end{bmatrix} \begin{bmatrix} 1 & 2 \\ -3 & 4 \end{bmatrix} = \begin{bmatrix} 5 & 0 \\ -13 & 4 \end{bmatrix}.$$

Aside from these two negative results, matrix multiplication does satisfy an analogous property to multiplication of real numbers, given in the following theorem.

Theorem 2

■ Consider matrices A, B, and C for which their sizes are such that the products $A(BC)$ and $(AB)C$ are defined. Then

$$A(BC) = (AB)C \qquad \text{(associative property)} \qquad ■$$

EXAMPLES

5. Verify the associative property of matrix multiplication using the matrices

$$A = \begin{bmatrix} 2 & 1 & 0 \\ -1 & 3 & 4 \end{bmatrix}, \qquad B = \begin{bmatrix} 0 & 1 & 2 \\ -1 & 3 & -2 \\ 2 & 3 & -4 \end{bmatrix}, \qquad \text{and}$$

$$C = \begin{bmatrix} 1 & 3 \\ 5 & 2 \\ 4 & 1 \end{bmatrix}.$$

$$A(BC) = \begin{bmatrix} 2 & 1 & 0 \\ -1 & 3 & 4 \end{bmatrix} \begin{bmatrix} 13 & 4 \\ 6 & 1 \\ 1 & 8 \end{bmatrix} = \begin{bmatrix} 32 & 9 \\ 9 & 31 \end{bmatrix}$$

and

$$(AB)C = \begin{bmatrix} -1 & 5 & 2 \\ 5 & 20 & -24 \end{bmatrix} \begin{bmatrix} 1 & 3 \\ 5 & 2 \\ 4 & 1 \end{bmatrix} = \begin{bmatrix} 32 & 9 \\ 9 & 31 \end{bmatrix}.$$

6. Consider again the transition matrix

Changes to

$$P = \begin{matrix} & X & Y & Z \\ X \\ Y \\ Z \end{matrix} \begin{bmatrix} 0.60 & 0.20 & 0.20 \\ 0.10 & 0.70 & 0.20 \\ 0.40 & 0.20 & 0.40 \end{bmatrix}$$

of example 3. Let S be a distribution vector and let us find the distribution of shoppers who will be using the different brands after two months.

Recall that SP is the distribution of shoppers who will be using the different brands after one month. Therefore, the distribution of shoppers after two months is $(SP)P$. However, by the associative property, $(SP)P = S(PP)$. Thus an initial distribution vector of shoppers times the matrix

Changes in two months to

$$P^2 = PP = \begin{array}{c} X \\ Y \\ Z \end{array} \begin{array}{ccc} X & Y & Z \\ \begin{bmatrix} 0.46 & 0.30 & 0.24 \\ 0.21 & 0.55 & 0.24 \\ 0.42 & 0.30 & 0.28 \end{bmatrix} \end{array}$$

yields the distribution of the shoppers using the three brands after two months. For example, if $S = [400 \quad 200 \quad 400]$, then the number of shoppers using the different brands after two months is $S(P^2) = [394 \quad 350 \quad 256]$. If $S = [500 \quad 300 \quad 200]$ then we have $S(P^2) = [377 \quad 375 \quad 248]$.

The application of the associative property in example 6 will be useful in our study of Markov chains in chapter 10.

Matrix multiplication, together with addition, satisfies another analogous property of arithmetic of real numbers.

Theorem 3   ■ Consider matrices A, B, and C for which their sizes are such that the sums and products in each of the following are defined. Then

a. $A(B + C) = AB + AC$ (distributive property)
b. $(B + C)A = BA + CA$ (distributive property) ■

EXAMPLE 7   Verify the distributive property $A(B + C) = AB + AC$ using the matrices

$$A = \begin{bmatrix} 4 & 5 & 2 \\ 3 & -2 & 0 \\ -1 & 0 & 1 \end{bmatrix}, \quad B = \begin{bmatrix} -3 & 6 \\ 0 & 7 \\ 1 & 2 \end{bmatrix}, \quad \text{and} \quad C = \begin{bmatrix} 5 & -4 \\ 2 & 0 \\ 7 & -6 \end{bmatrix}.$$

We have that

$$A(B + C) = \begin{bmatrix} 4 & 5 & 2 \\ 3 & -2 & 0 \\ -1 & 0 & 1 \end{bmatrix} \begin{bmatrix} 2 & 2 \\ 2 & 7 \\ 8 & -4 \end{bmatrix} = \begin{bmatrix} 34 & 35 \\ 2 & -8 \\ 6 & -6 \end{bmatrix}$$

and

$$AB + AC = \begin{bmatrix} -10 & 63 \\ -9 & 4 \\ 4 & -4 \end{bmatrix} + \begin{bmatrix} 44 & -28 \\ 11 & -12 \\ 2 & -2 \end{bmatrix} = \begin{bmatrix} 34 & 35 \\ 2 & -8 \\ 6 & -6 \end{bmatrix}.$$

The next theorem gives properties of multiplication by a scalar.

Theorem 4 ■ Let A and B be two matrices of sizes for which the sums and products are defined in the following. Let s and t be two scalars. Then

a. $s(A + B) = sA + sB$,

b. $(s + t)A = sA + tA$,

c. $s(AB) = (sA)B = A(sB)$,

d. $(st)A = s(tA) = t(sA)$. ■

The properties of matrix operations can be helpful when doing computations, as we see in the next example.

EXAMPLE 8 | 8. Compute the product

$$AB = \begin{bmatrix} \frac{6}{5} & -\frac{4}{5} \\ \frac{2}{5} & \frac{8}{5} \end{bmatrix}\begin{bmatrix} \frac{5}{3} & \frac{4}{3} \\ -\frac{1}{3} & \frac{2}{3} \end{bmatrix},$$

using property (c) of theorem 4.
We have

$$AB = \frac{2}{5}\begin{bmatrix} 3 & -2 \\ 1 & 4 \end{bmatrix}\frac{1}{3}\begin{bmatrix} 5 & 4 \\ -1 & 2 \end{bmatrix} = \frac{2}{15}\begin{bmatrix} 3 & -2 \\ 1 & 4 \end{bmatrix}\begin{bmatrix} 5 & 4 \\ -1 & 2 \end{bmatrix}$$

$$= \frac{2}{15}\begin{bmatrix} 17 & 8 \\ 1 & 12 \end{bmatrix} = \begin{bmatrix} \frac{34}{15} & \frac{16}{15} \\ \frac{2}{15} & \frac{24}{15} \end{bmatrix}.$$

EXERCISES 5.3

In exercises 1 and 2 assume $A, B, C,$ and D are matrices of sizes $3 \times 3, 2 \times 3, 4 \times 3,$ and $3 \times 4,$ respectively. State whether or not the indicated product is defined and give the size of the product matrix, if it is defined.

1. a. BA b. AB c. CD d. DC
2. a. CA b. BD c. BC d. AC

For exercises 3–6 let

$$A = \begin{bmatrix} 1 & -1 \\ 2 & 3 \\ 0 & 4 \end{bmatrix}, \quad B = \begin{bmatrix} 2 & 3 \\ -1 & 5 \\ -2 & 0 \end{bmatrix}, \quad C = \begin{bmatrix} 1 & -2 & 5 & -3 \\ 2 & -1 & 4 & -2 \end{bmatrix}, \quad \text{and}$$

$$D = \begin{bmatrix} 1 & 0 & 0 \\ 1 & 1 & 0 \\ 0 & 1 & 1 \\ 1 & 0 & 1 \end{bmatrix},$$

and compute the indicated product, if it is defined.

3. DB 4. AC 5. AB 6. CD

For exercises 7–10 let

$$A = \begin{bmatrix} 3 & 0 \\ 0 & 4 \end{bmatrix}, \quad B = \begin{bmatrix} 1 & -2 & 4 \\ -5 & 0 & 3 \end{bmatrix}, \quad C = \begin{bmatrix} 2 & 4 & 5 \\ 6 & 2 & 1 \end{bmatrix}, \quad \text{and}$$

$$D = \begin{bmatrix} 1 & 0 \\ 0 & 1 \\ 1 & 1 \end{bmatrix},$$

and compute the indicated product if it is defined.

7. AB 8. AD 9. DC 10. C^tB

11. Let

$$A = \begin{bmatrix} 1 & -2 \\ -3 & 4 \end{bmatrix}, \quad B = \begin{bmatrix} 0 & 5 \\ 6 & -2 \end{bmatrix}, \quad \text{and} \quad C = \begin{bmatrix} 7 & -3 \\ 8 & 6 \end{bmatrix}.$$

Verify that $A(BC) = (AB)C$.

12. Let

$$A = \begin{bmatrix} 1 & 0 & 2 \\ 0 & 1 & 1 \\ 1 & 0 & 3 \end{bmatrix}, \quad B = \begin{bmatrix} -1 & -2 & 0 \\ 2 & 0 & 1 \\ 3 & -1 & -4 \end{bmatrix}, \quad \text{and} \quad C = \begin{bmatrix} 2 & 5 & 1 \\ -1 & 1 & -2 \\ 3 & 0 & 1 \end{bmatrix}.$$

Verify that $A(BC) = (AB)C$.

13. Let

$$A = \begin{bmatrix} 1 & -1 \\ 2 & 3 \\ 0 & 4 \end{bmatrix}, \quad B = \begin{bmatrix} 2 & 3 \\ -1 & 5 \\ -2 & 0 \end{bmatrix}, \quad \text{and} \quad C = \begin{bmatrix} 7 & -3 \\ 8 & 6 \end{bmatrix}.$$

Verify that $(A + B)C = AC + BC$.

14. Let A, B, and C be the matrices given in exercise 12. Verify that $(A + B)C = AC + BC$.

15. Let

$$A = \begin{bmatrix} 1 & -1 \\ 2 & 3 \end{bmatrix}, \quad B = \begin{bmatrix} -1 & 5 \\ -2 & 0 \end{bmatrix}, \quad s = \frac{1}{2}, \quad \text{and} \quad t = 3.$$

For these matrices and real numbers, verify that:

a. $s(A + B) = sA + sB$
b. $(s + t)A = sA + tA$
c. $s(AB) = (sA)B = A(sB)$
d. $(st)A = s(tA) = t(sA)$

16. Repeat exercise 15 for the matrices

$$A = \begin{bmatrix} 1 & 0 & 2 \\ 0 & 1 & 1 \\ 1 & 0 & 3 \end{bmatrix} \quad \text{and} \quad B = \begin{bmatrix} -1 & -2 & 0 \\ 2 & 0 & 1 \\ 3 & -1 & -4 \end{bmatrix}$$

and the real numbers $s = -\frac{2}{3}$ and $t = -2$.

In exercises 17 and 18 do the matrix computation using theorem 4 on page 180.

17. $3\begin{bmatrix} 1 & -2 \\ -3 & 4 \end{bmatrix} + 4\begin{bmatrix} 1 & -2 \\ -3 & 4 \end{bmatrix}$

18. $7\begin{bmatrix} 2 & -1 & -2 \\ 3 & 5 & 0 \end{bmatrix} - 2\begin{bmatrix} 2 & -1 & -2 \\ 3 & 5 & 0 \end{bmatrix}$

In exercises 19 and 20 do the matrix computation using theorem 4 on page 180.

19. $7\begin{bmatrix} \frac{3}{7} & -\frac{1}{7} \\ -\frac{5}{7} & \frac{4}{7} \end{bmatrix}\begin{bmatrix} -2 & 3 \\ 4 & 1 \end{bmatrix}$

20. $15\begin{bmatrix} \frac{1}{5} & -\frac{2}{5} & \frac{6}{5} \\ \frac{7}{5} & 0 & -\frac{1}{5} \end{bmatrix}\begin{bmatrix} -\frac{4}{3} & \frac{1}{3} \\ -\frac{2}{3} & \frac{5}{3} \\ \frac{7}{3} & -\frac{4}{3} \end{bmatrix}$

In exercises 21 and 22 do the matrix computation using the distributive property $AB + AC = A(B + C)$.

21. $\begin{bmatrix} 1 & -1 \\ 2 & 3 \\ 0 & 4 \end{bmatrix}\begin{bmatrix} 0 & 5 \\ 6 & -2 \end{bmatrix} + \begin{bmatrix} 1 & -1 \\ 2 & 3 \\ 0 & 4 \end{bmatrix}\begin{bmatrix} 7 & -3 \\ 8 & 6 \end{bmatrix}$

22. $\begin{bmatrix} -1 & -2 & 0 \\ 2 & 0 & 1 \end{bmatrix}\begin{bmatrix} 2 & 3 \\ -1 & 5 \\ -2 & 0 \end{bmatrix} + \begin{bmatrix} -1 & -2 & 0 \\ 2 & 0 & 1 \end{bmatrix}\begin{bmatrix} 1 & 0 \\ -2 & -3 \\ 4 & 5 \end{bmatrix}$

23. Suppose each of three meals consists of three foods: A, B, and C. The number of ounces of each food eaten at each of the three meals is given in the following matrix.

Ounces of

		A	B	C
eaten	1	8	6	4
at	2	4	0	7
meal	3	9	3	5

One ounce of A contains 2.5 units of protein and 35 calories, one ounce of B contains 4 units of protein and 50 calories, and one ounce of C contains 3.5 units of protein and 70 calories. Express by a matrix product the units of protein and the number of calories that are contained in each meal.

24. Imperial Industries manufactures metal file cabinets and metal desks. Labor is needed in the manufacture of each. One file cabinet requires 1.5 man-hours of labor, and one desk requires 2 man-hours of labor. The number of file cabinets and desks produced on each of five days is given in the following matrix.

Produced on day

	Mon.	Tues.	Wed.	Thurs.	Fri.
number of cabinets	20	10	12	24	15
number of desks	15	20	13	18	14

Express by a matrix product the amount of labor needed on each of the five days.

25. The Glass Works company has three production plants, each of which produces one-quart, one-pint, and half-pint canning jars.

a. During one hour of operation, plant I makes 1500 quart jars, 1500 pint jars, and 2000 half-pint jars. Plant II can produce in one hour 2000 quart, 4000 pint, and 1500 half-pint jars. Plant III produces 3000 quart, 2500 pint, and 1000 half-pint jars. Express this information in a matrix, labeling rows and columns appropriately.

b. During the first week of May, plant I operated 35 hours, plant II operated 30 hours, and plant III operated 40 hours. During the second week of May, plants I, II, and III operated 40 hours, 60 hours, and 30 hours, respectively. Express this information in a matrix, labeling rows and columns appropriately.

c. Express by a matrix product the number of jars of each size produced by the Glass Works Company during each of the first two weeks of May.

26. Refer to exercise 24. In addition to labor, steel and power are used to make the file cabinets and desks. One file cabinet requires 35 pounds of steel and 0.1 kilowatt of power, whereas one desk requires 50 pounds of steel and 0.3 kilowatt of power. Express by a matrix product the amount of labor, steel, and power needed on each of the five days to produce the file cabinets and desks. Use the matrix given in exercise 24.

27. A salesperson for Snapp-Off Tools recorded in one week sales on three types of tool kits, Standard, Deluxe and Industrial, according to table 5.1

TABLE 5.1

	Mon.	Tues.	Wed.	Thurs.	Fri.
			Sales on day		
standard	0	13	11	8	5
deluxe	9	0	4	7	10
industrial	4	2	10	11	0

The person received a commission of $15.00 on each Standard kit, $18.00 on each Deluxe kit, and $12.50 on each Industrial kit sold. Express by a matrix product the total commission on each of the five days.

5.4 Lines of Best Fit

We realize that one step in solving many problems is first to model a dependency of one quantity y on another quantity x. Furthermore, when it is known (or at least assumed) that the function is linear, the dependency can be described by an equation of the form $y = mx + b$. In this case, as seen in chapter 2, it is sufficient to have two data points (x_1, y_1) and (x_2, y_2) available in order to write the linear equation. When the two points do not lie on a vertical line $(x_1 \neq x_2)$, m and b can then be found by solving the system of linear equations

$$b + x_1 m = y_1$$
$$b + x_2 m = y_2.$$

In many practical situations the data points may be obtained through experimentation, and therefore more than two could be gathered. For a line $y = mx + b$ to pass through n data points $(x_1, y_1), (x_2, y_2), \ldots, (x_n, y_n)$, m and b must be a solution to the system of n equations

$$b + x_1 m = y_1$$
$$b + x_2 m = y_2$$
$$\vdots \quad \vdots \quad \vdots$$
$$b + x_n m = y_n,$$

which in matrix form is $AW = Y$ with

$$A = \begin{bmatrix} 1 & x_1 \\ 1 & x_2 \\ \vdots & \vdots \\ 1 & x_n \end{bmatrix} \qquad W = \begin{bmatrix} b \\ m \end{bmatrix}, \qquad \text{and} \qquad Y = \begin{bmatrix} y_1 \\ y_2 \\ \vdots \\ y_n \end{bmatrix}.$$

For example, consider the following situation.

A laboratory testing the energy efficiency of a particular electric motor wishes to express the dependency of watt-hours h used on the running time t. Suppose the laboratory finds through measurement that when $t = 10$ hours, $h = 19.8$; when $t = 15$ hours, $h = 29.8$; and when $t = 20$ hours, $h = 39.2$. The laboratory assumes that the trend in the data is linear and wishes to find a linear equation $h = mt + b$ that "fits" the data points. If there exists a line that goes through the three data points $(10, 19.8)$, $(15, 29.8)$, and $(20, 39.2)$, then m and b must be a solution to the system

$$\begin{bmatrix} 1 & 10 \\ 1 & 15 \\ 1 & 20 \end{bmatrix} \begin{bmatrix} b \\ m \end{bmatrix} = \begin{bmatrix} 19.8 \\ 29.8 \\ 39.2 \end{bmatrix}.$$

Setting up the augmented matrix and using Gauss–Jordan reduction, we find that the system does not have a solution. Therefore there is no line that goes through all three data points.

This situation is not unusual. In fact it is a lot to expect that data points gathered through measurement would all lie on one line. However, it would be reasonable to assume that the data points are scattered close to some line. Therefore our problem is to find a line that "best fits" the data points in some sense.

The line $h = mt + b$ that best fits the data is one in which the differences between the values of h determined from the equation at each t and the actual values of h are small. As seen in figure 5.1, these errors are the differences in the ordinates. In particular, the errors in using $h = mt + b$ are

$$e_1 = (10m + b) - 19.8,$$
$$e_2 = (15m + b) - 29.8,$$
$$e_3 = (20m + b) - 39.2.$$

Thus the line that best fits the data is one in which m and b are chosen such that the total error, as measured by

$$e_1^2 + e_2^2 + e_3^2,$$

is as small as possible. We squared the individual error terms e_i to avoid the possible cancellation of a negative error with a positive error when adding.

Since higher level mathematics is needed to demonstrate the existence of a unique line of best fit, we simply state here the main ideas and follow them with some examples.

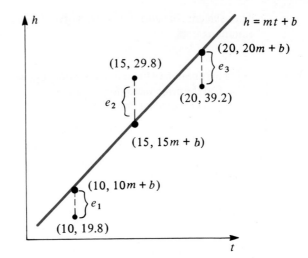

FIGURE 5.1

Definition   ■   Suppose we are given two or more data points $(x_1, y_1), (x_2, y_2), \ldots, (x_n, y_n)$, where no two lie on the same vertical line. A line $y = mx + b$ is a **line of best fit** to the data points when m and b are chosen such that

$$E^t E = e_1^2 + e_2^2 + \cdots + e_n^2$$

is minimal, where

$$E = \begin{bmatrix} e_1 \\ e_2 \\ \vdots \\ e_n \end{bmatrix} = \begin{bmatrix} (mx_1 + b) - y_1 \\ (mx_2 + b) - y_2 \\ \vdots \\ (mx_n + b) - y_n \end{bmatrix}.$$

E is called the **error vector** and the square root of $E^t E$ is called the **size of the error vector.**   ■

Theorem 5   ■   If $(x_1, y_1), (x_2, y_2), \ldots, (x_n, y_n)$ are two or more data points, where no two lie on the same vertical line, then there exists a unique line of best fit $y = mx + b$. Furthermore, the line is obtained by solving the system of linear equations

$$(A^t A)W = A^t y$$

for m and b, where

$$A = \begin{bmatrix} 1 & x_1 \\ 1 & x_2 \\ \vdots & \vdots \\ 1 & x_n \end{bmatrix}, \qquad Y = \begin{bmatrix} y_1 \\ y_2 \\ \vdots \\ y_n \end{bmatrix}, \qquad \text{and} \qquad W = \begin{bmatrix} b \\ m \end{bmatrix}.$$

The system of equations to solve for m and b are called the **normal equations.** ∎

EXAMPLE 1

Find the line of best fit for our laboratory example by solving the normal equations. Also find the error vector for this line.

For the data points $(10, 19.8)$, $(15, 29.8)$, and $(20, 39.2)$

$$A = \begin{bmatrix} 1 & 10 \\ 1 & 15 \\ 1 & 20 \end{bmatrix} \quad \text{and} \quad Y = \begin{bmatrix} 19.8 \\ 29.8 \\ 39.2 \end{bmatrix}.$$

Therefore,

$$A^t = \begin{bmatrix} 1 & 1 & 1 \\ 10 & 15 & 20 \end{bmatrix},$$

and the normal equations are

$$\begin{bmatrix} 1 & 1 & 1 \\ 10 & 15 & 20 \end{bmatrix} \begin{bmatrix} 1 & 10 \\ 1 & 15 \\ 1 & 20 \end{bmatrix} \begin{bmatrix} b \\ m \end{bmatrix} = \begin{bmatrix} 1 & 1 & 1 \\ 10 & 15 & 20 \end{bmatrix} \begin{bmatrix} 19.8 \\ 29.8 \\ 39.2 \end{bmatrix}.$$

Solving this 2×2 system

$$\begin{bmatrix} 3 & 45 \\ 45 & 725 \end{bmatrix} \begin{bmatrix} b \\ m \end{bmatrix} = \begin{bmatrix} 88.8 \\ 1429 \end{bmatrix},$$

we have $m = 1.94$ and $b = 0.5$, and the line of best fit is $h = 1.94t + 0.5$. The error vector we obtain for the line of best fit is

$$E = \begin{bmatrix} [1.94(10) + 0.5] - 19.8 \\ [1.94(15) + 0.5] - 29.8 \\ [1.94(20) + 0.5] - 39.2 \end{bmatrix} = \begin{bmatrix} 0.1 \\ -0.2 \\ 0.1 \end{bmatrix},$$

and $E^t E = 0.01 + 0.04 + 0.01 = 0.06$.

Note that the line of best fit in example 1 does not go through any of the data points since none of the entries of the error vector E is zero (see figure 5.1). The error vector for the line $h = 1.94t + 0.4$ that goes through two of the data points is

$$E = \begin{bmatrix} [1.94(10) + 0.4] - 19.8 \\ [1.94(15) + 0.4] - 29.8 \\ [1.94(20) + 0.4] - 39.2 \end{bmatrix} = \begin{bmatrix} 0 \\ -0.3 \\ 0 \end{bmatrix},$$

and $E^t E = 0.09$. However, the size of the error vector for $h = 1.94t + 0.4$ is larger than for the line of best fit, even though it goes through two data points.

As in example 1, we see that the line of best fit need not go through any of the data points. However, in the case when all the points lie on a line, the line of best fit is that line, as we see in the next example.

EXAMPLES

2. Consider the data points $(1, 1), (2, 2.1), (3, 3.2),$ and $(4, 4.3)$. Determine, by solving the normal equations, the line of best fit and the error vector.

For the given points,

$$A = \begin{bmatrix} 1 & 1 \\ 1 & 2 \\ 1 & 3 \\ 1 & 4 \end{bmatrix} \quad \text{and} \quad Y = \begin{bmatrix} 1.0 \\ 2.1 \\ 3.2 \\ 4.3 \end{bmatrix}.$$

Therefore the normal equations are

$$\begin{bmatrix} 1 & 1 & 1 & 1 \\ 1 & 2 & 3 & 4 \end{bmatrix} \begin{bmatrix} 1 & 1 \\ 1 & 2 \\ 1 & 3 \\ 1 & 4 \end{bmatrix} \begin{bmatrix} b \\ m \end{bmatrix} = \begin{bmatrix} 1 & 1 & 1 & 1 \\ 1 & 2 & 3 & 4 \end{bmatrix} \begin{bmatrix} 1.0 \\ 2.1 \\ 3.2 \\ 4.3 \end{bmatrix}$$

or

$$\begin{bmatrix} 4 & 10 \\ 10 & 30 \end{bmatrix} \begin{bmatrix} b \\ m \end{bmatrix} = \begin{bmatrix} 10.6 \\ 32.0 \end{bmatrix}.$$

Solving this 2×2 system by Gaussian elimination with back substitution, we have $m = 1.1$ and $b = -0.1$; the line of best fit is $y = 1.1x - 0.1$.

The error vector for the data points and the line $y = 1.1x - 0.1$ is

$$E = \begin{bmatrix} [1.1(1) - 0.1] - 1.0 \\ [1.1(2) - 0.1] - 2.1 \\ [1.1(3) - 0.1] - 3.2 \\ [1.1(4) - 0.1] - 4.3 \end{bmatrix} = \begin{bmatrix} 0 \\ 0 \\ 0 \\ 0 \end{bmatrix}.$$

Since each entry of E is zero, the line of best fit goes through each of the points.

3. The price deflator index for the gross national product for certain years is given in table 5.2. Also included in the table is the percent increase in the index over the previous year's index.

TABLE 5.2

t	Index	Percent increase
0	110.9	1.9
1	113.9	2.7
2	117.6	3.2
3	122.3	4.0
4	128.2	4.8

If we plot the index against time, we see that the points are not "close" to a line, whereas if we do the same for the percent increase, we see that the graph is *almost* a line (see figure 5.2). Let us find the line that best fits the data, percent increase versus time.

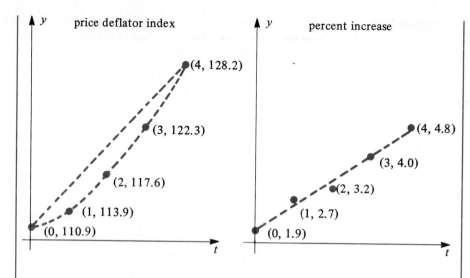

FIGURE 5.2

The normal equations are

$$
\begin{bmatrix} 1 & 1 & 1 & 1 & 1 \\ 0 & 1 & 2 & 3 & 4 \end{bmatrix}
\begin{bmatrix} 1 & 0 \\ 1 & 1 \\ 1 & 2 \\ 1 & 3 \\ 1 & 4 \end{bmatrix}
\begin{bmatrix} b \\ m \end{bmatrix}
=
\begin{bmatrix} 1 & 1 & 1 & 1 & 1 \\ 0 & 1 & 2 & 3 & 4 \end{bmatrix}
\begin{bmatrix} 1.9 \\ 2.7 \\ 3.2 \\ 4.0 \\ 4.8 \end{bmatrix}
$$

from which we obtain

$$
\begin{bmatrix} 5 & 10 \\ 10 & 30 \end{bmatrix}
\begin{bmatrix} b \\ m \end{bmatrix}
=
\begin{bmatrix} 16.6 \\ 40.3 \end{bmatrix}.
$$

Solving these equations we have $b = 1.9$ and $m = 0.71$. The equation of the line of best fit is $y = 0.71t + 1.9$.

Let us use our model $y = 0.71t + 1.9$ to find the percent increase and the index for the next year. Setting $t = 5$ in the equation, we have $y = 0.71(5) + 1.9 = 5.45$. Thus the index is $128.2 + 0.0545(128.2) = 135.2$. These results compare favorably to the actual values, which were 5.5 for the percent increase and 135.2 for the index.

Sometimes the tedious calculations needed to find a line of best fit emphasize the value of using a computer program. When we use a computer in problem solving, our role is to pose the problem correctly, collect appropriate data, and interpret the results. Let us look at this process in the next example.

EXAMPLE 4

The annual percent change in the consumer price indexes for the years 1976–1980, as published by the Bureau of Census, is given in table 5.3. The information points to an upward trend in the data. Even though the changes from year to year are not constant, let us use a line to model the trend in the data.

TABLE 5.3

t	Year	Percent change
0	1976	5.8
1	1977	6.5
2	1978	7.7
3	1979	11.3
4	1980	13.5

We have used a computer program to find the line of best fit, as shown in figure 5.3. We read from the computer printout that the line of best fit is

$$p = 2.02t + 4.92.$$

Let us use this model to project the percent change into the next year, 1981.

FIGURE 5.3

```
How many data points are there?
!5
Enter a data point (x,y) at each prompt.
!0, 5. 8
!1, 6. 5
!2, 7. 7
!3, 11. 3
!4, 13. 5


THE NORMAL EQUATIONS ARE:
=== ====== ========= ===

5B + 10M = 44. 8
10B + 30M = 109. 8

THE LINE OF BEST FIT, Y = MX + B, IS
=== ==== == ==== === ========== ==

M = 2. 02            B = 4. 919999999999
```

Replacing t by 5 in the equation, we obtain

$$p = 2.02(5) + 4.92 = 15.02.$$

It happens that this result is far from the actual value. In 1981, the percent change in the consumer price index over 1980 was 10.4. In fact, there was even a greater decline in 1982 with a value of 6.7 percent.

Example 4 indicates a very important consideration. The computer is a very useful tool for finding a model but, as problem solvers, we must be aware of any limitations on the model. If the trend in the data did continue, it would be reasonable to use the model for predictions. However it is not difficult to realize that trends sometimes do not continue, such as in the consumer price index.

EXERCISES 5.4

For exercises 1–8 obtain the line of best fit for the given points. Also determine the error vector E and state which points, if any, lie on the line of best fit.

1. $(0,0)$, $(1,2)$, and $(2,1)$
2. $(0,1)$, $(1,1)$, $(2,3)$, and $(3,2)$
3. $(0,1.2)$, $(1,2.4)$, and $(2,3.6)$
4. $(1,11.4)$, $(2,8.3)$, $(3,5.2)$, and $(4,2.1)$
5. $(0,0.1)$, $(1,0.4)$, $(2,0.3)$, and $(3,0.5)$
6. $(0,6)$, $(1,7.2)$, $(2,8.6)$, and $(3,9.2)$
7. $(1,2.7)$, $(2,4.7)$, $(4,6.0)$, $(5,8.3)$, and $(8,5.3)$
8. $(1,12.7)$, $(2,8.7)$, $(4,10.0)$, $(5,8.3)$, and $(8,12.3)$
9. The reported sales of a major manufacturer for each of five years 1980–1984 are as given in table 5.4.

TABLE 5.4

Year	1980	1981	1982	1983	1984
Sales (in millions)	22.1	21.3	24.7	25.6	27.8

   a. Let t represent time with $t = 0$ meaning 1980 and let S represent sales in millions of dollars. Find the equation of the line that best fits the reported sales.

   b. Use the line found in part (a) to predict the sales for each of the years 1985 and 1986.

10. An environmental protection agency has been monitoring the presence of phosphates in the Kinzua Creek. Over several months of taking samples from the creek, the agency found the units of phosphates per 1000 as given in table 5.5.

TABLE 5.5

Month	1	2	3	4	5	6	7	8
Units of phosphates	10.2	10.7	12.3	14.0	15.6	16.1	17.2	17.2

   a. Find the equation of the line that best fits this data.

   b. Use the line determined in part (a) to predict the units of phosphates per 1000 in Kinzua Creek in the ninth month ($t = 9$) and in the twelfth month ($t = 12$).

11. Consider the data in table 5.6.

TABLE 5.6

Year	Purchasing power of the dollar
0	1.00
1	0.98
2	0.94
3	0.90
4	0.88

This data can be interpreted as the amount of goods a dollar can purchase in comparison to what it could purchase in the base year (year zero). For example, in year 3, a dollar could purchase only 90 percent of what it could in the base year.

a. Determine the line that best fits the data given in the table.

b. Use the line determined in part (a) to predict the purchasing power of the dollar in year 5. What conclusion can be drawn from your answer?

12. Table 5.7 gives the population of the United States for the years indicated, as published by the Bureau of the Census. Observe that the change in the population over each 10-year period is almost constant.

TABLE 5.7

Year	t	Population (in millions)
1890	0	62.9
1900	1	75.9
1910	2	91.9
1920	3	105.7
1930	4	122.7

a. Find the line of best fit $P = mt + b$ for the data.

b. Use the model found in part (a) to determine the population of the United States for the year 1950. (Note that the actual population in 1950 was 150.6 million.)

13. Consider the data in table 5.8, which compares the increase in the amount spent for weapons y to the average amount spent x. (Lewis Richardson, a British meteorologist and mathematician, used these data to test his model of the arms race between two alliances prior to World War I.)

TABLE 5.8

x	y
202.0	5.6
209.8	10.1
226.8	23.8
263.8	50.3

a. Plot the points and observe that they are close to a line.

b. Determine the line that best fits the points.

14. Consider the data in table 5.9, which compares the retail price of flour to the production of flour in the United States for selected years. Determine the line that best fits the data.

TABLE 5.9

Production of flour (in millions of barrels)	Retail price per unit (in cents)
125.2	59.6
129.1	58.9
129.7	58.1
132.8	57.0

5.5 Square Matrices and Inverses

In this section our attention is focused on systems of linear equations $AX = B$ in which the number of equations equals the number of unknowns. For such a system the coefficient matrix A has as many rows as columns.

Any matrix in which the number of rows is the same as the number of columns is said to be **square**. Several types of square matrices that will be useful in later discussions are defined.

Definition

■ The main diagonal entries of a square $n \times n$ matrix $A = [a_{ij}]$ are $a_{11}, a_{22}, \ldots, a_{n-1,n-1}$, and a_{nn}. A square matrix is **diagonal** if all entries *off* the main diagonal are zeros. ■

We are interested in a special type of diagonal matrix. When multiplying real numbers, the number 1 is the identity; specifically, for any real number a, $1 \cdot a = a \cdot 1 = a$. Let us now investigate the concept of an identity for matrix multiplication.

As in the product

$$[0 \quad 1 \quad 0] \begin{bmatrix} 1 & 2 \\ 3 & 4 \\ 5 & 6 \end{bmatrix} = [3 \quad 4]$$

we see that multiplying on the left by the 1×3 row matrix $[0 \quad 1 \quad 0]$ "picks off" the second row of the matrix on the right. In general, consider the $1 \times m$ row matrix

$$[0 \quad \cdots \quad 0 \quad \underset{\substack{\uparrow \\ i\text{th}}}{1} \quad 0 \quad \cdots \quad 0],$$

where the ith entry is 1 and all other entries are zeros. If we multiply an $m \times n$ matrix A on the left by this $1 \times m$ row matrix, the product is the ith row of A. Therefore the matrix needed to multiply on the left of an $m \times n$ matrix A so as to obtain A is the $m \times m$ diagonal matrix

$$\begin{bmatrix} 1 & 0 & 0 & \cdots & 0 \\ 0 & 1 & 0 & \cdots & 0 \\ 0 & 0 & 1 & \cdots & 0 \\ \vdots & \vdots & \vdots & & \vdots \\ 0 & 0 & 0 & \cdots & 1 \end{bmatrix}.$$

For example,

$$\begin{bmatrix} 1 & 0 & 0 \\ 0 & 1 & 0 \\ 0 & 0 & 1 \end{bmatrix} \begin{bmatrix} 1 & 2 \\ 3 & 4 \\ 5 & 6 \end{bmatrix} = \begin{bmatrix} 1 & 2 \\ 3 & 4 \\ 5 & 6 \end{bmatrix}.$$

An **identity** matrix is any diagonal matrix with entries on the main diagonal being 1. Each of the following is an identity matrix.

$$\begin{bmatrix} 1 & 0 \\ 0 & 1 \end{bmatrix} \qquad \begin{bmatrix} 1 & 0 & 0 \\ 0 & 1 & 0 \\ 0 & 0 & 1 \end{bmatrix} \qquad \begin{bmatrix} 1 & 0 & 0 & 0 \\ 0 & 1 & 0 & 0 \\ 0 & 0 & 1 & 0 \\ 0 & 0 & 0 & 1 \end{bmatrix}.$$

$$2 \times 2 \qquad\qquad 3 \times 3 \qquad\qquad\quad 4 \times 4$$

Multiplying an $m \times n$ matrix A on the right by the $n \times n$ identity matrix will also yield A. For example,

$$\begin{bmatrix} 1 & 2 \\ 3 & 4 \\ 5 & 6 \end{bmatrix} \begin{bmatrix} 1 & 0 \\ 0 & 1 \end{bmatrix} = \begin{bmatrix} 1 & 2 \\ 3 & 4 \\ 5 & 6 \end{bmatrix}.$$

We shall use I to denote an identity matrix. The size of I will be implied by the context in which it appears.

Each real number, except zero, has a multiplicative inverse. For each real number $b \neq 0$, $1/b$ is its multiplicative inverse since $1/b \cdot b = b \cdot 1/b = 1$. The inverse of a matrix is also an important concept. Matrix inverses are used in a diversity of applications such as economic modeling and deciphering codes.

Definition ■ A square matrix A is **invertible** if there is a matrix B for which $AB = BA = I$. If such a matrix B exists, it is called an **inverse** of A. ■

We see from the definition that if an $n \times n$ matrix A has an inverse B, the size of B is also $n \times n$.

EXAMPLES

1. Let

$$A = \begin{bmatrix} -2 & -3 \\ 5 & 7 \end{bmatrix} \qquad \text{and} \qquad B = \begin{bmatrix} 7 & 3 \\ -5 & -2 \end{bmatrix}.$$

Since

$$\begin{bmatrix} -2 & -3 \\ 5 & 7 \end{bmatrix} \begin{bmatrix} 7 & 3 \\ -5 & -2 \end{bmatrix} = \begin{bmatrix} 1 & 0 \\ 0 & 1 \end{bmatrix}$$

and

$$\begin{bmatrix} 7 & 3 \\ -5 & -2 \end{bmatrix} \begin{bmatrix} -2 & -3 \\ 5 & 7 \end{bmatrix} = \begin{bmatrix} 1 & 0 \\ 0 & 1 \end{bmatrix},$$

A and B are both invertible. Furthermore, B and A are inverses of each other.

2. Find, if possible, an inverse for the matrix

$$A = \begin{bmatrix} 2 & -4 \\ -1 & 2 \end{bmatrix}.$$

If
$$B = \begin{bmatrix} a & b \\ c & d \end{bmatrix}$$

is to be an inverse of A, it is necessary that

$$\begin{bmatrix} 2 & -4 \\ -1 & 2 \end{bmatrix} \begin{bmatrix} a & b \\ c & d \end{bmatrix} = \begin{bmatrix} 1 & 0 \\ 0 & 1 \end{bmatrix}.$$

We see from this product that two of the four equations that the entries of B must satisfy are

$$2a - 4c = 1 \tag{1}$$

and
$$-a + 2c = 0. \tag{2}$$

Since the equation $-a + 2c = 0$ is equivalent to the equation $2a - 4c = 0$, it follows that there cannot exist numbers a and c that satisfy *both* equations (1) and (2). Therefore it is impossible for the matrix A to have an inverse.

Example 2 points out that there are non-zero matrices that are not invertible. The next example indicates a procedure for finding the inverse of a matrix if it exists.

EXAMPLE 3

Find, if it exists, the inverse of the matrix

$$A = \begin{bmatrix} 1 & 2 \\ 3 & 4 \end{bmatrix}.$$

If the inverse
$$B = \begin{bmatrix} b_{11} & b_{12} \\ b_{21} & b_{22} \end{bmatrix}$$

exists, its first column must satisfy the matrix equation

$$\begin{bmatrix} 1 & 2 \\ 3 & 4 \end{bmatrix} \begin{bmatrix} b_{11} \\ b_{21} \end{bmatrix} = \begin{bmatrix} 1 \\ 0 \end{bmatrix} \tag{3}$$

and its second column must satisfy

$$\begin{bmatrix} 1 & 2 \\ 3 & 4 \end{bmatrix} \begin{bmatrix} b_{12} \\ b_{22} \end{bmatrix} = \begin{bmatrix} 0 \\ 1 \end{bmatrix}. \tag{4}$$

To solve systems (3) and (4), Gauss–Jordan reduction can be performed on the augmented matrices

$$\begin{bmatrix} 1 & 2 & | & 1 \\ 3 & 4 & | & 0 \end{bmatrix} \quad \text{and} \quad \begin{bmatrix} 1 & 2 & | & 0 \\ 3 & 4 & | & 1 \end{bmatrix}.$$

Since A is the coefficient matrix of both systems, the row operations performed on the first augmented matrix in the Gauss–Jordan reduction will be the same as those performed on the second augmented matrix. Therefore we can take the two augmented matrices together and perform Gauss–Jordan reduction on the one matrix

$$\begin{bmatrix} 1 & 2 & | & 1 & 0 \\ 3 & 4 & | & 0 & 1 \end{bmatrix}.$$

Performing Gauss–Jordan reduction on this 2×4 matrix, we have

$$\begin{bmatrix} 1 & 2 & | & 1 & 0 \\ 3 & 4 & | & 0 & 1 \end{bmatrix} \xrightarrow{-3R_1 + R_2} \begin{bmatrix} 1 & 2 & | & 1 & 0 \\ 0 & -2 & | & -3 & 1 \end{bmatrix}$$

$$\xrightarrow{-\frac{1}{2}R_2} \begin{bmatrix} 1 & 2 & | & 1 & 0 \\ 0 & 1 & | & \frac{3}{2} & -\frac{1}{2} \end{bmatrix}$$

$$\xrightarrow{R_1 - 2R_2} \begin{bmatrix} 1 & 0 & | & -2 & 1 \\ 0 & 1 & | & \frac{3}{2} & -\frac{1}{2} \end{bmatrix}.$$

By the interpretation of the rows of the last matrix, we see that the solutions to systems (3) and (4) are, respectively,

$$\begin{bmatrix} -2 \\ \frac{3}{2} \end{bmatrix} \quad \text{and} \quad \begin{bmatrix} 1 \\ -\frac{1}{2} \end{bmatrix}.$$

Therefore,

$$B = \begin{bmatrix} -2 & 1 \\ \frac{3}{2} & -\frac{1}{2} \end{bmatrix}$$

where $AB = I$. It is easy to verify that also $BA = I$. Thus A is invertible and B is an inverse of A.

As seen in example 3, if we find a matrix B such that $AB = I$, then it will also be true that $BA = I$. Since this will be the case in general, we only need to determine a matrix B for which $AB = I$. Furthermore, as seen in example 3, when finding the matrix B we solve several systems of linear equations; the augmented matrices are combined in one matrix $[A \,|\, I]$. By performing Gauss–Jordan reduction on this matrix, we see that each system has a unique solution if A is row equivalent to the identity matrix. Therefore A is invertible and has a unique inverse when A is row equivalent to the identity matrix. We shall denote the inverse by A^{-1}.

EXAMPLE 4

Find the inverse of

$$A = \begin{bmatrix} 3 & 2 \\ 7 & 5 \end{bmatrix}.$$

As in example 3, we perform Gauss–Jordan reduction on the matrix

$$\begin{bmatrix} 3 & 2 & | & 1 & 0 \\ 7 & 5 & | & 0 & 1 \end{bmatrix}$$

and obtain the matrix

$$\begin{bmatrix} 1 & 0 & | & 5 & -2 \\ 0 & 1 & | & -7 & 3 \end{bmatrix}.$$

Therefore A is row equivalent to I and the inverse of A is

$$A^{-1} = \begin{bmatrix} 5 & -2 \\ -7 & 3 \end{bmatrix}.$$

The situation that occurs in examples 3 and 4 can be extended to other square matrices.

Theorem 6 ■ Let A be an $n \times n$ matrix. A is row equivalent to the $n \times n$ identity matrix I if and only if A is invertible and has a unique inverse A^{-1}. ■

With the preceding examples, theorem, and comments in mind, we see that a procedure for finding the inverse is as follows.

Procedure for Finding the Inverse of an Invertible Matrix

1. Perform Gauss–Jordan reduction on the $n \times 2n$ matrix $[A \mid I]$.
2. If A is invertible, the $n \times 2n$ matrix that is derived in reduced row-echelon form is $[I \mid A^{-1}]$.
3. If A is not invertible, then A is not row equivalent to I in which case the $n \times 2n$ matrix derived in reduced row-echelon form is $[B \mid C]$ where $B \neq I$.

EXAMPLES

5. Use the method described to find the inverse of

$$A = \begin{bmatrix} 1 & -2 & -3 & 0 \\ 0 & 2 & 6 & -2 \\ -1 & 2 & 4 & -2 \\ 3 & -5 & -6 & 0 \end{bmatrix}$$

We perform Gauss–Jordan reduction on the 4×8 matrix

$$\left[\begin{array}{cccc|cccc} 1 & -2 & -3 & 0 & 1 & 0 & 0 & 0 \\ 0 & 2 & 6 & -2 & 0 & 1 & 0 & 0 \\ -1 & 2 & 4 & -2 & 0 & 0 & 1 & 0 \\ 3 & -5 & -6 & 0 & 0 & 0 & 0 & 1 \end{array}\right]$$

as follows.

$$\left[\begin{array}{cccc|cccc} 1 & -2 & -3 & 0 & 1 & 0 & 0 & 0 \\ 0 & 2 & 6 & -2 & 0 & 1 & 0 & 0 \\ -1 & 2 & 4 & -2 & 0 & 0 & 1 & 0 \\ 3 & -5 & -6 & 0 & 0 & 0 & 0 & 1 \end{array}\right]$$

$$\xrightarrow[R_1 + R_3]{\frac{1}{2}R_2} \left[\begin{array}{cccc|cccc} 1 & -2 & -3 & 0 & 1 & 0 & 0 & 0 \\ 0 & 1 & 3 & -1 & 0 & \frac{1}{2} & 0 & 0 \\ 0 & 0 & 1 & -2 & 1 & 0 & 1 & 0 \\ 0 & 1 & 3 & 0 & -3 & 0 & 0 & 1 \end{array}\right]$$

$$\xrightarrow[2R_2 + R_1]{-R_2 + R_4} \left[\begin{array}{cccc|cccc} 1 & 0 & 3 & -2 & 1 & 1 & 0 & 0 \\ 0 & 1 & 3 & -1 & 0 & \frac{1}{2} & 0 & 0 \\ 0 & 0 & 1 & -2 & 1 & 0 & 1 & 0 \\ 0 & 0 & 0 & 1 & -3 & -\frac{1}{2} & 0 & 1 \end{array}\right]$$

Continuing with the process we obtain the matrix

$$\left[\begin{array}{cccc|cccc} 1 & 0 & 0 & 0 & 10 & 3 & -3 & -4 \\ 0 & 1 & 0 & 0 & 12 & 3 & -3 & -5 \\ 0 & 0 & 1 & 0 & -5 & -1 & 1 & 2 \\ 0 & 0 & 0 & 1 & -3 & -\frac{1}{2} & 0 & 1 \end{array}\right].$$

We see from this 4×8 row equivalent matrix that A is invertible and that

$$A^{-1} = \left[\begin{array}{cccc} 10 & 3 & -3 & -4 \\ 12 & 3 & -3 & -5 \\ -5 & -1 & 1 & 2 \\ -3 & -\frac{1}{2} & 0 & 1 \end{array}\right].$$

6. Find the inverse, if it exists, of

$$A = \left[\begin{array}{ccc} 2 & -5 & 2 \\ -4 & 11 & 1 \\ 2 & -8 & -13 \end{array}\right].$$

We perform Gauss–Jordan reduction on the 3×6 matrix obtaining

$$\left[\begin{array}{ccc|ccc} 2 & -5 & 2 & 1 & 0 & 0 \\ -4 & 11 & 1 & 0 & 1 & 0 \\ 2 & -8 & -13 & 0 & 0 & 1 \end{array}\right]$$

$$\xrightarrow[\substack{\frac{1}{2}R_1}]{\substack{2R_1 + R_2 \\ -R_1 + R_3}} \left[\begin{array}{ccc|ccc} 1 & -\frac{5}{2} & 1 & \frac{1}{2} & 0 & 0 \\ 0 & 1 & 5 & 2 & 1 & 0 \\ 0 & -3 & -15 & -1 & 0 & 1 \end{array}\right]$$

$$\xrightarrow{3R_2 + R_3} \left[\begin{array}{ccc|ccc} 1 & -\frac{5}{2} & 1 & \frac{1}{2} & 0 & 0 \\ 0 & 1 & 5 & 2 & 1 & 0 \\ 0 & 0 & 0 & 5 & 3 & 1 \end{array}\right].$$

At this point we see that there is no need to continue with the reduction since the matrix A is row equivalent to a 3×3 matrix with a row of zeros, and therefore A cannot be row equivalent to the identity. Thus A is not invertible.

We now turn to square systems—systems in which the number of equations is the same as the number of unknowns. With real numbers the multiplicative inverse can be used to solve equations of the type $ax = b$. For example, the solution to the equation $\frac{2}{3}x = 5$ is $x = \frac{3}{2}(5)$, is obtained by multiplying each side of the equation by the inverse $\frac{3}{2}$. A similar situation occurs for an $n \times n$ system of linear equations $AX = B$. If the coefficient matrix A is invertible, then $X = A^{-1}B$ is a solution to the system since $A(A^{-1}B) = (AA^{-1})B = IB = B$. Furthermore, if A is invertible, $A^{-1}B$ is the only solution to $AX = B$. In the next theorems, we summarize this result along with other conditions for which a matrix is invertible.

Theorem 7   ■ Let A be an $n \times n$ matrix. A is invertible if and only if $AX = B$ has exactly one solution $X = A^{-1}B$. ■

EXAMPLE 7   Solve the system $AX = B$ given by

$$\begin{bmatrix} 3 & 2 \\ 7 & 5 \end{bmatrix} \begin{bmatrix} x \\ y \end{bmatrix} = \begin{bmatrix} 1 \\ -3 \end{bmatrix}.$$

In example 4 we found the inverse of A to be

$$A^{-1} = \begin{bmatrix} 5 & -2 \\ -7 & 3 \end{bmatrix}.$$

By theorem 7, the desired solution is

$$X = \begin{bmatrix} 5 & -2 \\ -7 & 3 \end{bmatrix} \begin{bmatrix} 1 \\ -3 \end{bmatrix} = \begin{bmatrix} 11 \\ -16 \end{bmatrix}.$$

Theorem 8   ■ Let A be an $n \times n$ matrix. If any one of the following statements is true for A, then each of the others must also be true.

a. A is row equivalent to the $n \times n$ identity matrix I.

b. A is invertible.

c. For each $n \times 1$ column B, the system $AX = B$ has one and only one solution.

d. $AX = 0$ has only the trivial solution. ■

EXAMPLE 8   Consider the matrix

$$A = \begin{bmatrix} 2 & -1 & 5 & 6 \\ 0 & 1 & -2 & 4 \\ 0 & 0 & 3 & -5 \\ 0 & 0 & 0 & 4 \end{bmatrix}$$

and the homogeneous system $AX = 0$. Since the system is in triangular form and there are no zeros on the main diagonal of A, we can uniquely solve for each unknown. Furthermore, since the system is homogeneous, the unique solution is the trivial one. Therefore, by theorem 8, A^{-1} exists.

The matrix in example 8 is in a special form called **upper triangular**. A square matrix is upper triangular if all entries below the main diagonal are zeros. As seen in example 8 any upper-triangular matrix with non-zero entries on the main diagonal is invertible.

We saw in this section that the inverse of the coefficient matrix can be used to solve a square linear system $AX = B$, provided that A^{-1} exists. When solving a single system, this method is more time consuming than using Gaussian elimination with back substitution. Suppose, though, we need to solve several systems, each having the same coefficient matrix A; for example, we wish to solve each of the systems $AX = B_1, AX = B_2, \ldots, AX = B_k$. Such a situation occurs in an application discussed in section 5.6. Instead of performing Gaussian elimination with back substitution on each of the augmented matrices $[A \mid B_i]$, an alternate method would be to find the inverse of A once, provided it exists, and obtain the solutions as the products $A^{-1}B_i$. The choice of which method to use, Gaussian elimination or finding the inverse, can be based on a comparison of the number of computations needed.

EXAMPLE 9

Solve each of the systems $AX = B_1$, $AY = B_2$, and $AW = B_3$, where

$$A = \begin{bmatrix} 1 & 2 \\ 3 & 4 \end{bmatrix}, \quad B_1 = \begin{bmatrix} 1 \\ 1 \end{bmatrix}, \quad B_2 = \begin{bmatrix} 0 \\ 0 \end{bmatrix}, \quad \text{and} \quad B_3 = \begin{bmatrix} 2 \\ -4 \end{bmatrix}.$$

We found in example 3 the inverse of A.

$$A^{-1} = \begin{bmatrix} -2 & 1 \\ \frac{3}{2} & -\frac{1}{2} \end{bmatrix}.$$

Thus each system has exactly one solution. The solutions are

$$X = \begin{bmatrix} -2 & 1 \\ \frac{3}{2} & -\frac{1}{2} \end{bmatrix} \begin{bmatrix} 1 \\ 1 \end{bmatrix} = \begin{bmatrix} -1 \\ 1 \end{bmatrix},$$

$$Y = \begin{bmatrix} -2 & 1 \\ \frac{3}{2} & -\frac{1}{2} \end{bmatrix} \begin{bmatrix} 0 \\ 0 \end{bmatrix} = \begin{bmatrix} 0 \\ 0 \end{bmatrix},$$

and

$$W = \begin{bmatrix} -2 & 1 \\ \frac{3}{2} & -\frac{1}{2} \end{bmatrix} \begin{bmatrix} 2 \\ -4 \end{bmatrix} = \begin{bmatrix} -8 \\ 5 \end{bmatrix}.$$

EXERCISES 5.5

In exercises 1 and 2 compute the indicated matrix product.

1. a.
$$\begin{bmatrix} 1 & 0 & 0 \end{bmatrix} \begin{bmatrix} 3 & 4 \\ -2 & 6 \\ 5 & 0 \end{bmatrix}$$

   b.
$$\begin{bmatrix} 0 & 1 & 0 \end{bmatrix} \begin{bmatrix} 3 & 4 \\ -2 & 6 \\ 5 & 0 \end{bmatrix}$$

   c.
$$\begin{bmatrix} 0 & 0 & 1 \end{bmatrix} \begin{bmatrix} 3 & 4 \\ -2 & 6 \\ 5 & 0 \end{bmatrix}$$

   d.
$$\begin{bmatrix} 1 & 0 & 0 \\ 0 & 1 & 0 \\ 0 & 0 & 1 \end{bmatrix} \begin{bmatrix} 3 & 4 \\ -2 & 6 \\ 5 & 0 \end{bmatrix}$$

2. a.
$$\begin{bmatrix} 0 & 1 & 0 & 0 \end{bmatrix} \begin{bmatrix} -2 & 5 & 4 & -7 \\ 0 & 1 & 0 & 0 \\ 8 & -9 & 3 & -6 \\ 0 & 1 & 1 & 0 \end{bmatrix}$$

b.
$$[0 \quad 0 \quad 0 \quad 1] \begin{bmatrix} -2 & 5 & 4 & -7 \\ 0 & 1 & 0 & 0 \\ 8 & -9 & 3 & -6 \\ 0 & 1 & 1 & 0 \end{bmatrix}$$

c.
$$\begin{bmatrix} 1 & 0 & 0 & 0 \\ 0 & 1 & 0 & 0 \\ 0 & 0 & 1 & 0 \\ 0 & 0 & 0 & 1 \end{bmatrix} \begin{bmatrix} -2 & 5 & 4 & -7 \\ 0 & 1 & 0 & 0 \\ 8 & -9 & 3 & -6 \\ 0 & 1 & 1 & 0 \end{bmatrix}$$

In exercises 3–20 determine whether or not the given matrix is invertible; if it is invertible, then find its inverse.

3. $\begin{bmatrix} 4 & 5 \\ 7 & 9 \end{bmatrix}$

4. $\begin{bmatrix} 7 & 6 \\ 8 & 7 \end{bmatrix}$

5. $\begin{bmatrix} 2 & -1 \\ -6 & 2 \end{bmatrix}$

6. $\begin{bmatrix} \frac{1}{3} & -\frac{5}{6} \\ -\frac{3}{5} & \frac{3}{2} \end{bmatrix}$

7. $\begin{bmatrix} 2 & 0 & 0 \\ 0 & -3 & 0 \\ 0 & 0 & 4 \end{bmatrix}$

8. $\begin{bmatrix} \frac{2}{3} & 0 & 0 \\ 0 & \frac{1}{2} & 0 \\ 0 & 0 & -5 \end{bmatrix}$

9. $\begin{bmatrix} -2 & 1 & 3 \\ 2 & 4 & -1 \\ 3 & 0 & -4 \end{bmatrix}$

10. $\begin{bmatrix} 3 & -2 & 5 \\ 0 & 5 & 3 \\ 1 & 0 & 2 \end{bmatrix}$

11. $\begin{bmatrix} 1 & -2 & 3 \\ -2 & -2 & 0 \\ 4 & -5 & 6 \end{bmatrix}$

12. $\begin{bmatrix} -2 & 0 & 1 \\ 4 & 3 & -1 \\ 0 & 6 & 3 \end{bmatrix}$

13. $\begin{bmatrix} 1 & 7 & 6 \\ 4 & -2 & 5 \\ 0 & 0 & 0 \end{bmatrix}$

14. $\begin{bmatrix} 1 & 0 & -2 \\ 3 & 0 & 4 \\ 2 & 0 & -5 \end{bmatrix}$

15. $\begin{bmatrix} 2 & 3 & -4 & 5 \\ 0 & 1 & 5 & 2 \\ 0 & 0 & -3 & 6 \\ 0 & 0 & 0 & 1 \end{bmatrix}$

16. $\begin{bmatrix} 2 & 3 & -4 & 5 \\ 0 & 0 & 5 & 2 \\ 0 & 0 & -3 & 6 \\ 0 & 0 & 0 & 1 \end{bmatrix}$

17. $\begin{bmatrix} 3 & 4 & -2 & 1 \\ 1 & 0 & 2 & 0 \\ 0 & 5 & 4 & -1 \\ -2 & 2 & 1 & -1 \end{bmatrix}$

18. $\begin{bmatrix} 8 & 0 & 5 & 12 \\ 2 & 1 & -6 & 0 \\ 0 & 1 & 0 & 3 \\ 1 & 0 & 0 & 1 \end{bmatrix}$

19. $\begin{bmatrix} 2 & -3 & 4 & 1 \\ 3 & -4 & 6 & 2 \\ 1 & 1 & 4 & 1 \\ -1 & 0 & -2 & -1 \end{bmatrix}$

20. $\begin{bmatrix} 1 & 0 & -1 & 1 \\ -2 & 3 & 3 & 7 \\ 2 & 0 & -6 & 0 \\ -1 & 1 & 2 & 2 \end{bmatrix}$

21. Let A be the 2×2 matrix given in exercise 3 and let $B = [2 \quad 5]^t$. Compute $A^{-1}B$ and verify that it is the solution to the system $AX = B$.

22. Let A be the 2×2 matrix given in exercise 5 and let $B = [10 \quad -5]^t$. Compute $A^{-1}B$ and verify that it is the solution to the system $AX = B$.

23. Let A be the 3×3 matrix given in exercise 9 and let $B = [1 \quad 1 \quad -1]^t$. Compute $A^{-1}B$ and verify that it is the solution to the system $AX = B$.

24. Let A be the 4×4 matrix given in exercise 20 and let $B = [1 \quad 0 \quad -1 \quad 2]^t$. Compute $A^{-1}B$ and verify that it is the solution to the system $AX = B$.

25. Let

$$A = \begin{bmatrix} 1 & -2 & 4 \\ 0 & 1 & -5 \\ 0 & 0 & 1 \end{bmatrix}.$$

First find A^{-1}. Then, using A^{-1} and theorem 7 on page 198, obtain the solution to the system $AX = B$ where:

a. $B = \begin{bmatrix} 9 \\ -7 \\ 4 \end{bmatrix}$ b. $B = \begin{bmatrix} 1 \\ 1 \\ 1 \end{bmatrix}$ c. $B = \begin{bmatrix} 0 \\ 3 \\ -2 \end{bmatrix}$ d. $B = \begin{bmatrix} 0 \\ 0 \\ 0 \end{bmatrix}$

26. Let

$$A = \begin{bmatrix} 1 & 0 & -1 & 1 \\ 0 & 3 & 3 & 7 \\ 0 & 0 & -6 & 0 \\ 0 & 0 & 0 & 2 \end{bmatrix}.$$

First find A^{-1}. Then, using A^{-1} and theorem 7 on page 198, obtain the solution to the system $AX = B$ where:

a. $B = \begin{bmatrix} 1 \\ 0 \\ 1 \\ 0 \end{bmatrix}$ b. $B = \begin{bmatrix} 4 \\ -2 \\ -3 \\ 5 \end{bmatrix}$ c. $B = \begin{bmatrix} -1 \\ 2 \\ 0 \\ 6 \end{bmatrix}$ d. $B = \begin{bmatrix} 0 \\ 0 \\ 0 \\ 0 \end{bmatrix}$

Exercises 27–33 indicate further theorems on the invertibility of a matrix.

27. a. Verify that the inverse of

$$A = \begin{bmatrix} 4 & 0 & 0 \\ 0 & -3 & 0 \\ 0 & 0 & 2 \end{bmatrix} \quad \text{is} \quad A^{-1} = \begin{bmatrix} \frac{1}{4} & 0 & 0 \\ 0 & -\frac{1}{3} & 0 \\ 0 & 0 & \frac{1}{2} \end{bmatrix}.$$

b. Find the inverse of the matrix

$$A = \begin{bmatrix} 3 & 0 & 0 \\ 0 & \frac{1}{2} & 0 \\ 0 & 0 & -5 \end{bmatrix}.$$

c. A diagonal matrix with non-zero entries on the main diagonal is invertible. Let

$$A = \begin{bmatrix} a_{11} & 0 & \cdots & 0 \\ 0 & a_{22} & \cdots & 0 \\ \vdots & \vdots & & \vdots \\ 0 & 0 & \cdots & a_{nn} \end{bmatrix}$$

where each entry a_{ii} is not zero, and verify that the matrix

$$\begin{bmatrix} \frac{1}{a_{11}} & 0 & \cdots & 0 \\ 0 & \frac{1}{a_{22}} & \cdots & 0 \\ \vdots & \vdots & & \vdots \\ 0 & 0 & \cdots & \frac{1}{a_{nn}} \end{bmatrix}$$

times A yields the identity I.

28. Let

$$A = \begin{bmatrix} 1 & 2 & 3 \\ 0 & 4 & -5 \\ 0 & 0 & 2 \end{bmatrix}.$$

a. Verify that the system $AX = 0$ has only the trivial solution.
b. Explain why the matrix A is invertible. (*Hint:* Use theorem 8 on page 198.)
c. Repeat parts (a) and (b) for the matrix

$$A = \begin{bmatrix} 2 & 0 & 3 & 4 \\ 0 & 1 & -2 & 6 \\ 0 & 0 & -3 & 7 \\ 0 & 0 & 0 & 1 \end{bmatrix}.$$

d. Explain why a matrix of the form

$$A = \begin{bmatrix} a_{11} & a_{12} & a_{13} \\ 0 & a_{22} & a_{23} \\ 0 & 0 & a_{33} \end{bmatrix},$$

where $a_{11} \neq 0$, $a_{22} \neq 0$, and $a_{33} \neq 0$, is invertible.

29. Let

$$A = \begin{bmatrix} 2 & -1 & 3 \\ 0 & 0 & 0 \\ 4 & 5 & -2 \end{bmatrix} \quad \text{and} \quad B = \begin{bmatrix} 0 \\ 1 \\ 0 \end{bmatrix}.$$

a. Show that the system $AX = B$ has no solution.
b. Using part (a) and theorem 8 on page 198, explain why A is not invertible.
c. Repeat parts (a) and (b) for the matrices

$$A = \begin{bmatrix} 0 & 1 & 2 & 4 \\ 2 & 0 & 5 & -3 \\ 0 & 0 & 0 & 0 \\ 1 & 3 & 0 & 6 \end{bmatrix} \quad \text{and} \quad B = \begin{bmatrix} 0 \\ 0 \\ 1 \\ 0 \end{bmatrix}.$$

d. Explain why a square matrix with a row of zeros is not invertible.

30. Let

$$A = \begin{bmatrix} 2 & -1 \\ -3 & 2 \end{bmatrix} \quad \text{and} \quad B = \begin{bmatrix} -1 & 0 & 1 \\ 4 & -1 & -3 \\ 1 & 2 & -2 \end{bmatrix}.$$

a. Find $(A^{-1})^t$.

b. Compute $(A^t)(A^{-1})^t$ and verify that the inverse of A^t is the transpose of A^{-1}; that is, $(A^t)^{-1} = (A^{-1})^t$.

c. Repeat parts (a) and (b) for the matrix B.

d. Explain why a matrix is invertible *if and only if* its transpose is invertible.

31. Let

$$A = \begin{bmatrix} 2 & 0 & 2 \\ -1 & 0 & 5 \\ 3 & 0 & -2 \end{bmatrix}.$$

a. Determine A^t.

b. Explain why A^t is not invertible. (*Hint:* See exercise 21.)

c. Explain why A is not invertible. (*Hint:* See exercise 22.)

d. Repeat parts (a), (b), and (c) for the matrix

$$A = \begin{bmatrix} 1 & 4 & 0 & -5 \\ 0 & 2 & 0 & 6 \\ 3 & 0 & 0 & -1 \\ 7 & -2 & 0 & 3 \end{bmatrix}.$$

e. Explain why a square matrix with a column of zeros is not invertible.

32. Let

$$A = \begin{bmatrix} 2 & 3 \\ 4 & 6 \end{bmatrix} \quad \text{and} \quad C = \begin{bmatrix} 2 & 3 \\ 0 & 0 \end{bmatrix}.$$

a. Verify that A is row equivalent to C.

b. Explain why A is not row equivalent to the 2×2 identity matrix I.

c. Explain why A is not invertible. (*Hint:* Use part (b) and theorem 8 on page 198.)

d. Repeat parts (a) through (c) for the matrices

$$A = \begin{bmatrix} 4 & -2 & 3 \\ 1 & 5 & -4 \\ -8 & 4 & -6 \end{bmatrix} \quad \text{and} \quad C = \begin{bmatrix} 4 & -2 & 3 \\ 1 & 5 & -4 \\ 0 & 0 & 0 \end{bmatrix}.$$

e. Explain why a matrix of the form

$$\begin{bmatrix} a_{11} & a_{12} & a_{13} \\ ta_{11} & ta_{12} & ta_{13} \\ a_{31} & a_{32} & a_{33} \end{bmatrix},$$

where t is any real number, is not invertible.

33. Let

$$A = \begin{bmatrix} 2 & -1 \\ -3 & 2 \end{bmatrix} \quad \text{and} \quad B = \begin{bmatrix} 5 & 2 \\ 2 & 1 \end{bmatrix}.$$

a. Find A^{-1} and B^{-1}.

b. Find $(AB)^{-1}$ and verify that $(AB)^{-1} = B^{-1}A^{-1}$.

c. Repeat parts (a) and (b) for the matrices

$$A = \begin{bmatrix} 1 & 2 & 1 \\ 0 & -3 & -1 \\ -1 & 1 & 2 \end{bmatrix} \quad \text{and} \quad B = \begin{bmatrix} -1 & 0 & 1 \\ 0 & -1 & 1 \\ 1 & 2 & -3 \end{bmatrix}.$$

5.6 Linear Production Models

The material in this section is based on ideas first considered by Wassily Leontief, who received the Nobel prize for economics in 1974. He developed a tool to study an economy, called *input–output analysis*, and constructed input–output tables for the United States.

Consider an economic system consisting of several *industries (sectors)* where the output of *goods* by one industry is dependent on the input of goods from other industries in the system. For example, manufacturing needs energy in order to produce machinery. In addition to supplying goods to other industries in the system, each industry "exports" goods outside the system to consumers. How much should each industry produce so as to satisfy the needs of others in the system and also to provide enough goods to meet the outside demand? To measure the flow of goods between industries in the system and for export, we use a standard monetary unit. For example, we can use for a United States economy the dollar value of the goods. Let us consider an example to illustrate this discussion.

Suppose that the United States economy is divided into three sectors— manufacturing (M), agriculture (A), and energy (E). To produce $1 worth of manufactured goods, manufacturing needs $0.40 worth of manufactured goods, $0.10 worth of agricultural goods, and $0.40 worth of energy. On the other hand, to produce $1 worth of agricultural goods, agriculture needs $0.20 worth of manufactured goods, $0.30 worth of agricultural goods, and $0.20 worth of energy. Energy needs $0.20 of manufactured goods, $0.10 worth of agricultural goods, and $0.30 worth of energy in order to produce $1 worth of energy.

Note that the value of goods a sector consumes in its production process depends on its total production. For example, as the manufacturing sector's production increases, it consumes more of its own goods and this rate of consumption is $0.40 for every $1 produced. To represent the consumption of goods within the system, let

$$x_1 = \text{the total value of goods produced by manufacturing}$$

$$x_2 = \text{the total value of goods produced by agriculture}$$

$$x_3 = \text{the total value of goods produced by energy.}$$

Thus, for example, the value of manufactured goods used by manufacturing, agriculture and energy are $0.4x_1, 0.2x_2$, and $0.2x_3$. The total value of manufactured goods consumed within the system is

$$0.4x_1 + 0.2x_2 + 0.2x_3.$$

Also, the total value of agricultural goods and the total value of energy used within the system are represented by

$$0.1x_1 + 0.3x_2 + 0.1x_3$$

and $$0.4x_1 + 0.2x_2 + 0.3x_3.$$

We can depict this flow of the value of goods within the system as shown in figure 5.4.

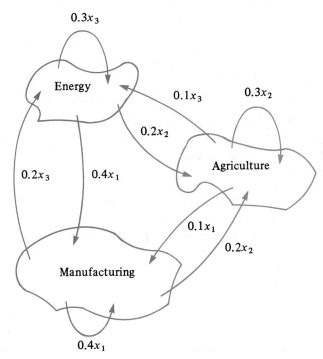

FIGURE 5.4

The value of each sector's goods consumed within the system is represented in the matrix product

$$\begin{bmatrix} 0.4 & 0.2 & 0.2 \\ 0.1 & 0.3 & 0.1 \\ 0.4 & 0.2 & 0.3 \end{bmatrix} \begin{bmatrix} x_1 \\ x_2 \\ x_3 \end{bmatrix}.$$

The matrix

Supplied to
sector

$$
\begin{array}{c}
\text{from}\\
\text{sector}
\end{array}
\begin{array}{c}
\\
M\\
A\\
E
\end{array}
\begin{array}{ccc}
M & A & E\\
\left[\begin{array}{ccc}
0.4 & 0.2 & 0.2\\
0.1 & 0.3 & 0.1\\
0.4 & 0.2 & 0.3
\end{array}\right]
\end{array}
$$

is called an **input–output matrix**.

Now suppose that the "export" demand, in millions of dollars, for manufactured goods, agricultural goods, and energy are 105, 95, and 70, respectively. Our problem is to determine for the given **demand vector**

$$
D = \begin{bmatrix} 105 \\ 95 \\ 70 \end{bmatrix}
$$

an **intensity vector**

$$
X = \begin{bmatrix} x_1 \\ x_2 \\ x_3 \end{bmatrix}
$$

so that the system's needs are met and the "export" demand is satisfied. Since the flow of the value of goods within the system and for export must be in balance with the total value of goods produced by each sector, our problem becomes one of solving the system of equations

$$
x_1 = 0.4x_1 + 0.2x_2 + 0.2x_3 + 105
$$

$$
x_2 = 0.1x_1 + 0.3x_2 + 0.1x_3 + 95
$$

$$
x_3 = 0.4x_1 + 0.2x_2 + 0.3x_3 + 70.
$$

First let us summarize our discussion. Let

a_{ij} = the value of goods industry j needs from industry i to produce one monetary unit worth of goods.

For example, if $a_{12} = \$0.05$, then, for each dollar's worth of goods industry 2 produces, it uses $\$0.05$ worth of goods from industry 1. The rates of input are given in an $n \times n$ input–output matrix:

Supplied to
industry

$$
A = \begin{array}{c}
\\
\text{from}\\
\text{industry}
\end{array}
\begin{array}{c}
\\
1\\
2\\
\vdots\\
n
\end{array}
\begin{array}{cccc}
1 & 2 & \cdots & n\\
\left[\begin{array}{cccc}
a_{11} & a_{12} & \cdots & a_{1n}\\
a_{21} & a_{22} & \cdots & a_{2n}\\
\vdots & \vdots & & \vdots\\
a_{n1} & a_{n2} & \cdots & a_{nn}
\end{array}\right]
\end{array}
$$

Note that each entry of A is non-negative ($a_{ij} \geq 0$).

Further, let
$$x_j = \text{the total output of industry } j.$$

Therefore, $a_{ij}x_j$ represents the value of goods from industry i needed by industry j and
$$a_{i1}x_1 + a_{i2}x_2 + \cdots + a_{in}x_n$$

represents the total value of goods from industry i needed within the system.

Now, let
$$d_i = \text{the external demand for goods produced by industry } i.$$

Therefore our question can be posed as to how large should $x_1, x_2, \ldots, x_n$ be so that
$$x_1 = a_{11}x_1 + a_{12}x_2 + \cdots + a_{1n}x_n + d_1$$
$$x_2 = a_{21}x_1 + a_{22}x_2 + \cdots + a_{2n}x_n + d_2$$
$$\vdots \qquad \vdots \qquad \vdots \qquad \qquad \vdots \qquad \vdots$$
$$x_n = a_{n1}x_1 + a_{n2}x_2 + \cdots + a_{nn}x_n + d_n.$$

This system of linear equations can be written in the matrix form
$$X = AX + D, \tag{1}$$

where A is the input-output matrix, D is the given demand vector, and the solution to the system X is the intensity vector.

Since $IX = X$, we can rewrite (1) as
$$IX - AX = D$$
and
$$(I - A)X = D. \tag{2}$$

We know that there is a solution X to the system (2) if $(I - A)^{-1}$ exists and this solution is
$$X = (I - A)^{-1}D. \tag{3}$$

However, we want a solution in which each entry is non-negative. Since each entry of D is non-negative, each entry of the solution $(I - A)^{-1}D$ is guaranteed to be non-negative whenever each entry of $(I - A)^{-1}$ is non-negative.

Definition ■ An input–output matrix A is **productive** if $(I - A)^{-1}$ exists and each entry of $(I - A)^{-1}$ is non-negative. ■

We can, therefore, find an intensity vector X with non-negative entries whenever the input-output matrix A is productive. The next theorem, proved in more advanced courses, gives a convenient method for checking if an input–output matrix is productive.

Theorem 9 ■ An input–output matrix A is productive if the sum of the entries in each row is less than one. ■

The key point we should keep in mind is that if the input–output matrix for a system is productive, then the system can operate at a unique intensity X to produce the given demand D.

EXAMPLE 1

Consider again the input–output matrix

$$A = \begin{bmatrix} 0.4 & 0.2 & 0.2 \\ 0.1 & 0.3 & 0.1 \\ 0.4 & 0.2 & 0.3 \end{bmatrix}.$$

Find, if possible, the intensity vector $X = [x_1 \quad x_2 \quad x_3]^t$ for the given demand vector $D = [105 \quad 95 \quad 70]^t$.

Since the sum of the entries in the rows are 0.8, 0.5, and 0.9, and since each sum is less than one, the matrix A is productive. Therefore there exists a unique intensity vector X for the demand vector D.

Using Gauss–Jordan reduction, we find that the inverse of

$$I - A = \begin{bmatrix} 0.6 & -0.2 & -0.2 \\ -0.1 & 0.7 & -0.1 \\ -0.4 & -0.2 & 0.7 \end{bmatrix}$$

is

$$(I - A)^{-1} = \begin{bmatrix} 2.35 & 0.90 & 0.80 \\ 0.55 & 1.70 & 0.40 \\ 1.50 & 1.00 & 2.00 \end{bmatrix}.$$

Thus, for the given demand vector, we have

$$X = (I - A)^{-1} \begin{bmatrix} 150 \\ 95 \\ 70 \end{bmatrix} = \begin{bmatrix} 494 \\ 272 \\ 460 \end{bmatrix}$$

where the entries represent millions of dollars worth of goods; that is, manufacturing should produce $494 million worth of goods, agriculture should produce $272 million worth of goods, and energy should produce $460 million worth of goods.

Generally, the input–output matrix for a system would remain the same while the demand vector would vary. Therefore, once the inverse $(I - A)^{-1}$ has been found, we can find for each new demand vector D the intensity vector X by using the product $(I - A)^{-1}D$. For instance, if the demand vector in our example should change to $D = [200 \quad 100 \quad 80]^t$, then the new intensity vector would be

$$X = \begin{bmatrix} 2.35 & 0.90 & 0.80 \\ 0.55 & 1.70 & 0.40 \\ 1.50 & 1.00 & 2.00 \end{bmatrix} \begin{bmatrix} 200 \\ 100 \\ 80 \end{bmatrix} = \begin{bmatrix} 624 \\ 312 \\ 560 \end{bmatrix}.$$

EXAMPLE 2

An economic system consists of three industrial sectors

I—primary and fabricated metals,

II—petroleum and coal products,

III—chemicals and chemical products.

Each sector's production is dependent on products produced within its own sector and those produced by the other two sectors. For sector I to produce $1 worth of primary and fabricated metals, it requires $0.29 worth of its own products, $0.01 worth of products produced by section II and $0.02 worth of products produced by sector III. For sector II to produce $1 worth of petroleum and coal products, it requires $0.7 worth of its own products, $0.01 worth of products produced by sector I and $0.03 worth of products produced by sector III. Finally, for sector III to produce $1 worth of chemicals and chemical products, it requires $0.22 worth of its own products, $0.03 worth of products produced by sector I and $0.05 worth of products produced by sector II. The demand, in millions of dollars, from outside this system is $60,510 for sector I products, $22,863 for sector II products, and $33,295 for sector III products. (These data were obtained from an input–output table published by the Bureau of Economic Analysis; the input coefficients were rounded off to two decimal places.)

The entries in a column of the input–output matrix are the requirements of the sector corresponding to the column. For example, column *I* is

$$\begin{array}{c} \\ \text{I} \\ \text{II} \\ \text{III} \end{array} \begin{array}{c} \text{I} \\ \begin{bmatrix} 0.29 \\ 0.01 \\ 0.02 \end{bmatrix} \end{array}.$$

The complete input–output matrix is

$$A = \begin{array}{c} \\ \text{I} \\ \text{II} \\ \text{III} \end{array} \begin{array}{c} \begin{array}{ccc} \text{I} & \text{II} & \text{III} \end{array} \\ \begin{bmatrix} 0.29 & 0.01 & 0.03 \\ 0.01 & 0.07 & 0.05 \\ 0.02 & 0.03 & 0.22 \end{bmatrix} \end{array}.$$

The given demand vector is

$$D = \begin{bmatrix} 60{,}510 \\ 22{,}863 \\ 33{,}295 \end{bmatrix}.$$

Thus our problem is to determine the intensity vector X, using the equation

$$X = (I - A)^{-1}D.$$

Now that the problem has been formulated, only the computations remain to be done. For this the computer program described in chapter 11 was used. The result of the computer run is given in figure 5.5. We see from the computer

```
THE INPUT-OUTPUT MATRIX IS
=== ============ ====== ==
 .29   .01   .03
 .01   .07   .05
 .02   .03   .22

THE DEMAND VECTOR IS
=== ====== ====== ==
 60510
 22863
 33295

THE INTENSITY VECTOR IS
=== ========= ====== ==
 87563.6
 27998.9
 46008
```

FIGURE 5.5

printout that the total production of the primary and fabricating metals sector must be $87,563.6 million, the total production of the petroleum and coal products sector must be $27,998.9 million, and the total production of the chemicals and chemical products sector must be $46,008 million.

Even though we used a computer to solve the problem in example 2, these same computations could have been done with pencil and paper. When doing hand computations, the following fact on the inverse of a matrix may be useful.

Consider two $n \times n$ matrices B and C and suppose there is a scalar $k \neq 0$ such that $C = kB$. C is invertible if and only if B is invertible and $C^{-1} = (1/k)B^{-1}$. This fact can be shown by the multiplication

$$CC^{-1} = (kB)\left(\frac{1}{k}B^{-1}\right) = \left(k\frac{1}{k}\right)(BB^{-1}) = I.$$

For example, the matrix $I - A$ from example 2 is

$$I - A = \begin{bmatrix} 0.71 & -0.01 & -0.03 \\ -0.01 & 0.93 & -0.05 \\ -0.02 & -0.03 & 0.78 \end{bmatrix}.$$

Instead of computing the inverse of $I - A$ directly, we can compute the inverse of

$$100(I - A) = \begin{bmatrix} 71 & -1 & -3 \\ -1 & 93 & -5 \\ -2 & -3 & 78 \end{bmatrix}$$

and then multiply this result by $\frac{1}{100}$ to obtain $(I - A)^{-1}$.

Even if a computer is used to handle the sometimes tedious calculations, we must still formulate the problem correctly and set up the data correctly for input to the computer.

EXERCISES 5.6

---

1. Consider a two-sector economic system consisting of manufacturing and energy and consider the following input–output matrix.

$$A = \begin{array}{cc} & \begin{array}{cc} \text{Supplied to} \\ \text{sector} \end{array} \\ & \begin{array}{cc} \text{M} & \text{E} \end{array} \\ \begin{array}{c} \text{from} \quad \text{M} \\ \text{sector} \quad \text{E} \end{array} & \begin{bmatrix} 0.4 & 0.2 \\ 0.6 & 0.1 \end{bmatrix} \end{array}$$

a. What is the value of manufactured goods required by energy so that it can produce $1 worth of energy?

b. What is the value of energy required by manufacturing so that it can produce $1 worth of manufactured goods?

c. Let x_1 be the total value of manufactured goods, and x_2 be the total value of energy produced. Draw a diagram similar to figure 5.4 to depict the flow of the value of goods within the system.

d. Let d_1 be the external demand for manufactured goods, and d_2 be the external demand for energy. Give the system of linear equations $X = [x_1 \quad x_2]^t$ must satisfy so that the needs of the system are met and the external demand is satisfied.

e. Find the intensity vector X for each of the following demand vectors.

$$D = \begin{bmatrix} 150 \\ 90 \end{bmatrix} \quad \text{and} \quad D = \begin{bmatrix} 250 \\ 150 \end{bmatrix}$$

2. Consider an economic system consisting of three sectors I, II, and III and consider the following input–output matrix.

$$A = \begin{array}{c} \\ \text{I} \\ \text{II} \\ \text{III} \end{array} \begin{array}{ccc} \text{I} & \text{II} & \text{III} \\ \begin{bmatrix} 0 & 0.4 & 0.2 \\ 0.1 & 0 & 0.3 \\ 0.5 & 0.2 & 0 \end{bmatrix} \end{array}$$

a. What is the value of goods produced by sector II and required by sector III so that sector III can produce $1 worth of goods?

b. What is the value of goods produced by sector III and required by sector II so that sector II can produce $1 worth of goods?

c. Observe that the entries on the main diagonal of the input–output matrix are zeros. What do these mean with respect to the input–output of the sectors?

d. Let x_i be the total value of goods produced by sector i. Draw a diagram similar to figure 3.3 to depict the flow of the value of goods within the system.

e. Let d_i be the external demand for goods produced by sector i. Give the system of linear equations that $X = [x_1 \quad x_2 \quad x_3]^t$ must satisfy so that the needs of the system are met and the external demand is satisfied.

f. Find the intensity vector X for each of the following demand vectors.

$$D = \begin{bmatrix} 200 \\ 100 \\ 200 \end{bmatrix} \quad \text{and} \quad D = \begin{bmatrix} 100 \\ 200 \\ 100 \end{bmatrix}$$

In exercises 3–8 an input–output matrix and corresponding demand vectors are given. First, use theorem 9 on page 207 to determine if the matrix is productive. If the matrix is productive, find the intensity vector for each of the demand vectors.

3. $\begin{bmatrix} 0.1 & 0.4 \\ 0.6 & 0.2 \end{bmatrix} \quad D_1 = \begin{bmatrix} 30 \\ 50 \end{bmatrix}, \quad D_2 = \begin{bmatrix} 60 \\ 100 \end{bmatrix}, \quad D_3 = \begin{bmatrix} 15 \\ 25 \end{bmatrix}$

4. $\begin{bmatrix} 0.2 & 0.4 \\ 0.5 & 0.3 \end{bmatrix} \quad D_1 = \begin{bmatrix} 100 \\ 200 \end{bmatrix} \quad \text{and} \quad D_2 = \begin{bmatrix} 50 \\ 40 \end{bmatrix}$

5. $\begin{bmatrix} 0.5 & 0.1 \\ 0.5 & 0.4 \end{bmatrix} \quad D_1 = \begin{bmatrix} 100 \\ 200 \end{bmatrix}, \quad D_2 = \begin{bmatrix} 150 \\ 250 \end{bmatrix}, \quad D_3 = \begin{bmatrix} 50 \\ 150 \end{bmatrix}$

6. $\begin{bmatrix} 0.75 & 0.25 \\ 0.75 & 0 \end{bmatrix}$ $D_1 = \begin{bmatrix} 300 \\ 150 \end{bmatrix}$, $D_2 = \begin{bmatrix} 120 \\ 100 \end{bmatrix}$, $D_3 = \begin{bmatrix} 90 \\ 50 \end{bmatrix}$

7. $\begin{bmatrix} 0.4 & 0 & 0.1 \\ 0 & 0.5 & 0.4 \\ 0.4 & 0 & 0.4 \end{bmatrix}$ $D_1 = \begin{bmatrix} 80 \\ 70 \\ 90 \end{bmatrix}$, $D_2 = \begin{bmatrix} 150 \\ 200 \\ 300 \end{bmatrix}$, $D_3 = \begin{bmatrix} 140 \\ 220 \\ 210 \end{bmatrix}$

8. $\begin{bmatrix} 0.6 & 0.4 & 0 \\ 0 & 0.6 & 0 \\ 0.2 & 0.2 & 0.6 \end{bmatrix}$ $D_1 = \begin{bmatrix} 1000 \\ 2000 \\ 3500 \end{bmatrix}$ and $D_2 = \begin{bmatrix} 2500 \\ 1500 \\ 1200 \end{bmatrix}$

9. Consider again the input–output matrix in exercise 6. We should have observed when doing the exercise that theorem 9 could not be used to conclude that the matrix is productive.

   a. However, verify that the matrix A is productive by showing that $(I - A)^{-1}$ exists and its entries are non-negative.

   b. Now determine the intensity vector X for each of the demand vectors given in exercise 6.

10. Consider again the input–output matrix in exercise 8. We should have observed when doing the exercise that theorem 9 could not be used to conclude that the matrix is productive.

    a. However, verify that the matrix A is productive by showing that $(I - A)^{-1}$ exists and its entries are non-negative.

    b. Now determine the intensity vector X for each of the demand vectors given in exercise 8.

11. Suppose that the United States economy is divided into two sectors, agriculture and manufacturing. Suppose that for each $1 of agricultural goods produced, agriculture uses $0.25 worth of their own products and $0.15 worth of manufactured goods. Also, when producing $1 worth of manufactured goods, manufacturing uses $0.05 worth of agricultural goods and $0.40 worth of their own goods. Finally, suppose that the consumer sector requires $15 billion worth of agricultural goods and $100 billion worth of manufactured goods.

    a. Write the input–output matrix.

    b. Find the intensity vector that will meet these requirements.

12. Refer to exercise 11. Suppose, in addition to the demand by United States consumers, there is foreign demand for our agricultural and manufactured products. If foreign demand is for $5 billion worth of grain and $10 billion worth of machinery, with what intensity must the United States economy operate at in order to meet these added demands?

13. Consider a society that consists of three groups of workers–shelter builders, clothing makers, and food producers. These groups of workers exchange goods among themselves. For the sake of convenience, we take $1 as one monetary unit in measuring the value of goods. In particular, the shelter builders need $0.15 worth of shelter, $0.10 worth of clothing, and $0.10 worth of food for each $1 worth of shelter built. The clothing makers need $0.25 worth of shelter, $0.30 worth of clothing, and $0.20 worth of food for each $1 worth of clothing made. Lastly, the food producers need $0.55 worth of shelter, $0.35 worth of clothing, and $0.30 worth of food for each $1 worth of food produced.

a. Give the input–output matrix for this three-sector economic system.

b. Suppose that this society also has a group of leaders; the leaders do not work but require shelter, clothing, and food. If the leaders demand $800 worth of shelter, $280 worth of clothing, and $370 worth of food, then how much must each of the three groups of workers produce so as to meet their own needs and to satisfy the demands of the leaders?

14. A carpenter, an electrician, and a plumber each own a home construction firm. When any one of them has a contract to build a home, he or she uses the other two for their respective skills. The carpenter uses $0.35 worth of his own time, $0.20 worth of the electrician's time, and $0.30 worth of the plumber's time for each $1 worth of home construction he does. For each $1 worth of home construction the electrician does, she uses $0.25 worth of the carpenter's time, $0.10 worth of her own time, and $0.25 worth of the plumber's time. Finally, for each $1 worth of home construction the plumber does, he uses $0.25 worth of the carpenter's time, $0.15 worth of the electrician's time, and $0.30 worth of his own time.

a. Give the input–output matrix for this system.

b. Suppose the carpenter has a contract for a $55,000 home, the electrician has a contract for a $45,000 home, and the plumber has a contract for a $50,000 home. What must be the dollar value of the work done by each of the three to satisfy these requirements?

5.7 The Number of Solutions of a Linear System

We have seen examples of systems that have no solutions, a unique solution, or infinitely many solutions. In fact, these are the only possibilities for the solution set of a system of linear equations. In other words, it is not possible for a system to have, for example, exactly two solutions. We have also seen in theorem 8, section 5.5, that a square system has a unique solution only when the coefficient matrix is invertible. In fact, the only systems that may have a unique solution are the square systems (ones with the same number of unknowns as equations). If a system has more unknowns than equations, the system will have either no solution or infinitely many solutions. The purpose of this section is to convey some feeling as to why this is so. The section is given simply for your reading, and therefore there are no exercises at the end of the section.

We begin our discussion by considering the linear system

$$\begin{bmatrix} 2 & -3 \\ -4 & 6 \end{bmatrix} \begin{bmatrix} x \\ y \end{bmatrix} = \begin{bmatrix} 1 \\ -2 \end{bmatrix}$$

that we denote by $AX = B$. It is not difficult to verify that each of the column matrices

$$S = \begin{bmatrix} 2 \\ 1 \end{bmatrix} \quad \text{and} \quad T = \begin{bmatrix} -1 \\ -1 \end{bmatrix}$$

is a solution to the system—that is, $AS = B$ and $AT = B$. Therefore we have

$$A(S - T) = AS - AT$$
$$= B - B$$
$$= 0.$$

This tells us that the column matrix $S - T$ is a solution to the homogeneous system $AX = 0$.

Our discussion points to a general statement that can be made about linear systems.

Theorem 10   ■ If S and T are solutions to an $m \times n$ system $AX = B$, then the difference $S - T$ is a solution to the homogeneous system $AX = 0$. ■

A homogeneous system $AX = 0$ will always have at least one solution, namely, the trivial solution $X = 0$. The trivial solution may be the only solution to the system. However, if a homogeneous system has one non-trivial solution, the system will have infinitely many solutions. This is true since all we need to do in order to obtain other non-trivial solutions is to multiply the non-trivial solution by non-zero scalars, as we see next.

Let

$$A = \begin{bmatrix} -10 & 11 & 0 & 5 \\ 1 & -2 & -3 & -2 \\ -4 & 5 & 2 & 3 \end{bmatrix} \quad \text{and} \quad C = \begin{bmatrix} 2 \\ 5 \\ 2 \\ -7 \end{bmatrix}$$

It is easily verified that C is a solution to the homogeneous system $AX = 0$; that is, $AC = 0$. Now let t be any real number. Since

$$A(tC) = t(AC) = t(0) = 0,$$

we have that

$$tC = \begin{bmatrix} 2t \\ 5t \\ 2t \\ -7t \end{bmatrix}$$

is a solution to $AX = 0$. Therefore the system $AX = 0$ has an infinite number of solutions.

Theorem 11   ■ If an $m \times n$ homogeneous system $AX = 0$ has a non-trivial solution C, the system has infinitely many solutions of the form tC, where t is any real number. Thus a homogeneous system $AX = 0$ has either

a. the trivial solution as its only one solution, or

b. infinitely many solutions. ■

We turn our attention now to non-homogeneous systems $AX = B$, where $B \neq 0$. We already realize that such a system may have no solutions. Suppose, on the other hand, $AX = B$ has a solution S. If the system has another solution T, where $S \neq T$, then the system will have infinitely many solutions. To see this, first note from theorem 10 that $S - T$ is a non-trivial solution to the homogeneous system $AX = 0$. Thus the system $AX = 0$ has infinitely many solutions of the form $t(S - T)$. Finally, $S + t(S - T)$, for any real number t, is a solution to $AX = B$ since

$$A(S + t(S - T)) = AS + A(t(S - T))$$
$$= B + 0$$
$$= B.$$

Therefore, if $AX = B$ has more than one solution, it must have infinitely many. This discussion of non-homogeneous systems is summarized in the next theorem.

Theorem 12 ■ An $m \times n$ system of linear equations $AX = B$ has either

a. no solutions,

b. one and only one solution, or

c. infinitely many solutions.

Furthermore, if the homogeneous system $AX = 0$ has a non-trivial solution W and if the non-homogeneous system $AX = B$ has a solution T, then for any real number t, $T + tW$ is a solution to $AX = B$. ■

Our final consideration is of systems in which there are more unknowns than equations. Consider, for example, the homogeneous system $AX = 0$ where

$$A = \begin{bmatrix} 1 & 0 & 0 & -3 \\ 0 & 1 & 0 & 1 \\ 0 & 0 & 1 & 2 \end{bmatrix}.$$

The matrix A is in reduced row-echelon form, and therefore we see the solutions to $AX = 0$ are given by the parametric equations

$$x_1 = \quad 3t$$
$$x_2 = \quad -t$$
$$x_3 = \quad -2t$$
$$x_4 = \quad t, \quad \text{any real number.}$$

The solutions can be written in the matrix form

$$tC = t\begin{bmatrix} 3 \\ -1 \\ -2 \\ 1 \end{bmatrix}.$$

Now consider the non-homogeneous system $AX = B$ where

$$B = \begin{bmatrix} -8 \\ 1 \\ 2 \end{bmatrix}.$$

It is easily verified that

$$T = \begin{bmatrix} 1 \\ -2 \\ -4 \\ 3 \end{bmatrix}$$

is a solution to $AX = B$. Therefore other solutions to $AX = B$ are

$$T + tC = \begin{bmatrix} 1 \\ -2 \\ -4 \\ 3 \end{bmatrix} + t \begin{bmatrix} 3 \\ -1 \\ -2 \\ 1 \end{bmatrix} = \begin{bmatrix} 1 + 3t \\ -2 - t \\ -4 - 2t \\ 3 + t \end{bmatrix}.$$

As in this last example, a homogeneous system with more unknowns than equations will always have a non-trivial solution. This is true since the coefficient matrix, when brought into reduced row-echelon form, will have at least one non-leading variable. This last comment, and our discussion thus far, points to the following important statement.

Theorem 13   ■ Consider an $m \times n$ non-homogeneous system $AX = B$ that has more unknowns than equations $(m < n)$. If the system has one solution, it must have infinitely many. ■

The statement in theorem 13 is pertinent to solving linear programming problems algebraically, which is the topic of our next chapter. In particular, in solving linear programming problems, we will be working with systems of linear equations in which there are more unknowns than equations.

SUMMARY OF TERMS

Operations with Matrices
  Row, Column Product
  Matrix Multiplication
  Matrix Addition
  Matrix Subtraction
  Multiplication by a Scalar
  Transpose of a Matrix

Properties of Operations
  Associative
  Commutative
  Additive Identity
  Additive Inverse
  Distributive

Types of Matrices
  Coefficient Matrix
  Equal Matrices
  Zero Matrix
  Identity Matrix
  Square Matrix
  Diagonal Matrix
  Upper-Triangular Matrix
  Invertible Matrix
  Inverse of a Matrix
  Price Vector
  Transition Matrix
  Distribution Vector

Lines of Best Fit
  Normal Equations
  Error Vector
  Size of the Error Vector

Linear Production Models
  Input–Output Matrix
  Productive Matrix
  Demand Vector
  Intensity Vector

REVIEW EXERCISES (CH. 5)

For exercises 1–12 do the indicated computations, if defined, using the following matrices.

$$A = \begin{bmatrix} 0 & 1 & -2 \\ 3 & 4 & 5 \\ -2 & -9 & 0 \end{bmatrix} \qquad B = \begin{bmatrix} 5 & -6 & 0 & 3 \\ 0 & 0 & -2 & 1 \\ 4 & -3 & -1 & 7 \end{bmatrix}$$

$$C = \begin{bmatrix} 1 & 7 & -8 & 4 \\ 0 & 5 & 6 & -1 \\ 2 & -3 & 0 & 9 \end{bmatrix} \qquad D = \begin{bmatrix} -0.1 & 0.3 & -0.4 \\ 0.5 & 0 & 0.2 \\ -0.7 & 0.5 & 0.1 \\ 0.2 & -0.9 & 0.3 \end{bmatrix}$$

$$E = \begin{bmatrix} 1 \\ 1 \\ 1 \end{bmatrix} \qquad F = \begin{bmatrix} 0.8 \\ -0.6 \\ -0.4 \\ 0.1 \end{bmatrix}$$

$$G = \begin{bmatrix} 1.7 & -2.3 & 0.5 \end{bmatrix} \qquad H = \begin{bmatrix} 2 & 3 & 4 & 5 \end{bmatrix}$$

1. $B + C$
2. $C - 2B$
3. GE
4. FH
5. HC
6. GA
7. AE
8. BF
9. AB
10. DC
11. $A + BD$
12. $B^t C$

For exercises 13–16 find the inverse, if it exists, for the given matrix.

13. $\begin{bmatrix} 2 & 0 & -2 \\ 0 & -3 & -4 \\ 1 & 2 & 3 \end{bmatrix}$

14. $\begin{bmatrix} -\frac{3}{5} & 0 & 0 \\ 0 & \frac{8}{7} & 0 \\ 0 & 0 & \frac{5}{6} \end{bmatrix}$

15. $\begin{bmatrix} 0.2 & 0 & -0.4 & 0.2 \\ -1.2 & 0.1 & 1.9 & -1 \\ -0.1 & 0 & -0.3 & -0.3 \\ 0 & 0.2 & -1 & 0.3 \end{bmatrix}$

16. $\begin{bmatrix} 5 & 0 & -2 & 1 \\ 0 & -3 & 4 & 0 \\ 0 & 0 & 1 & 3 \\ 0 & 0 & 0 & 2 \end{bmatrix}$

17. Let A be the matrix given in exercise 13. Using A^{-1}, obtain the solution to the system $AX = B$ where:

a.
$$B = \begin{bmatrix} 16 \\ -4 \\ 8 \end{bmatrix}$$

b.
$$B = \begin{bmatrix} -1 \\ 1 \\ -1 \end{bmatrix}$$

c.
$$B = \begin{bmatrix} 0 \\ -1 \\ 0 \end{bmatrix}$$

18. Let A be the matrix given in exercise 16. Using A^{-1}, obtain the solution to the system $AX = B$ where:

a.
$$B = \begin{bmatrix} 1 \\ 1 \\ 1 \\ 1 \end{bmatrix}$$

b.
$$B = \begin{bmatrix} 2 \\ -3 \\ 4 \\ -1 \end{bmatrix}$$

c.
$$B = \begin{bmatrix} 0 \\ 0 \\ 0 \\ 0 \end{bmatrix}$$

19. The Dayton Electronics Company has two plants, I and II, and each plant makes a hand calculator and a microcomputer.

    a. During one week of operation in May, plant I produced 465 hand calculators and 25 microcomputers. During the same week, plant II produced 364 hand calculators and 57 microcomputers. Express this information in a matrix.

    b. During the second week of operation in May, plant I made 359 hand calculators and 34 microcomputers, and plant II made 401 hand calculators and 48 microcomputers. Express by matrix addition the total number of hand calculators and microcomputers produced by each plant during the two weeks of operation.

20. The ABC Company manufactures widgets and gidgets. Labor, steel, and electrical power are needed to make these items. One widget requires 1.5 man-hours of labor, 0.5 pound of steel, and 0.1 kilowatt of power, whereas one gidget requires 2 man-hours of labor, 1 pound of steel, and 0.3 kilowatts of power. Express by a matrix product the total amount of labor, total pounds of steel, and the total kilowatts used to make 1500 widgets and 1250 gidgets.

For exercises 21 and 22 obtain the line of best fit for the given data points. Also determine the error vector E.

21. (0, 17.2),   (1, 16.4),   (2, 15.4),   and   (3, 14.8)
22. (1, 8.1),   (3, 10.5),   (5, 12.4),   and   (7, 15.0)

23. Consider the following productive input–output matrix.

$$A = \begin{bmatrix} 0.7 & 0.1 & 0.1 \\ 0.05 & 0.65 & 0.05 \\ 0.2 & 0.1 & 0.65 \end{bmatrix}$$

Determine the intensity vector X for this three-industry economy when the demand vector is

$$D = \begin{bmatrix} 150 \\ 200 \\ 300 \end{bmatrix}.$$

CHAPTER 6

Linear Programming: The Simplex Method

Introduction

In this chapter we discuss an algebraic approach, called the **simplex method**. This approach is used for solving linear programming problems in n variables. In chapter 3 we considered problems in two variables and their solution by the graphical method. The method we present here is not a generalization of that method. However, reference to problems in two variables and their solution by the graphical method can provide understanding of the simplex method.

In the graphical method we determine the vertices of the convex polygonal set of feasible solutions and determine which one, if any, gives the optimal value for the objective function. A vertex is a special type of feasible solution called **basic**. For a problem in more than two variables, as in the case for problems in two variables, it happens that if the problem has an optimal solution, then it is a **basic feasible solution**. Furthermore, for a problem in n variables, there are only a finite number of basic feasible solutions. Therefore we might be tempted to seek an algebraic procedure for finding all of them and evaluating the objective function at each, as done in the graphical method. However, this procedure is not practical since the number of basic feasible solutions could grow quite large with the number of variables and constraints. Instead we start at a basic feasible solution and determine whether or not the value of the objective function would improve by moving to another basic feasible solution.

The idea of moving from basic feasible solution to basic feasible solution, checking for improvement in the value of the objective function, is fundamental to the simplex algorithm. George Dantzig, a leader in an Air Force research group working on military logistics problems during World War II, developed the simplex method in the late 1940s. Our discussion in this chapter is based on material appearing in his book, *Linear Programming and Extensions*.

6.1 Introducing Slack Variables

We saw in chapter 3 how to solve linear programming problems in two unknowns by the graphical method. In this section and the next we develop an algebraic method for solving standard maximization problems. We consider other types of problems in sections 6.3 and 6.4.

A **standard maximization problem** is a linear programming problem in the following form.

$$\text{Maximize} \quad z = c_1 x_1 + c_2 x_2 + \cdots + c_n x_n$$

$$\text{subject to} \quad a_{11} x_1 + a_{12} x_2 + \cdots + a_{1n} x_n \leq b_1, \quad b_1 \geq 0$$

$$a_{21} x_1 + a_{22} x_2 + \cdots + a_{2n} x_n \leq b_2, \quad b_2 \geq 0$$

$$\vdots \qquad\qquad\qquad \vdots$$

$$a_{m1} x_1 + a_{m2} x_2 + \cdots + a_{mn} x_n \leq b_m, \quad b_m \geq 0$$

$$x_1 \geq 0, \quad x_2 \geq 0, \quad \ldots \quad x_n \geq 0.$$

An example of a standard maximization problem is the production problem introduced in chapter 3.

$$\text{Maximize} \quad z = 20x + 18y \qquad \text{(profit function)}$$

$$\text{subject to} \quad x + 2y \leq 575 \qquad \text{(labor constraint)}$$

$$10x + 15y \leq 5000 \qquad \text{(gas constraint)}$$

$$x \geq 0 \quad \text{and} \quad y \geq 0. \qquad \text{(non-negativity)}$$

The first step of the algebraic method is to convert the inequality **constraints** of the problem to equalities. In doing so we can find feasible solutions by solving a system of linear equations. To make the conversion we introduce additional variables into the problem, which we illustrate with our production problem.

For a production program to be feasible, the difference between the available labor (575 man-hours) and the amount used in production ($x + 2y$) must be non-negative. This difference, called **slack**, can be represented by a new variable s_1, where

$$s_1 = 575 - (x + 2y).$$

Thus, by introducing a **slack variable** s_1 into the problem, the inequality constraint $x + 2y \leq 575$ is converted to the equation

$$x + 2y + s_1 = 575.$$

Also the constraint on natural gas is converted to an equation by introducing a second slack variable s_2. Specifically, from the representation of s_2,

$$s_2 = 5000 - (10x + 15y),$$

we obtain the linear equation

$$10x + 15y + s_2 = 5000.$$

Observe that slack variables are meaningful variables in the problem; they represent the unused resource. Also slack variables must be non-negative. If a slack variable is negative, the production program will need more of that resource than what is available and the program would not be feasible. For example, if $s_1 = -10$ in the equation $x + 2y + s_1 = 575$, then $x + 2y = 585$ and the available man-hours would be exceeded.

The slack variables are also considered in the objective function. However, since slack is left-over resources, each unit of slack contributes nothing to the profit, and therefore its coefficient in the profit expression is zero. For our production problem, the profit function is

$$z = 0s_1 + 0s_2 + 20x + 18y = 20x + 18y.$$

Thus, by introducing slack variables, our production problem now can be viewed as finding a solution to the system of linear equations

$$x + 2y + s_1 \qquad = 575$$
$$10x + 15y \qquad + s_2 = 5000$$

for which $x \geq 0$, $y \geq 0$, $s_1 \geq 0$, $s_2 \geq 0$, and $z = 20x + 18y$ is maximum.

When solving this problem we not only want to solve the system of equations for the variables $s_1, s_2, x,$ and y, but we also need to determine the value of z. Instead of determining z by substituting a solution for the other variables, we can include the equation

$$-20x - 18y + z = 0$$

as the last equation of the system and obtain the value for z along with the other variables. Thus our production problem can be stated as: Find a solution to the system of equations

$$x + 2y + s_1 \qquad\qquad = 575$$
$$10x + 15y \qquad + s_2 \qquad = 5000$$
$$-20x - 18y \qquad\qquad + z = 0$$

for which $x \geq 0$, $y \geq 0$, $s_1 \geq 0$, $s_2 \geq 0$, and z is maximum.

We wish to standardize the type of problems to which we apply the simplex algorithm; we use the form obtained for our production problem, which results in the following definition.

Definition ■ A linear programming problem is said to be **a problem in canonical form** when we are to find a solution to the system of $(m + 1)$ linear equations in

$(m + 1 + n)$ variables

$$
\begin{array}{lll}
a_{11}x_1 + a_{12}x_2 + \cdots + a_{1n}x_n + s_1 & = b_1 & (b_1 \geq 0) \\
a_{21}x_1 + a_{22}x_2 + \cdots + a_{2n}x_n \quad\quad + s_2 & = b_2 & (b_2 \geq 0) \\
\quad\quad\quad \vdots & \quad \vdots & \\
a_{m1}x_1 + a_{m2}x_2 + \cdots + a_{mn}x_n \quad\quad + s_m & = b_m & (b_m \geq 0) \\
-c_1x_1 - c_2x_2 \quad - \cdots - c_nx_n \quad\quad\quad\quad\;\; + z = z_0 &
\end{array}
\tag{1}
$$

that satisfies the **non-negative conditions**

$$
x_1 \geq 0, x_2 \geq 0, \ldots, x_n \geq 0; s_1 \geq 0, s_2 \geq 0, \ldots, s_m \geq 0;
\tag{2}
$$

and for which z is **maximum**.

A solution to the system of equations (1) that satisfies the non-negative conditions (2) is called **feasible**. A feasible solution at which the value of z is maximum is called an **optimal solution**. The maximum value of z is called the **optimal value**. ∎

A standard maximization problem is easily converted to canonical form. First we add a slack variable to each constraint to change the inequality to an equality. We also include in the system of linear equations the equation of the objective function. Therefore, the number of linear equations for the problem in canonical form is $m + 1$ when the original problem has m linear inequality constraints. Furthermore, since the dependent variable of the objective function is included along with each slack variable as variables in the problem, the problem in canonical form has $m + 1 + n$ unknowns when the original problem has n unknowns.

EXAMPLE 1

Convert the following linear programming problem to canonical form.

$$
\begin{array}{ll}
\text{Maximize} & z = 3x_1 + 7x_2 + 5x_3 \\
\text{subject to} & 2x_1 \quad\quad + x_3 \leq 4 \\
& x_1 + 4x_2 + 6x_3 \leq 8 \\
& x_1 \geq 0, \quad x_2 \geq 0, \quad x_3 \geq 0.
\end{array}
$$

We introduce a slack variable in each of the two constraints. Also we include the equation of the objective function in the system. The coefficients of the slack variables are zeros. Thus the problem written in canonical form is: Find a solution to the 3×6 system

$$
\begin{array}{ll}
2x_1 \quad\quad + x_3 + s_1 \quad\quad\quad = 4 \\
x_1 + 4x_2 + 6x_3 \quad\quad + s_2 \quad = 8 \\
-3x_1 - 7x_2 - 5x_3 \quad\quad\quad\quad + z = 0
\end{array}
$$

that satisfies the non-negative conditions $x_1 \geq 0, x_2 \geq 0, x_3 \geq 0; s_1 \geq 0, s_2 \geq 0;$ and for which z is maximum.

The following points on the description of a problem in canonical form should be kept in mind:

1. The system of linear equations to be solved contains the equation for the objective function.
2. Each equation in the system contains a variable with coefficient 1 and this is the only equation containing that variable with a non-zero coefficient.
3. The right-hand sides of the equations, except possibly the equation for the objective function, are non-negative.
4. The problem is to determine a solution for which the value of the objective function is maximum.

The $(m + 1)$ by $(m + 1 + n)$ system of linear equations (1) for a problem in canonical form has at least one solution, specifically,

$$x_1 = 0, x_2 = 0, \ldots, x_n = 0$$
$$s_1 = b_1, s_2 = b_2, \ldots, s_m = b_m \tag{3}$$
$$z = z_0.$$

Furthermore this solution is feasible since the right-hand sides b_i are non-negative. Recall that a feasible solution is one in which the values of all the variables are non-negative.

The solution (3) is not the only solution to the equations (1). As discussed in section 5.7, we know that a system with more unknowns than equations will have infinitely many solutions if it has one solution. This is the case for a problem in canonical form. Therefore the system (1) has infinitely many solutions, one of which is the feasible solution (3). Some other solutions, however, may not be feasible. Our primary interest is, of course, in feasible solutions. We have a further interest in feasible solutions that are also basic.

Definition   ■ Consider a linear system $AX = B$ of $(m + 1)$ equations in $(m + 1 + n)$ unknowns. A **basic solution** to the system is obtained by setting n of the unknowns equal to zero and solving for the remaining $m + 1$ unknowns. The n unknowns set equal to zero are called **non-basic variables** and the other $m + 1$ unknowns are called **basic variables** in the solution.   ■

The solution (3) for a problem in canonical form is basic, where the basic variables in the solution are the slack variables $s_1, s_2, \ldots, s_m$ and the variable z. The non-basic variables in the solution are the original unknowns of the problem $x_1, x_2, \ldots, x_n$. As we have already noted, this basic solution is feasible.

The fundamental idea of the simplex algorithm is to start with a basic feasible solution and move, if possible, to another basic feasible solution for which the value of z is larger. Therefore the algorithm starts with an **initial basic feasible solution**. We will always take the basic feasible solution (3) as the initial one. Thus the algorithm will always start from the solution obtained by setting the original unknowns of the problem $x_1, x_2, \ldots, x_n$ equal to zeros. In this case the basic variables in the solution will be the slack variables and z, the values of which are easily obtained since each will appear in one and only one equation with coefficient 1.

EXAMPLES

2. Consider the 3×6 linear system

$$2x_1 \quad\quad + x_3 + s_1 \quad\quad\quad = 4$$
$$x_1 + 4x_2 + 6x_3 \quad\quad + s_2 \quad = 8$$
$$-3x_1 - 7x_2 - 5x_3 \quad\quad\quad + z = 0$$

which was obtained in example 1. Give the initial basic feasible solution.

    We obtain the initial basic feasible solution by setting the original unknowns of the problem, x_1, x_2, and x_3, equal to zeros. In doing so, and due to the form of the equations, the right-hand sides 4, 8, and 0 become the values of the basic variables s_1, s_2, and z, respectively. Thus we can easily read from the equations the initial basic feasible solution

$$x_1 = 0, \quad x_2 = 0, \quad x_3 = 0, \quad s_1 = 4, \quad s_2 = 8, \quad \text{and} \quad z = 0.$$

3. Consider the 3×5 system

$$x + 2y + s_1 \quad\quad\quad = 575$$
$$10x + 15y \quad + s_2 \quad = 5000$$
$$-20x - 18y \quad\quad + z = \quad 0$$

which is the system for our production problem. Let us give the initial basic feasible solution and find, if possible, the basic solution with basic variables x, s_1, and z.

    We obtain the initial basic feasible solution by setting $x = y = 0$ and solving for the basic variables s_1, s_2, and z. In particular $x = 0, y = 0, s_1 = 575$, $s_2 = 5000$, and $z = 0$. Observe that this solution corresponds to the vertex $A(0, 0)$ of the region of feasible solutions determined by the inequalities

$$x + 2y \leq 575$$
$$10x + 15y \leq 5000$$
$$x \geq 0 \quad \text{and} \quad y \geq 0$$

(see figure 6.1).

    We can also obtain a basic solution to the system with basic variables x, s_1, and z. In particular, setting $s_2 = 0$ and $y = 0$, we obtain the 3×3 system

$$x + s_1 \quad\quad = 575$$
$$10x \quad\quad = 5000$$
$$-20x \quad + z = \quad 0.$$

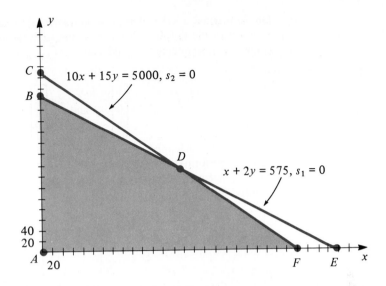

Point	x	y	s_1	s_2	z
A	0	0	575	5000	0
B	0	$\frac{575}{2}$	0	$\frac{1375}{2}$	5175
C	0	$\frac{1000}{3}$	$-\frac{275}{3}$	0	6000
D	275	150	0	0	8200
E	575	0	0	-750	11,500
F	500	0	75	0	10,000

FIGURE 6.1

This system has a solution since the coefficient matrix

$$\begin{bmatrix} 1 & 1 & 0 \\ 10 & 0 & 0 \\ -20 & 0 & 1 \end{bmatrix}$$

is row equivalent to the 3×3 invertible matrix

$$\begin{bmatrix} 0 & 1 & 0 \\ 1 & 0 & 0 \\ 0 & 0 & 1 \end{bmatrix}.$$

To solve for x, s_1, and z, let us perform the necessary row operations on the augmented matrix

$$A_1 = \begin{array}{ccccc} x & y & s_1 & s_2 & z \\ \begin{bmatrix} 1 & 2 & 1 & 0 & 0 \\ 10 & 15 & 0 & 1 & 0 \\ -20 & -18 & 0 & 0 & 1 \end{bmatrix} & & & & \begin{array}{c} 575 \\ 5000 \\ 0 \end{array} \end{array}$$

for the original 3×5 system. For convenience, we label the columns of the matrix with the variables they correspond to in the system. Observe that the columns corresponding to s_1 and z are

$$s_1 \qquad\qquad z$$
$$\begin{bmatrix} 1 \\ 0 \\ 0 \end{bmatrix} \quad \text{and} \quad \begin{bmatrix} 0 \\ 0 \\ 1 \end{bmatrix}.$$

Since s_1 and z each appear in one and only one equation with coefficient 1, we need only perform row operations so that x appears in one and only one equation with coefficient 1. Therefore we perform elimination of variables on A_1 so as to transform the column

$$x \qquad\qquad x$$
$$\begin{bmatrix} 1 \\ 10 \\ -20 \end{bmatrix} \quad \text{to} \quad \begin{bmatrix} 0 \\ 1 \\ 0 \end{bmatrix}.$$

First, we divide the second row of A_1 by 10. After performing this row operation, we subtract the resulting second row from the first row and add 20 times the second row to the third row. Performing this sequence of row operations, we have the following augmented matrix, which is row equivalent to A_1.

$$A_2 = \begin{bmatrix} x & y & s_1 & s_2 & z \\ 0 & \frac{1}{2} & 1 & -\frac{1}{10} & 0 & | & 75 \\ 1 & \frac{3}{2} & 0 & \frac{1}{10} & 0 & | & 500 \\ 0 & 12 & 0 & 2 & 1 & | & 10{,}000 \end{bmatrix}.$$

Upon setting $y = 0$ and $s_2 = 0$, we read from A_2 the basic solution with basic variables x, s_1, and z. Therefore we have the basic solution $x = 500$, $y = 0$, $s_1 = 75$, $s_2 = 0$, and $z = 10{,}000$. This solution is feasible and corresponds to the point F in figure 6.1. Furthermore, the value of z in this solution is greater than in the basic solution with basic variables s_1, s_2, and z.

As shown in example 3, we can obtain a basic solution with a given set of basic variables if the columns corresponding to the variables can be transformed to those of the identity matrix. The number of basic variables in the solution is the same as the number of equations. Furthermore, each basic variable appears in one and only one equation with coefficient 1.

EXAMPLE 4

Consider again the system and its augmented matrix A_1 in example 3. Let us obtain, if possible, the basic solution with basic variables y, s_1, and z and the basic solution with basic variables y, s_2, and z.

$$A_1 = \begin{bmatrix} x & y & s_1 & s_2 & z \\ 1 & 2 & 1 & 0 & 0 & | & 575 \\ 10 & 15 & 0 & 1 & 0 & | & 5000 \\ -20 & -18 & 0 & 0 & 1 & | & 0 \end{bmatrix}$$

To obtain the basic solution with basic variables y, s_1, and z, we perform row operations on A_1 to transform the column corresponding to y to the column $[0 \quad 1 \quad 0]^t$. To obtain the basic solution with basic variables y, s_2, and z, we transform the column corresponding to y to the column $[1 \quad 0 \quad 0]^t$. Performing the necessary row operations on A_1 we have

$$
\begin{array}{ccccc}
x & y & s_1 & s_2 & z \\
\end{array}
$$
$$
\begin{bmatrix}
1 & 2 & 1 & 0 & 0 & 575 \\
10 & \textcircled{15} & 0 & 1 & 0 & 5000 \\
-20 & -18 & 0 & 0 & 1 & 0
\end{bmatrix}
$$

$$
\xrightarrow[\substack{-2R_2+R_1 \\ 18R_2+R_3}]{\frac{1}{15}R_2}
$$

$$
\begin{array}{ccccc}
x & y & s_1 & s_2 & z \\
\end{array}
$$
$$
\begin{bmatrix}
-\frac{1}{3} & 0 & 1 & -\frac{2}{15} & 0 & -\frac{275}{3} \\
\frac{2}{3} & 1 & 0 & \frac{1}{15} & 0 & \frac{1000}{3} \\
-8 & 0 & 0 & \frac{6}{5} & 1 & 6000
\end{bmatrix}
$$

Basic Solution and Variables

$$[0 \quad \tfrac{1000}{3} \quad -\tfrac{275}{3} \quad 0 \quad 6000]^t$$

$$y, s_1, z$$

not feasible

and

$$
\begin{array}{ccccc}
x & y & s_1 & s_2 & z \\
\end{array}
$$
$$
\begin{bmatrix}
1 & \textcircled{2} & 1 & 0 & 0 & 575 \\
10 & 15 & 0 & 1 & 0 & 5000 \\
-20 & -18 & 0 & 0 & 1 & 0
\end{bmatrix}
$$

$$
\xrightarrow[\substack{-15R_1+R_2 \\ 18R_1+R_3}]{\frac{1}{2}R_1}
$$

$$
\begin{array}{ccccc}
x & y & s_1 & s_2 & z \\
\end{array}
$$
$$
\begin{bmatrix}
\frac{1}{2} & 1 & \frac{1}{2} & 0 & 0 & \frac{575}{2} \\
\frac{5}{2} & 0 & -\frac{15}{2} & 1 & 0 & \frac{1375}{2} \\
-11 & 0 & 9 & 0 & 1 & 5175
\end{bmatrix}
$$

Basic Solution and Variables

$$[0 \quad \tfrac{575}{2} \quad 0 \quad \tfrac{1375}{2} \quad 5175]^t$$

$$y, s_2, z$$

feasible

Observe that the basic solution

$$[0 \quad \tfrac{575}{2} \quad 0 \quad \tfrac{1375}{2} \quad 5175]^t$$

with basic variables y, s_2, and z is feasible and corresponds to the vertex B of the region of feasible solutions in figure 6.1. Furthermore, the value of z in this basic solution is smaller than in the basic solution with basic variables x, s_1, and z obtained in example 3. The basic solution with basic variables y, s_1, and z is not feasible and corresponds to the point C.

It is possible that the value of a basic variable, other than z, is zero. Such a basic solution is called **degenerate**.

EXAMPLE 5

The system of linear equations with augmented matrix

$$
A = \begin{bmatrix}
x & y & s_1 & s_2 & s_3 & z & \\
5 & 2 & 1 & 0 & 0 & 0 & 20 \\
6 & 3 & 0 & 1 & 0 & 0 & 12 \\
11 & 4 & 0 & 0 & 1 & 0 & 16 \\
-5 & -12 & 0 & 0 & 0 & 1 & 0
\end{bmatrix}
$$

is for a linear programming problem in canonical form. Consider the basic feasible solution

$$[0 \quad 0 \quad 20 \quad 12 \quad 16 \quad 0]^t.$$

Determine each of the three basic solutions obtained by replacing one of the basic variables s_1, s_2, or s_3 with the variable y.

We perform row operations on A and read from the derived row equivalent augmented matrix the basic solution. Specifically, we have the following three equivalent matrices and corresponding basic solutions.

$$
\begin{bmatrix}
x & y & s_1 & s_2 & s_3 & z & \\
5 & 2 & 1 & 0 & 0 & 0 & 20 \\
6 & 3 & 0 & 1 & 0 & 0 & 12 \\
11 & 4 & 0 & 0 & 1 & 0 & 16 \\
-5 & -12 & 0 & 0 & 0 & 1 & 0
\end{bmatrix}
$$

$$
\begin{array}{c}
\frac{1}{2}R_1 \\
-3R_1 + R_2 \\
\hline
-4R_1 + R_3 \\
12R_1 + R_4
\end{array}
\begin{bmatrix}
x & y & s_1 & s_2 & s_3 & z & \\
\frac{5}{2} & 1 & \frac{1}{2} & 0 & 0 & 0 & 10 \\
-\frac{3}{2} & 0 & -\frac{3}{2} & 1 & 0 & 0 & -18 \\
1 & 0 & -2 & 0 & 1 & 0 & -24 \\
25 & 0 & 6 & 0 & 0 & 1 & 120
\end{bmatrix}
$$

Basic Solution and Basic Variables

$$[0 \quad 10 \quad 0 \quad -18 \quad -24 \quad 120]^t$$

$$y, s_2, s_3, z$$

not feasible

$$
\begin{bmatrix}
x & y & s_1 & s_2 & s_3 & z & \\
5 & 2 & 1 & 0 & 0 & 0 & 20 \\
6 & 3 & 0 & 1 & 0 & 0 & 12 \\
11 & 4 & 0 & 0 & 1 & 0 & 16 \\
-5 & -12 & 0 & 0 & 0 & 1 & 0
\end{bmatrix}
$$

$$
\begin{array}{c}
\frac{1}{3}R_2 \\
-2R_2 + R_1 \\
\hline
-4R_2 + R_3 \\
12R_2 + R_4
\end{array}
\begin{bmatrix}
x & y & s_1 & s_2 & s_3 & z & \\
1 & 0 & 1 & -\frac{2}{3} & 0 & 0 & 12 \\
2 & 1 & 0 & \frac{1}{3} & 0 & 0 & 4 \\
3 & 0 & 0 & -\frac{4}{3} & 1 & 0 & 0 \\
19 & 0 & 0 & 4 & 0 & 1 & 48
\end{bmatrix}
$$

Basic Solution and Basic Variables

$$[0 \quad 4 \quad 12 \quad 0 \quad 0 \quad 48]^t$$

$$y, s_1, s_3, z$$

feasible, but degenerate since $s_3 = 0$

$$
\begin{array}{cccccc}
x & y & s_1 & s_2 & s_3 & z \\
\end{array}
$$
$$
\left[
\begin{array}{cccccc|c}
5 & 2 & 1 & 0 & 0 & 0 & 20 \\
6 & 3 & 0 & 1 & 0 & 0 & 12 \\
11 & 4 & 0 & 0 & 1 & 0 & 16 \\
-5 & -12 & 0 & 0 & 0 & 1 & 0 \\
\end{array}
\right]
$$

$$
\begin{array}{cccccc}
x & y & s_1 & s_2 & s_3 & z \\
\end{array}
$$
$$
\begin{array}{c}
\frac{1}{4}R_3 \\
-2R_3 + R_1 \\
\xrightarrow{\hspace{1.5cm}} \\
-3R_3 + R_2 \\
12R_3 + R_4 \\
\end{array}
\left[
\begin{array}{cccccc|c}
-\frac{1}{2} & 0 & 1 & 0 & -\frac{1}{2} & 0 & 12 \\
-\frac{9}{4} & 0 & 0 & 1 & -\frac{3}{4} & 0 & 0 \\
\frac{11}{4} & 1 & 0 & 0 & \frac{1}{4} & 0 & 4 \\
28 & 0 & 0 & 0 & 3 & 1 & 48 \\
\end{array}
\right]
$$

Basic Solution and Basic Variables

$$[0 \quad 4 \quad 12 \quad 0 \quad 0 \quad 48]^t$$

$$y, s_1, s_2, z$$

feasible, but degenerate since $s_2 = 0$

As discussed in this section we can go from one basic solution to another by changing a non-basic variable to a basic one. This non-basic variable is called the **entering variable**. Of course, we must also pick one of the basic variables to make non-basic. This variable is called the **departing variable**. Since we want to obtain a new basic solution that is feasible and a value of z that is larger, the entering variable and departing variable cannot be chosen arbitrarily, as seen in the examples. In the next section we discuss criteria for selecting the departing and entering variables.

EXERCISES 6.1

For exercises 1–6 convert the linear programming problem to canonical form. Do not attempt to solve these problems.

1. Maximize $z = 5x + 2y$
   subject to $3x + 4y \le 26$
   $4x + 3y \le 23$
   $x \ge 0$ and $y \ge 0$.

2. Maximize $z = 0.3x + 0.4y$
   subject to $7x + 5y \le 43$
   $2x + 9y \le 35$
   $x \ge 0$ and $y \ge 0$.

3. Maximize $z = 4x + 5y$
   subject to $x + 3y \leq 26$
   $$4x + 3y \leq 44$$
   $$2x + 3y \leq 28$$
   $$x \geq 0 \quad \text{and} \quad y \geq 0.$$

4. Maximize $z = 3x_1 + 2x_2 + x_3$
   subject to $x_1 + 4x_2 + 5x_3 \leq 32$
   $$4x_1 + 2x_2 + 8x_3 \leq 39$$
   $$x_1 \geq 0, \quad x_2 \geq 0, \quad \text{and} \quad x_3 \geq 0.$$

5. Maximize $z = 1.5x_1 + 0.75x_2 + 0.50x_3$
   subject to $5x_1 + 3x_2 \qquad \leq 30$
   $$4x_1 + \;\; x_2 + 7x_3 \leq 20$$
   $$5x_2 + 8x_3 \leq 32$$
   $$x_1 \geq 0, \quad x_2 \geq 0, \quad \text{and} \quad x_3 \geq 0.$$

6. Maximize $z = 15x_1 + 8x_2 + 12x_3$
   subject to $3x_1 + 2x_2 + 6x_3 \leq 18$
   $$x_1 \qquad + 7x_3 \leq 21$$
   $$2x_1 + 4x_2 + 5x_3 \leq 32$$
   $$9x_1 + 7x_2 \qquad \leq 56$$
   $$x_1 \geq 0, \quad x_2 \geq 0, \quad \text{and} \quad x_3 \geq 0.$$

For exercises 7–10 is the problem in canonical form? If not, state why.

7. Determine a solution to

$$
\begin{aligned}
2x + 5y + s_1 \qquad\qquad &= 20 \\
5x + 2y \qquad + s_2 \qquad &= 20 \\
7x - 2y \qquad\qquad + z &= 0
\end{aligned}
$$

for which $x \geq 0, y \geq 0, s_1 \geq 0, s_2 \geq 0$, and z is maximum.

8. Determine a solution to

$$
\begin{aligned}
3x + 4y + s_1 \qquad\qquad &= 16 \\
5x + 2y \qquad - s_2 \qquad &= 25 \\
-8x - 9y \qquad\qquad + z &= 0
\end{aligned}
$$

for which $x \geq 0, y \geq 0, s_1 \geq 0, s_2 \geq 0$, and z is maximum.

9. Determine a solution to

$$
\begin{aligned}
2x + 4y - s_1 \qquad\qquad &= 20 \\
5x + 3y \qquad - s_2 \qquad &= 16 \\
-7x - 2y \qquad\qquad + z &= 0
\end{aligned}
$$

for which $x \geq 0, y \geq 0, s_1 \geq 0, s_2 \geq 0$, and z is minimum.

10. Determine a solution to

$$
\begin{aligned}
9x + 5y + s_1 \qquad\qquad &= 90 \\
8x + 9y \qquad + s_2 \qquad &= -72 \\
-2x - 3y \qquad\qquad + z &= 0
\end{aligned}
$$

for which $x \geq 0, y \geq 0, s_1 \geq 0, s_2 \geq 0$, and z is maximum.

For exercises 11–14 read directly from the augmented matrix a basic solution, give the basic variables, and state whether or not the solution is feasible.

11.

$$
\begin{array}{ccccc}
x & y & s_1 & s_2 & z
\end{array}
$$
$$
\begin{bmatrix}
5 & -2 & 1 & 0 & 0 & 6 \\
-4 & 3 & 0 & 1 & 0 & 8 \\
3 & -1 & 0 & 0 & 1 & 0
\end{bmatrix}
$$

12.

$$
\begin{array}{cccccc}
x & y & s_1 & s_2 & s_3 & z
\end{array}
$$
$$
\begin{bmatrix}
1 & 4 & -2 & 0 & 0 & 0 & -3 \\
0 & -1 & 3 & 1 & 0 & 0 & 5 \\
0 & 2 & -1 & 0 & 1 & 0 & 9 \\
0 & -3 & 2 & 0 & 0 & 1 & 16
\end{bmatrix}
$$

13.

$$
\begin{array}{ccccc}
x_1 & x_2 & x_3 & s_1 & s_2 & z
\end{array}
$$
$$
\begin{bmatrix}
-3 & 1 & 0 & 2 & -1 & 0 & 5 \\
4 & 0 & 1 & -3 & 2 & 0 & 6 \\
7 & 0 & 0 & 2 & 5 & 1 & 24
\end{bmatrix}
$$

14.

$$
\begin{array}{cccccc}
x_1 & x_2 & x_3 & s_1 & s_2 & s_3 & z
\end{array}
$$
$$
\begin{bmatrix}
0 & 1 & -3 & -1 & 0 & 2 & 0 & 5 \\
0 & 0 & 4 & 2 & 1 & -4 & 0 & 7 \\
1 & 0 & 2 & -1 & 0 & 3 & 0 & -4 \\
0 & 0 & -6 & 3 & 0 & -8 & 1 & 15
\end{bmatrix}
$$

15. Consider the system of linear equations

$$
\begin{aligned}
3x + 6y + s_1 \quad\quad &= 24 \\
4x + 3y \quad + s_2 \quad &= 36 \\
-5x - 2y \quad\quad + z &= 0
\end{aligned}
$$

and the initial basic feasible solution

$$
x = y = 0, \quad s_1 = 24, \quad s_2 = 36, \quad \text{and} \quad z = 0.
$$

a. Determine the basic solution with basic variables x, s_2, and z.

b. Determine the basic solution with basic variables x, s_1, and z.

c. Are the solutions in parts (a) and (b) both feasible?

16. Consider the system of linear equations in exercise 15.

a. Determine the basic solution with basic variables y, s_1, and z.

b. Determine the basic solution with basic variables y, s_2, and z.

c. Are the solutions in parts (a) and (b) both feasible?

17. Consider the system of linear equations

$$
\begin{aligned}
5x_1 \quad\quad + 3x_3 + s_1 \quad\quad\quad &= 30 \\
4x_1 + x_2 + 7x_3 \quad + s_2 \quad\quad &= 28 \\
5x_2 + 8x_3 \quad\quad + s_3 &= 40 \\
-7x_1 - 3x_2 - 5x_3 \quad\quad\quad + z &= 0
\end{aligned}
$$

and the initial basic feasible solution

$$x_1 = x_2 = x_3 = 0, \quad s_1 = 30, \quad s_2 = 28, \quad s_3 = 40, \quad \text{and} \quad z = 0.$$

a. Determine the basic solution with basic variables x_1, s_2, s_3, and z.

b. Determine the basic solution with basic variables x_1, s_1, s_3, and z.

c. Are the solutions determined in parts (a) and (b) both feasible?

18. Consider the system of linear equations in exercise 17.

a. Determine the basic solution with basic variables x_3, s_1, s_3, and z.

b. Determine the basic solution with basic variables x_3, s_2, s_3, and z.

c. Are the solutions determined in parts (a) and (b) both feasible?

d. Which of the variables s_2 or s_3 should be basic in a basic feasible solution in which the other basic variables are x_3, s_1, and z?

19. Consider the system of linear equations with augmented matrix

$$
\begin{array}{ccccc}
x & y & s_1 & s_2 & z \\
\end{array}
$$
$$
\left[
\begin{array}{ccccc|c}
8 & 4 & 1 & 0 & 0 & 48 \\
7 & 14 & 0 & 1 & 0 & 56 \\
-6 & -8 & 0 & 0 & 1 & 0 \\
\end{array}
\right]
$$

and the basic feasible solution

$$x = y = 0, \quad s_1 = 48, \quad s_2 = 56, \quad \text{and} \quad z = 0.$$

a. Determine, if possible, a basic feasible solution obtained by changing y to a basic variable and s_2 to a non-basic variable in the solution.

b. Determine, if possible, a basic feasible solution obtained by changing y to a basic variable and s_1 to a non-basic variable in the solution.

c. To obtain a basic feasible solution by changing y to a basic variable, which of the two variables s_1 or s_2 should be changed to non-basic?

20. Consider the system of linear equations with augmented matrix

$$
\begin{array}{cccccc}
x_1 & x_2 & x_3 & s_1 & s_2 & z \\
\end{array}
$$
$$
\left[
\begin{array}{cccccc|c}
0 & 2 & 1 & 1 & 8 & 0 & 5 \\
1 & 3 & 6 & 0 & 9 & 0 & 12 \\
0 & -4 & -4 & 0 & 7 & 1 & 3 \\
\end{array}
\right]
$$

and the basic feasible solution

$$x_2 = x_3 = s_2 = 0, \quad x_1 = 12, \quad s_1 = 5, \quad \text{and} \quad z = 3.$$

a. Determine the basic feasible solution obtained from changing x_2 from a non-basic variable to a basic variable.

b. Determine the basic feasible solution obtained from changing x_3 from a non-basic variable to a basic variable.

21. a. Graph the system of linear inequalities

$$4x + 3y \leq 48$$

$$x + 2y \leq 22$$

$$x \geq 0 \quad \text{and} \quad y \geq 0.$$

b. Determine all the basic solutions to the linear system

$$4x + 3y + s_1 \qquad\quad = 48$$

$$x + 2y \quad\;\; + s_2 \qquad = 22$$

$$-x - \;\; y \qquad\qquad + z = \;\; 0$$

and identify the points on the graph of part (a) that correspond to these basic solutions.

22. a. Graph the system of linear inequalities

$$3x + 4y \le 44$$

$$3x + 2y \le 28$$

$$x \ge 0 \quad\text{and}\quad y \ge 0.$$

b. Determine all the basic solutions to the linear system

$$3x + 4y + s_1 \qquad\quad = 44$$

$$3x + 2y \quad\;\; + s_2 \qquad = 28$$

$$-5x - \;\; y \qquad\qquad + z = \;\; 0$$

and identify the points on the graph of part (a) that correspond to these basic solutions.

6.2 Solving Standard Maximization Problems

In the previous section we saw how to perform row operations to change from one basic solution to another. The non-basic variable that becomes basic in this process is called the **entering variable**. The basic variable that becomes, in turn, non-basic is called the **departing variable**. The examples in the previous section also pointed out that these variables cannot be chosen arbitrarily. We discuss in this section how the selections are made for the entering variable and the departing variable. With these criteria we will have completed the simplex algorithm for solving problems in canonical form.

Let us first add more notation with the augmented matrix to form a **simplex tableau** (see tableau 6.1). As before, the names of the variables appear above their respective columns. These labels stay the same as we do row operations on the augmented matrix. The augmented matrix appears in the tableau within matrix brackets ([ ]). However, we draw a dotted line above the last row, the row for the equation of the objective function, to make this row stand out from the others. Finally, each basic variable in the solution appears in one and only one equation with coefficient 1, and its value is obtained from that equation. Therefore we write the names of the basic variables to the left of the augmented matrix, each on the same line as the equation in which it appears. The simplex tableau is the augmented matrix with this notation added. Note, however, that we will be doing row operations only on the augmented matrix portion of the tableau.

TABLEAU 6.1   *Initial Simplex Tableau for a Problem in Canonical Form*

	x_1	x_2	$\cdots$	x_n	s_1	s_2	$\cdots$	s_m	z	
s_1	a_{11}	a_{12}	$\cdots$	a_{1n}	1	0	$\cdots$	0	0	b_1
s_2	a_{21}	a_{22}	$\cdots$	a_{2n}	0	1	$\cdots$	0	0	b_2
$\vdots$		$\vdots$					$\ddots$			$\vdots$
s_m	a_{m1}	a_{m2}	$\cdots$	a_{mn}	0	0	$\cdots$	1	0	b_m
z	$-c_1$	$-c_2$	$\cdots$	$-c_m$	0	0	$\cdots$	0	1	z_0

EXAMPLE 1

Consider the following problem in canonical form: Find a solution to

$$3x_1 + 5x_2 + 2x_3 + s_1 \qquad\qquad = 20$$
$$6x_2 + 3x_3 \qquad + s_2 \qquad\quad = 12$$
$$11x_1 \qquad - 4x_3 \qquad\qquad + s_3 \quad = 30$$
$$-4x_1 - 7x_2 - 8x_3 \qquad\qquad\quad + z = 0$$

for which $x_i \geq 0$, $i = 1, 2, 3$; $s_j \geq 0$, $j = 1, 2$; and z is maximum.

Give the simplex tableau for this problem.

From the equations it can be seen that the basic variables are s_1, s_2, s_3, and z, appearing in the first through fourth equations, respectively. Therefore tableau 6.2 is the desired initial simplex tableau.

TABLEAU 6.2

	x_1	x_2	x_3	s_1	s_2	s_3	z	
s_1	3	5	2	1	0	0	0	20
s_2	0	6	3	0	1	0	0	12
s_3	11	0	-4	0	0	1	0	30
z	-4	-7	-8	0	0	0	1	0

We now discuss the criterion for selecting the entering variable. Consider again our production problem for the glassware manufacturer:
Find a solution to

$$x + 2y + s_1 \qquad\qquad = 575$$
$$10x + 15y \qquad + s_2 \qquad = 5000$$
$$-20x - 18y \qquad\qquad + z = 0$$

for which $x \geq 0$, $y \geq 0$, $s_1 \geq 0$, $s_2 \geq 0$, and z is maximum.

The initial simplex tableau for the problem is tableau 6.3.

TABLEAU 6.3

	x	y	s_1	s_2	z	
s_1	1	2	1	0	0	575
s_2	10	15	0	1	0	5000
z	-20	-18	0	0	1	0

From tableau 6.3 we read the initial basic feasible solution $x = 0$, $y = 0$, $s_1 = 575$, $s_2 = 5000$, and $z = 0$. This solution corresponds to the vertex A in figure 6.2. We wish to see how the value of z changes when we move away from the vertex A to points on the positive x-axis or to points on the positive y-axis. Therefore consider the equation of the objective function, which is written as

$$z - 20x - 18y = 0.$$

Since the coefficient of x in this equation is negative, we see that z must increase from its value of zero at A to a positive value at solutions represented by points on the positive x-axis ($x > 0$ and $y = 0$). Also since the coefficient of y is negative, the value of z must also increase by moving away from $A(0, 0)$ to points on the positive y-axis ($x = 0$ and $y > 0$). However, since the coefficient of x in the equation of the objective function is more negative than the coefficient of y, a greater increase in the value of z will occur per unit increase in x rather than per unit increase in y. Therefore let us obtain a basic solution when x is changed from non-basic to basic and y is left as non-basic; that is, let us obtain a basic solution represented by a point on the positive x-axis.

To change x to a basic variable, we must change one of the variables s_1, s_2, or z to non-basic. Since z must always be a basic variable in the solution, we are left with two choices, $s_1 = 0$ or $s_2 = 0$. However, observe from figure 6.2 that $s_1 = 0$ at the point E and $s_2 = 0$ at the point F. Since the basic solution must be feasible, we see from figure 6.2 that we can move along the positive x-axis only as far as the vertex F. Therefore we obtain the basic solution with x as a basic variable by changing s_2 to a non-basic variable. This was done in example 3 in section 6.1. The augmented matrix that is row equivalent to the augmented matrix in tableau 6.3 is repeated here in tableau 6.4.

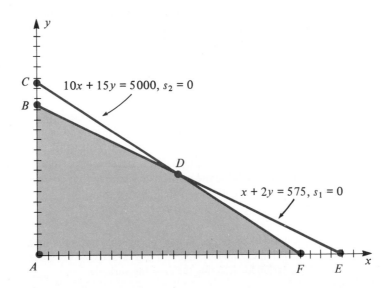

FIGURE 6.2

TABLEAU 6.4

	x	y	s_1	s_2	z	
s_1	0	$\frac{1}{2}$	1	$-\frac{1}{10}$	0	75
x	1	$\frac{3}{2}$	0	$\frac{1}{10}$	0	500
z	0	12	0	2	1	10,000

We have the basic feasible solution $x = 500$, $y = 0$, $s_1 = 75$, $s_2 = 0$, and $z = 10,000$. As before, we wish to determine whether the value of z can be larger at any other feasible solution (basic or not) than its present value 10,000.

Consider any solution to the system of equations in which y and/or s_2 are positive. Such solutions must satisfy the last equation of the system $12y + 2s_2 + z = 10,000$, which we rewrite as

$$(z - 10,000) + 12y + 2s_2 = 0.$$

Since the coefficients of y and s_2 are positive, $(z - 10,000)$ must be negative for any solution to the equation in which $y > 0$ and/or $s_2 > 0$; that is, z must be smaller than 10,000. Therefore the optimal value of z is 10,000 and it occurs when $y = 0$ and $s_2 = 0$. Thus we have a basic feasible solution that is optimal.

With our discussion in mind, we should be able to accept the following theorem on optimality.

Theorem

■ *Test for an Optimal Solution:*
Consider a simplex tableau and the basic feasible solution determined from the tableau. Suppose all entries in the last row of the tableau corresponding to non-basic variables are non-negative. (Observe that the coefficients corresponding to the basic variables are zero.) Then the z-value of the basic feasible solution is the optimal value for z and the basic feasible solution is an optimal solution for the problem. On the other hand, suppose the entry in the last row corresponding to a non-basic variable is negative. Then the basic feasible solution is not optimal, and the z-value is larger at another feasible solution. ■

Using the preceding theorem, we can now state the following:

■ *Criterion for Selecting the Entering Variable:*
Suppose we have a basic feasible solution that is not optimal, and suppose the entry in the last row of the simplex tableau for at least one-basic variable is negative. Then, to obtain a new basic solution with larger z-value, select as the entering variable the non-basic variable that corresponds to the most negative entry. If there are several such variables, pick any one of them. ■

EXAMPLE 2

Consider the simplex tableau 6.5 which is the initial simplex tableau for the problem given in example 1. Apply to this tableau the test for an optimal solution and the criterion for selecting the entering variable.

TABLEAU 6.5

	x_1	x_2	x_3	s_1	s_2	s_3	z	
s_1	3	5	2	1	0	0	0	20
s_2	0	6	3	0	1	0	0	12
s_3	11	0	−4	0	0	1	0	30
z	−4	−7	−8	0	0	0	1	0

Since there are negative entries in the last row of the tableau corresponding to non-basic variables, the initial basic feasible solution $x_1 = x_2 = x_3 = 0$, $s_1 = 20$, $s_2 = 12$, $s_3 = 30$, and $z = 0$ is not optimal. Furthermore, x_3 is the entering variable since it corresponds to the most negative of the entries.

Once we have determined an entering variable, we must also select a departing variable, one of the basic variables to make non-basic. Since we want to obtain a new basic solution that is feasible, we cannot select the departing variable arbitrarily. We now consider a method for determining algebraically the departing variable.

Consider again tableau 6.5. As determined in example 2, x_3 is the entering variable. We want to determine which one of the three basic variables $s_1, s_2,$ or s_3 should be replaced by x_3. To this end, consider the first three equations in the tableau. Keeping $x_1 = x_2 = 0$ and solving for the basic variables in terms of x_3, we have

$$s_1 = 20 - 2x_3$$

$$s_2 = 12 - 3x_3$$

$$s_3 = 30 + 4x_3.$$

Since we want to obtain a feasible solution, we must pick $x_3 \geq 0$ such that $s_1 \geq 0$, $s_2 \geq 0$, and $s_3 \geq 0$. Therefore we must pick $x_3 \geq 0$ such that

$$20 - 2x_3 \geq 0 \qquad \text{or} \qquad x_3 \leq \frac{20}{2} = 10$$

and

$$12 - 3x_3 \geq 0 \qquad \text{or} \qquad x_3 \leq \frac{12}{3} = 4$$

and

$$30 + 4x_3 \geq 0 \qquad \text{or} \qquad x_3 \geq -\frac{30}{4}.$$

These three conditions are satisfied by choosing x_3 equal to the smaller of the two non-negative ratios, that is, $x_3 = \frac{12}{3} = 4$. With this choice for x_3, the departing variable is s_2, since $s_2 = 12 - 3(4) = 0$. Observe that the departing variable s_2 is the basic variable of the equation from which we obtained the smallest non-negative ratio.

It is interesting to note that if we take $x_3 = 10$ instead, then $s_1 = 20 - 2(10) = 0$, but $s_2 = 12 - 3(10) = -18$. Therefore for this choice, we would not obtain a feasible solution. Also it is clear that we cannot take $x_3 = -\frac{30}{4}$.

Based on our discussion, we have the following:

Theorem

■ *Criterion for Selecting the Departing Variable:*
Let x_j be the entering variable and let

$$\begin{bmatrix} a_{1j} \\ a_{2j} \\ \vdots \\ a_{mj} \end{bmatrix} \quad \text{and} \quad \begin{bmatrix} b_1 \\ b_2 \\ \vdots \\ b_m \end{bmatrix}$$

be, with the entry in the z row deleted, the column corresponding to x_j and the last column of the augmented matrix, respectively. For each $a_{kj} > 0$, compute the ratio b_k / a_{kj}. The basic variable of the row from which we obtain the smallest non-negative ratio is the departing variable. If the smallest non-negative ratio is the ratio for several basic variables, pick any one of them as the departing variable. If the smallest non-negative ratio is 0, the value of x_j will be 0 in the new basic feasible solution and this solution will be degenerate. (Later in this section we will discuss the significance of a degenerate solution.) If none of the entries a_{kj} is positive, the problem has an unbounded solution. (We will also discuss this point later in the section.) ■

The entry a_{kj} of the augmented matrix in the column of the entering variable and the row of the departing variable is called the **pivot**. Observe that the pivot will always be positive. Also the new basic feasible solution is obtained by performing those row operations on the simplex tableau that convert the pivot to 1 and all entries above and below the pivot to 0. We call the necessary row operations **pivot operations**.

EXAMPLE 3

Solve the problem given in examples 1 and 2 by applying to simplex tableau 6.5 the test for an optimal solution and the criterion for selecting the entering variable and the departing variable if the solution is not optimal.

As we saw in example 2, the initial basic feasible solution

$$x_1 = x_2 = x_3 = 0, \quad s_1 = 20, \quad s_2 = 12, \quad s_3 = 30, \quad \text{and} \quad z = 0$$

is not optimal, and the entering variable is x_3. We also saw that s_2 is the departing variable. Thus the pivot is 3, which is in the column of x_3 and the row of s_2. We now perform row operations on the augmented matrix in tableau 6.5 to change the column of x_3 from

$$\begin{bmatrix} 2 \\ ③ \\ -4 \\ -8 \end{bmatrix} \quad \text{to} \quad \begin{bmatrix} 0 \\ 1 \\ 0 \\ 0 \end{bmatrix}$$

By performing the necessary pivot operations on tableau 6.5, we obtain the simplex tableau 6.6.

TABLEAU 6.6

	x_1	x_2	x_3	s_1	s_2	s_3	z	
s_1	3	5	2	1	0	0	0	20
s_2	0	6	③	0	1	0	0	12
s_3	11	0	-4	0	0	1	0	30
z	-4	-7	-8	0	0	0	1	0

$\dfrac{1}{3}R_2$
$-2R_2 + R_1$
$\xrightarrow{\hspace{2cm}}$
$4R_2 + R_3$
$8R_2 + R_4$

	x_1	x_2	x_3	s_1	s_2	s_3	z	
s_1	3	1	0	1	$-\frac{2}{3}$	0	0	12
x_3	0	2	1	0	$\frac{1}{3}$	0	0	4
s_3	11	8	0	0	$\frac{4}{3}$	1	0	46
z	-4	9	0	0	$\frac{8}{3}$	0	1	32

The new basic feasible solution is

$$x_1 = x_2 = 0, \quad x_3 = 4, \quad s_1 = 12, \quad s_2 = 0, \quad s_3 = 46, \quad \text{and} \quad z = 32.$$

However, this solution is not optimal since there is a negative entry below a variable in the last row of tableau 6.6, namely, -4 below x_1. Thus we can obtain a larger value for z by changing x_1 to a basic variable.

We find the departing variable with x_1 as the entering variable using the ratios determined from the two columns

$$\begin{matrix} s_1 \\ x_3 \\ s_3 \end{matrix} \begin{bmatrix} 3 \\ 0 \\ 11 \end{bmatrix} \quad \text{and} \quad \begin{bmatrix} 12 \\ 4 \\ 46 \end{bmatrix}.$$

Since the smallest of the non-negative ratios $\frac{12}{3} = 4$ and $\frac{46}{11} = 4\frac{2}{11}$ is in the row of s_1, the departing variable is s_1 and the pivot is 3.

Performing the necessary pivot operations on tableau 6.6 to change the column of x_1 from

$$\begin{bmatrix} ③ \\ 0 \\ 11 \\ -4 \end{bmatrix} \quad \text{to} \quad \begin{bmatrix} 1 \\ 0 \\ 0 \\ 0 \end{bmatrix}$$

we have tableau 6.7.

TABLEAU 6.7

	x_1	x_2	x_3	s_1	s_2	s_3	z	
x_1	1	$\frac{1}{3}$	0	$\frac{1}{3}$	$-\frac{2}{9}$	0	0	4
x_3	0	2	1	0	$\frac{1}{3}$	0	0	4
s_3	0	$\frac{13}{3}$	0	$-\frac{11}{3}$	$\frac{34}{9}$	1	0	2
z	0	$\frac{31}{3}$	0	$\frac{4}{3}$	$\frac{16}{9}$	0	1	48

Since the entries below the variables in the last row of tableau 6.7 are non-negative, by the test for optimality we have that the new basic feasible solution $x_1 = 4, x_2 = 0, x_3 = 4, s_1 = 0, s_2 = 0, s_3 = 2$, and $z = 48$ is optimal.

We now summarize our discussion with the steps to be performed when solving a linear programming problem written in canonical form.

The Simplex Algorithm

Step 1: Write the initial simplex tableau for the problem that is in canonical form.

Step 2: Check the entries below variables in the last row of the tableau. If each entry is non-negative, the basic feasible solution determined from the tableau is an optimal solution. Otherwise, go to step 3.

Step 3: Determine the entering variable by applying **the criterion for selecting the entering variable**. In particular, the entering variable is a non-basic variable that appears above the most non-negative entry in the last row.

Step 4: If all entries in the column of the entering variable and above the last row are non-positive, the problem has an unbounded solution. Otherwise, go to step 5.

Step 5: Determine the departing variable by applying **the criterion for selecting the departing variable**. In particular, for each positive entry a_k in the column of the entering variable and above the last row, compute the ratio b_k/a_k. The departing variable is the basic variable corresponding to the row from which we obtain the smallest non-negative ratio. If the smallest non-negative ratio is the same for several basic variables, pick any one as the departing variable.

Step 6: Derive the new simplex tableau by performing the necessary pivot operations, where the pivot is the entry in the column of the entering variable and the row of the departing variable. In particular, convert the pivot to 1 and all other entries in its column to 0's by using row operations on the augmented matrix. Relabel the row of the pivot with the name of the new basic variable; that is, replace the name of the departing variable to the left of the augmented matrix with the name of the entering variable. Return to step 2.

EXAMPLE 4

Convert the linear programming problem:

$$\text{Maximize} \quad z = 3x_1 + 8x_2 + 5x_3$$

$$\text{subject to} \quad 2x_1 \qquad + \ x_3 \leq 4$$

$$x_1 + 4x_2 + 6x_3 \leq 8$$

$$x_i \geq 0, \quad i = 1, 2, 3$$

to canonical form and solve the problem using the simplex algorithm.

We convert the problem to canonical form by introducing two slack variables s_1 and s_2. The converted problem is: Find a solution to

$$2x_1 \qquad + \ x_3 + s_1 \qquad\qquad = 4$$

$$x_1 + 4x_2 + 6x_3 \qquad + s_2 \qquad = 8$$

$$-3x_1 - 8x_2 - 5x_3 \qquad\qquad + z = 0$$

for which $x_i \geq 0, i = 1, 2, 3; s_j \geq 0, j = 1, 2;$ and z is maximum.

Tableau 6.8 are the initial simplex tableau and the sequence of tableaux derived from performing the simplex algorithm. We see from the tableaux that we pass through the steps of the algorithm two times until we arrive at an optimal solution $x_1 = 2$, $x_2 = \frac{3}{2}$, $x_3 = 0$, $s_1 = 0$, $s_2 = 0$, and $z = 18$.

TABLEAUX 6.8

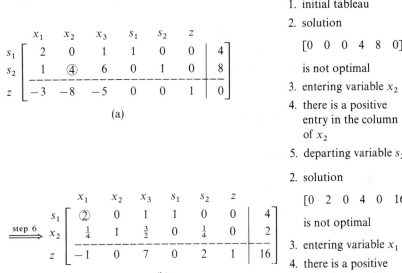

Steps of Algorithm

1. initial tableau
2. solution

$$[0 \quad 0 \quad 0 \quad 4 \quad 8 \quad 0]^t$$

   is not optimal
3. entering variable x_2
4. there is a positive entry in the column of x_2
5. departing variable s_2

2. solution

$$[0 \quad 2 \quad 0 \quad 4 \quad 0 \quad 16]^t$$

   is not optimal
3. entering variable x_1
4. there is a positive entry in the column of x_1
5. departing variable s_1

2. the solution

$$[2 \quad \tfrac{3}{2} \quad 0 \quad 0 \quad 0 \quad 18]^t$$

   is optimal

We can solve problems in two variables with greater ease if we use the simplex algorithm rather than the graphical method. With the graphical method we need to determine coordinates of vertices and evaluate the objective function at these vertices. This is also accomplished with the simplex tableau. However, we must determine which points are vertices of the convex polygonal set by graphing and determine the coordinates of all these vertices. This need not be done when applying the simplex algorithm.

EXAMPLE 5

Consider the problem in two variables:

$$\text{Maximize} \quad z = x + 2y$$

$$\text{subject to} \quad 3x + \ y \le 27$$

$$x + 2y \le 24$$

$$x \ge 0 \quad \text{and} \quad y \ge 0.$$

This problem was solved in chapter 3 (example 2, page 104) using the graphical method. Let us now solve the problem using the simplex algorithm. The problem written in canonical form is: Find a solution to

$$3x + y + s_1 \qquad\qquad = 27$$
$$x + 2y \qquad + s_2 \qquad = 24$$
$$-1x - 2y \qquad\qquad + z = 0$$

for which $x \geq 0$, $y \geq 0$, $s_1 \geq 0$, $s_2 \geq 0$, and z is maximum.

The initial simplex tableau and the tableau derived from applying the simplex algorithm are given as tableaux 6.9. We see that we need only pass through the steps of the algorithm once to obtain an optimal solution.

TABLEAUX 6.9

	x	y	s_1	s_2	z	
s_1	3	1	1	0	0	27
s_2	1	②	0	1	0	24
z	-1	-2	0	0	1	0

(a)

Steps of Algorithm

1. initial tableau
2. solution $[0\ \ 0\ \ 27\ \ 24\ \ 0]^t$ is not optimal
3. entering variable y
4. there is a positive entry in the column of y
5. departing variable s_2

	x	y	s_1	s_2	z	
s_1	$\frac{5}{2}$	0	1	$-\frac{1}{2}$	0	15
y	$\frac{1}{2}$	1	0	$\frac{1}{2}$	0	12
z	0	0	0	1	1	24

step 6 $\Longrightarrow$

(b)

2. the solution

$$[0\ \ 12\ \ 15\ \ 0\ \ 24]^t$$

is optimal

Let us examine more closely the solution $x = 0$, $y = 12$, $s_1 = 15$, $s_2 = 0$, and $z = 24$ in example 5. The non-basic variables are x and s_2, and the last row of the final tableau corresponds to the equation

$$0x + s_2 + z = 24.$$

Since the coefficient of the non-basic variable x is 0 in the equation, any other feasible solution to the other equations in the system

$$\frac{5}{2}x + s_1 - \frac{1}{2}s_2 = 15 \qquad \text{and} \qquad \frac{1}{2}x + y + \frac{1}{2}s_2 = 12$$

in which $x > 0$ and $s_2 = 0$ is also an optimal solution. Observe that for such a solution the value of z will remain at the optimal value 24. In particular, any solution for which

$$s_2 = 0, \quad x + 2y = 24, \quad s_1 = 15 - \frac{5}{2}x \geq 0, \quad x > 0, \quad y \geq 0$$

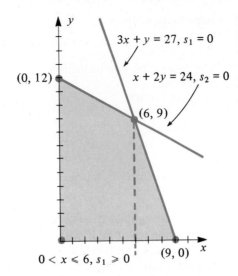

FIGURE 6.3

is an optimal solution. Compare this result to the graph in figure 6.3. These observations on the non-uniqueness of an optimal solution are true in general.

Step 4 of the simplex algorithm states that if all the entries in the column of the entering variable are non-positive, the problem has an unbounded solution. Let us see why with the following example.

EXAMPLE 6

Consider the problem:

$$\text{Maximize} \quad z = 2x + y$$

$$\text{subject to} \quad -2x + y \le 1$$

$$- x + y \le 2$$

$$x \ge 0 \quad \text{and} \quad y \ge 0.$$

This problem was solved in chapter 3 (example 4, page 106), and it was found that the problem has an unbounded solution. Let us now solve the problem using the simplex algorithm.

Converting the problem to canonical form by introducing slack variables s_1 and s_2, we have the initial simplex tableau 6.10.

TABLEAU 6.10

	x	y	s_1	s_2	z	
s_1	-2	1	1	0	0	1
s_2	-1	1	0	1	0	2
z	-2	-1	0	0	1	0

By examining the last row of tableau 6.10, we see that the entering variable is x. Keeping $y = 0$ and solving the first two equations of the system for s_1 and s_2, respectively, we have

$$s_1 = 1 + 2x \qquad \text{and} \qquad s_2 = 2 + x.$$

Therefore for any positive number A assigned to x we can obtain a feasible solution in the form

$$x = A, \quad y = 0, \quad s_1 = 1 + 2A, \quad s_2 = 2 + A$$

in which case,

$$z = 2A.$$

Therefore z can have any large value and the problem has an unbounded solution.

A final comment is in order on applying the simplex algorithm to problems written in canonical form. If, with each pass through the steps of the algorithm, the value of z increases, we will find in a finite number of passes either an optimal or an unbounded solution. However, there is the possibility that we have a basic feasible solution that is degenerate, which is a solution in which one of the basic variables is 0. In this case, if the solution is not optimal, we make another pass through the steps, but the value of z may not increase (for an example, see exercise 22). Such a situation is called **degeneracy**. Even though a problem may have an optimal solution, degeneracy raises the possibility of cycling through the steps over and over again without reaching the solution. We should be aware of this possibility, even though methods for avoiding it are beyond the scope of this text.

EXAMPLE 7

A Production Problem. A small fabricating company makes three products, A, B, and C. The products go through three phases of production—cutting, assembling, and painting. The amount of time needed in each phase to produce one unit of each item and the total amount of time available for each phase are given in table 6.1.

TABLE 6.1

	Man-hours required per unit			Man-hours available per day
	A	B	C	
cutting	$\frac{1}{12}$	$\frac{1}{15}$	$\frac{1}{9}$	19
assembling	$\frac{1}{5}$	$\frac{1}{10}$	$\frac{1}{3}$	42
painting	$\frac{1}{8}$	0	$\frac{1}{5}$	19

If the company realizes a profit of \$8 on each unit of product A, \$4 on each unit of product B, and \$13 on each unit of product C, how many units of each product should the company plan to make in a day in order to achieve the largest profit? We start our model by defining the unknowns:

$$A = \text{the number of units of product A to be made}$$

$$B = \text{the number of units of product B to be made}$$

$$C = \text{the number of units of product C to be made}.$$

The goal of the company is to obtain a production program for which the profit

$$z = 8A + 4B + 13C$$

is maximum.

For each phase let us write the time used as a function of the production levels A, B, and C. We read from the table that

$$\text{cutting time used} = \frac{1}{12}A + \frac{1}{15}B + \frac{1}{9}C$$

$$\text{assembly time used} = \frac{1}{5}A + \frac{1}{10}B + \frac{1}{3}C$$

$$\text{painting time used} = \frac{1}{8}A + 0B + \frac{1}{5}C.$$

Since each of these times must be no more than what is available for that phase, we have the constraints

$$\frac{1}{12}A + \frac{1}{15}B + \frac{1}{9}C \leq 19$$

$$\frac{1}{5}A + \frac{1}{10}B + \frac{1}{3}C \leq 42$$

$$\frac{1}{8}A + 0B + \frac{1}{5}C \leq 19.$$

We can simplify this system of linear inequalities by clearing the fractions. We multiply the first, second, and third inequalities by 180, 30, and 40, respectively. Our linear programming problem is therefore:

$$\text{Maximize} \quad z = 8A + 4B + 13C$$
$$\text{subject to} \quad 15A + 12B + 20C \leq 3420$$
$$6A + 3B + 10C \leq 1260$$
$$5A + 0B + 8C \leq 760$$
$$A \geq 0 \quad B \geq 0 \quad \text{and} \quad C \geq 0.$$

Now that the problem has been formulated, we can obtain the solution. This can be done with pencil and paper once the problem is written in the canonical form: Find a solution to

$$15A + 12B + 20C + s_1 \qquad\qquad = 3420$$
$$6A + 3B + 10C \qquad + s_2 \qquad = 1260$$
$$5A \qquad + 8C \qquad\quad + s_3 \quad = 760$$
$$-8A - 4B - 13C \qquad\qquad + z = 0$$

for which $A \geq 0$, $B \geq 0$, $C \geq 0$; $s_i \geq 0$, $i = 1, 2, 3$; and z is maximum.

Now applying the simplex algorithm we will find the optimal solution $A = 80$, $B = 110$, $C = 45$, and $z = 1665$. You are asked in exercises 6.2 to derive the solution to this problem.

EXERCISES 6.2

Write each of the problems in exercises 1 and 2 in canonical form and give the initial simplex tableau.

1. Maximize $\quad z = 3x + 2y$
   subject to $\quad 5x + 3y \le 41$
   $$x + y \le 11$$
   $$2x + y \le 18$$
   $$x \ge 0 \quad \text{and} \quad y \ge 0.$$

2. Maximize $\quad z = 12x_1 + 7x_2 + 4x_3$
   subject to $\quad 1.5x_1 + x_2 + 0.25x_3 \le 10$
   $$1.25x_2 + 0.75x_3 \le 20$$
   $$x_1 \ge 0, \quad x_2 \ge 0, \quad \text{and} \quad x_3 \ge 0.$$

For exercises 3–6 determine from the simplex tableau the basic feasible solution and whether or not it is optimal. If the solution is not optimal, determine the entering variable, the departing variable, and the new basic feasible solution.

3.

	x_1	x_2	s_1	s_2	z	
s_1	1	3	1	0	0	24
s_2	4	3	0	1	0	42
z	-2	-4	0	0	1	0

4.

	x_1	x_2	x_3	s_1	s_2	z	
s_1	-2	1	1	1	0	0	5
s_2	6	-2	3	0	1	0	21
z	-5	-5	-3	0	0	1	0

5.

	x_1	x_2	x_3	x_4	s_1	s_2	s_3	z	
s_1	3	0	-2	0	1	1	-1	0	9
x_4	-4	0	2	1	0	1	2	0	17
x_2	5	1	-1	0	0	1	1	0	15
z	-7	0	-6	0	0	5	8	1	35

6.

	x_1	x_2	x_3	s_1	s_2	s_3	z	
x_3	0	-2	1	7	0	-3	0	12
s_2	0	0	0	-2	1	4	0	9
x_1	1	-3	0	-1	0	5	0	10
z	0	-5	0	3	0	2	1	21

The problems in exercises 7 and 8 are in canonical form. For each of these problems, give the initial simplex tableau, the initial basic feasible solution, and then solve the problem.

7. Find a solution to

$$5x_1 \qquad + 2x_3 + s_1 \qquad = 21$$
$$2x_1 + 3x_2 \qquad + s_2 \qquad = 11$$
$$-7x_1 - 10x_2 - 8x_3 \qquad + z = 0$$

for which $x_i \geq 0$, $i = 1, 2, 3$; $s_j \geq 0$, $j = 1, 2$; and z is maximum.

8. Find a solution to

$$\frac{1}{2}x_1 - x_2 + \frac{1}{3}x_3 + s_1 \qquad\qquad = 15$$
$$\frac{5}{3}x_1 \qquad -\frac{1}{4}x_3 \quad + s_2 \qquad = 7$$
$$x_1 - 2x_2 + 3x_3 \qquad + s_3 \quad = 8$$
$$-11x_1 - 9x_2 - 7x_3 \qquad\qquad + z = 0$$

for which $x_i \geq 0$, $i = 1, 2, 3$; $s_j \geq 0$, $j = 1, 2, 3$; and z is maximum.

In exercises 9–20 solve each linear programming problem using the simplex method.

9. Maximize $z = x + 2y$
   subject to $3x + 5y \leq 45$
   $6x + 5y \leq 60$
   $x \geq 0$ and $y \geq 0$.

10. Maximize $z = 3x + 2y$
    subject to $x + 2y \geq 20$
    $8x + 7y \leq 88$
    $x \geq 0$ and $y \geq 0$.

11. Maximize $z = 2x + 3y$
    subject to $x + 2y \leq 18$
    $x + y \leq 11$
    $5x - 2y \leq 20$
    $x \geq 0$ and $y \geq 0$.

12. Maximize $z = 5x + 4y$
    subject to $-x + 2y \leq 6$
    $x + 3y \leq 14$
    $3x + 2y \leq 21$
    $x \geq 0$ and $y \geq 0$.

13. Maximize $z = 10x_1 + 8x_2 + 4x_3$
    subject to $x_1 + 3x_2 + 4x_3 \leq 100$
    $2x_1 + x_2 + 2x_3 \leq 100$
    $x_1 \geq 0$, $x_2 \geq 0$,   and $x_3 \geq 0$.

14. Maximize $z = 2x_1 + 3x_2 + x_3$
    subject to $2x_1 + x_2 + 3x_2 \leq 10$
    $-x_1 + 2x_2 + x_3 \leq 12$
    $x_1 \geq 0$, $x_2 \geq 0$,   and $x_3 \geq 0$.

15. Maximize $z = x_1 + 3x_2 + 2x_3$
    subject to $2x_1 + 4x_2 - x_3 \leq 2$
    $2x_1 + 5x_2 + x_3 \leq 6$
    $x_1 \geq 0$, $x_2 \geq 0$,   and $x_3 \geq 0$.

16. Maximize $z = 2x_1 + 10x_2 + 6x_3$
    subject to $2x_1 + 3x_2 + 6x_3 \leq 30$
    $3x_1 + 5x_2 + x_3 \leq 60$
    $5x_1 + 5x_2 + x_3 \leq 10$
    $x_1 \geq 0$, $x_2 \geq 0$,   and $x_3 \geq 0$.

17. Maximize $\quad z = x_1 + x_2 + 4x_3 + 6x_4$
    subject to $\quad x_1 + 3x_2 + 2x_3 + \;\; x_4 \le 12$
    $\qquad\qquad\;\; 4x_1 + 2x_2 + 3x_3 + 2x_4 \le 15$
    $\qquad\qquad\;\; x_i \ge 0, \quad i = 1, 2, 3, 4.$

18. Maximize $\quad z = x_1 + x_2 + 5x_3 + 3x_4$
    subject to $\quad -2x_1 + \;\; x_2 + 2x_3 + 3x_4 \le 11$
    $\qquad\qquad\quad x_1 + 2x_2 \qquad\quad + \;\; x_4 \le 12$
    $\qquad\qquad\;\; 2x_1 - \;\; x_2 + \;\; x_3 + 2x_4 \le 4$
    $\qquad\qquad\;\; x_i \ge 0, \quad i = 1, 2, 3, 4.$

19. Maximize $\quad z = x_1 + 2x_2 + 3x_3 + 6x_4 + 3x_5$
    subject to $\quad x_1 + 2x_2 - 2x_3 + 3x_4 + \;\; x_5 \le 12$
    $\qquad\qquad\;\; 2x_1 + 3x_2 + 3x_3 + 4x_4 + 4x_5 \le 33$
    $\qquad\qquad\;\; x_i \ge 0, \quad i = 1, 2, 3, 4, 5.$

20. Maximize $\quad z = 4x_1 + 3x_2 + \;\; x_3 + 2x_4 + 6x_5$
    subject to $\quad x_1 + 2x_2 - \;\; x_3 - 4x_4 \qquad\qquad \le 12$

    $\qquad\qquad\;\; 2x_1 - \dfrac{1}{2}x_2 + 2x_3 + 2x_4 + \;\; x_5 \le 8$

    $\qquad\qquad\;\; 3x_1 - 2x_2 + \;\; x_3 + 7x_4 + 2x_5 \le 18$
    $\qquad\qquad\;\; x_i \ge 0, \quad i = 1, 2, 3, 4, 5.$

21. Consider the problem:

    Maximize $\quad z = 4x + 5y$
    subject to $\quad -2x + \;\; y \le 5$
    $\qquad\qquad\;\; -\;\; x + 2y \le 16$
    $\qquad\qquad\;\; x \ge 0 \qquad$ and $\qquad y \ge 0.$

    Show that the problem has an unbounded solution by
    a. The graphical method
    b. The simplex algorithm.

22. Consider the simplex tableau

	x	y	s_1	s_2	z	
s_1	1	0	1	−3	0	0
y	2	1	0	−5	0	4
z	−2	0	0	3	1	48

From the tableau we read the basic feasible solution $x = 0$, $y = 4$, $s_1 = 0$, $s_2 = 0$, and $z = 48$. Observe that this solution is degenerate since the basic variables $s_1 = 0$. Also the solution is not optimal. Determine the new basic feasible solution and observe that the z-value in this new solution remains at the previous value 48.

For exercises 23–30 model each as a linear programming problem. Make sure to state what each variable in the problem represents. Then solve the problem using the simplex algorithm.

23. The Spangle Confection Company makes three kinds of lollipops; Licorice Delight, Centre Surprise, and All-Day Sucker. The three lollipops are produced by two machines, I and II. The amount of time, in hours, needed on each machine to make one dozen of each kind of lollipop is given in table 6.2.

TABLE 6.2

	Licorice Delight	Centre Surprise	All-Day Sucker
machine I	0.1	0.1	0.2
machine II	0.4	0.2	0.2

Machine I is operated at most 10 hours a day and machine II is operated at most 12 hours a day. If the profits are $2 per dozen Licorice Delights, $3 per dozen Centre Surprises, and $4 per dozen All-Day Suckers, how many dozen of each kind of lollipop should be made each day so as to maximize the daily total profit? What is the total maximum daily profit?

24. Woodhaven Development Corporation has a 240-acre tract of land that it plans to subdivide into 1-acre, 2-acre, and 3-acre building lots. It costs Woodhaven $1500, $2500, and $3000 to develop a 1-acre lot, a 2-acre lot, and a 3-acre lot, respectively. The corporation has available $300,000 to cover these development costs. Woodhaven anticipates the following profit on the sale of each lot: $600 for each 1-acre lot, $1000 for each 2-acre lot, and $1500 for each 3-acre lot. How many lots of each size should be allotted so that Woodhaven will realize the maximum profit? What is the maximum profit?

25. Starmont Products manufactures three video games: Zapman, Snakepit, and Monkey Kong. The manufacturing of each game involves three machines: I, II, and III. The amount of time, in hours, needed on each machine to make a carton of each game is given in table 6.3.

TABLE 6.3

Machine	Zapman	Snakepit	Monkey Kong
I	$\frac{3}{4}$	1	$2\frac{1}{4}$
II	$\frac{1}{2}$	1	1
III	1	$\frac{1}{2}$	$\frac{1}{4}$

Each day, machines I, II, and III can be used at most 10, 8, and 7 hours, respectively. Starmont makes a profit of $100, $120, and $110 on each carton of Zapman, Snakepit, and Monkey Kong, respectively. How many cartons of each game should be made each day to maximize the total profit? What is the maximum total profit?

26. Refer to exercise 24. Suppose in addition to the limitations on the amount of land available and the amount of money available to cover the development costs, Woodhaven wants the number of 1-acre lots to be less than or equal to the number of 2- and 3-acre lots combined. With this additional constraint, how many lots of each size should now be planned to maximize the profit? What is the maximum profit?

27. The Grendal Corporation makes three kinds of painkillers: Extra-Strength, Regular, and Placebo Plus. Each type requires three machines in its manufacture. Each batch of Extra-Strength requires $\frac{1}{2}$ hour on machine A, $\frac{1}{2}$ hour on machine B, and $\frac{3}{2}$ hours on machine C. Each batch of Regular requires 1 hour on machine A, 1 hour on machine B, and $\frac{1}{2}$ hour on machine C. Each batch of Placebo Plus requires $\frac{1}{2}$ hour on machine A, $\frac{1}{4}$ hour on machine B, and 1 hour on machine C. Machines A, B, and C are available at most $6\frac{1}{2}$, 6, and 5 hours a day, respectively. If the corporation makes $2000, $3000, and $1500 on each batch of Extra-Strength, Regular, and Placebo Plus, respectively, how many batches of

each kind of painkiller should be made each day to yield the largest profit? What is the largest profit?

28. Lyn has up to $50,000 available to invest in three different types of securities: one that yields 8% annually, another that yields 10% annually, and a third that yields 13% annually. Her goal is to achieve the largest possible annual income; however, she has restrictions on how the money is to be invested. The amounts invested at 10% and 13% combined can be no more than $25,000, and the amount invested at 13% can be no more than the amounts invested at 8% and 10% combined. How much should Lyn invest in each type of investment to maximize the annual income? What is the maximum annual income?

29. The Imperial Cigar Company makes four brands of cigars: The Lady Jane, El Macho, Slim, and Mascot. Each brand requires four machines in its production: blender, chopper, liquefier, and binder. The number of hours required on each machine for a case of each brand, the hours available on each machine per day, and the profit per case (in dollars) of each brand are given in table 6.4.

TABLE 6.4

	Lady Jane	El Macho	Slim	Mascot	Hours Available
Blender	$\frac{1}{2}$	2	1	3	17
Chopper	$\frac{1}{4}$	1	1	1	9
Liquefier	$\frac{1}{2}$	1	2	2	19
Binder	1	$\frac{1}{2}$	1	2	12
Profit per case	30	30	40	10	

How many cases of each brand should be made each day to maximize the total profit? What is the maximum profit?

30. Saddlebrook Farms has 400 acres available for planting wheat, corn, soybeans, alfalfa, or a combination of them. It costs the farm $90 per acre to plant wheat, $120 per acre to plant corn, $100 per acre to plant soybeans, and $70 per acre to plant alfalfa. The farm has $60,000 available to cover its planting costs. Saddlebrook Farms wants to plant no more than 150 acres in corn or alfalfa or a combination of them. The farm anticipates income from the crops as follows: $300 per acre of wheat, $450 per acre of corn, $400 per acre of soybeans, and $250 per acre of alfalfa. How many acres of each crop should the farm plant to achieve the greatest income? How much is this maximum income?

6.3 Standard Minimization Problems: Duality

Knowing how to solve standard maximization problems, we proceed in this section to the solution of standard minimization problems.

A standard minimization problem is a linear programming problem in which an objective function with non-negative coefficients is to be minimized over a set of feasible solutions. A solution $(x_1, x_2, \ldots, x_n)$ is feasible in this case if all of its entries are non-negative and satisfy inequality constraints of the form

$$a_{i1}x_1 + a_{i2}x_2 + \cdots + a_{in}x_n \geq b_i.$$

Observe that we do not require that the right-hand sides of the constraints be non-negative, but we do require that the coefficients of the objective function be non-negative. An example of a standard minimization linear programming problem is as follows:

$$\begin{aligned}
\text{Minimize} \quad & Z = 4x_1 + 8x_2 \\
\text{subject to} \quad & 2x_1 + x_2 \geq 3 \\
& \qquad\quad 4x_2 \geq 8 \\
& \quad x_1 + x_2 \geq 5 \\
& x_1 \geq 0 \quad \text{and} \quad x_2 \geq 0.
\end{aligned}$$

There is a very interesting relationship, called **duality**, between standard minimization problems and standard maximization problems. For every minimization problem there is a corresponding maximization problem such that by solving the latter problem we can find the solution to the former problem. Thus our method for solving a standard minimization problem, called a **primal** problem, is first to obtain the corresponding standard maximization problem, called the **dual** of the primal problem. Next, we solve the dual problem by the simplex algorithm of section 6.2. Finally, upon deriving the final simplex tableau for the dual problem, we can determine from this tableau the solution to the primal problem. With this strategy for solving a standard minimization problem, our first goal is to see how to write its dual problem.

We begin by introducing convenient matrix notation for writing a linear programming problem. By the **matrix inequality**

$$A \geq B$$

we mean that the two matrices A and B are of the same size and each entry of A is greater than or equal to the corresponding entry of B. In particular, if $A = [a_{ij}]$ and $B = [b_{ij}]$ are two $m \times n$ matrices, then $A \geq B$ provided that $a_{ij} \geq b_{ij}$ for every i and j. A similar meaning can be given for the inequality $A \leq B$. For example

$$\begin{bmatrix} 9 & 5 & 3 \\ 2 & 4 & 7 \end{bmatrix} \geq \begin{bmatrix} 6 & 1 & 3 \\ 0 & 4 & 5 \end{bmatrix}$$

whereas the matrices

$$A = \begin{bmatrix} 9 & 5 & 3 \\ 2 & 4 & 7 \end{bmatrix} \quad \text{and} \quad B = \begin{bmatrix} 6 & 7 & 3 \\ 0 & 4 & 5 \end{bmatrix}$$

cannot be compared—neither $A \geq B$ nor $A \leq B$.

1. Rewrite, using matrix notation, the linear programming problem

$$\text{Minimize} \quad Z = 4x_1 + 8x_2$$

$$\text{subject to} \quad 2x_1 + x_2 \geq 3$$

$$4x_2 \geq 8$$

$$x_1 + 6x_2 \geq 5$$

$$x_1 \geq 0 \quad \text{and} \quad x_2 \geq 0.$$

First, let

$$X = \begin{bmatrix} x_1 \\ x_2 \end{bmatrix}.$$

As in using matrix multiplication for writing systems of linear equations, by letting

$$A = \begin{bmatrix} 2 & 1 \\ 0 & 4 \\ 1 & 6 \end{bmatrix}, \quad B = \begin{bmatrix} 3 \\ 8 \\ 5 \end{bmatrix}, \quad \text{and} \quad 0 = \begin{bmatrix} 0 \\ 0 \end{bmatrix},$$

we can write the linear inequalities as

$$AX \geq B$$

$$X \geq 0.$$

Finally, we see that with $C = \begin{bmatrix} 4 & 8 \end{bmatrix}$

$$Z = CX.$$

Therefore our linear programming problem can be given as

$$\text{Minimize} \quad Z = CX$$

$$\text{subject to} \quad AX \geq B$$

$$X \geq 0.$$

2. Let

$$A' = \begin{bmatrix} 2 & 0 & 1 \\ 1 & 4 & 6 \end{bmatrix}, \quad C' = \begin{bmatrix} 4 \\ 8 \end{bmatrix}, \quad B' = \begin{bmatrix} 3 & 8 & 5 \end{bmatrix}, \quad \text{and} \quad Y = \begin{bmatrix} y_1 \\ y_2 \\ y_3 \end{bmatrix}.$$

Rewrite the standard maximization problem

$$\text{Maximize} \quad z = B'Y$$

$$\text{subject to} \quad A'Y \leq C'$$

$$Y \geq 0$$

without matrices.

Performing the matrix multiplication and using the meaning of matrix inequality, we have the linear programming problem

$$\text{Maximize} \quad z = 3y_1 + 8y_2 + 5y_3$$

$$\text{subject to} \quad 2y_1 \quad + \; y_2 \leq 4$$

$$y_1 + 4y_2 + 6y_3 \leq 8$$

$$y_1 \geq 0, \quad y_2 \geq 0, \quad \text{and} \quad y_3 \geq 0.$$

It so happens that the dual of the minimization problem given in example 1 is the maximization problem given in example 2. To see how to write the dual problem, observe that the coefficient matrix A' of the linear constraints of the dual problem is the transpose of the coefficient matrix A of the linear constraints of the primal problem; that is,

$$A' = A^t.$$

Also the relationships between the other matrices B' and C' in the dual problem and the matrices B and C in the primal problem are

$$B' = B^t \qquad \text{and} \qquad C' = C^t;$$

that is, the coefficients of the objective function in the dual are the right-hand sides of the constraints in the primal and the right-hand sides of the constraints in the dual are the coefficients of the objective function in the primal problem. Finally, the number of unknowns in the dual problem is the same as the number of constraints in the primal problem. We now state these observations in the following definition.

Definition ■ Assume a standard minimization problem

$$\text{Minimize} \quad Z = CX, \quad C \geq 0$$

$$\text{subject to} \quad AX \geq B$$

$$X \geq 0.$$

The dual of this primal problem is the standard maximization problem

$$\text{Maximize} \quad z = B^t Y$$

$$\text{subject to} \quad A^t Y \leq C^t$$

$$Y \geq 0. \quad ■$$

EXAMPLE 3 Write the dual problem for

$$\text{Minimize} \quad Z = 16x_1 + 24x_2 + 20x_3$$

$$\text{subject to} \quad 4x_1 + 3x_2 + 2x_3 \geq 72$$

$$x_1 + 2x_2 + 2x_3 \geq 36$$

$$5x_2 + 4x_3 \geq 68$$

$$x_i \geq 0, \quad i = 1, 2, 3$$

in matrix form.

The matrix form of the primal problem is

$$\text{Minimize} \quad Z = CX$$

$$\text{subject to} \quad AX \geq B$$

$$X \geq 0$$

where

$$X = \begin{bmatrix} x_1 \\ x_2 \\ x_3 \end{bmatrix}, \quad A = \begin{bmatrix} 4 & 3 & 2 \\ 1 & 2 & 2 \\ 0 & 5 & 4 \end{bmatrix}, \quad B = \begin{bmatrix} 72 \\ 36 \\ 68 \end{bmatrix}, \quad \text{and} \quad C = [16 \quad 24 \quad 20].$$

Since there are three constraints in the primal problem, the dual problem has three unknowns; thus, let

$$Y = \begin{bmatrix} y_1 \\ y_2 \\ y_3 \end{bmatrix}.$$

Finally, since

$$A^t = \begin{bmatrix} 4 & 1 & 0 \\ 3 & 2 & 5 \\ 2 & 2 & 4 \end{bmatrix}, \quad C^t = \begin{bmatrix} 16 \\ 24 \\ 20 \end{bmatrix}, \quad \text{and} \quad B^t = [72 \quad 36 \quad 68],$$

the dual problem, in matrix form, is

$$\text{Maximize} \quad z = [72 \quad 36 \quad 68] \begin{bmatrix} y_1 \\ y_2 \\ y_3 \end{bmatrix}$$

$$\text{subject to} \quad \begin{bmatrix} 4 & 1 & 0 \\ 3 & 2 & 5 \\ 2 & 2 & 4 \end{bmatrix} \begin{bmatrix} y_1 \\ y_2 \\ y_3 \end{bmatrix} \geq \begin{bmatrix} 16 \\ 24 \\ 20 \end{bmatrix}$$

$$Y \geq 0.$$

Knowing how to form the dual problem, we will now see how it can be used to find the solution to the primal problem. Consider again the dual problem of example 2 with its initial simplex tableau given as tableau 6.11.

TABLEAU 6.11

	y_1	y_2	y_3	s_1	s_2	z	
s_1	2	0	1	1	0	0	4
s_2	1	4	6	0	1	0	8
z	-3	-8	-5	0	0	1	0

Recall that the optimality test of the simplex algorithm tells us that a solution has been reached when all the entries in the last row of the tableau below the variables are non-negative. Therefore the simplex algorithm can be restated as:

Perform row operations on the simplex tableau in order to change the last row to a row with non-negative entries below the variables.

Observe how the last row changes while applying the simplex algorithm. In particular, performing pivot operations on the simplex tableau 6.11, we have the sequence of tableau in tableau 6.12.

TABLEAU 6.12

$$
\begin{array}{c}
\begin{array}{ccccccc}
\quad & y_1 & y_2 & y_3 & s_1 & s_2 & z \\
\end{array} \\
\begin{array}{c}
s_1 \\
s_2 \\
z
\end{array}
\left[
\begin{array}{cccccc|c}
2 & 0 & 1 & 1 & 0 & 0 & 4 \\
1 & ④ & 6 & 0 & 1 & 0 & 8 \\
\hline
-3 & -8 & -5 & 0 & 0 & 1 & 0
\end{array}
\right]
\end{array}
$$

$$
\xrightarrow{2R_2 + R_3}
\begin{array}{c}
\begin{array}{ccccccc}
\quad & y_1 & y_2 & y_3 & s_1 & s_2 & z \\
\end{array} \\
\begin{array}{c}
s \\
s^2 \\
z
\end{array}
\left[
\begin{array}{cccccc|c}
② & 0 & 1 & 1 & 0 & 0 & 4 \\
\frac{1}{4} & 1 & \frac{3}{2} & 0 & \frac{1}{4} & 0 & 2 \\
\hline
-1 & 0 & 7 & 0 & 2 & 1 & 16
\end{array}
\right]
\end{array}
$$

$$
\xrightarrow{\frac{1}{2}R_1 + R_3}
\begin{array}{c}
\begin{array}{ccccccc}
\quad & y_1 & y_2 & y_3 & s_1 & s_2 & z \\
\end{array} \\
\begin{array}{c}
y_1 \\
y_2 \\
z
\end{array}
\left[
\begin{array}{cccccc|c}
1 & 0 & \frac{1}{2} & \frac{1}{2} & 0 & 0 & 2 \\
0 & 1 & \frac{11}{8} & -\frac{1}{8} & \frac{1}{4} & 0 & \frac{3}{2} \\
\hline
0 & 0 & \frac{15}{2} & \frac{1}{2} & 2 & 1 & 18
\end{array}
\right]
\end{array}
$$

We see in tableaux 6.12 that the last row of the final tableau is derived by

1. adding $\frac{1}{2}$ times row 1 to row 3, and
2. adding 2 times row 2 to row 3.

Observe that these scalars, $\frac{1}{2}$ and 2, also appear in the last row below the slack variables. In particular, $\frac{1}{2}$ appears below s_1, the slack variable of row 1, and 2 appears below s_2, the slack variable of row 2. Furthermore, $(\frac{1}{2}, 2)$ is a feasible solution to the primal problem, which can be seen in the following way.

Let $r_1 = \frac{1}{2}$ and $r_2 = 2$. Using scalar multiplication and matrix addition, the sequence of changes to row three

$$
(2R_2 + R_3) + \frac{1}{2}R_1 = \frac{1}{2}R_1 + 2R_2 + R_3
$$

can be written as

$$
\begin{array}{rcccccccc}
r_1 [& 2 & 0 & 1 & 1 & 0 & 0 & 4 &] \\
+ r_2 [& 1 & 4 & 6 & 0 & 1 & 0 & 8 &] \\
+ \ [& -3 & -8 & -5 & 0 & 0 & 1 & 0 &] \\
= & [2r_1 + r_2 - 3 & 4r_2 - 8 & r_1 + 6r_2 - 5 & r_1 & r_2 & 1 & 4r_1 + 8r_2]. &
\end{array}
\tag{1}
$$

Therefore an optimal solution to the dual problem will be obtained when the row multipliers r_1 and r_2 are such that

$$2r_1 + r_2 \geq 3$$

$$4r_2 \geq 8$$

$$r_1 + 6r_2 \geq 5$$

$$r_1 \geq 0 \quad \text{and} \quad r_2 \geq 0.$$

Recalling the constraints of the primal problem, we are saying that $(r_1, r_2) = \left(\frac{1}{2}, 2\right)$ is a feasible solution to that problem.

Let us now look at the last entry of (1), namely,

$$4r_1 + 8r_2.$$

From the optimality criterion, this is the optimal value for the dual problem when each of the other entries in (1) is non-negative. But observe that $4r_1 + 8r_2$ is the value of the objective function in the primal problem. It can be shown, even though we will not be able to do it here, that the optimal values of the dual and primal problems are equal. Therefore, by using the simplex algorithm to solve the dual problem, we will also derive the solution to the primal problem.

Let us now summarize our discussion with the following theorem and procedure for solving a standard minimization problem. First, given a standard minimization problem

$$\begin{aligned} \text{Minimize} \quad & Z = CX, \quad \text{where} \quad C \geq 0 \\ \text{subject to} \quad & AX \geq B \\ & X \geq 0, \end{aligned} \qquad (2)$$

its dual problem is

$$\begin{aligned} \text{Maximize} \quad & z = B^t Y \\ \text{subject to} \quad & A^t Y \leq C^t \\ & Y \geq 0. \end{aligned} \qquad (3)$$

Theorem

■ *Duality:*

If the dual problem (3) for a given primal problem (2) has a finite optimal solution, the primal problem has an optimal solution. Furthermore the optimal value for the primal problem is equal to the optimal value for the dual problem. Also, if the dual problem has an unbounded solution, the primal problem has no feasible solutions let alone an optimal one. ■

Procedure For Solving A Standard Minimization Problem

Step 1: Given a standard minimization problem (2), write the initial simplex tableau for the dual problem (3).

Step 2: Using the simplex algorithm of section 6.2, solve the dual problem with the simplex tableau obtained in step 1.

Step 3: If we obtain a solution in step 2 to the dual problem, the primal problem has a solution and its optimal value is equal to the optimal value of the dual problem. Furthermore the optimal solution for the primal problem appears below the slack variables in the last row of the final tableau derived in step 2. If, on the other hand, we find in step 2 that the dual problem has an unbounded solution, the primal problem does not have a solution and, in fact, has no feasible solutions.

EXAMPLE 4

Solve the standard minimization problem

$$\text{Minimize} \quad Z = 32x_1 + 48x_2 + 11x_3$$

$$\text{subject to} \quad x_1 + 4x_2 \qquad \geq 1$$

$$4x_1 + 2x_2 + x_3 \geq 2$$

$$2x_1 \qquad + x_3 \geq 6$$

$$x_1 \geq 0, \quad x_2 \geq 0, \quad \text{and} \quad x_3 \geq 0$$

by solving its dual problem.

The given problem in matrix form is

$$\text{Minimize} \quad Z = \begin{bmatrix} 32 & 48 & 11 \end{bmatrix} \begin{bmatrix} x_1 \\ x_2 \\ x_3 \end{bmatrix}$$

$$\text{subject to} \quad \begin{bmatrix} 1 & 4 & 0 \\ 4 & 2 & 1 \\ 2 & 0 & 1 \end{bmatrix} \begin{bmatrix} x_1 \\ x_2 \\ x_3 \end{bmatrix} \geq \begin{bmatrix} 1 \\ 2 \\ 6 \end{bmatrix}$$

$$X \geq 0.$$

The dual problem in matrix form is, therefore

$$\text{Maximize} \quad z = \begin{bmatrix} 1 & 2 & 6 \end{bmatrix} \begin{bmatrix} y_1 \\ y_2 \\ y_3 \end{bmatrix}$$

$$\text{subject to} \quad \begin{bmatrix} 1 & 4 & 2 \\ 4 & 2 & 0 \\ 0 & 1 & 1 \end{bmatrix} \begin{bmatrix} y_1 \\ y_2 \\ y_3 \end{bmatrix} \leq \begin{bmatrix} 32 \\ 48 \\ 11 \end{bmatrix}$$

$$Y \geq 0.$$

Applying the simplex algorithm to the initial tableau, we obtain the sequence of tableaux in tableaux 6.13.

We see from the last row of the final tableau that the dual problem has an optimal solution, and therefore the primal problem also has an optimal solution.

Furthermore the optimal value of either problem is 76. Finally, the optimal solution to the primal problem is found in the last row under the slack variables. Specifically, we have the optimal solution to the primal problem

$$x_1 = 1, \qquad x_2 = 0, \qquad \text{and} \qquad x_3 = 4.$$

TABLEAU 6.13

	y_1	y_2	y_3	s_1	s_2	s_3	z	
s_1	1	4	2	1	0	0	0	32
s_2	4	2	0	0	1	0	0	48
s_3	0	1	1	0	0	1	0	11
z	-1	-2	-6	0	0	0	1	0

$\xrightarrow{\begin{array}{c}-2R_3+R_1\\6R_3+R_4\end{array}}$

	y_1	y_2	y_3	s_1	s_2	s_3	z	
s_1	1	2	0	1	0	-2	0	10
s_2	4	2	0	0	1	0	0	48
y_3	0	1	1	0	0	1	0	11
z	-1	4	0	0	0	6	1	66

$\xrightarrow{\begin{array}{c}-4R_1+R_2\\R_1+R_4\end{array}}$

	y_1	y_2	y_3	s_1	s_2	s_3	z	
y_1	1	2	0	1	0	-2	0	10
s_2	0	-6	0	-4	1	8	0	8
y_3	0	1	1	0	0	1	0	11
z	0	6	0	1	0	4	1	76

With practice, we find it easier to write the initial simplex tableau of the dual problem directly from the primal problem. We only need to keep in mind the following points. The right-hand sides in the primal problem become the coefficients in the objective function of the dual problem; the coefficients in the objective function of the primal problem become the right-hand sides in the dual problem; and the transpose of the coefficient matrix of the system of linear constraints in the primal problem is the coefficient matrix in the dual.

EXAMPLE 5

Solve the linear programming problem

$$\text{Minimize} \quad Z = x_1 + x_2 + x_3$$
$$\text{subject to} \quad 2x_1 + 3x_2 - x_3 \geq 5$$
$$x_1 + \frac{3}{2}x_2 - 2x_3 \geq 12$$
$$x_1 \geq 0, \qquad x_2 \geq 0, \qquad \text{and} \qquad x_3 \geq 0.$$

We write the initial simplex tableau for the dual problem directly from this primal problem as tableau 6.14.

TABLEAU 6.14

$$\begin{array}{c|cccccc|c} & y_1 & y_2 & s_1 & s_2 & s_3 & z & \\ \hline s_1 & 2 & 1 & 1 & 0 & 0 & 0 & 1 \\ s_2 & 3 & \frac{3}{2} & 0 & 1 & 0 & 0 & 1 \\ s_3 & -1 & -2 & 0 & 0 & 1 & 0 & 1 \\ \hline z & -5 & -12 & 0 & 0 & 0 & 1 & 0 \end{array}$$

One pass through the simplex algorithm is sufficient to obtain tableau 6.15 and an optimal solution. Reading from tableau 6.15, we have the optimal solution to the primal problem $x_1 = 0$, $x_2 = 8$, and $x_3 = 0$ and the optimal value 8.

TABLEAU 6.15

$$\begin{array}{c|cccccc|c} & y_1 & y_2 & s_1 & s_2 & s_3 & z & \\ \hline s_1 & 0 & 0 & 1 & -\frac{2}{3} & 0 & 0 & \frac{1}{3} \\ y_2 & 2 & 1 & 0 & \frac{2}{3} & 0 & 0 & \frac{2}{3} \\ s_3 & 3 & 0 & 0 & \frac{4}{3} & 1 & 0 & \frac{7}{3} \\ \hline z & 19 & 0 & 0 & 8 & 0 & 1 & 8 \end{array}$$

As stated in the procedure for solving a standard minimization problem, if the dual problem has an unbounded solution, the primal problem has no solution. This is the case in the next example.

EXAMPLE 6

Solve the linear programming problem

$$\text{Minimize} \quad Z = 3x + y$$
$$\text{subject to} \quad -2x + y \geq -2$$
$$-x - y \geq \quad 2$$
$$x \geq 0 \quad \text{and} \quad y \geq 0.$$

Note that this is a standard minimization problem even with the right-hand side of the first constraint being negative. The initial simplex tableau for the dual problem is given as tableau 6.16.

TABLEAU 6.16

$$\begin{array}{c|cccc|c} & x' & y' & s_1 & s_2 & z & \\ \hline s_1 & -2 & -1 & 1 & 0 & 0 & 3 \\ s_2 & 1 & -1 & 0 & 1 & 0 & 1 \\ \hline z & 2 & -2 & 0 & 0 & 1 & 0 \end{array}$$

Considering the column y' in tableau 6.16, we immediately conclude that the dual problem has an unbounded solution. Therefore the primal problem has no solution. Comparing this result to the graph of the system of inequalities

$$-2x + y \geq -2$$
$$-x - y \geq \quad 2$$

as given in figure 6.4, we see that the set of feasible solutions for the primal problem is empty.

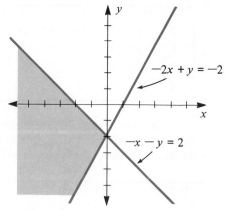

FIGURE 6.4

Graph of the system $-2x + y \geqslant -2$ and $-x - y \geqslant 2$

We conclude this section with two applications that give rise to solving standard minimization problems.

7. *A Diet Problem.* The daily diet for a pregnant woman requires at least 12 units of iron, 15 units of protein, and 8 units of calcium. Suppose a mixture of three types of food is used to obtain a diet that meets these requirements. The amounts of iron, protein, and calcium supplied by a unit of each type of food are given in table 6.5.

TABLE 6.5

	Supplied by one unit of			*Minimum*
	A	*B*	*C*	*requirement*
iron	4	2	2	12
protein	4	2	3	15
calcium	2	4	1	8

If type A food costs $1.00 per unit, type B food costs $0.50 per unit, and type C food costs $0.75 per unit, then how many units of each type of food should be used in the daily diet so that the cost is minimal?

We see that the problem is to determine three numbers, represented as

$$x_1 = \text{units of type A food used in the diet}$$

$$x_2 = \text{units of type B food used in the diet}$$

$$x_3 = \text{units of type C food used in the diet.}$$

The objective is to minimize the cost of the diet, which we represent as

$$Z = 1.00x_1 + 0.50x_2 + 0.75x_3.$$

Since we require a diet in which the iron, protein, and calcium are no less than the minimum requirements, our problem is modeled by the linear programming problem:

$$\text{Minimize} \quad Z = 1.00x_1 + 0.50x_2 + 0.75x_3 \quad \text{(cost)}$$

$$\text{subject to} \quad 4x_1 + 2x_2 + 2x_3 \geq 12 \quad \text{(iron)}$$

$$4x_1 + 2x_2 + 3x_3 \geq 15 \quad \text{(protein)}$$

$$2x_1 + 4x_2 + x_3 \geq 8 \quad \text{(calcium)}$$

$$x_i \geq 0 \quad i = 1, 2, 3.$$

The initial simplex tableau for the dual problem is given as tableau 6.17. In the exercises you will be asked to apply the simplex algorithm to this tableau and, therefore, find the solution to this diet problem.

TABLEAU 6.17

	y_1	y_2	y_3	s_1	s_2	s_3	z	
s_1	4	4	2	1	0	0	0	1.00
s_2	2	2	4	0	1	0	0	0.50
s_3	2	3	1	0	0	1	0	0.75
z	-12	-15	-8	0	0	0	1	0

8. *A Stock-Cutting Problem.* The Wood Lumber Company has as part of its standard stock boards that are 18 inches wide by 8 feet long. Suppose a customer needs boards 8 feet long but in widths of 7 inches and 8 inches. Suppose the customer needs at least 20 of the 7-inch-wide boards, and at least 40 of the 8-inch-wide boards. To fill this order, the 18-in-wide boards are to be "ripped" lengthwise. Since the customer must pay for the entire 18-inch-wide board, and since he has use only for boards 7 inches and 8 inches wide, the trimmings (pieces less than 7 inches wide) are waste. In cutting an 18-inch-wide board there are three options, given in table 6.6.

TABLE 6.6

	Number of 7" boards	Number of 8" boards	Amount of waste
option 1	2	0	4"
option 2	1	1	3"
option 3	0	2	2"

How many 18-inch-wide boards should be cut under each option so as to minimize the waste?

In this problem we must determine three numbers that can be represented as follows:

x_1 = number of 18" wide boards to be cut at option 1

x_2 = number of 18" wide boards to be cut at option 2

x_3 = number of 18" wide boards to be cut at option 3.

We can see that the total waste is represented by $4x_1 + 3x_2 + 2x_3$. Thus, the objective is to minimize the number

$$Z = 4x_1 + 3x_2 + 2x_3.$$

The constraints arise from the minimum number of 7-inch and 8-inch boards that are needed. Since the number of 7-inch boards obtained is $2x_1 + x_2$, and since the number of 8-inch boards obtained is $x_2 + 2x_3$, we require $2x_1 + x_2 \geq 20$ and $x_2 + 2x_3 \geq 40$.

Imposing the natural conditions on x_1, x_2, and x_3, our stock-cutting problem can now be modeled by the linear programming problem:

$$\text{Minimize} \qquad Z = 4x_1 + 3x_2 + 2x_3$$

$$\text{subject to} \qquad 2x_1 + \ x_2 \geq 20$$

$$x_2 + 2x_3 \geq 40$$

$$x_1 \geq 0, \qquad x_2 \geq 0, \qquad \text{and} \qquad x_3 \geq 0.$$

In the exercises you will be asked to complete the solution of this stock-cutting example by setting up and solving its dual problem.

EXERCISES 6.3

For exercises 1–4 rewrite the linear programming problem using matrix notation.

1. Minimize $\quad Z = 4x + 5y$
   subject to
   $$5x + 9y \geq \ 90$$
   $$15x + 9y \geq 180$$
   $$x \geq 0 \qquad \text{and} \qquad y \geq 0.$$

2. Minimize $\quad Z = 10x_1 + 12x_2 + 8x_3$
   subject to
   $$-x_1 + 2x_2 + 5x_3 \geq 2$$
   $$2x_1 + \ x_2 + 3x_3 \geq 4$$
   $$x_1 - \ x_2 + \ x_3 \geq 5$$
   $$x_1 \qquad - \ 2x_3 \geq 6$$
   $$x_1 \geq 0, \qquad x_2 \geq 0, \qquad \text{and} \qquad x_3 \geq 0.$$

3. Maximize $\quad z = 2x + 3y$
   subject to
   $$-4x + 3y \leq \ 1$$
   $$14x + 6y \leq 35$$
   $$x \geq 0 \qquad \text{and} \qquad y \geq 0.$$

4. Maximize $\quad z = 10x_1 + 8x_2 + 4x_3$
   subject to
   $$x_1 + 3x_2 + 4x_3 \leq 100$$
   $$2x_1 + \ x_2 + 2x_3 \leq 100$$
   $$x_1 \geq 0, \qquad x_2 \geq 0, \qquad \text{and} \qquad x_3 \geq 0.$$

5. Give the dual problem for the standard minimization problem of exercise 1.

6. Give the dual problem for the standard minimization problem of exercise 2.

7. Give a standard minimization problem for which its dual problem is the standard maximization problem of exercise 3.

8. Give a standard minimization problem for which its dual problem is the standard maximization problem of exercise 4.

For exercises 9–20 solve the standard minimization problem by solving its dual problem.

9. The problem of exercise 1.

10. The problem of exercise 2.

11. Minimize
    subject to
    $$Z = 12x + 6y$$
    $$-x + 2y \geq 1$$
    $$3x - y \geq 2$$
    $$x \geq 0 \quad \text{and} \quad y \geq 0.$$

12. Minimize
    subject to
    $$Z = x + y$$
    $$3x + 5y \geq 60$$
    $$6x + 5y \geq 90$$
    $$x \geq 0 \quad \text{and} \quad y \geq 0.$$

13. Minimize
    subject to
    $$Z = 3x + y$$
    $$5x + 4y \geq 52$$
    $$7x + 3y \geq 52$$
    $$3x + 5y \geq 39$$
    $$x \geq 0 \quad \text{and} \quad y \geq 0.$$

14. Minimize
    subject to
    $$Z = 5x + 3y$$
    $$x + y \geq 9$$
    $$18x - 8y \geq 45$$
    $$x - y \geq 0$$
    $$x \geq 0 \quad \text{and} \quad y \geq 0.$$

15. Minimize
    subject to
    $$Z = x_1 + x_2 + x_3$$
    $$2x_1 + 3x_2 - x_3 \geq 5$$
    $$x_1 + \frac{3}{2}x_2 - 2x_3 \geq 12$$
    $$x_1 \geq 0, \quad x_2 \geq 0, \quad \text{and} \quad x_3 \geq 0.$$

16. Minimize
    subject to
    $$Z = 6x_1 + 12x_2 + 2x_3$$
    $$2x_1 + 3x_2 + 5x_3 \geq 2$$
    $$3x_1 + 5x_2 + 5x_3 \geq 10$$
    $$6x_1 + x_2 + x_3 \geq 6$$
    $$x_1 \geq 0, \quad x_2 \geq 0, \quad \text{and} \quad x_3 \geq 0.$$

17. Minimize
    subject to
    $$Z = 7x_1 + 19x_2 + 11x_3$$
    $$3x_1 - x_2 + x_3 \geq 4$$
    $$-x_1 + 3x_2 + 11x_3 \geq 1$$
    $$-x_1 - 3x_2 + 2x_3 \geq 2$$
    $$x_1 + 2x_2 \geq 6$$
    $$x_1 \geq 0, \quad x_2 \geq 0, \quad \text{and} \quad x_3 \geq 0.$$

18. Minimize
    subject to
    $$Z = 7x_1 + 9x_2 + 15x_3$$
    $$2x_1 + x_2 - x_3 \geq 5$$
    $$2x_1 - x_2 + 3x_3 \geq 4$$
    $$x_1 + 2x_3 \geq 6$$
    $$-x_1 + 3x_2 + x_3 \geq 2$$
    $$x_1 + 20x_2 + 15x_3 \geq 15$$
    $$x_1 \geq 0, \quad x_2 \geq 0, \quad \text{and} \quad x_3 \geq 0.$$

19. Minimize
    subject to
    $$Z = 20x_1 + 10x_2 + 5x_3 + 2x_4$$
    $$8x_1 + 4x_2 + 2x_3 \geq 20$$
    $$2x_1 + 3x_2 \geq 6$$
    $$3x_1 + x_2 + x_4 \geq 8$$
    $$x_i \geq 0, \quad i = 1, 2, 3, 4.$$

20. Minimize $Z = 14x_1 + 10x_2 + 8x_3 + 10x_4$
    subject to $2x_1 + 4x_2 + 2x_3 + 10x_4 \geq 40$
                   $8x_1 + 8x_2 + 4x_3 + 1x_4 \geq 80$
                   $2x_1 + 2x_2 + 8x_3 + 1x_4 \geq 50$
                   $6x_1 \qquad + 4x_3 + 6x_4 \geq 20$
                   $x_i \geq 0, \quad i = 1, 2, 3, 4.$

21. Complete the solution to the diet problem given in example 7.

22. Joe is planning a meal of salad, steak, and potato. He wants the meal to supply at least 16 grams of protein and 1500 calories. Each unit of salad supplies 1 gram of protein and 20 calories. Each unit of steak has 3 grams of protein and 200 calories, whereas each unit of potato has 300 calories and no protein. Joe wants the meal to contain at least 2 units of steak. If each unit of salad costs $0.30, each unit of steak costs $1.50, and each unit of potato costs $0.10, how many units of each food should Joe plan for the meal so that the requirements of protein and calories are met or exceeded but the cost is minimal? What is the minimum cost?

23. The Mayfair Feed Company wants each bag of its chicken feed to contain at least 50 kilograms of vitamin D, 25 kilograms of vitamin C, and 20 kilograms of fish oil. Two ingredients, A and B, are used in making the feed. Ingredient A contains 25% vitamin D, 30% vitamin C, and 10% fish oil, whereas ingredient B contains 35% vitamin D, 20% vitamin C, and 15% fish oil. If A costs $0.50 per kilogram and B costs $0.75 per kilogram, how many kilograms of each ingredient should a bag of feed contain to satisfy the requirements in vitamins and fish oil but the cost is minimal?

24. A dietician is planning a low carbohydrate meal in which the food will be selected from among four types, A, B, C, and D. Each unit of type A contains 2 units of carbohydrates. Each unit of types B, C, and D contains 4 units, 1 unit, and 3 units of carbohydrates, respectively. The foods must be selected so that the meal contains at least 20 units of protein. Each unit of type A food contains 3 units of protein. Each unit of types B, C, and D contains 2 units, 2 units, and 4 units of protein, respectively. Also the dietician requires that the combined number of units of types A and C food be no less than the combined number of units of types B and D foods. How many units of each type of food should be used to meet the requirements so that the number of units of carbohydrates is minimal?

25. Solve by duality the stock-cutting problem presented in example 8.

26. The Forest Paper Company produces rolls of paper which are 20 inches wide. The company has orders for rolls of paper in widths of 6 inches, $8\frac{1}{2}$ inches, and 11 inches. In order to meet these orders, a 20-inch-wide roll is cut into these smaller widths. However, rolls of paper that are less than 6 inches wide are waste. The possible options for cutting a 20-inch-wide roll are given in table 6.7.

TABLE 6.7

	Number of 6″ rolls	Number of $8\frac{1}{2}$″ rolls	Number of 11″ rolls	Amount of waste
option 1	3	0	0	2″
option 2	1	1	0	$5\frac{1}{2}$″
option 3	1	0	1	3″
option 4	0	2	0	3″
option 5	0	1	1	$\frac{1}{2}$″

If Forest Paper has orders for at least 96 six-inch rolls, 44 eight-and-one-half-inch rolls, and 35 eleven-inch rolls, then how many twenty-inch rolls should be cut under each possible option to minimize waste?

27. Refer to exercise 26. Suppose Forest Paper has rolls of paper 15 inches wide from which are cut 4-inch, 6-inch, and $8\frac{1}{2}$-inch-wide rolls. Suppose the company has orders for at least 50 four-inch rolls, 45 six-inch rolls, and 56 eight-and-one-half-inch rolls. Give a table of the possible cutting options and determine the number of fifteen-inch rolls that should be cut under each option so as to minimize waste.

28. A carpenter can buy a particular kind of lumber in 8-foot and 12-foot length boards. He is working on a job and needs 45 pieces of wood in 4-foot lengths, 20 pieces in 5-foot lengths, and 15 pieces in 7-foot lengths. Give a table of possible cutting options and determine the number of 8-foot length and 12-foot length boards the carpenter should buy in order to minimize the waste. Also, how many 8-foot length boards should be cut under each option, and how many 12-foot length boards should be cut under each of the options?

29. The Sun Tile Company is planning to discontinue two of its styles of tiles, style A and style B. However, the company still has orders that it must fill for these tiles. In particular, the company needs 2500 units of style A and 1800 units of style B. Sun Tile has three plants located in Bristol, Lewistown, and Acton. In one hour of operation the Bristol plant can produce 100 units of style A and 50 units of style B. The plant at Lewistown can only produce style B tile at the rate of 75 units per each hour of operation. In one hour the Acton plant can produce 50 units of style A and 80 units of style B. Each hour of operation of the Bristol, Lewistown, and Acton plants costs $200, $125 and $150, respectively. How many units of each style of tile should be produced at each plant so that Sun Tile Company produces the number of units needed but at a minimal cost?

6.4 Linear Programming Problems in General: The Two-Phase Method and Computer Solutions

We have seen in sections 6.2 and 6.3 how to solve standard linear programming problems. We develop here a general method that can be applied to any type of problem, including the standard ones. This method, called the **two-phase method**, can be used on problems with mixed constraints and on problems in which the system of constraints includes equations. Also there are no restrictions on either the coefficients of the objective function, as in the case of a standard minimization problem, or on the right-hand sides of the constraints, as in the case of a standard maximization problem.

In addition to being a general procedure, the two-phase method is used as the basis for a computer algorithm. It is not difficult to realize the value of a computer as an aid to problem solving, especially when the necessary computations become long and tedious. However, using a computer does not diminish our role as a problem solver. We must still know how to describe the problem, set up the necessary data, and interpret the results. We need, therefore, to have a fundamental understanding of the computer algorithm being used, which in this case is the two-phase method.

One of the key ideas in the two-phase method is expressing the problem in canonical form. Once in canonical form, we can apply the simplex algorithm of section 6.2 to obtain the desired results. We saw in section 6.1 that a standard maximization problem can be converted to canonical form by introducing a slack variable for each constraint. We now consider techniques for converting any type of linear programming problem to canonical form.

Let us begin our development with a standard minimization problem. Consider, for example, the following diet problem. We wish to select x units of cereal and y units of nuts to satisfy recommended minimum requirements of calories and protein while keeping the cost

$$z = 0.05x + 0.12y$$

to a minimum. The constraints of the problem are

$$2x + y \geq 50 \quad \text{(calories)}$$

$$x + 3y \geq 45 \quad \text{(protein)}.$$

To derive a problem in canonical form, we first change the problem from a minimization problem to a maximization one. The conversion is based on the idea that the smallest number in a set of numbers

$$\{a_1, a_2, \ldots, a_k\}$$

is the negative of the largest number in the set

$$\{-a_1, -a_2, \ldots, -a_k\}.$$

Therefore our problem can be given as maximization of the negative of the objective function. In particular, we wish to

$$\text{maximize} \quad z' = -z = -0.05x - 0.12y.$$

We next convert the inequality constraints to equations. A feasible diet (x, y) may be one where there is a **surplus** of calories in the diet; that is, the calories $2x + y$ in the diet may be greater than 50. Therefore we introduce a **surplus variable** S_1 by setting $S_1 = 2x + y - 50$. Observe that the surplus must be non-negative, otherwise the amount of calories in the diet would be less than 50. We also introduce into the problem a surplus variable S_2 to represent the difference between the protein $x + 3y$ in the diet and the minimum requirement 45. Thus we set $S_2 = x + 3y - 45$ and require that S_2 be non-negative. Finally, since surplus contributes nothing to cost, the objective function remains unchanged. We now restate our diet problem as follows: Find a solution to the system

$$2x + y - S_1 \qquad = 50$$
$$x + 3y \quad - S_2 \quad = 45 \qquad \text{(1)}$$
$$0.05x + 0.12y \qquad + z' = 0$$

for which $x \geq 0$, $y \geq 0$, $S_1 \geq 0$, $S_2 \geq 0$, and z' is maximum.

We have not yet obtained a problem in canonical form, since the first two equations of the system (1) do not contain a basic variable, a variable with coefficient 1 in one equation and coefficient 0 in the other. This situation can be easily rectified by simply adding to these equations other variables $A_1 \geq 0$ and

$A_2 \geq 0$. In doing so, our system of equations becomes

$$
\begin{aligned}
2x + \quad y - S_1 \quad\quad + A_1 \quad\quad\quad\quad &= 50 \\
x + \quad 3y \quad\quad -S_2 + A_2 \quad\quad\quad &= 45 \\
0.05x + 0.12y \quad\quad\quad\quad\quad + z' &= 0.
\end{aligned}
\tag{2}
$$

Each equation in system (2) now has a basic variable. However, this was accomplished by introducing **artificial variables** A_1 and A_2 that have no meaningful interpretation with respect to a feasible diet. Therefore, if the only solutions to the system are those in which $A_1 \neq 0$ or $A_2 \neq 0$, the diet problem has no feasible solution, let alone an optimal one. Thus our first concern is to determine, if possible, solutions to system (2), satisfying non-negative conditions, and for which $A_1 = 0$ and $A_2 = 0$. Let us see how to phrase this as a linear programming problem.

Let $w = -(A_1 + A_2)$. Since we require the artificial variables A_1 and A_2 to be non-negative, w must be less than or equal to 0. Therefore a solution to system (2) in which $A_1 = 0$ and $A_2 = 0$ is found when the maximum value of w is 0. Thus, as the linear programming problem for eliminating the artificial variables from the set of basic variables, we have: Find a solution to the system of equations

$$
\begin{aligned}
2x + \quad y - S_1 \quad\quad + A_1 \quad\quad\quad\quad\quad\quad &= 50 \\
x + \quad 3y \quad\quad - S_2 \quad\quad + A_2 \quad\quad\quad\quad &= 45 \\
0.05x + 0.12y \quad\quad\quad\quad\quad\quad + z' \quad\quad &= 0 \\
A_1 + A_2 \quad\quad + w &= 0
\end{aligned}
$$

for which $x \geq 0,\ y \geq 0,\ S_1 \geq 0,\ S_2 \geq 0,\ A_1 \geq 0,\ A_2 \geq 0$, and w is maximum.

This problem is called the **phase I** problem. Our phase I problem is not in canonical form since the basic variables A_1 and A_2 in the first two equations of the system also appear in the w equation. However, by subtracting the first and the second equations from the last equation, we have the phase I problem in canonical form: Find a solution to the system

$$
\begin{aligned}
2x + \quad y - S_1 \quad\quad + A_1 \quad\quad\quad\quad\quad\quad &= \quad 50 \\
x + \quad 3y \quad\quad - S_2 \quad\quad + A_2 \quad\quad\quad\quad &= \quad 45 \\
0.05x + 0.12y \quad\quad\quad\quad\quad\quad + z' \quad\quad &= \quad 0 \\
-3x - \quad 4y + S_1 + S_2 \quad\quad\quad\quad\quad + w &= -95
\end{aligned}
$$

for which $x \geq 0,\ y \geq 0,\ S_1 \geq 0,\ S_2 \geq 0,\ A_1 \geq 0,\ A_2 \geq 0$, and w is maximum.

We apply the simplex algorithm to this problem. If we find an optimal solution to the phase I problem in which the w-value is 0 (meaning $A_1 = 0$ and $A_2 = 0$), we have found a basic feasible solution to our original diet problem. Thus we can begin the simplex algorithm for the original problem starting with this basic feasible solution. This is called **phase II** of the procedure.

In solving the phase I problem, we should note that z' would never be eligible for a departing variable since we want to return to the problem of maximizing z'. The initial simplex tableau for our phase I problem is given as tableau 6.18.

TABLEAU 6.18

	x	y	S_1	S_2	A_1	A_2	z'	w	
A_1	2	1	-1	0	1	0	0	0	50
A_2	1	3	0	-1	0	1	0	0	45
z'	0.05	0.12	0	0	0	0	1	0	0
w	-3	-4	1	1	0	0	0	1	-95

We apply the simplex algorithm to tableau 6.18. With two passes through the steps of the algorithm we obtain tableau 6.19.

TABLEAU 6.19

	x	y	S_1	S_2	A_1	A_2	z'	w	
x	1	0	$-\frac{3}{5}$	$\frac{1}{5}$	$\frac{3}{5}$	$-\frac{1}{5}$	0	0	21
y	0	1	$\frac{1}{5}$	$-\frac{2}{5}$	$-\frac{1}{5}$	$\frac{2}{5}$	0	0	8
z'	0	0	0.006	0.038	0.006	-0.038	1	0	-2.01
w	0	0	0	0	1	1	0	1	0

We see from the w-row of tableau 6.19 that we have obtained an optimal solution and phase I is complete. Furthermore, the w-value in this optimal solution is 0. Therefore, $x = 21$, $y = 8$, $S_1 = 0$, $S_2 = 0$, and $z' = -2.01$ is a basic feasible solution to the system of equations in our original problem (1).

Since we want to continue the simplex algorithm and find, if necessary, other basic feasible solutions to the original problem, we will always want to keep $A_1 = 0$ and $A_2 = 0$, in which case w is also 0. Therefore, when proceeding to phase II, we can drop from further consideration the variables A_1, A_2, and w. Thus, for phase II we apply the simplex algorithm to tableau 6.20, which was obtained from tableau 6.19 by deleting the w-row and the columns for w, A_1, and A_2.

TABLEAU 6.20

	x	y	S_1	S_2	z'	
x	1	0	$-\frac{3}{5}$	$\frac{1}{5}$	0	21
y	0	1	$\frac{1}{5}$	$-\frac{2}{5}$	0	8
z'	0	0	0.006	0.038	1	-2.01

Since all entries in the last row below variables are non-negative, phase II is complete without a complete pass through the algorithm and we have the optimal solution $x = 21$, $y = 8$, $S_1 = 0$, $S_2 = 0$, and $z = -z' = 2.01$.

In our diet problem the right-hand sides of the inequalities are positive. If we have a problem in which a right-hand side is negative, we first multiply that constraint by -1. In doing so, we should recall that

$$-a \geq -b \quad \text{is equivalent to} \quad a \leq b.$$

For example,

$$2x - y \geq -2 \quad \text{is equivalent to} \quad -2x + y \leq 2.$$

We now summarize our discussion as follows:

Converting Problems to Canonical Form and the Two-Phase Method

1. If the problem is a minimization one, multiply the objective function by -1, set $z' = -z$, and take the problem to be the maximization of z'. If this maximization problem has a solution, this solution with z' replaced by $-z'$ is an optimal solution to the minimization problem.

2. If the right-hand side of a constraint is negative, multiply through the constraint by -1. (Note that the direction of the inequality will change.)

3. Convert inequality constraints (as changed by step 2 if necessary) to equalities. If the inequality is $\leq$, add a slack variable. If the inequality is $\geq$, subtract a surplus variable and add an artificial variable.

4. If the problem has no artificial variables, apply the simplex algorithm to this problem as done in section 6.2. On the other hand, if the problem has artificial variables $A_1, A_2, \ldots, A_k$, include the equation $A_1 + A_2 + \cdots + A_k + w = 0$ in the system and consider the problem of finding a solution to this system for which w is maximum, which is the phase I problem.

5. (Phase I): Bring the phase I problem into canonical form by subtracting from the w equation the equations in the system for which $A_1, A_2, \ldots, A_k$ are basic variables. Obtain an optimal solution to the phase I problem, written in canonical form, by applying the simplex algorithm to its initial simplex tableau. Observe that in applying the algorithm z' will always be a basic variable.

6. If the w-value in the optimal solution obtained in step 5 is negative, the original problem has no feasible solutions and the process stops. On the other hand, if the w-value in the optimal solution is 0, the original problem has feasible solutions. In this case, proceed to phase II.

7. (Phase II): Drop from the tableau the columns of all variables with positive entry in the w-row and drop the w-row. Continue the simplex algorithm with this tableau. We find with the completion of this phase either an optimal solution or an unbounded solution for the original problem.

We shall refer to the process of converting a problem to canonical form and applying the two-phase method as the **simplex method**.

EXAMPLE 1

Solve the following problem using the simplex method:

$$\text{Minimize} \quad z = 2x + y$$
$$\text{subject to} \quad x + y \geq 13$$
$$x + 4y \geq 31$$
$$5x + 8y \leq 95$$
$$x \geq 0 \quad \text{and} \quad y \geq 0.$$

We change the problem from minimizing z to maximizing $z' = -2x - y$. We skip step 2 since the right-hand sides of the constraints are non-negative.

Next, the first two constraints are converted to equalities by subtracting surplus variables $S_1 \geq 0$ and $S_2 \geq 0$ and adding artificial variables A_1 and A_2. The third constraint is changed to an equation by adding a slack variable $s \geq 0$. In doing so we have the equations

$$
\begin{array}{rcl}
x + y - S_1 \qquad\quad + A_1 \qquad\qquad &=& 13 \\
x + 4y \qquad - S_2 \qquad + A_2 &=& 31 \\
5x + 8y \qquad\qquad\qquad\quad + s &=& 95.
\end{array}
$$

Since the problem contains artificial variables, we include the equation $A_1 + A_2 = w$ in the system. The phase I problem written in canonical form is: Find a solution to the system

$$
\begin{array}{rcl}
x + y - S_1 \qquad\quad + A_1 \qquad\qquad\qquad\quad &=& 13 \\
x + 4y \qquad - S_2 \qquad + A_2 \qquad\qquad\qquad &=& 31 \\
5x + 8y \qquad\qquad\qquad\qquad + s \qquad\qquad &=& 95 \\
2x + y \qquad\qquad\qquad\qquad\qquad + z' \qquad &=& 0 \\
-2x - 5y + S_1 + S_2 \qquad\qquad\qquad\qquad + w &=& -44
\end{array}
$$

for which $x \geq 0$, $y \geq 0$, $S_1 \geq 0$, $S_2 \geq 0$, $A_1 \geq 0$, $A_2 \geq 0$, $s \geq 0$, and w is maximum. The initial simplex tableau for our phase I problem is tableau 6.21.

TABLEAU 6.21

	x	y	S_1	S_2	A_1	A_2	s	z'	w	
A_1	1	1	-1	0	1	0	0	0	0	13
A_2	1	4	0	-1	0	1	0	0	0	31
s	5	8	0	0	0	0	1	0	0	95
z'	2	1	0	0	0	0	0	1	0	0
w	-2	-5	1	1	0	0	0	0	1	-44

Applying the simplex algorithm to tableau 6.21, we obtain tableau 6.22, which completes phase I. We read from tableau 6.22 the optimal solution $x = 7$, $y = 6$, $S_1 = 0$, $S_2 = 0$, $A_1 = 0$, $A_2 = 0$, $s = 12$, $z' = -20$, and $w = 0$. Since the optimal value for w is 0, we can drop from consideration A_1, A_2, and w and begin phase II with tableau 6.23.

With one pass through the steps of the simplex algorithm, we obtain tableau 6.24 and the optimal solution for the original problem $x = 3$, $y = 10$, $S_1 = 0$, $S_2 = 12$, $s = 0$, and $z = -z' = 16$.

TABLEAU 6.22

	x	y	S_1	S_2	A_1	A_2	s	z'	w	
x	1	0	$-\frac{4}{3}$	$\frac{1}{3}$	$\frac{4}{3}$	$-\frac{1}{3}$	0	0	0	7
y	0	1	$\frac{1}{3}$	$-\frac{1}{3}$	$-\frac{1}{3}$	$\frac{1}{3}$	0	0	0	6
s	0	0	4	1	-4	-1	1	0	0	12
z'	0	0	$\frac{7}{3}$	$-\frac{1}{3}$	$-\frac{7}{3}$	$\frac{1}{3}$	0	1	0	-20
w	0	0	0	0	1	1	0	0	1	0

TABLEAU 6.23

	x	y	S_1	S_2	s	z'	
x	1	0	$-\frac{4}{3}$	$\frac{1}{3}$	0	0	7
y	0	1	$\frac{1}{3}$	$-\frac{1}{3}$	0	0	6
s	0	0	4	1	1	0	12
z'	0	0	$\frac{7}{3}$	$-\frac{1}{3}$	0	1	-20

TABLEAU 6.24

	x	y	S_1	S_2	s	z'	
x	1	0	$-\frac{8}{3}$	0	$-\frac{1}{3}$	0	3
y	0	1	$\frac{5}{3}$	0	$\frac{1}{3}$	0	10
S_2	0	0	4	1	1	0	12
z'	0	0	$\frac{11}{3}$	0	$\frac{1}{3}$	1	-16

When we obtain an optimal solution to the phase I problem in which the w-value is negative (meaning at least one artificial variable has a non-zero value in the solution), the original problem has no feasible solutions. An example of such a problem was solved in example 6 of section 6.3 using duality. Let us see how the same conclusion arises using the two-phase method; in particular, we consider this problem again in the next example.

EXAMPLE 2

Solve the following problem using the simplex method

$$\text{Minimize} \quad z = 3x + y$$
$$\text{subject to} \quad -2x + y \geq -2$$
$$-x - y \geq \quad 2$$
$$x \geq 0 \quad \text{and} \quad y \geq 0.$$

We change this minimization problem to a maximization one by setting $z' = -z = -3x - y$. Next, we multiply the first constraint by -1 to obtain an inequality with a non-negative right-hand side. With these changes, our problem becomes:

$$\text{Maximize} \quad z' = -3x - y$$
$$\text{subject to} \quad 2x - y \leq 2$$
$$-x - y \geq 2$$
$$x \geq 0 \quad \text{and} \quad y \geq 0.$$

We now convert the inequalities to equalities by adding a slack variable s to the first inequality and subtracting a surplus variable S and adding an artificial variable A to the second inequality. Since we now have an artificial variable in the problem, we have the phase I problem: Find a solution to the system

$$2x - y + s \qquad\qquad\qquad = 2$$
$$-x - y \qquad - S + A \qquad\qquad = 2$$
$$3x + y \qquad\qquad + z' \quad = 0$$
$$A \qquad + w = 0$$

for which $x \geq 0$, $y \geq 0$, $s \geq 0$, $S \geq 0$, $A \geq 0$, and w is maximum.

This phase I problem is written in canonical form by subtracting the second equation of the system from the last equation. Thus the initial simplex tableau for our phase I problem is tableau 6.25.

TABLEAU 6.25

	x	y	s	S	A	z'	w	
s	2	-1	1	0	0	0	0	2
A	-1	-1	0	-1	1	0	0	2
z'	3	1	0	0	0	1	0	0
w	1	1	0	1	0	0	1	-2

We see from the last row of tableau 6.25 that we have an optimal solution for the phase I problem, $x = 0$, $y = 0$, $s = 2$, $S = 0$, $A = 2$, $z' = 0$, and $w = -2$. Since the value of w is not zero, the original problem has no feasible solutions.

As noted, standard maximization problems and standard minimization problems can be solved using the simplex method. For example, the production problem of section 6.2 and the diet problem and stock-cutting problem of section 6.3 could be solved this way and, therefore, could be solved with the aid of a computer program. Instead of showing the computer solution for any of these problems, we conclude this chapter with two other applications and show the computer solution for one of them.

EXAMPLES

3. *A Transportation Problem.*   The Forest Paper Company has two pulp mills, one at Johnsonburg and the other at Machias. The company has also three paper finishing plants located at Warren Glen, Riegelsville, and Milford. The cost of shipping one ton of unfinished paper from each mill to each plant is given in table 6.8. Also included in the table is the maximum number of tons of unfinished paper that each mill can supply to each finishing plant and the minimum requirement by each finishing plant.

TABLE 6.8

	Destination			
Source	*Warren Glen*	*Riegelsville*	*Milford*	*Supply*
Johnsonburg	$8	$3	$5	100
Machias	$4	$8	$7	150
demand	80	60	90	

One concern of Forest Paper is to determine the amount of unfinished paper to ship from each pulp mill to each finishing plant so that the total shipping cost is minimal. We can model this optimization problem as a linear programming problem in the following way.

Let x_{ij} be the amount to be shipped from the ith mill to the jth plant. For example, x_{11} represents the amount shipped from Johnsonburg to Warren Glen and x_{23} represents the amount to be shipped from Machias to Milford. There are six unknowns to be determined, namely, x_{11}, x_{12}, x_{13}, and x_{21}, x_{22}, x_{23}.

The total amount shipped from Johnsonburg is $x_{11} + x_{12} + x_{13}$, and for Machias it is $x_{21} + x_{22} + x_{23}$. The amounts received by the finishing plants at Warren Glen, Riegelsville, and Milford are $x_{11} + x_{21}, x_{12} + x_{22}$, and $x_{13} + x_{23}$, respectively.

Considering now the limitations on supply and the requirements by the plants, we have the constraints

$$x_{11} + x_{12} + x_{13} \leq 100 \qquad \text{(supply by Johnsonburg)}$$

$$x_{21} + x_{22} + x_{23} \leq 150 \qquad \text{(supply by Machias)}$$

$$x_{11} + x_{21} \geq 80 \qquad \text{(demand by Warren Glen)}$$

$$x_{12} + x_{22} \geq 60 \qquad \text{(demand by Riegelsville)}$$

$$x_{13} + x_{23} \geq 90 \qquad \text{(demand by Milford)}$$

Thus our linear programming problem is to determine a feasible solution to the system of constraints that minimizes the total shipping cost

$$z = 8x_{11} + 3x_{12} + 5x_{13} + 4x_{21} + 8x_{22} + 7x_{23}.$$

Now that the problem has been described, we complete the solution with the aid of a computer program. The choice of using a computer to obtain numbers is due primarily to the size of the problem. The data necessary to the program includes the coefficient matrix of the constraints, the right-hand sides, and the coefficients in the objective function. The latter of these is easily read from the description of the linear programming problem. The coefficient matrix may not be as easy to read from the constraints because many variables are missing. Therefore we give the coefficient matrix as follows:

$$
\begin{array}{cccccc}
x_{11} & x_{12} & x_{13} & x_{21} & x_{22} & x_{23}
\end{array}
$$
$$
\begin{bmatrix}
1 & 1 & 1 & 0 & 0 & 0 \\
0 & 0 & 0 & 1 & 1 & 1 \\
1 & 0 & 0 & 1 & 0 & 0 \\
0 & 1 & 0 & 0 & 1 & 0 \\
0 & 0 & 1 & 0 & 0 & 1
\end{bmatrix}
$$

Now running the program with this data, we obtain the printout in figure 6.5. Reading the solution from this printout, we have that Forest Paper will achieve the minimum cost of $1050 if it ships unfinished paper from the pulp mills to the finishing plants according to the schedule given in table 6.9.

```
            Is this a MAX problem or a MIN problem?
            !MIN
            Enter the number of constraints of type <=, >=, and =.
            !2,3,0
            Enter the number of activities.
            !6

            Enter the coefficient matrix of the constraints.
            !1,1,1,0,0,0
            !0,0,0,1,1,1
            !1,0,0,1,0,0
            !0,1,0,0,1,0
            !0,0,1,0,0,1
            Enter the right-hand sides of the contraints.
            !100,150,80,60,90
            Enter the coefficients of the objective function.
            !8,3,5,4,8,7
```

```
        VARIABLES
        =========
        ACTIVITIES    1 - 6
        SLACK         7 - 8
        SURPLUS       9 - 11
        ARTIFICIAL    12 - 14


                        INITIAL PHASE I TABLEAU
                        ======= ===== = =======


    X(1)     X(2)     X(3)     X(4)     X(5)     X(6)
    1        1        1        0        0        0
    0        0        0        1        1        1
    1        0        0        1        0        0
    0        1        0        0        1        0
    0        0        1        0        0        1
    Z - ROW
    8        3        5        4        8        7
    W - ROW
    -1       -1       -1       -1       -1       -1

    X(7)     X(8)     X(9)     X(10)    X(11)    X(12)
    1        0        0        0        0        0
    0        1        0        0        0        0
    0        0        -1       0        0        1
    0        0        0        -1       0        0
    0        0        0        0        -1       0
    Z - ROW
    0        0        0        0        0        0
    W - ROW
    0        0        1        1        1        0

    X(13)    X(14)    RIGHT SIDE
    0        0        100
    0        0        150
    0        0        80
    1        0        60
    0        1        90
    Z - ROW
    0        0        0
    W - ROW
    0        0        -230
```

FIGURE 6.5

PHASE II STARTED AFTER 5 ITERATIONS

INITIAL PHASE II TABLEAU
======= ===== == ========

X(1)	X(2)	X(3)	X(4)	X(5)	X(6)
1	1	0	0	0	-1
0	0	0	0	0	0
0	0	1	0	0	1
1	0	0	1	0	0
-1	0	0	0	1	1

Z - ROW

| 9 | 0 | 0 | 0 | 0 | -3 |

X(7)	X(8)	X(9)	X(10)	X(11)	RIGHT SIDE
1	0	0	0	1	10
1	1	1	1	1	20
0	0	0	0	-1	90
0	0	-1	0	0	80
-1	0	0	-1	-1	50

Z - ROW

| 5 | 0 | 4 | 8 | 10 | -1200 |

FINAL TABLEAU
===== =======

X(1)	X(2)	X(3)	X(4)	X(5)	X(6)
0	1	0	0	1	0
0	0	0	0	0	0
1	0	1	0	-1	0
1	0	0	1	0	0
-1	0	0	0	1	1

Z - ROW

| 6 | 0 | 0 | 0 | 3 | 0 |

X(7)	X(8)	X(9)	X(10)	X(11)	RIGHT SIDE
0	0	0	-1	0	60
1	1	1	1	1	20
1	0	0	1	0	40
0	0	-1	0	0	80
-1	0	0	-1	-1	50

Z - ROW

| 2 | 0 | 4 | 5 | 7 | -1050 |

OPTIMAL SOLUTION FOUND AFTER 1 PHASE II ITERATION(S)

BASIC VARIABLE		VALUE
===== ========		=====
X(2)	=	60
X(3)	=	40
X(4)	=	80
X(6)	=	50
X(8)	=	20

FIGURE 6.5 (*cont.*)   Z = 1050

TABLE 6.9

Source	Destination Warren Glen	Riegelsville	Milford	Supply
Johnsonburg	0 tons	60 tons	40 tons	100 tons
Machias	80 tons	0 tons	50 tons	130 tons
demand	80 tons	60 tons	90 tons	

We also see in this solution that the pulp mill at Machias would have a slack of 20 tons.

4. *A Supply Problem.* The Leather Goods Company uses two grades of leather, high and medium, supplied from two tanneries, one at Brockton and the other at Lowell. Leather Goods needs exactly 200 pounds of high-grade and 150 pounds of medium-grade leather. The tannery at Brockton can supply at most 100 pounds of medium-grade leather, and at most 175 pounds of the two grades combined. The tannery at Lowell can supply at most 100 pounds of high-grade leather, and at most 200 pounds of the two grades combined. The cost per pound of each grade leather is given in table 6.10.

TABLE 6.10

Supplier	Cost per pound High-grade	Medium-grade
Brockton	$6	$3
Lowell	$5	$4

Leather Goods wishes to determine the number of pounds of each grade leather it should obtain from each tannery so that the cost is minimal.

First, we need to model the problem mathematically. To begin, let the unknowns be denoted by

x_{11} = pounds of high-grade supplied by Brockton

x_{12} = pounds of medium-grade supplied by Brockton

x_{21} = pounds of high-grade supplied by Lowell

x_{22} = pounds of medium-grade supplied by Lowell.

The objective function is given by the cost

$$z = 6x_{11} + 3x_{12} + 5x_{21} + 4x_{22},$$

which we wish to minimize.

There are two types of constraints. One type of constraint comes from the amounts needed by Leather Goods, and these are equations. The second type of constraint is on the amounts that can be supplied by the tanneries, and these are inequalities. In particular, the constraints of the problem are as follows:

$$x_{11} + x_{21} = 200$$

$$x_{12} + x_{22} = 150$$

$$x_{12} \leq 100$$

$$x_{11} + x_{12} \leq 175$$
$$x_{21} \leq 100$$
$$x_{21} + x_{22} \leq 200.$$

Therefore we have the linear programming problem of minimizing cost subject to the preceding constraints. Note that an artificial variable must be included for each of the two equations, if we were to solve this by hand. On the other hand, the computer program will automatically take care of the artificial variables.

You will be asked in the exercises to complete the solution to this problem.

EXERCISES 6.4

For exercises 1–8 solve the linear programming problem using the two-phase method.

1. Minimize $z = 3x + 2y$
   subject to $x + 5y \geq 41$
   $$4x + y \geq 31$$
   $$x \geq 0 \quad \text{and} \quad y \geq 0.$$

2. Maximize $z = 3x + 2y$
   subject to $7x - 5y \geq -10$
   $$7x - 3y \geq 5$$
   $$2x - 3y \geq -10$$
   $$x \geq 0 \quad \text{and} \quad y \geq 0.$$

3. Maximize $z = 22x + 7y$
   subject to $7x + 3y \leq 21$
   $$3x + 2y \geq 6$$
   $$y \leq 4$$
   $$x \geq 0 \quad \text{and} \quad y \geq 0.$$

4. Maximize $z = 2x + y$
   subject to $x + y \geq 13$
   $$5x + 8y \leq 95$$
   $$x + 4y \geq 31$$
   $$x \geq 0 \quad \text{and} \quad y \geq 0.$$

5. Maximize $z = 9x_1 + 4x_2 + 5x_3$
   subject to $x_1 + x_2 + x_3 \leq 20$

   $$-\frac{1}{2}x_1 - \frac{1}{2}x_2 + x_3 \geq 0$$

   $$x_1 \leq 4$$
   $$x_2 \geq 7$$
   $$x_1 \geq 0, \quad x_2 \geq 0, \quad \text{and} \quad x_3 \geq 0.$$

6. Minimize $z = x_1 + x_2 + x_3$
   subject to $2x_1 + 3x_2 + 4x_3 \le 14$
                $4x_1 + 7x_2 \qquad\quad \le 24$
                $2x_1 + \ x_2 - 2x_3 \ge 16$
                $x_1 \ge 0, \qquad x_2 \ge 0, \qquad \text{and} \qquad x_3 \ge 0.$

7. Minimize $z = 8x_1 + 7x_2 + 15x_3$
   subject to $x_1 \qquad\quad - \ x_3 \ge 10$
                $2x_1 + 3x_2 + \ x_3 \le 18$
                    $4x_2 + 3x_3 \ge 12$
                $x_1 \ge 0, \qquad x_2 \ge 0, \qquad \text{and} \qquad x_3 \ge 0.$

8. Maximize $z = 3x_1 + x_2 + 4x_3 + 2x_4$
   subject to $x_1 + 3x_2 + 2x_3 + 3x_4 = 10$
          $-3x_1 + \ x_2 \qquad\quad + 2x_4 \le 7$
            $2x_1 + \ x_2 + \ x_3 + 3x_4 \ge 4$
            $x_i \ge 0, \quad i = 1, 2, 3, 4.$

9. Bradford Beef has two meat-packing plants, one at Bradford and the other at Lewis Run. Bradford Beef supplies two supermarket chains, Topps and Marketville. The cost for shipping one ton of beef from each plant to each of the markets is given in table 6.11. Also the table contains the largest amount that each plant can supply and the minimum amount required by each market. Determine the amount of beef that should be shipped from each plant to each market to minimize the total shipping cost.

TABLE 6.11

	Destination		
	Topps	Marketville	Supply
Bradford	$60	$50	9
Lewis Run	$30	$40	13
demand	10	12	

10. Refer to exercise 9. Suppose that the demand for beef has changed. Topps now requires at least 14 tons; Marketville requires at least 15 tons. To satisfy these increased requirements, Bradford Beef opens another meat packing plant at Westline. The Westline plant can supply at most 7 tons of beef. The shipping cost per ton from Westline to Topps is $50, whereas the cost per ton to Marketville is $60. How many tons should now be shipped from each plant to each market so that the total shipping cost is minimal?

11. The Micro Company uses two types of floppy disks, single-sided and double-sided. Micro has two suppliers of the disks, CompuDisk and PCDisk. Micro Company needs at least 1500 single-sided disks and 1200 double-sided disks. CompuDisk can supply at most 900 single-sided disks and 1300 combined, whereas PCDisk can supply at most 1100 of the double-sided disks and 1700 combined. The cost per disk from each supplier is given in table 6.12.

TABLE 6.12

	Cost per disk	
	Single-sided	Double-sided
CompuDisk	$3.00	$3.00
PCDisk	$2.50	$4.00

How many disks of each type should be bought from each supplier so that Micro's total cost is minimal?

12. Complete the solution of the supply problem given in example 4.

13. Derby Feed Company guarantees at least 12 pounds of protein, 15 pounds of fat, and 9.75 pounds of carbohydrates in each bag of its feed. Three ingredients, I, II, and III, are used to make the feed. Ingredient I contains 40% protein, 45% fat, and 15% carbohydrates. Ingredient II contains 20% protein, 30% fat, and 50% carbohydrates. Ingredient III contains 30% protein, 15% fat, and 30% carbohydrates. Derby's recipe requires that the combined amount of ingredients I and II used be at least as much as the amount of ingredient III used. If ingredient I costs 50 cents per pound, ingredient II costs 30 cents per pound, and ingredient III costs 30 cents per pound, how many pounds of each ingredient should be used in the mixture so that the requirements of protein, fat, and carbohydrates are met but the total cost is minimal?

14. Lobo Kennels has been feeding its dogs food that is 22% protein and 9% fat; however, they want to use a food that is at least 24.5% protein and 11% fat. Three types of food, High-Protein, Regular, and Low-Protein, can be blended to obtain the desired food. The percent of protein and fat and the cost per pound for each type of food is given in table 6.13.

TABLE 6.13

	High-Protein	Regular	Low-Protein
protein (%)	26	22	20
fat (%)	12	9	8
cost/lb	20¢	16¢	14¢

How many pounds of each type food should be used to obtain 50 pounds of mixture that is at least 24.5% protein and 11% fat and gives the smallest cost?

15. An investor has up to $20,000 available to invest in three types of investments, a low-risk investment that yields 9% annually, a medium-risk investment that yields 15% annually, and a high-risk investment that yields 20% annually. The combined amount the investor wishes to put into the medium- and high-risk investments should be no more than the amount put into the low-risk investment. Also the investor wishes to put no more than $5000 in the high-risk investment and at least $10,000 in the low-risk investment. How much should the person put into each type of investment so that the annual income is maximum?

16. Refer to exercise 15. Suppose the person invests the entire $20,000. How much should be put into each type of investment so that the investment requirements stated in exercise 15 are met and the annual income is maximum?

SUMMARY OF TERMS

Linear Programming Problems in n Variables
    Problem in Canonical Form
    Slack
    Slack Variables
    Constraints

    Non-Negative Conditions
    Basic Solution
    Feasible Solutions
    Basic Feasible Solutions

Basic Variables in Solution
Non-Basic Variables in Solution
Optimal Solution

Simplex Algorithm
Standard Maximization Problem
Simplex Tableau
Initial Basic Feasible Solution
Entering Variable
Departing Variable
Pivot
Pivot Operations
Degeneracy

Optimal Value
Degenerate Solution

Duality
Standard Minimization Problem
Matrix Inequality
Primal Problem
Dual Problem

Two-Phase Method, Simplex Method
Surplus
Surplus Variables
Artificial Variables
Phase I Problem
Phase II Procedure

REVIEW EXERCISES (CH. 6)

Convert the linear programming problems in exercises 1 and 2 to canonical form.

1. Maximize $z = 8x_1 + 6x_2 + 5x_3$
   subject to
   $$2x_1 + x_2 + 6x_3 \leq 20$$
   $$2x_1 + 5x_2 + 7x_3 \leq 22$$
   $$x_i \geq 0 \qquad i = 1, 2, 3.$$

2. Maximize $z = 5x_1 + 18x_2 + 2.5x_3$
   subject to
   $$1.5x_1 + 6.5x_2 + x_3 \leq 28$$
   $$5x_1 + 16x_2 + 2x_3 \leq 71.5$$
   $$2x_1 + 7x_2 + x_3 \leq 31$$
   $$x_i \geq 0 \qquad i = 1, 2, 3.$$

3. Consider the system of linear equations

$$3x_1 + 4x_2 + 5x_3 + x_4 + s_1 \qquad\qquad = 18$$
$$x_1 + 2x_2 + 7x_3 + 2x_4 \qquad + s_2 \qquad\qquad = 5$$
$$4x_1 + 5x_2 + 3x_3 + 8x_4 \qquad\qquad + s_3 \qquad = 30$$
$$-7x_1 - 8x_2 - 5x_3 - 9x_4 \qquad\qquad\qquad + z = 0$$

and the initial basic feasible solution $x_1 = x_2 = x_3 = x_4 = 0$; $s_1 = 18$, $s_2 = 5$, $s_3 = 30$; and $z = 0$.

a. Determine the basic solution with basic variables x_1, s_1, s_3, and z and state whether or not this solution is feasible.

b. Determine the basic solution with basic variables x_1, s_1, s_2, and z and state whether or not this solution is feasible.

4. Consider the system of linear equations and the initial basic feasible solution given in exercise 3.

a. Determine the basic solution with basic variables x_4, s_2, s_3, and z and state whether or not this solution is feasible.

b. Determine the basic solution with basic variables $x_4, s_1, s_3,$ and z and state whether or not this solution is feasible.

In exercises 5 and 6 determine from the simplex tableau the basic feasible solution and whether or not it is optimal. If the solution is not optimal, determine the entering variable, the departing variable, and the new basic feasible solution.

5.

	x_1	x_2	x_3	s_1	s_2	z	
s_1	$\frac{1}{2}$	$\frac{1}{3}$	$\frac{1}{4}$	1	0	0	$\frac{5}{2}$
s_2	$\frac{3}{4}$	$\frac{1}{6}$	$\frac{1}{2}$	0	1	0	$\frac{13}{2}$
z	-8	-6	-10	0	0	1	0

6.

	x_1	x_2	x_3	s_1	s_2	s_3	z	
s_1	3	4	2	1	0	0	0	21
s_2	5	7	3	0	1	0	0	22
s_3	2	1	6	0	0	1	0	13
z	-5.5	-4	-3.5	0	0	0	1	0

7. Using the simplex algorithm solve the linear programming problem given in exercise 1.

8. Using the simplex algorithm solve the linear programming problem given in exercise 2.

For exercises 9–12 solve the standard maximization problem by the simplex algorithm.

9. Maximize $z = 3x + y$
   subject to
   $$x + 3y \le 24$$
   $$7x + 2y \le 35$$
   $$x \ge 0 \quad \text{and} \quad y \ge 0.$$

10. Maximize $z = x_1 + 5x_2 + 2x_3$
    subject to
    $$x_1 + x_2 + 6x_3 \le 14$$
    $$x_1 + x_2 - 2x_3 \le 6$$
    $$x_1 \ge 0, \quad x_2 \ge 0, \quad \text{and} \quad x_3 \ge 0.$$

11. Maximize $z = 3x_1 + 12x_2 + 5x_3 + 8x_4$
    subject to
    $$10x_1 + 7x_2 + 3x_3 + 17x_4 \le 124$$
    $$7x_1 + 5x_2 + 2x_3 + 12x_4 \le 87$$
    $$x_i \ge 0, \quad i = 1, 2, 3, 4.$$

12. Maximize $z = 25x_1 + 9x_2 + 2x_3 + 6x_4$
    subject to
    $$10x_1 + 3x_2 + x_3 + 14x_4 \le 57$$
    $$2x_1 + x_2 \quad\quad + 3x_4 \le 13$$
    $$13x_1 + 5x_2 + x_3 + 19x_4 \le 79$$
    $$x_i \ge 0, \quad i = 1, 2, 3, 4.$$

For exercises 13 and 14 solve the standard minimization problem by solving its dual problem.

13. Minimize $z = 21x + 12y$
    subject to
    $$3x + 2y \ge 68$$
    $$5x + 6y \ge 132$$
    $$13x + 2y \ge 154$$
    $$x \ge 0 \quad \text{and} \quad y \ge 0.$$

14. Minimize $\quad z = 16x_1 + 24x_2 + 20x_3$
    subject to $\quad 4x_1 + 3x_2 + 2x_3 \geq 72$
               $x_1 + 2x_2 + 2x_3 \geq 36$
                    $5x_2 + 4x_3 \geq 68$
             $x_i \geq 0, \quad i = 1, 2, 3.$

For exercises 15 and 16 solve the linear programming problem using the two-phase method.

15. Maximize $\quad z = 3x + 2y$
    subject to $\quad -7x + 5y \leq 10$
                $7x - 3y \geq 5$
               $-2x + 3y \leq 10$
               $x \geq 0 \quad$ and $\quad y \geq 0.$

16. Minimize $\quad z = 5x + 3y$
    subject to $\quad 7x + 3y \geq 30$
                $2x + 3y \geq 25$
                $8x + 5y \leq 40$
               $x \leq 0 \quad$ and $\quad y \geq 0.$

17. Fox Run Estates plans to construct a maximum of 80 new houses in one subdivision. For each house the developer will use one of the four basic designs: ranch, split-level, modern, and Cape Cod. The developer will have the foundation and framing done by a masonry contractor and a framing contractor. The number of days needed and the profit for each of the designs are given in table 6.14.

TABLE 6.14

	Ranch	Split-level	Modern	Cape Cod
foundation	1	2	2	1
framing	3	6	5	4
profit	$5,000	$7,500	$7,000	$6,000

If the developer allows at most 130 days for the masonry contractor to complete the foundations for all the houses and at most 387 days for the framing contractor to complete the framing for all the houses, how many of each basic design should the developer plan to build in order to maximize the profit? What is the maximum profit?

18. Three items, A, B, and C, are produced using two machines, I and II. The amount of time in hours needed on each machine for one unit of each item is given in table 6.15.

TABLE 6.15

	Hours needed for one unit of		
	Item A	Item B	Item C
machine I	0.03	0.02	0.02
machine II	0.02	0.04	0.02

If each machine is operated for at most 30 hours, and if the profits are $0.10 for one unit of item A, $0.12 for one unit of item B, and $0.08 for one unit of item C, how many units of each item should be produced so as to maximize total profit?

CHAPTER 7

Mathematics of Finance

Introduction

The use of credit cards is widespread in modern American society. Even such things as luggage, watches, and cameras can be purchased by mail with a gasoline company's credit card. In the past, Americans wanted to be debt free. Today it is not unusual for a person to buy an automobile on credit and trade it in for a newer model before the payments on the first car are completed. The person doing this thinks of the process as "renting" an automobile.

Buying on credit does not come without cost. Whenever a person buys an item and pays for it over a period of time, that person is borrowing money, and thus will pay **interest**, a fee for the use of the borrowed money. Borrowing money is not restricted to the individual consumer; many institutions borrow money. Companies borrow money to meet payrolls and to make capital improvements. Banks even "borrow" money from their depositors.

The decision to purchase an item on credit, to borrow money, or to put money into a particular type of savings account depends on individual circumstances. To make a good decision, a borrower should know how to weigh the alternatives. This chapter considers some aspects of borrowing (or lending) money.

7.1 Simple Interest and Bank Discount

In the world of finance, the value of money changes with time. Disregarding inflation, a dollar today is worth more in the future since it can be loaned, and interest will be added to the value of the dollar. Therefore we must keep in mind that money has a **present value** and a **future value**. In a loan transaction, the amount of money borrowed is the present value of the loan, called the **principal** of the loan. The total amount of money repaid is the future value of the loan, called the **amount** of the loan. The difference between the amount and the principal is the **interest**. Letting

$$P = \text{principal}$$

$$A = \text{amount}$$

$$I = \text{interest}$$

we have $\qquad\qquad\qquad\qquad A = P + I.$ $\qquad\qquad\qquad\qquad$ (1)

EXAMPLE 1

A person borrows $1000 and at the end of one year repays $1100 and satisfies the debt. What is the principal, amount, and interest?

The principal is the money borrowed, $1000. The amount is the money repaid, $1100. The interest is the difference, $100.

One of two questions is generally associated with any loan transaction:

1. Knowing the principal, what is the amount (the future value of the principal)?
2. Knowing the amount, what is the principal (the present value of the amount)?

From equation (1) we see that either question can be easily answered by knowing the interest. Thus we must concern ourselves with methods of computing interest. The three methods are **simple interest**, **bank discount**, and **compound interest**. We consider in this section the first two methods, leaving compound interest for the next section.

Definition

■ **Simple interest** is the interest computed on the original principal. Simple interest is given as a fixed percentage of the principal for one unit of time. The percentage is called the **rate of interest**. ■

Rates of interest can be stated as 6% per year, which means that the interest owed is 0.06 of the original principal for each year the loan is outstanding. A rate of 1% per month means the interest is 0.01 of the original principal for each month the loan is outstanding.

EXAMPLE 2

What is the total interest and the amount when $100 is borrowed at 6% per year and the loan is repaid at the end of 2 years?

Each year the loan remains unpaid, the interest is 0.06 of the principal $100. Since the loan is outstanding for 2 years, the interest is

$$I = (100)(0.06)(2) = \$12.$$

Therefore the amount is

$$A = 100 + 100(0.06)(2) = \$112.$$

For simple interest, the formulas for computing the interest and the amount are

$$I = Prt \tag{2}$$

$$A = P + I$$
$$= P(1 + rt) \tag{3}$$

where I = the interest in dollars

P = the principal in dollars

A = the amount due in dollars

r = percentage of the principal charged per each unit of time (rate of interest)

t = units of time consistent with those stated in the rate of interest.

EXAMPLE 3

A person borrows $1050 for 2 months at 10% per year. What is the interest and the amount?

Since the rate of interest is per year, the time must be converted to years, so, $t = \frac{2}{12}$ of a year. Using formulas (2) and (1), we have

$$I = (1050)(0.10)\left(\frac{2}{12}\right) = \$17.50$$

$$A = 1050 + 17.50 = \$1067.50.$$

Many times the stated rate of interest is per year, but the term of the loan is for several months or days. As in example 3, we must convert the time to years. If the term of the loan is for a number of days, we use in the conversion either 360 days for one year, called an **ordinary interest year**, or 365 days for one year, called an **exact interest year**. Unless otherwise stated, we will use 360 days. The rate of interest is often stated without a time unit; in this case, it will be understood that it is **per annum** (per year).

EXAMPLES

4. What is the amount of a loan of $1500 at 12% due in 45 days?

To apply formula (3), we must convert 45 days to a part of a year; it is understood that the rate of interest is per annum. We use ordinary interest, and therefore, 45 days is equivalent to $\frac{45}{360} = \frac{1}{8}$ year. The amount is

$$A = 1500\left[1 + 0.12\left(\frac{1}{8}\right)\right] = \$1522.50.$$

5. On credit card purchases where an amount owed is carried from one month to the next, a bank charges $1\frac{1}{2}\%$ per month on the amount owed during the month. If a person owes $500 during a month and makes a payment of $50, how much of the payment is applied to reducing the debt of $500?

A portion of the payment is used to pay the interest; the remainder of the payment is applied to the debt. Using formula (2), we have

$$I = 500(0.015)(1) = \$7.50.$$

Therefore the debt is reduced $50.00 - 7.50 = \$42.50$.

6. Every 3 months an investor receives \$750 interest on an investment which pays at 12%. How much money is invested?

We use formula (2), with $I = 750$, $r = 0.12$, and $t = \frac{1}{4}$, giving

$$750 = P(0.12)\left(\frac{1}{4}\right).$$

Solving for P, we have that the amount invested is

$$P = \frac{750}{0.03} = \$25,000.$$

7. A company advertises a sale of color television sets. The buyer can try the set at home for two months, at which time he will be billed \$366.68; or, if the purchaser desires, he can pay \$356 now. What annual rate of interest is the buyer charged if he chooses to use the television for two months before paying for it?

From formula (1), we have that the interest is $I = \$366.68 - \$356.00 = \$10.68$. We now use formula (2) to obtain r. Setting $I = 10.68$, $P = 356.00$, and $t = \frac{1}{6}$, we have

$$10.68 = 356\left(\frac{1}{6}\right)r$$

and

$$r = \frac{10.68}{356(\frac{1}{6})} = 0.18.$$

Thus the annual rate of interest is 18%.

8. A company has to pay \$5000 in one month for equipment it has purchased. In anticipation of satisfying this future debt, the company puts money in a savings account that pays $6\frac{3}{4}\%$. How much money must be put in the account so that there will be \$5000 in one month?

We can view the problem as lending P dollars to the bank so as to obtain the amount \$5000 in one month. We need to find the present value of \$5000 where interest is at the rate of $6\frac{3}{4}\%$. Therefore we can use formula (3) with $A = 5000$, $r = 0.0675$, and $t = \frac{1}{12}$ and solve for P. Using these values in the formula, we have

$$5000 = P\left[1 + (0.0675)\left(\frac{1}{12}\right)\right]$$

and

$$P = \frac{5000}{1 + (0.0675)(\frac{1}{12})} = \$4972.03.$$

Thus the present value of the \$5000 due in one month at $6\frac{3}{4}\%$ is \$4972.03 rounded to the nearest cent.

In examples 2, 3, and 4, we used formula (3) to move money forward to obtain A, the future value of P. In example 8, we used the formula

$$P = \frac{A}{1 + rt} \tag{4}$$

to move money backward to obtain P, the present value of A. As we observed, the value of money changes with time. In a problem where we wish to compare alternatives, we must move money forward or backward to a common date, called the **focal date**, to make the comparison. A graph, called a **time diagram**, can be used to display the situation (see figure 7.1).

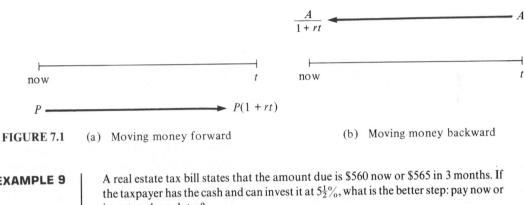

FIGURE 7.1   (a) Moving money forward   (b) Moving money backward

EXAMPLE 9   A real estate tax bill states that the amount due is $560 now or $565 in 3 months. If the taxpayer has the cash and can invest it at $5\frac{1}{2}\%$, what is the better step: pay now or invest and pay later?

To compare the alternatives—pay the $560 now or invest the money and pay $565 in 3 months—we must pick a focal date and move all money to this date. We shall take 3 months hence as the focal date, in which case we must move the $560 forward at $5\frac{1}{2}\%$. The time diagram in figure 7.2 displays the situation.

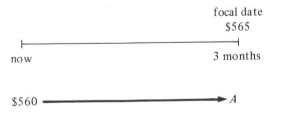

FIGURE 7.2

The future value A of 560 at $5\frac{1}{2}\%$ is $A = 560[1 + (0.055)(\frac{1}{4})] = \$567.70$. Therefore the better alternative is to invest $560 and pay the tax bill in 3 months. This way, the taxpayer will save $2.70 in 3 months.

EXAMPLE 10   A company contracts to have a $24,000 tractor delivered in 3 months and a $20,000 forklift delivered in 6 months. Instead of paying for each item when it is delivered, the company wishes to make a single payment now for both items. The company and supplier agree to use $9\frac{1}{2}\%$ interest. How much should the single payment be?

We need to move the two amounts, $24,000 and $20,000, due in the future to the present using $9\frac{1}{2}\%$ (see figure 7.3).

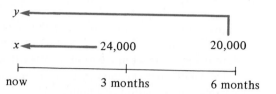

FIGURE 7.3

We use the present-value formula to determine x and y. We have

$$x = \frac{24{,}000}{1 + (0.095)(\frac{1}{4})} = 23{,}443.22 \quad \text{and} \quad y = \frac{20{,}000}{1 + (0.095)(\frac{1}{2})} = 19{,}093.08.$$

Therefore the single payment should be

$$x + y = \$42{,}536.30$$

In some transactions, the lender deducts the fee from the amount due. For this type of transaction it is common to call the fee **bank discount** or **discount** rather than interest. The discount is given as a fixed percentage of the **amount** due for one unit of time. The percentage is called the **discount rate**. The money the borrower receives, the difference between the amount due and the discount, is called the **proceeds**. For example, a borrower signs a note for the amount of $1000 due in 1 year where the lender discounts the amount at the rate of 8%. The proceeds of this note are $1000 - (0.08)(1000) = \$920$, which the borrower receives now.

For bank discount, the formulas for computing the discount and the proceeds are

$$D = Sdt \tag{5}$$

$$P = S - D$$
$$= S(1 - dt) \tag{6}$$

where D = the discount in dollars

S = the amount due

P = the proceeds

d = percentage of the amount charged per each unit of time (discount rate)

t = units of time consistent with those stated in the discount rate.

The proceeds P can be viewed as the present value of the amount S.

EXAMPLES

11. A borrower obtains a loan from a bank for $900 due in 6 months where the bank uses a discount rate of 12%. How much is the discount and what are the proceeds?

   We use formula (5) to obtain the discount D.

$$D = 900(0.12)\left(\frac{1}{2}\right) = \$54.$$

   Thus the discount is $54, and the proceeds are $900 − $54 = $846.

12. Suppose the borrower in example 11 misunderstood the terms of the loan and wants to receive $900 now. What is the amount due in 6 months if the proceeds are $900?

   We use the values in formula (6) and solve for S.

$$900 = S\left[1 - (0.12)\left(\frac{1}{2}\right)\right]$$

$$S = \frac{900}{1 - (0.12)(\frac{1}{2})} = \$957.45$$

As in example 12, when given the proceeds, we find the amount due by the formula

$$S = \frac{P}{1 - dt}. \tag{7}$$

Simple interest and bank discount are two distinct procedures. If the discount rate equals the simple interest rate and if the proceeds equals the principal, the amounts due for the two transactions will be different. This is illustrated in the next example.

EXAMPLE 13

Suppose the person in example 12 can borrow $900 from a different lender, due in 6 months, at a simple interest rate of 12%. What is the amount for this loan transaction?

   Using the formula for simple interest, we have

$$A = 900\left[1 + (0.12)\left(\frac{1}{2}\right)\right]$$

$$= \$954.$$

Comparing this result to that of example 12, we see that simple interest at 12% is better for the borrower than bank discount at 12%.

As seen in example 13, when the rates are equal the amounts due are different. If, however, the discount rate and the simple interest rate result in the same amount, we say that the two rates are **equivalent**.

EXAMPLE 14

Consider example 12 again. Let us find the equivalent simple interest rate. From example 12, the amount due for the discount transaction is

$$\frac{900}{[1 - (0.12)(\frac{1}{2})]}.$$

Since the equivalent simple interest rate r must yield this amount and since the amount for the simple interest transaction is $900[1 + \frac{1}{2}r]$, we have that r must satisfy the equation

$$900\left[1 + \frac{1}{2}r\right] = \frac{900}{1 - (0.12)(\frac{1}{2})}.$$

From this equation we have

$$\frac{1}{2}r = \frac{1}{1 - (0.12)(\frac{1}{2})} - 1$$

$$= \frac{(0.12)(\frac{1}{2})}{1 - (0.12)(\frac{1}{2})}$$

and

$$r = \frac{0.12}{1 - (0.12)(\frac{1}{2})} = 0.1277.$$

The general formula for the equivalent interest rate r, when given a discount rate d and time t, is

$$r = \frac{d}{1 - dt}. \tag{8}$$

This formula can be obtained using computations similar to those in example 14.

As seen in example 14, the equivalent interest rate will be larger than the discount rate.

EXAMPLE 15

A person wishes to obtain $1000 now and repay the loan in 120 days. One lender uses bank discount at the rate of 12%; a second lender uses simple interest at the rate of $12\frac{1}{2}\%$. Which of the two terms is better for the borrower?

We can choose the better of the two alternatives either by comparing the amounts due or by comparing the simple interest rate to the equivalent discount rate. We choose the latter method.

The equivalent rate is obtained by formula (8). To convert 120 days to a fraction of a year, we use an ordinary interest year. Thus $t = \frac{120}{360} = \frac{1}{3}$ and we have

$$r = \frac{0.12}{1 - (0.12)(\frac{1}{3})} = 0.125.$$

Therefore the terms of the two possible lenders are equivalent since the second lender's interest rate is equivalent to the first lender's discount rate.

EXERCISES 7.1

In exercises 1–4 what is the principal, amount, and interest for each of the loans?

1. A person borrows $600 and at the end of 8 months repays $632 to satisfy the debt.
2. A person borrows $1500 and at the end of $1\frac{1}{2}$ years repays $1871.25 to satisfy the debt.
3. A person repays $5020.68 ninety days after borrowing $5000.
4. A repayment of $3924.80 is made to satisfy a $3200 loan 18 months after the loan was taken out.

In exercises 5–12 find the interest and amount for each of the loans.

5. $1400 borrowed for 2 years at 8% per year.
6. $750 borrowed for $1\frac{1}{2}$ years at 15% per year.
7. A loan of $1500 for 9 months at $1\frac{1}{3}$% per month.
8. A loan of $675 for 5 months at $1\frac{3}{4}$% per month.
9. A person borrows $850 for 8 months at 18.5% per year.
10. A person borrows $2075 for 1 year and 4 months at 15.75% per year.
11. $1450 borrowed for 90 days at 14% per year (use an ordinary interest year).
12. $2000 borrowed for 150 days at 18.6% per year (use an ordinary interest year).

In exercises 13–18 determine the annual simple interest rate for the stated amount, principal, and time.

13. $903 paid in 6 months to satisfy a debt of $860.
14. $12,281.25 paid in 15 months to satisfy a debt of $10,000.
15. A person borrows $2500 for $1\frac{1}{2}$ years and repays $2968.75.
16. A person borrows $1800 for 39 weeks and repays $1914.75.
17. A loan of $600 is repaid in 135 days with $620.70 (use an ordinary interest year).
18. A loan of $7100 is repaid in 105 days with $7377.49 (use an ordinary interest year).

In exercises 19–24 determine the principal to be deposited in an account to obtain the amounts shown by using simple interest.

19. $1000 in 8 months where the account pays 6% interest per year.
20. $3500 in $1\frac{1}{4}$ years where the account pays 8% interest per year.
21. $750 in 3 months into an account that pays $\frac{3}{4}$% interest per month.
22. $2100 in 13 weeks into an account that pays $5\frac{1}{4}$% interest per year.
23. $1100 in 120 days into an account that pays 7.8% interest per year (use an ordinary interest year).
24. $15,000 in 6 months into an account that pays 9.425% interest per year.
25. On cash advances obtained with one of its credit cards, a bank charges $1\frac{1}{4}$% interest per month on the balance. If a person uses a credit card to borrow $1000, what finance charge for this cash advance will appear on the person's next monthly statement from the bank?
26. On credit card purchases, a major oil company charges $1\frac{3}{4}$% interest per month on amounts past due. A person receives a statement from the oil company for $175.50 due on

credit card purchases. If the person does not make the payment when it is due, what finance charge on the past due amount will appear on the next monthly statement?

27. Refer to exercise 25. Suppose the person makes a payment of $400 when the first monthly statement following the cash advance is received. How much will the second monthly statement following the cash advance show as due, including finance charges if no other charges or cash advances are made?

28. On credit card purchases, a bank charges $1\frac{1}{2}\%$ interest per month on the amount owed during the month. Suppose a person uses a credit card to purchase an item costing $450. When each of the next two monthly statements following the purchase is received from the bank, the person makes a payment of $100. How much will the third monthly statement following the purchase show as due, including finance charges?

29. A stereo system costs $695. An advertisement states that the system can be obtained by putting $125 down and in one month paying $578.55. What is the annual rate used to compute the finance charges?

30. A video recorder and camera system cost $1250. Instead of paying cash, an advertisement states that the buyer can put $300 down and pay $974.54 in one month. What is the interest charged and what is the annual rate?

31. A company owes $10,000 due in 60 days and $3000 due in 180 days on equipment it ordered. Instead of making separate payments when the amounts are due, the company arranges with the supplier to make a single payment in 240 days. If the company and the supplier agree to use a 9% interest rate, how much should the single payment be?

32. A loan company holds notes for $6000 due in 2 months and $4000 due in 5 months. The loan company wishes to sell the notes to a bank and receive a single amount now for the notes. If the loan company and the bank agree to use a 14% simple interest note, how much should the bank pay now for the notes?

33. A bank holds notes from a company for $10,000 due in 2 months and $15,000 due in 8 months. Instead of making separate payments, the company wishes to make a single payment in 6 months and satisfy both debts. If the bank and the company agree to use a 9.5% simple interest note, how much should the single payment in 6 months be?

34. A company orders a drill press for $25,000 to be delivered in 2 months, a grinder for $12,000 to be delivered in 3 months, and a lathe for $8000 to be delivered in 6 months. Instead of making separate payments when the equipment is delivered, the company arranges to make a single payment for all three pieces of equipment when the grinder is delivered. If the company and the equipment supplier agree to use a $12\frac{1}{4}\%$ simple interest rate, how much should the single payment be?

In exercises 35–38 find the discount and proceeds for each of the loans.

35. A loan of $1000 due in 1 year using a discount rate of 15%.

36. A loan of $3500 due in 45 days using a discount rate of 10%.

37. A loan of $5000 due in 4 months using a discount rate of 14.7%.

38. A loan of $800 due in 39 weeks using a discount rate of 12.4%.

39. Determine the amount due in $1\frac{1}{2}$ years when the proceeds are $1500 and the discount rate is 10%.

40. Determine the amount due in 3 months when the proceeds are $880 and the discount rate is 8.2%.

For each of the loan transactions in exercises 41 and 42, determine the better rate for the borrower.

41. A loan for 3 months using a discount rate of 10.25% or using a simple interest rate of 10.5%.

42. A loan for 45 days using a discount rate of 9% or using a simple interest rate of 9.25%.

7.2 Compound Interest

Simple interest is a linear process since the interest remains constant for each time period. For example, if the rate of interest is 6% and the principal is $100, the interest for each year is $6.00. In contrast, our second method of computing interest, the method of compound interest, is not linear since the interest for one time period is computed on the sum of the principal and interest from the previous period. For example, suppose a person deposits $100 in a savings account where the bank compounds the interest each year at 6%. The interest for the first year is $6, which is the same as for simple interest. However, $6 is added to the principal and the interest for the second year is 6% of this amount. Therefore the interest for the second year is $(0.06)(106) = \$6.36$ if compound interest is used, whereas the interest remains at $6 per year if simple interest is used.

The method of **compound interest** applied to a loan that extends over several time periods is to convert the principal at the end of each period to a new principal by adding on the interest. The new principal is then used in the interest computation for the next period. The length of time between conversions is called a **conversion period** or an **interest period**. Any financial transaction using compound interest must include as part of its terms a **frequency of conversion** (e.g., annually, semi-annually, quarterly, monthly, or daily). For example, if interest is compounded quarterly, the conversions are done at the end of every three-month period.

The interest for a given conversion period is obtained by taking a fixed percentage of the principal for that period. This fixed percentage is called the **conversion period interest rate**. However, in many transactions an annual rate, called a **nominal rate**, is given even though the frequency of conversion is something other than annual. Therefore the conversion period interest rate must be obtained from the nominal rate by dividing the number of conversion periods per year into the nominal rate. In particular,

$$i = \frac{r}{m} \tag{9}$$

where

i = the conversion period interest rate

r = the nominal rate

m = the number of conversion periods per year.

A financial transaction may state that interest is 6% compounded monthly. In this case, $r = 0.06$, $m = 12$, and $i = 0.005$. Therefore the fixed percentage used every month in the interest computations is $\frac{1}{2}\%$.

A final aspect that must be considered in a compound interest transaction is the number of conversion periods.

EXAMPLE 1

An investor puts \$10,000 in a certificate of deposit that pays 6% compounded quarterly. What will the value of the investment be at the end of $1\frac{1}{2}$ years?

The conversion period interest rate is

$$i = \frac{0.06}{4} = 0.015.$$

The number of conversion periods is $n = 6$. The interest for the first conversion period is $(0.015)(10,000) = \$150$. The principal is converted to the new principal \$10,150, which is the value of the investment at the end of the first period. Observe that

$$10,150 = 10,000 + (0.015)10,000 = 10,000(1 + 0.015).$$

The interest for the second period is computed on the converted principal, \$10,150. The value of the investment at the end of the second period is again obtained by moving the converted principal forward one quarter at the rate of 1.5% per quarter. Thus the value of the investment at the end of two conversion periods, and the principal for the third period, is

$$10,150 + (0.015)10,150 = 10,150(1 + 0.015)$$
$$= \$10,302.25.$$

We continue for the remaining periods, moving forward the value of the investment at the end of the previous period. Since the value of the investment at the end of a period is the principal for the period plus the interest for the period, we can move the principal forward over the period with a multiplication by the factor $(1 + 0.015)$. For example, the value of the investment at the end of the third period is

$$10,302.25 + (0.015)(10,302.25) = 10,302.25(1 + 0.015)$$
$$= \$10,456.78.$$

The results of these conversions for the six periods are displayed in table 7.1.

TABLE 7.1

Conversion period n	Principal for period	Converted principal (value of investment at end of period)
1	10,000.00	$10,000.00(1 + 0.015) = 10,150.00$
2	10,150.00	$10,150.00(1 + 0.015) = 10,302.25$
3	10,302.25	$10,302.25(1 + 0.015) = 10,456.78$
4	10,456.78	$10,456.78(1 + 0.015) = 10,613.63$
5	10,613.63	$10,613.63(1 + 0.015) = 10,772.83$
6	10,772.83	$10,772.83(1 + 0.015) = 10,934.42$

The value of the investment at the end of the conversion periods is called the **compound amount** or **amount**. The compound amount in example 1 for 6 conversion periods is $10,934.42.

Let us make an important observation on example 1. The amount $10,934.42 was obtained with a repeated multiplication by the factor $(1 + 0.015)$. Observe that a repeated multiplication by a factor can be expressed using exponent notation. For example, we can write the value of the investment at the end of two periods as

$$10,302.25 = 10,150(1 + 0.015) = 10,000(1 + 0.015)(1 + 0.015)$$
$$= 10,000(1 + 0.015)^2$$

and the value at the end of three periods can be written as

$$10,456.78 = 10,302.25(1 + 0.015) = 10,000(1 + 0.015)^2(1 + 0.015)$$
$$= 10,000(1 + 0.015)^3.$$

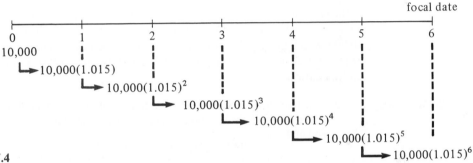

FIGURE 7.4

We can continue this idea for the six periods as shown in figure 7.4. Therefore, in finding the amount for six periods, we can combine the factors as $(1 + 0.015)^6$ and multiply the original principal $10,000 by this number. The compound amount can be expressed as $10,000(1 + 0.015)^6$. The idea is used in general and gives the following formula:

$$S = P(1 + i)^n \qquad \textbf{(10)}$$

where S = the compound amount (value of the investment at end of the conversion periods)

P = the original principal

i = the conversion period interest rate

n = the number of conversion periods.

Combining formulas (9) and (10) into one formula we have

$$S = P\left(1 + \frac{r}{m}\right)^{tm} \qquad \textbf{(11)}$$

where $\quad r =$ the nominal rate

$m =$ the number of conversion periods per year

$t =$ the number of years.

The number $(1 + i)^n$ in formula (10) is called the **accumulation factor** or the **amount of \$1**. It would be a laborious task to evaluate this factor for given values of i and n by repeated multiplication. However, the factor can be easily computed on hand calculators with an exponent key, for example, a key y^x. For example, to obtain the value of $(1 + 0.015)^6$ with such a calculator, enter 1.015 and raise it to the power 6 by using the exponent key. On obtaining the value of the accumulation factor, we multiply it by the given principal.

Tables of accumulation factors can also be used, especially when a hand calculator with an exponent key is not available. A set of such tables is given in appendix C for selected values of i and n. We find, for example, the accumulation factor

$$(1 + 0.015)^6 = 1.09344326$$

in the table labeled $i = 1\frac{1}{2}\%$, under column $(1 + i)^n$, and in row $n = 6$. Therefore the compound amount for $P = \$10,000$ is derived from multiplying this factor by 10,000, which gives \$10,934.43 rounded off to the nearest cent. The answer given in example 1 differs from this amount by 0.01 because of rounding off after each multiplication.

EXAMPLES

2. A person invests \$10,000 in a certificate of deposit that pays 6% interest converted monthly. What is the compound amount at the end of 2 years? At the end of 4 years?

Since the frequency of conversion is monthly, the conversion period rate is $i = \frac{6}{12}\%$. Furthermore, the number of conversion periods in 2 years is $n = 24$, and in 4 years is $n = 48$. Using the accumulation factors found in the table $i = \frac{6}{12}\%$, we have

$$S = 10,000(1 + 0.005)^{24} = 10,000(1.12715978)$$
$$= \$11,271.60$$

at the end of 2 years and

$$S = 10,000(1 + 0.005)^{48} = 10,000(1.27048916)$$
$$= \$12,704.89$$

at the end of 4 years.

3. Suppose the person in example 2 continues the investment until the end of 6 years. What is the compound amount?

The number of conversion periods in 6 years is $n = 72$. We observe that the table $i = \frac{6}{12}\%$ does not contain the accumulation factor $(1 + 0.005)^{72}$. However, we can obtain the answer in the following way: move \$10,000 forward $n = 50$ periods, obtain an amount, and then move this amount forward 22 periods. The compound amount at the end of 6 years is

$$S = [10,000(1 + 0.005)^{50}](1 + 0.005)^{22} = 10,000(1.28322581)(1.11597216)$$
$$= \$14,320.44.$$

Example 3 illustrates how the accumulation factor table can be extended for values of n larger than those in the table. In particular, for $n = l + k$,

$$(1 + i)^{l+k} = (1 + i)^l(1 + i)^k.$$

As another example, the compound amount at the end of 8 years ($n = 96$) for the investment of example 2 is

$$S = 10,000(1 + 0.005)^{50}(1 + 0.005)^{46} = 10,000(1.28322581)(1.25787892)$$
$$= \$16,141.43.$$

It is interesting to compare the effect of compound interest with that of simple interest. If the person in example 2 were to receive simple interest, he would receive less interest, as seen in table 7.2.

TABLE 7.2

Number of years	Amount simple interest	Amount compound interest	Difference in interest
2	11,200	11,271.60	71.60
4	12,400	12,704.89	304.89
6	13,600	14,320.44	720.44
8	14,800	16,141.43	1341.43
10	16,000	18,193.97	2193.97
12	17,200	20,507.51	3307.51

From this example and the graph in figure 7.5, we see the dramatic effect of compound interest.

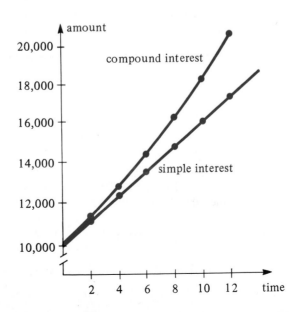

FIGURE 7.5

The choice of using the tables or a hand calculator to obtain an accumulation factor depends on the situation. Example 3 illustrates how one type of problem is handled when the accumulation factor does not appear directly in the table. In this case, we have a table for the desired conversion rate, but we must extend the table for a value of n larger than what appears in the table. However, sometimes we have a problem involving a conversion period rate i for which no table is available. In this case, we must use formula (10) and carry out the computations on a calculator. For example, the amount of $1000 at the end of 2 years, converted monthly at 7.2% interest, is found by multiplying 1000 by the accumulation factor $(1 + i)^{24}$, where $i = 0.072/12$. Since no table is available for this conversion period rate, we resort to using a calculator with an exponent key to obtain

$$S = 1000(1 + 0.072/12)^{24} = 1000(1.006)^{24}$$
$$= 1000(1.154387)$$
$$= \$1154.39,$$

rounded to the nearest cent.

EXAMPLE 4

A person wishes to invest $10,000 in a certificate of deposit. Each of three banks pays 6% interest, but they use different frequencies of conversion: semiannual, quarterly, and monthly. What is the amount at the end of 5 years using each frequency of conversion?

The necessary numbers and computations are given in table 7.3.

TABLE 7.3

Frequency of conversion	Conversion period rate i	Number of conversions n	Accumulation factor $(1 + i)^n$	Compound amount S
Bank 1 semiannually	3%	10	1.34391638	$13,439.16
Bank 2 quarterly	1.5%	20	1.34685501	$13,468.55
Bank 3 monthly	$\frac{1}{2}$%	60	1.34885015	$13,488.50

Example 4 shows that the frequency of conversion has a significant effect on the amount and the interest. More frequent conversions result in larger interest payments.

Although banks may pick any method of conversion for savings accounts, many banks use daily compounding. The amount converted daily can be found by formula (11) where we use either $m = 360$ (ordinary interest year) or $m = 365$ (exact interest year). For example, the value of the investment in example 4, using daily compounding and an ordinary interest year, is

$$S = 10,000\left(1 + \frac{0.06}{360}\right)^{5(360)} = 10,000(1.34982506)$$
$$= \$13,498.25.$$

Note that a hand calculator was used to compute the accumulation factor since a table for the conversion period interest rate $i = 0.06/360$ was not available.

Some banks even use **continuous compounding**. Continuous compounding means that the principal is converted at every instant of time. The reader may find this difficult to understand. One difficulty is how to determine the number of conversion periods per year. The number of conversion periods m increases as the frequency of conversion increases. Therefore we can view the amount, converted continuously, as the result of letting m become "very" large in formula (11)

$$S = P\left(1 + \frac{r}{m}\right)^{tm}.$$

It can be shown (using calculus) that this result yields the formula

$$S = Pe^{rt} \tag{12}$$

where $e = 2.71828$, rounded to five places.

The amount of the investment of example 4 converted continuously is $S = 10{,}000e^{0.30} = \$13{,}498.59$. Observe that the difference between compounding continuously and compounding daily is not as significant as the difference between compounding monthly and compounding daily.

At times we are confronted with alternative investments and we need to determine which one is better. As our discussion points out, if the nominal rates are the same but have different frequencies of conversion, the investment converted more frequently will produce the larger amount. However, sometimes we are confronted with alternatives in which not only are the frequencies of conversion different but also the nominal rates are different. In this case, we can base our comparison of the alternatives on the effective rate of interest.

The **effective rate of interest** r_E for an investment is the rate converted annually that would produce the same interest in one year as would the investment.

For example, on an investment of P dollars at 8% interest compounded quarterly, the interest at the end of one year is the amount less the principal; in particular, we have

$$I = P\left(1 + \frac{0.08}{4}\right)^4 - P$$

$$= P\left[\left(1 + \frac{0.08}{4}\right)^4 - 1\right].$$

On the other hand, the interest at the effective rate r_E is

$$I = P(1 + r_E) - P$$
$$= Pr_E.$$

Therefore, setting the interests equal, we have

$$Pr_E = P\left[\left(1 + \frac{0.08}{4}\right)^4 - 1\right]$$

and
$$r_E = \left(1 + \frac{0.08}{4}\right)^4 - 1 = 0.0824.$$

In general, the effective rate is given by the formula

$$r_E = \left(1 + \frac{r}{m}\right)^m - 1, \tag{13}$$

where r is the quoted nominal rate and m is the number of conversion periods per year. We see from formula (13) that the effective rate is the difference between the accumulation factor for one year and one. For continuous compounding, the effective rate is given by

$$r_E = e^r - 1. \tag{14}$$

EXAMPLE 5

Which is a better investment, 6% interest converted monthly or 6.2% interest converted annually?

The effective rate for the first investment is

$$r_E = \left(1 + \frac{0.06}{12}\right)^{12} - 1 = 1.06167781 - 1$$

$$= 6.17\%.$$

Since the conversion is annual for the second investment, its effective rate is the nominal rate 6.2%. Comparing these effective rates, we see that the second investment is slightly better and would produce a larger return.

We have seen thus far how to move money forward using compound interest. In some financial transactions we wish to move money backward to the present; that is, we wish to determine the **present value** P of some future amount S. Solving the compound amount formulas (10) and (11) for P, we have, respectively,

$$P = \frac{S}{(1 + i)^n}$$

$$= S(1 + i)^{-n} \tag{15}$$

and
$$P = \frac{S}{(1 + r/m)^{tm}}$$

$$= S(1 + r/m)^{-tm} \tag{16}$$

The factors $(1 + i)^{-n}$ and $(1 + r/m)^{-tm}$ are each called the **discount factor** or the **present value of $1**.

The factor $(1 + i)^{-n}$ can be obtained by using either tables or a hand calculator with an exponent key. Tables are given in appendix C for selected values of i and n.

EXAMPLE 6

What is the present value of $1000 due in 2 years using 7% interest compounded quarterly? That is, what must be invested now so as to have $1000 in 2 years where interest is at 7% compounded quarterly?

Since $i = 1\frac{3}{4}\%$, we find the discount factor in that table under $(1 + i)^{-n}$ in row $n = 8$.

$$P = 1000(1 + 0.0175)^{-8} = 1000(0.87041157)$$
$$= \$870.41.$$

This means that if $870.41 is invested today at 7% interest converted quarterly, the investment will have a value of nearly $1000 in 2 years. (We say "nearly" because of rounding the present value).

If no table is available for a given conversion rate i, the computations can be done easily on a calculator with an exponent key using formula (16). Also, if n is too large for a particular table, we can write n as $l + k$, move the amount backward l periods, and then move this result backward k periods. For $n = l + k$,

$$S(1 + i)^{-(l+k)} = S(1 + i)^{-l}(1 + i)^{-k}. \tag{17}$$

EXAMPLES

7. How much money must be invested now at 6% interest converted monthly so that in 8 years the value of the investment is $10,000?

We see that the discount factor for $n = 96$ does not appear in the table $i = \frac{6}{12}\%$. However we set $n = 50 + 46$ and use formula (17).

$$P = 10,000(1 + 0.005)^{-50}(1 + 0.005)^{-46}$$
$$= 10,000(0.77928607)(0.79498907)$$
$$= \$6195.24.$$

8. A person has a debt of $3000 due in 1 year. In anticipation of having this amount, she now deposits $1000 in a savings account and plans to make another deposit in 6 months. If the savings account pays 5% interest converted monthly, what must be the size of the deposit in 6 months?

To find the answer, we first move $1000 forward 1 year and subtract this future value from $3000. Next, we move the remaining balance backward 6 months. A time diagram is useful here (see figure 7.6).

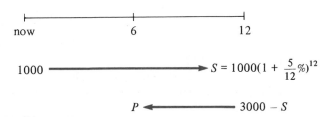

FIGURE 7.6

The value of $1000 in 1 year is

$$S = 1000\left(1 + \frac{5}{12}\%\right)^{12} = 1000(1.05116190)$$

$$= \$1051.16.$$

Now we move the balance $3000 - 1051.16 = 1948.84$ backward, yielding the size of the deposit

$$P = 1948.84\left(1 + \frac{5}{12}\%\right)^{-6} = 1948.84(0.97536057)$$

$$= \$1900.82.$$

9. A company orders a machine from a supplier. The supplier wants $2000 down and $8000 when the machine is delivered in 6 months. However, the company would rather make two equal payments, one in 3 months and the other in 6 months. The supplier will agree to this arrangement provided that the company accepts that money is worth 7% interest converted monthly. Accepting these terms, what is the size of the two payments?

A time diagram is helpful. We put the focal date at 6 months and move all money to this date (see figure 7.7). We let x represent the size of each payment. As the time diagram shows, the value of the company's payments at the focal date is

$$x\left(1 + \frac{7}{12}\%\right)^{3} + x = x\left[\left(1 + \frac{7}{12}\%\right)^{3} + 1\right]$$

$$= x(2.01760228).$$

This value must equal the sum of the amount of $2000 at the focal date and $8000. Therefore we have

$$x(2.01760228) = 2000(1.0355144) + 8000$$

$$= 10,071.03.$$

Therefore

$$x = \frac{10,071.03}{2.01760228}$$

$$= \$4991.58.$$

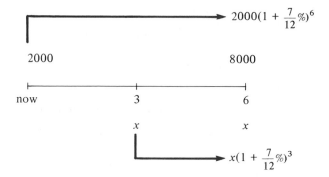

FIGURE 7.7

EXERCISES 7.2

Factors necessary to carry out the computations cannot be obtained from the tables for those problems marked with a calculator symbol (▦). In these cases the formulas of the section can be used with the aid of a calculator to derive the answers.

In exercises 1–10 find the compound amount at the end of the stated term for the stated principal and rate.

1. $7500 invested for 3 years at 5% interest converted annually.
2. $975 invested for 8 months at $1\frac{1}{2}$% interest per month converted monthly.
3. $3750 invested for 18 months at 7% interest converted monthly.
4. $5000 invested for 3 years at 8% interest converted semiannually.
5. $10,000 invested for 6 years at 9% interest converted monthly.
6. $8500 invested for 15 years at 12% interest converted quarterly.
7. $3200 invested for 1 year at 9.7% interest converted monthly.
8. $11,000 invested for 15 months at 11.2% interest converted monthly.
9. $5500 invested for 1 year at 7% interest converted daily.
10. $5500 invested for 1 year at 7% interest converted continuously.

In exercises 11–18 determine the effective rate for the stated nominal rate.

11. 6% interest converted quarterly.
12. 6% interest converted monthly.
13. 8% interest converted monthly.
14. 8% interest converted semiannually.
15. 10% interest converted quarterly.
16. 9.2% interest converted monthly.
17. 8% interest converted continuously.
18. 8% interest converted daily.

In exercises 19–22 which of the two given rates would yield the larger amount in one year?

19. 7% interest converted monthly or 7.3% interest converted annually.
20. 8% interest converted quarterly or 8.2% interest converted annually.
21. 10% interest converted semiannually or 9.8% interest converted continuously.
22. 11% interest converted monthly or 10.8% interest converted daily.

In exercises 23–30 find the present value of the stated amount.

23. $5000 due in 3 years at 5% interest converted annually.
24. $1500 due in 20 months at $1\frac{1}{4}$% interest per month converted monthly.
25. $10,000 due in 6 years at 8% interest converted quarterly.
26. $20,000 due in 4 years at 12% interest converted monthly.
27. $15,000 due in 7 years at 11% interest converted monthly.
28. $75,000 due in 15 years at 12% interest converted quarterly.
29. $2000 due in 2 years at 5.2% interest converted monthly.
30. $1000 due in 1 year at 9.4% interest converted daily.

31. Instead of buying new furniture now, a couple puts $1500 in a bank account that pays 6% interest converted monthly. How much will be in their account in 8 months?

32. To have $3000 in 2 years to use toward the purchase of a car, Joe decides to put money aside now in a money market account that pays 9% interest converted monthly. How much must Joe now put into the account so he will have $3000 in the account in 2 years?

33. The Farmer's Bank holds a note from a company for $20,000 due in 8 months. If the Farmer's Bank sells the note to the First National Bank and if the two banks agree to use 10% interest converted monthly to determine the present value of the note, how much will Farmer's Bank receive from the First National Bank for the note?

34. Louise establishes a trust fund for her son on the son's twelfth birthday. The fund pays 8% interest converted monthly. How much money must she put into the fund initially if the value of the fund is to be $5000 on the son's eighteenth birthday?

35. Three deposits of $500 each are made now, in 3 months, and in 6 months in an account that pays 7% interest converted monthly. How much will be in the account in 6 months, just after the last $500 deposit is made?

36. A man plans to take two vacation trips next year, one in 12 months and the other in 18 months. In anticipation of having money for these trips, he puts $5000 in a money management account that pays 9% interest converted monthly. In 12 months he plans to withdraw $2500 from the account for the first trip. How much will be in the account in 18 months when he plans to take the second trip?

37. A trucking company orders a truck costing $60,000 to be delivered in 3 months and a second truck costing $75,000 to be delivered in 8 months. Instead of making separate payments when the trucks are delivered, the trucking company arranges with the supplier to make a single payment in 5 months. If the trucking company and the supplier agree to use 6% interest converted monthly, how much should the single payment be?

38. A homeowner will need $895 in 9 months to pay real estate taxes and $200 in 12 months to pay home insurance. In anticipation of these expenses, the owner plans to make three deposits in a savings account that pays 7% interest converted monthly. If $450 is deposited now and another $450 in 3 months, how much should be deposited in 6 months so that there is enough in the account to cover the taxes and the insurance when each is due?

39. Inflation tends to diminish the purchasing power of the dollar. An item that costs $100 today would cost more several years from now due to inflation. Suppose a house is worth $70,000 today and suppose the value of homes went up on the average of 8% per year over the last 5 years. What was the value of this $70,000 house 5 years ago?

40. Refer to exercise 39. Suppose a new automobile costs $8700 today and suppose the price of new automobiles is increasing at the rate of 12% per year. How much will it cost 2 years from now to purchase the automobile that costs $8700 today?

7.3 Annuities

It is a common practice to borrow money and repay the loan in equal installments. For example, a person may take out a loan to purchase a car or a house and make equal monthly payments on the loan.

A financial transaction in which equal payments are made at equal time periods is called an **annuity**. The size of the payment is called the **periodic payment** or

periodic rent, and the length of the period between payments is called the **payment period** or **rent period**.

Installment loans are not the only type of annuities. A person can pay into a retirement fund and upon retirement receive equal monthly payments from the fund. If an annuity specifies that payments continue until the death of the person, it is called a **contingent annuity** since the date at which the payments begin or end (or both) is contingent on some event.

We shall consider only annuities where the payment periods begin and end at specified dates, called **certain annuities**. The length of time between the beginning of the first payment period and the end of the last payment period is called the **term** of the annuity.

Installment loans and periodic deposits to Christmas clubs are examples of certain annuities. Another example is a home mortgage where the first payment is made one month after the money is obtained and payments are made each month thereafter.

An annuity in which the payments are made at the end of each payment period is called an **ordinary annuity**. If the payments are made at the beginning of each period, it is called an **annuity due**. We will develop formulas for ordinary annuities.

The **future value**, **final value**, or **amount** of an ordinary annuity is the sum of all periodic payments moved forward to the end of the term using compound interest.

EXAMPLE 1

A person plans to deposit $300 in a savings account at the end of each month for the next 4 months. What would be the amount in the account after the fourth deposit is made if the account pays 5% interest converted monthly?

The focal date is the end of the term of the annuity, so we move all deposits forward to this date as depicted in figure 7.8. As the time diagram shows, the amount is the sum

$$S = 300 + 300\left(1 + \frac{5}{12}\%\right) + 300\left(1 + \frac{5}{12}\%\right)^2 + 300\left(1 + \frac{5}{12}\%\right)^3$$

$$= 300\left[1 + \left(1 + \frac{5}{12}\%\right) + \left(1 + \frac{5}{12}\%\right)^2 + \left(1 + \frac{5}{12}\%\right)^3\right]$$

$$= 300[1 + (1.004167) + (1.008351) + (1.012552)]$$

$$= \$1207.52.$$

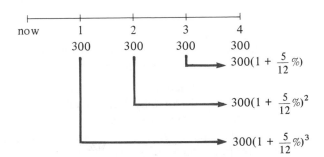

FIGURE 7.8

Example 1 indicates the general method for finding the amount of an ordinary annuity. We see from figure 7.9 that the amount is given by

$$S = R[1 + (1 + i) + (1 + i)^2 + \cdots + (1 + i)^{n-1}].$$

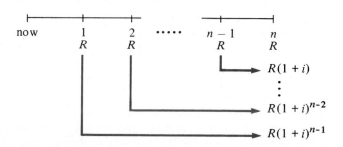

FIGURE 7.9

Instead of finding each accumulation factor $(1 + i)^k$ and adding them, we use the concept of a geometric progression to obtain the amount in a more convenient form. We multiply both sides of

$$S = R + R(1 + i) + R(1 + i)^2 + \cdots + R(1 + i)^{n-1}$$

by $(1 + i)$ to obtain

$$(1 + i)S = R(1 + i) + R(1 + i)^2 + \cdots + R(1 + i)^{n-1} + R(1 + i)^n.$$

Subtracting the first of these two equations from the second, we have

$$(1 + i)S - S = R(1 + i)^n - R$$

and

$$iS = R[(1 + i)^n - 1]$$

Thus,

$$S = R\left[\frac{(1 + i)^n - 1}{i}\right]$$

$$= Rs_{\overline{n}|i} \qquad (18)$$

where

$$s_{\overline{n}|i} = \frac{(1 + i)^n - 1}{i} \qquad (19)$$

S = the amount of the ordinary annuity

R = the periodic payment

n = the number of payment periods

i = the conversion period interest rate.

The factor $s_{\overline{n}|i}$, read "s, n bracket i" is called the **amount of $1 per period**. Tables of $s_{\overline{n}|i}$ for selected values of i and n are included in appendix C.

Let us consider example 1 again, but this time we use the tables for $s_{\overline{n}|i}$. We find the necessary factor in table $i = \frac{5}{12}\%$, under column $s_{\overline{n}|i}$ and in row $n = 4$. In particular, we find

$$s_{\overline{4}|\frac{5}{12}\%} = 4.02506952.$$

Now, by formula (18),

$$S = 300(4.02506952)$$
$$= \$1207.52.$$

EXAMPLES

2. To save for next year's vacation, Susan decides to deposit $200 per month for the next 12 months in a savings account that pays 6% interest converted monthly. How much will be in the account when the twelfth deposit is made?

In the table $i = \frac{6}{12}\%$ we find the needed factor in column $s_{\overline{n}|i}$, row $n = 12$. In particular, $s_{\overline{12}|\frac{1}{2}\%} = 12.33556237$ and the amount in the account will be

$$S = 200(12.33556237)$$
$$= \$2467.11.$$

3. At the end of each quarter, Harry deposits $1000 in a certificate of deposit that pays 7% converted quarterly. If he has done this for the last 5 years, how much has accumulated when the last deposit was made?

The periodic deposits at the end of each quarter form an ordinary annuity. In this case, $i = 1\frac{3}{4}\%$ and $n = 20$. The needed factor $s_{\overline{n}|i}$ is found in the tables. Therefore the amount accumulated is

$$S = 1000(23.70161119)$$
$$= \$23,701.61.$$

In some financial transactions, an amount due in the future is known and an account is established in which periodic deposits are made to the account so as to accumulate to this amount. Such an account is called a **sinking fund**. In this type of transaction, we wish to find the size of each deposit R. Solving formulas (18) and (19) we have

$$R = S \frac{1}{s_{\overline{n}|i}} \qquad (20)$$

where

$$\frac{1}{s_{\overline{n}|i}} = \frac{i}{(1 + i)^n - 1}. \qquad (21)$$

The factor $1/s_{\overline{n}|i}$ is the **periodic deposit needed to accumulate to $1 in n payments**. Tables of $1/s_{\overline{n}|i}$ for selected values of i and n are given in appendix C.

EXAMPLE 4

A company anticipates that it will need to replace one of its machines in 2 years. The replacement will cost $15,000. The company establishes a sinking fund that pays 8% interest converted quarterly. What must the size of each quarterly deposit be?

We find the size of each deposit using formula (20). The needed factor is found in table $i = 2\%$, column $1/s_{\overline{n}|i}$ and row $n = 8$. In particular, $1/s_{\overline{8}|2\%} = 0.1165098$ and the size of each deposit must be

$$R = 15,000(0.1165098)$$
$$= \$1747.65.$$

As indicated before, tables may not be available for the desired conversion period rate i. This may be true also when we need to determine either of the factors $s_{\overline{n}|i}$ or $1/s_{\overline{n}|i}$. In such cases we must resort directly to the formulas (19) or (21) and carry out the computations using a calculator with an exponent key. Therefore it is important to remember the availability of these formulas and how to use them.

The **present value** of an ordinary annuity is the sum of all periodic payments moved backward to the beginning of the term using compound interest. Applying the present value formula (15) to each payment R, we see from figure 7.10 that the present value of the annuity is given by the following formula.

$$A = R[(1 + i)^{-1} + (1 + i)^{-2} + \cdots + (1 + i)^{-n}].$$

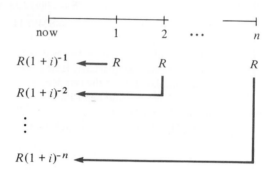

FIGURE 7.10

Using the concept of a geometric progression, it can be shown that the present value is given more conveniently by the formula

$$A = Ra_{\overline{n}|i} \tag{22}$$

where

$$a_{\overline{n}|i} = \frac{1 - (1 + i)^{-n}}{i} \tag{23}$$

$A =$ the present value of the ordinary annuity

$R =$ the periodic payment

$n =$ the number of payment periods

$i =$ the conversion period interest rate.

The factor $a_{\overline{n}|i}$, read "a, n bracket i," is called the **present value of $1 per period**. Tables of $a_{\overline{n}|i}$ for selected values of i and n are given in appendix C.

EXAMPLES

5. Nancy obtains an automobile loan on which she makes 36 monthly payments of $200 each, beginning one month after she obtained the loan. If the bank charges 12% interest converted monthly, what was the size of the loan?

The size of the loan is the present value of the annuity, which we can find using formula (22). The needed factor is found in table $i = 1\%$, column $a_{\overline{n}|i}$ and row $n = 36$. In particular, $a_{\overline{36}|1\%} = 30.107505$ and the size of the loan is

$$A = 200(30.10750504)$$
$$= \$6021.50.$$

6. David plans that in 3 years he will take a year-long vacation. He anticipates that he will need $1500 per month for 12 months and he establishes a savings account in which he will make monthly deposits. One month after he makes his last deposit, he will begin drawing $1500 per month from the account. If the account pays 5% interest converted monthly, what must the size of each deposit be?

There are two ordinary annuities: the first is periodic deposits x to the account and the second is periodic payments of $1500 from the account. We place the focal date at the end of the first annuity and the beginning of the second (see figure 7.11).

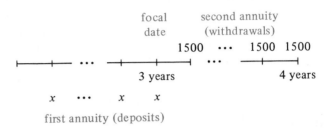

FIGURE 7.11

First we find the present value at the focal date of the second annuity. This is the amount that must be accumulated in the account. Therefore,

$$A = 1500_{\overline{12}|\frac{5}{12}\%}$$
$$= 1500(11.68122200) = \$17,521.83.$$

The first annuity is a sinking fund that must accumulate to $A = \$17,521.83$ at the focal date. Therefore the periodic deposit must be

$$X = A\frac{1}{s_{\overline{36}|\frac{5}{12}\%}}$$
$$= 17,521.83(0.02580423)$$
$$= \$452.14.$$

In many loan transactions the present value A is known, but we wish to obtain the periodic payment R. Solving formulas (22) and (23), we have

$$R = A\frac{1}{a_{\overline{n}|i}} \tag{24}$$

where

$$\frac{1}{a_{\overline{n}|i}} = \frac{i}{1 - (1 + i)^{-n}}. \tag{25}$$

The factor $1/a_{\overline{n}|i}$ is the **periodic payment needed to pay \$1 in n payments**. Tables of $1/a_{\overline{n}|i}$ for selected values of i and n are given in appendix C.

EXAMPLE 7

Kathy obtains a loan of $1000 which is to be repaid with 6 equal monthly payments. If the lender charges 15% converted monthly, what is the size of each payment?

The size of the payment is found using formula (24). The needed factor is found in table $i = 1\frac{1}{4}\%$, column $1/a_{\overline{n}|i}$ and row $n = 6$. In particular,

$1/a_{\overline{m}|i} = 0.17403381$ and

$$R = 1000(0.17403381)$$
$$= \$174.04.$$

(In finding the periodic payment, it is general practice to round up any fraction of a cent.)

The repayment of many loans, such as a home mortgage, is accomplished as an ordinary annuity. The periodic payment R is found by formula (24). As payments are made, the **loan balance** or the **outstanding principal** is reduced. Part of the payment is used to pay the interest for the period, and the remainder of the payment is applied to reducing the outstanding principal. In this case, we say the loan is **amortized**.

EXAMPLE 8

Consider the loan described in example 7. What portion of the first payment is applied to reducing the debt?

Since the outstanding principal during the first month is the original debt of $1000 and since the lender charges $1\frac{1}{4}\%$ interest per month, the interest for the first month is $(0.0125)(1000) = \$12.50$. This interest is deducted from the first payment and the remainder, $174.04 - 12.50 = \$161.54$, is applied to the debt. Therefore the outstanding principal during the second month is $838.46.

For an amortized loan, it is important to the borrower and the lender to know how much interest has been paid and what the outstanding principal is. This information is displayed in an **amortization schedule**. The schedule contains the portion of each payment that goes to interest, the portion applied to reducing the debt, and the outstanding principal.

The computation of an amortization schedule lends itself nicely to the use of a computer. This is due to the repetitious nature of the calculations. The amortization schedule for the loan described in example 7 was obtained from a computer program. The computer printout, and thus the amortization schedule, is shown in figure 7.12.

Instead of computing the entire amortization schedule for a loan, the two parties may only want to know the outstanding principal after the kth payment is made. The outstanding principal can be viewed as the present value of an annuity consisting of the remaining $n - k$ payments. For example, the outstanding principal for the loan of example 7 after the fourth payment is

$$B = 174.04 a_{\overline{2}|\,1\frac{1}{4}\%}$$
$$= 174.04(1.96311538)$$
$$= \$341.66.$$

Compare this result to the amortization schedule in figure 7.12.

Therefore, for a loan that is amortized, we have

$$B = R a_{\overline{n-k}|\,i} \tag{26}$$

This program computes an amortization schedule for
the given input principal, nominal rate, and term
in years. The conversion period used is monthly.
Enter the data at the appropriate prompt.

PRINCIPAL: !1000
NOMINAL RATE (percentage): !15
TERM IN YEARS: !0.5

AMORTIZATION SCHEDULE

Payment Number	Size of Payment	Portion to Interest	Portion to Principal	Outstanding Principal 1000
1	174.04	12.5	161.54	838.46
2	174.04	10.49	163.55	674.91
3	174.04	8.44	165.6	509.31
4	174.04	6.37	167.67	341.64
5	174.04	4.28	169.76	171.88
6	174.04	2.16	171.88	0

FIGURE 7.12

where B = the outstanding principal after the kth payment is made

R = the periodic payment

n = the entire number of payment periods

i = the conversion period rate.

EXAMPLE 9

A loan of $20,000 is taken out for 4 years at 9% interest converted monthly. What is the outstanding balance after the twelfth monthly payment is made?
First, we find that the periodic payment is

$$R = 20,000\frac{1}{a_{\overline{48}|\frac{3}{4}\%}} = 20,000(0.02488504) = 497.70.$$

The number of payments remaining is 36. The outstanding principal is the present value of the remaining payments. Therefore,

$$B = 497.70a_{\overline{36}|\frac{3}{4}\%} = 497.70(31.44680525) = \$15,651.07.$$

The debt has been reduced by $4348.93. However, the payments amount to $5972.40. Therefore $1623.47 went to interest during the first year of the loan.

Home mortgages usually are taken for a large number of years. It is interesting to observe the effect the term of the loan has on the total interest paid. Furthermore, a difference of even 1% interest in the nominal rate can have a significant effect on the total interest. A mortgage of $50,000 was computed for various terms and interest rates with the results given in table 7.4. Since the values of n exceeded those in the tables of appendix C, the present value factors and the periodic payment factors were computed directly from formulas (23) and (25) with the aid of a calculator.

TABLE 7.4   *Mortgage of $50,000*

Rate	Term (years)	Number of payments	Periodic payment (monthly)	Total paid	Total interest paid	Outstanding balance after 1 year
11%	15	180	568.30	102,294.00	52,294.00	48,611.95
	20	240	516.10	123,864.00	73,864.00	49,271.39
	25	300	490.06	147,018.00	97,018.00	49,599.88
12%	15	180	600.09	108,016.20	58,016.20	48,731.17
	20	240	550.55	132,132.00	82,132.00	49,359.71
	25	300	526.62	157,986.00	107,986.00	49,663.24
13%	15	180	632.63	113,873.40	63,873.40	48,841.80
	20	240	585.79	140,589.60	90,589.60	49,438.02
	25	300	563.92	169,176.00	119,176.00	49,716.70

EXERCISES 7.3

Factors necessary to carry out the computations cannot be obtained from the tables for those problems marked with a calculator symbol (▦). In these cases the formulas of the section can be used with the aid of a calculator to derive the answer.

In exercises 1–6 determine the amount that will be in the account when the last deposit is made for the stated periodic deposit, rate, and term.

1. $500 deposited at the end of each month for 15 months in an account that pays 7% interest converted monthly.

2. $800 deposited at the end of each quarter for 18 quarters in an account that pays 12% interest converted quarterly.

3. $100 deposited at the end of each month for 24 months in an account that pays $\frac{3}{4}$% interest per month converted monthly.

4. $3000 deposited at the end of each 6-month period for 3 years in an account that pays 10% interest converted semiannually.

5. $50 deposited at the end of each month for 5 years in an account that pays 6% interest converted monthly.

▦ 6. $200 deposited at the end of each month for 12 months in an account that pays 5.75% interest converted monthly.

For each of the sinking funds in exercises 7–12 determine the periodic deposit that must be made to the fund at the end of each conversion period so as to accumulate to the stated amount in the stated time.

7. The amount must be $5000 at the end of 8 months, where the sinking fund pays 6% interest converted monthly.

8. The amount must be $7500 at the end of 10 quarters, where the sinking fund pays 8% interest converted quarterly.

9. The amount must be $10,000 at the end of $1\frac{1}{2}$ years, where the sinking fund pays 1.25% per month converted monthly.

10. The amount must be $35,000 at the end of 5 years, where the sinking fund pays 6% interest converted quarterly.

11. The amount must be $20,000 at the end of 15 years, where the sinking fund pays 7% interest converted quarterly.

12. The amount must be $8000 at the end of 3 years, where the sinking fund pays $6\frac{1}{4}$% interest converted monthly.

In exercises 13–18 for the stated periodic payment, term, and rate, find the present value of the ordinary annuity.

13. $750 paid quarterly for 6 years, where the rate is 12% interest converted quarterly.

14. $306.54 paid monthly for 36 months, where the rate is 10% interest converted monthly.

15. $179.50 paid monthly for 4 years, where the rate is $1\frac{1}{4}$% interest per month converted monthly.

16. $276.13 paid monthly for 5 years, where the rate is 11% interest converted monthly.

17. $199.80 paid monthly for 20 years, where the rate is 7% interest converted monthly.

18. $167.07 paid monthly for 15 years, where the rate is 9.2% interest converted monthly.

For the loans in each of the exercises 19–24 determine the periodic payment needed to repay the loan in the stated time and at the stated rate.

19. $7000 to be repaid in equal monthly payments in 36 months at the rate of 18% interest.

20. $8500 to be repaid in equal quarterly payments in 4 years at the rate of 15% interest.

21. $1500 to be repaid in equal monthly payments in 1 year at the rate of 9% interest.

22. $2500 to be repaid in equal monthly payments in $1\frac{1}{2}$ years at the rate of $1\frac{3}{4}$% interest per month.

23. $26,000 to be repaid in equal monthly payments in 25 years at the rate of 11.75% interest.

24. $20,000 to be repaid in equal monthly payments in 20 years at the rate of 13% interest.

Determine the outstanding principal in each of the exercises 25–28 using the stated periodic payment, term, and rate.

25. $325.17 to be paid monthly for 3 years at 10% interest after 24 payments are made.

26. $659.25 to be paid quarterly for 20 quarters at 16% interest after 10 payments are made.

27. $261.52 to be paid monthly for 5 years at 12% interest after payments are made for 1 year.

28. $179.80 to be paid monthly for 15 years at 9% interest after payments are made for 11 years.

29. A loan of $3000 is obtained for 6 months at 18% interest converted monthly. Determine the monthly payment and construct an amortization schedule for this loan.

30. A loan of $9500 is obtained for 2 years at 10% interest converted quarterly. Determine the quarterly payment and construct an amortization schedule for this loan.

31. Dave buys a new car for which he must finance $7595 of the cost. The terms of the financing are 3 years at 15% interest converted monthly. What monthly payment must Dave make?

32. Carol buys a new car costing $15,640. The dealer allows her $4500 for her trade-in. Carol finances the remaining balance for 4 years at 13% interest converted monthly. What monthly payment must Carol make?

33. An advertisement states that a stereo outfit can be purchased by paying $100 cash and 18 monthly payments of $51.87 each. If the finance charges are computed at the rate of $1\frac{3}{4}\%$ interest per month, what is the original cost of the stereo outfit?

34. An appliance store advertises a sale of color television sets. The store states that for $125 down and 12 monthly payments of $38.96 each, a person can buy one of these sets. If the store uses 18% interest converted monthly, what is the original cost of the television set?

35. A homeowner's mortage is for 15 years at 9% interest. After making monthly payments of $167 for 11 years, the homeowner is selling her house. How much should she sell it for in order to have $15,000 left after the outstanding principal on the mortgage is paid?

36. John receives $1200 per quarter on his share of ownership in an oil-producing well. It is estimated that the oil reserve in the well is sufficient to generate this periodic payment for the next 8 years. The return on his original investment is 12% interest. If John sells his interest in the well after he receives his twentieth quarterly payment, what should he sell it for?

37. Arthur establishes a trust fund for his daughter one month after her twelfth birthday. The fund pays 9% interest converted monthly. If Arthur wants the value of the fund to be $5000 on his daughter's sixteenth birthday and when he makes the last deposit to the fund, how much must he deposit at the end of each month?

38. A couple want to save toward the down-payment on a house, which they plan to buy in 3 years. They anticipate that they will need $9000 for the down-payment. The couple plan to make equal monthly deposits to a money management account that pays 9% interest converted monthly. How much must they deposit each month?

39. Mr. and Mrs. Smith plan to sell their home and put the proceeds of the sale in a certificate of deposit that pays 10% interest converted quarterly. After 8 years, when they are retired, they plan to withdraw an equal amount at the end of each 3-month period for 12 years. How much will they withdraw if the proceeds from the sale of their home amount to $50,000?

40. Jean plans to take two vacation trips next year, one in 12 months and the other in 18 months. To have enough money for these trips, she plans to deposit $200 per month for 12 months in an account that pays 7% interest converted monthly. When Jean makes her twelfth deposit, she plans to withdraw $1500 from the account for her first trip and leave the balance in the account. How much will be in the account when she is ready to take her second trip?

41. A trucking company orders a truck costing $75,000 to be delivered in 3 months and a second truck costing $60,000 to be delivered in 8 months. Instead of making separate payments when the trucks are delivered, the trucking company arranges with the supplier to make 12 equal monthly payments beginning in one month. If the trucking company and the supplier agree to use 9% interest converted monthly, how much should each monthly payment be?

42. After making monthly payments of $233.70 for 12 years on their mortgage, a couple sold their house. The mortgage was for 15 years at 7% interest converted monthly. They bought the house 12 years ago for $32,000. However, they estimated that the house appreciated in value at the rate of 5% per year. If they sold their house for the appreciated value, how much did they have left from the sale after paying the outstanding principal on the mortgage?

SUMMARY OF TERMS AND FORMULAS

Simple Interest and Bank Discount
   Principal
   Present Value
   Future Value
   Focal Date
   Time Diagram
   Ordinary Interest Year
   Exact Interest Year
   Per Annum
   Rate of Interest
   Discount Rate
   Equivalent Rate
   Proceeds

Compound Interest
   Conversion Period (Interest Period)
   Frequency of Conversion
   Conversion Period Interest Rate
   Nominal Rate
   Effective Rate of Interest
   Continuous Compounding
   Amount of $1 (Accumulation Factor)
   Present Value
   Present Value of $1 (Discount Factor)

Annuities
   Periodic Payment (Periodic Rent)
   Payment Period (Rent Period)
   Contingent Annuity
   Certain Annuity
   Term of Annuity
   Ordinary Annuity
   Annuity Due
   Amount of Annuity
     (Future Value, Final Value)
   Amount of $1 Per Period
   Sinking Fund
   Present Value
   Loan Balance
   Outstanding Principal
   Amortized
   Amortization Schedule

Formulas

The **amount** A that the principal P will be worth in time t at the simple interest rate r

$$A = P(1 + rt).$$

The **principal** P that is needed to be worth the amount A in time t at the simple interest rate r

$$P = \frac{A}{1 + rt}.$$

The **proceeds** P determined from the amount S due in time t at discount rate d

$$P = S(1 - dt).$$

The **equivalent simple interest rate** r to the bank discount rate d for time t

$$r = \frac{d}{1 - dt}.$$

The **compound amount** S obtained from the principal P for n conversion periods at the rate i per period

$$S = P(1 + i)^n.$$

The **amount** S obtained from the principal P through continuous compounding at the nominal rate r for t years

$$S = Pe^{rt}.$$

The **principal** P necessary to obtain the amount S through compound interest at the rate i for n conversion periods

$$P = S(1 + i)^{-n}.$$

The **effective rate** r_E for a given nominal rate r with m conversion periods per year

$$r_E = \left(1 + \frac{r}{m}\right)^m - 1.$$

The **effective rate** r_E for a given nominal rate r with continuous compounding

$$r_E = e^r - 1.$$

The **amount** S that n periodic payments R will grow to with interest i per payment period

$$S = R\frac{(1 + i)^n - 1}{i}$$

$$= Rs_{\overline{n}|i}.$$

The **periodic payment** R necessary to grow to the amount S in n periods at the interest rate i per period

$$R = S\frac{i}{(1 + i)^n - 1}$$

$$= S\frac{1}{s_{\overline{n}|i}}.$$

The **present value** A of an ordinary annuity with n payments R at the interest rate i per payment period

$$A = R\frac{1 - (1 + i)^{-n}}{i}$$

$$= Ra_{\overline{n}|i}.$$

The **periodic payment** R necessary to repay a loan A for n periods at the interest rate i per payment period

$$R = A\frac{i}{1 - (1 + i)^{-n}}$$

$$= A\frac{1}{a_{\overline{n}|i}}.$$

The **outstanding principal** B on a loan with periodic payment R for n periods at interest rate i after the kth payment is made

$$B = R\frac{1 - (1 + i)^{-n+k}}{i}$$

$$= Ra_{\overline{n-k}|i}.$$

REVIEW EXERCISES (CH. 7)

In exercises 1–4 determine the annual simple interest rate for the stated amount, principal, and time.

1. $6360 paid in 8 months to satisfy a debt of $6000.

2. $879.42 paid in 12 weeks to satisfy a debt of $850.

3. $518.23 paid in 105 days to satisfy a debt of $500. (Use an ordinary interest year.)

4. $1106.75 paid in $1\frac{1}{2}$ years to satisfy a debt of $950.

5. A major oil company charges $1\frac{1}{4}\%$ interest per month on past due bills. If a customer's bill is $155.40 and the customer is one month late in paying this bill, what is the total amount the customer must pay when the next bill is received?

6. A department store charges its credit card customers $1\frac{1}{2}\%$ interest per month on an unpaid balance. If a person owes $425 during one month and makes a payment of $150 at the end of the month, how much will the next monthly statement following this payment show is due, including finance charges?

7. A bank holds notes from a company for $2000 due now, $5000 due in 6 weeks and $3000 due in 2 months. The company wishes to make a single payment now to satisfy all the debts. If the bank and the company agree to use 13% simple interest, how much should this single payment be?

8. A company owes a supplier $4000 due in 120 days and $5000 due in 300 days. Instead of making separate payments when the amounts are due, the company arranges with the supplier to make a single payment in 240 days. If the company and the supplier agree to use 15% simple interest, how much should the single payment be? (Use an ordinary interest year.)

9. Find the discount and proceeds for a loan of $3500 due in 8 months using a discount rate of 12% interest.

10. Find the discount and proceeds for a loan of $2000 due in 270 days using a discount rate of 14.5% interest. (Use an ordinary interest year.)

In exercises 11–14 find the compound amount at the end of the stated term for the stated principal and rate.

11. $15,000 invested for 6 months at 9% interest converted monthly.

12. $12,500 invested for 5 years at 7% interest converted monthly.

13. $2000 invested for 2 years at 6% interest converted quarterly.

14. $7500 invested for 8 months at 14.2% interest converted monthly.

15. What is the effective rate when the nominal rate is 7% interest converted monthly?

16. What is the effective rate when the nominal rate is 5% interest converted continuously?

17. A man establishes a trust fund for his son on his son's tenth birthday. The fund pays 10% interest converted monthly. How much must be put into the fund now if the value of the fund is to be $10,000 on the son's eighteenth birthday?

18. A couple wants to save toward a vacation that they plan to take in $1\frac{1}{2}$ years. They open a

savings account with a deposit of $500. The account pays 6% interest converted monthly. The couple plans to make the following additional deposits to the account: $400 in 6 months, $300 in 9 months, $500 in 12 months, and $600 in 15 months. How much will they have in their account at the end of $1\frac{1}{2}$ years?

In exercises 19 and 20 use the stated periodic deposit, rate, and time to determine the amount that will be in the account when the last deposit is made.

19. $250 deposited at the end of each month for 2 years in an account that pays 8% interest converted monthly.

20. $650 deposited at the end of each quarter for 5 years in an account that pays 6% interest converted quarterly.

In exercises 21 and 22 determine the periodic deposit that must be made to a fund at the end of each conversion period to accumulate the stated amount in the stated time.

21. The amount must be $5000 at the end of 18 months, where the sinking fund pays 9% interest converted monthly.

22. The amount must be $12,000 at the end of 4 years, where the sinking fund pays 8.4% interest converted monthly.

23. A person obtains a car loan of $7500 to be repaid by equal monthly payments in 3 years at the rate of 15% interest.

   a. What is the size of the periodic payment needed to repay this loan?

   b. What will the outstanding balance be on the loan after the eighteenth payment is made?

24. A couple obtains a home mortgage of $50,000 to be repaid by equal monthly payments in 20 years at the rate of 13% interest.

   a. What is the size of the periodic payment needed to repay this mortgage?

   b. What will the outstanding principal be after the first year?

CHAPTER 8

Probability

Introduction

Uncertainty is something we all have to live with. For example, no one can predict with absolute certainty what tomorrow's weather will be. A meteorologist may state, however, what the chances are that a particular weather condition may occur. At the beginning of a semester, a student cannot say what grade he or she will receive for a given course; however, the student knows that the grade will be A, B, C, D, or F. If we play roulette, or any other game of chance, we cannot say with certainty whether we will win or lose. The possible outcomes of playing roulette are known, but our own actual outcome cannot be predicted.

Mathematicians have devised a model for a chance process — a process in which we know what could happen but do not know what will happen prior to performing the process. The name given to such a process is an **experiment**, and the model is called a **probability model**.

8.1 What Is Probability?

The first step in building a mathematical model of an experiment is to describe the set of possible outcomes, called the **sample space**, for the experiment.

EXAMPLES

1. Suppose an experiment consists of rolling a six-sided die and observing the upturned face. The six possible outcomes are given in the sample space $S = \{1, 2, 3, 4, 5, 6\}$.

2. An urn contains 2 red, 3 blue, and 4 green balls. An experiment consists of drawing a ball at random from the urn and observing its color. The sample space

S consists of the three outcomes: draw a red ball, draw a blue ball, or draw a green ball. For brevity, we give the sample space as $S = \{R, B, G\}$.

3. A market survey is made to determine the buying preference of shoppers with respect to brands of cereal. An experiment consists of selecting shoppers in a market and asking whether they buy brand X, brand Y, brand Z, or some other brand. The sample space for this experiment can be given as $S = \{X, Y, Z, Other\}$.

4. A survey of 100 males shows that 30 percent shave with an electric razor, 60 percent shave with a safety razor, and 10 percent do not shave. An experiment consists of selecting a person at random from the 100 in the survey and observing the category to which he belongs. The sample space for the experiment consists of three possible outcomes: the man selected either uses an electric razor, or uses a safety razor, or does not shave.

5. Consider an experiment that consists of flipping a coin repeatedly and observing the number of flips it takes for the first head to appear. The set of possible outcomes is infinite, and we give the sample space as $S = \{1, 2, 3, \ldots\}$.

Note that the sample space depends on the observation being made. For example, if we are to observe whether the number appearing on the roll of a six-sided die is even or odd, the sample space is $S = \{even, odd\}$, which is different from the sample space given in example 1.

Even though there are experiments with an infinite number of possible outcomes, such as given in example 5, in this text we will only consider experiments with a finite sample space.

One of our interests in experiments is to determine the likelihood that an event may occur. Since an event must consist of possible outcomes for the experiment, we define an **event** as any subset of the sample space. The empty set is a subset of any sample space; this subset is considered to be an event and is called the **impossible event** or **empty event**. Also, since the sample space is a subset of itself, the sample space is one of the events for an experiment and is called the **certain event**.

EXAMPLE 6

Consider an experiment that consists of flipping a coin. The sample space is $S = \{H, T\}$. There are four possible events: the empty set $\varnothing$, $\{H\}$, $\{T\}$, and S. The subsets $\{H\}$ and $\{T\}$ can be described as the event of obtaining a head and the event of obtaining a tail, respectively.

In addition to giving descriptions for the events $\{H\}$ and $\{T\}$, we can also give properties that determine the event $\varnothing$. For example, $\varnothing$ can be described as the event of obtaining both a head and a tail on the flip of the coin. This description is not unique, since $\varnothing$ can also be described as the event of obtaining a 2 on the flip of the coin.

We can also give a description for S. For example, S is the event of obtaining either a head or a tail.

As we see in example 6, a description can be given for each subset of the sample space. In addition, each verbally described event defines a subset of the sample space.

EXAMPLE 7

Consider an experiment that consists of rolling a six-sided die and that has the sample space $S = \{1, 2, 3, 4, 5, 6\}$. The property of obtaining an odd number determines the event $\{1, 3, 5\}$. The event of rolling an even number less than 5 is the subset $\{2, 4\}$. The event of rolling a 3 and a 5 is the empty event $\varnothing$.

In addition to a sample space and an event, a mathematical model includes a method of measuring the degree of certainty, called the **probability**, that an event may occur. On the one hand, a meteorologist may say that there is a 90 percent chance of rain tomorrow. We can interpret this statement to mean that it is almost certain to rain. On the other hand, if the forecast were a 10 percent chance of rain, we would consider that it is not very likely to rain tomorrow. If an event is absolutely certain, such as obtaining a number less than 7 when rolling a six-sided die, we can say that the event has 100 percent chance of occurring. As the likelihood that an event may occur diminishes, the probability of the event becomes smaller. If an event is impossible, such as obtaining an 8 on the single roll of a six-sided die, then the probability of the event is zero. Instead of using percentages to describe the chance that an event may occur, we will use fractional equivalents of percents in this text.

Before we can determine the probability of an event, we first assign a probability to each outcome in the sample space. The guideline for such an assignment of probabilities is given next.

Definition

■ *Assignment of Probabilities:*

Consider an experiment with n possible outcomes where $S = \{e_1, e_2, \ldots, e_n\}$. We assign to each outcome e_k in S a number called the **probability of** e_k, denoted by $Pr(e_k)$. This assignment of probabilities for S must satisfy the following two properties.

1. The probability of each outcome in S is a number between 0 and 1, inclusive; that is, $0 \leq Pr(e_k) \leq 1$, for each outcome e_k.

2. The sum of the probabilities of all the outcomes in S equals one; that is, $Pr(e_1) + Pr(e_2) + \cdots + Pr(e_n) = 1$. ■

The definition of a probability assignment gives only a general framework. It does not tell us how to assign probabilities for a particular experiment. However, since we wish to make a model of the experiment, we select the assignment that not only obeys the properties of the definition, but also describes the experiment most accurately.

EXAMPLE 8

Consider an urn that contains 3 red balls, 3 blue balls, and 3 green balls. An experiment consists of picking a ball at random from the urn and observing its color. The sample space for the experiment is $S = \{R, B, G\}$. Let us assign probabilities to S.

Consider the distribution of colors. We see that each color has the same chance of occurring. We want our assignment to reflect this; therefore, we must have $Pr(R) = Pr(B) = Pr(G)$. Furthermore, by the definition we must also have $Pr(R) + Pr(B) + Pr(G) = 1$. Thus we assign probabilities

$$Pr(R) = \frac{1}{3}, \quad Pr(B) = \frac{1}{3}, \quad \text{and} \quad Pr(G) = \frac{1}{3}.$$

Note that in example 8 the appropriate assignment for each color can be obtained from the proportion of the number of balls of that color to the total number of balls. In particular,

$$Pr(R) = \frac{\text{number of red balls}}{\text{total number of balls}} = \frac{3}{9} = \frac{1}{3}$$

$$Pr(B) = \frac{\text{number of blue balls}}{\text{total number of balls}} = \frac{3}{9} = \frac{1}{3}$$

$$Pr(G) = \frac{\text{number of green balls}}{\text{total number of balls}} = \frac{3}{9} = \frac{1}{3}.$$

EXAMPLES

9. Consider an urn that contains 3 red balls, 6 blue balls, and 9 green balls. An experiment consists of selecting a ball at random from the urn and observing its color. The sample space for this experiment is the same as that in example 8, namely, $S = \{R, B, G\}$. However, for this experiment the chance of obtaining a blue ball is twice that of obtaining a red one. Also the chance of obtaining a green ball is three times that of obtaining a red one. Therefore the model in example 8 would not be appropriate for this experiment. Let us determine an assignment of probabilities that characterizes the experiment.

Let us assign to each color the proportion of the number of balls of that color to the total number of balls. Thus we have

$$Pr(R) = \frac{\text{number of red balls}}{\text{total number of balls}} = \frac{3}{18} = \frac{1}{6}$$

$$Pr(B) = \frac{\text{number of blue balls}}{\text{total number of balls}} = \frac{6}{18} = \frac{1}{3}$$

$$Pr(G) = \frac{\text{number of green balls}}{\text{total number of balls}} = \frac{9}{18} = \frac{1}{2}.$$

Not only does this assignment meet the conditions that each probability be between 0 and 1 and that the sum of all the probabilities be 1, but also it reflects that $Pr(B) = 2Pr(R)$ and $Pr(G) = 3Pr(R)$.

10. Let us determine which assignment of probabilities for the outcomes of the

experiment given in example 4 will be the most accurate model for the experiment.

Recall that the sample space consists of the following three outcomes:

e_1 = the man selected uses an electric razor

e_2 = the man selected uses a safety razor

e_3 = the man selected does not shave.

Also recall that 30 percent of the men in the survey use an electric razor, 60 percent use a safety razor, and 10 percent do not shave. Since a man is selected at random, the chance that the outcome is e_2 is twice the chance of e_1 occurring and six times the chance of e_3 occurring. The assignment that most accurately models this is obtained from the percentages of men in each category, that is, $Pr(e_1) = 0.3$, $Pr(e_2) = 0.6$, and $Pr(e_3) = 0.1$.

The probabilities assigned in the examples thus far have been based on physical characteristics of the experiment. Let us now consider a fact in probability theory that will give us another way to determine a probability assignment.

The experiment in example 8 was simulated on a computer, and the printout is given in figure 8.1. The right-hand column is the ratio of the number of times the outcome occurred to the number of times the experiment was performed. Each of these ratios is called a **relative frequency**.

NUMBER OF TRIALS OF EXPERIMENT	OUTCOME	NUMBER OF TIMES OUTCOME OCCURRED	RELATIVE FREQUENCY
100	RED	33	.33
	BLUE	37	.37
	GREEN	30	.3
500	RED	167	.334
	BLUE	147	.294
	GREEN	186	.372
1000	RED	351	.351
	BLUE	301	.301
	GREEN	348	.348
2000	RED	693	.3465
	BLUE	626	.313
	GREEN	681	.3405
10000	RED	3327	.3327
	BLUE	3325	.3325
	GREEN	3348	.3348
50000	RED	16583	.33166
	BLUE	16728	.33456
	GREEN	16689	.33378

FIGURE 8.1

Definition   ■ Consider an experiment with a sample space S and suppose the experiment is performed a number of times. The **relative frequency** for an outcome in S is the ratio

$$\frac{\text{number of times the outcome occurred}}{\text{number of times the experiment was performed}} \,. \quad ■$$

Observe that there is no single relative frequency for a particular outcome; the relative frequency can change with the number of times the experiment is performed. For example, in figure 8.1 the relative frequencies for the outcome red are

$$0.33, \quad 0.334, \quad 0.351, \quad 0.3465, \quad 0.3327, \quad 0.33166.$$

Observe that the relative frequencies for a particular outcome do not change much as the number of performances of the experiment increases. Furthermore, it appears that they remain around $\frac{1}{3}$, which is the probability assigned in example 8. In fact, a theorem in probability theory states that this is always the case.

Theorem   ■ *Law of Large Numbers:*
Consider an experiment that can be repeated many times. As the number of performances of the experiment increases, the relative frequencies of an outcome for the experiment become and remain close to the probability of the outcome. ■

The probability of an outcome can be thought of as the long-term relative frequency of the outcome. Thus, for certain experiments, the probability assignment can be based on past behavior using relative frequency.

EXAMPLES   11. Suppose we observe the day-to-day performance of a particular stock listed on the New York Stock Exchange. We observe whether the price of the stock goes up, goes down, or remains the same. We make 100 observations and obtain the following results. The price of the stock went up 57 times, went down 32 times, and remained the same 11 times. Considering this situation as an experiment, let us give a probability model for it.

   The sample space for the experiment is given by $S = \{$Up, Down, Same$\}$. Let us assume that the model should reflect the past performance of the stock. Therefore we use relative frequency as a basis for the assignment of probability. Since the stock went up on 57 of the 100 observations, we can assign to the outcome Up the probability $\frac{57}{100}$. We can use similar fractions for the number of times each of the other possible outcomes were observed. Thus our assignment of probabilities is $Pr(\text{Up}) = \frac{57}{100}$, $Pr(\text{Down}) = \frac{32}{100}$, and $Pr(\text{Same}) = \frac{11}{100}$.

   12. Suppose the Weather Bureau compiled the data given in table 8.1 on rainfall (in inches) during the month of April over the last 80 years.

TABLE 8.1

Outcome		Frequency of occurrence
e_1	less than 1 inch	6 years
e_2	1 or more inches but less than 2 inches	26 years
e_3	2 or more inches but less than 3 inches	31 years
e_4	3 or more inches but less than 4 inches	14 years
e_5	4 or more inches	3 years

Using this information, give a model for the experiment of observing the amount of rainfall during the month of April.

The sample space is the set of five outcomes $S = \{e_1, e_2, e_3, e_4, e_5\}$. To assign probabilities that reflect the information obtained over 80 years, we assign to each outcome the relative frequency consisting of the number of years the outcome occurred divided by 80:

$$Pr(e_1) = \frac{6}{80}, \quad Pr(e_2) = \frac{26}{80}, \quad Pr(e_3) = \frac{31}{80}, \quad Pr(e_4) = \frac{14}{80}, \quad \text{and} \quad Pr(e_5) = \frac{3}{80}.$$

Sometimes the assignment of probabilities for an experiment is obtained through educated guessing. For example, suppose the experiment is the Superbowl game between two football teams, A and B. The sample space is {A wins, B wins}. What probability that team A will win would accurately reflect the situation? Many factors could be considered: the personnel on each team, the location of the game, past performances against common opponents during the regular season, and so forth. Based on perceived relevant factors, we then assign probabilities. Clearly this assignment is **subjective** and may differ from person to person.

EXAMPLES

13. Eight horses are entered in a race. Tom feels that only three of the horses, which we call A, B, and C, have any chance of winning. He believes that horse A is twice as likely to win as horse B and that horse C is four times as likely to win as horse B. He also feels that the chance of a tie is negligible. Based on his feelings, what would Tom assign as probabilities for the three possible outcomes: A wins, B wins, and C wins?

Let $x = Pr(\text{B wins})$. Then $Pr(\text{A wins}) = 2x$ and $Pr(\text{C wins}) = 4x$. Since Tom feels that the probability that any one of the other five horses might win is 0, and since the sum of all the probabilities must be 1, we have

$$Pr(\text{A wins}) + Pr(\text{B wins}) + Pr(\text{C wins}) = 1$$

that is,

$$2x + x + 4x = 1.$$

Solving for x, we obtain $7x = 1$ and $x = \frac{1}{7}$. Therefore the assignment that reflects Tom's intuition is $Pr(\text{A wins}) = \frac{2}{7}$, $Pr(\text{B wins}) = \frac{1}{7}$, $Pr(\text{C wins}) = \frac{4}{7}$, and the probability of another horse winning is 0.

14. Diane is considering her grade on the next test in her mathematics course. She feels that her chance for a C is four times that of a D, her chance for a B is not as high as that for a C but still is three times that for a D, and her chance for an A is

the same as that for a B. She has not ruled out the possibility of failing, but she feels that the chance of it happening is only 1 out of 100 (0.01). What assignment for the set of possibilities $S = \{A, B, C, D, F\}$ would reflect her intuition?

The chance of each of the outcomes A, B, and C is given in terms of the chance of D occurring. Therefore let $Pr(D) = x$. The probabilities of the other outcomes are $Pr(A) = Pr(B) = 3x$, $Pr(C) = 4x$, and $Pr(F) = 0.01$. Since the sum of all the probabilities must be 1, we must choose x such that

$$3x + 3x + 4x + x + 0.01 = 1.$$

Therefore, $\qquad\qquad 11x + 0.01 = 1, \quad x = 0.09.$

The assignment of probabilities is $Pr(A) = 0.27$, $Pr(B) = 0.27$, $Pr(C) = 0.36$, $Pr(D) = 0.09$, and $Pr(F) = 0.01$.

We have seen three ways in which a probability assignment can be obtained. These are based on

1. The physical characteristics of the experiment (examples 8, 9, and 10).
2. Past behavior using relative frequency (examples 11 and 12).
3. Subjective judgments (examples 13 and 14).

Now that we have discussed ways to obtain a probability assignment for a sample space S, we need a method for determining the probability of an event. This method is given in the following definition.

Definition   ■ *Probability of Events:*
Consider an experiment with a sample space $S = \{e_1, e_2, \ldots, e_n\}$ and an assignment of probabilities $Pr(e_1)$, $Pr(e_2)$, $\ldots$, $Pr(e_n)$. Let E be an event; that is, E is a subset of S. If E is the empty event, then the probability of E is zero. Symbolically, if $E = \varnothing$, then $Pr(E) = Pr(\varnothing) = 0$. If E is not the empty event, then the probability of E is the sum of the probabilities of the outcomes belonging to E. (For example, if $E = \{e_1, e_3, e_n\}$, then $Pr(E) = Pr(e_1) + Pr(e_3) + Pr(e_n)$.)   ■

Note (from the definition) that if E is the entire sample space S, then the probability of E is one. We therefore have two extremes: if the event cannot happen ($E = \varnothing$), then its probability is zero; if the event must happen ($E = S$), then its probability is one. Conversely, if the probability of an event is zero, the event cannot happen; if the probability of an event is one, the event must happen. The probability of any event, other than $\varnothing$ and S, must be strictly between 0 and 1. (See figure 8.2.)

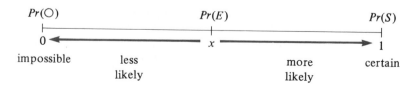

FIGURE 8.2

15. Consider an experiment that consists of rolling an unbiased six-sided die. The sample space is $S = \{1, 2, 3, 4, 5, 6\}$. Let E be the event of rolling an odd number, F be the event of rolling a number less than 5, and G be the event of rolling a number greater than 6. What is the probability of each of these events?

To answer these questions, we must first have a probability assignment for S. Since the die is unbiased, each of the six outcomes is equally likely to occur. Therefore the assignment that models this is $Pr(1) = Pr(2) = Pr(3) = Pr(4) = Pr(5) = Pr(6) = \frac{1}{6}$.

Next we identify the outcomes in each of the events and apply the definition for the probability of an event. In particular, $E = \{1, 3, 5\}$ and

$$Pr(E) = Pr(1) + Pr(3) + Pr(5) = \frac{1}{6} + \frac{1}{6} + \frac{1}{6} = \frac{3}{6}.$$

Also, $F = \{1, 2, 3, 4\}$ and

$$Pr(F) = 4\left(\frac{1}{6}\right) = \frac{4}{6}.$$

Finally, $G = \varnothing$ and therefore, $Pr(G) = 0$.

16. Consider the experiment in example 13 and the probability assignment obtained in that example. What is the probability that either horse A wins or horse B wins?

The event of interest is $E = \{A \text{ wins}, B \text{ wins}\}$. The probability of E is given by

$$Pr(E) = Pr(A \text{ wins}) + Pr(B \text{ wins}) = \frac{2}{7} + \frac{1}{7} = \frac{3}{7}.$$

17. Consider the experiment in example 11. Let us compute the probability that the price of the stock will not remain the same.

The event of interest is the subset $E = \{\text{Up, Down}\}$. Thus, using the probability assignment given in example 11, we have

$$Pr(E) = Pr(\text{Up}) + Pr(\text{Down}) = \frac{57}{100} + \frac{32}{100} = \frac{89}{100}.$$

18. Consider the experiment and the sample space $S = \{e_1, e_2, e_3, e_4, e_5\}$ in example 12. Recall from example 12 the meaning of each of these outcomes and the probability assignment for S. Let E be the event of less than 2 inches of rainfall during April, F be the event of 2 or more inches of rainfall during April, and G be the event of no less than 3 inches of rainfall during April. Find the probability of each of these events.

Since $E = \{e_1, e_2\}$, we have that the probability of less than 2 inches of rainfall during April is

$$Pr(e_1) + Pr(e_2) = \frac{6}{80} - \frac{26}{80} = \frac{32}{80}.$$

The outcomes in event F are e_3, e_4, and e_5. Thus,

$$Pr(F) = \frac{31}{80} + \frac{14}{80} + \frac{3}{80} = \frac{48}{80}.$$

Finally, event G is the same as the event of 3 or more inches of rainfall during April; therefore,

$$Pr(G) = Pr(e_4) + Pr(e_5) = \frac{14}{80} + \frac{3}{80} = \frac{17}{80}.$$

EXERCISES 8.1

1. Consider the experiment of rolling a die. Which of the following sets are sample spaces for this experiment?

   a. $S = $ {even number, odd number}.

   b. $S = $ The set of counting numbers greater than one and less than 6.

   c.

   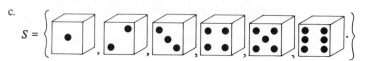

2. Consider the experiment of flipping two coins. Which of the following sets are sample spaces for this experiment?

   a. $S = $ {no heads, one head, two heads}.

   b. $S = $ {two heads, at least one tail}.

   c. $S = $ {no heads, one tail}.

3. In an experiment a chimpanzee pulls either a blue or a purple lever. Past observations indicate that the purple lever is pulled four times as frequently as the blue lever.

   a. Give a sample space for the experiment.

   b. Give an appropriate assignment of probabilities.

4. Consider the experiment of a mouse running a T-maze, where the mouse can go either right or left. Past observations indicate that the mouse will turn left five times as frequently as turning right.

   a. Give a sample space for the experiment.

   b. Give an appropriate assignment of probabilities.

5. In a market survey of 150 people, it was found that 52 people preferred drinking soda, 65 preferred drinking beer, 27 preferred drinking iced tea, and 5 preferred drinking none of the three. A person is selected at random from the 150 in the survey.

   a. Give a sample space for the experiment.

   b. Give an appropriate assignment of probabilities.

   c. What is the probability that the person selected prefers to drink either beer or soda?

6. In a survey group it was found that 41 percent of the people brush their teeth with Fluorsafe, 36 percent of the people brush with Xtra-Bright, 17 percent of the people brush with Dentfree, and 6 percent of the group do not brush their teeth. A person is selected at random from the survey group.

    a. Give a sample space for this experiment.

    b. Give an appropriate assignment of probabilities.

    c. What is the probability that the person selected does not brush with Xtra-Bright?

7. A survey of the number of cars waiting in line at a drive-in window at the Tri-City Bank was conducted. The survey consisted of 200 observations; the results are given in table 8.2.

TABLE 8.2

Number waiting in line	0	1	2	3	4 or more
Number of times observed	32	75	55	18	20

Consider the experiment of observing the number of cars waiting in line at this drive-in window.

    a. Give a sample space for this experiment.

    b. Give an appropriate assignment of probabilities.

    c. What is the probability that two or more cars are waiting in line?

    d. What is the probability that at most three cars are waiting in line?

8. The data in table 8.3 gives the number of minutes in a single hour that a worker on a production line is idle. This information was gathered over five 40-hour weeks.

TABLE 8.3

Minutes idle during one hour	Number of one-hour periods when this occurred
0	73
5	81
10	17
15	17
more than 15	12
	200

    a. Assign probabilities to the outcomes in $S = \{0, 5, 10, 15, \text{more than } 15\}$ that reflect the results in the table. Use this assignment to answer the questions in parts (b)–(d).

    b. What is the probability that a worker will be idle 5 minutes or 10 minutes in a 1-hour period?

    c. What is the probability that a production worker will be idle 15 or more minutes in a 1-hour period?

    d. What is the probability that a worker will not be idle more than 5 minutes in a 1-hour period?

9. A coin is weighted so that the probability of a head is $\frac{1}{3}$. What is the probability of a tail?

10. A coin is weighted so that the probability of a head is three times the probability of a tail. What is the probability of a head? What is the probability of a tail?

11. Consider the spinner in figure 8.3. It is divided into four equal spaces labeled 1, 2, 3, and 4. Consider the experiment of spinning the metal arrow and observing the space in which it stops. (If the arrow lands on a boundary line, it is spun again.)

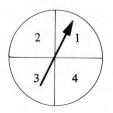

FIGURE 8.3

   a. Give a sample space for this experiment.

   b. Give an appropriate assignment of probabilities.

   c. What is the probability that the arrow lands in either the space labeled 1 or the one labeled 3?

12. Consider the spinner in figure 8.4. It is divided into three spaces colored red, blue, and green.

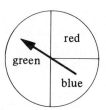

FIGURE 8.4

Consider the experiment of spinning the metal arrow and observing the space in which it stops. (If the arrow lands on a boundary line, it is spun again.)

   a. Give a sample space for this experiment.

   b. Give an appropriate assignment of probabilities.

   c. What is the probability that the arrow lands in either the blue or green space?

13. Consider an experiment with a sample space

$$S = \{e_1, e_2, e_3, e_4\}.$$

Suppose $Pr(e_1) = 3Pr(e_2)$, $Pr(e_2) = (\frac{1}{6})Pr(e_3)$, and $Pr(e_3) = Pr(e_4)$.

   a. Find the assignment of probabilities for S.

   b. If $E = \{e_1, e_4\}$, find $Pr(E)$.

   c. Describe an experiment for which S and the assignment of probabilities given in part (a) would be an appropriate model.

14. Consider an experiment with a sample space

$$S = \{e_1, e_2, e_3, e_4\}.$$

Suppose $Pr(e_1) = (\frac{1}{2})Pr(e_3)$, $Pr(e_3) = 4Pr(e_4)$, and $Pr(e_3) = (\frac{2}{3})Pr(e_2)$.

   a. Find the assignment of probabilities for S.

   b. If $F = \{e_3, e_2\}$, find $Pr(F)$.

   c. Describe an experiment for which S and the assignment of probabilities given in part (a) would be an appropriate model.

15. Erica is studying for a test in calculus. Based on her preparation, her performance on previous tests, and the information the professor gave to the class, she feels that she will pass the test. Furthermore, she feels that the probability of getting an A is twice that of getting a B, the probability of getting a B is three times the probability of getting a C, and the probability of getting a C is the same as the probability of getting a D.

    Based on these feelings, Erica finds the probability of getting a C or D. If it is larger than 0.1, she will do more studying; if it is less than 0.1, she will go to a party. What is Erica's decision?

16. In a tennis tournament there are four possible winners: Andrew, Barry, Charles, and Dean. The probability of Andrew winning is three times that of Dean winning. The probability of Barry winning is four times that of Dean winning. Barry and Charles each have the same chance of winning.

    a. Give an appropriate sample space for the experiment of running the tournament.

    b. Find an assignment of probability that reflects the information about the winners in the tournament.

    c. What is the probability that either Charles or Dean will win the tournament?

17. Labyrinth Games has 1200 cartridges in stock of which 52 are defective. A cartridge is selected at random from the stock and tested.

    a. Give an appropriate sample space for this experiment.

    b. Give an appropriate assignment of probabilities for the sample space of part (a).

18. In a group of 25 people, 4 of them have a cold. A person is selected at random. What is the probability that this person has a cold? What is the probability that this person does not have a cold?

19. In a group of 18 teenagers, 5 of them smoke cigarettes. A person is selected at random from this group. What is the probability that this person smokes cigarettes? What is the probability that this person does not smoke cigarettes?

20. A mathematics class has 12 males and 8 females. The names of the students are written on slips of paper, placed in a hat, and then mixed. The experiment is to draw one name at random from the hat. Give the assignment of probabilities for the sample space $S = \{\text{male, female}\}$.

21. Give a sample space for each experiment.

    a. A die is rolled and then a coin is flipped.

    b. A coin is flipped until a tail appears or three flips have been made.

22. An unbiased coin is flipped three times.

    a. Give a sample space for this experiment.

    b. Give an appropriate assignment of probabilities.

8.2 Properties of Probability Models

For a given experiment, we saw that any event E concerning the experiment is a subset of the sample space S. Thus we may think of S as the **universal set** and use the set theory to obtain properties of a probability model. Let us start by interpreting the set operations in terms of events. Let E and F be two events (subsets) in a sample space (universal set) S.

The **union** of the sets E and F

$$E \cup F = \{x \mid x \in E \quad \text{or} \quad x \in F\}$$

is the event "E or F (or both) happen."

The **intersection** of E and F

$$E \cap F = \{x \mid x \in E \quad \text{and} \quad x \in F\}$$

is the event "both E and F happen."

The **difference** of F from E

$$E - F = \{x \mid x \in E \quad \text{and} \quad x \notin F\}$$

is the event "E but not F happens."

The **complement** of E

$$E' = \{x \mid x \notin E\}$$

is the event "E does not happen."

Venn diagrams, as in figure 8.5 are useful in visualizing these concepts.

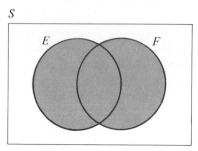

$E \cup F$

"E or F (or both) happen"

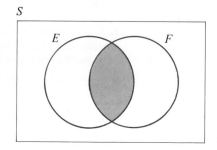

$E \cap F$

"both E and F happen"

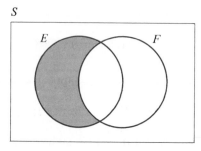

$E - F$

"E but not F happens"

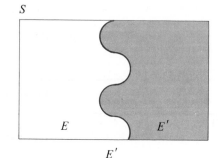

E'

"E does not happen"

FIGURE 8.5

EXAMPLE 1

Consider the experiment of drawing a card from a standard deck of 52 cards. The sample space S is given as

$$S = \begin{cases} A\clubsuit, K\clubsuit, Q\clubsuit, J\clubsuit, 10\clubsuit, \ldots, 2\clubsuit & \text{(clubs, black)} \\ A\diamondsuit, K\diamondsuit, Q\diamondsuit, J\diamondsuit, 10\diamondsuit, \ldots, 2\diamondsuit & \text{(diamonds, red)} \\ A\heartsuit, K\heartsuit, Q\heartsuit, J\heartsuit, 10\heartsuit, \ldots, 2\heartsuit & \text{(hearts, red)} \\ A\spadesuit, K\spadesuit, Q\spadesuit, J\spadesuit, 10\spadesuit, \ldots, 2\spadesuit & \text{(spades, black)} \end{cases}$$

Consider the events

E: the card drawn is red.

F: the card drawn is a black 2.

G: the card drawn is a face card (A, K, Q, or J).

Let us describe each of the following events

(a) $E \cap F$ (b) $E \cup F$ (c) $E \cap G$ (d) $G - E$ (e) E'.

It is clear that $E \cap F$ is empty and thus impossible. $E \cup F$ is the event of "drawing either a black 2 or a red card." $E \cap G$ is the event of "drawing a red face card." $G - E$ is the event of "drawing a face card that is not red." Finally, E' is the event of "drawing a black card."

For the events E and F in example 1, we have $E \cap F = \varnothing$; that is, they have no outcomes in common. As sets, we say that E and F are **disjoint**; as events, we say that E and F are **mutually exclusive** (see figure 8.6).

Mutually Exclusive Events

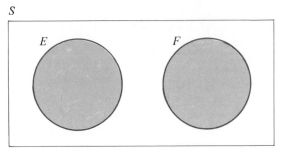

$Pr(E \cup F) = Pr(E) + Pr(F)$
$Pr(E$ or F happening$) = Pr(E$ happening$) + Pr(F$ happening$)$

FIGURE 8.6

Let us determine the probability of E happening or F happening (i.e., $Pr(E \cup F)$) when the two events E and F are mutually exclusive. Since E and F have no outcomes in common, the union of E and F is obtained simply by putting the outcomes of E with the outcomes of F. Further, since the probability of the event $E \cup F$ is the sum of the probabilities of the outcomes in $E \cup F$, we have

$$Pr(E \cup F) = Pr(E) + Pr(F); \tag{1}$$

that is,

$$Pr(E \text{ or } F \text{ happening}) = Pr(E \text{ happening}) + Pr(F \text{ happening}).$$

EXAMPLE 2

Consider the experiment of rolling an unbiased die once. What is the probability of obtaining 4 or an odd number?

The probability of obtaining 4 is $\frac{1}{6}$, and the probability of obtaining an odd number is $\frac{1}{2}$. Since the two events are mutually exclusive, the probability of obtaining 4 or an odd number is the sum of these two probabilities, $\frac{1}{6} + \frac{1}{2} = \frac{2}{3}$.

Observe that any event E and its complement E' are mutually exclusive. Thus, from formula (1), we have $Pr(E \cup E') = Pr(E) + Pr(E')$. Furthermore, since the union of E and E' is the sample space S, $Pr(E \cup E') = Pr(S) = 1$ (see figure 8.7). Therefore we have

$$Pr(E) = 1 - Pr(E') \tag{2}$$

that is,

$$Pr(E \text{ happening}) = 1 - Pr(E \text{ not happening}).$$

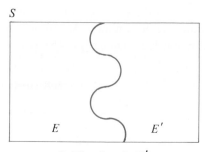

FIGURE 8.7

$$Pr(E) = 1 - Pr(E')$$
$$Pr(E \text{ happening}) = 1 - Pr(E \text{ not happening})$$

EXAMPLE 3

A survey was taken of the number of cars waiting for gasoline at a service station during a period of one week. Once each hour, 10 hours a day, an observer recorded the number of cars waiting in line. The results are given in table 8.4.

TABLE 8.4

Number of cars waiting in line	0	1	2	3	4	5	6	7	8	9	10 or more
Number of times observed	1	3	3	7	9	14	11	8	5	4	5

What is the probability that 3 or more cars will be waiting for gasoline?

The complement of this event is that 2 or fewer cars are waiting for gasoline; its probability is $\frac{7}{70}$. Therefore, by equation (2), the probability that three or more cars are waiting for gasoline is $1 - \frac{7}{70} = \frac{63}{70} = 0.9$.

Consider two events E and F that are not mutually exclusive, as in figure 8.8. Since in this case there are outcomes common to both events, the probability of E added to the probability of F "counts" twice the probability of both happening. Therefore the probability of E or F happening is equal to the sum of the probability of E and the probability of F minus the probability of both happening; that is

$$Pr(E \cup F) = Pr(E) + Pr(F) - Pr(E \cap F). \tag{3}$$

Not Mutually Exclusive

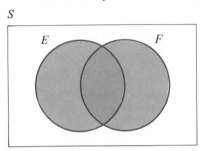

$$Pr(E \cup F) = Pr(E) + Pr(F) - Pr(E \cap F)$$
$$Pr(E \text{ or } F \text{ happening}) = Pr(E \text{ happening}) + Pr(F \text{ happening})$$
$$-Pr(\text{both happening})$$

FIGURE 8.8

EXAMPLES

4. Consider the experiment of drawing a card at random from a standard deck of 52 cards given in example 1. Let E be the event of drawing an ace and let F be the event of drawing a heart. Let us use formula (3) to find the probability of drawing either an ace or a heart.

Recall from example 1 the sample space of 52 outcomes. Since each card has the same chance of being drawn and the sum of the probabilities of these 52 outcomes must be 1, we assign to each outcome the probability $\frac{1}{52}$. Now, the probability of drawing either an ace or a heart is $Pr(E \cup F)$. Since the events E and F have the outcome ace of hearts in common, the events are not mutually exclusive and

$$Pr(E \cup F) = Pr(E) + Pr(F) - Pr(E \cap F).$$

Since there are 4 outcomes in E, $Pr(E) = 4(\frac{1}{52}) = \frac{4}{52}$. Similarly there are 13 outcomes in F, and therefore $Pr(F) = \frac{13}{52}$. Thus $Pr(\text{ace or heart}) = \frac{4}{52} + \frac{13}{52} - \frac{1}{52} = \frac{16}{52}$.

5. A middle-level executive in a large company feels that the probability she will be transferred to another division within the company is 0.5 and that the probability she will be promoted to assistant vice-president is 0.6. She believes that the probability of being either transferred or promoted to assistant vice-president is 0.8. What is the probability that she will be transferred and promoted? Let E be the event that she is transferred and F be the event that she is promoted to assistant vice-president. From the information given.

$$Pr(E) = 0.5 \qquad Pr(F) = 0.6 \qquad \text{and} \qquad Pr(E \cup F) = 0.8.$$

We want to find $Pr(E \cap F)$. Solving formula (3) for $Pr(E \cap F)$, we have

$$Pr(E \cap F) = Pr(E) + Pr(F) - Pr(E \cup F).$$

Therefore,

$$Pr(\text{transferred and promoted}) = 0.5 + 0.6 - 0.8 = 0.3.$$

We conclude this section with two additional relationships for two events E and F. These relationships are called De Morgan's laws.

$$(E \cup F)' = E' \cap F' \qquad \text{first De Morgan law}$$

$$(E \cup F)' = E' \cup F' \qquad \text{second De Morgan law}$$

Both laws can be verified by using Venn diagrams. The first law is verified by figure 8.9, and the verification of the second law is left as an exercise.

We can express De Morgan's laws in words as follows:

> E or F does not happen
> means the same as   first De Morgan law
> neither E nor F happens

and

> E and F does not happen
> means the same as   (second De Morgan law)
> either E does not happen or F does not happen

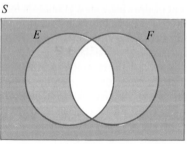

E' and F'

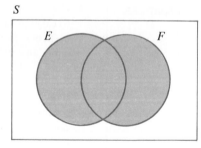

$E \cup F$

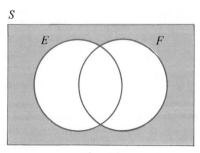

$E' \cap F'$

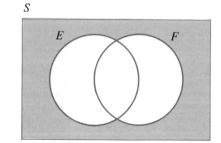

$(E \cup F)'$

$E' \cap F' = (E \cup F)'$

FIGURE 8.9

Since the probabilities are the same for two equal events, we have from De Morgan's laws that

$$Pr(E \text{ or } F \text{ does not happen}) = Pr(\text{neither } E \text{ nor } F \text{ happens})$$

and

$$Pr(E \text{ and } F \text{ does not happen}) = Pr(\text{either } E \text{ does not happen or } F \text{ does not happen}).$$

EXAMPLE 6

Two college freshmen, Scott and Ralph, try out for the basketball team. Scott and Ralph feel that the likelihood of neither of them making the team is $\frac{1}{3}$. What is the probability of at least one of them making the team?

Let E be the event that Scott makes the team and F be the event that Ralph makes the team. From the information given, $Pr(E' \cap F') = \frac{1}{3}$ and, therefore, from the first De Morgan law

$$Pr((E \cup F)') = Pr(E' \cap F') = \tfrac{1}{3}.$$

Thus,

$$Pr(\text{at least one makes the team}) = Pr(E \cup F)$$
$$= 1 - \tfrac{1}{3}$$
$$= \tfrac{2}{3}.$$

EXERCISES 8.2

---

1. Consider the experiment of selecting a person at random from a set of registered voters. Let E be the event that the person selected makes more than \$30,000 a year and let F be the event that the person selected is a Republican. Express in words the following events.

   a. $E \cap F$ b. $E \cup F$ c. F' d. $F - E$ e. $E \cap F'$ f. $E' - F$

   g. $(E \cup F)'$ h. $E' \cap F'$ i. $E' - F'$ j. $F - E'$

2. David is taking a computer science course and a history course during the spring term. Let E be the event that he passes the computer science course and F be the event that he passes the history course. Express in words the following events.

   a. $E \cup F$ b. $E \cap F$ c. E' d. $E - F$ e. $E' \cup F$ f. $F - E$

   g. $F - E'$ h. $(E \cap F)'$ i. $E' \cup F'$ j. $(E - F)'$

3. Let E and F be two events. Symbolize the following events in terms of E and F.

   a. At least one of the two events occurs.

   b. Both events occur.

   c. Either E or F occurs.

   d. Neither E nor F occurs.

4. Let E and F be two events. Symbolize the following events in terms of E and F.
   a. F, but not E, occurs.
   b. Either E occurs or F does not occur.
   c. It is not the case that both events occur.
   d. Exactly one of the two events occurs.

5. Consider the experiment of selecting a student at random from Midwest University. Let E be the event that the student selected smokes cigarettes and F be the event that the student selected drinks beer. For these two events, state in words each of De Morgan's laws.

6. Consider the experiment of drawing a card at random from a standard deck of 52 cards. Let E be the event that the card drawn is an ace and let F be the event that the card drawn is a spade. For these two events, state in words each of De Morgan's laws.

7. Consider the experiment of selecting a student at random from Clarion University. Let

$$E = \text{The student selected owns a car.}$$

$$F = \text{The student selected owns a bicycle.}$$

   Are E and F mutually exclusive? Why?

8. A card is drawn at random from a standard deck of 52 cards. Which of the following pairs of events are mutually exclusive?
   a. "The card is black" and "The card is a spade."
   b. "The card is a diamond" and "The card is not red."

9. Let E and F be two mutually exclusive events.
   a. If $Pr(E) = 0.31$ and $Pr(F) = 0.22$, find $Pr(E \cup F)$.
   b. If $Pr(E) = \frac{1}{3}$ and $Pr(E \cup F) = \frac{5}{12}$, find $Pr(F)$.

10. Let E and F be two mutually exclusive events.
   a. If $Pr(E) = \frac{1}{4}$ and $Pr(F) = \frac{1}{20}$, find $Pr(E \cup F)$.
   b. If $Pr(E \cup F) = 0.751$ and $Pr(F) = 0.369$, find $Pr(E)$.

11. Let A and B be two events of a sample space S.
   a. If $Pr(A) = 0.35$, $Pr(B) = 0.50$, and $Pr(A \cap B) = 0.15$, find $Pr(A \cup B)$.
   b. If $Pr(A) = \frac{4}{21}$, $Pr(B) = \frac{13}{42}$, and $Pr(A \cup B) = \frac{17}{42}$, find $Pr(A \cap B)$.

12. Let A and B be two events of a sample space S.
   a. If $Pr(A) = \frac{4}{9}$, $Pr(B) = \frac{1}{6}$, and $Pr(A \cap B) = \frac{1}{9}$, find $Pr(A \cup B)$.
   b. If $Pr(A) = 0.75$, $Pr(B) = 0.90$, and $Pr(A \cup B) = 0.96$, find $Pr(A \cap B)$.

13. An item that is mass produced must pass two tests, A and B, before it is put on the market. From past experience, the probability that an item will pass test A is 0.87, and the probability that an item will pass test B is 0.77. If the probability that an item will pass at least one of the tests is 0.91, find the probability that an item will pass both tests.

14. Cindy feels that the probability of passing business mathematics is 0.6 and that the probability of passing accounting is 0.5. She feels that her chances of passing at least one of the two courses is 0.8. What is the probability that she will pass both courses?

15. Use the information given in example 3 to find
   a. The probability that between 4 and 7 cars, inclusive, are waiting for gasoline.

b. The probability that 8 or more cars are waiting for gasoline.

c. The probability that 7 or fewer cars are waiting for gasoline.

16. Use the information given in example 3 to find

   a. The probability that between 2 and 6 cars, inclusive, are waiting for gasoline.

   b. The probability that 7 or more cars are waiting for gasoline.

   c. The probability that 6 or fewer cars are waiting for gasoline.

17. Consider an experiment with a sample space S and two events E and F.

   a. If $Pr(E' \cap F') = \frac{2}{7}$, find $Pr(E \cup F)$.

   b. If $Pr(E' \cup F') = 0.723$, find $Pr(E \cap F)$.

18. Consider an experiment with a sample space S and two events E and F.

   a. If $Pr(E' \cap F') = 0.391$, find $Pr(E \cup F)$.

   b. If $Pr(E' \cup F') = \frac{5}{13}$, find $Pr(E \cap F)$.

19. Consider an experiment with a sample space S and two events E and F. Suppose $Pr(E) = 0.58$, $Pr(F) = 0.40$, and $Pr(E' \cap F') = 0.25$. Find

   a. $Pr(E \cup F)$
   b. $Pr(E \cap F)$.

20. Let E and F be two events of a sample space S. Suppose $Pr(E) = \frac{5}{11}$, $Pr(F) = \frac{6}{11}$, and $Pr(E' \cap F') = \frac{2}{11}$. Find

   a. $Pr(E \cup F)$
   b. $Pr(E \cap F)$.

21. Let E and F be two events of a sample space S. Suppose $Pr(E) = \frac{4}{15}$, $Pr(F) = \frac{1}{5}$, and $Pr(E' \cup F') = \frac{14}{15}$. Find ·

   a. $Pr(E \cap F)$
   b. $Pr(E \cup F)$.

22. Consider an experiment with a sample space S and two events E and F. Suppose $Pr(E) = 0.620$, $Pr(F) = 0.698$, and $Pr(E' \cup F') = 0.539$. Find

   a. $Pr(E \cap F)$
   b. $Pr(E \cup F)$.

23. Carole and Maria try out for the girls' tennis team. The probability that at least one of them will make the team is $\frac{2}{3}$. What is the probability that neither of them will make the team?

24. The weather forecast for the greater Beaver Valley metropolitan area is 85 percent chance of rain or snow during the next 24 hours. What is the likelihood of neither rain nor snow?

25. Paul and Beth each apply for a job at Crossfire Enterprises. The probability that both will be offered jobs is 0.8. What is the probability that at least one of them will not be offered a job?

26. Richard applies for two jobs, one with Gas Products and the other with Digital Supplies. After interviewing with both companies, he feels that there is $\frac{1}{3}$ chance of either Gas Products not offering him a job or Digital Supplies not offering him a job. Find the probability that both companies will offer Richard a job.

27. The probability that a university student selected at random from the University population smokes cigarettes is 0.5 and the probability that the student drinks beer is 0.9. Furthermore, the probability that the student neither smokes cigarettes nor drinks beer is 0.02. What is the probability that the student smokes cigarettes and drinks beer?

28. An advertising firm has produced two different commercials for Dentfree toothpaste. They conducted a survey to obtain the following information. The probability that a person will like commercial A is 0.39. The probability that a person will like commercial B

is 0.59. The probability that a person will like neither commercial is 0.31. Find the probability that a person will like both commercials.

29. Let E and F be two events of a sample space S. Show that

$$Pr(E \cap F') = Pr(E) - Pr(E \cap F).$$

(*Hint:* Refer to figure 8.10.)

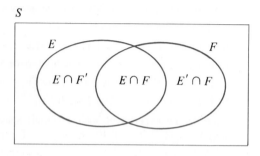

FIGURE 8.10

30. Let E and F be two events of a sample space S.
    a. If $Pr(E) = \frac{5}{13}$ and $Pr(E \cap F) = \frac{2}{13}$, find $Pr(E \cap F')$.
    b. If $Pr(F) = 0.53$ and $Pr(F \cap E') = 0.41$, find $Pr(E \cap F)$.
    (*Hint:* Refer to exercise 29.)

31. Let E and F be two events of a sample space S.
    a. If $Pr(E) = 0.62$ and $Pr(E \cap F) = 0.27$, find $Pr(E \cap F')$.
    b. If $Pr(F) = \frac{2}{3}$ and $Pr(F \cap E') = \frac{7}{15}$, find $Pr(E \cap F)$.
    (*Hint:* Refer to exercise 29.)

32. Use the information in example 5 to find the probability that the executive is transferred but not promoted.
    (*Hint:* Refer to exercise 29.)

33. The probability that a college student chosen at random from the student population at State University is taking a computer science course is 0.652. The probability that the student is taking a computer science course but not a business course is 0.286. Find the probability that the student is taking both a computer science course and a business course. (*Hint:* Refer to exercise 29.)

34. Let A, B, and C be three events associated with an experiment. Symbolize the following events.
    a. C and A, but not B, occur.
    b. B and C do not occur.
    c. At least one of the three events occurs.

35. Let E_1, E_2, and E_3 be three events associated with an experiment. Symbolize the following events.
    a. All three events occur.
    b. None of the three events occurs.
    c. Exactly one of the three events occurs.

36. Let E and F be two events of a sample space S. Show that
    a. $Pr(E \cup F) \le Pr(E) + Pr(F)$. (*Hint:* Refer to equation (3).)
    b. $Pr(E \cap F) \le Pr(E)$. (*Hint:* Refer to exercise 29.)

37. Let E, F, and G be three events where E and F are mutually exclusive. Show that $E \cap G$ and $F \cap G$ are mutually exclusive.

8.3 Equiprobable Models: The Principle of Counting and Permutations

We have seen experiments (examples 8 and 15 in section 8.1 and example 4 in section 8.2) in which all the outcomes have the same chance of occurring. In this case we say that the outcomes are **equally likely** or **equiprobable**. For example, the experiment of drawing a card at random from a standard deck of 52 cards has 52 equally likely outcomes. For experiments with equally likely outcomes there is a special formula for the probability of an event, which we now introduce.

Consider a sample space S of an experiment with equally likely outcomes. To reflect that the outcomes have the same chance of occurring, we assign to each outcome the same number. In particular, let $N(S)$ denote the **number of elements of** S. Since there are $N(S)$ outcomes and since the sum of the probabilities of all these outcomes must be 1, we must assign to any outcome e in S the probability

$$Pr(e) = \frac{1}{N(S)} \tag{4}$$

Now, let E be an event in S. Since the probability of E is the sum of the probabilities of the outcomes in E, the probability of E is obtained by adding the common probability $1/N(S)$ as many times as there are outcomes in E: that is,

$$Pr(E) = \underbrace{\frac{1}{N(S)} + \frac{1}{N(S)} + \cdots + \frac{1}{N(S)}}_{N(E) \text{ terms}} = \frac{N(E)}{N(S)}. \tag{5}$$

Thus, for an experiment with equally likely outcomes, the probability of an event is the number of ways it can happen, $N(E)$, divided by the total number of possibilities, $N(S)$.

EXAMPLE 1

Consider the experiment of rolling a pair of fair dice. An outcome in the sample space S is an ordered pair (a, b), where a is the number on the upturned face of the first die and b is the number on the upturned face of the second die. This set of ordered pairs can be denoted by the **Cartesian product** $\{1, 2, 3, 4, 5, 6\} \times \{1, 2, 3, 4, 5, 6\}$ and displayed as shown in table 8.5.

TABLE 8.5

	1	2	3	4	5	6
1	(1,1)	(1,2)	(1,3)	(1,4)	(1,5)	(1,6)
2	(2,1)	(2,2)	(2,3)	(2,4)	(2,5)	(2,6)
3	(3,1)	(3,2)	(3,3)	(3,4)	(3,5)	(3,6)
4	(4,1)	(4,2)	(4,3)	(4,4)	(4,5)	(4,6)
5	(5,1)	(5,2)	(5,3)	(5,4)	(5,5)	(5,6)
6	(6,1)	(6,2)	(6,3)	(6,4)	(6,5)	(6,6)

Since table 8.5 has 6 rows and 6 columns, there are 36 entries. Furthermore, since the outcomes in S are equiprobable, each outcome in S has probability $\frac{1}{36}$.

Let E be the event "the sum of the numbers on the upturned faces is eleven." E contains the two outcomes $(5,6)$ and $(6,5)$; therefore, by formula (5), we have $Pr(E) = \frac{2}{36}$.

The number of outcomes can be found by listing and counting them. However, this can be tedious when the experiment has a large number of outcomes. Furthermore, often we are only interested in the size of the sample space. Therefore we need counting techniques for finding the number of outcomes.

Our first counting technique was illustrated in example 1. Consider two sets $A = \{a_1, a_2, \ldots, a_m\}$ and $B = \{b_1, b_2, \ldots, b_n\}$. The **Cartesian product** $A \times B$ is the set of all ordered pairs (a_i, b_j). The elements of $A \times B$ can be displayed in an array with $N(A)$ rows and $N(B)$ columns as shown in table 8.6. Since there are mn entries in table 8.6, the number of elements in the Cartesian product $A \times B$ is

$$N(A \times B) = N(A) \cdot N(B) = m \cdot n. \tag{6}$$

TABLE 8.6

	b_1	b_2	$\cdots$	b_n	
a_1	(a_1,b_1)	(a_1,b_2)	$\cdots$	(a_1,b_n)	
a_2	(a_2,b_1)	(a_2,b_2)	$\cdots$	(a_2,b_n)	$N(A)$ rows
$\vdots$		$\vdots$			
a_m	(a_m,b_1)	(a_m,b_2)	$\cdots$	(a_m,b_n)	

$N(B)$ columns

EXAMPLE 2

Consider an experiment that consists of flipping a coin followed by rolling a die. Let S be the sample space for this experiment and determine $N(S)$.

An outcome in S is an ordered pair (a, b) where a is the result of flipping the coin and b is the result of rolling the die. Thus, setting $A = \{H, T\}$ and $B = \{1, 2, 3, 4, 5, 6\}$, we have $S = A \times B$ and $N(S) = N(A)N(B) = 2 \cdot 6 = 12$ (see table 8.7).

TABLE 8.7

	1	2	3	4	5	6
H	(H,1)	(H,2)	(H,3)	(H,4)	(H,5)	(H,6)
T	(T,1)	(T,2)	(T,3)	(T,4)	(T,5)	(T,6)

The outcomes for the experiment in example 2 can also be viewed as the result of performing two tasks in succession. The first task is to pick a result for the flip of the coin, and the second task is to pick a result for the roll of the die. In this case we can also display the outcomes in S by a **tree diagram** as shown in figure 8.11.

To construct the tree diagram, we choose a point **start** and draw from this point a **branch** for each way the first task can be accomplished. At the end of each of these branches, we then draw branches that correspond to the possible ways of performing the second task. An outcome in S is determined by a **path** in the tree. A path is a sequence of branches beginning at start and finishing at an endpoint of the tree. The number of outcomes in S is the same as the number of paths in the tree diagram. We can use formula (6) to count the number of paths in the tree diagram.

first task | second task | outcome

	1	(H, 1)
	2	(H, 2)
	3	(H, 3)
H	4	(H, 4)
	5	(H, 5)
	6	(H, 6)
START	1	(T, 1)
	2	(T, 2)
	3	(T, 3)
T	4	(T, 4)
	5	(T, 5)
	6	(T, 6)

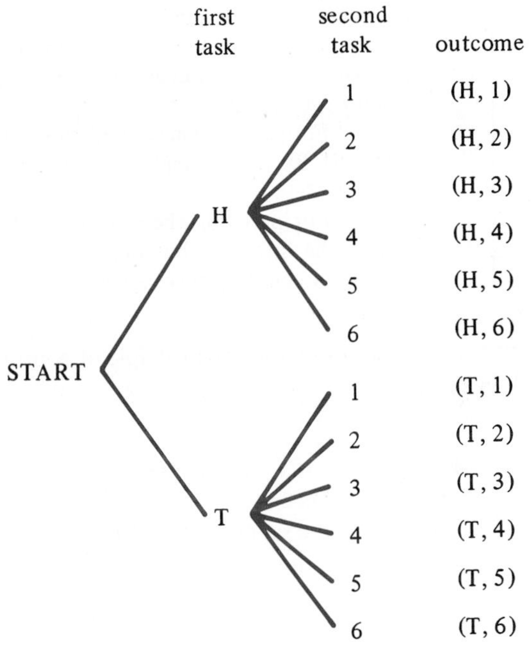

FIGURE 8.11

Definition

■ *Principle of Counting:*

Suppose two tasks are to be performed in succession. Let A_1 be the set of results of performing the first task and A_2 be the set of results of performing the second task. The outcomes of performing the two tasks are given in $A_1 \times A_2$, and the number of ways to perform the two tasks is

$$N(A_1) \cdot N(A_2). \quad ■ \tag{7}$$

EXAMPLES

3. There are 5 major roads from Allentown to Bethlehem and 4 major roads from Bethlehem to Easton. How many routes from Allentown to Easton are available to a bus company with a stop in Bethlehem?

We can answer this question by applying the principle of counting. A possible route can be constructed by performing two tasks in succession. The first task is to choose a major road from Allentown to Bethlehem. The second task is to choose a major road from Bethlehem to Easton. The first task can be done in 5 ways and the second task can be done in 4 ways. Therefore, by the principle of counting, there are $5 \cdot 4 = 20$ routes available to the bus company.

4. How many two-digit numbers can be formed using the six digits in the set $\{1, 2, 3, 4, 5, 6\}$ where

    a. Digits can be used more than once?

    b. Digits cannot be used more than once?

We answer both (a) and (b) by using the principle of counting. To construct a two-digit number, we perform two tasks in succession. The first task is to select a digit for the first position in the number and the second task is to select a digit for the second position in the number. We count the number of ways of doing each task.

For part (a), where digits can be repeated, there are 6 ways to perform each task. By the principle of counting, there are $6 \cdot 6 = 36$ such two-digit numbers.

For part (b), where digits cannot be repeated, there are 6 ways to perform the first task, but only 5 ways to perform the second task. By the principle of counting, there are $6 \cdot 5 = 30$ such two-digit numbers.

We can extend the principle of counting to any number of tasks performed in succession.

Definition

■ *Principle of Counting (General):*
Suppose n tasks are to be performed in succession. Let $A_1, A_2, \ldots, A_n$ be, respectively, the set of results of performing the first task, the second task, and so forth. The number of ways to perform the n tasks in succession is the product

$$N(A_1)N(A_2) \cdots N(A_n). \quad ■ \qquad (8)$$

EXAMPLES

5. The research and development division of an automobile company is designing an electric car. The car must be competitively priced and mass-produced. Within these parameters, there are 12 lightweight metal alloys possible for the chassis, 8 rechargeable battery systems, and 5 aerodynamic designs available for the body of the car. With this information, how many design options does the company have in developing an electric car?

The answer to this question can be obtained by using the (general) principle of counting. There are three tasks to be performed in succession. The first task can be done in 12 ways, the second task in 8 ways, and the third task in 5 ways. By the principle of counting, there are $12 \cdot 8 \cdot 5 = 480$ design options.

6. Consider the experiment of flipping 4 coins. Let us find the number of elements in the sample space S for this experiment.

We can construct an outcome in S by performing four tasks in succession, namely, select either heads (H) or tails (T) for each of the 4 coins. Thus, by the principle of counting, there are $2 \cdot 2 \cdot 2 \cdot 2 = 2^4 = 16$ possible outcomes. The outcomes for this experiment are given by the tree diagram in figure 8.12.

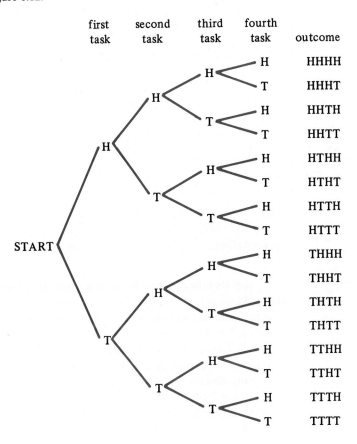

FIGURE 8.12

The outcomes for some experiments consist of different arrangements of objects in a list.

EXAMPLE 7

In a history test a student is given three historical events and asked to list them in chronological order, from the earliest to the most recent. If the student simply guesses, how many different answers are possible?

Let the historical events be designated as 1, 2, and 3. Each possible answer is an arrangement of the three events. We can view the problem of arranging the three events as performing three tasks in succession. The first task is to pick an historical event for the first position in the list, which can be accomplished in 3 ways. The second task is to select from among the remaining 2 historical events one for the second position in the list, which can be done in 2 ways. The final task is to place the remaining event in the list, which can be done in 1 way. Using the principle of counting, we see that the number of possible answers is $3 \cdot 2 \cdot 1 = 6$. (See figure 8.13.)

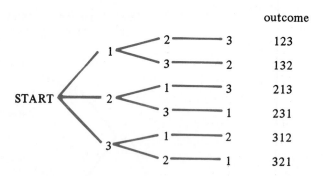

FIGURE 8.13

Definition

■ Given a set of n distinct objects, an arrangement of these n objects in a definite order where no object is repeated in the arrangement, is called a **permutation** of the set of n objects. ■

Let us now obtain a formula for counting the number of permutations of n objects. As we saw in example 7, we can view the problem of arranging objects in a list as performing tasks in succession. To count the permutations of the objects in the set $\{a, b, c, d\}$, we can count the number of ways to perform four tasks in succession. The first task is to select one of the objects for the first position in the list, which can be accomplished 4 ways. The second task is to select from among the 3 remaining objects one for the second position. The third task is to select one of the remaining 2 objects for the third position, and the fourth task is to place the 1 remaining object in the last position. The number of ways to perform these four tasks is $4 \cdot 3 \cdot 2 \cdot 1 = 24$. With example 7 and the preceding discussion in mind, there should be no difficulty in accepting the following formula for counting permutations.

Definition

■ *Formula for Counting Permutations:*
The number of permutations of n distinct objects is given by

$$n! = n(n - 1)(n - 2) \cdots (2)(1) \tag{9}$$

$n!$ is read "*n* factorial." ■

The number of permutations of n objects is the product of n integers starting with n and decreasing to 1. Also we define $0! = 1$. Observe from the definition of $n!$ and table 8.8 that, for example, $7! = 7(6!)$ and, in general, $n! = n[(n-1)!]$. Thus,

$$8! = 8(7!) = 8(5040) = 40,320.$$

TABLE 8.8   **Table of Factorials**

$n!$
$0! = 1$
$1! = 1$
$2! = 2(1) = 2$
$3! = 3(2)(1) = 6$
$4! = 4(3)(2)(1) = 24$
$5! = 5(4)(3)(2)(1) = 120$
$6! = 6(5)(4)(3)(2)(1) = 720$
$7! = 7(6)(5)(4)(3)(2)(1) = 5040$

EXAMPLE 8

Consider an experiment that tests a monkey's learning ability. The monkey is required to pull five levers in a particular order to receive a reward. If the sequence in which the levers are pulled is different from the required one, no reward is received. The experiment is performed a large number of times. The relative frequency of the outcome of picking the correct sequence is computed and compared to the probability of picking this sequence at random (by guessing). If these two numbers are close, we can say that little or no learning took place. What is the probability that the monkey will pull the levers at random in the correct order?

Suppose the levers are labeled 1, 2, 3, 4, and 5. The sample space S consists of all possible arrangements of these five numbers. If the monkey pulls the five levers in a random fashion, the outcomes in the sample space are equiprobable and the probability of any one of them is $1/N(S)$. It is clear that $N(S) = 5! = 120$. Thus the probability of pulling the five levers at random and obtaining the correct arrangement is $\frac{1}{120} = 0.008$.

In some counting problems involving arrangements, we do not wish to arrange all the objects from a set.

EXAMPLE 9

In an ESP experiment a person is asked to select and arrange 2 cards from a set of 5 cards labeled A, B, C, D, and E. Without seeing the cards, a second person is asked to give the arrangement he thinks he perceives. Let us determine the number of possible responses by the second person if he simply guesses.

We can think of an arrangement of the 2 cards selected from the 5 cards as the result of performing two tasks in succession. The first task is to pick one of the 5 cards for the first position of the arrangement, which can be done in 5 ways. Since a card cannot be repeated, the second task is to pick from the 4 remaining cards one for the second position of the arrangement, which can be done in 4 ways. The possible arrangements of the 2 cards selected from the 5 are given by the tree diagram in figure 8.14.

By the counting principle, the number of possible arrangements of 2 cards chosen from the set of 5 cards is $5 \cdot 4 = 20$.

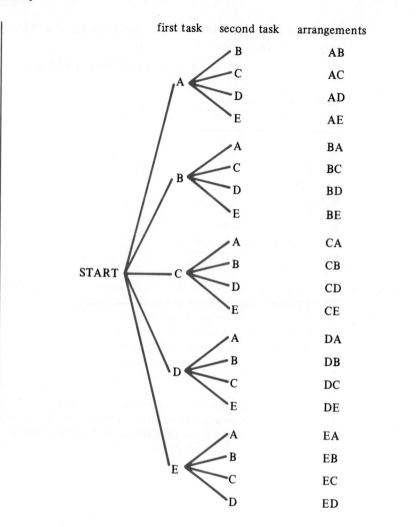

first task   second task   arrangements

A — B   AB
A — C   AC
A — D   AD
A — E   AE

B — A   BA
B — C   BC
B — D   BD
B — E   BE

C — A   CA
C — B   CB
C — D   CD
C — E   CE

D — A   DA
D — B   DB
D — C   DC
D — E   DE

E — A   EA
E — B   EB
E — C   EC
E — D   ED

FIGURE 8.14

Definition   ■ An arrangement of r objects that are chosen from a set of n distinct objects is called a **permutation of n objects taken r at a time**. The number of all such permutations is denoted by $P[n, r]$. ■

With example 9 in mind, let us now develop the general formula for $P[n, r]$, the number of permutations of n objects taken r at a time. As in example 9, to construct such a permutation we must perform r tasks in succession. The first task is to pick one of the n objects for the first position of the arrangement, which can be done in n ways. Since we cannot repeat an object, the second task is to select from among the $(n - 1)$ remaining objects one for the second position, which can be done in $(n - 1)$ ways. Similarly, after performing the first two tasks, we pick one of the $(n - 2)$ remaining objects for the third position of the arrangement. This process continues until we have performed the rth task.

There is a pattern in table 8.9 and we can say that there are $n - (r - 1) = n - r + 1$ ways to perform the rth task.

TABLE 8.9

Task	Number of ways to perform this task
first	n
second	$n - 1$
third	$n - 2$
fourth	$n - 3$
$\vdots$	$\vdots$
rth	$n - (r - 1)$

Definition

■ *Formula for P[n, r]:*

The number of permutations of n objects taken r at a time is given by

$$P[n, r] = \underbrace{n(n - 1)(n - 2) \cdots (n - r + 1)}_{r \text{ factors}}. \quad ■ \qquad (10)$$

Note that r must always be less than or equal to n. An easy way to remember the right-hand side of formula (10) is to start with n and "count down" r factors. For example,

$$P[n, 1] = n$$
$$1 \text{ factor}$$

$$P[5, 1] = 5$$

$$P[n, 2] = n(n - 1)$$
$$2 \text{ factors}$$

$$P[4, 2] = 4(3) = 12$$

$$P[n, 3] = n(n - 1)(n - 2)$$
$$3 \text{ factors}$$

$$P[7, 3] = 7(6)(5) = 210$$

$$P[n, 4] = n(n - 1)(n - 2)(n - 3)$$
$$4 \text{ factors}$$

$$P[10, 4] = 10(9)(8)(7) = 5040$$

In the case where $r = n$, formula (10) becomes

$$P[n, n] = n(n - 1)(n - 2) \cdots (2)(1) = n!$$

which is the number of permutations of all n objects.

EXAMPLES

10. Consider the set of digits $\{1, 2, 3, 4, 5\}$. Let us count the number of three-digit numbers that can be formed from the given set where no digit can be repeated.

Each three-digit number corresponds to a permutation of the 5 objects taken 3 at a time. Therefore there are $P[5, 3] = 5(4)(3) = 60$ such three-digit numbers.

11. There are 8 candidates for 4 seats on an executive board. A voter is asked to cast a preferential ballot in which he must select and rank 4 of the 8 candidates. How many different ballots can be cast?

   When a voter marks his ballot, he must select 4 candidates and write them down in the order of first choice, second choice, and so forth. Therefore the number of different ballots is the number of permutations of 8 objects taken 4 at a time and the answer is $P[8, 4] = 8(7)(6)(5) = 1680$.

12. On a literature test, the names of 3 novels and 5 authors are given. The student is asked to pick 3 of the 5 authors and match them with the 3 given novels. What is the probability of a student answering the question entirely correctly if he only guesses?

   The number of possible responses to the question is the number of permutations of 5 things taken 3 at a time, $P[5, 3] = 60$. If the student purely guesses then any one of these possible responses is equally likely. Furthermore, since only one of the possible responses is the correct answer to the question, the probability of obtaining the correct one by guessing is $\frac{1}{60}$.

EXERCISES 8.3

---

1. Bluegrass Farms has 14 mares and 5 stallions. How many different possible ways can they be mated?

2. Allen Printing Company offers a choice of 13 colors and 5 different sizes of paper for making pamphlets. How many different types (color and size) of pamphlets are possible?

3. Tony, who is buying a car, has a choice of 4 body styles, 5 engines, and 25 accessory packages. How many different cars does he have to choose from?

4. The city of Bethlehem has seven telephone exchanges: 691, 694, 861, 865, 866, 867, and 868. How many different telephone numbers (a telephone exchange followed by any four digits) are possible for this city?

5. Biltmore Research Laboratories uses identification numbers that contain one letter followed by three digits selected from $1, 2, 3, \ldots, 9$.

   a. If digits may be repeated, how many different identification numbers are possible?

   b. If digits are not repeated, how many different identification numbers are possible?

6. How many different license plate numbers can be made using two letters followed by four digits selected from the digits 0 through 9, if

   a. Letters and digits may be repeated?

   b. Letters may be repeated, but digits are not repeated?

   c. Neither letters nor digits may be repeated?

7. How many outcomes are there for the experiment of flipping two coins? three coins? four coins? n coins?

8. How many outcomes are there for the experiment of rolling two dice? three dice? four dice? n dice?

9. Count the permutations of the following sets.

   a. $\{5, 8, 9, 2\}$ b. $\{a, b, c, 1, 2, 4\}$

10. Count the permutations of the following sets.

    a. $\{A, B, C, a, b\}$ b. $\{2, 4, 6, C, T, B\}$

11. Use a tree diagram to list all the permutations of the set $\{a, h, m, t\}$.

12. Use a tree diagram to list all the permutations of the set $\{e, f, g\}$.

13. a. Use a tree diagram to list and then count all the permutations of $\{A, B, C, D\}$ taken two at a time.

    b. Compute $P[4, 2]$ to check your answer in part (a).

14. a. Use a tree diagram to list and then count all the permutations of $\{0, 6, 9\}$ taken two at a time.

    b. Compute $P[3, 2]$ to check your answer in part (a).

15. A hat contains 4 one-dollar bills, 5 ten-dollar bills, and 1 one-hundred-dollar bill. A bill is drawn at random, its value is noted, then the bill is replaced. A second bill is then drawn at random and its value is noted. Use a tree diagram to list the possible outcomes for this experiment.

16. An urn contains two white balls, three green balls, and one red ball. A ball is drawn at random, its color is recorded, then the ball is replaced in the urn. A second ball is then drawn at random and its color is recorded. Use a tree diagram to list the possible outcomes for this experiment.

17. A hat contains 1 five-hundred-dollar bill, 6 one-hundred-dollar bills, and 12 twenty-dollar bills. A bill is drawn at random and its value is noted. Without replacing the first bill, a second bill is drawn at random and its value is recorded. Use a tree diagram to list the possible outcomes for this experiment.

18. An urn contains one white ball, two red balls, and two blue balls. A ball is drawn at random and its color is recorded. Without replacing the first ball, a second ball is drawn at random and its color recorded. Use a tree diagram to list the possible outcomes for this experiment.

19. Find each of the following.

    a. $1!$ b. $5!$ c. $9!$

    d. $P[5, 2]$ e. $P[7, 4]$ f. $P[34, 5]$

20. Find each of the following.

    a. $0!$ b. $6!$ c. $10!$

    d. $P[19, 2]$ e. $P[8, 5]$ f. $P[89, 3]$

21. How many four-letter code words can be formed from the set $\{s, f, g, h, j, t\}$ if

    a. Repetitions are not allowed?

    b. Repetitions are allowed?

22. How many three-digit numbers can be formed from the set $\{4, 5, 7, 8, 9\}$ if

    a. Repetitions are not allowed?

    b. Repetitions are allowed?

23. In exacta horse race betting, the player must correctly choose the first and second place winners of the race. If eleven horses are entered in a race, how many possible ways can there be a first and second place winner?

24. Twenty pigs are to be judged at the Rockford County Fair. Ribbons are to be awarded for first, second, third, and fourth places. If no pig can be awarded more than one ribbon, how many different possibilities must the judges consider?

25. Joe watches five television shows and is asked to rate them according to preference. How many rankings are possible?

26. The characteristics of a protein molecule depend on the arrangement of its amino acids in its molecular structure. How many different characteristics are possible for a protein molecule that contains eight amino acids?

27. From seven brands of wine, four are selected at random for a taste test. What is the probability that the four wines are correctly identified by guessing?

28. Troy claims that he can distinguish among various brands of colas. From five brands of colas, four are selected at random for a taste test. What is the probability that Troy correctly identifies the four colas, if he is only guessing?

29. In a group of four people, what is the probability that at least two have the same birthday? (*Hint:* Find the probability of the complement.)

8.4 Equiprobable Models: Combinations

In example 11 of section 8.3 it is not enough for a voter to select four of the eight candidates. In addition, the four selected must be arranged on the ballot according to first, second, third, and fourth preference. Therefore the voter must give an **ordered** subset of four candidates. If the voter were asked to pick four candidates but not to rank them according to preference, then the order that the voter writes the names on the ballot would not matter and the voter would be picking an **unordered** subset of four candidates from the eight. An *ordered* subset is called a **permutation**, whereas an unordered subset is called a **combination**.

Definition   ■   An unordered subset (or simply a subset) of r objects selected from a set of n distinct objects is a **combination of n objects taken r at a time**. The number of all such subsets is denoted by $C[n,r]$.   ■

In counting problems we must decide whether or not the order in which the objects are selected is important. In particular,

if the order is important, count permutations (ordered subsets);

otherwise,

if the order is not important, count combinations (unordered subsets).

To realize a formula for $C[n,r]$, we should observe the relationship between $P[n,r]$ and $C[n,r]$. To this end, consider the set of 4 objects $\{a,b,c,d\}$, and let us determine how $P[4,3]$ is related to $C[4,3]$. With each combination (unordered subset of 3 objects) we may associate $3! = 6$ permutations (ordered subsets of 3 objects), as indicated in table 8.10. Therefore the number of permutations of 4 objects taken 3 at a time is 6 times the number of combinations of 4 objects taken 3 at a time; that is,

$$P[4,3] = 3!\,C[4,3]$$

and therefore

$$C[4,3] = \frac{P[4,3]}{3!}.$$

TABLE 8.10

Combinations	Permutations
$\{a,b,c\}$	abc, acb, bac, bca, cab, cba
$\{a,b,d\}$	abd, adb, bad, bda, dab, dba
$\{a,c,d\}$	acd, adc, cad, cda, dac, dca
$\{b,c,d\}$	bcd, bdc, cbd, cdb, dbc, dcb

This idea may be generalized to give the desired formula.

Definition

■ *Formula for $C[n,r]$:*
The number of combinations of n objects taken r at a time is given by

$$C[n,r] = \frac{P[n,r]}{r!}. \quad ■ \tag{11}$$

By formula (11), to find $C[n,r]$ we first determine $P[n,r]$ and then divide by $r!$. For example,

$$C[3,2] = \frac{P[3,2]}{2!} = \frac{3\cdot 2}{2} = 3$$

$$C[5,4] = \frac{P[5,4]}{4!} = \frac{5\cdot 4\cdot 3\cdot 2}{4\cdot 3\cdot 2\cdot 1} = 5$$

$$C[10,3] = \frac{P[10,3]}{3!} = \frac{10\cdot 9\cdot 8}{3\cdot 2\cdot 1} = 120$$

$$C[10,7] = \frac{P[10,7]}{7!} = \frac{10\cdot 9\cdot 8\cdot 7\cdot 6\cdot 5\cdot 4}{7\cdot 6\cdot 5\cdot 4\cdot 3\cdot 2\cdot 1} = 120.$$

Note that $C[10,7] = C[10,3]$. This is no coincidence. For each subset E of 7 elements taken from a set S of 10 elements, there is a corresponding subset of 3 elements, namely, $S - E$. Therefore the number of subsets with 7 elements is equal to the number of subsets with 3 elements. In general,

$$C[n,r] = C[n,n-r]. \tag{12}$$

Formula (12) is useful when r is larger than one-half of n. For example, to find $C[13,11]$, we use formula (12) to obtain

$$C[13,11] = C[13,2] = \frac{13\cdot 12}{2} = 78.$$

EXAMPLES

1. A 4-member committee is to be selected from a group of 12 people. How many different committees are possible?

   To answer this question we observe that a committee of 4 is an unordered subset of the 12 people. Thus we count combinations and the answer is

$$C[12,4] = \frac{12 \cdot 11 \cdot 10 \cdot 9}{4 \cdot 3 \cdot 2 \cdot 1} = 495.$$

2. Find the number of different 5-card hands that can be dealt from a standard deck of 52 cards.

   In this situation, the order in which the cards are dealt is not important. For example, consider the 5-card hand

$$\{2\blacklozenge, A\spadesuit, 10\spadesuit, 8\heartsuit, J\blacklozenge\}.$$

   It does not matter, for instance, whether the ace of spades is dealt on the second round or on the fifth round; the same 5-card hand results. Therefore the number of 5-card hands is equal to the number of combinations of 52 cards taken 5 at a time. The answer is $C[52,5] = 2{,}598{,}960$.

3. What is the probability of being dealt a 5-card hand in which all the cards are of the same suit?

   If the cards are well shuffled and the dealer is honest, we can assume that all 5-card hands are equiprobable. Thus we can use formula (5), $Pr(E) = N(E)/N(S)$, to determine the answer.

   The number of elements in the sample space S (see example 2) is $N(S) = C[52,5]$.

   To count the number of outcomes in the event E, we can view an outcome in E as the result of performing two tasks in succession. First, we pick a suit, which can be accomplished in 4 ways. Next, we select 5 cards from the 13 cards in the suit, which can be done in $C[13,5]$ ways.

   Therefore the probability of being dealt 5 cards of the same suit is

$$\frac{4C[13,5]}{C[52,5]} = \frac{4(13 \cdot 12 \cdot 11 \cdot 10 \cdot 9)}{52 \cdot 51 \cdot 50 \cdot 49 \cdot 48} = \frac{11 \cdot 3}{4 \cdot 17 \cdot 5 \cdot 49} = \frac{33}{16{,}660} = 0.002.$$

4. What is the probability of obtaining exactly 3 heads in 7 trials with an unbiased coin?

   We can view the outcomes for this experiment as sequences of H's and T's. For example,

$$\text{THHTTHT} \qquad \text{TTHTHTH} \qquad \text{TTTHHTH} \qquad (13)$$

   are three such outcomes. By the principle of counting, there are $2^7 = 128$ outcomes in the sample space. Next we count the number of outcomes that give us exactly 3 heads. Each sequence in (13) has exactly 3 H's and therefore belongs to the event of obtaining exactly 3 heads. For example, the first sequence is the outcome of obtaining a head on the second, third, and sixth flip of the coin. This outcome corresponds to the selection of the subset $\{2,3,6\}$ from the set $\{1,2,3,4,5,6,7\}$. Similarly the second sequence in (13) corresponds to the subset $\{3,5,7\}$, and the third sequence corresponds to the subset $\{4,5,7\}$. Therefore the number of sequences with exactly 3 H's is the number of subsets with 3 elements taken from a set of 7 elements; that is, $C[7, 3]$.

Since the outcomes are equiprobable, we have

$$Pr(\text{exactly 3 heads}) = \frac{C[7, 3]}{128} = \frac{35}{128}.$$

5. What is the probability of obtaining at least 8 heads in 10 trials with an unbiased coin?

The number of equally likely outcomes when flipping an unbiased coin 10 times is $2^{10} = 1024$. Of these possibilities, we now count those in the event E of obtaining at least 8 heads.

The number of ways to obtain exactly 8 heads in the 10 trials is the same as the number of ways of picking 8 of the 10 trials to show heads, which is $C[10, 8]$. Likewise, the number of ways to obtain exactly 9 heads on the 10 trials is $C[10, 9]$. Finally, the number of ways for all 10 trials to show heads is $C[10, 10]$. Thus,

$$N(E) = C[10, 8] + C[10, 9] + C[10, 10] = C[10, 2] + C[10, 1] + 1$$

$$= 45 + 10 + 1 = 56.$$

Since the possible outcomes are equally likely, we have $Pr(E) = \frac{56}{1024}$.

6. A 13-card hand is dealt from a well-shuffled 52-card deck. What is the probability that the hand has exactly 6 red cards and 7 black cards?

The sample space is the set of all 13-card hands, the number of which is $C[52, 13]$. We are assuming equiprobable outcomes, so we next count the number of hands with 6 red cards and 7 black cards.

To construct such a hand, we perform two tasks.

Task 1: Choose 6 red cards from the available 26 red cards.

Task 2: Choose 7 black cards from the available 26 black cards.

The first task can be done in $C[26, 6]$ ways, and the second task can be done in $C[26, 7]$ ways. By the principle of counting, there are $C[26, 6] \cdot C[26, 7]$ hands with 6 red cards and 7 black cards. Therefore,

$$Pr(\text{6 red cards and 7 black cards}) = \frac{C[26, 6] \cdot C[26, 7]}{C[52, 13]} = 0.239.$$

Example 6 illustrates a technique that can be used in problems other than counting card hands.

EXAMPLES

7. A store has 25 radios in its inventory, 5 of which are defective. If 8 radios are selected at random, what is the probability that exactly 2 of them are defective?

Using the deck of cards analogy, we have a "deck" of 25 radios (cards) of which 5 are defective (red) and 20 are not defective (black). The number of 8-radio samples corresponds to the number of 8-card hands possible from a deck of 25 cards, which is $C[25, 8]$. The number of samples with exactly 2 defective radios corresponds to the number of hands with exactly 2 red cards and 6 black cards, which is $C[5, 2] \cdot C[20, 6]$. Therefore,

$$Pr(\text{exactly 2 defective radios}) = \frac{C[5, 2] \cdot C[20, 6]}{C[25, 8]}$$

$$= 0.358.$$

8. A bin contains 72 lightbulbs of which 10 are defective and 62 are perfect. A sample of 8 bulbs is drawn at random from the bin. Find:

a. The probability that exactly one defective bulb is drawn.

b. The probability that at least one defective bulb is drawn.

The total number of possible samples of 8 bulbs taken from 72 bulbs is $C[72, 8]$. For part (a), constructing a sample with exactly one defective bulb requires two tasks.

Task 1:   Select 1 defective bulb from the 10 defective bulbs.

Task 2:   Select 7 perfect bulbs from the 62 perfect bulbs.

Task 1 can be done in $C[10, 1]$ ways and task 2 can be done in $C[62, 7]$ ways. By the principle of counting, there are $C[10, 1] \cdot C[62, 7]$ samples with exactly 1 defective bulb. Therefore,

$$Pr(\text{exactly 1 defective bulb}) = \frac{C[10, 1] \cdot C[62, 7]}{C[72, 8]}$$

$$= 0.411.$$

For part (b), it is easier to find the probability of the complement and subtract from 1. Therefore,

$$Pr(\text{at least 1 defective bulb}) = 1 - Pr(\text{all bulbs are perfect})$$

$$= 1 - \frac{C[62, 8]}{C[72, 8]}$$

$$= 1 - 0.282$$

$$= 0.718.$$

EXERCISES 8.4

1. Find each of the following.
   a. $C[3, 2]$ b. $C[7, 5]$ c. $C[23, 0]$ d. $C[56, 55]$

2. Find each of the following.
   a. $C[6, 3]$ b. $C[20, 2]$ c. $C[9, 4]$ d. $C[201, 199]$

3. List all combinations of $\{A, B, C, D, E\}$ taken 3 at a time. Use formula (11) to check the number of combinations you listed.

4. List all combinations of $\{1, 2, 3, 4\}$ taken 2 at a time. Use formula (11) to check the number of combinations you listed.

5. Ariel Rental Agency has a fleet of 28 cars. If four cars are to be tested for exhaust emissions, how many choices are possible?

6. Rita is suffering from an allergy of an unknown cause. There are 15 possible sites on her back for the doctor to test 5 different allergens. How many different ways can the 5 sites be selected?

7. Next semester, David wants to take 3 business courses, 2 mathematics courses, and 1 economics course. There are 6 business courses, 5 mathematics courses, and 6 economics courses from which to choose. How many different schedules are possible?

8. Dr. Jones plans to buy 3 of 6 different stocks and 9 of 12 different bonds. How many possibilities must she consider?

9. The Pennsylvania lotto game requires the player to choose 6 numbers from 40. (The order does not matter.) How many ways can this be done?

10. The New York lotto game requires the player to choose 6 numbers from 44. (The order does not matter.) How many ways can this be done?

11. Fifteen boys are trying out for the high school basketball team. If 2 of them are centers, 6 are forwards, and 7 are guards, how many different teams are possible? (A team has 1 center, 2 forwards, and 2 guards.)

12. A college committee of 8 is to be selected from 12 people.
    a. How many possible ways can this be done?
    b. If 5 are faculty and 7 are students, how many possible committees are there consisting of 3 faculty and 5 students?
    c. If 3 are deans, 2 are faculty, and 7 are students, how many possible committees are there consisting of 1 dean, 2 faculty, and 5 students?

13. The meet and discuss committee at State University is comprised of 5 administrators and 5 faculty members. If there are 7 administrators and 11 faculty members to choose from, how many different committees are possible?

14. Consider the set of all 6-letter code words comprised of D's and R's. How many different words have three D's and three R's?

Exercises 15–35 use counting techniques such as permutations, combinations, and the principle of counting.

15. Fifty tickets are sold in a lottery. There are three $1000 prizes to be won. How many possible outcomes are there?

16. How many ways can 7 people stand in line at a ticket window?

17. Fifty tickets are sold in a lottery. A $1000 prize, a $500 prize, and a $200 prize are to be won. How many possible outcomes are there?

18. The Internal Revenue Service plans to audit 8 out of 11 people. How many ways can this be done?

19. What is the probability of obtaining exactly 5 heads when flipping an unbiased coin 6 times?

20. What is the probability of obtaining exactly 4 heads when flipping an unbiased coin 7 times?

21. What is the probability of obtaining at most 5 heads when flipping an unbiased coin 6 times?

22. What is the probability of obtaining at least 2 heads when flipping an unbiased coin 6 times?

23. Jodi is to take 5 different drugs in a definite order. She has forgotten what the order is, but takes the drugs anyway. What is the probability that she took them in the correct order?

24. A numbers game consists of the random selection of a three-digit number. (The possibilities range from 000 to 999.) Lyn bets $10 on the number 411 as she was born on April 11. What is the probability that her number will win the game?

25. Dave is to take a 6-question true-false quiz. He is totally unprepared and decides to answer each question by flipping a coin. If heads turns up, he writes true and if tails turns up, he writes false.

    a. What is the probability that Dave answers all the questions correctly?

    b. What is the probability that Dave answers exactly 5 questions correctly?

    c. If a passing score on the quiz is at least 5 correct answers, what is the probability that Dave will pass the quiz?

26. Jane claims she can identify 4 different brands of beer by taste. She is given 4 glasses of beer, one of each brand, and asked to identify correctly the brand in each glass. What is the probability of her getting all 4 correct if she cannot distinguish the taste and only guesses?

27. From 6 brands of instant coffee, 3 are selected at random for a taste test. What is the probability that the 3 brands are correctly identified by guessing?

28. A 6-card hand is dealt from a standard 52-card deck. What is the probability that the hand has exactly 2 diamonds and 4 black cards?

29. A 5-card hand is dealt from a standard 52-card deck. What is the probability that the hand has exactly 2 hearts and 3 spades?

30. There are 8 girls and 7 boys in a mathematics class of 15 students. If 4 students are selected at random to put problems on the board, what is the probability that 2 are girls and 2 are boys?

31. A manufacturer of blank cassette tapes sells them in boxes of 12. If a particular box has exactly 3 defective tapes, what is the probability that a random selection of 6 cassettes from that box will have exactly 1 defective tape?

32. A manufacturer sells electric heating coils in lots of 25. One purchaser randomly selects 6 coils from each lot and tests them. If none of the 6 is defective, the lot is accepted. A rejected lot is returned to the manufacturer. If a lot has exactly 5 defective coils, what is the probability that the lot is accepted?

33. The bakery section of a supermarket has 15 loaves of bread, 7 of which are stale and 8 of which are fresh. If Jason buys 5 loaves selected at random, what is the probability that at least 4 loaves are fresh?

34. The meat section of a supermarket has 25 identical packages of hamburger on display — 12 packages contain fresh meat, whereas the other 13 contain old meat. (The old hamburger has been treated with chemicals to preserve the red color.) If Lyn buys 4 packages of hamburger selected at random, what is the probability that all 4 packages contain fresh meat?

35. Three dice are rolled. What is the probability that all three numbers are different?

8.5 Conditional Probability

Suppose that we construct a probability model for an experiment, but subsequently we obtain further information on the situation. With this additional information the probabilities of events may change. For example, a person reports to a medical center suffering from nausea and dizziness. The initial evaluation by the doctor is that there are three possible causes and all are equally likely. The doctor proceeds to ask the patient questions about his medical history and any unusual happenings. From this additional information, the doctor would reevaluate the likelihood of each of the possible causes.

Additional information may sometimes allow us to rule out completely some of the possibilities in the original sample space. For example, consider the experiment of rolling a pair of unbiased dice. In example 1 in section 8.3 (page 341), the sample space S consists of 36 ordered pairs of integers between 1 and 6 inclusive (see figure 8.15). Let E be the event that the sum of the numbers showing on the two dice is 7. Prior to performing the experiment, we know that the probability of E is $\frac{1}{6}$. Now, suppose the dice have been rolled without our seeing the outcome, but we are told that at least one of the dice shows a 2. How does this additional information affect the probability of E?

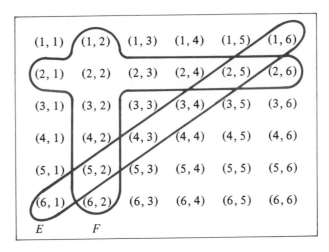

FIGURE 8.15

Let F be the event that at least one of the two dice shows a 2. The event of obtaining a 7 knowing that one of the dice shows as 2 is $E \cap F$. With the information that a 2 occurred, we can exclude from consideration all outcomes in the original sample space S that are not in F and, therefore, we can consider F as a new sample space. Futhermore, the remaining outcomes in F are still equiprobable. Therefore the probability of E given that F has occurred is equal to the number of outcomes in $E \cap F$ divided by the number of outcomes in F; that is,

$$Pr(E \text{ given } F) = \frac{N(E \cap F)}{N(F)} = \frac{2}{11}.$$

EXAMPLE 1

An insurance company conducted a survey of 100 drivers concerning drinking and accidents. The results of this survey are given in table 8.11.

TABLE 8.11

	F Drinkers	F' Nondrinkers	Totals
E had an accident	42	12	54
E' did not have an accident	18	28	46
totals	60	40	100

Consider the experiments of selecting at random a person in the survey. Let E be the event that the person selected had an accident, and let F be the event that the person selected drinks. Let us find the probability that the person selected had an accident given that he drinks; that is, let us find $Pr(E$ given $F)$.

Since we are given that the event F occurred, we can eliminate from consideration the nondrinkers in the survey and consider only the first column of the table. Since 42 out of the 60 drinkers had an accident, the probability that the person had an accident given that he drinks is

$$Pr(E \text{ given } F) = \frac{N(E \cap F)}{N(F)} = \frac{42}{60} = 0.7.$$

Let us consider another approach to finding $Pr(E$ given $F)$. If we divide $N(S)$ into the numerator and denominator of the fraction $(N(E \cap F))/(N(F))$, we obtain

$$Pr(E \text{ given } F) = \frac{N(E \cap F)/N(S)}{N(F)/N(S)} = \frac{Pr(E \cap F)}{Pr(F)}.$$

For example, we see from table 8.11 that $N(E \cap F) = 42$, so $Pr(E \cap F) = \frac{42}{100} = 0.42$. Likewise, $N(F) = 60$ and $Pr(F) = \frac{60}{100} = 0.6$. Therefore,

$$Pr(E \text{ given } F) = \frac{Pr(E \cap F)}{Pr(F)} = \frac{0.42}{0.60} = 0.7.$$

Note that the probability that a person selected from the survey had an accident is 0.54. With the additional information that the person drinks, the probability that he had an accident increases to 0.7.

This alternate approach to finding $Pr(E$ given $F)$ serves as the general definition for any probability model.

Definition

■ Let E and F be two events of a sample space S with $Pr(F) \neq 0$. The **conditional probability of E given that F has occurred**, $Pr(E$ given $F)$, which we now denote by $Pr(E|F)$, is

$$Pr(E|F) = \frac{Pr(E \cap F)}{Pr(F)} \qquad (14)$$

If $Pr(F) = 0$, then F is the impossible event, and it is not meaningful to talk about the conditional probability of E given that the impossible event has occurred. ∎

EXAMPLES

2. Let E and F be two events such that $Pr(E \cap F) = 0.6$ and $Pr(F) = 0.8$. Find $Pr(E|F)$.

From formula (14), we have that

$$Pr(E|F) = \frac{0.6}{0.8} = 0.75.$$

3. An executive feels that the probability that he will be transferred to another division in the company is 0.8 and that the probability that he will be transferred and promoted is 0.7. What is the probability that he will be promoted given that he is transferred to another division?

Let E be the event that he is promoted and F be the event that he is transferred. We use formula (14) to find $Pr(E|F)$. From the information given, $Pr(F) = 0.8$ and $Pr(E \cap F) = 0.7$. Therefore,

$$Pr(E|F) = \frac{Pr(E \cap F)}{Pr(F)} = \frac{0.7}{0.8} = 0.875.$$

4. Consider example 3 and the information given there. Let us now find the probability that the executive is not promoted given that he is transferred.

Letting E and F be the events as described in example 3, we want to find $Pr(E'|F)$. By formula (14),

$$Pr(E'|F) = \frac{Pr(E' \cap F)}{Pr(F)}.$$

From the information given in example 3, $Pr(F) = 0.8$ and $Pr(E \cap F) = 0.7$. A Venn diagram is helpful in finding $Pr(E' \cap F)$. From the diagram in figure 8.16 we see that F is the union of two disjoint events $E \cap F$ and $E' \cap F$. Therefore,

$$Pr(F) = Pr(E \cap F) + Pr(E' \cap F).$$

Solving for $Pr(E' \cap F')$, we have

$$Pr(E' \cap F) = Pr(F) - Pr(E \cap F)$$
$$= 0.8 - 0.7 = 0.1.$$

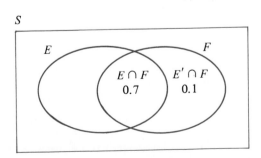

FIGURE 8.16

Finally,

$$Pr(E'|F) = \frac{0.1}{0.8} = 0.125.$$

From examples 3 and 4, we have, respectively, $Pr(E|F) = 0.875$ and $Pr(E'|F) = 0.125$. Observe that the sum of these two probabilities is 1. This property is true in general. In particular, if E and F are two events of a sample space S, then $Pr(E|F) + Pr(E'|F) = 1$ and therefore

$$Pr(E'|F) = 1 - Pr(E|F). \tag{15}$$

Sometimes we may know the probability of an event F as well as the conditional probability of E given F. From this information we can determine the probability of both E and F happening.

EXAMPLE 5

The Electron Company has two production lines, I and II. Each line produces the same type of transistor. The company's production comes 65 percent from line I and 35 percent from line II. The quality control department has determined, based on past performance, that 2 percent of the transistors produced on line I are defective, whereas 1 percent of the transistors produced on line II are defective. Suppose a transistor is selected at random from a day's production. What is the probability that the transistor is defective and was produced on line I?

Let D be the event that the transistor is defective and I be the event that the transistor was produced on line I. We wish to find $Pr(D \cap I)$. We know that $Pr(I) = 0.65$. Furthermore, we know that the probability that a transistor is defective given that it was produced on line I is 0.02; that is, $Pr(D|I) = 0.02$. Solving the conditional probability formula (14) for $Pr(D \cap I)$, we have

$$Pr(D \cap I) = Pr(I)Pr(D|I)$$
$$= (0.65)(0.02) = 0.013.$$

Rule

■ *Product Rule for Two Events:*
Let E and F be events of a sample space S. Then

$$Pr(E \cap F) = Pr(F) \cdot Pr(E|F). \tag{16}$$

Therefore, the probability of E and F occurring is the product of the probability of F occurring times the conditional probability of E occurring given that F has already occurred. ■

EXAMPLES

6. Let E and F be two events such that $Pr(F) = 0.7$ and $Pr(E|F) = 0.9$. Find $Pr(E \cap F)$.

From formula (16) we have

$$Pr(E \cap F) = Pr(F)Pr(E|F) = (0.7)(0.9) = 0.63.$$

7. A specialist administers a diabetes test to all new patients. The test is not 100 percent accurate, but it is known to detect diabetes in 97 percent of those people who do indeed have diabetes. The specialist estimates from her past medical records that 75 percent of her new patients are diabetic. What is the probability that a new patient who has a positive test is diabetic?

Let E be the event that a new patient has a positive test and let F be the event that a new patient is diabetic. We want to find $Pr(E \cap F)$. From the information, we have the probability of E knowing F; in particular, $Pr(E|F) = 0.97$. Furthermore, $Pr(F) = 0.75$. Therefore, using formula (16),

$$Pr(E \cap F) = Pr(F)Pr(E|F) = (0.75)(0.97) = 0.7275.$$

Consider the experiment of flipping two coins. Let E be the event that the second coin turns up tails and F be the event that the first coin turns up heads. If we know that event F occurred, does this additional information affect the probability of E? Intuitively, we say it does not. To verify our answer mathematically, we show that $Pr(E|F) = Pr(E)$.

For the sample space $\{HH, HT, TH, TT\}$, we have $E = \{HT, TT\}$ and $F = \{HH, HT\}$. Therefore, $Pr(E|F) = \frac{1}{2}$ and $Pr(E) = \frac{1}{2}$. Thus the event that the second coin turns up tails does not depend on the event that the first coin turned up heads.

Definition ■ Let E and F be two events of an experiment. We say that E and F are **independent events** if

$$Pr(E|F) = Pr(E). \tag{17}$$

When events E and F are independent, it will also be true that

$$Pr(F|E) = Pr(F).$$

Events that are not independent are called **dependent events**. ■

Thus two events E and F are independent provided that the probability of E occurring given the condition that F has already occurred is the same as the unconditional probability that E occurs.

EXAMPLE 8 Recall example 1 where E is the event that a driver selected from the survey had an accident and F is the event that the selected driver drinks. We saw that $Pr(E|F) = 0.7$ and $Pr(E) = 0.54$. Since $Pr(E|F) \neq Pr(E)$, then E and F are dependent. Note that the fact that the driver drinks increases the probability that he had an accident.

Let us consider again the experiment of flipping two coins. It can be shown that the outcome that results from flipping the first coin does not affect the chances of heads or tails when flipping the second coin. This observation is true for any two independent events. Whether or not one event occurred does not change the probability of the other event occurring.

Since $Pr(E|F) = Pr(E)$ for two independent events, then by formula (16),

$$Pr(E \cap F) = Pr(F)Pr(E|F)$$
$$= Pr(F)Pr(E).$$

Conversely it can shown that $Pr(E \cap F) = Pr(E)Pr(F)$ implies $Pr(E|F) = Pr(E)$. We therefore have the following rule.

Rule   ■ *Product Rule for Independent Events:*
Two events E and F are independent if and only if

$$Pr(E \cap F) = Pr(E)Pr(F). \quad ■ \qquad (18)$$

The product rule for independent events gives us a method to determine if two events are independent when we know $Pr(E)$, $Pr(E)$, and $Pr(E \cap F)$.

EXAMPLES   9. Let E and F be two events such that $Pr(E) = \frac{1}{4}$, $Pr(F) = \frac{2}{3}$, and $Pr(E \cap F) = \frac{1}{6}$. Since

$$Pr(E)Pr(F) = \left(\frac{1}{4}\right)\left(\frac{2}{3}\right)$$

$$= \frac{1}{6}$$

$$= Pr(E \cap F)$$

by the product rule for independent events, E and F are independent events.

10. Let E and F be two events such that $Pr(E) = 0.3$, $Pr(F) = 0.7$, and $Pr(E \cap F) = 0.2$. Since

$$Pr(E)Pr(F) = (0.3)(0.7)$$

$$= 0.21$$

$$\neq Pr(E \cap F)$$

by the product rule for independent events, E and F are dependent events.

The product rule (18) can be used to find the probability of the intersection of two independent events if we know the probability of each one.

EXAMPLES   11. If E and F are two independent events with $Pr(E) = 0.5$ and $Pr(F) = 0.7$, then by product rule (18), we have

$$Pr(E \cap F) = Pr(E)Pr(F)$$

$$= (0.5)(0.7)$$

$$= 0.35.$$

12. A company puts together minicomputrs using two components, a mainframe and a terminal. These components are supplied independently to the company

by two different manufacturers. From past experience, it has been found that 0.2 percent of the mainframes are defective and 0.5 percent of the terminals are defective. What is the probability that the company puts together a mini-computer in which both the mainframe and the terminal are defective?

Let E be the event that the mainframe is defective and let F be the event that the terminal is defective. We want to find $Pr(E \cap F)$. Since the two events E and F are independent, using product rule (18), we have

$$Pr(E \cap F) = Pr(E)Pr(F)$$
$$= (0.002)(0.005) = 0.00001.$$

The concept of independence can be applied to more than two events. For example, three events E, F, and G are independent if pairwise they obey the product rule (18).

$$Pr(E \cap F) = Pr(E)Pr(F)$$

$$Pr(F \cap G) = Pr(F)Pr(G)$$

$$Pr(E \cap G) = Pr(E)Pr(G),$$

as well as $\qquad Pr(E \cap F \cap G) = Pr(E)Pr(F)Pr(G).$

Independence of four or more events is defined in an analogous manner. Given n independent events, we have the following product rule.

Rule ■ *Product Rule for Independent Events (General):*
If $E_1, E_2, E_3, \ldots, E_n$ are n independent events, then

$$Pr(E_1 \cap E_2 \cap E_3 \cap \cdots \cap E_n) = Pr(E_1)Pr(E_2)Pr(E_3) \cdots Pr(E_n). \quad ■ \quad (19)$$

For example, if $n = 5$, equation (19) is

$$Pr(E_1 \cap E_2 \cap E_3 \cap E_4 \cap E_5) = Pr(E_1)Pr(E_2)Pr(E_3)Pr(E_4)Pr(E_5).$$

EXAMPLES

13. Let E, F, and G be three independent events where $Pr(E) = \frac{1}{2}$, $Pr(F) = \frac{7}{8}$, and $Pr(G) = \frac{6}{7}$. Then by the product rule for independent events, we have

$$Pr(E \cap F \cap G) = Pr(E)Pr(F)Pr(G)$$

$$= \left(\frac{1}{2}\right)\left(\frac{7}{8}\right)\left(\frac{6}{7}\right)$$

$$= \frac{3}{8}.$$

14. Let E_1, E_2, E_3, and E_4 be four independent events with $Pr(E_1) = 0.2$, $Pr(E_2) = 0.53$, $Pr(E_3) = 0.9$, and $Pr(E_4) = 0.5$. Then by equation (19),

$$Pr(E_1 \cap E_2 \cap E_3 \cap E_4) = Pr(E_1)Pr(E_2)Pr(E_3)Pr(E_4)$$

$$= (0.2)(0.53)(0.9)(0.5)$$

$$= 0.0477.$$

We close this section with some applications involving the concept of independence. It is sometimes difficult to determine if events are independent. For example, suppose statistics show that a basketball player makes 45 percent of his shots. In today's game, however, suppose that he misses his first 6 shots. As a result, he may lose his confidence, thereby decreasing the probability that he scores on his next attempt. There are two other possibilities. He may increase his concentration, resulting in a higher probability that he scores the next time; or he may be unaffected by the fact that he missed his first 6 attempts. Therefore, in many real-world problems, we rely on intuition and simply *assume* that the events in question are independent.

EXAMPLES

15. A basketball player makes 80 percent of his free throws. If he attempts 4 free throws during the first quarter, what is the probability, assuming independence, that he makes all of them?

　　　We consider each free throw as an event and these events are assumed to be independent. Therefore,

$$Pr(\text{he makes all 4 free throws}) = (0.8)(0.8)(0.8)(0.8)$$
$$= 0.4096.$$

16. An enemy code is sent to three experts, E, F, and G, to decipher. If the probability that E cracks the code is 0.4, the probability that F cracks the code is 0.3, and the probability that G cracks the code is 0.6, find the probability that all three decipher the code. Assume independence.

　　　From the information given,

$$Pr(\text{E cracks the code}) = 0.4$$
$$Pr(\text{F cracks the code}) = 0.3$$
$$Pr(\text{G cracks the code}) = 0.6.$$

Since we are assuming independence,

$$Pr(\text{all three decipher the code}) = (0.4)(0.3)(0.6)$$
$$= 0.072.$$

17. In a certain mass-production process, it is estimated that 10 percent of the items produced are defective. If five items are independently selected at random from the assembly line, find

a. The probability that all five items are defective.

b. The probability that all five items are not defective.

From the information given,

$$Pr(\text{an item selected is defective}) = 0.1.$$

If we let

$$E_i = \text{the } i\text{th item selected is defective,}$$

$i = 1, 2, 3, 4, 5$, then for part (a)

Pr(all five items are defective)

$$= Pr(E_1 \cap E_2 \cap E_3 \cap E_4 \cap E_5)$$
$$= Pr(E_1)Pr(E_2)Pr(E_3)Pr(E_4)Pr(E_5)$$
$$= (0.1)^5$$
$$= 0.00001.$$

For part (b),

$$E'_i = \text{the } i\text{th item selected is not defective,}$$

$i = 1, 2, 3, 4, 5$. Now, for independent events, their complements are also independent. Therefore,

Pr(all five items are not defective)

$$= Pr(E'_1 \cap E'_2 \cap E'_3 \cap E'_4 \cap E'_5)$$
$$= Pr(E'_1)Pr(E'_2)Pr(E'_3)Pr(E'_4)Pr(E'_5)$$
$$= (0.9)^5$$
$$= 0.59049.$$

Note in example 17 that we used the fact that the complements of independent events are independent. Furthermore, if E and F are independent events, then E and F' are independent. (See exercise 45.) This idea can be generalized to three or more independent events.

EXERCISES 8.5

1. Consider the experiment of selecting a person at random from the Fox River Valley. Let

    $E = $ The person selected likes to hunt.

    $F = $ The person selected is in favor of stricter gun control laws.

    Express in words the following conditional probabilities.
    a. $Pr(E|F)$ b. $Pr(F|E)$ c. $Pr(E'|F)$ d. $Pr(F'|E')$

2. Consider the experiment of selecting a person at random from a set of registered voters. Let

    $E = $ The person selected is a Republican.

    $F = $ The person selected is unemployed.

    Express in words the following conditional probabilities.
    a. $Pr(F|E)$ b. $Pr(E|F')$ c. $Pr(E'|F)$ d. $Pr(E'|F')$

3. Consider the experiment of rolling a die. Let

$$E = \text{The number showing is even.}$$

$$F = \text{The number showing is less than four.}$$

Symbolize each of the following conditional probability statements in terms of E and F using set notation.

a. The probability that the number showing is even, given that it is less than 4.

b. The probability that the number showing is odd, given that it is not less than 4.

c. The probability that the number showing is greater than or equal to 4, given that it is even.

4. Consider the experiment of flipping three coins. Let

$$E = \text{At least one coin was heads.}$$

$$F = \text{The second coin was tails.}$$

Symbolize each of the following conditional probability statements in terms of E and F using set notation.

a. The probability that at least one coin was heads, given that the second coin was not tails.

b. The probability that the second coin was tails, given that at least one coin was heads.

c. The probability that the second coin was not tails, given that at least one coin was heads.

5. Three coins were flipped and it is known that at least one was heads. Using this information,

a. Find the probability that all three coins were heads.

b. Find the probability that exactly two coins were heads.

c. Find the probability that all three coins were tails.

d. Find the probability that at least one coin was tails.

6. Two dice are rolled. What is the probability that the sum of the numbers showing is either 7 or 11, given that at least one of the dice shows a 5?

7. For the experiment in example 1, find the probability that a driver selected from the survey did not have an accident given that he is a drinker.

8. For the experiment in example 1, find the probability that a driver selected from the survey is a nondrinker, given that he did not have an accident.

9. A survey of students at State University concerning alcohol and grade point average (GPA) is given in table 8.12.

TABLE 8.12

	F *The student has a drinking problem*	F' *The student does not have a drinking problem*
E GPA is between 2.00 and 4.00 inclusive	120	270
E' GPA is under 2.00	250	360

Let E be the event that a student selected from the survey has a GPA between 2.00 and 4.00 inclusive. Let F be the event that the student selected from the survey has a drinking problem. Find

a. $Pr(E|F)$ b. $Pr(F|E)$ c. $Pr(E'|F)$

d. $Pr(E|F')$ e. $Pr(E'|F')$ f. $Pr(F'|E')$

10. A survey was taken of 200 people and the results are given in table 8.13.

TABLE 8.13

	F Income higher than $30,000 per year	F' Income no more than $30,000 per year
E college degree	52	43
E' no college degree	38	67

Let E be the event that a person selected from the survey has a college degree. Let F be the event that a person selected from the survey makes more than $30,000 per year. Find

a. $Pr(E|F)$ b. $Pr(E'|F)$ c. $Pr(F|E)$

d. $Pr(F|E')$ e. $Pr(F'|E')$ f. $Pr(E'|F')$

11. Let E and F be events of an experiment such that $Pr(F) = 0.4$ and $Pr(E \cap F) = 0.18$. Find $Pr(E|F)$.

12. Let E and F be events of an experiment such that $Pr(E) = \frac{3}{5}$ and $Pr(E \cap F) = \frac{11}{20}$. Find $Pr(F|E)$.

13. Let E and F be events of an experiment such that $Pr(F) = \frac{6}{17}$ and $Pr(E|F) = \frac{2}{3}$. Find $Pr(E \cap F)$.

14. Let E and F be events of an experiment such that $Pr(F) = 0.56$ and $Pr(E|F) = 0.41$. Find $Pr(E \cap F)$.

15. Let E and F be events of an experiment with $Pr(E \cap F) = \frac{1}{4}$, $Pr(E|F) = \frac{1}{2}$, and $Pr(F|E) = \frac{3}{4}$. Find

a. $Pr(E)$ b. $Pr(F)$

16. Let E and F be events of an experiment with $Pr(E \cap F) = \frac{1}{10}$, $Pr(E|F) = \frac{1}{3}$, and $Pr(F|E) = \frac{1}{7}$. Find

a. $Pr(E)$ b. $Pr(F)$

17. Let E and F be events of an experiment with $Pr(F|E) = \frac{1}{3}$, $Pr(E \cap F) = \frac{1}{5}$, and $Pr(F) = \frac{2}{5}$. Find $Pr(E \cup F)$.

18. Let E and F be events of an experiment with $Pr(E|F) = 0.75$, $Pr(E) = 0.5$, and $Pr(E \cap F) = 0.3$. Find $Pr(E \cup F)$.

19. The probability that a person selected at random from a certain group reads both $NEWS$ magazine and $AWARE$ magazine is 0.6. The probability that he reads $NEWS$ given that he reads $AWARE$ is 0.8. What is the probability that he reads $AWARE$ magazine?

20. Patricia feels that the probability that she will get a raise and be transferred is 0.4. She also thinks that the probability she will get a raise given that she is transferred is 0.8. What is the probability that she will be transferred?

21. Refer to example 5. Find the probability that if a transistor is selected at random from the day's production, the transistor is not defective and it was produced on line I.

22. Refer to example 5. Find the probability that a transistor selected at random from the day's production is defective and was produced on line II.

23. At State University, the probability that a student chosen at random owns a car is 0.387. If the probability that the student owns a bicycle given that he or she owns a car is 0.195, find the probability that the student owns both a car and a bicycle.

24. Past records indicate the probability that a student who passed English composition at State University had completed the assignments regularly is 0.892. Furthermore, the probability that a student passes English composition and completes the assignments regularly is 0.726. Find the probability that a student taking English composition passes the course.

25. If a survey of the Lehigh Valley, it was found that the probability that a person selected at random is unemployed and has a college degree is 0.216. If the probability that a person selected at random has a college degree is 0.493, find the probability that the person is unemployed given that the person has a college degree.

26. Dr. Barrington conducted a survey with the following results. The probability that a person selected at random had a sinus problem was 0.309. Furthermore, the probability that a person with a sinus problem smoked cigarettes was 0.695. Find the probability that a person selected at random smoked cigarettes and had a sinus problem.

27. Let E and F be two events such that $Pr(E) = \frac{2}{21}$, $Pr(F) = \frac{7}{16}$, and $Pr(E \cap F) = \frac{1}{26}$. Are E and F independent?

28. Let E and F be two events such that $Pr(E) = 0.34$, $Pr(F) = 0.61$, and $Pr(E \cap F) = 0.2740$. Are E and F independent?

29. Let E and F be two independent events with $Pr(E) = \frac{7}{13}$ and $Pr(F) = \frac{26}{35}$. Find $Pr(E \cap F)$.

30. Let E and F be two independent events with $Pr(E) = 0.192$ and $Pr(F) = 0.865$. Find $Pr(E \cap F)$.

31. Let E_1, E_2, E_3, and E_4 be four independent events with $Pr(E_1) = 0.73$, $Pr(E_2) = 0.29$, $Pr(E_3) = 0.95$, and $Pr(E_4) = 0.64$. Find $Pr(E_1 \cap E_2 \cap E_3 \cap E_4)$.

32. Let E, F, and G be three independent events with $Pr(E) = \frac{3}{7}$, $Pr(F) = \frac{11}{27}$, and $Pr(G) = \frac{9}{22}$. Find $Pr(E \cap F \cap G)$.

33. Let E and F be independent events with $Pr(E) = 0.291$ and $Pr(F) = 0.613$. Find
    a. $Pr(E \cap F')$ b. $Pr(E' \cap F)$ c. $Pr(E' \cap F')$

34. Let E and F be independent events with $Pr(E) = \frac{2}{3}$ and $Pr(F) = \frac{7}{8}$. Find
    a. $Pr(E \cap F')$ b. $Pr(E' \cap F)$ c. $Pr(E' \cap F')$

35. A candidate believes that $\frac{2}{3}$ of the registered voters in his district will vote for him in the next election. If two registered voters are independently selected at random, what is the probability that
    a. Both of them will vote for him in the next election?
    b. Neither will vote for him in the next election?
    c. Exactly one of them will vote for him in the next election?

36. The probability that a child of diabetic parents will be diabetic is $\frac{1}{4}$. Assuming independence, what is the probability that a diabetic couple will have two children such that
    a. Both children are diabetic?
    b. Neither child is diabetic?
    c. Exactly one child is diabetic?

37. A new skin medication cures acne 80% of the time. If 4 randomly selected people use this creme, assuming independence, what is the probability that
   a. All 4 are cured of acne?
   b. All 4 are not cured of acne?

38. A particular type of sweet pea plant has either red or white blossoms. From genetic theory, the probability that the plant has red blossoms is $\frac{1}{4}$. If 5 seeds for this plant are randomly selected and planted, assuming independence, what is the probability that
   a. All 5 will grow into red-blossomed plants?
   b. All 5 will not grow into red-blossomed plants?

39. The recovery rate from a certain disease is 0.7. If 5 people have this disease, what is the probability (assuming independence) that
   a. All will recover?
   b. Exactly 3 will recover?
   c. At least 2 will recover?

40. Past experience shows that 65% of the applicants for a job will be qualified and 35% will not be qualified. If 4 people apply for the job, what is the probability, assuming independence, that
   a. All are qualified?
   b. Exactly 2 are qualified?
   c. At least 1 is qualified?

41. Suppose E and F are two events that are mutually exclusive and independent. Show that at least one of them cannot happen.

42. Suppose E, F, and G are 3 events. If E and F are mutually exclusive, show that
$$Pr[(E \cup F)|G] = Pr(E|G) + Pr(F|G).$$
   (*Hint:* $(E \cup F) \cap G = (E \cap G) \cup (F \cap G)$)

43. An urn contains 3 white and 5 blue balls. A ball is drawn at random from the urn. Without replacing the first ball, a second ball is drawn from the urn. Find the probability of
   a. A blue ball on the second draw, given a white ball on the first draw.
   b. A white ball on the second draw, given a white ball on the first draw.
   c. A blue ball on the second draw, given a blue ball on the first draw.

44. Let E and F be independent events with $Pr(E) = 4Pr(F)$, and $Pr(E \cup F) = \frac{23}{50}$. Find $Pr(F)$.

45. If E and F are independent events, show that E and F' are independent events.

46. If E, F, and G are three events, show that
$$Pr(E \cap F \cap G) = Pr(E)Pr(F|E)Pr(G|E \cap F)$$

47. Three cards are drawn one at a time without replacement from a standard 52-card deck. Let

$$E = \text{The first card drawn is a king.}$$

$$F = \text{The second card drawn is a king.}$$

$$G = \text{The third card drawn is a king.}$$

Find the probability that all three cards drawn are kings. (*Hint:* Use the result of exercise 46.)

8.6 Sequences of Experiments

In this section we develop a model for a sequence of experiments, called a **stochastic process**. The sequence will contain only a finite number of experiments or **stages**. The sequence may consist of the same experiment repeated several times (flipping a coin five times), or the experiments in the sequence may be unrelated (rolling a die followed by drawing a ball from an urn). Furthermore, the outcomes of one experiment in the sequence may or may not affect the probabilities of the outcomes of the following experiments.

A probability model for a stochastic process consists of a sample space S of possible outcomes for the process and probabilities assigned to the outcomes in S. The probability model for a stochastic process can be represented by a tree diagram, as we see in the next example.

EXAMPLE 1

An urn contains 2 white balls and 3 red ones. Consider the experiment of drawing at random a ball from the urn and noting its color. The experiment is performed twice **without replacement**; that is, a ball is drawn from the urn and is not put back, and a second ball is then drawn from the urn. Let us construct a probability model for this stochastic process of two stages.

We use a tree diagram to display the outcomes for this process (see figure 8.17). Each of the four paths of the tree gives a possible outcome for the process. For example, if a white ball (W) is drawn first and a red ball (R) is drawn second, this outcome corresponds to the path labeled WR. Thus a sample space for this process is

$$S = \{WW, WR, RW, RR\}.$$

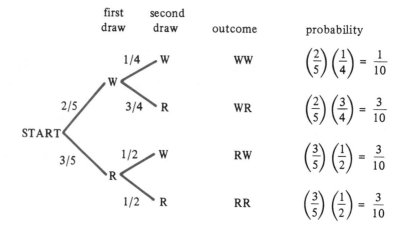

first draw	second draw	outcome	probability
	1/4 W	WW	$\left(\frac{2}{5}\right)\left(\frac{1}{4}\right) = \frac{1}{10}$
W 2/5	3/4 R	WR	$\left(\frac{2}{5}\right)\left(\frac{3}{4}\right) = \frac{3}{10}$
START			
R 3/5	1/2 W	RW	$\left(\frac{3}{5}\right)\left(\frac{1}{2}\right) = \frac{3}{10}$
	1/2 R	RR	$\left(\frac{3}{5}\right)\left(\frac{1}{2}\right) = \frac{3}{10}$

FIGURE 8.17

The next step in our development of a model is to assign probabilities to the four outcomes in the sample space. To make this assignment, we use the product rule for two events, $Pr(E \cap F) = Pr(F)Pr(E|F)$, given as formula (16) in section 8.5 (page 362).

For example, WW is the intersection of two events, drawing a white ball on the first draw and drawing a white ball on the second draw. Using the product rule for two events, we have that $Pr(WW)$ is the probability of a white ball on the first draw times the probability of a white ball on the second draw, given that a white ball was drawn first; that is

$$Pr(WW) = Pr(W \text{ on 1st draw})Pr(W \text{ on 2nd draw}|W \text{ on 1st draw}).$$

Since on the first draw, 2 of the 5 balls are white,

$$Pr(W \text{ on 1st draw}) = \frac{2}{5}.$$

Given that a white ball was drawn first, 1 of the remaining 4 balls is white. Therefore,

$$Pr(W \text{ on 2nd}|W \text{ on 1st}) = \frac{1}{4}$$

and

$$Pr(WW) = \left(\frac{2}{5}\right)\left(\frac{1}{4}\right) = \frac{1}{10}.$$

We find the probabilities for the other three outcomes WR, RW, and RR using the same method. We can use the tree diagram in figure 8.17 to display the results.

The path that corresponds to the outcome WW has two branches. The first branch corresponds to getting a white ball on the first draw. The probability of doing this is $\frac{2}{5}$. We write $\frac{2}{5}$ over this branch as in figure 8.17. The second branch of the path corresponds to getting a white ball on the second draw given that a white ball was drawn first. The probability of this occurring is $\frac{1}{4}$. We write $\frac{1}{4}$ over the branch. The probability of the path WW is now the product of the probabilities of its two branches, $Pr(WW) = \left(\frac{2}{5}\right)\left(\frac{1}{4}\right)$.

Observe that the event of obtaining a white ball on the second step of the process depends on what happened on the first step.

The remainder of the tree can be completed in the same way as for the outcome WW, resulting in what is called a **probability tree** for the stochastic process. The assignment of probabilities to the outcomes in the sample space

$$\{WW, WR, RW, RR\}$$

is given in the column at the extreme right of the probability tree (see figure 8.17).

The method employed to assign probabilities in example 1 can be used for any stochastic process.

Definition   ■ *Assignment of Probabilities for a Stochastic Process:*
Consider a stochastic process and its **probability tree**, where the probability that an outcome occurs at a given stage of the process is indicated on the corresponding branch of the tree. The probability that the process follows a

particular path in the tree is the product of the probabilities indicated on the branches of that path. This is also the probability of the result represented by that path. ∎

As seen in example 1, the probabilities of the outcomes on the second draw depend on what occurred on the first draw. This is because the first ball is not replaced before the second ball is drawn. If the process is done **with replacement**, what occurs on the second draw is independent of what happened on the first draw.

EXAMPLE 2

Consider again the urn in example 1 and the two-stage stochastic process of drawing two balls in succession from the urn. However, this time the process is done with replacement; that is, the first ball is replaced prior to performing the experiment a second time. Give a probability model for the process using a probability tree.

The probability tree in figure 8.18 is similar to the one given in example 1 but differs in the conditional probabilities given on the branches for the second step of the process. To obtain these conditional probabilities, again let W be the event of drawing a white ball and R be the event of drawing a red ball.

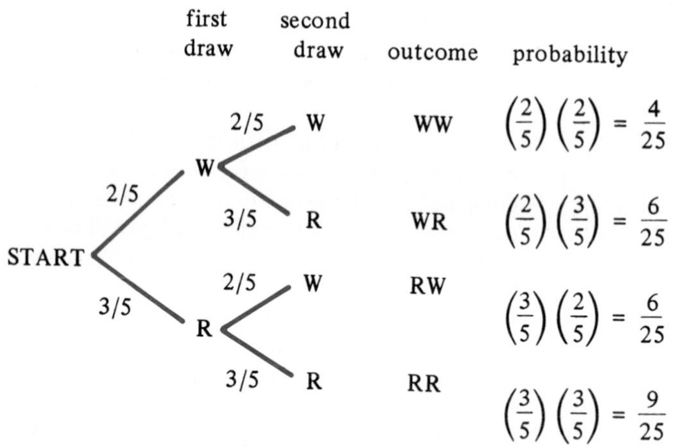

FIGURE 8.18

When the experiment is performed the first time, $Pr(W) = \frac{2}{5}$ and $Pr(R) = \frac{3}{5}$. In example 1, we saw that $Pr(R|W) \neq Pr(R)$ since the process was done without replacement. However, in this case the events are independent since the process is done *with* replacement. For example, the probability of obtaining a red ball on the second draw, given that a white ball was drawn first, is the same as the probability of drawing a red ball the first time; that is

$$Pr(R|W) = Pr(R) = \frac{3}{5}.$$

Similarly we can see that

$$Pr(R|R) = Pr(R) = \frac{3}{5}.$$

Also, $Pr(W|R) = Pr(W) = \frac{2}{5}$ and $Pr(W|W) = Pr(W) = \frac{2}{5}$. Writing these conditional probabilities on the branches for the second step of the process, we have the probability tree in figure 8.18.

Finally, using the principle of assigning probabilities for a stochastic process, we obtain the probabilities in the column to the right of the tree as in figure 8.18.

There are many experiments that can be considered as **urn-like experiments**. For example, the "urn" may be a collection of 12 transistor radios (balls) where 3 of them are defective (red) and 9 of them are perfect (blue). The experiment is to draw 2 radios from the "urn" at random without replacement. In the area of survey taking, the "urn" may consist of the people (balls) living in a certain area. Some of the people are in favor of a tax proposal (red balls), some of the people are against the tax proposal (white balls), and some of them are indifferent (blue balls). The experiment is to conduct a survey by drawing 250 "balls" from the "urn" and to observe the colors of the balls drawn.

When a stochastic process is to perform an urn-like experiment several times in succession, it can be done with or without replacement.

EXAMPLES

3. The Metal Fabrication Company has a machine that produces metal parts. Past performance of the machine shows that 90 percent of the parts have no type of defect. Each morning the first 50 parts produced on the machine are collected in a bin. If 3 parts are selected at random from the 50, what is the probability that all 3 have no type of defect? Also, what is the probability that at least 1 of the 3 has some type of defect?

We can view the random selection of the 3 parts from the bin of 50 as a stochastic process. Consider the experiment of selecting at random 1 part from the bin. The stochastic process is to perform this experiment 3 times without replacement. The tree diagram that displays the possibilities for the process is given in figure 8.19, where G means the part has no defects and D means that it has some type of defect. To answer the first question, we need to find the probability of the outcome GGG.

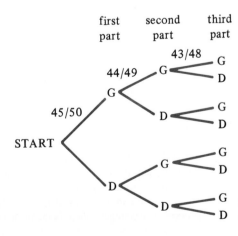

FIGURE 8.19

Let us assume that the machine is operating as usual and 90 percent of the 50 parts have no defects. Therefore 45 parts in the bin are of type G. Also, since the stochastic process is without replacement, the outcome at any step of the process is dependent on the previous outcome. Using the principle of assigning probabilities to a stochastic process, we have that

$$Pr(GGG) = \frac{45}{50} \cdot \frac{44}{49} \cdot \frac{43}{48} = 0.724.$$

(It is interesting to note that this answer is the same as $C[45, 3]/C[50, 3]$.)
Since all outcomes other than GGG satisfy the condition that at least 1 of the 3 parts has some type of defect, the answer to the second question is

$$1 - 0.724 = 0.276.$$

4. Whenever the Opinion Poll Company wants to take a poll it selects the sample from a very large set of people. Of those in this set of people, 45 percent are registered Democrats, 35 percent are registered Republicans, and the remaining 20 percent are not registered in any political party. If a sample of 4 is selected at random from Opinion Poll's set of people, what is the probability that either all 4 are registered Democrats, all 4 are registered Republicans, or all 4 are not registered in a political party?

Observe that the random selection of the sample of 4 from the set is an urn-like experiment done 4 times without replacement. Let D, R, and N represent, respectively, the events that a person selected is a Democrat, a person selected is a Republican, and a person selected is not registered in a political party. We want to find the probability of {DDDD, RRRR, NNNN}.

Even though the experiment is performed without replacement, we can assume that the events D, R, and N are independent since the set from which the sample is selected is very large. A draw does not significantly affect the proportion of Democrats, Republicans, and unregistered voters. Therefore we can assume that $Pr(D|D) = Pr(D)$, $Pr(R|R) = Pr(R)$, and $Pr(N|N) = Pr(N)$. Therefore, since 45 percent of those in the set are registered Democrats, we have $Pr(DDDD) = (0.45)^4$. Likewise, $Pr(RRRR) = (0.35)^4$ and $Pr(NNNN) = (0.20)^4$. Finally, the probability that all 4 are of the same type is

$$Pr(DDDD) + Pr(RRRR) + Pr(NNNN) = (0.45)^4 + (0.35)^4 + (0.20)^4$$
$$= 0.0410 + 0.0150 + 0.0016$$
$$= 0.0576.$$

In contrast to a stochastic process in which an urn-like experiment is repeated, we have other types of stochastic processes.

EXAMPLES

5. Consider again example 12 of section 8.5 (page 364), in which a company assembles minicomputers by putting together a mainframe and a terminal. It was found that 0.2 percent of the mainframes used were defective and 0.5 percent of the terminals used were defective. What is the probability that an assembled minicomputer has exactly one defective component?

We can think of the assembly of a minicomputer as a stochastic process of two stages, where N means a component is not defective and D means a component is defective. The probability tree for this process is given in figure 8.20. We read from the probability tree that

$$Pr(\text{exactly one component is defective}) = Pr(ND) + Pr(DN)$$
$$= (0.998)(0.005) + (0.002)(0.995)$$
$$= 0.00698.$$

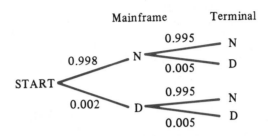

FIGURE 8.20

6. The United States population is divided into two sectors, urban (U) and nonurban (N). Let us assume that over a one-year period, 4 percent of the urban population moves into nonurban areas. Thus, if a person is chosen at random from the urban population, the probability of the person living in a nonurban area at the end of the year is 0.04. Also, let us assume that over a one-year period, 1 percent of the nonurban population moves into urban areas. If a person is selected at random from an urban area, what is the probability that the person will be living in a nonurban area in two years?

The set of possibilities for the stochastic process of moving from one sector to another (or remaining in a sector) is given by the tree diagram in figure 8.21. We have written on the branches the appropriate probabilities of the process following that branch. Observe that we are starting the process with a person living in an urban area.

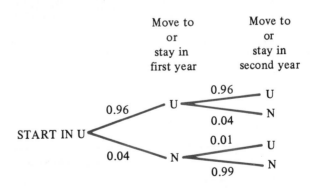

FIGURE 8.21

The answer to our question is $Pr(UN) + Pr(NN)$. From the probability tree in figure 8.21 we have

$$Pr(UN) + Pr(NN) = (0.96)(0.04) + (0.04)(0.99)$$
$$= 0.078.$$

7. Suppose Alice competes in a best-of-three tennis match; that is, the first person to win two sets wins the match. Suppose the chance of Alice winning the first set is $\frac{1}{2}$. Also, suppose that her behavior pattern is such that if she loses a set, her chance of winning the next set is 50 percent of what it was prior to her loss. If she wins a set, her chance of winning the next set is the same as what it was prior to her win. Let us find the probability that Alice wins the match.

The model for the stochastic process is given by the probability tree in figure 8.22.

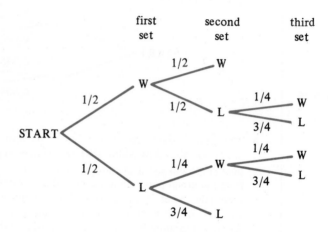

FIGURE 8.22

The event of interest is the set $E = \{WW, WLW, LWW\}$. Reading from the probability tree we have

$$Pr(WW) = \left(\frac{1}{2}\right)\left(\frac{1}{2}\right) = \frac{1}{4}$$

$$Pr(WLW) = \left(\frac{1}{2}\right)\left(\frac{1}{2}\right)\left(\frac{1}{4}\right) = \frac{1}{16}.$$

$$Pr(LWW) = \left(\frac{1}{2}\right)\left(\frac{1}{4}\right)\left(\frac{1}{4}\right) = \frac{1}{32}.$$

Thus Alice's chance of winning the match is $Pr(E) = \frac{1}{4} + \frac{1}{16} + \frac{1}{32} = \frac{11}{32}$.

EXERCISES 8.6

1. According to transactional analysis, there are three states—parent, adult, and child— existing in each person. When one person talks to another, the communication can come from any one of the three states of the talker and can be directed to any one of the three states of the listener. Draw a tree diagram to illustrate the possible ways a communication can be transmitted from a talker to a listener.

2. Professor Johnson is making a survey concerning socioeconomic conditions in the Delaware Valley. He classifies people according to sex and income level (low, middle, high).

   a. Draw a tree diagram to illustrate the possibilities with respect to an individual in the survey. (The tree has two stages and six paths.)

   b. In addition to sex and income level, a person is also classified according to educational level achieved (grade school, high school, college). Add to the tree diagram of part (a) branches to depict this additional category. How many paths does the tree diagram have with this third category of classification?

3. An urn contains 4 red balls, 5 blue balls, and 9 green balls. An experiment is to draw a ball at random from the urn and observe its color. Consider the stochastic process in which this experiment is performed two times with replacement.

   a. Draw a tree diagram for this process with the appropriate probabilities indicated on each branch.

   b. What is the probability that both balls drawn are green?

   c. What is the probability that the first ball drawn is a color other than red and the next ball drawn is red?

   d. What is the probability that a red ball is drawn both times?

4. An urn contains 2 red balls, 7 blue balls, and 3 green balls. An experiment is to draw one ball at random from the urn and observe its color. Consider the stochastic process in which this experiment is performed two times without replacement.

   a. Draw a tree diagram for this process with the appropriate probabilities indicated on each branch.

   b. What is the probability that both balls drawn are green?

   c. What is the probability that the first ball drawn is a color other than red and the next ball drawn is red?

   d. What is the probability that a red ball is drawn both times?

5. An urn contains 3 red balls and 4 blue balls. Consider the experiment of selecting at random a ball from the urn and observing its color. This experiment is performed without replacement until a blue ball is obtained.

   a. Draw a probability tree for this stochastic process.

   b. What is the probability that exactly 3 balls must be drawn to obtain a blue ball?

   c. What is the probability that at least 3 balls must be drawn to obtain a blue ball?

6. A storage room contains 6 slot machines from a production line. Three of the machines are defective. Consider the experiment of selecting a machine at random from the room and testing it. This experiment is performed without replacement until a defective machine is obtained.

   a. Draw a probability tree for this stochastic process.

   b. What is the probability that exactly 2 machines must be taken from the room to obtain a defective one?

   c. What is the probability that at least 3 machines must be taken from the room to obtain a defective one?

7. Consider the best-of-three tennis match discussed in example 7. Suppose that if Alice wins a set, her chance of winning the next set is 50 percent more than what it was prior to

winning. Except for this change, assume that everything remains the same as described in example 7.

a. Draw a tree diagram for this stochastic process with the appropriate probabilities indicated on each branch.

b. What is the probability that Alice wins the match?

c. What is the probability that the match will go to three sets?

8. In a government survey of teenage smoking, it was found that 4 percent of boys aged 12 to 14 smoked regularly, 18 percent of boys aged 15 to 16 smoked regularly, and 31 percent of boys aged 17 to 18 smoked regularly. Suppose a boy is selected at random from a very large group of teenagers in which each age category is equally represented.

a. Find the probability that the boy selected is in age group 17 to 18 and smokes regularly.

b. Find the probability that the boy selected is between 12 to 16 years old and smokes regularly.

c. Find the probability that the boy selected is in age group 12 to 14 and does not smoke regularly.

9. Consider the experiment of flipping a weighted coin, where the probability of obtaining a head is $\frac{3}{4}$. Suppose we perform this experiment three times, resulting in a stochastic process.

a. Draw a probability tree for this process.

b. List the possible outcomes of this process.

c. Make an appropriate assignment of probabilities for the sample space found in part (b).

d. What is the probability of obtaining at least two heads?

10. Consider the experiment of flipping an unbiased coin and the stochastic process consisting of performing this experiment three times.

a. Draw a probability tree for this process.

b. List the possible results of this process.

c. Assign a probability to each possible result of the process.

d. What is the probability of obtaining exactly two tails?

e. What is the probability of obtaining at least one head?

11. A test contains 3 true-false questions. Suppose Pete takes the test and guesses on each question. Thus the result of answering one question (right or wrong) is independent of the result of answering a previous question, and the probability of Pete answering any question correctly is $\frac{1}{2}$.

a. What is the probability of Pete answering just the first question correctly?

b. What is the probability of Pete answering only one of the 3 questions correctly?

c. What is the probability of Pete answering all 3 questions correctly?

12. A test contains 3 multiple-choice questions in which each question has 4 choices and only 1 of them is correct. Suppose Sue takes the test and guesses on each question.

a. What is the probability of Sue answering only the first question correctly?

b. What is the probability of Sue answering only 1 of the 3 questions correctly?

c. What is the probability of Sue answering all 3 correctly?

13. Past records show that the probability that a person fails an actuarial exam on the first try is 0.6. The probability that a person who failed the first time fails the second time is 0.3. What is the probability that a person passes on either the first or second try?

14. From past records, the probability that a mathematics major at State University passes the comprehensive examinations on the first try is 0.7. The probability of passing the exams on the second try, given a failure on the first try, is 0.6. What is the probability that a mathematics major does not pass the exams?

15. The manufacture of briar pipes involves two phases, carving and finishing. After a piece of briar is carved, it can be discarded or passed on to finishing. After finishing, the pipe can be discarded, sold as second quality, or sold as first quality. Suppose that 95 percent of the carved pieces of briar are passed on to finishing. Suppose also that 75 percent of the finished pipes are sold as first quality, and 23 percent are sold as second quality.

    a. What is the probability of a piece of briar becoming a pipe of first quality?

    b. What is the probability of a piece of briar being discarded?

16. In the manufacture of pottery, a molded piece of clay goes through the following three phases: drying, first firing, and second firing. After either of the first two phases, a piece can be either discarded or passed on to the next phase. After the third phase, the piece can be discarded, sold as second quality, or sold as first quality. Suppose the pottery company has obtained the following information through past experience. The probability of a molded piece being discarded after drying is 0.04. The probability of a dried piece being discarded after the first firing is 0.25. The probability of a fired piece being discarded after the second firing is 0.11. The probability of the piece being sold as second quality is 0.22.

    a. What is the probability of a molded piece being discarded prior to going to the second firing?

    b. What is the probability of a molded piece being discarded?

    c. What is the probability of a molded piece being sold as first quality?

    d. What is the probability of a molded piece being sold as either first quality or second quality?

17. Tom has a class at 8 o'clock in the morning and the instructor observes the following behavior pattern in the student. Whenever Tom is present (P) one day, there is only a 50 percent chance that he will be present the next day. However, if he is absent (A) one day, he makes a greater effort to attend class and his chance of being present the next day increases to 70 percent.

    a. Find the probability that Tom will be absent on the second day following one in which he is present.

    b. Find the probability that Tom will be absent on the second day following one on which he was absent.

18. Consider two urns A and B. Urn A contains 2 white and 2 green balls, whereas urn B contains 3 white and 4 green balls. Consider the experiment of selecting an urn at random and drawing a ball at random from that urn.

    a. Draw a tree diagram for this two-stage stochastic process with the appropriate probabilities indicated on each branch.

    b. Find the probability that a green ball is drawn.

    c. Find the probability that a white ball is drawn.

19. The diagram shown in figure 8.23 is a map of Judy's neighborhood, where each square indicates a block. Judy is located at the lower left corner, labeled *J*. She performs the following experiment. She flips a coin. If it is heads, she moves one block north; if it is tails, she moves one block east. She continues this process, block by block, flipping the coin and moving either north or east. What is the probability that she will reach intersection *F*?

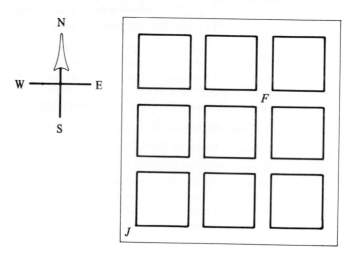

FIGURE 8.23

8.7 Bayes's Formula

In many problems, an experimenter wants to find the probable cause of an event. For example, if the demand for microchips has surged, we may want to know the probability that the electronic game business is the cause. The event that has occurred may be in the form of a test result. For example, in medicine, the results of a blood test determine the probability that a patient has high cholesterol. In quality control, the sampling of items from an assembly line determines the probability that the production process is operating efficiently.

A method used to find the probability of a particular cause when the result of a test is given was developed by Thomas Bayes (1702–1763). Before we derive this method and look at some applications, we first consider the concept of **partitioning** (dividing) a sample space into two or more disjoint events.

Suppose there are two urns A and B. Urn A contains 3 white balls and 4 red balls and urn B contains 2 white balls and 1 red ball. Consider the experiment of selecting an urn at random and drawing a ball. What is the probability that the ball drawn is white?

Let *A* be the event that the ball is drawn from urn A and let *B* be the event that the ball is drawn from urn B. Let *E* be the event that the ball drawn is white. To find the probability of *E*, we refer to the Venn diagram in figure 8.24.

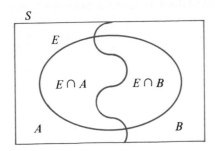

FIGURE 8.24

We have that A and B are mutually exclusive events and their union is the entire sample space S. In this case we say that A and B form a partition of S.

Definition   ■   Let S be a sample space for an experiment. Let A and B be two events of S. We say that A and B form a **partition** of S if

1. $A \cap B = \emptyset$, that is, A and B are mutually exclusive events; and
2. $A \cup B = S$, that is, the union of A and B is S.

To distinguish the events A and B of the partition from other events, we shall refer to A and B as **states**.   ■

Returning to our urn example, we have, from the Venn diagram in figure 8.24, that

$$E = (E \cap A) \cup (E \cap B).$$

Furthermore, $E \cap A$ and $E \cap B$ are mutually exclusive. Therefore,

$$Pr(E) = Pr(E \cap A) + Pr(E \cap B).$$

Apply formula (15), from page 362

$$Pr(E \cap F) = Pr(F)Pr(E|F),$$

we have the following rule for the probability of E.

Rule   ■   *Expansion Rule:*
Given a partition of a sample space S into two states A and B, then for any event E

$$Pr(E) = Pr(A)Pr(E|A) + Pr(B)Pr(E|B).$$ ■   (20)

Let us use the expansion rule for our urn example. Since an urn is selected at random, $Pr(A) = \frac{1}{2}$ and $Pr(B) = \frac{1}{2}$. The probability that a white ball is drawn given that urn A was selected is $\frac{3}{7}$, that is $Pr(E|A) = \frac{3}{7}$. Likewise, $Pr(E|B) = \frac{2}{3}$. Using this information we find the probability of E.

$$Pr(E) = \left(\frac{1}{2}\right)\left(\frac{3}{7}\right) + \left(\frac{1}{2}\right)\left(\frac{2}{3}\right)$$

$$= \frac{23}{42} = 0.548.$$

EXAMPLE 1

The accounting department of a bank has divided the people with mortgages with the bank into two groups. Group A are those people who 90 percent of the time pay their monthly installment when due. Group B are those people who 70 percent of the time pay their monthly installment when due. Eightly percent of all those having mortgages with the bank are in group A and 20 percent are in group B. Now, suppose the bank grants a mortgage. Let E be the event that the mortgagee pays the first installment when due. Let us find the probability of E.

Let A be the event that the person belongs to group A, and let B be the event that the person belongs to group B. Since A and B form a partition of the sample space, we may use formula (20) to find the probability of E.

From the information given, we have

$$Pr(A) = 0.8 \qquad Pr(B) = 0.2 \qquad Pr(E|A) = 0.9 \qquad \text{and} \qquad Pr(E|B) = 0.7.$$

Therefore,

$$Pr(E) = Pr(A)Pr(E|A) + Pr(B)Pr(E|B)$$
$$= (0.8)(0.9) + (0.2)(0.7)$$
$$= 0.86.$$

Let us return to the urn example at the beginning of this section and consider the following situation. Suppose that we do not see the experiment being performed, but we are told that the ball drawn is white. What is the probability that the ball came from urn A?

As before, let A be the event that the ball is drawn from urn A, B the event that the ball is drawn from urn B, and E the event that the ball drawn is white. We want to find the probability of A given E, that is, $Pr(A|E)$. Using the formula for conditional probability (formula 14 on page 360),

$$Pr(A|E) = \frac{Pr(A \cap E)}{Pr(E)}.$$

But, $Pr(A \cap E) = Pr(A) \cdot Pr(E|A)$. Therefore,

$$Pr(A|E) = \frac{Pr(A) \cdot Pr(E|A)}{Pr(E)}.$$

From the information given, we saw that $Pr(A) = \frac{1}{2}$ and $Pr(E|A) = \frac{3}{7}$. Furthermore, $Pr(E) = \frac{23}{42}$ was found by using the expansion rule. Therefore,

$$Pr(A|E) = \frac{\left(\frac{1}{2}\right)\left(\frac{3}{7}\right)}{\frac{23}{42}} = 0.391.$$

This illustrates Bayes's formula for a partition of the sample space into two states A and B.

Rule

■ *Bayes's Formula:*
Let events A and B form a partition of a sample space S. Then for any event E of S, we have

$$Pr(A|E) = \frac{Pr(A) \cdot Pr(E|A)}{Pr(E)} \tag{21}$$

where, by the expansion rule, $Pr(E) = Pr(A) \cdot Pr(E|A) + Pr(B) \cdot Pr(E|B)$. ■

Note that $Pr(B|E)$ can also be found by using formula (21) with A replaced by B. There is an alternate way to find it. Given E, either A occurs or B occurs, so $Pr(A|E)$ and $Pr(B|E)$ are related by $Pr(A|E) + Pr(B|E) = 1$.

EXAMPLES

2. Refer to example 1. Suppose that a person who has been granted a mortgage by the bank pays the first installment when due. What is the probability that he belongs to group A? What is the probability that he belongs to group B? We let A, B, and E denote the same events as in example 1. The conditional probabilities $Pr(E|A)$ and $Pr(E|B)$, as given in example 1, are determined from bank records of previous clients. We want to use this information to find $Pr(A|E)$; that is, we want to find the chances that a new client belongs to group A given that he has made his first payment on time. Also, we wish to find the chances that this person belongs to group B.

We use Bayes's formula to find $Pr(A|E)$; in particular,

$$Pr(A|E) = \frac{Pr(A)Pr(E|A)}{Pr(E)}$$

$$= \frac{(0.8)(0.9)}{0.86}$$

$$= 0.837.$$

To find $Pr(B|E)$, we use the fact that $Pr(A|E) + Pr(B|E) = 1$. Therefore,

$$Pr(B|E) = 1 - Pr(A|E)$$

$$= 1 - 0.837$$

$$= 0.163.$$

3. A criminal lawyer estimates that 70 percent of her clients are guilty. She feels that her guilty clients are acquitted 60 percent of the time and that her innocent clients are acquitted 95 percent of the time. Suppose a particular client has been acquitted. Let us find the probability that the client is innocent.

Let A be the state that the client is innocent, and let B be the state that the client is guilty. Let E be the event that the client has been acquitted. We want to find $Pr(A|E)$. From the information given,

$$Pr(A) = 0.3 \qquad Pr(B) = 0.7 \qquad Pr(E|A) = 0.95 \qquad Pr(E|B) = 0.6.$$

By the expansion rule,

$$Pr(E) = (0.3)(0.95) + (0.7)(0.6)$$
$$= 0.705.$$

By Bayes's formula,

$$Pr(A|E) = \frac{(0.3)(0.95)}{0.705}$$
$$= 0.404.$$

Now that we have seen Bayes's formula for two states, we consider the general formula for n states. We begin with the definition of a partition of a sample space into n states.

Definition

■ Let S be a sample space for an experiment. Let $A_1, A_2, \ldots, A_n$ be n events of S. We say that $A_1, A_2, \ldots, A_n$ form a **partition** of S if,

1. $A_i \cap A_j = \varnothing$, for $i \neq j$, that is, the events in the partition are **pair-wise mutually exclusive**; and if
2. $A_1 \cup A_2 \cup \cdots A_n = S$, that is, the union of these events is S.

As for two states, we shall also refer to the n events $A_1, A_2, \ldots, A_n$ of the partition as **states**. ■

Rule

■ *Expansion Rule (General):*
Let $A_1, A_2, \ldots, A_n$ form a partition of a sample space S and let E be any event of S, then

$$Pr(E) = Pr(A_1)Pr(E|A_1) + Pr(A_2)Pr(E|A_2) + \cdots + Pr(A_n)Pr(E|A_n). \quad ■ \ (22)$$

EXAMPLE 4

A truck manufacturer gets transmissions from four different suppliers A_1, A_2, A_3, and A_4. From past experience, the company estimates that 4 percent of the transmissions from supplier A_1, 5 percent of the transmissions from supplier A_2, 3 percent of the transmissions from supplier A_3, and 3 percent of the transmissions from supplier A_4 are defective. The company receives 30 percent of its stock of transmissions from supplier A_1, 40 percent from supplier A_2, 10 percent from supplier A_3, and 20 percent from supplier A_4. If a transmission is selected at random from the company's stock, what is the probability that the transmission is defective?

Let E be the event that the transmission is defective. Also, let A_1, A_2, A_3, and A_4 be the events that the transmission came from supplier A_1, A_2, A_3, and A_4, respectively. Since a transmission must be in one and only one of the events A_1, A_2, A_3, and A_4, these four events form a partition of the sample space. We find the probability of the event E by using the expansion rule (22) with $n = 4$.

$$Pr(E) = Pr(A_1)Pr(E|A_1) + Pr(A_2)Pr(E|A_2) + Pr(A_3)Pr(E|A_3) + Pr(A_4)Pr(E|A_4).$$

From the given information

$$Pr(A_1) = 0.3 \quad Pr(A_2) = 0.4 \quad Pr(A_3) = 0.1 \quad Pr(A_4) = 0.2,$$

$$Pr(E|A_1) = 0.04 \quad Pr(E|A_2) = 0.05 \quad Pr(E|A_3) = 0.03 \quad Pr(E|A_4) = 0.03$$

we have

$$Pr(E) = (0.3)(0.04) + (0.4)(0.05) + (0.1)(0.03) + (0.2)(0.03)$$
$$= 0.041.$$

We now state Bayes's formula for a partition of the sample space into n states.

Definition   ■ *Bayes's Formula (General):*
Let $A_1, A_2, \ldots, A_n$ form a partition of a sample space S. Then for any event E of S, we have

$$Pr(A_i|E) = \frac{Pr(A_i)Pr(E|A_i)}{Pr(E)} \qquad i = 1, 2, \ldots, n \tag{23}$$

where $Pr(E)$ is found by the expansion rule (20)

$$Pr(E) = Pr(A_1)Pr(E|A_1) + Pr(A_2)Pr(E|A_2) + \cdots + Pr(A_n)Pr(E|A_n). \quad ■$$

EXAMPLE 5   Refer to example 4. Suppose a transmission is selected at random from the company's stock and it is found to be defective. What is the probability that it came from supplier A_3?
If we let A_1, A_2, A_3, A_4, and E denote the same events as in example 4, then we want to find $Pr(A_3|E)$.
By Bayes's formula (equation (23) with $i = 3$), we have

$$Pr(A_3|E) = \frac{Pr(A_3)Pr(E|A_3)}{Pr(E)}$$

$$= \frac{(0.1)(0.03)}{(0.041)}$$

$$= 0.073.$$

EXERCISES 8.7

1. Let a sample space S be partitioned into two states A and B with $Pr(A) = \frac{1}{3}$ and $Pr(B) = \frac{2}{3}$. Let E be an event such that $Pr(E|A) = \frac{3}{4}$ and $Pr(E|B) = \frac{6}{11}$. Find
   a. $Pr(E)$ b. $Pr(A|E)$ c. $Pr(B|E)$

2. Let a sample space S be partitioned into two states A and B with $Pr(A) = 0.65$ and $Pr(B) = 0.35$. Let E be an event such that $Pr(E|A) = 0.21$ and $Pr(E|B) = 0.64$. Find
   a. $Pr(E)$ b. $Pr(A|E)$ c. $Pr(B|E)$

3. Let a sample space S be partitioned into two states A and B with $Pr(A) = 0.46$ and $Pr(B) = 0.54$. Let E be an event with $Pr(E|A) = 0.92$ and $Pr(E|B) = 0.75$. Find

    a. $Pr(E)$ b. $Pr(A|E)$ c. $Pr(B|E)$

4. Let a sample space S be partitioned into two states A and B with $Pr(A) = \frac{5}{12}$ and $Pr(B) = \frac{7}{12}$. Let E be an event with $Pr(E|A) = \frac{3}{5}$ and $Pr(E|B) = \frac{6}{7}$. Find

    a. $Pr(E)$ b. $Pr(A|E)$ c. $Pr(B|E)$

5. Urn A contains 2 white and 2 red balls; urn B contains 3 white and 5 red balls. An experiment is to select an urn and draw a ball from that urn. Let E be the event that the ball drawn is red. Suppose that the probability of selecting urn A is $\frac{1}{3}$ and the probability of selecting urn B is $\frac{2}{3}$. Find

    a. $Pr(E)$ b. $Pr(A|E)$ c. $Pr(B|E)$

6. Urn A contains 2 blue balls and 3 red balls; urn B contains 1 blue ball and 2 red balls. An experiment is to select an urn at random and draw a ball from that urn. Let E be the event that the ball drawn is blue. Find

    a. $Pr(E)$ b. $Pr(A|E)$ c. $Pr(B|E)$

7. A study is made involving a group of people, 40 percent of whom work at a nuclear power facility. It is found that 0.02 percent of those people who work at the facility have a high number of leukocytes in their blood and 0.005 percent of those people who do not work at the facility have a high number of leukocytes in their blood. A person is selected at random from the people in the study.

    a. What is the probability that the person has a high number of leukocytes in his blood?

    b. Given that the person selected has a high number of leukocytes in his blood, what is the probability that he works at the nuclear power facility?

8. Berkshire Aircraft Company has two storerooms that contain engine pressure guages. Suppose that 12% of the guages in storeroom A are defective and 15% of the guages in storeroom B are defective. A storeroom is picked at random and a guage selected at random from that storeroom.

    a. What is the probability that the guage selected is defective?

    b. Suppose that the guage selected is defective. What is the probability that it came from storeroom B?

9. A do-it-yourself pregnancy detection kit has been tested for accuracy. The test is 95 percent accurate if the woman is pregnant and 75 percent accurate if the woman is not pregnant. It is estimated that 80 percent of the women who use the kit are pregnant.

    a. What is the probability that a woman who uses the kit has a test showing pregnancy (a positive test)?

    b. What is the probability that a woman who uses the kit has a negative test?

10. A test for diabetes is 90 percent accurate if the person has diabetes and 70 percent accurate if the person does not have diabetes. It is estimated that 80 percent of the people who take the test have diabetes.

    a. What is the probability that a person who takes the test has a positive result? (The test indicates the person has diabetes.)

    b. What is the probability that a person who takes the test has a negative result?

11. Considering the information given in exercise 9, suppose a woman who uses the kit has a test showing pregnancy.

a. What is the probability that the woman is pregnant?

b. What is the probability that the woman is not pregnant?

12. Considering the information given in exercise 10, suppose a person who takes the test has a positive result?

    a. What is the probability that the person has diabetes?

    b. What is the probability that the person does not have diabetes?

13. Let a sample space S be partitioned into three states A_1, A_2, and A_3, with $Pr(A_1) = 0.23$, $Pr(A_2) = 0.47$, and $Pr(A_3) = 0.3$. Let E be an event with $Pr(E|A_1) = 0.68$, $Pr(E|A_2) = 0.31$, and $Pr(E|A_3) = 0.92$. Find

    a. $Pr(E)$ b. $Pr(A_1|E)$ c. $Pr(A_2|E)$ d. $Pr(A_3|E)$

14. Let a sample space S be partitioned into three states A_1, A_2, and A_3 with $Pr(A_1) = \frac{1}{3}$, $Pr(A_2) = \frac{1}{2}$, and $Pr(A_3) = \frac{1}{6}$. Let E be an event with $Pr(E|A_1) = \frac{2}{3}$, $Pr(E|A_2) = \frac{3}{4}$, and $Pr(E|A_3) = \frac{1}{3}$. Find

    a. $Pr(E)$ b. $Pr(A_1|E)$ c. $Pr(A_2|E)$ d. $Pr(A_3|E)$

SUMMARY OF TERMS AND FORMULAS

Basic Probability
  Experiment
  Probability Model
  Sample Space
  Event
  Assignment of Probability
  Relative Frequency
  Law of Large Numbers
  Probability of Events

Properties of Probability Models
  Universal Set
  Union
  Intersection
  Difference
  Complement
  Mutually Exclusive Events

Equiprobable Models: Counting and Permutations
  Cartesian Product
  Tree Diagram
  Principle of Counting
  Permutation
  n Factorial

Equiprobable Models: Combinations
  Combination

Conditional Probability
  Independent Events
  Dependent Events

Stochastic Process
  Probability Tree
  Urn-like Experiments

Bayes's Formula
  Partition of a Sample Space
  States
  Pair-Wise Mutually Exclusive Events

Formulas

Let E and F be events, then

$$Pr(E \cup F) = Pr(E) + Pr(F) - Pr(E \cap F).$$

$$Pr(E \cup F) = Pr(E) + Pr(F) \quad \text{if } E \text{ and } F \text{ are mutually exclusive.}$$

$$Pr(E) = 1 - Pr(E').$$

1. $(E \cup F)' = E' \cap F'$

2. $(E \cap F)' = E' \cup F'$ (De Morgan's laws).

$$Pr(E|F) = \frac{Pr(E \cap F)}{Pr(F)}$$

$$Pr(E \cap F) = Pr(F)Pr(E|F) \quad \text{(product rule).}$$

$$Pr(E_1 \cap E_2 \cap \cdots \cap E_n) = Pr(E_1)Pr(E_2)\cdots Pr(E_n)$$
$$\text{if } E_1, E_2, \ldots, E_n \text{ are independent events.}$$

$$n! = n(n-1)(n-2)\cdots 2 \cdot 1 \quad\quad C[n,r] = \frac{P[n,r]}{r!}.$$

$$P[n,r] = n(n-1)\cdots(n-r+1) \quad\quad C[n,r] = C[n, n-r].$$

Let A and B form a partition of a sample space and E be any event then

$$Pr(E) = Pr(A)Pr(E|A) + Pr(B)Pr(E|B) \quad \text{(expansion rule).}$$

$$Pr(A|E) = \frac{Pr(A)Pr(E|A)}{Pr(E)}, \quad Pr(B|E) = \frac{Pr(B)Pr(E|B)}{Pr(E)} \quad \text{(Bayes's formulas).}$$

Let $A_1, A_2, \ldots, A_n$ form a partition of a sample space and let E be any event, then

$$Pr(E) = Pr(A_1)Pr(E|A_1) + Pr(A_2)Pr(E|A_2) + \cdots + Pr(A_n)Pr(E|A_n)$$
$$\text{(general expansion rule).}$$

$$Pr(A_i|E) = \frac{Pr(A_i)Pr(E|A_i)}{Pr(E)} \quad i = 1, 2, \ldots, n \quad \text{(general Bayes's formula).}$$

REVIEW EXERCISES (CH. 8)

---

1. An urn contains 2 red, 6 white, 1 green, and 4 blue balls. A single ball is drawn at random from the urn, and its color is noted.

   a. Give the sample space for the experiment.

   b. Give an appropriate assignment of probabilities for the sample space.

   c. What is the probability that either a white or a blue ball is drawn?

2. In a market survey of tobacco users, it was discovered that 52 percent preferred smoking cigarettes, 18 percent preferred smoking cigars, 27 percent preferred smoking pipes, and 3

percent preferred chewing tobacco. A person is selected at random from the survey group.

a. What is the probability that the person selected prefers smoking cigars?

b. What is the probability that the person selected prefers either smoking cigarettes or chewing tobacco?

c. What is the probability that the person selected does not prefer smoking cigarettes?

3. For a certain group of car owners, the probability that a person selected at random owns a compact car is 0.55, the probability that the person earns less than $20,000 a year is 0.67, and the probability that the person either owns a compact car or earns less than $20,000 a year is 0.75. Find the probability that the person owns a compact car and earns less than $20,000 a year.

4. Let E and F be two events of a sample space S.

a. If E and F are mutually exclusive, $Pr(E) = 0.89$, and $Pr(F) = 0.03$, find $Pr(E \cup F)$.

b. If $Pr(E) = \frac{4}{11}$, $Pr(F) = \frac{5}{11}$, and $Pr(E \cap F) = \frac{1}{11}$, find $Pr(E \cup F)$.

5. Let E and F be two events of a sample space S.

a. If $Pr(E' \cap F') = 0.476$, find $Pr(E \cup F)$.

b. If $Pr(E' \cup F') = \frac{4}{21}$, find $Pr(E \cap F)$.

6. David plans to buy a television component system. He has a choice of 7 monitors, 3 receivers, 5 VCR's, and 6 types of speakers. How many different systems are possible?

7. Find each of the following.

a. $P[4, 4]$ b. $P[3, 2]$ c. $C[7, 2]$ d. $C[60, 58]$

8. How many three-digit numbers can be formed by taking the first digit from the set $\{3, 5, 6\}$, the second digit from the set $\{4, 1\}$, and the third digit from the set $\{0, 3, 6, 9\}$?

9. An unbiased coin is flipped 5 times.

a. Find the probability of obtaining exactly 4 heads.

b. Find the probability of 2 or 3 heads.

c. Find the probability of at most 4 heads.

10. Amazon Discount Store has 10 hot-air corn poppers in its inventory of which 3 are defective. If 4 poppers are selected at random, what is the probability that at least 1 of them is defective?

11. Byzantine Meat Packing Company has 20 electronic insect eliminators in a store room of which 6 are defective. If 5 of these machines are selected at random, find the probability that at least 2 of them are defective.

12. Let E and F be two events of a sample space S.

a. If $Pr(E) = \frac{3}{5}$ and $Pr(F|E) = \frac{20}{21}$, find $Pr(E \cap F)$.

b. If $Pr(E \cap F) = 0.795$ and $Pr(F) = 0.821$, find $Pr(E|F)$.

13. A survey was taken of 100 people and the results are given in table 8.14.

TABLE 8.14

	F Single	F' Married	Totals
E: income is under $15,000	27	10	37
E: income is $15,000 or more	32	31	63
totals	59	41	100

Let E be the event that a person selected at random makes under $15,000 per year. Let F be the event that a person selected at random is single. Find:

a. $Pr(E|F)$ b. $Pr(E'|F)$ c. $Pr(F|E')$ d. $Pr(E'|F')$.

14. Let E, F, and G be three independent events of a sample space S. If $Pr(E) = 0.45$, $Pr(F) = 0.39$, and $Pr(G) = 0.81$, find $Pr(E \cap F \cap G)$.

15. It is estimated that 65 percent of the people in the Lower Glennwood District are in favor of increased spending on solar energy development. Five people are selected at random from this district. Assuming independence, what is the probability that

a. All 5 favor increased spending on solar energy?

b. Exactly 3 favor increased spending on solar energy?

c. At most 4 favor increased spending on solar energy?

16. A salesman must travel from his home office to each of three cities A, B, and C. If he travels to city A first, the probability of traveling to city B next is $\frac{1}{2}$. If he travels to city B first, the probability of traveling to city A next is $\frac{1}{3}$. If he travels to city C first, the probability of traveling to city A next is $\frac{1}{4}$. The probability that he travels to city A first is $\frac{1}{2}$, whereas the probability that he travels to city B first is $\frac{1}{4}$.

a. What is the probability that the salesman travels to city B last?

b. What is the probability that the salesman travels to city C last?

c. What is the probability that the salesman travels to city C before traveling to city A?

17. An urn contains 4 red, 1 white, and 1 blue ball. An experiment is to draw a ball at random from the urn and observe its color. Consider the stochastic process in which this experiment is performed 3 times without replacement.

a. Draw a probability tree for the stochastic process.

b. What is the probability of drawing exactly 2 red balls?

18. A box contains 4 lightbulbs of which 3 are defective. Lightbulbs are drawn without replacement until the good one is drawn.

a. Draw a probability tree for this experiment.

b. What is the probability that 4 draws are necessary to obtain the good lightbulb?

19. Mr. Koehler estimates that 20 percent of his soybeans have root rot. A representative sample of 4 plants is independently drawn from the field.

a. What is the probability that all 4 plants have root rot?

b. What is the probability that exactly 2 out of the 4 plants have root rot?

20. Consider two urns A and B. Urn A contains 2 white balls and 3 red balls and urn B contains 1 white ball and 2 red balls. An urn is selected at random and a ball is drawn. Let A and B be the events that urns A and B are chosen, and E be the event that the ball drawn is red.

a. Find $Pr(E)$.   b. Find $Pr(A|E)$.

21. Let A_1, A_2, and A_3 be a partition of a sample space S with

$$Pr(A_1) = \frac{1}{3}, \qquad Pr(A_2) = \frac{1}{2}, \qquad \text{and} \qquad Pr(A_3) = \frac{1}{6}.$$

Let E be an event such that

$$Pr(E|A_1) = \frac{1}{4}, \qquad Pr(E|A_2) = \frac{3}{4}, \qquad \text{and} \qquad Pr(E|A_3) = \frac{1}{2}.$$

a. Find $Pr(E)$.   b. Find $Pr(A_2|E)$.

22. An IRS agent estimates that 60% of the people she audits have cheated on their taxes. She feels that 55% of the cheaters pay more tax after the audit, whereas 35% of the people who did not cheat pay more tax after the audit. Suppose that a particular person whose tax return has been audited by the agent must pay more tax. What is the probability that this person did not cheat on his taxes?

23. The Digitized Corporation makes compact disks on three machines I, II, and III. Machines I, II, and III account for 40%, 36%, and 24%, respectively, of the daily production. It is estimated that 3% of the disks from machine I, 4% of the disks from machine II, and 6% of the disks from machine III are defective. A compact disk is selected at random from the corporation's output and it is found to be defective. Find

a. The probability that the disk was made on machine I.

b. The probability that the disk was made on machine II.

c. The probability that the disk was made on machine III.

d. The probability that the disk was not made on machine II.

CHAPTER 9

Statistics

Introduction

Many of us associate the word *statistics* with numerical data. We read about stock market averages, the statistics of a football game, and changes in the consumer price index. However, when statistical techniques are used to solve problems, collecting numerical data is only one part of the process.

In addition to collecting data, the statistician is interested in

1. describing the data—applying measures to the data such as the average of the numbers and the spread of the numbers in the data;
2. drawing conclusions—inferring answers from the data.

Often the collection of data is obtained through experimentation, and therefore probability theory is an integral part of statistical inquiry.

In this chapter we touch on the descriptive aspect of statistics and consider probability models that are useful in our study. Inferential statistics is introduced in section 9.7, Hypothesis Testing, which uses the descriptive statistics and probability models developed throughout the chapter.

9.1 Random Variables

Observation of what occurs is an important aspect of probability experiments. Frequently the experimenter associates a numerical value with each possible outcome and observes the numerical values that occur. For example, in the experiment of rolling two dice, 36 things could happen, each of which can be denoted by an ordered pair (a, b). However, as experimenters, we may not be so interested in the particular pair that occurs, but we may wish to observe instead the sum of the numbers showing. Therefore we would assign to each of the 36 possible outcomes one of the integers 2 through 12.

In the area of quality control, a group of items may be selected from an assembly line to test the state of the production process. For example, suppose the quality control department selects without replacement three items from a bin containing the previous hour's production. The items in the bin can be either defective (D) or normal (N). The sample space S is the triples of D's and N's.

$$S = \{DDD, DDN, DND, NDD, DNN, NDN, NND, NNN\}.$$

The quality control department may be interested only in the number of defective items in the sample of three. In this case, quality control would assign to each of the eight possible outcomes in S the number of defective items in the sample. For example, the number 2 would be assigned to each of the outcomes DDN, DND, and NDD.

The assignment of numerical values to the possible outcomes of an experiment is given a special name.

Definition
■ A **random variable** is a rule for assigning a number to each outcome in the sample space of an experiment. ■

Random variables are usually denoted by capital letters such as X, Y, and Z. The numbers obtained from the rule are called the **values of the random variable**. The values of a random variable are denoted by lower-case letters, possibly with subscripts. For example, we may say that the values of the random variable X are $x_1, x_2,$ and x_3. We may now illustrate a variety of random variables in the following examples.

EXAMPLES
1. Consider, as previously discussed, the experiment of selecting without replacement a sample of three items from a bin containing defective (D) and normal (N) items. Consider the random variable that assigns to each possible sample the number of defective items in the sample, which we will call X. There are four values that X can have, $x_1 = 0, x_2 = 1, x_3 = 2,$ and $x_4 = 3$. The outcomes in the sample space and the corresponding values of X are given as follows.

outcome	DDD	DDN	DND	NDD	DNN	NDN	NND	NNN
value of X	3	2	2	2	1	1	1	0

2. A club runs a lottery to raise money. The club sells 1000 tickets for $1 each. The prizes are as follows: a first prize worth $400, two second-place prizes worth $100 each, and four third-place prizes worth $25 each. The experiment of buying a ticket has a sample space with the four possible outcomes

$$e_1 = \text{the ticket wins the first-place prize}$$
$$e_2 = \text{the ticket wins a second-place prize}$$
$$e_3 = \text{the ticket wins a third-place prize}$$
$$e_4 = \text{the ticket wins nothing.}$$

Let X be the random variable that assigns the net amount won or lost to each outcome; that is, X assigns to each outcome the **payoff** for the outcome. We should observe that if a person wins a prize, the net amount won is $1 less than the actual prize since it cost a dollar to buy the ticket. We can construct the following table of outcomes with the corresponding payoffs.

outcome	e_1	e_2	e_3	e_4
value of X	$399	$99	$24	−$1

3. An urn contains 2 red balls and 1 white ball. Balls are drawn from the urn with replacement until a white ball is obtained. Let the random variable Y be the number of balls drawn. Since the balls are drawn with replacement (the ball is put back into the urn before the next ball is drawn), there are infinitely many values for Y, as shown in the following table.

outcome	W	RW	RRW	RRRW	$\cdots$
value of Y	1	2	3	4	$\cdots$

4. Suppose a can of peas is selected at random from the daily production of a food processing plant. Let the random variable V be the volume of water in the can. Suppose the radius of the can is $\frac{3}{2}$ inches and the height is 4 inches. Then the can holds a maximum of 9π cubic inches of water. Assuming that the volume can be measured with complete accuracy, the possible values of V are any real numbers between 0 and 9π, inclusive. This set of values for V can be displayed as an interval in the real number line (see figure 9.1).

FIGURE 9.1

$$0 \leqslant V \leqslant 9\pi$$

A random variable is classified according to its set of values. As seen in examples 1, 2, and 3, the set of distinct values can be of the type.

$$\{x_1, x_2, x_3, \ldots, x_n\} \quad \text{or} \quad \{x_1, x_2, x_3, \ldots \quad \}.$$

In this case we say that the random variable is **discrete**. Further, the random variable is **finite** discrete or **infinite** discrete according to whether the set of values is finite or infinite. The random variables of examples 1 and 2 are finite discrete, whereas the random variable of example 3 is infinite discrete. In example 4, the set of values for the random variable is an interval in the line of real numbers. This set of values cannot be arranged in a sequence as can discrete random variables. When the set of values is an interval, we say that the random variable is **continuous**. Some typical examples of continuous random variables are the height or weight of a person, the life of a lightbulb, and the production time needed to complete an item on an assembly line. We will discuss continuous random variables in more detail later. For now, we restrict our attention to finite discrete random variables.

Just as probabilities are assigned to the outcomes of an experiment, we consider the probabilities that the values of a random variable may occur. Consider again the experiment of rolling a pair of unbiased dice and the random variable X, where the sum of the numbers showing is assigned to the outcome. Since we are assuming that the dice are unbiased, the 36 outcomes are equally likely and the probability of each is $\frac{1}{36}$. Now consider, for example, the value 3 of the random variable X. The chance of obtaining this value when the experiment is performed is the same as the probability of obtaining a pair whose sum is 3; that is, the probability of obtaining the value 3 is the same as the probability of the event $E = \{(1, 2), (2, 1)\}$, or $\frac{2}{36}$. As another example, the probability of obtaining the value 7 is the same as the probability of the event $E\{(1, 6), (2, 5), (3, 4), (4, 3), (5, 2), (6, 1)\}$, or $\frac{6}{36}$. This idea of assigning probabilities to the values of the random variable is generalized in the following definition.

Definition

■ Consider an experiment with sample space S and let X be a finite discrete random variable. Let x be a value of the random variable and let E be the event of those outcomes in S which are assigned the value x by the random variable X. Then the probability of the value x occurring is the probability of E occurring, which we denote by

$$Pr(X = x) = Pr(E). \quad ■$$

EXAMPLE 5

Consider the experiment of flipping three unbiased coins. The sample space is given as

$$S = \{HHH, HHT, HTH, HTT, THH, THT, TTH, TTT\}.$$

Let X be the random variable that assigns to each outcome the number of heads appearing. The values of the random variable are 0, 1, 2, and 3. Let us determine the probability of each of these values occurring.

Since the coins are unbiased, the probability for each of the eight outcomes in S is $\frac{1}{8}$. With this probability assignment for S, we have

$$Pr(X = 0) = Pr\{TTT\} = \frac{1}{8}$$

$$Pr(X = 1) = Pr\{HTT, THT, TTH\} = \frac{3}{8}$$

$$Pr(X = 2) = Pr\{HHT, HTH, THH\} = \frac{3}{8}$$

and

$$Pr(X = 3) = Pr\{HHH\} = \frac{1}{8}.$$

We summarize this in the following table.

value of $X(x)$	0	1	2	3
$Pr(X = x)(p)$	$\frac{1}{8}$	$\frac{3}{8}$	$\frac{3}{8}$	$\frac{1}{8}$

Given a probability model of an experiment and a random variable X, we can assign to each value x the probability p that X takes on the value x. These ordered pairs (x, p) can be given by a table, as in example 5. Another method for displaying the pairs (x, p) is to sketch them on a graph in a Cartesian coordinate system where the horizontal axis is used for the value x of the random variable and the vertical axis is used for the probability p. For example, the ordered pairs $(0, \frac{1}{8})$, $(1, \frac{3}{8})$, $(2, \frac{3}{8})$, and $(3, \frac{1}{8})$ obtained in example 5 are shown on the graph in figure 9.2a.

When the values of a random variable with their associated probabilities are shown on a graph, it is common practice to connect each point in the graph to the x-axis with a vertical line. The resulting graph is called a **bar graph**. The bar graph for the pairs of example 5 is given in figure 9.2b. The graph for a random variable X shows how the probabilities of the possible values of X are distributed. In general, the **probability distribution** of a random variable is any table, graph, or formula that gives the pairs (x, p) where x is any value of X and p is $Pr(X = x)$.

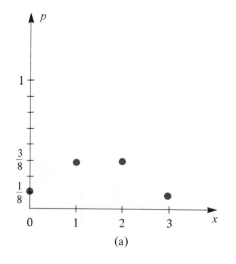

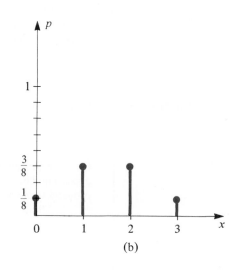

FIGURE 9.2 (a) (b)

EXAMPLES

6. Consider the experiment of rolling a pair of unbiased dice and the random variable X, where the sum of the numbers showing is assigned to the 36 possible outcomes. Let us give the probability distribution of X by a table and a bar graph.

The following table gives the probability distribution.

value of $X(x)$	2	3	4	5	6	7	8	9	10	11	12
$Pr(X = x)(p)$	$\frac{1}{36}$	$\frac{2}{36}$	$\frac{3}{36}$	$\frac{4}{36}$	$\frac{5}{36}$	$\frac{6}{36}$	$\frac{5}{36}$	$\frac{4}{36}$	$\frac{3}{36}$	$\frac{2}{36}$	$\frac{1}{36}$

The bar graph of the probability distribution is given in figure 9.3.

7. Let X be a random variable for an experiment where the probability distribution of X is given by the bar graph in figure 9.4. From this bar graph, make a table showing the probability distribution.

There are five possible values for X: $x = 1, 2, 3, 4, 5$. The probability that X takes on a particular value x, $Pr(X = x)$, is obtained by measuring the

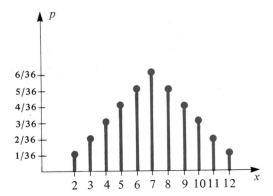

FIGURE 9.3

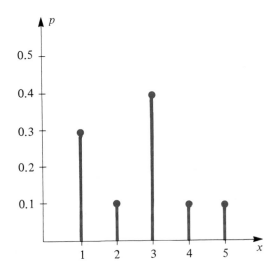

FIGURE 9.4

height of the vertical bar associated with x. This height is found from the scale on the p-axis. The results are given in the following table.

value of $X(x)$	1	2	3	4	5
$Pr(X = x)(p)$	0.3	0.1	0.4	0.1	0.1

8. A company that produces phonograph records knows from past experience that 10 percent of all records produced are defective. Consider the experiment of selecting 2 records without replacement from a warehouse that contains a very large number of records. Let X denote the number of defective records in a sample of 2. Give the probability distribution for X.

   We first construct the sample space for the experiment. Letting D mean that the record is defective and N mean that the record is normal, the sample space is the set of four outcomes

   $$S = \{DD, DN, ND, NN\}.$$

   Let us assume that the number of records in the warehouse is large enough that we can consider the selection of the second record to be independent of the first. We are assuming, for example, that $Pr(D|N) = Pr(D)$. With this assumption, we have

$$Pr(DD) = Pr(D)Pr(D) = (0.1)(0.1) = 0.01$$

$$Pr(DN) = Pr(D)Pr(N) = (0.1)(0.9) = 0.09$$

$$Pr(ND) = Pr(N)Pr(D) = (0.9)(0.1) = 0.09$$

and $\qquad Pr(NN) = Pr(N)Pr(N) = (0.9)(0.9) = 0.81.$

The values of the random variable are 0, 1, and 2. We find, by the definition, the probabilities that X takes on each of these three values.

$$Pr(X = 2) = Pr(DD) \qquad\qquad = 0.01$$

$$Pr(X = 1) = Pr(DN) + Pr(ND) = 0.18$$

and $\qquad Pr(X = 0) = Pr(NN) \qquad\qquad = 0.81.$

We display this probability distribution by the bar graph in figure 9.5.

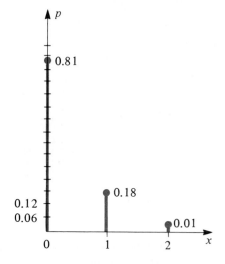

FIGURE 9.5

EXERCISES 9.1

1. Decide whether each of the following describes a discrete or continuous random variable.

   a. The number of cars waiting in line at a drive-in window of the Tri-city bank.

   b. The volume of coffee drunk each day by a mathematics professor.

   c. The amount of money won or lost by playing the daily number.

   d. The amount of time spent by a student each day doing calculus homework.

   e. The moisture content in an ear of corn.

   f. The number of defective radios sold each day by a department store chain.

2. Decide whether each of the following describes a discrete or continuous random variable.

    a. The number of defective tires produced by a company in a year.

    b. The weight of a nickel.

    c. The diameter of an oil drum.

    d. The Dow–Jones daily average.

    e. The amount of money won or lost in a football lottery.

    f. The number of days it rained during the month of April.

3. A bin contains 6 electric starters of which 3 are defective. Two starters are selected at random without replacement. Let X denote the number of defective starters selected.

    a. Construct a table that gives the probability distribution of X.

    b. Draw a bar graph for the probability distribution of X.

4. An urn contains 3 red balls and 2 blue balls. Two balls are drawn at random without replacement. Let X denote the number of blue balls drawn.

    a. Construct a table that gives the probability distribution of X.

    b. Draw a bar graph for the probability distribution of X.

5. An urn contains 1 red ball, 2 white balls, and 3 blue balls. Two balls are drawn at random in succession with replacement. Let X be the number of white balls drawn.

    a. Construct a table that gives the probability distribution of X.

    b. Draw a bar graph for the probability distribution of X.

6. An urn contains 3 red balls and 2 blue balls. Two balls are drawn at random with replacement. Let X be the number of blue balls drawn.

    a. Construct a table that gives the probability distribution of X.

    b. Draw a bar graph for the probability distribution of X.

7. A coin is weighted so that the probability of a head occurring when the coin is flipped is 0.7. Consider the experiment of flipping the coin 2 times. Let X denote the number of heads that appeared.

    a. Construct a table that gives the probability distribution of X.

    b. Draw a bar graph for the probability distribution of X.

8. A coin is weighted so that the probability of a head occurring when the coin is flipped is $\frac{3}{4}$. Consider the experiment of flipping the coin 3 times. Let X denote the number of heads, that appeared.

    a. Construct a table that gives the probability distribution of X.

    b. Draw a bar graph for the probability distribution of X.

9. Let X be a random variable for an experiment where the probability distribution of X is given by the bar graph in figure 9.6. Construct the probability distribution table for X.

10. Let X be a random variable for an experiment where the probability distribution of X is given by the bar graph in figure 9.7. Construct the probability distribution table for X.

11. Sabre Appliances estimates that 15% of all food processors that they produce are defective. Three food processors are selected at random without replacement from a large quantity of food processors. If X denotes the number of defective food processors selected, construct the probability distribution table for X.

12. In a large truckload of apples, it is estimated that 20% of the apples are bruised. Four apples are selected at random without replacement from this large shipment. If X denotes the number of bruised apples, construct the probability distribution table for X.

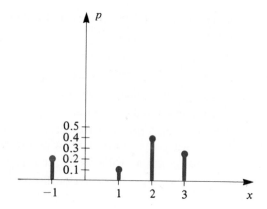

FIGURE 9.6

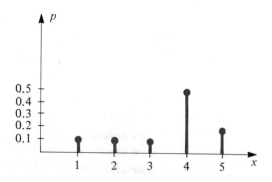

FIGURE 9.7

13. A box contains three slips of paper, numbered 0, 6, and 9, respectively. A slip of paper is drawn at random, the number on it is noted, then it is replaced in the box. A second slip of paper is then drawn at random and the number on it noted. Let X denote the sum of the numbers on the two slips of paper that were drawn. Construct the probability distribution table for X.

14. A speculator estimates that the price of each of three commodities is equally likely (independently of the other two) to increase, remain the same, or decrease by the end of trading tomorrow. Let X denote the number of commodities that increase in price by the end of trading tomorrow. Construct the probability distribution table for X.

15. Suppose X is a random variable with the following distribution table.

value of $X(x_i)$	0	1	2	3
$Pr(X = x)$ (p_i)	k	$\frac{1}{2}k$	$\frac{3}{4}k$	$\frac{5}{8}k$

Find the value of k, where k is a constant.

16. Suppose X is a random variable with the following distribution table.

value of $X(x_i)$	0	1	2	3	4
$Pr(X = x)$ (p_i)	$k \cdot C[4,0]$	$k \cdot C[4,1]$	$k \cdot C[4,2]$	$k \cdot C[4,3]$	$k \cdot C[4,4]$

Find the value of k, where k is a constant.

9.2 Expected Value

When we consider a collection of data, we often wish to describe certain of its characteristics, such as the location of a central point in the data. Probably the best known **measure of central tendency** of a collection of numbers is the arithmetic average. Some examples are the class average on a test, the batting average of a baseball player, or the average miles per gallon of an automobile.

Test scores and a batter's performance over a number of games are common situations in which data are collected. A collection of data can also be generated by the repeated performance of an experiment. In this case, the numbers in the collection are the values of a random variable. In this section, we consider the arithmetic average of a collection of data as well as the "long-term" average of a random variable.

Definition ■ Given a collection of n numbers $x_1, x_2, \ldots, x_n$, the **arithmetic average** or **mean** of the collection is

$$\bar{x} = \frac{x_1 + x_2 + \cdots + x_n}{n} \qquad ■ \qquad (1)$$

EXAMPLE 1 Suppose that the results on a 10-point quiz, given to 15 students in a class, are as follows:

$$6, 7, 8, 7, 9, 7, 7, 8, 10, 9, 10, 10, 9, 9, 7$$

Then the class average for the quiz is

$$\bar{x} = \frac{6 + 7 + 8 + 7 + 9 + 7 + 7 + 8 + 10 + 9 + 10 + 10 + 9 + 9 + 7}{15}$$

$$= \frac{123}{15}$$

$$= 8.2.$$

Although no one in the class actually obtained 8.2 on the quiz, we can think of it as the central point of the scores.

In the preceding example, all the scores except the 6 occurred more than once. By observing the number of times a score occurred in the data, called the **frequency** of the score, we can simplify the calculations of the mean. The test scores in the example are given with their corresponding frequencies in the following table.

quiz score (x_i)	6	7	8	9	10
frequency (f_i)	1	5	2	4	3

By grouping the score and using the frequencies of the scores, we can compute the mean in example 1 as follows:

$$\bar{x} = \frac{6 + (7 + 7 + 7 + 7 + 7) + (8 + 8) + (9 + 9 + 9 + 9) + (10 + 10 + 10)}{15}$$

$$= \frac{1(6) + 5(7) + 2(8) + 4(9) + 3(10)}{15}$$

$$= 8.2.$$

Thus the mean can be obtained by multiplying each of the distinct test scores by its frequency, adding the resulting terms, and then dividing by the total number of test scores.

Definition ∎ Consider a collection of n numbers containing k distinct values $x_1, x_2, \ldots, x_k$. Suppose the number of times a value x_i occurs in the data is given by a **frequency table**

value (x_i)	x_1	x_2	$\cdots$	x_k
frequency (f_i)	f_1	f_2	$\cdots$	f_k

Then the **mean** $\bar{x}$ of the collection of numbers is

$$\bar{x} = \frac{f_1 x_1 + f_2 x_2 + \cdots + f_k x_k}{n}. \tag{2}$$

Note that the sum of the frequencies is equal to n; that is,

$$f_1 + f_2 + \cdots + f_k = n. \quad ∎$$

EXAMPLE 2 Consider the experiment of rolling an unbiased die. Let X be the random variable

$$X = \text{the number appearing on the upturned face.}$$

Suppose the experiment was performed 20 times with the results given in the following table.

value of X	1	2	3	4	5	6
frequency	3	5	5	1	4	2

Let us use equation (2) to compute the mean for this collection of data.

$$\bar{x} = \frac{3(1) + 5(2) + 5(3) + 1(4) + 4(5) + 2(6)}{20}$$

$$= \frac{64}{20} = 3.2.$$

In example 2, the die was rolled 20 times. Suppose that the die is rolled n times, where n is a large number, and the resulting values of the random variable are collected in a frequency table. Can we make a prediction as to the approximate value for the mean of this collection of data? The answer to the question is yes, as we now see.

Consider an experiment with a finite discrete random variable X. Suppose X has six distinct values x_1, x_2, x_3, x_4, x_5, and x_6, with the probability distribution given in the following table.

value of $X(x_i)$	x_1	x_2	x_3	x_4	x_5	x_6
$Pr(X = x_i)(p_i)$	p_1	p_2	p_3	p_4	p_5	p_6

Suppose further that the experiment is performed n times with the results collected in the following frequency table.

value of $X(x_i)$	x_1	x_2	x_3	x_4	x_5	x_6
frequency (f_i)	f_1	f_2	f_3	f_4	f_5	f_6

By using formula (2), the mean $\bar{x}$ is given by

$$\bar{x} = \frac{f_1 x_1 + f_2 x_2 + f_3 x_3 + f_4 x_4 + f_5 x_5 + f_6 x_6}{n}$$

which can be rewritten as

$$\bar{x} = \left(\frac{f_1}{n}\right)x_1 + \left(\frac{f_2}{n}\right)x_2 + \left(\frac{f_3}{n}\right)x_3 + \left(\frac{f_4}{n}\right)x_4 + \left(\frac{f_5}{n}\right)x_5 + \left(\frac{f_6}{n}\right)x_6.$$

Recall from chapter 8 that the probability of an outcome can be thought of as the long-term relative frequency of the outcome. Since f_i/n is the relative frequency of x_i, it is approximately equal to the probability p_i, which we write as

$$\frac{f_i}{n} \approx p_i.$$

Therefore, for large n, a good approximation of $\bar{x}$ is given as

$$\bar{x} \approx p_1 x_1 + p_2 x_2 + p_3 x_3 + p_4 x_4 + p_5 x_5 + p_6 x_6.$$

Definition
■ Let X be a finite discrete random variable whose probability distribution is given by the following table.

value of $X(x_i)$	x_1	x_2	x_3	$\cdots$	x_k
$Pr(X = x_i)(p_i)$	p_1	p_2	p_3	$\cdots$	p_k

Then the **expected value** of X is

$$E(X) = p_1 x_1 + p_2 x_2 + p_3 x_3 + \cdots + p_k x_k. \quad\blacksquare \qquad (3)$$

EXAMPLE 3
Let us return to the experiment of rolling an unbiased die where X is the number appearing on the upturned face. Let us find the expected value of X.
The probability distribution for X is given as follows.

value of $X(x_i)$	1	2	3	4	5	6
$Pr(X = x_i)(p_i)$	$\frac{1}{6}$	$\frac{1}{6}$	$\frac{1}{6}$	$\frac{1}{6}$	$\frac{1}{6}$	$\frac{1}{6}$

Therefore, by equation (3), the expected value of X is

$$E(X) = \frac{1}{6}(1) + \frac{1}{6}(2) + \frac{1}{6}(3) + \frac{1}{6}(4) + \frac{1}{6}(5) + \frac{1}{6}(6)$$

$$= \frac{21}{6}$$

$$= 3.5.$$

Thus, if the experiment is performed a large number of times, we would expect the average of the numbers appearing on the upturned face to be close to 3.5.

The experiment of rolling an unbiased die n times, for $n = 100, 500, 1000,$ and 2000, was simulated on a computer with the results given in figure 9.8. For each n, the computer generated a frequency table and computed the mean for the results. Note from the table that the means are "close" to 3.5, the expected value of the random variable X.

We observe, then, that the expected value of a random variable can be interpreted as a long-term average. Suppose an experiment with a random variable X is performed many times. The frequencies at which the values of X occurred can be collected, and in turn the mean $\bar{x}$ for this collection of data can be computed. If the number of times the experiment was performed is large, then $\bar{x}$ and $E(X)$ will be close. Therefore, instead of performing an experiment many times, we can use the expected value $E(X)$ as an approximation to the mean $\bar{x}$ of the data that would have been collected. For this reason, the expected value of a random variable X is also called the **mean of** X. However, we must keep in mind that the mean of X and the mean $\bar{x}$ of a collection of data are not the same, even though they can be close in value.

For the remainder of this section we illustrate the varied applications of expected value with several examples.

EXAMPLES

4. Let us find the expected value for the random variable X in example 2 of the previous section (page 395).

We first construct the probability distribution table for X. Since there is one ticket that will win the first prize and since the tickets are drawn at random, we have

$$Pr(X = \$399) = \frac{1}{1000}.$$

There are 2 tickets that will win a second-place prize; therefore

$$Pr(X = \$99) = \frac{2}{1000}.$$

There are 4 tickets that will win a third-place prize and we have

$$Pr(X = \$24) = \frac{4}{1000}.$$

```
DIE ROLLED  100 TIMES

            VALUE OF X      FREQUENCY
            ===== == =      =========
               1               16
               2               11
               3               21
               4               14
               5               22
               6               16

            MEAN =   363 / 100  =  3.63


DIE ROLLED  500 TIMES

            VALUE OF X      FREQUENCY
            ===== == =      =========
               1               89
               2               73
               3               87
               4               83
               5               84
               6               84

            MEAN =   1752 / 500  =  3.504


DIE ROLLED  1000 TIMES

            VALUE OF X      FREQUENCY
            ===== == =      =========
               1              171
               2              147
               3              177
               4              167
               5              172
               6              166

            MEAN =   3520 / 1000  =  3.52


DIE ROLLED  2000 TIMES

            VALUE OF X      FREQUENCY
            ===== == =      =========
               1              339
               2              301
               3              354
               4              334
               5              337
               6              335

            MEAN =   7034 / 2000  =  3.517
```

FIGURE 9.8

Finally, there are 993 tickets that will win no prize, and therefore,

$$Pr(X = -\$1) = \frac{993}{1000}.$$

We summarize these results in the probability distribution table for X.

value of $X(x_i)$	$399	$99	$24	$-$1
$Pr(X = x_i)(p_i)$	$\frac{1}{1000}$	$\frac{2}{1000}$	$\frac{4}{1000}$	$\frac{993}{1000}$

Using equation (3) with $n = 4$ and the numbers from the probability distribution table for X, we have

$$E(X) = \frac{1}{1000}(399) + \frac{2}{1000}(99) + \frac{4}{1000}(24) + \frac{993}{1000}(-1)$$

$$= -\frac{300}{1000}$$

$$= -\$0.30.$$

Applying the idea that the expected value is a long-term average, we can interpret $E(X) = -\$0.30$ in the following way. Suppose the lottery is run on a daily basis and suppose Sam buys one ticket each day. If he records his net winnings or loss each day and divides the total by the number of days he plays the lottery, this daily average will be close to a loss of $0.30 after many days of playing.

5. From past experience, the customer relations department at Metro Power and Light has categorized telephone calls into five types, e_1, e_2, e_3, e_4, and e_5. The lengths of the five types of calls and the probabilities of the types of calls occurring are given in table 9.1. Using this information, determine the expected length of a call to customer relations.

TABLE 9.1

Type of call	Average length of call	Probability
e_1	5 minutes	0.25
e_2	6 minutes	0.32
e_3	4 minutes	0.13
e_4	7 minutes	0.16
e_5	10 minutes	0.14

Let X be the random variable that assigns to each type of call the length of the call. The probability distribution table for X is as follows.

value of $X(x_i)$	5	6	4	7	10
$Pr(X = x_i)(p_i)$	0.25	0.32	0.13	0.16	0.14

Using the formula for expected value of X, the expected length of a telephone call to customer relations is

$$E(X) = 0.25(5) + 0.32(6) + 0.13(4) + 0.16(7) + 0.14(10)$$
$$= 6.21 \text{ minutes.}$$

We can interpret the expected value of 6.21 minutes as follows. If the customer relations department made a record of the length of a large number of future calls and then found the average length per call, this average should be close to 6.21 minutes.

6. A jogger runs 3 miles a day in good weather and runs 1.5 miles a day in bad weather. During the month of March, the probability of bad weather on a given day is 40 percent. What is the average number of miles per day the jogger can expect to run in March?

We can think of the experiment in this situation as one of observing the weather conditions on a given day in March. The sample space consists of the two outcomes e_1 = good weather and e_2 = bad weather. (We shall assume no other possibilities exist.) Now, $Pr(e_1) = 0.6$ and $Pr(e_2) = 0.4$.

Let the random variable X be the number of miles the jogger runs in a day and we have the values $x_1 = 3$ miles and $x_2 = 1.5$ miles. Using formula (3) for expected value, the average number of miles per day that the jogger can expect to run during March is

$$E(X) = 0.6(3) + 0.4(1.5) = 2.4 \text{ miles.}$$

7. A roulette wheel consists of 37 equally spaced slots labeled 0 through 36. (Some wheels have also a double zero slot.) One of the games of roulette is as follows. The wheel is spun and a ball comes to rest in one of the slots. If a player bets on a slot and the ball lands in that slot, he receives 36 times his bet. If the ball lands in any other slot, the player loses his bet. Let the random variable X be the player's winnings (or loss) per play. Find the expected value of X, if the player bets $1 per game.

The sample space for one play consists of the outcomes

$$e_1 = \text{the player wins}$$

$$e_2 = \text{the player loses.}$$

Since there is only one way to win and 36 equally likely ways to lose, we have $Pr(e_1) = \frac{1}{37}$ and $Pr(e_2) = \frac{36}{37}$.

The payoff for winning is $36 less the original bet of $1, whereas the payoff for losing is $-\$1$. Therefore the random variable X has the two values $x_1 = \$35$ and $x_2 = -\$1$. The expected value for X is

$$E(X) = \frac{1}{37}(35) + \frac{36}{37}(-1) = -\$0.027.$$

We say that the expected value for this game with a bet of $1 is $-\$0.027$.

The significance of the negative expected value in example 7 is that the player will lose, in the long run, an average $0.027 per play. In general, when the expected value of a game, from the player's viewpoint, is negative, the game is unfavorable to the player. When the expected value is positive, the game favors the player. If the expected value of the game is zero, the game is said to be **fair**. Therefore we can use expected value to evaluate the worth of playing the game. For the most part, games of chance and lotteries have a negative expected value for the player.

Another application of expected value is in the field of insurance.

EXAMPLE 8

Life Insurance Premium. An insurance company plans to sell 5-year term life insurance policies to men 60 years old. If the value of the policy is $10,000, what should the company charge as the one-time premium in order to make an expected profit of $10 per policy?

To answer this question, the company's actuarial department finds that the probability that a 60-year-old man will live to age 65 is 0.883. We can think of this situation as the experiment of selecting a 60-year-old man at random from the set of 60-year-old men in the United States. The sample space has two possible outcomes: e_1 = the man lives to age 65 and e_2 = the man does not live to age 65. The probability of e_1 is 0.883 and the probability of e_2 is 0.117. Let x be the amount (in dollars) of the premium. If e_1 occurs, the payoff for the insurance company is x. If e_2 occurs, the insurance company loses $10,000 but still retains the premium x. Therefore the payoff for e_2 is $x - 10,000$. The expected value from the company's viewpoint is $E = (x)(0.883) + (x - 10,000)(0.117)$. The insurance company wants this expected value to be $10, so we set $E = 10$ and solve for x in

$$(x)(0.883) + (x - 10,000)(0.117) = 10.$$

Thus, to achieve an "average" profit of $10 per policy, the insurance company should charge a premium of $1180.

As shown in example 8, expected value can be used in a decision-making process. In that example the decision was about what to charge for a policy. The next example gives another problem in decision making.

EXAMPLE 9

Sun Bakery must purchase a large quantity of flour either today or tomorrow. Today's price is $155 per unit. The bakery believes that tomorrow's price will be either $160 per unit, with a probability of 0.7, or $148 per unit, with a probability of 0.3. Should the bakery buy the flour today or wait until tomorrow?

To answer this question, we compute the expected price per unit of flour tomorrow. This expected value is given by

$$(\$160)(0.7) + (\$148)(0.3) = \$156.40.$$

Compared with today's certain price of $155 per unit, it is better to buy the flour today.

EXERCISES 9.2

1. Jason has kept a monthly record of the cost of gasoline for his car. He obtained the following data during the past 12 months.

    $92, $98, $100, $109, $100, $60, $88, $170, $165, $150, $180, $205

    Find the (arithmetic) average monthly cost of gasoline for the past 12 months.

2. The following temperatures (Celsius) were recorded on a daily basis for two weeks in the month of February.

    6, 8, −3, 3, 2, 1, 7, −4, 1, 0, 1, 3, 4, 6

    Find the mean of these temperatures.

3. A survey of age distribution at a certain college is summarized in the following frequency table.

age (x_i)	17	18	19	20	21	22	23
frequency (f_i)	105	451	325	212	176	80	62

Find the mean age of the students in the survey.

4. The speed of motorists was checked by radar on Interstate 78 and the data collected in the following frequency table.

miles per hour (x_i)	45	50	55	60	65	70	75
frequency (f_i)	1	10	12	20	38	15	4

Find the (arithmetic) average speed.

5. Consider the experiment of rolling a die. Let X denote the number appearing on the upturned face. Suppose that the experiment was performed 25 times with the results given in the following frequency table.

value of X	1	2	3	4	5	6
frequency	6	4	5	2	7	1

Compute the mean for this collection of data.

6. Consider the experiment of rolling a die. Let X denote the number appearing on the upturned face. Suppose the experiment was performed 15 times with the results given in the following frequency table.

value of X	1	2	3	4	5	6
frequency	2	1	5	3	0	4

Computer the mean for this collection of data.

7. Consider an experiment with a finite discrete random variable X with the probability distribution of X given in the following table.

value of $X(x_i)$	315	-217	-58	2001	296
$Pr(X = x_i)(p_i)$	0.17	0.31	0.41	0.01	0.10

Find the expected value of X.

8. Consider an experiment with a finite discrete random variable X with the probability distribution of X given in the following table.

value of $X(x_i)$	-37	41	-18
$Pr(X = x_i)(p_i)$	0.12	0.37	0.51

Find the expected value of X.

For exercises 9–16 find the expected value of the random variable X described in the indicated exercise of section 9.1.

9. exercise 3

10. exercise 4

11. exercise 9

12. exercise 10

13. exercise 11

14. exercise 12

15. exercise 13

16. exercise 14

17. Sharon plays the following game. A hat contains 7 one-dollar bills, 9 five-dollar bills, and 4 ten-dollar bills. Sharon draws a bill at random from the hat and is permitted to keep the bill that is drawn. Let X denote the amount of money Sharon won. Find the expected value of X. How much would Sharon have to pay to play this game in order to make it a fair game?

18. Jeffrey plays the following game. A coin is weighted so that the probability of a head is 0.4. The coin is flipped three times and Jeffrey receives $50 for each head that appears. How much should Jeffrey pay each time he plays to make it a fair game?

19. An insurance company plans to sell 10-year term life insurance policies to women 25 years old. The probability that a 25-year-old woman lives to age 35 is 0.979. What should the company charge as the one-time premium in order to make an expected profit of $95 per policy if the value of the policy is $25,000?

20. An insurance company plans to sell 5-year term life insurance policies to men 40 years old. The probability that a 40-year-old man lives to age 45 is 0.996. What should the company charge as the one-time premium in order to make an expected profit of $15 per policy if the value of the policy is $10,000?

21. Experience has shown that attendance at a football game averages 50,000 people if the weather is good and 35,000 people if the weather is bad. The probability for good weather for Sunday's game is 0.7. What is the expected attendance for the game?

22. The express check-out line in a grocery store is reserved for shoppers buying 10 or fewer items. The store manager wants to operate the express line in such a way that (a) people need not wait too long in line and (b) there are enough people using the express line to warrant the cost of maintaining it. He conducts a survey of the express line and the results are given in the following table.

number of people waiting in line	0	1	2	3	4	5	6	7
probability	0.1	0.1	0.17	0.16	0.15	0.1	0.12	0.1

From the table, the manager finds the expected number of people waiting in the express line. If this number is larger than 3, he will decrease from 10 to 7 the maximum number of items a shopper may have in order to use the express line. What should be his decision?

23. A professor must decide either to write a textbook or to start a consulting business. She does not have the time to do both, so she makes a decision based on expected value. She feels that the success of both ventures depends on economic conditions during the next three years. She classifies the economic outlook as belonging to the set {poor, average, good}. By educated guessing, she assigns probabilities: $Pr(\text{poor}) = 0.3$, $Pr(\text{average}) = 0.5$, and $Pr(\text{good}) = 0.2$. She also estimates the payoffs for both the book and the consulting business for the three outcomes of the economic outlook as given in the following table.

	Economic outlook		
	Poor	Average	Good
book	$3000	$5000	$7000
business	$1000	$4000	$8000

Should the professor write a text book or start a consulting business?

24. An oil company must decide to drill an exploratory well at one of two sites, A or B. The company estimates that the probability of finding oil at site A is 0.15 and the probability of finding oil at site B is 0.2. It will cost $120,000 to drill at site A and $340,000 to drill at site B. If oil is present at site A, its estimated worth is $3,900,000, whereas if oil is present at site B, its estimated worth is $4,920,000. Based on expected value, which site should the company choose for the exploratory well?

25. a. Mr. Blackwell is producing an outdoor rock concert. He estimates that he will make $250,000 if it does not rain and make $50,000 if it does rain. There is a 65% chance of rain. What are Mr. Blackwell's expected earnings from the concert?

    b. Mr. Blackwell can buy an insurance policy for $140,000 that will pay him $250,000 if it rains. If he buys this policy, what are his expected earnings from the concert?

    c. Based on expected earnings, should Mr. Blackwell buy an insurance policy?

26. a. The student government at State University is sponsoring an outdoor concert. They estimate a profit of $20,000 if it does not rain and a profit of $5000 if it does rain. There is a 45% chance of rain. What is the student government's expected profit from the concert?

    b. The student government can buy an insurance policy for $5750 that will pay them $15,000 if it rains. If they buy this policy, what is the expected profit from the concert?

    c. Based on expected profit, should the student government buy an insurance policy?

27. A coin is weighted so that the probability of a head is $\frac{2}{3}$. Consider the experiment of flipping the coin until a head or three tails occur. Find the expected number of tails.

28. Consider the experiment of simultaneously flipping two unbiased coins. This experiment is performed four times or until two heads simultaneously occur, whichever comes first. Find the expected number of times the experiment is performed.

29. Three cards are dealt to Charles from a standard 52-card deck. If all three cards are of the same suit, Charles wins $425. How much should Charles pay per play to make this a fair game?

30. Suppose we subtract 15 from each value x_i of the random variable X in exercise 7 and let these be the values of a new random variable Y.

    a. Find the expected value of Y.
    b. What is the relationship between $E(X)$ and $E(Y)$?

31. Suppose we multiply each value x_i of the random variable X in exercise 8 by 10 and let these be the values of a new random variable Y.

    a. Find the expected value of Y.
    b. What is the relationship between $E(X)$ and $E(Y)$?

32. Consider an experiment with a finite discrete random variable X with the probability distribution of X given in the following table.

value of $X(x_i)$	x_1	x_2	$\cdots$	x_n
$Pr(X = x_i)(p_i)$	p_1	p_2	$\cdots$	p_n

Let b be a constant and consider the random variable $Y = X + b$ defined by the probability distribution table

value of $Y(y_i)$	$x_1 + b$	$x_2 + b$	$\cdots$	$x_n + b$
$Pr(Y = y_i)(p_i)$	p_1	p_2	$\cdots$	p_n

Show that

$$E(Y) = E(X) + b,$$

that is, show that

$$E(X + b) = E(X) + b.$$

33. Consider the experiment with the random variable X given in exercise 32. Let m be a constant and consider the random variable $Z = mX$ defined by the probability distribution table

value of $Z(z_i)$	mx_1	mx_2	$\cdots$	mx_n
$Pr(Z = z_i)(p_i)$	p_1	p_2	$\cdots$	p_n

Show that

$$E(Z) = mE(X),$$

that is, show that

$$E(mX) = mE(X).$$

34. Suppose, in exercise 17, that Sharon is required to pay $5.50 each time she plays the game. Let Y denote the amount of money Sharon won or lost. Find the expected value of Y. (*Hint:* $Y = X - \$5.50$, where X is the random variable defined in exercise 17; then use exercise 32.)

35. Suppose, in exercise 18, the player receives ten dollars each time a head appears. How much should it cost to play in order to make this a fair game? (*Hint:*   Use exercise 33.)

9.3 Bernoulli Trials: Binomial Random Variables

A given experiment may be repeated a number of times to generate a **sequence of trials**. Rolling dice, playing roulette, and dealing cards are typical examples from gambling. There are situations other than games of chance where an experiment is performed repeatedly.

We can think of the manufacture of an item as an experiment that is repeated possibly thousands of times a day. This experiment has two outcomes: either the item produced is defective or it is perfect. Another example is testing an experimental serum on 100 animals, where the outcome on each trial is either a positive or a negative result.

In this section we will study sequences of trials where the trials are **independent**. This means (see Chapter 8) that the probability of an outcome on a trial of the experiment does not depend on the outcome on the previous trial of the experiment. For example, if a coin is flipped 10 times, the probability of getting a head (or a tail) remains the same for each flip.

Whenever we have a sequence of trials of an experiment, the experiment may have a number of possible outcomes. The observation we wish to make may allow us to divide the outcomes into two groups arbitrarily labeled success (s) and failure (f). For example, the experiment of rolling a die has the possible outcomes 1, 2, 3, 4, 5, and 6. If we are interested in the number of times a 6 appears on repeated trials of the experiment, we can consider the possibilities on each trial as success (rolling a 6) or failure (not rolling a 6). Furthermore, if the die is unbiased, the probability of success is $\frac{1}{6}$ and the probability of failure is $\frac{5}{6}$ on each roll.

Definition

■ Consider an experiment and the sequence of trials obtained by repeated performance of the experiment. If there are only two possible outcomes, s and f, on each trial and if the outcomes are independent, then the sequence is called a **sequence of Bernoulli trials.** ■

Note that a sequence of Bernoulli trials is a special type of stochastic process, however, not every stochastic process can be considered a sequence of Bernoulli trials. For example, consider an urn containing red balls and white balls and the experiment of selecting a ball at random from the urn. If the experiment is performed three times with replacement, then we have a sequence of Bernoulli trials. If the experiment is performed three times without replacement, then the trials are not independent and the process is not Bernoulli.

A process that is a sequence of Bernoulli trials is called a **binomial experiment**. A random variable that is associated with a binomial experiment is

$$X = \text{the number of successes.}$$

In this case, X is called a **binomial random variable**.

We now develop a method for obtaining a probability model for a binomial experiment and, in turn, the probability distribution for the random variable X.

Consider a binomial experiment consisting of three trials where on any given trial the outcomes are s and f. The sequence sfs, for example, represents the outcome of obtaining a success on the first and third trial and a failure on the second trial. The sample space S for this binomial experiment has eight outcomes:

$$S = \{\text{sss, ssf, sfs, fss, ffs, fsf, sff, fff}\}.$$

The binomial random variable X that assigns to each outcome the number of successes has the four values

$$X = 0, \quad X = 1, \quad X = 2, \quad X = 3.$$

To give the probability distribution of X, we set $Pr(\text{s}) = p$ and $Pr(\text{f}) = q$. Since there are only two possible outcomes, we observe that $p + q = 1$. Let us first find $Pr(X = 0)$.

The event "no successes" contains the one outcome fff. Therefore, $Pr(X = 0) = Pr(\text{fff})$. Since the trials are independent,

$$Pr(\text{fff}) = Pr(\text{f})Pr(\text{f})Pr(\text{f})$$
$$= q \cdot q \cdot q$$
$$= q^3.$$

Therefore,
$$Pr(X = 0) = q^3.$$

Next, let us find $Pr(X = 1)$. The event "1 success" contains the three outcomes sff, fsf, ffs. Therefore,

$$Pr(X = 1) = Pr(\text{sff}) + Pr(\text{fsf}) + Pr(\text{ffs})$$
$$= Pr(s)Pr(f)Pr(f) + Pr(f)Pr(s)Pr(f) + Pr(f)Pr(f)Pr(s)$$
$$= pqq + qpq + qqp$$
$$= 3pq^2.$$

In a like manner, we obtain

$$Pr(X = 2) = 3p^2q$$

and
$$Pr(X = 3) = p^3.$$

These results can also be shown on a probability tree as in figure 9.9.

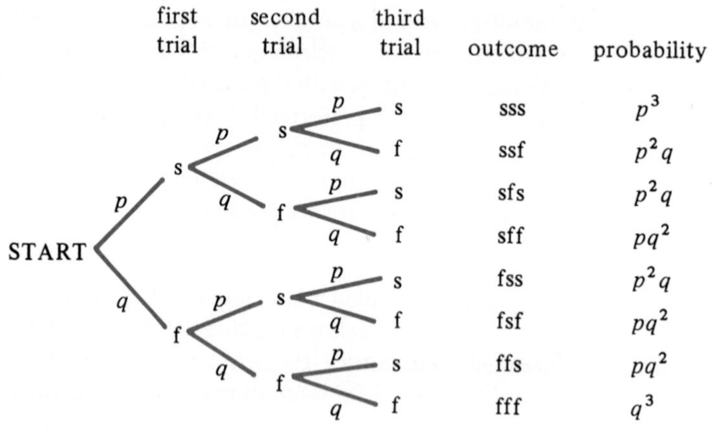

	first trial	second trial	third trial	outcome	probability
			p — s	sss	p^3
		p s	q — f	ssf	p^2q
	p s		p — s	sfs	p^2q
		q f	q — f	sff	pq^2
START			p — s	fss	p^2q
	q	p s	q — f	fsf	pq^2
	f		p — s	ffs	pq^2
		q f	q — f	fff	q^3

FIGURE 9.9   Binomial experiment with 3 trials

EXAMPLE 1

Consider the binomial experiment of rolling an unbiased die three times, where success means obtaining a 6 on a roll. Let us find the probability distribution for the random variable

$$X = \text{the number of successes.}$$

Using the results for a binomial experiment of three trials with $p = \frac{1}{6}$ and $q = \frac{5}{6}$, we have

$$Pr(X = 0) = \left(\frac{5}{6}\right)^3 = \frac{125}{216}$$

$$Pr(X = 1) = 3\left(\frac{1}{6}\right)\left(\frac{5}{6}\right)^2 = \frac{75}{216}$$

$$Pr(X = 2) = 3\left(\frac{1}{6}\right)^2\left(\frac{5}{6}\right) = \frac{15}{216}$$

$$Pr(X = 3) = \left(\frac{1}{6}\right)^3 = \frac{1}{216}.$$

When the number of trials in a binomial experiment is large, the method we used to find the probability distribution for three trials is not the best. It would be tedious to write out the entire sample space and collect those outcomes assigned the same value by X. It is necessary to find a general pattern for binomial experiments, which we now demonstrate.

Consider a binomial experiment consisting of eight trials where the probability of success on any trial is p and the probability of failure on any trial is q. Let X be the binomial random variable

$$X = \text{the number of successes.}$$

Let us find the probability of exactly three successes; that is, $Pr(X = 3)$.

Some ways of obtaining exactly three successes in eight trials are

$$\text{sffsfsff, sssfffff, ffsffssf, fssfffffs.}$$

Note that each of these eight-letter sequences has exactly 3 s's and 5 f's. Therefore the probability of such a sequence is the product of 3 p's and 5 q's, which can be written as $p^3 q^5$. Thus, to find $Pr(X = 3)$, it suffices to count the number of sequences with 3 s's and 5 f's and then multiply this number by $p^3 q^5$.

A sequence is composed of eight positions with three of the positions filled with s's. Therefore the total number of such sequences is the number of ways of selecting 3 positions from 8 positions. This can be done in $C[8, 3]$ ways. Therefore,

$$Pr(X = 3) = C[8, 3] p^3 q^5$$

$$= \frac{8 \cdot 7 \cdot 6}{3!} p^3 q^5$$

$$= 56 p^3 q^5.$$

With this discussion in mind, we now state the general formula.

Theorem 1　■　Consider a binomial experiment with n trials where p is the probability of success and q is the probability of failure on any given trial. Consider also the binomial random variable

$$X = \text{the number of successes.}$$

X can take on the values $0, 1, 2, 3, \ldots, n$. Let x be one of these values. Then the probability of exactly x successes on the n trials is given by

$$Pr(X = x) = C[n, x] p^x q^{n-x}. \quad ■ \tag{4}$$

EXAMPLE 2　Find the probability of exactly 5 successes for a binomial experiment with 12 trials, where the probability of success on any given trial is 0.4.

By formula (4), with $p = 0.4$ and $q = 0.6$, we have

$$Pr(X = 5) = C[12, 5](0.4)^5 (0.6)^7$$

$$= \frac{12 \cdot 11 \cdot 10 \cdot 9 \cdot 8}{5!} (0.4)^5 (0.6)^7$$

$$= 0.227.$$

As shown in example 2, the computation of $Pr(X = 5)$ is tedious even with the use of a calculator. For this reason, table I in appendix B gives $Pr(X = x)$ for certain values of n and p. We show how this table is used with the next examples.

EXAMPLES

3. Use table I in appendix B to find $Pr(X = 5)$ for the binomial experiment given in example 2.

First we locate table $n = 12$. Next, we find the row labeled $x = 5$ and the column labeled $p = 0.4$. The value for $Pr(X = 5)$ is the number at the intersection of that row and column, $Pr(X = 5) = 0.227$.

4. Consider a binomial experiment consisting of 15 trials where the probability of success is 0.75 on any given trial. Use table I in appendix B to find the probability of exactly 10 successes; that is, $Pr(X = 10)$.

First locate table $n = 15$. Next, look in row $x = 10$. There are no values listed for $p = 0.75$. This is no problem, for we can change our point of view. Obtaining exactly 10 successes in 15 trials is equivalent to obtaining exactly 5 failures in 15 trials. Therefore,

$$Pr(10 \text{ successes in 15 trials}) = Pr(5 \text{ failures in 15 trials}).$$

Recall that $p + q = 1$, so the probability of failure on any trial is

$$q = 1 - 0.75 = 0.25.$$

We now need to find the probability of obtaining exactly 5 "successes" in 15 trials with the probability of "success" on any given trial being 0.25. Thus, to find the answer to the problem, we look in row $x = 5$ and column $p = 0.25$, giving us 0.1651.

We conclude this section with some examples of situations that can be modeled as binomial experiments. When considering applications, we must keep in mind that the binomial experiment is an appropriate model only when the trials are independent.

EXAMPLES

5. *Genetics.* A particular type of sweet pea plant has either red or white blossoms. From genetic theory, the probability that the plant has red blossoms is $\frac{1}{4}$. If 10 seeds for this plant are randomly selected and planted, what is the probability that the number of red-blossomed plants will be (a) exactly 6? (b) at least 7? (c) less than 7?

We model this situation with a binomial experiment of 10 trials where success on any given trial means growing a red-blossomed plant. The probability of success is $\frac{1}{4}$. Let X be the number of red-blossomed plants.

For part (a), we want to find $Pr(X = 6)$. Using table I in appendix B, with $n = 10$, $x = 6$, and $p = 0.25$, we have $Pr(X = 6) = 0.0162$.

For part (b), we want to find $Pr(X \geq 7)$.

$$Pr(X \geq 7) = Pr(X = 7) + Pr(X = 8) + Pr(X = 9) + Pr(X = 10)$$
$$= 0.0031 + 0.0004 + 0.0000 + 0.0000$$
$$= 0.0035.$$

For part (c), we want to find $Pr(X < 7)$. Recall that for any event E, $Pr(E) = 1 - Pr(E')$. Therefore, if we let E be the event of growing fewer than 7 red-blossomed plants,

$$Pr(X < 7) = 1 - Pr(X \geq 7)$$
$$= 1 - 0.0035$$
$$= 0.9965.$$

6. *Opinion Polls.* Suppose that 70 percent of the voters in a city oppose an increase in the school tax. If 9 voters are selected at random, what is the probability that a majority of them will favor an increase in the school tax?

This opinion poll can be modeled with a binomial experiment of 9 trials. Success is that a person favors an increase in the school tax. Therefore the probability of success is 0.3. Letting X denote the number of successes, we want to find the probability that X is greater than or equal to 5. Using table I in appendix B, we have

$$Pr(X \geq 5) = Pr(X = 5) + Pr(X = 6) + Pr(X = 7) + Pr(X = 8) + Pr(X = 9)$$
$$= 0.0735 + 0.02100 + 0.0039 + 0.0004 + 0.0000$$
$$= 0.0988.$$

This example illustrates how poll taking can be misleading.

7. *Test-Taking by Guessing.* In Dave's history class, the instructor gave an unannounced 20-question true-false test. Unfortunately Dave was totally unprepared and decided to guess on each question. If a passing score on the test is 14 or more correct answers, what is the probability that Dave passed the test?

We can view this situation as a binomial experiment consisting of 20 trials. Since Dave guessed on each question, the probability of a success on any one of the questions is $\frac{1}{2}$. Therefore the probability that Dave passed this test is $Pr(X \geq 14)$, which we find in the tables to be

$$0.0370 + 0.0148 + 0.0046 + 0.0011 + 0.0002 + 0.0000 + 0.0000 = 0.0577.$$

8. *Drug Company Claim.* A drug company claims that their new skin creme cures acne 80% of the time. Dr. Snyder tested the new product on 12 of his patients. What is the probability that the number of patients cured of acne is less than or equal to 6, given that the company's claim is correct?

The tests performed on the 12 patients can be viewed as a binomial experiment. If we assume that the company's claim is correct, the probability that a patient is cured by the skin creme is 0.8. As in example 4, we must change our point of view since the number $p = 0.8$ does not appear in the tables. Thus we must find the probability of 7 or more failures, with the probability that the skin creme fails being 0.2. Using the tables for $n = 12$, $x \geq 7$, and $p = 0.2$, we have the desired answer

$$0.0033 + 0.0005 + 0.0001 + 0.0000 + 0.0000 + 0.0000 = 0.0039.$$

Consider example 8 again. Suppose after administering the tests, Dr. Snyder found that exactly 6 of his patients were actually cured of acne. As we saw in the example that, if the company's claim is correct, the probability of this happening would be even less than 0.0039. This result caused Dr. Snyder to doubt the validity of the drug company's claim. Therefore, he tested the skin creme on a larger group of 100 of his patients with the result that 70 or fewer are cured. Let us now find the probability that the number of patients cured of acne is less than or equal to 70 if the company's claim is correct.

The tests can be regarded as a binomial experiment of 100 trials where the probability of success is 0.8. If X is the number of successes, we want to find the probability that X is less than or equal to 70.

$$Pr(X \leq 70) = Pr(X = 70) + Pr(X = 69) + Pr(X = 68) + \cdots + Pr(X = 0).$$

Table I in appendix B cannot be used because $n = 100$ and the table only goes up to $n = 20$. Furthermore, using the formula

$$Pr(X = x) = C[100, x](0.8)^x(0.2)^{100-x}$$

71 times is not practical. Fortunately there is a method to approximate $Pr(X \leq 70)$, which gives us that $Pr(X \leq 70)$ is approximately equal to 0.0087. (The method used to obtain this answer will be explained in section 9.6.)

EXERCISES 9.3

1. Use formula (4) to find $Pr(X = x)$ for the following values of n, x, and p.
   a. $n = 3$, $x = 2$, $p = \frac{3}{4}$
   b. $n = 7$, $x = 4$, $p = 0.45$
   c. $n = 8$, $x = 3$, $p = 0.57$
   d. $n = 6$, $x = 3$, $p = 0.65$

2. Use formula (4) to find $Pr(X = x)$ for the following values of n, x, and p.
   a. $n = 5$, $x = 4$, $p = 0.7$
   b. $n = 3$, $x = 2$, $p = \frac{3}{7}$
   c. $n = 6$, $x = 2$, $p = 0.62$
   d. $n = 7$, $x = 4$, $p = 0.48$

3. Use table I in appendix B to find the following binomial probabilities.
   a. $Pr(X = 4)$, $n = 5$, $p = 0.30$
   b. $Pr(X = 7)$, $n = 12$, $p = 0.20$
   c. $Pr(X = 8)$, $n = 17$, $p = 0.60$
   d. $Pr(X \leq 5)$, $n = 13$, $p = 0.20$
   e. $Pr(X \geq 7)$, $n = 20$, $p = 0.70$
   f. $Pr(X \leq 4)$, $n = 7$, $p = 0.75$

4. Use table I in appendix B to find the following binomial probabilities.
   a. $Pr(X = 9)$, $n = 10$, $p = 0.50$
   b. $Pr(X = 2)$, $n = 3$, $p = 0.25$

   c. $Pr(X = 7)$, $n = 11$, $p = 0.75$

   d. $Pr(X \leq 6)$, $n = 17$, $p = 0.30$

   e. $Pr(X \geq 9)$, $n = 14$, $p = 0.60$

   f. $Pr(X \geq 14)$, $n = 19$, $p = 0.80$

5. A coin is weighted so that the probability of getting a head when the coin is flipped is 0.48. If the coin is flipped 5 times, find the probability of obtaining

   a. exactly 2 heads.   b. at least 1 head.

   c. at most 3 tails.   d. more than 3 tails.

6. An unbiased die is rolled 4 times. Find the probability of getting a 1 on the upturned face the following number of times.

   a. exactly twice.   b. more than twice.

   c. at least once.   d. exactly once.

7. The Atlantis Corporation feels that 20% of its water pumps being made on assembly line A are defective. If 20 water pumps are selected at random from this assembly line, what is the probability that

   a. exactly 3 will be defective?

   b. more than 13 will be defective?

   c. at most 5 will be defective?

   d. at least 4 will be defective?

8. It is estimated that 40% of all cars in a particular city would fail an exhaust emissions test. If 20 cars are randomly selected for testing, what is the probability that

   a. exactly 8 will fail?

   b. at least 10 will fail?

   c. at least 12 will pass?

   d. less than 12 will pass?

9. Suppose that 80% of the people in Lehigh County favor increased financial aid to their community college. If 20 people are selected at random, what is the probability that 11 or more of them will oppose an increase in financial aid to the college?

10. Suppose that 75% of the people in a city oppose an increase in the property tax. If 18 people are selected at random, what is the probability that 10 or more of them will favor an increase in the property tax?

11. A new drug used to combat PMS is effective 88% of the time. If five women use this drug, find the probability that

   a. exactly three found the drug to be effective.

   b. more than two found the drug not to be effective.

   c. all five found the drug to be effective.

12. From records of past performances, the probability that a student passes Professor Johnson's statistics course is 0.593. Four students are selected at random from Professor Johnson's class. Find the probability that

   a. all four pass the course.

   b. exactly three pass the course.

   c. at least two pass the course.

13. From past records, 60% of freshmen computer science majors at State University finish the program. If 18 computer science majors are selected at random, find the probability that

    a. at most 8 will complete the program.

    b. at least 6 will complete the program.

    c. more than 11 will not complete the program.

14. It is estimated that 30% of all students that take a mathematics achievement test get scores that require them to take a remedial mathematics course. If 15 students take the test, what is the probability that

    a. at most 6 must take a remedial mathematics course?

    b. at least 10 will not be required to take a remedial mathematics course?

    c. more than 9 must take a remedial mathematics course?

15. It is estimated that 92% of migraine headaches are caused by chronic tension. From a study group of people having migraine headaches, six people are selected at random. Find the probability that

    a. exactly four have migraine headaches caused by chronic tension.

    b. all six have migraine headaches caused by chronic tension.

    c. at least four have migraine headaches caused by chronic tension.

16. From past records, the probability that a trainee completes the managerial course at National Steel is 0.793. If four trainees are selected at random, find the probability that

    a. all four complete the managerial course.

    b. exactly two do not complete the managerial course.

    c. at most three do not complete the managerial course.

17. The Green Thumb Garden Shop estimates that 80% of the tulip bulbs that they sell will bloom. If Jason buys 18 tulip bulbs at this shop, what is the probability that

    a. at least 12 will bloom?

    b. exactly 15 will bloom?

    c. at most 4 will not bloom?

18. The Viking Drug Company makes a decongestant that produces unwanted side effects in 1 out of every 10 people who use it. If 17 people use the decongestant, what is the probability that

    a. at least 10 will have unwanted side effects?

    b. at most 7 will have unwanted side effects?

    c. more than 12 will not have unwanted side effects?

19. A do-it-yourself pregnancy detection kit is 70% accurate. Fifteen women used the kit and obtained the indication that they are pregnant. Find the probability that

    a. at least 12 of them are pregnant.

    b. at most 10 of them are pregnant.

    c. at most 8 of them are not pregnant.

20. A hormone called PK, given to women needing induced labor because of complications in pregnancy, works 90% of the time. If this hormone is administered to 12 such pregnant women, what is the probability that all 12 women will go into labor?

21. A study indicates that 50% of all marriages end in divorce. Out of 16 marriages, find the probability that

    a. at least 8 end in divorce.

    b. at least 10 do not end in divorce.

    c. at most 6 end in divorce.

22. In David's history class, the instructor gave an unannounced 20-question multiple-choice test. David is totally unprepared and decides to guess on each question. Each question has a choice of 5 answers. What is the probability that

    a. David will answer at most 9 questions correctly?

    b. David will answer at least 12 questions incorrectly?

    c. David will answer no more than 5 questions correctly?

23. Consider a binomial experiment with three trials. Let

$$X = \text{the number of successes}$$

where p is the probability of success and q is the probability of failure on any given trial.

    a. Construct the probability distribution table of X.

    b. Show that $E(X) = 3p$.

9.4 Variance and Standard Deviation

We wish to obtain a method to approximate the probability distribution of a binomial random variable when the number of trials is large. To do this, we must first consider two additional concepts—another method of characterizing a collection of data and the concept of a normal random variable. The first of these concepts is covered in this section, and the second is covered in the next section.

From our work in section 9.2, we know that the mean of a collection of numbers gives a "central" point for the data. The mean does not give any information on the spread of the numbers. We illustrate this with the following example.

Suppose two groups of 6 students each were given a 10-point quiz with the following results:

$$\text{scores in group 1:} \quad 4, 6, 5, 6, 5, 4$$

$$\text{scores in group 2:} \quad 10, 9, 0, 1, 8, 2$$

The average for both groups is $\bar{x} = 5$, but we observe that the two groups have a different "spread" of scores. We now consider methods, called **measures of dispersion**, which give this characteristic of a collection of data.

For a collection of numbers with mean $\bar{x}$ and for a number x in the collection, the **deviation from the mean for x** is $\bar{x} - x$.

The deviations from the mean for the quiz scores in our example are given in table 9.2.

TABLE 9.2

Group 1		Group 2	
x	$\bar{x} - x$	x	$\bar{x} - x$
4	1	10	-5
6	-1	9	-4
5	0	0	5
6	-1	1	4
5	0	8	-3
4	1	2	3

Suppose we want to obtain a single number that measures the dispersion for the collection of data. We might consider adding the deviations from the mean and taking the average for each of the two groups. We would have

$$\text{group 1: } \quad \text{average deviation} = \frac{2(1) + 2(-1) + 2(0)}{6} = 0$$

$$\text{group 2: } \quad \text{average deviation} = \frac{-5 + (-4) + 5 + 4 + (-3) + 3}{6} = 0.$$

We see that this method is of no value since the average deviation is zero for both groups. In fact, this is true for any collection data. The sum of the deviations from the mean is always zero.

To avoid cancellation of the deviations when we add them, we first square each deviation to obtain a nonnegative number and then take the average. This measure of dispersion is given a special name.

Definition

■ Consider a collection of n numbers containing k distinct values x_1, $x_2, \ldots, x_k$. Let f_i be the frequency of the value x_i in the collection and let $\bar{x}$ be the mean of the collection. Then the **variance** s^2 is given by

$$s^2 = \frac{f_1(\bar{x} - x_1)^2 + f_2(\bar{x} - x_2)^2 + \cdots + f_k(\bar{x} - x_k)^2}{n} \tag{5}$$

where we observe that $f_1 + f_2 + \cdots + f_k = n$. ■

For each of the two groups of quiz scores, we obtain the variance as follows:

$$\text{group 1: } \quad s^2 = \frac{2(1)^2 + 2(-1)^2 + 2(0)^2}{6} = \frac{4}{6} = 0.67$$

$$\text{group 2: } \quad s^2 = \frac{(-5)^2 + (-4)^2 + 5^2 + 4^2 + (-3)^2 + 3^2}{6} = \frac{100}{6} = 16.67$$

We see that the two computed variances, especially the variance for the second group of scores, do not give a good indication of an "average" deviation from the mean. For the second group of scores, no score is more than 5 units from the mean but the variance is 16.67. To compensate for having distorted the deviations by squaring them, we must take the square root of the variance. We have $\sqrt{4/6} = 0.82$ for the first group of scores and $\sqrt{100/6} = 4.08$ for the second group of scores. Thus the square root of the variance should give us a better feeling for the spread of the scores in the collection. As the square root of the variance becomes smaller, we can say that the scores become more clustered about the mean. This important measure of dispersion is given a special name.

Definition

■ Consider a collection of n numbers containing k distinct values x_1, $x_2, \ldots, x_k$. Let f_i be the frequency of the value x_i in the collection and let $\bar{x}$ be the mean of the collection. The **standard deviation** s of the collection is the square root of the variance; that is,

$$s = \sqrt{\frac{f_1(\bar{x} - x_1)^2 + f_2(\bar{x} - x_1)^2 + \cdots + f_k(\bar{x} - x_k)^2}{n}}. \quad \blacksquare \quad (6)$$

EXAMPLE 1

A die was rolled 20 times with the results given in table 9.3. Also included in the table are the deviations from the mean. Let us find the variance and the standard deviation for this collection of data. (Note that the mean $\bar{x} = 3.2$ for the data was obtained in example 2 of section 9.2 on page 404.)

TABLE 9.3

Value x_i	Frequency f_i	Deviation about $\bar{x}$ $(\bar{x} - x_i)$
1	3	2.2
2	5	1.2
3	5	0.2
4	1	−0.8
5	4	−1.8
6	2	−2.8

Using equation (5) and the values in the table, we have for the variance

$$s^2 = \frac{3(2.2)^2 + 5(1.2)^2 + 5(0.2)^2 + 1(-0.8)^2 + 4(-1.8)^2 + 2(-2.8)^2}{20}$$

$$= \frac{51.2}{20}$$

$$= 2.56.$$

Thus the standard deviation is

$$s = \sqrt{2.56} = 1.6.$$

Using equation (5) to calculate the variance s^2 involves k subtractions. The following alternate formula for computing the variance can be derived from (5) by applying certain algebraic manipulations.

$$s^2 = \frac{f_1 x_1^2 + f_2 x_2^2 + f_3 x_3^2 + \cdots + f_k x_k^2}{n} - \bar{x}^2 \quad (7)$$

This formula is more convenient to use.

Consider a collection of data generated by n repetitions of an experiment and consider a random variable X for the experiment. We saw in section 9.2 that expected value $E(X)$ of X can be used as an approximation to the mean $\bar{x}$ of the data when n is large. We now consider the variance and standard variation of the random variable that will also be good approximations to the variance and standard variation of the data when n is large.

Consider an experiment with a random variable X having, for example, six values x_1, x_2, x_3, x_4, x_5, and x_6 and the probability distribution given in the following table.

value of $X(x_i)$	x_1	x_2	x_3	x_4	x_5	x_6
$Pr(X = x_i)(p_i)$	p_1	p_2	p_3	p_4	p_5	p_6

Suppose the experiment is performed n times with the result that the values x_i occur with respective frequencies f_i. Let $\bar{x}$ be the mean of this collection of data. From equation (7) we have the variance of the data given by

$$s^2 = \frac{f_1}{n}(x_1)^2 + \frac{f_2}{n}(x_2)^2 + \frac{f_3}{n}(x_3)^2 + \frac{f_4}{n}(x_4)^2 + \frac{f_5}{n}(x_5)^2 + \frac{f_6}{n}(x_6)^2 - \bar{x}^2.$$

Recall that the probability of an outcome can be thought of as the long-term relative frequency of the outcome. We have that for each value x_i, the relative frequency f_i/n is approximately equal to the probability p_i; that is,

$$\frac{f_i}{n} \approx p_i.$$

Furthermore, the mean (expected value) of X, which we now write for convenience as $\mu = E(X)$, is approximately equal to $\bar{x}$. Therefore,

$$s^2 \approx p_1(x_1)^2 + p_2(x_2)^2 + p_3(x_3)^2 + p_4(x_4)^2 + p_5(x_5)^2 + p_6(x_6)^2 - \mu^2$$

where the right-hand side of this expression is the variance of X. We summarize this discussion as follows.

Definition

■ Let X be a finite discrete random variable with values $x_1, x_2, x_3, \ldots, x_k$ and let $Pr(X = x_i) = p_i$. Also, let μ be the mean (expected value) of X. Then the **variance of X** is given by

$$\sigma^2 = p_1(x_1)^2 + p_2(x_2)^2 + \cdots + p_k(x_k)^2 - \mu^2. \tag{8}$$

Furthermore, the **standard deviation σ of X** is the square root of the variance of X. ■

EXAMPLE 2

Consider the experiment of rolling an unbiased die and the random variable X having values 1, 2, 3, 4, 5, and 6. For each of these values x, we have the probability $Pr(X = x) = \frac{1}{6}$. We computed the mean of X in example 3, section 9.2 on page 405 and found it to be $\mu = 3.5$. Let us now find the variance and standard deviation of X.

Using formula (8), we find the variance of X to be

$$\sigma^2 = \frac{1}{6}(1)^2 + \frac{1}{6}(2)^2 + \frac{1}{6}(3)^2 + \frac{1}{6}(4)^2 + \frac{1}{6}(5)^2 + \frac{1}{6}(6)^2 - (3.5)^2$$

$$= \frac{91}{6} - 12.25$$

$$= 2.92.$$

Taking the square root of the variance, we have the standard deviation

$$\sigma = \sqrt{2.92}$$

$$= 1.71.$$

EXERCISES 9.4

For exercises 1–4 find the variance and standard deviation for the collection of data given in the indicated exercises of section 9.2.

1. exercise 1
2. exercise 2
3. exercise 3
4. exercise 4

5. Consider an experiment with a random variable X where the probability distribution of X is given in the following table.

value of $X(x_i)$	10	15	20	30	-20
$Pr(X = x_i)(p_i)$	0.20	0.25	0.10	0.15	0.30

Find the mean, variance, and standard deviation of X.

6. Consider an experiment with a random variable X where the probability distribution of X is given in the following table.

value of $X(x_i)$	0	-5	3	2
$Pr(X = x_i)(p_i)$	0.1	0.2	0.4	0.3

Find the mean, variance, and standard deviation of X.

7. Consider an experiment with a random variable X where the probability distribution of X is given in the following table.

value of $X(x_i)$	105	-215	316	-46	569
$Pr(X = x)(p_i)$	0.325	0.176	0.191	0.215	0.093

Find the mean, variance, and standard deviation of X.

8. Consider an experiment with a random variable X where the probability distribution of X is given in the following table.

value of $X(x_i)$	467	451	-899	-582	-907
$Pr(X = x)(p_i)$	0.315	0.101	0.015	0.437	0.132

Find the mean, variance, and standard deviation of X.

For exercises 9–16 find the variance and standard deviation of the random variable X described in the indicated exercises of section 9.1.

9. exercise 3

10. exercise 4

11. exercise 9

12. exercise 10

13. exercise 11

14. exercise 12

15. exercise 13

16. exercise 14

17. Let X be a finite random variable. If b is a constant, show that the variance of $X + b$ is the same as the variance of X. (*Hint*: See exercise 32 in section 9.2.)

18. Let X be a finite random variable. If m is a constant, show that the variance of mX is equal to m^2 times the variance of X. (*Hint*: See exercise 33 in section 9.2.)

9.5 Normal Random Variables: Normal Curve

In this section we discuss continuous random variables, focusing our attention on normal random variables. The bell-shaped curve that many people associate with statistics is related to normal random variables.

Many real-world variables can be modeled with normal random variables. Some examples are the heights of people in a certain group, the life spans for the lightbulbs in a collection, the percentage of alcohol per bottle of beer, and the amount of nicotine per cigarette. Before we consider the varied applications of normal random variables, we must first develop the necessary tools.

We have discussed finite discrete random variables. This type of random variable is associated with an experiment whose sample space contains a finite number of outcomes. The probability distribution for the random variable is then obtained from the probability model of the experiment.

A continuous random variable is associated with an experiment whose sample space consists of a continuum of outcomes. For example, consider the experiment of selecting a point at random from the line segment between 0 and 1. The sample space S for this experiment is the set of all points Q on the real number line with coordinate q between 0 and 1, as shown in figure 9.10. Let X denote the coordinate associated with a point in S. Our goal is to obtain a probability distribution for the continuous random variable X.

FIGURE 9.10

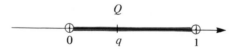

Recall that in a probability model with a finite sample space S, each outcome was assigned a real number that satisfied the following two conditions:

1. For each $e \in S, 0 \le Pr(e) \le 1$.
2. The sum of the probabilities of all the elements in S is equal to 1.

In the case where the sample space contains a continuum of outcomes, the procedure of assigning probabilities to individual outcomes no longer works. For example, let us attempt to assign probabilities to the individual points of the line segment between 0 and 1.

Since the experiment is to select a point at random from S, each point would be assigned the same probability. Therefore, for some positive number k, we set $Pr(Q) = k$ for each Q in S. However, no matter how small k is, condition 2 for a probability model could not be satisfied; the sum of all the $Pr(Q)$ would exceed 1.

One way around this problem is as follows. Instead of attempting to assign probabilities to individual outcomes, we assign probabilities to *sets* of outcomes. If E is a line segment between a and b contained in S, the probability of E is given by

$$Pr(E) = b - a.$$

Observe that $Pr(E)$ is the length of the line segment between a and b in figure 9.11.

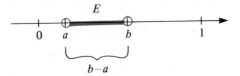

FIGURE 9.11

This assignment of probabilities seems to be a reasonable way to model this experiment. For example, any two line segments in S of the same length are assigned the same probability. If E is the line segment between 0.1 and 0.3 and F is the line segment between 0.75 and 0.95, then $Pr(E) = Pr(F) = 0.2$. In other words, the probability that the point selected at random lies between 0.1 and 0.3 is the same as the probability that the point lies between 0.75 and 0.95.

Now, given the random variable X and a line segment between a and b contained in S, we set

$$Pr(a < X < b) = b - a. \tag{9}$$

For example $Pr(0.2 < X < 0.5) = 0.3$ and $Pr(0.64 < X < 0.65) = 0.01$. We can visualize (9) geometrically in the following way. We draw a rectangle of unit height above the line segment between a and b (see figure 9.12). Then the area of the rectangle is given by

$$\text{area} = \text{height} \times \text{width}$$
$$= 1 \times (b - a)$$
$$= b - a$$
$$= Pr(a < X < b).$$

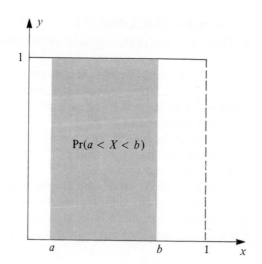

FIGURE 9.12

Therefore the probability distribution for X is given by areas under the line $y = 1$ and above the x-axis (see figure 9.12). The line segment $y = 1$, $0 < x < 1$ is called the **distribution curve** of X. Note that the total area under the curve from $x = 0$ to $x = 1$ is 1. We now generalize these ideas.

Let X be a continuous random variable. Associated with X is a **distribution curve** (see figure 9.13).

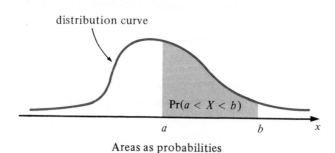

FIGURE 9.13

Areas as probabilities

Properties of a Distribution Curve

1. The total area under the curve and above the x-axis is one.
2. $Pr(a < X < b) =$ the area under the curve from a to b.

One feature of the probability distribution for a continuous random variable that differs from the discrete case is the following. Since probabilities are given in terms of areas under a curve, then for a single value x we have $Pr(X = x) = 0$. This can be seen from the fact that the area under a curve and above a single point x on the x-axis is zero. Therefore, *for continuous random variables, we only consider probabilities where X takes a value from some interval of values.*

For the remainder of this chapter, we will discuss the particular class of continuous random variables called **normal** random variables. Normal random variables with their associated **normal distribution curves**, or simply **normal curves**, have many applications. Before considering applications, let us first look at some characteristics of a normal curve.

A normal curve is symmetrical and bell shaped (see figure 9.14). The curve is symmetrical with respect to the line perpendicular to the x-axis and passing through the highest point on the curve. This line is called the **center of symmetry**. The curve tapers off in both directions, becoming closer and closer to the x-axis but never touching it. Nevertheless, the total area under the curve is one.

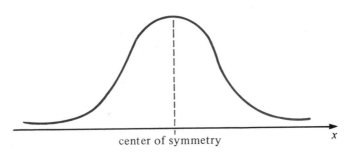

center of symmetry

x

FIGURE 9.14

A normal curve

As for the discrete case, a continuous random variable possesses a mean that measures central tendency and a standard deviation that measures dispersion. As before, we denote the mean and standard deviation by μ and σ. It is beyond the scope of this book actually to compute them.

The mean and standard deviation have geometrical significance. The mean of a normal random variable is the point on the x-axis that lies on the center of symmetry. The standard deviation determines the "spread" of the curve. In advanced statistics courses it is shown that the area under the curve within one standard deviation of μ (from $\mu - \sigma$ to $\mu + \sigma$) is approximately 68.3% of the total area under the curve. Also, 95.4% of the total area lies within two standard deviations of μ (from $\mu - 2\sigma$ to $\mu + 2\sigma$) and 99.7% of the total area lies within three standard deviations of μ (from $\mu - 3\sigma$ to $\mu + 3\sigma$). (See figure 9.15.)

A normal curve is uniquely determined by its mean and standard deviation. The mean μ locates the center of symmetry on the x-axis and the standard deviation σ dictates the shape of the curve. Figure 9.16 gives three normal curves with the same mean but with three different standard deviations. As shown in figure 9.16, the larger the standard deviation, the flatter and wider the curve.

For each μ and σ there is a unique normal curve. If we change μ or σ (or both), we obtain a different normal curve. There is one particular normal curve, used

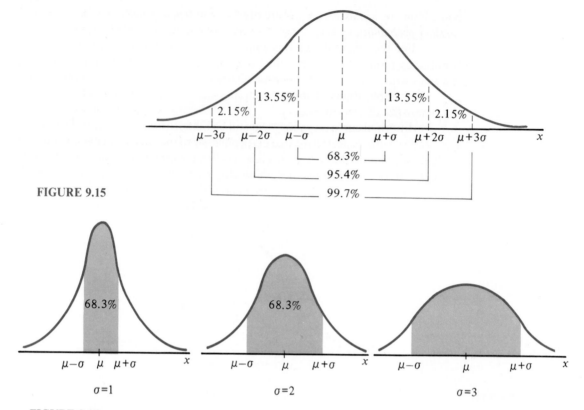

FIGURE 9.15

FIGURE 9.16

as a comparison for all other normal curves, called the **standard normal curve**. The standard normal curve is the normal curve with mean $\mu = 0$ and standard deviation $\sigma = 1$. The corresponding random variable is called the **standard normal random variable** and is usually denoted by Z. The graph of the standard normal curve is shown in figure 9.17.

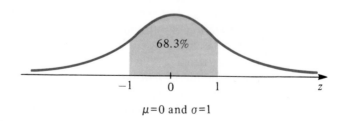

FIGURE 9.17

Our goal for the remainder of this section is to find probabilities of normal random variables. In particular, given a random variable X, we want to find the following probabilities.

Probabilities of Interest

1. $Pr(a < X < b)$
2. $Pr(X < c)$ **(10)**
3. $Pr(X > c)$

The three probabilities in (10) can be represented as areas under the normal curve of X. Referring to figure 9.18, we have

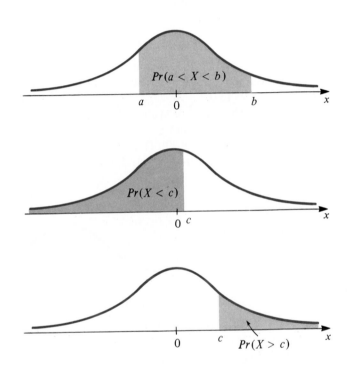

FIGURE 9.18

Probabilities of a Normal Random Variable

1. $Pr(a < X < b)$ = the area under the normal curve between a and b
2. $Pr(X < c)$ = the area under the normal curve to the left of c
3. $Pr(X > c)$ = the area under the normal curve to the right of c

Therefore, to achieve our goal of finding probabilities, it suffices to find areas under the normal curve of X.

Let us begin by considering the standard normal random variable Z. To help us find areas under the standard normal curve, table II in appendix B gives the area under the curve between 0 and z for values of Z from 0 to 2.99. Figure 9.19 indicates such an area.

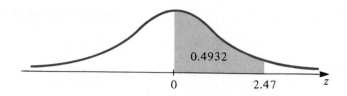

FIGURE 9.19

To find the area of the region between 0 and $z = 2.47$, shown in figure 9.19, we find the row labeled 2.4 and the column labeled 7 in table II. The entry in the table at the intersection of the row and column gives the desired area. The area under the curve between 0 and $z = 2.47$ is 0.4932. Therefore,

$$Pr(0 < Z < 2.47) = 0.4932.$$

To find the area under the curve for a negative z, we make use of the symmetry property of normal curves. Let us find the area of the region from $z = -1.75$ to 0.

The center of symmetry for the standard normal curve is the y-axis, so the area of the region from $z = -1.75$ to 0 is the same as the area of the region from 0 to $z = 1.75$ (see figure 9.20). Therefore, using table II with $z = 1.75$ we see that the area under the curve between -1.75 and 0 is 0.4599. Thus,

$$Pr(-1.75 < Z < 0) = 0.4599.$$

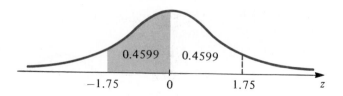

FIGURE 9.20

Now, suppose we want to find the area of the region from $z = z_1$ to $z = z_2$ with $z_1 < z_2$. For example, suppose z_1 and z_2 are both positive (see figure 9.21). From the figure we see that

area from z_1 to z_2 = (area from 0 to z_2) − (area from 0 to z_1).

For example, let $z_1 = 0.46$ and $z_2 = 2.11$. From table II we have

area from 0 to 2.11 = 0.4826

area from 0 to 0.46 = 0.1772.

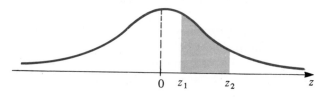

FIGURE 9.21

area from z_1 to z_2 = (area from 0 to z_2) - (area from 0 to z_1)

Therefore,

$$\text{area from } 0.46 \text{ to } 2.11 = 0.4826 - 0.1772$$
$$= 0.3054.$$

Another type of area problem comes in two forms:

form 1: Find the area to the right of z.

form 2: Find the area to the left of z.

An example of form 1 is to find the area to the right of $z = 0.93$ (see figure 9.22). From the figure we see that

$$\text{area to the right of } 0.93 = (\text{area to the right of } 0) - (\text{area from } 0 \text{ to } 0.93).$$

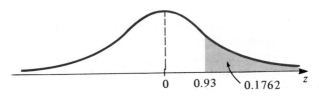

FIGURE 9.22   area to the right of $0.93 = 0.5000 - (\text{area from } 0 \text{ to } 0.93)$

Since the total area under the curve is one, by symmetry the area to the right of 0 is 0.5000. Therefore,

$$\text{area to the right of } 0.93 = 0.5000 - 0.3238$$
$$= 0.1762.$$

An example of form 2 is to find the area to the left of $z = -2.55$ (see figure 9.23). From the figure we see that

$$\text{area to the left of } -2.55 = \text{area to the right of } 2.55$$
$$= 0.5000 - 0.4946$$
$$= 0.0054.$$

Whenever we want to determine the area of a region, it is best first to draw a figure showing the region. Then we may express the area of the region, using symmetry when appropriate, in such a way that table II can be used.

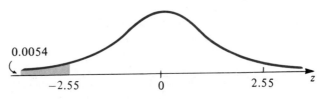

FIGURE 9.23   area to the left of -2.55 = area to the right of 2.55

EXAMPLE 1

Let us find the area of the region to the right of $z = -2.40$. First we draw the indicated region (see figure 9.24). From the figure, we see that

$$\text{area to the right of } -2.40 = (\text{area from } -2.40 \text{ to } 0) + 0.5000$$
$$= 0.4918 + 0.5000$$
$$= 0.9918.$$

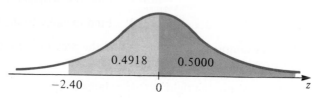

FIGURE 9.24

area to the right of $-2.40 = (\text{area from } -2.40 \text{ to } 0) + 0.5000$

We now know how to solve area problems with respect to the standard normal variable Z. In general, given a normal random variable X with mean μ and standard deviation σ, we can convert area problems with respect to X to area problems with respect to Z.

Rule

■ *Conversion Rule:*

Let X be a normal random variable with mean μ and standard deviation σ. Then the area under the normal curve of X between μ and x_1 is the same as the area under the standard normal curve between 0 and z_1, where

$$z_1 = \frac{x_1 - \mu}{\sigma}. \quad ■$$

Figure 9.25 illustrates the conversion rule.

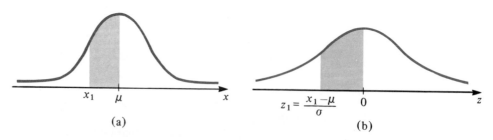

(a)   (b)

FIGURE 9.25

area between x_1 and μ in (a) = area between z_1 and 0 in (b)

EXAMPLE 2

Let X be a normal random variable with $\mu = 25$ and $\sigma = 6$. Find the probability that X lies between 21 and μ.

The probability that X lies between 21 and μ is equal to the area under the normal curve of X between $x_1 = 21$ and μ (see figure 9.26). By the conversion rule, this area is the same as the area under the standard normal curve between

$$z_1 = \frac{21 - 25}{6} = -0.67 \quad \text{and} \quad 0.$$

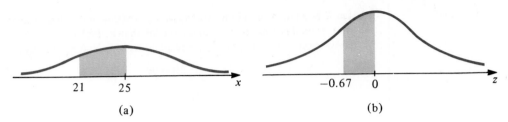

FIGURE 9.26 $Pr(21 < X < 25)$ = area under the standard normal curve between $z = -0.67$ and 0

(see figure 9.26). Therefore,

$$Pr(21 < X < 25) = \text{area between } z = -0.67 \text{ and } 0$$
$$= 0.2486.$$

The conversion rule can be used to find probabilities with respect to any normal random variable X. The next example illustrates this technique.

EXAMPLES   3. Let X be a normal random variable with $\mu = 4$ and $\sigma = 0.39$. Let us find the probability that X lies between $x_1 = 3.5$ and $x_2 = 4.7$.
    Using the conversion rule with z_1 and z_2 corresponding to x_1 and x_2, respectively, we have

$$z_1 = \frac{3.5 - 4}{0.39} = -1.28$$

$$z_2 = \frac{4.7 - 4}{0.39} = 1.79.$$

Thus the probability that X lies between 3.5 and 4.7 is equal to the area under the standard normal curve between -1.28 and 1.79 (see figure 9.27). We can find the area indicated in figure 9.27 by using table II in appendix B. Therefore,

$$Pr(3.5 < X < 4.7) = (\text{area between } 0 \text{ and } z = 1.28)$$
$$+ (\text{area between } 0 \text{ and } z = 1.79)$$
$$= 0.3997 + 0.4633$$
$$= 0.8630.$$

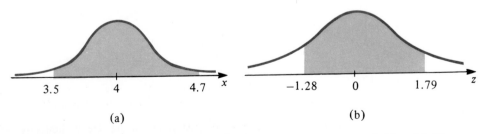

FIGURE 9.27 $Pr(3.5 < X < 4.7)$ = area under standard normal curve between -1.28 and 1.79

4. Let X be a normal random variable with $\mu = 90$ and $\sigma = 5$. Find the probability that X lies between $x_1 = 10$ and $x_2 = 80$, that is, $Pr(10 < X < 80)$. Using the conversion rule with z_1 and z_2 corresponding to x_1 and x_2, we have

$$z_1 = \frac{10 - 90}{5} = -16$$

$$z_2 = \frac{80 - 90}{5} = -2.$$

Thus the probability that X lies between 10 and 80 is equal to the area under the standard normal curve between -16 and -2.

Our next step would be to use table II in appendix B. However, in this case, the table cannot be used with $z_1 = -16$. This is not a problem as z_1 is 16 standard deviations from the mean and therefore the area to the left of z_1 is negligible. (Recall that 99.7 percent of the area under the curve is within three standard deviations of the mean.) Therefore,

$$Pr(10 < X < 80) = \text{area to the left of } z_2 = -2$$
$$= 0.5000 - 0.4773$$
$$= 0.0227.$$

We conclude this section with applications. Few real-world variables behave precisely as normal random variables; however, normal variables are adequate mathematical models for many real-world situations. Thus in our applications we will assume that the variable in question is normally distributed; that is, we model the real-world variable with a normal random variable.

EXAMPLES

5. *Selecting from a Machine's Production.* Bottles of cola are filled by a machine so that the weight X per bottle has a mean value of 16 ounces. If we assume (based on past experience) that X is normally distributed with standard deviation $\sigma = 0.24$ ounces, let us find the probability that a bottle selected at random weighs less than 15.5 ounces.

We want to find $Pr(X < 15.5)$. Using the conversion rule, we convert $x = 15.5$ to $z = (15.5 - 16)/0.24 = -2.08$. Therefore, from figure 9.28 we have

$$Pr(X < 15.5) = \text{area to the left of } z = -2.08$$
$$= 0.5000 - 0.4812$$
$$= 0.0188.$$

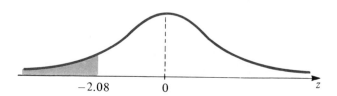

FIGURE 9.28

6. *The Life of a Manufactured Item.* If the life in hours of STAY-BRITE lightbulbs is normally distributed with a mean of 1200 hours and a standard

deviation of 80 hours, let us find the probability that a lightbulb selected at random will last

a. more than 1300 hours.

b. between 1025 and 1150 hours.

c. less than 1000 hours.

Let X be the life in hours of STAY-BRITE lightbulbs. Then X has a mean $\mu = 1200$ and a standard deviation $\sigma = 80$.

For part (a), we want to find $Pr(X > 1300)$. Using the conversion rule, $x = 1300$ corresponds to $z = (1300 - 1200)/80 = 1.25$. Therefore, from figure 9.29a, we have

$$Pr(X > 1300) = \text{area to the right of } z = 1.25$$
$$= 0.5000 - 0.3944$$
$$= 0.1056.$$

For part (b), we want to find $Pr(1025 < X < 1150)$. Using the conversion rule, $x = 1025$ corresponds to $z = (1025 - 1200)/80 = -2.19$, and $x = 1150$ corresponds to $z = (1150 - 1200)/80 = -0.63$. Therefore, from figure 9.29b, we have

$$Pr(1025 < X < 1150) = \text{area between } z = -2.19 \text{ and } z = -0.63$$
$$= 0.4857 - 0.2357$$
$$= 0.2500.$$

For part (c), we want to find $Pr(X < 1000)$. Using the conversion rule, $x = 1000$ corresponds to $z = (1000 - 1200)/80 = -2.50$. Therefore, from figure 9.29c, we have

$$Pr(X < 1000) = \text{area to the left of } z = -2.50$$
$$= 0.5000 - 0.4938$$
$$= 0.0062.$$

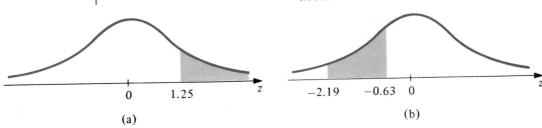

(a)

(b)

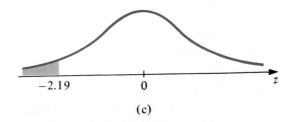

(c)

FIGURE 9.29

1. Use table II in appendix B to find the following areas under the standard normal curve.

   a. between $z = 0$ and $z = 1.95$

   b. to the right of $z = 2.06$

   c. to the right of $z = -0.21$

   d. to the right of $z = 0.79$

   e. to the left of $z = -1.02$

   f. between $z = 0.89$ and $z = 1.93$

   g. between $z = -2.01$ and $z = -1.87$

   h. between $z = -1.09$ and $z = 0.63$

2. Use table II in appendix B to find the following areas under the standard normal curve.

   a. between $z = 0$ and $z = 1.18$

   b. to the right of $z = 1.13$

   c. to the right of $z = -2.11$

   d. to the left of $z = -1.98$

   e. to the left of $z = 0.93$

   f. between $z = 0.09$ and $z = 0.92$.

   g. between $z = -2.17$ and $z = -0.99$

   h. between $z = -1.57$ and $z = 1.86$

3. Let Z denote the standard normal random variable. Find the following probabilities.

   a. $Pr(Z > 1.33)$ b. $Pr(\text{either } Z > 0.20 \text{ or } Z < -1.39)$

   c. $Pr(-1.76 < Z < -0.65)$ d. $Pr(1.20 < Z < 2.57)$

4. Let Z denote the standard normal random variable. Find the following probabilities.

   a. $Pr(Z < 0.52)$

   b. $Pr(-1.26 < Z < 1.23)$

   c. $Pr(\text{either } Z > 2.65 \text{ or } Z < 0.06)$

   d. $Pr(1.99 < Z < 2.43)$

5. Let X be a normal random variable with $\mu = 476$ and $\sigma = 37$. Find the following probabilities.

   a. $Pr(X > 443)$

   b. $Pr(X < 521)$

   c. $Pr(450 < X < 535)$

   d. $Pr(\text{either } X < 421 \text{ or } X > 543)$

6. Let X be a normal random variable with $\mu = 0.51$ and $\sigma = 0.23$. Find the following probabilities.

   a. $Pr(X < 0.65)$

   b. $Pr(X > 0.09)$

  c. $Pr(0.65 < X < 0.73)$

  d. $Pr($either $X < 0.19$ or $X > 0.82)$

7. If the life in hours of Destiny refrigerator lights is normally distributed with a mean of 900 hours and a standard deviation of 55 hours, find the probability that a refrigerator light selected at random will last

  a. less than 789 hours.

  b. between 925 and 1000 hours.

  c. more than 850 hours.

8. The time required for a machinist to produce a compression cylinder is assumed to be a normal random variable with a mean of 25 minutes and a standard deviation of 3 minutes. What is the probability that the time required to make a compression cylinder is

  a. less than 20 minutes?

  b. between 23 and 29 minutes?

  c. greater than 30 minutes?

9. The amount of snowfall in the Lehigh Valley in January is normally distributed with a mean of 15 inches and a standard deviation of 4 inches. Find the probability that the amount of snowfall next January in the Lehigh Valley will be

  a. more than 22 inches.

  b. less than 10 inches.

  c. between 6 and 9 inches.

10. The number of calories in a bottle of light beer is normally distributed with a mean of 96 calories and a standard deviation of 2.5 calories. Find the probability that a bottle of light beer selected at random will have

  a. less than 92.5 calories.

  b. between 93 and 100 calories.

  c. more than 100 calories.

11. The average miles per gallon in city driving of the new XLT is normally distributed with a mean of 19 miles per gallon and a standard deviation of 2.8 miles per gallon. Find the probability that a new XLT, selected at random, has a city driving average mileage that. is

  a. greater than 21.5 miles per gallon.

  b. between 17 and 21 miles per gallon.

  c. less than 15 miles per gallon.

12. If the height of students at Orange Community College is normally distributed with a mean of 67 inches and a standard deviation of 3.5 inches, find

  a. the percentage of students whose height is less than 62 inches.

  b. the percentage of students whose height is greater than 6 feet.

  c. the percentage of students whose height is within one standard deviation of the mean.

13. The Horbocker Brewery plans to introduce its new Horbocker LA beer with the percentage of alcohol in a bottle having a mean of 1.4 and a standard deviation of 0.18%. Find the probability that a bottle of Horbocker LA beer selected at random will have

a. less than 1.2% alcohol.

b. between 1.35% and 1.89% alcohol.

c. more than 1.56% alcohol.

14. The amount of nicotine per cigarette of a particular brand is normally distributed with a mean of 0.4 mg (milligrams) and a standard deviation of 0.02 mg. Find the probability that a cigarette selected at random will have

a. less than 0.424 mg of nicotine.

b. between 0.408 and 0.458 mg of nicotine.

c. more than 0.379 mg of nicotine.

15. Let X be a normal random variable with $\mu = 82$ and $\sigma = 6$. Find x_0 such that

$$Pr(X > x_0) = 0.1711.$$

16. Let X be a normal random variable with mean μ and standard deviation σ. Find a positive constant k such that

$$Pr(\mu - k\sigma < X < \mu + k\sigma) = 0.8472.$$

9.6 Normal Approximation of a Binomial Random Variable

In example 8 of section 9.3 (page 419), a drug company claims that their new skin creme cures acne 80% of the time. Dr. Snyder tested the new product on 100 of his patients. We modeled this sequence of 100 tests as a binomial experiment where success on a given test (trial) was 0.8. If X denotes the number of successes, we wanted to find the probability of 70 or less successes—that is, $Pr(0 \leq X \leq 70)$. We saw that finding $Pr(0 \leq X \leq 70)$ would involve 71 tedious calculations.

In this section we will develop a method to approximate probabilities associated with binomial experiments by using areas under a normal curve. Before starting this development, we digress for a moment.

Recall that any random variable has a mean and a standard deviation. Consider a binomial experiment of n trials where the probability of success is p and the probability of failure is q. Let X denote the (discrete) random variable, $X =$ the number of successes in n trials. Although we will not give proofs, the following theorem states formulas for the mean μ and standard deviation σ of X in terms of the number of trials n, the probability of success p, and the probability of failure q.

Theorem ■ Consider a binomial experiment of n trials with

$$p = \text{probability of a success on any trial}$$

$$q = \text{probability of a failure on any trial.}$$

If X denotes the number of successes in n trials, then the mean μ and standard deviation σ of X are given by

$$\mu = np$$
$$\sigma = \sqrt{npq}. \quad \blacksquare$$

(11)

EXAMPLE 1

Let us return to the binomial experiment of testing the new skin creme on 100 patients. The random variable X in this case has a mean of $\mu = 100(0.8) = 80$ and a standard deviation of $\sigma = \sqrt{100(0.8)(0.2)} = 4$.

We now develop a method to *approximate* binomial probabilities with areas under the standard normal curve. As in section 9.1, the probability distribution of a discrete random variable can be visualized as a bar graph. Let us construct the bar graph of the variable $X =$ the number of successes, for a binomial experiment of 3 trials, with $p = \frac{1}{3}$ as the probability of success on any given trial. Using formula (4) (page 417),

$$Pr(X = 0) = 0.2963 \qquad Pr(X = 1) = 0.4444$$
$$Pr(X = 2) = 0.2222 \qquad Pr(X = 3) = 0.0370.$$

The bar graph for X is given in figure 9.30a. From the bar graph, we can draw a **histogram** for X. For each value of x on the x-axis, we draw a rectangle centered at x of width one and with height $h = Pr(X = x)$. The histogram for the binomial variable X with $n = 3$ and $p = \frac{1}{3}$ is shown in figure 9.30b.

Histograms have the property that the areas of the rectangles can be interpreted as probabilities. For example, consider the rectangle centered at $x = 1$ in figure 9.30b. The area A of this rectangle is

$$A = \text{width} \times \text{height}$$
$$= 1 \times 0.4444$$
$$= 0.4444$$
$$= Pr(X = 1).$$

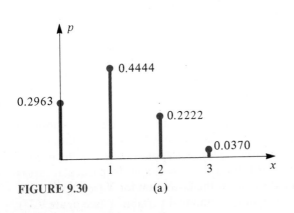

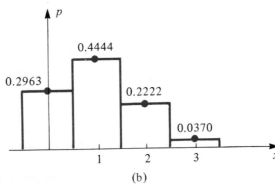

FIGURE 9.30 (a) (b)

Let us now increase the number of trials and observe the shapes of the resulting histograms. Figure 9.31 shows histograms where $n = 5, 10, 15,$ and 20 for a binomial variable with $p = 0.3$.

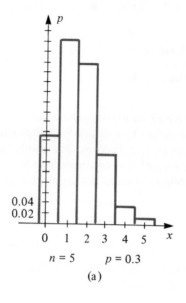

n = 5   p = 0.3

(a)

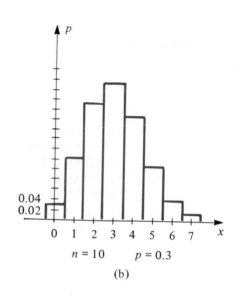

n = 10   p = 0.3

(b)

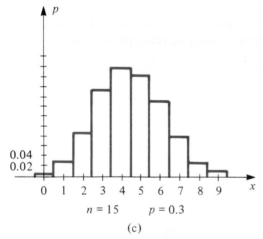

n = 15   p = 0.3

(c)

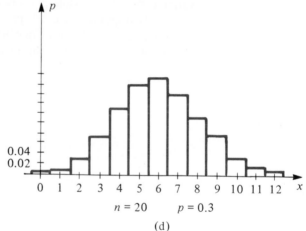

n = 20   p = 0.3

(d)

FIGURE 9.31

It is not hard to imagine that for large n the resulting histogram resembles a bell-shaped curve—namely, a normal curve (see figure 9.31). In fact, a theorem in statistics states that for large n, the histogram for X is approximated by the normal curve with mean $\mu = np$ and standard deviation $\sigma = \sqrt{npq}$. In particular, areas under the normal curve approximate areas of the histogram for X (see figure 9.32). For example, consider a fixed value x of our binomial variable X (see figure 9.33).

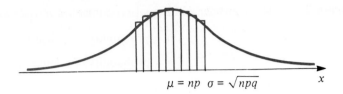

FIGURE 9.32

Then

$$Pr(X = x) = \text{area of the rectangle centered at } x$$

$$\approx \text{area under the normal curve between } x - \frac{1}{2} \quad \text{and} \quad x + \frac{1}{2}$$

where "$\approx$" means "approximately equal to."

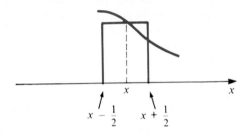

FIGURE 9.33

We may extend this idea to two or more values of X as shown in figure 9.34. Then,

$$Pr(a \le X \le b) \approx \text{area under the curve between } a - \frac{1}{2} \text{ and } b + \frac{1}{2}.$$

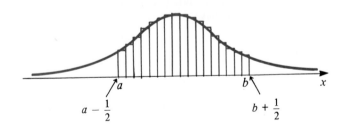

FIGURE 9.34

Thus we have that probabilities of the binomial variable X can be approximated by areas under the normal curve with mean $\mu = np$ and standard deviation $\sigma = \sqrt{npq}$. The final step is to use the conversion rule (see page 436 in section 9.5) to convert areas under the approximating normal curve to areas under the standard normal curve.

We summarize the procedure for approximating binomial probabilities with areas under the standard normal curve in the following theorem.

Theorem 3   ■   Consider a binomial experiment of n trials with

$$p = \text{probability of a success on any trial}$$
$$q = \text{probability of a failure on any trial.}$$

Let X denote the number of successes in n trials. If a and b are two values of X with $a < b$, then

$$Pr(a \leq X \leq b) \approx \text{area under the standard normal curve}$$

$$\text{between} \quad z_1 = \frac{\left(a - \frac{1}{2}\right) - \mu}{\sigma} \quad \text{and} \quad z_2 = \frac{\left(b + \frac{1}{2}\right) - \mu}{\sigma} \quad (12)$$

where $\mu = np$ and $\sigma = \sqrt{npq}$.   ■

EXAMPLE 2   Let X be a binomial variable with $n = 250$ and $p = 0.6$. Let us use theorem 3 to approximate the probability that X lies between 140 and 165, inclusive.
　　We first compute the mean μ and standard deviation σ. Using formula (11), we have

$$\mu = 250(0.6)$$
$$= 150$$

and

$$\sigma = \sqrt{(250)(0.6)(0.4)} = 7.75.$$

$$Pr(140 \leq X \leq 165) \approx \text{area under the standard normal}$$
$$\text{curve between } z_1 \text{ and } z_2$$

where

$$z_1 = \frac{139.5 - 150}{7.75} = -1.35 \quad \text{and} \quad z_2 = \frac{165.5 - 150}{7.75} = 2.$$

Using the methods described in section 9.5 to find areas under the standard normal curve, we have

$$Pr(140 \leq X \leq 165) \approx 0.8888.$$

We now know how to approximate binomial probabilities using the standard normal curve. Note that this method should be used only when the number of trials n is large enough so that $n \geq 25$, $np \geq 5$, and $n(1 - p) \geq 5$.

EXAMPLES   3. Recall again example 8 of section 9.3 (page 419), discussed at the beginning of this section, in which Dr. Snyder tested a new skin creme on 100 patients. Let us approximate the probability that 70 or fewer patients are cured of acne.
　　As before, let X denote the number of successes. We approximate $Pr(0 \leq X \leq 70)$ by using formula (12). The mean of X is $\mu = 100(0.8) = 80$, and the standard deviation of X is $\sigma = \sqrt{100(0.8)(0.2)} = 4$. Therefore,

$$Pr(0 \leq X \leq 70) \approx \text{area between } z_1 \text{ and } z_2$$

where

$$z_1 = \frac{-0.5 - 80}{4} = -20.13 \quad \text{and} \quad z_2 = \frac{70.5 - 80}{4} = -2.38.$$

Since z_1 is more than 20 standard deviations from the mean, the area under the curve to the left of z_1 is negligible. Therefore,

$$Pr(0 \le X \le 70) \approx \text{area to the left of } z_2$$
$$= 0.5000 - 0.4913$$
$$= 0.0087.$$

4. The admissions office at Slippery Rock University knows from past experience that 60% of all high school applicants will enroll as freshmen. If 1200 high school students apply for admissions for next year, what is the probability that (a) no more than 750 will enroll? (b) at least 700 will enroll?

For part (a),

$$Pr(\text{no more than 750 will enroll}) = Pr(0 \le X \le 750).$$

We use formula (12) to approximate this probability. The mean of X is $\mu = (1200)(0.6) = 720$ and the standard deviation of X is $\sigma = \sqrt{1200(0.6)(0.4)} = 16.97$.

$$Pr(0 \le X \le 750) \approx \text{area between } z_1 \text{ and } z_2$$

where

$$z_1 = \frac{-0.5 - 720}{16.97} = -42.46 \quad \text{and} \quad z_2 = \frac{750.5 - 720}{16.97} = 1.80.$$

Since z_1 is more than 42 standard deviations from the mean, the area under the curve to the left of z_1 is negligible. Therefore,

$$Pr(0 \le X \le 750) \approx \text{area to the left of } z_2$$
$$= 0.5000 + 0.4641$$
$$= 0.9641.$$

For part (b),

$$Pr(\text{at least 700 will enroll}) = Pr(700 \le X \le 1200).$$

We use formula (12) to approximate this probability.

$$Pr(700 \le X \le 1200) \approx \text{area between } z_1 \text{ and } z_2$$

where

$$z_1 = \frac{699.5 - 720}{16.97} = -1.21 \quad \text{and} \quad z_2 = \frac{1200.5 - 720}{16.97} = 28.31.$$

Since z_2 is more than 28 standard deviations from the mean, the area under the curve to the right of z_2 is negligible. Therefore,

$$Pr(700 \le X \le 1200) \approx \text{area to the right of } z_1$$
$$= 0.5000 + 0.3869$$
$$= 0.8869.$$

EXERCISES 9.6

1. Find the mean and standard deviation of the binomial random variable X for the given values of n and p.

   a. $n = 625$, $p = \dfrac{3}{5}$

   b. $n = 517$, $p = 0.53$

   c. $n = 136$, $p = 0.471$

   d. $n = 4121$, $p = 0.769$

2. Find the mean and standard deviation of the binomial random variable X for the given values of n and p.

   a. $n = 35$, $p = \dfrac{2}{3}$

   b. $n = 726$, $p = 0.492$

   c. $n = 319$, $p = 0.759$

   d. $n = 20{,}517$, $p = 0.718$

3. Draw the histogram for the binomial random variable X for the given values of n and p.

   a. $n = 2$, $p = 0.6$

   b. $n = 5$, $p = 0.4$

   c. $n = 6$, $p = 0.7$

4. Draw the histogram for the binomial random variable X for the given values of n and p.

   a. $n = 2$, $p = \dfrac{1}{2}$

   b. $n = 4$, $p = 0.3$

   c. $n = 8$, $p = 0.4$

5. Let X be a binomial random variable with $n = 1650$ and $p = 0.55$. Use a normal approximation to find the following probabilities.

   a. $Pr(X \geq 928)$

   b. $Pr(X \leq 882)$

   c. $Pr(890 \leq X \leq 938)$

   d. $Pr(\text{either } X \leq 879 \text{ or } X \geq 947)$

6. Let X be a binomial random variable with $n = 528$ and $p = 0.852$. Use a normal approximation to find the following probabilities.

   a. $Pr(X \leq 460)$

   b. $Pr(X \geq 435)$

   c. $Pr(440 \leq X \leq 460)$

   d. $Pr(\text{either } X \leq 430 \text{ or } X \geq 470)$

In exercises 7–15 find the binomial probabilities by using a normal approximation.

7. A coin is weighted so that the probability of getting a head when the coin is flipped is $\frac{2}{3}$. If the coin is flipped 250 times, find the probability of

   a. obtaining at most 157 heads.

   b. obtaining at least 170 heads.

   c. obtaining between 160 and 165 heads, inclusive.

   d. obtaining 225 or more heads.

8. An unbiased die is rolled 300 times. Let X denote getting a one on the upturned face. Find the following binomial probabilities.

   a. $Pr(X \geq 60)$

   b. $Pr(X \leq 50)$

   c. $Pr(45 \leq X \leq 55)$

   d. $Pr(\text{either } X \leq 48 \text{ or } X \geq 62)$

9. Suppose that 15% of all items produced by an assembly line are defective. If 700 items are selected at random from this assembly line, use a normal approximation to find the probability that

    a. 84 or fewer items are defective.

    b. between 100 and 110 items, inclusive, are defective.

    c. 125 or more items are defective.

10. Suppose that 60% of the people in a large city oppose an increase in the property tax. If 2500 people are selected at random, what is the probability that 950 or more of them will favor an increase in the property tax?

11. It is estimated that 30% of all cars in a particular city would fail an exhaust emissions test. If 250 cars are randomly selected for testing, what is the probability that 170 or more will pass the test?

12. The number of birds in a batch of 2000 chicks that will be suitable for market when full grown is a binomial random variable with $n = 2000$ and $p = 0.8$. What is the probability that 1650 or more of the chicks will be suitable for market?

13. It is estimated that 15% of the electric motors produced at a certain factory are defective. If 1200 electric motors are selected at random from this factory, what is the probability that

    a. 185 or fewer are defective?

    b. between 170 and 180, inclusive, are defective?

    c. 600 or more are defective?

14. Suppose that 10% of all items produced by an assembly line are defective. If 100 items are selected at random from this assembly line, what is the probability that

    a. 80 or fewer items are defective?

    b. between 85 and 95 items, inclusive, are defective?

    c. 200 or more items are defective?

15. From past experience, it is known that 12% of all microwave ovens of a particular brand will not outlast the warranty. If a store sells 600 of these microwave ovens, use a normal approximation to find the probability that the store will eventually replace

    a. between 60 and 80 ovens, inclusive.

    b. no more than 55 ovens.

    c. 90 or more ovens.

16. Consider a binomial experiment with three trials. Let

$$X = \text{the number of successes}$$

    where p is the probability of success and q is the probability of failure on any given trial. Show that the variance of X is $3pq$.

17. Consider a binomial experiment with n trials. Let

$$X = \text{the number of successes}$$

    where p is the probability of success and q is the probability of failure on any given trial. If the expected value of X is 18 and the variance of X is 4.5, find p and q.

9.7 Hypothesis Testing

Hypothesis testing is a part of statistical inference that has applications in areas such as market studies, surveys, and quality control. Companies may make market studies to determine what percentage of buyers of a product prefer the company's brand. A politician may take a survey to determine the percentage of people who will vote for him in the next election. A manufacturer may select at random a sample from the production line to determine the quality of the manufacturing process.

We introduce hypothesis testing with the following problem. Suppose we have a coin that comes up heads more often than tails when flipped several times. This seems unusual and we question the "honesty" of the coin. We want to conduct an experiment to test whether the coin is unbiased. We are suspicious about the coin, so our position on the matter is that the coin is biased. In fact, we conjecture that the coin favors heads. There are two positions on this coin, one of which is an alternative to the other.

Conjecture 1

■　The coin that we want to test is unbiased. This conjecture is called the **null hypothesis** and is denoted by H_0. Furthermore, this hypothesis is equivalent to saying that the probability of obtaining a head on any given flip of the coin is $\frac{1}{2}$, which we denote by $H_0 : p = \frac{1}{2}$. ■

Conjecture 2

■　The coin favors heads, which is our position as the experimenter, and which is the alternative to conjecture 1. This conjecture is called the **alternate hypothesis** and is denoted by H_1. Furthermore, this hypothesis is equivalent to saying that the probability of obtaining a head on any given flip of the coin is greater than $\frac{1}{2}$, which we denote by $H_1 : p > \frac{1}{2}$. ■

Let us now consider our test of the null hypothesis H_0. We flip the coin 100 times and record the number of heads that appear. This number is called a **test statistic**. We want to use this statistic to arrive at a decision about H_0: should we reject the null hypothesis H_0 in favor of our position H_1 or should we accept H_0 over H_1?

As the tester, we should be neutral even though we have taken a position contrary to H_0. It would not be proper for us to reject H_0 after obtaining the results of the experiment. Furthermore, we want to feel comfortable with our final decision as to the honesty of the coin. With these considerations in mind, we seek a criterion for making our decision. We start with the following remark.

We will reject H_0 in favor of H_1 if the probability of obtaining the test statistic, given that the coin is unbiased, is small.

Taking a closer look at this tentative criterion, we find that we could not be comfortable with it because the probability of obtaining exactly a certain number of heads on 100 flips of a coin is small to begin with. For example, the probability of obtaining exactly 50 heads on 100 flips of an unbiased coin is 0.0796. Thus we consider a different criterion.

Since our position is that the coin favors heads, we should be comfortable with the following criterion.

Let E be the event of obtaining at least the test statistic on the 100 trials. Then we will reject H_0 in favor of H_1 if the probability of E happening, given that H_0 is true, is small; that is, we will reject H_0 in favor of H_1 if $Pr(E|H_0)$ is small.

With this criterion, we need to consider one final question: what do we mean by "small"?

We choose a small number α and check whether or not $Pr(E|H_0)$ is less than α. 100α percent is called the **level of significance**. Values commonly used for α in hypothesis testing are 0.05 and 0.01. We will take $\alpha = 0.05$. Therefore the criterion on which to base our decision is

if $Pr(E|H_0) < 0.05$, we will reject H_0 in favor of H_1.

Of course,

if $Pr(E|H_0) \geq 0.05$, we will accept H_0 over H_1.

With the basis for making the decision now established, let us proceed with the test.

Suppose the coin is flipped 100 times and we obtain 59 heads as the test statistic. Therefore E is the event of obtaining 59 or more successes in a binomial experiment of 100 trials where success means getting a head. Consider the binomial random variable

$$X = \text{the number of successes.}$$

Since we want to obtain $Pr(E|H_0)$, the binomial distribution of X is determined by taking $p = \frac{1}{2}$. Therefore,

$$Pr(E|H_0) = Pr(59 \leq X \leq 100).$$

Let us now compute $Pr(59 \leq X \leq 100)$.

Since $n = 100$ and $p = \frac{1}{2}$, we have $n \geq 25$, $np \geq 5$, and $n(1 - p) \geq 5$. Therefore we approximate $Pr(59 \leq X \leq 100)$ using the standard normal curve, as discussed in section 9.6.

The mean and standard deviation of X are $\mu = (100)(\frac{1}{2}) = 50$ and $\sigma = \sqrt{(100)(\frac{1}{2})(\frac{1}{2})} = 5$. Since $(100.5 - 50)/5 = 10.1$, we have

$$Pr(59 \leq X \leq 100) \approx \text{area under the curve to the right of } z$$

where
$$z = \frac{58.5 - 50}{5} = 1.7,$$

as shown in figure 9.35. Therefore,

$$Pr(59 \leq X \leq 100) \approx 0.5000 - 0.4554$$
$$= 0.0446.$$

We conclude that $Pr(E|H_0) < 0.05$. Thus we reject the null hypothesis that the coin

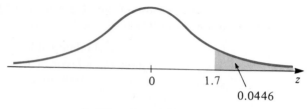

FIGURE 9.35

$$Pr(59 < X < 100) \approx 0.0446$$

is unbiased in favor of the alternate hypothesis that the coin favors heads, at the 5% level of significance.

The preceding example illustrates just one special type of hypothesis testing. We will not go into the general theory, but will limit our discussions to binomial experiments.

Let us now summarize the preceding example.

Structure for Hypothesis Testing

1. Determine the hypothesis to be tested, which is called the **null hypothesis** and is denoted by H_0.

2. Determine the hypothesis of the experimenter from his suspicions and H_0, which is called the **alternate hypothesis** and is denoted by H_1.

3. Choose a **level of significance** 100α percent. Two common choices are $\alpha = 0.05$ and $\alpha = 0.01$.

4. Obtain a **test statistic** from the performance of a binomial experiment.

5. Determine the event E from H_1 and the test statistic.

Criteria for Making the Decision about H_0

1. If $Pr(E|H_0) < \alpha$, reject H_0 in favor of H_1.

2. If $Pr(E|H_0) \geq \alpha$, accept H_0 over H_1.

EXAMPLE 1

Recall the examples concerning a drug company's claims that its new skin creme cures acne 80% of the time. Dr. Snyder has some doubts about the validity of this claim and suspects that the new creme cures acne less than 80% of the time. Let us make a structure for a test of the company's claim.

Dr. Snyder wishes to test the claim that the creme cures acne 80% of the time, so we take as the null hypothesis

$$H_0 : p = 0.8.$$

Dr. Snyder's position is that the creme is less than 80% effective; therefore, the alternate hypothesis is

$$H_1 : p < 0.8.$$

A level of significance must be chosen; let us pick $\alpha = 0.05$.

With these preliminaries established, Dr. Snyder can proceed to testing the creme on his patients. Suppose he tries the creme on 115 patients and finds that 84 of the 115 are cured of acne. This test statistic and Dr. Snyder's position that the creme is not as effective as claimed give us the event E of 84 or fewer patients cured by the creme. Thus, the criterion used in the test is

if $Pr(E|H_0) = Pr(0 \leq X \leq 84)$ is less than 0.05, then Dr. Snyder can reject the company's claim in favor of his hypothesis.

We use the standard normal curve to approximate $Pr(0 \leq X \leq 84)$ and obtain $Pr(0 \leq X \leq 84) = 0.0401$. Since $Pr(E|H_0) < 0.05$, Dr. Snyder rejects the company's claim.

We now develop an alternate way of approximating a probability $Pr(a \leq X \leq b)$ as used in testing a hypothesis. Consider again $Pr(59 \leq X \leq 100)$ from the example concerning the biased coin.

From table II in appendix B, we see that the area under the standard normal curve to the right of $z = 1.64$ is 5% of the total area. (More precisely, it is 5.05% of the total area, but for our purposes, we round it to 5%.) Therefore, as shown in figure 9.36,

$$Pr(59 \leq X \leq 100) < 0.05$$

if and only if

$$z_1 = \frac{58.5 - 50}{5} > 1.64.$$

Since $z_1 = 1.7$, we can conclude that $Pr(59 \leq X \leq 100) < 0.05$, and therefore we reject H_0. This method is called a **one-tailed test** as it involves one "tail" of the standard normal curve.

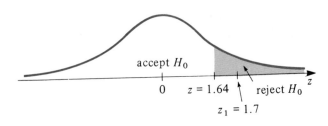

One-tailed test for $a = 0.05$

FIGURE 9.36

Let us change the coin example slightly. We still want to test the hypothesis $H_0 : p = \frac{1}{2}$ at the 5% level of significance, but suppose we have no suspicion this time as to whether the coin favors heads or tails. In this case it is appropriate to change the alternate hypothesis to

$$H_1 : p \neq \frac{1}{2}.$$

Suppose the test statistic is the same as before—59 heads were obtained when the coin was flipped 100 times.

The formulation of the event E must take our new alternate hypothesis into account. Assuming the coin is unbiased, then 59 heads is 9 from the expected number $\mu = 50$. Therefore E is the event of deviating 9 or more from the mean of 50 heads. In terms of the binomial variable $X =$ the number of heads, we have

$$Pr(E|H_0) = Pr(X \leq 41 \text{ or } X \geq 59).$$

To determine if $Pr(E|H_0) < 0.05$, we apply a **two-tailed test**, which we describe for $\alpha = 0.05$.

As shown in figure 9.37, the area to the left of $z = -1.96$ plus the area to the right of $z = 1.96$ is 5% of the area under the standard normal curve. Setting

$$z_1 = \frac{58.5 - 50}{5} = 1.7 \quad \text{and} \quad z_2 = \frac{41.5 - 50}{5} = -1.7 = -z_1$$

we have the following two-tailed test:

$$Pr(E|H_0) = Pr(X \leq 41 \text{ or } X \geq 59) < 0.05$$

if and only if either $z_1 < -1.96$ or $z_1 > 1.96$. Since $z_1 = 1.7$, then $Pr(E|H_0) \geq 0.05$ and we accept H_0 over our new position H_1.

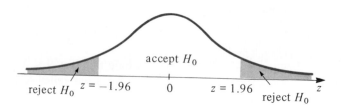

FIGURE 9.37

Two-tailed test for $\alpha = 0.05$

The choice to use either a one-tailed or two-tailed test depends entirely on the alternate hypothesis. In the coin problem, when H_1 was $p > \frac{1}{2}$, we used a one-tailed test; when H_1 was $p \neq \frac{1}{2}$, we used a two-tailed test.

In general, to test a hypothesis concerning the probability p_0 of success in a binomial experiment, we will take the alternate hypothesis to be one of three types:

1. $H_1 : p < p_0$, in which case we use a one-tailed test as in figure 9.38a.
2. $H_1 : p > p_0$, in which case we use a one-tailed test as in figure 9.38b.
3. $H_1 : p \neq p_0$, in which we use a two-tailed test as in figure 9.38c.

Furthermore, if we are testing a hypothesis at the 5% level of significance, then the shaded regions in figure 9.35a, b, and c are determined by $z = -1.64$, $z = 1.64$, and $z = \pm 1.96$, respectively.

We conclude this section with further examples of hypothesis testing as it relates to binomial experiments.

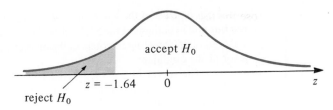

(a) One-tailed test for $\alpha = 0.05$

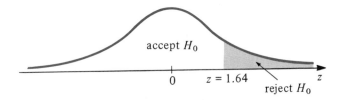

(b) One-tailed test for $\alpha = 0.05$

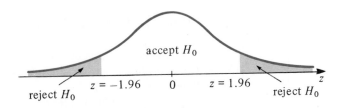

(c) Two-tailed test for $\alpha = 0.05$

FIGURE 9.38

EXAMPLES

2. Consider again Dr. Snyder's test as given in the previous example. Let us now use a one-tailed test to show that the company's claim should be rejected.

Since Dr. Snyder's alternate hypothesis is

$$H_1 : p < 0.8$$

we use the one-tailed test as shown in figure 9.38a.

Using the test statistic that 84 out of 115 of Dr. Snyder's patients were cured of acne when treated with the new skin creme, we have

$$z_1 = \frac{84.5 - (115)(0.8)}{\sqrt{(115)(0.8)(0.2)}}$$

$$= \frac{84.5 - 92}{4.29} = -1.75.$$

Since $z_1 < -1.64$, we reject the company's claim in favor of Dr. Snyder's position that the creme is less than 80% effective.

3. *Voter Preference.* A candidate believes that more than two-thirds of the registered voters in her district will vote for her in the next election. We can take the candidate's conjecture as an alternate hypothesis to the null hypothesis that

two-thirds or fewer of the voters in the district will vote for the candidate. Let us test this null hypothesis at the 5% level of significance, using the test statistic that 1702 out of 2500 registered voters in the district were found by a survey to be in favor of the candidate.

The survey of 2500 registered voters can be modeled as a binomial experiment with 2500 trials, where success on any given trial means the voter favors the candidate. If p denotes the probability of success on any given trial, then the null hypothesis can be stated as

$$H_0 : p \leq \frac{2}{3}$$

and the alternate hypothesis as

$$H_1 : p > \frac{2}{3}.$$

Thus we use the one-tailed test for $\alpha = 0.05$, as shown in figure 9.38b.

It suffices to take $p = \frac{2}{3}$ to test the null hypothesis. The mean and standard deviation for $X =$ the number of successes is

$$\mu = (2500)\left(\frac{2}{3}\right) = 1667 \quad \text{and} \quad \sigma = \sqrt{(2500)\left(\frac{2}{3}\right)\left(\frac{1}{3}\right)} = 23.57.$$

In terms of the random variable X, we need to determine $Pr(X \geq 1702)$. Applying the one-tailed test for $\alpha = 0.05$, as in figure 9.38b, we determine from

$$z_1 = \frac{1701.5 - 1667}{23.57} = 1.46 < 1.64$$

that the null hypothesis should be accepted over the candidate's hypothesis. Therefore, at the 5% level of significance, the candidate should now believe that two-thirds or fewer of the voters in her district will vote for her in the next election.

4. *Reliability of a Production Process.* A company that makes flashbulbs wants to check the quality of its production process. The company hopes that less than 7% of the flashbulbs produced are defective, which can be taken as an alternate to the hypothesis that 7% or more of the flashbulbs being produced are defective. At the 5% level of significance, should the null hypothesis be rejected if 300 flashbulbs are selected at random from the production line and it is found that 16 of them are defective?

The random selection of 300 flashbulbs can be modeled as a binomial experiment with success meaning that the flashbulb selected is defective. If p denotes the probability of success on any given trial, then we have for the null hypothesis

$$H_0 : p \geq 0.07.$$

Since the alternate hypothesis is

$$H_1 : p < 0.07$$

we use the one-tailed test for $\alpha = 0.05$ as shown in figure 9.38a.

We take $p = 0.07$ to test the null hypothesis. The mean and the standard variation for the random variable $X =$ the number of successes is

$$\mu = (300)(0.07) = 21 \quad \text{and} \quad \sigma = \sqrt{(300)(0.07)(0.93)} = 4.42.$$

Using the test statistic that 16 out of 300 flashbulbs were found to be defective and the alternate hypothesis $p < 0.07$, we have

$$Pr(E|H_0) = Pr(X \leq 16).$$

Since

$$z_1 = \frac{16.5 - 21}{4.42} = -1.02 > -1.64$$

we have from the one-tailed test that H_0 should be accepted at the 5% level of significance. Therefore the company decides that 7% or more of the flashbulbs being produced are defective.

EXERCISES 9.7

---

1. Joan wants to test the "honesty" of a coin, although she has no suspicion as to whether the coin favors heads or tails. Letting p denote the probability of getting a head on one flip of the coin, Joan tests the hypothesis $H_0 : p = \frac{1}{2}$. She flips the coin 125 times and records 52 heads. At the 5% significance level should Joan accept the null hypothesis?

2. Marty has a coin that he suspects favors tails when it is flipped. If p denotes the probability of getting a tail on one flip of the coin, Marty tests the hypothesis $H_0 : p = \frac{1}{2}$ where the alternate hypothesis is $H_1 : p > \frac{1}{2}$. He flips the coin 76 times and records 44 tails. At the 5% significance level should Marty accept the null hypothesis?

3. Mr. Koehler buys a new insecticide that is supposed to kill on contact at least 75% of the larva of a moth that feeds on the ears and stalks of corn. To test this claim, he sprays a section of his cornfield. Upon inspection, he counts 721 dead larva out of a total of 996. At the 5% significance level, should Mr. Koehler reject the claim?

4. Richard wants to test the "honesty" of a coin, although he has no suspicion as to whether the coin favors heads or tails. Letting p denote the probability of getting a head on one flip of the coin, Richard tests the hypothesis $H_0 : p = \frac{1}{2}$ where the alternate hypothesis is $H_1 : p \neq \frac{1}{2}$. He flips the coin 150 times and records 88 heads. At the 5% significance level should Richard accept the null hypothesis?

5. Mr. Ryan believes that more than 70% of the registered voters in his district will vote for him in the next election. Let p denote the probability that a voter selected at random favors the candidate. He wants to test the null hypothesis $H_0 : p \leq 0.7$ where the alternate hypothesis is $H_1 : p > 0.7$. A poll was conducted with the result that 2149 out of 3000 registered voters in the district favored the candidate. At the 5% significance level should Mr. Ryan accept the null hypothesis?

6. Professor Swann believes that less than half of the students at his college are moderate drinkers. Let p denote the probability that a student selected at random is a moderate drinker. The professor wants to test the null hypothesis $H_0 : p \geq 0.5$ where the alternate hypothesis is $H_1 : p < 0.5$. He conducts a poll with the result that 122 out of 280 students

indicate that they are moderate drinkers. At the 5% significance level should the professor reject the null hypothesis?

7. Four Seasons Nursery claims that at least 95% of its black raspberry bushes will survive transplanting. Bruce buys 80 black raspberry bushes and 8 do not survive. At the 5% significance level, should Bruce reject the nursery's claim?

8. Reflections Flower and Garden Center claims that at least 95% of its seeds will germinate. Jodi buys 118 seeds and 108 germinate. At the 5% significance level, should she reject the company's claim?

9. Quality Drug Company is testing a new decongestant that the company hopes will relieve stuffy noses at least 85% of the time. The new drug is tested on 2000 people, and 1724 indicate that it is effective. Is there evidence at the 5% significance level for the company to accept the position that the new decongestant is effective at least 85% of the time?

10. A drug company claims that its new antibiotic produces unwanted side effects in at most 1 out of 10 people. Dr. Long tests the new antibiotic on 150 patients, and 23 of them have unwanted side effects. At the 5% significance level, should Dr. Long reject the drug company's claim?

11. The Newhope Company claims that no more than 5% of its automatic citrus juicers are defective. A department store orders 300 juicers and finds that 23 of them are defective. At the 5% significance level, should the department store accept the company's claim?

12. The Alpine Corporation makes jackhammers and hopes that at most 8% of its product is defective. From the assembly line 1050 jackhammers are selected at random and it is determined that 67 of them are defective. At the 5% significance level, should the company accept the position that at most 8% of its product is defective?

13. The market analysis department of Horbocker brewery wants to determine if people can identify their beer from other brands. Two hundred beer drinkers are selected at random and each is given four glasses of beer, only one of which is Horbocker. After tasting all four drinks, 63 correctly identify the glass containing Horbocker beer. At the 5% significance level, are these results greater than expected by chance?

14. A drug company claims that their new product WORM-GONE is effective in eliminating tapeworms in cats at least 96% of the time. Dr. Wood tests the new product on 112 cats and finds the new product effective on 104 of them. At the 5% significance level, should Dr. Wood reject the company's claim?

15. Devise a one-tailed test for $\alpha = 0.01$ similar to the one-tailed test for $\alpha = 0.05$ as shown in figure 9.38b.

SUMMARY OF TERMS AND FORMULAS

Random Variable
  Values of a Random Variable
  Finite Random Variables
  Continuous Random Variables
  Bar Graph
  Payoff

Normal Random Variables
  Distribution Curves
  Normal Curves
  Center of Symmetry
  The Standard Normal Curve
  The Standard Normal Random Variable
  Conversion Rule

Expected Value
  Arithmetic Average (Mean)
  Frequency Table
  Probability Distribution
  Fair Game

Binomial Random Variable
  Bernoulli Trials
  Binomial Random Variable

Variable and Standard Deviation
  Measures of Dispersion

Normal Approximation of a Binomial Random Variable

Hypothesis Testing
  Null Hypothesis
  Alternate Hypothesis
  Test Statistic
  Level of Significance
  One-tailed Test
  Two-tailed Tests

Formulas
Consider a collection of numbers whose frequency table is given by

value x_i	x_1	x_2	$\cdots$	x_k
frequency f_i	f_1	f_2	$\cdots$	f_k

The **mean** of the collection of numbers is

$$\bar{x} = \frac{f_1 x_1 + f_2 x_2 + \cdots + f_k x_k}{n}$$

where $$f_1 + f_2 + \cdots + f_k = n.$$

The **variance** of the collection of numbers is

$$s^2 = \frac{f_1 x_1^2 + f_2 x_2^2 + \cdots + f_k x_k^2}{n} - \bar{x}^2.$$

The **standard deviation** of the collection of numbers is

$$s = \sqrt{s^2}.$$

Let X be a random variable whose probability distribution is given by

value of $X(x_i)$	x_1	x_2	$\cdots$	x_k
$Pr(X = x)(p_i)$	p_1	p_2	$\cdots$	p_k

The **expected value** (or **mean**) of X is

$$\mu = E(X) = p_1 x_1 + p_2 x_2 + \cdots + p_k x_k.$$

The **variance** of X is

$$\sigma^2 = p_1(x_1)^2 + p_2(x_2)^2 + \cdots + p_k(x_k)^2 - \mu^2.$$

The **standard deviation** of X is

$$\sigma = \sqrt{\sigma^2}.$$

If X is the number of successes in n trials of a binomial experiment where p is the probability of success and q is the probability of failure on any trial, then the probability of exactly x successes is given by

$$Pr(X = x) = C[n, x] p^x q^{n-x}.$$

REVIEW EXERCISES (CH. 9)

---

1. Give examples of three random variables that are discrete.

2. Give examples of three random variables that are continuous.

3. A box contains 11 fuses of which 6 are good and 5 are defective. Three fuses are drawn at random without replacement. Let X denote the number of defective fuses. Construct the probability distribution table of X.

4. An urn contains 2 red balls and 4 blue balls. Three balls are drawn at random in succession with replacement. Let X denote the number of red balls drawn.

   a. Construct a table that gives the probability distribution of X.

   b. Construct a bar graph for the probability distribution of X.

5. Consider an experiment with a finite discrete random variable X with the probability distribution of X given in the following table.

value of $X(x_i)$	250	-115	315	452	-23
$Pr(X = x_i)(p_i)$	0.150	0.321	0.209	0.010	0.310

   a. Find the mean of X.

   b. Find the variance of X.

   c. Find the standard deviation of X.

6. An unbiased coin is flipped three times. What is the expected number of heads?

7. Suppose we play the following game with another person. An unbiased die is rolled. If the number on the upturned face is even, our opponent pays us that amount in dollars. If the number on the upturned face is odd, we pay our opponent that amount in dollars. What is our expected value? Is the game fair?

8. Use formula (4) to find $Pr(X = x)$ for the following values of n, x, and p.

   a. $n = 5$, $x = 3$, $p = 0.6$

   b. $n = 7$, $x = 5$, $p = 0.85$

9. Use table I in appendix B to find the following binomial probabilities.

   a. $Pr(X = 12)$, $n = 15$, $p = 0.75$

   b. $Pr(X \geq 5)$, $n = 10$, $p = 0.4$

   c. $Pr(X \leq 7)$, $n = 18$, $p = 0.6$

10. Suppose that 20% of all items produced by an assembly line are defective. If 20 items are selected at random from this assembly line, what is the probability that

    a. exactly 4 will be defective?

    b. at most 4 will be defective?

    c. at most 15 will not be defective?

11. It is estimated that 80% of all zucchini squash seeds sold by a mail-order seed company will germinate. If Gene buys a packet of 15 zucchini squash seeds, what is the probability that

    a. at least 12 seeds will germinate?

    b. exactly 5 will not germinate?

12. Let X be a normal random variable with $\mu = 37$ and $\sigma = 4.56$. Find the following

probabilities.

a. $Pr(X < 29)$

b. $Pr(32 < X < 45)$

c. $Pr(X > 33)$

13. If the life in hours of Richmore floodlights is normally distributed with a mean of 800 hours and a standard deviation of 45 hours, find the probability that a floodlight selected at random will last

a. less than 725 hours.

b. between 750 and 900 hours.

c. more than 875 hours.

14. The amount (in milligrams) of tar per cigarette of a particular brand is normally distributed with a mean of 5 mg and a standard deviation of 0.1 mg. Find the probability that a cigarette selected at random will have

a. less than 4.85 mg of tar.

b. between 4.95 and 5.15 mg of tar.

c. more than 5.25 mg of tar.

15. Let X be a binomial random variable with $n = 1055$ and $p = 0.6$. Use a normal approximation to find the following probabilities.

a. $Pr(X \leq 615)$

b. $Pr(625 \leq X \leq 645)$

c. $Pr(\text{either } X \leq 610 \text{ or } X \geq 650)$

16. Suppose that 15% of all items produced by an assembly line are defective. If 700 items are selected at random from this assembly line, use a normal approximation to find the probability that

a. 84 or fewer items are defective.

b. between 100 and 110 items, inclusive, are defective.

c. 125 or more items are defective.

17. From past experience, it is known that 12% of all microwave ovens of a particular brand will not outlast the warranty. If a store sells 600 of these microwave ovens, use a normal approximation to find the probability that the store will eventually replace

a. between 60 and 80 ovens, inclusive.

b. no more than 55 ovens.

c. 90 or more ovens.

18. If the weight of pigs sent to the Midlands slaughterhouse is normally distributed with a mean of 205 pounds and standard deviation of 26 pounds, find

a. the percentage of pigs that weigh less than 200 pounds.

b. the percentage of pigs that weigh between 215 and 275 pounds.

c. the percentage of pigs that weigh more than 300 pounds.

19. Beth is playing dice with her friends. She feels that one of the dice is biased. She suspects that the number one shows on the upturned face more than it should if the die were unbiased. If p denotes the probability of getting a one on the upturned face of the die when it is rolled, Beth tests the hypothesis $H_0 : p = \frac{1}{6}$ where the alternate hypothesis is $H_1 : p > \frac{1}{6}$. She rolls the die 200 times and records 42 ones. At the 5% significance level, should Beth accept the null hypothesis?

CHAPTER 10

Decision Making: Markov Chains and Matrix Games

Introduction

In chapter 6 we studied the use of linear programming in decision-making situations. Linear programming models are deterministic in the sense that the consequence of the decision is completely known, given the assumptions that lead to the model. In chapters 8 and 9 we used probability models to analyze a decision-making situation under conditions of uncertainty. In this case the result of a decision is not certain, but it is predictable in a probabilistic sense. We saw how the expected value could be used to evaluate the result of a decision. In this chapter, we study two decision-making situations that involve matrices and probability. The first situation deals with systems in transition, and the second deals with situations that involve conflict of interest.

The systems in transition that we will study are called **Markov chains**. We live in a world of systems in transition. Many systems move among a finite number of describable states. For example, a particular machine can be either operative or inoperative at any moment. The number of cars waiting in line at a car wash changes as time progresses. In genetics, the percentage of offspring that retain a given characteristic changes from generation to generation. In the first two sections of this chapter, we will study these systems.

In the remaining two sections, we will study decision-making situations that involve conflict of interest. The mathematical term for this study is **game theory**. Conflict situations arise in many areas. For example, in the automobile industry there is competition among several companies. A particular company must make decisions regarding its product in hopes of increasing its share of the market in direct opposition to the objectives of the other companies. Recreational games such as poker and chess are also included in game theory.

An opponent need not be another person or a company. For example, the search for new energy sources can be thought of as a game with nature as the opponent. In competing against nature, a farmer makes decisions intended to increase crop yield, or a doctor makes decisions regarding the treatment of a patient.

In this chapter we consider Markov chains and matrix games where the structure can be studied using the mathematical tools developed in previous chapters.

10.1 Matrix Models of Markov Chains

A system that is moving from one state to another and in which future states can be predicted in a probabilistic sense can be viewed as a stochastic process. If, in addition, the probabilities of the next possible state of the system depend only on the present state of the system, and if these conditional probabilities do not change from one step of the process to the next, then the process is a **Markov chain**. We illustrate these points in the following situations.

A college is having difficulty with a new computer system. The system is either up and operative during an entire day, which we denote as state U, or it is down and inoperative sometime during the day, which we denote as state D. It is found that if the system is in state U one day, there is a 45% chance that it will be in state D the next day. If the computer system is in state D one day, there is a 65% chance that it will be in state U the next day.

The college's computer system is moving between the two states U and D from day to day, for which we have the conditional probabilities $Pr(D|U) = 0.45$ and $Pr(U|D) = 0.65$. The day-to-day transitions of the computer system can be viewed as a stochastic process. However, since the state of the computer system one day is dependent only on its state the previous day, and since the conditional probabilities are taken to be unchanging, we can say that the process is a Markov chain.

An example of a stochastic process that is not a Markov chain is found in example 7 of section 8.6 (page 378). Even though the tennis player in the example is moving between the states win (W) and lose (L) with each set, the conditional probabilities that she is in one of these states, given the previous state, change.

Returning to our computer system, we have, in addition to the conditional probabilities $Pr(D|U) = 0.45$ and $Pr(U|D) = 0.65$, the conditional probabilities

$$Pr(U|U) = 0.55 \quad \text{and} \quad Pr(D|D) = 0.35.$$

Both these probabilities are found by using the property

$$Pr(E|F) + Pr(E'|F) = 1.$$

We can display in a matrix the four conditional probabilities of the computer system. We label the rows of the matrix with the possible present states of the system and label the columns with the possible states of the system on the next day. Then the (U, D) entry of the matrix is the conditional probability that the system is in state D, given that it was in state U; that is, the (U, D) entry is $Pr(D|U)$. The complete matrix is

$$
\begin{array}{cc}
 & \begin{array}{cc} U & D \end{array} \\
\begin{array}{c} U \\ D \end{array} &
\begin{bmatrix} 0.55 & 0.45 \\ 0.65 & 0.35 \end{bmatrix}.
\end{array}
$$

We formalize these concepts in the following definition.

Definition

■ A **Markov chain** is a stochastic process in which a single experiment with possible outcomes $e_1, e_2, \ldots, e_m$, called **states**, is repeated several times. Furthermore, the probability that an outcome e_j occurs at a particular stage is dependent only on the outcome e_i that occurred in the immediately prior stage of the process, and this probability does not change from stage to stage. This conditional probability $Pr(e_j|e_i) = p_{ij}$ is called a **transition probability**. The transition probabilities can be conveniently displayed in the $m \times m$ matrix.

$$
\begin{array}{c}
\text{next state} \\
\begin{array}{c}
\begin{array}{c} \\ \text{present} \\ \text{state} \end{array}
\begin{array}{c} e_1 \\ e_2 \\ \vdots \\ e_m \end{array}
\begin{bmatrix}
\begin{array}{cccc} e_1 & e_2 & \cdots & e_m \end{array} \\
p_{11} & p_{12} & \cdots & p_{1m} \\
p_{21} & p_{22} & \cdots & p_{2m} \\
\vdots & & & \vdots \\
p_{m1} & p_{m2} & \cdots & p_{mm}
\end{bmatrix},
\end{array}
\end{array}
$$

called the **transition matrix** of the Markov chain. ■

Observe that an entry of a transition matrix is the probability of moving from a given state (row heading) to another state (column heading), so it is a number between zero and one, inclusive. Furthermore, since the only states that the system can move among are the possibilities $e_1, e_2, \ldots, e_m$, the sum of the entries in each row is 1.

EXAMPLES

1. *Behavioral Pattern.* Suppose a student has an 8 o'clock class and the instructor observes the following behavior pattern. Whenever the student is present (P) one day, there is only a 50% chance that he will be present the next day. However, if the student is absent (A) one day, he makes a greater effort to attend class and the chance of his being present the next day is 70%. Let us model this process as a Markov chain and give the transition matrix.

First, we view the student as a system moving between the two states P and A. Furthermore, the state of the student with respect to a day depends on his state on the previous day. The conditional probabilities remain unchanged from one day to the next. Thus we are dealing with a Markov chain.

From the information given, we have the conditional probabilities $Pr(P|P) = 0.50$ and $Pr(P|A) = 0.70$. From these probabilities we obtain

$$Pr(A|P) = 1 - Pr(P|P) = 0.50 \quad \text{and} \quad Pr(A|A) = 1 - Pr(P|A) = 0.30.$$

With these four conditional probabilities we give the transition matrix as

state the next day

$$
\begin{array}{cc}
 & \begin{array}{cc} P & A \end{array} \\
\begin{array}{c} \text{state} \\ \text{one day} \end{array} \begin{array}{c} P \\ A \end{array} & \begin{bmatrix} 0.50 & 0.50 \\ 0.70 & 0.30 \end{bmatrix}.
\end{array}
$$

2. *Brand Switching.* A survey is taken of homemakers to determine their buying habits with respect to three detergents, X, Y, and Z. The results of this survey indicate that 60% of the homemakers who used detergent X bought detergent X again the next month, whereas 20% changed to detergent Y and 20% changed to detergent Z. Furthermore, 10% of those who used detergent Y changed to detergent X, whereas 70% stayed with detergent Y and 20% changed to detergent Z. Finally, 40% of the users of detergent Z changed to detergent X, whereas 20% changed to detergent Y and 40% stayed with detergent Z. Let us give the transition matrix for this process.

First, let X denote the state of buying brand X one month, Y denote the state of buying brand Y, and Z denote the state of buying brand Z. We consider a homemaker selected at random as a system moving among the states X, Y, and Z from month to month.

From the information given, we have the conditional probabilities

$$Pr(X|X) = 0.60 \qquad Pr(Y|X) = 0.20 \qquad Pr(Z|X) = 0.20$$

$$Pr(X|Y) = 0.10 \qquad Pr(Y|Y) = 0.70 \qquad Pr(Z|Y) = 0.20$$

$$Pr(X|Z) = 0.40 \qquad Pr(Y|Z) = 0.20 \qquad Pr(Z|Z) = 0.40.$$

Therefore, the transition matrix is

brand next month

$$
\begin{array}{cc}
 & \begin{array}{ccc} X & Y & Z \end{array} \\
\begin{array}{c} \text{present} \\ \text{brand} \end{array} \begin{array}{c} X \\ Y \\ Z \end{array} & \begin{bmatrix} 0.60 & 0.20 & 0.20 \\ 0.10 & 0.70 & 0.20 \\ 0.40 & 0.20 & 0.40 \end{bmatrix}
\end{array}
$$

One application of Markov chains is to study the pattern of changing states of a system as time progresses. In particular, given initial states of the system, we wish to determine the probabilities of the possible future states of the system.

Definition ■ Consider a Markov chain with m possible states $e_1, e_2, \ldots, e_m$. Let $P_i^{(k)}$ denote the probability that the system is in state e_i at the kth stage of the chain. The $1 \times m$ matrix

$$S^{(k)} = [P_1^{(k)}, P_2^{(k)}, \ldots, P_m^{(k)}]$$

is called a **state distribution vector** at the kth stage. For $k = 0$, $P_i^{(0)}$ is the probability that the system starts from state e_i, and $S^{(0)}$ is called an **initial state distribution vector**. ■

EXAMPLE 3

Consider again the brand-switching situation described in example 2. Suppose for a certain group of homemakers 30% now use brand X, 40% use Y, and 30% use Z. Suppose a homemaker is selected at random from this group. Let us find the probabilities that this homemaker is in states X, Y, and Z after one month. Let us also find these probabilities after two months of transition.

If a homemaker is selected at random, the probability that this person is presently buying brand X is 0.30. The probabilities that this person is presently buying brand Y, and is presently buying brand Z are, respectively, 0.40 and 0.30. Thus the initial state distribution vector for this person selected at random is

$$\begin{array}{ccc} X & Y & Z \end{array}$$
$$S^{(0)} = [0.3 \quad 0.4 \quad 0.3].$$

The entries of $S^{(1)}$ are the probabilities that the person is in one of the three states after one month. To see how to obtain this state distribution vector, consider the probability tree in figure 10.1.

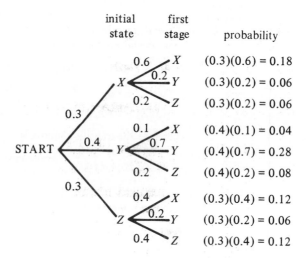

FIGURE 10.1

We can view the person as entering the chain through the point "START" in the tree. The person would then be at the initial state X, Y, or Z with probabilities 0.3, 0.4, or 0.3, respectively. The person then moves from the initial states. The probability that the person moves to state X through the initial state X is $(0.3)(0.6) = 0.18$. The probability that the person goes to state X from the initial state Y is $(0.4)(0.1) = 0.04$. Finally, the probability that the person moves to state X from Z is $(0.3)(0.4) = 0.12$. Therefore the probability that the person will be in state X after one month is the sum of these three probabilities. This sum can be expressed as follows:

$$[0.3 \quad 0.4 \quad 0.3]\begin{bmatrix} 0.6 \\ 0.1 \\ 0.4 \end{bmatrix} = 0.18 + 0.04 + 0.12 = 0.34. \tag{1}$$

Observe that the row–column product in (1) is the initial state distribution vector $S^{(0)}$ times the first column of the transition matrix P given in example 2.

In a like manner, we obtain the probability that the homemaker is in state Y after one month by the product

$$[0.3 \quad 0.4 \quad 0.3] \begin{bmatrix} 0.2 \\ 0.7 \\ 0.2 \end{bmatrix} = 0.06 + 0.28 + 0.06 = 0.40. \qquad (2)$$

This can be verified by considering the probability tree in figure 10.1.

Finally, the probability that the person is in state Z after one month is the vector $S^{(0)}$ times the Z column of the transition matrix P, as follows:

$$[0.3 \quad 0.4 \quad 0.3] \begin{bmatrix} 0.2 \\ 0.2 \\ 0.4 \end{bmatrix} = 0.26. \qquad (3)$$

We can combine the row–column products (1), (2), and (3) in the one matrix product

$$[0.3 \quad 0.4 \quad 0.3] \begin{bmatrix} 0.6 & 0.2 & 0.2 \\ 0.1 & 0.7 & 0.2 \\ 0.4 & 0.2 & 0.4 \end{bmatrix} = \begin{matrix} X & Y & Z \\ [0.34 & 0.40 & 0.26] \end{matrix}$$

which is the initial state distribution vector $S^{(0)}$ times the transition matrix P and which is the state distribution vector $S^{(1)}$. Thus we have

$$S^{(1)} = S^{(0)}P.$$

To obtain $S^{(2)}$, which gives the probabilities that the person is in one of the three states after two months, we can view $S^{(1)}$ as an initial state distribution vector and compute $S^{(2)}$ in a manner similar to that used to compute $S^{(1)}$. Therefore $S^{(2)} = S^{(1)}P$, and we have

$$S^{(2)} = [0.34 \quad 0.40 \quad 0.26] \begin{bmatrix} 0.6 & 0.2 & 0.2 \\ 0.1 & 0.7 & 0.2 \\ 0.4 & 0.2 & 0.4 \end{bmatrix} = \begin{matrix} X & Y & Z \\ [0.348 & 0.400 & 0.252]. \end{matrix}$$

As seen in example 3, if we start with a state distribution vector S, the state distribution vector after a transition is S times the transition matrix P. For example,

$$S^{(1)} = S^{(0)}P \qquad \text{and} \qquad S^{(2)} = S^{(1)}P.$$

Furthermore, by substituting $S^{(0)}P$ for $S^{(1)}$ and using the associative property of matrix multiplication, we have

$$S^{(2)} = S^{(1)}P = (S^{(0)}P)P = S^{(0)}P^2.$$

We can continue this reasoning to obtain

$$S^{(3)} = S^{(2)}P = (S^{(0)}P^2)P = S^{(0)}P^3.$$

We summarize these observations in the following theorem.

Theorem 1 ■ Consider a Markov chain with transition matrix P. If $S^{(0)}$ is the initial state distribution vector, the state distribution vector after one transition is

$$S^{(1)} = S^{(0)}P.$$

In general, the state distribution vector $S^{(k)}$ after k transitions is

$$S^{(k)} = S^{(k-1)}P$$
$$= S^{(0)}P^k$$

where P^k is the kth power of the transition matrix. ■

We indicate in the next two examples an important interpretation of the matrix P^k.

EXAMPLES

4. Consider again the student and his 8 o'clock class as described in example 1. Let us determine the state distribution vector for the second day following a day on which the student was present.

Since we start with a day on which the student was present, we consider that the initial state of the student is present, and this state has probability 1. Therefore we take as the initial state distribution vector $S^{(0)} = [1 \quad 0]$.

Now, starting at $S^{(0)}$ we wish to find the state distribution vector after 2 transitions. Applying theorem 1, we have

$$S^{(2)} = S^{(0)}P^2.$$

For the transition matrix P, given in example 1, we have

$$P^2 = \begin{bmatrix} 0.5 & 0.5 \\ 0.7 & 0.3 \end{bmatrix} \begin{bmatrix} 0.5 & 0.5 \\ 0.7 & 0.3 \end{bmatrix} = \begin{bmatrix} 0.60 & 0.40 \\ 0.56 & 0.44 \end{bmatrix}.$$

Thus,

$$[1 \quad 0] \begin{bmatrix} 0.60 & 0.40 \\ 0.56 & 0.44 \end{bmatrix} = \overset{\textstyle P \quad\ A}{[0.60 \quad 0.40]}$$

which is the first row of P^2.

5. We continue with example 4, but this time we determine the state distribution vector for the second day following a day on which the student was absent.

In this case, we consider that the initial state of the student is absent, and this state has probability 1. Therefore we take as the initial state distribution vector $S^{(0)} = [0 \quad 1]$. The state distribution vector after two transitions is, in this case,

$$[0 \quad 1] \begin{bmatrix} 0.60 & 0.40 \\ 0.56 & 0.44 \end{bmatrix} = \overset{\textstyle P \quad\ A}{[0.56 \quad 0.44]},$$

which is the second row of P^2.

Theorem 2 ■ If P is the transition matrix of a Markov chain, then the (i, j) entry of the matrix P^k, the kth power of P, is the probability that the system moves from state e_i to state e_j in k transitions. ■

EXAMPLE 6

Population Transition. The U.S. population is divided into two sectors, urban (U) and nonurban (N). Let us assume that over a single time period (one year) 15% of the urban population moves into nonurban areas. Thus, if a person is chosen at random from the urban population, the chance that he will move to a nonurban area is 0.15. Also, let us assume that over a single period 5% of the nonurban population moves into urban areas. We will find the probability that an individual lives in an urban area and the probability that an individual lives in a nonurban area after three years.

The transition matrix of the Markov chain is

$$P = \begin{matrix} U \\ N \end{matrix} \begin{bmatrix} 0.85 & 0.15 \\ 0.05 & 0.95 \end{bmatrix}.$$

It can be verified that

$$P^3 = \begin{matrix} U \\ N \end{matrix} \begin{bmatrix} 0.634 & 0.366 \\ 0.122 & 0.878 \end{bmatrix}.$$

By theorem 2 we can interpret the first row [0.634 0.366] of P^3 as the probabilities of the states a person can be in after three years if the person was initially living in an urban area. Observe that this matrix of probabilities is also obtained by the product

$$[1 \quad 0] \begin{bmatrix} 0.634 & 0.366 \\ 0.122 & 0.878 \end{bmatrix} = [0.634 \quad 0.366].$$

The probabilities of the states of a person who was initially living in a nonurban area are obtained from the product

$$[0 \quad 1] \begin{bmatrix} 0.634 & 0.366 \\ 0.122 & 0.878 \end{bmatrix} = [0.122 \quad 0.878].$$

We conclude this section with two major applications of Markov chains.

Queuing Theory

We now illustrate an application of Markov chains to queuing (waiting-in-line) theory. Many examples of queuing theory abound in everyday life. People wait in line at supermarket check-outs and at bank teller windows. People sit in doctors' and dentists' waiting rooms. We have simplified the assumptions in the following illustration, but it gives you an idea of the nature of queuing theory.

Suppose that customers arrive at a drive-in bank window at random. The approach lane can handle at most 4 cars at a time. Suppose that every 5 minutes at most one car arrives and at most one customer completes the transaction with the teller and leaves. Suppose also that the probability that a car arrives in a 5-minute time interval is $\frac{2}{3}$ and the probability that a customer finishes the transaction with the teller and leaves is $\frac{3}{4}$. We are assuming that the events of cars arriving and cars departing are independent.

We can model this situation as a Markov chain where the states of the process are the number of cars waiting in line. We observe the line at 5-minute intervals thereby generating a stochastic process. We now construct the transition matrix for this Markov chain.

There are five states for this chain: 0, 1, 2, 3, or 4 cars waiting in line. Therefore the transition matrix P has the form

$$
\begin{array}{c}
\text{next state} \\
\begin{array}{ccccc}
0 & 1 & 2 & 3 & 4
\end{array} \\
\begin{array}{c}
\text{present} \\
\text{state}
\end{array}
\begin{array}{c}
0 \\
1 \\
2 \\
3 \\
4
\end{array}
\begin{bmatrix}
p_{11} & p_{12} & p_{13} & p_{14} & p_{15} \\
p_{21} & p_{22} & p_{23} & p_{24} & p_{25} \\
p_{31} & p_{32} & p_{33} & p_{34} & p_{35} \\
p_{41} & p_{42} & p_{43} & p_{44} & p_{45} \\
p_{51} & p_{52} & p_{53} & p_{54} & p_{55}
\end{bmatrix}.
\end{array}
$$

The "next state" is determined by the observation of the number of cars waiting in line 5 minutes from the "present state."

We next find the entries in this transition matrix. The number p_{11} is the probability that the process has a (static) transition from state 0 to state 0. That is, no car is waiting in line at the present state and no car is waiting in line 5 minutes later. This can happen in exactly one way. Namely, no car arrives during the 5-minute time interval. The probability of this happening is $\frac{1}{3}$, since the complementary event is that one car arrives.

Next, p_{12} is the probability that the process has a transition from state 0 to state 1. This can happen in exactly one way. Namely, one car arrives in the 5-minute time interval. Therefore $p_{12} = \frac{2}{3}$.

The next entry p_{13} is the probability that the process has a transition from state 0 to state 2. Since we are assuming that at most one car arrives during a 5-minute interval, $p_{13} = 0$. Similarly, p_{14} and p_{15} are also zero.

The entry p_{21} is the probability that the process changes from one car waiting in line to no cars waiting in line. If we let

$$E = \text{No cars arrive}$$

$$F = \text{One car leaves},$$

then p_{21} is the probability of the intersection of E and F. Furthermore, E and F are independent. Therefore,

$$
\begin{aligned}
p_{21} &= Pr(E \cap F) \\
&= Pr(E) \cdot Pr(F) \\
&= \left(\frac{3}{4}\right) \cdot \left(\frac{1}{3}\right) = \frac{1}{4}.
\end{aligned}
$$

The entry p_{22} is the probability that the process remains at one car waiting in line from the present state to the next state. This can happen in exactly two mutually exclusive ways. Either no car arrives and no car leaves or one car arrives and one car leaves. Therefore,

$$p_{22} = Pr(\text{no car arrives and no car leaves})$$
$$+ Pr(\text{one car arrives and one car leaves})$$
$$= Pr(\text{no car arrives}) \cdot Pr(\text{no car leaves})$$
$$+ Pr(\text{one car arrives}) \cdot Pr(\text{one car leaves})$$

$$= \left(\frac{1}{3}\right)\left(\frac{1}{4}\right) + \left(\frac{2}{3}\right)\left(\frac{3}{4}\right) = \frac{7}{12}.$$

The rest of the entries are found in similar fashion. We will ask you to find some of them in the exercise problems (see exercises 15 and 16). The finished transition matrix follows.

$$
\begin{array}{c}
\quad\quad\quad\quad\quad \text{next state} \\
\quad\quad\quad\quad 0 \quad 1 \quad 2 \quad 3 \quad 4 \\
\begin{array}{cc}
 & \begin{array}{c} 0 \\ 1 \\ \text{present} \quad 2 \\ \text{state} \quad 3 \\ 4 \end{array}
\end{array}
\left[
\begin{array}{ccccc}
\frac{1}{3} & \frac{2}{3} & 0 & 0 & 0 \\
\frac{1}{4} & \frac{7}{12} & \frac{1}{6} & 0 & 0 \\
0 & \frac{1}{4} & \frac{7}{12} & \frac{1}{6} & 0 \\
0 & 0 & \frac{1}{4} & \frac{7}{12} & \frac{1}{6} \\
0 & 0 & 0 & \frac{3}{4} & \frac{1}{4}
\end{array}
\right]
\end{array}
$$

Inheritance Models: An Application to Genetics

Markov chains can be used to model certain breeding experiments. Johann Mendel (1822–1884) investigated the inherited trait of seed color, yellow and green, in a particular variety of peas. In the Mendelian model, the appearance of a certain trait is determined by a pair of genes of types A and a. Each individual carries a pair of genes of the form

$$\text{AA, \quad Aa (or aA), \quad and \quad aa}$$

called *genotypes*. When two individuals (parents) mate, each offspring inherits a single gene from each parent. We will assume that the offspring has equal probability of $\frac{1}{2}$ of receiving either one of the two genes from a given parent. For example, suppose the first parent has genotype Aa and the second parent also has genotype Aa. The mating of these two parents, denoted by Aa × Aa, will result in offspring that again possess one of the three genotypes. We can exhibit this situation by a probability tree. (See figure 10.2.)

From the probability tree, we obtain the following information.

$$Pr(\text{an offspring has genotype AA}) = \frac{1}{4}$$

$$Pr(\text{an offspring has genotype Aa}) = \frac{1}{2}$$

$$Pr(\text{an offspring has genotype aa}) = \frac{1}{4}$$

	first parent	second parent	outcome	probability
	1/2 A	1/2 A	AA	1/4
Aa cross Aa		1/2 a	Aa	1/4
	1/2 a	1/2 A	aA	1/4
		1/2 a	aa	1/4

FIGURE 10.2

Note that when we formed the probability tree, we assumed that the selection of a gene from the first parent and the selection of a gene from the second parent were independent events.

Now, consider the following sequence of matings. We require that the first parent for each mating is of genotype Aa. For the first mating, we select a second parent at random from a population in which all genotypes are present. From this mating, we select an offspring at random and mate it with a parent of genotype Aa. We continue this process, always mating an offspring selected at random with a parent of genotype Aa. This process is a Markov chain with three states: the offspring selected for mating is of genotype AA, Aa, or aa.

We now construct the transition matrix for this process. We can fill in the second row from the information obtained from the probability tree given in figure 10.2.

$$
\begin{array}{c}
\text{offspring} \\
\begin{array}{ccc} AA & Aa & aa \end{array} \\
\begin{array}{c} \text{parents} \\ \text{Aa cross} \end{array}
\begin{array}{c} AA \\ Aa \\ aa \end{array}
\left[\begin{array}{ccc}
 & & \\
\frac{1}{4} & \frac{1}{2} & \frac{1}{4} \\
 & &
\end{array}\right]
\end{array}
$$

To complete this transition matrix, we form two more probability trees for the two matings Aa × AA and Aa × aa. (See figure 10.3.)

	first parent	second parent	outcome	probability
Aa × AA	1/2 A	1 A	AA	1/2
		0 a	Aa	0
	1/2 a	1 A	aA	1/2
		0 a	aa	0

	first parent	second parent	outcome	probability
Aa × aa	1/2 A	0 A	AA	0
		1 a	Aa	1/2
	1/2 a	0 A	aA	0
		1 a	aa	1/2

FIGURE 10.3

Using the information obtained from the two probability trees in figure 10.3, we complete the following transition matrix for the Markov chain.

offspring

$$
\begin{array}{c}
\text{parents} \\
\text{Aa cross}
\end{array}
\quad
\begin{array}{c}
 \\
\text{AA} \\
\text{Aa} \\
\text{aa}
\end{array}
\begin{array}{ccc}
\text{AA} & \text{Aa} & \text{aa}
\end{array}
\begin{bmatrix}
\frac{1}{2} & \frac{1}{2} & 0 \\
\frac{1}{4} & \frac{1}{2} & \frac{1}{4} \\
0 & \frac{1}{2} & \frac{1}{2}
\end{bmatrix}
$$

EXAMPLE 7

Refer to the Markov chain with the preceding transition matrix. Suppose that the original population from which the individual was selected to start the process was composed of 30% genotype AA, 45% genotype Aa, and 25% genotype aa. Find $S^{(1)}$ and $S^{(2)}$.

From the information given, the initial state distribution vector is

$$S^{(0)} = [0.3, 0.45, 0.25].$$

To find $S^{(1)}$, we use theorem 1 with $k = 1$.

$$S^{(1)} = S^{(0)} p$$

$$= [0.3, \quad 0.45, \quad 0.25]
\begin{bmatrix}
\frac{1}{2} & \frac{1}{2} & 0 \\
\frac{1}{4} & \frac{1}{2} & \frac{1}{4} \\
0 & \frac{1}{2} & \frac{1}{2}
\end{bmatrix}$$

$$= [0.2625, \quad 0.5, \quad 0.2375].$$

To find $S^{(2)}$, we use theorem 1 with $k = 2$.

$$S^{(2)} = S^{(1)}P$$

$$= [0.2625, \quad 0.5, \quad 0.2375]
\begin{bmatrix}
\frac{1}{2} & \frac{1}{2} & 0 \\
\frac{1}{4} & \frac{1}{2} & \frac{1}{4} \\
0 & \frac{1}{2} & \frac{1}{2}
\end{bmatrix}$$

$$= [0.25625, \quad 0.5, \quad 0.24375].$$

Note that in example 7, $S^{(1)}$ gives the distribution of genotypes one generation from the initial mating and $S^{(2)}$ gives the distribution of genotypes two generations from the initial mating. In the next section we investigate the long-term behavior of a Markov chain.

EXERCISES 10.1

1. State which of the following matrices can be the transition matrix for a Markov chain. For those that cannot be a transition matrix, give a reason why.

a. $\begin{bmatrix} \frac{1}{2} & \frac{1}{3} & \frac{1}{6} \\ 0 & \frac{3}{4} & \frac{1}{4} \end{bmatrix}$

b. $\begin{bmatrix} \frac{1}{5} & 0 & \frac{3}{5} & 1 \\ 1 & 0 & 0 & 0 \\ \frac{1}{3} & 0 & 0 & \frac{2}{3} \end{bmatrix}$

c. $\begin{bmatrix} 0.25 & 0.75 \\ 0.05 & 0.05 \end{bmatrix}$ d. $\begin{bmatrix} 0.01 & 0.99 \\ 0.54 & 0.46 \\ 0.38 & 0.62 \end{bmatrix}$

2. State which of the following matrices can be the transition matrix for a Markov chain. For those that cannot be a transition matrix, give a reason why.

   a. $\begin{bmatrix} 1 & 0 & 0 \\ \frac{1}{5} & 0 & \frac{4}{5} \\ 0.2 & 0 & 0.8 \end{bmatrix}$ b. $\begin{bmatrix} 0 & 0.67 & 0.33 \\ 0 & -0.01 & 1.01 \\ 1 & 0 & 0 \end{bmatrix}$

   c. $\begin{bmatrix} 0.81 & 0.09 & \frac{1}{10} \\ 0.45 & 0.19 & 36 \end{bmatrix}$ d. $\begin{bmatrix} 0.51 & 0.15 & 0.33 \\ 0.05 & 0.81 & 0.14 \end{bmatrix}$

3. The transition matrix P of a Markov chain with states A, B, and C is the following.

$$P = \begin{array}{c} \\ A \\ B \\ C \end{array} \begin{array}{c} \begin{array}{ccc} A & B & C \end{array} \\ \begin{bmatrix} 0.51 & 0.24 & 0.25 \\ 0.35 & 0.19 & 0.46 \\ 0.01 & 0.69 & 0.30 \end{bmatrix} \end{array}$$

   a. Compute P^2 and P^3.

   b. What is the probability that the system moves from state C to state A in one transition?

   c. What is the probability that the system moves from state B to state B in two transitions?

   d. What is the probability that the system moves from state A to state C in three transitions?

4. The transition matrix P of a Markov chain with states A, B, and C is the following.

$$P = \begin{array}{c} \\ A \\ B \\ C \end{array} \begin{array}{c} \begin{array}{ccc} A & B & C \end{array} \\ \begin{bmatrix} 0.05 & 0.81 & 0.14 \\ 0.51 & 0.37 & 0.12 \\ 0.29 & 0.29 & 0.42 \end{bmatrix} \end{array}$$

   a. Compute P^2 and P^3.

   b. What is the probability that the system moves from state B to state A in one transition?

   c. What is the probability that the system moves from state A to state C in two transitions?

   d. What is the probability that the system moves from state C to state C in three transitions?

5. Blake is studying the learning behavior of a rat. He performs the following experiment. He places a rat in a T-maze. The rat receives food if it turns right (R) and receives an electric shock if it turns left (L). The conjecture is that if the rat receives food on a given trial, it will turn right on the next trial with probability $\frac{3}{5}$. If the rat receives an electric shock on a given trial, it will turn right on the next trial with probability $\frac{7}{10}$.

   a. Give the transition matrix P for the Markov chain.

   b. Compute P^2.

c. What is the probability that the rat will turn right on the second trial following one in which it turned right?

d. What is the probability that the rat will turn left on the second trial following one in which it turned left?

6. In a survey the height of males is categorized as tall or short. The probability that a son of a tall father is also tall is 0.65 and the probability that a son of a short father is also short is 0.48. Consider the Markov chain of transitions of heights from generation to generation.

a. Give the transition matrix P for the Markov chain.

b. Compute P^2 and P^3.

c. What is the probability that the grandson of a tall father is tall?

d. What is the probability that the great-grandson of a short father is tall?

e. What is the probability that the great-grandson of a tall father is short?

7. A survey indicated that 72% of the children who have at least one parent with a college degree graduate from college, whereas 33% of children who have parents without college degrees graduate from college.

a. Give the transition matrix P for this Markov chain.

b. Compute P^2 and P^3.

c. What is the probability that the grandchild of parents who do not have college degrees graduates from college?

d. What is the probability that the great-grandchild of a parent who has a college degree does not graduate from college?

e. What is the probability that the great-grandchild of parents who do not have college degrees does not graduate from college?

8. Consider the transition matrix of population movement given in example 6 on page 469.

a. Suppose the distribution of population between the two areas is initially 20% in urban areas and 80% in nonurban areas. Compute the state distribution vectors $S^{(1)}$, $S^{(2)}$, $S^{(3)}$, $S^{(4)}$, and $S^{(5)}$.

b. Repeat part (a) where there are initially 30% in urban areas and 70% in nonurban areas.

c. Repeat part (a) where there are initially 25% in urban areas and 75% in nonurban areas.

9. A survey of cola drinkers showed that 58% of those who bought Black Star Cola one week bought it again the next week, whereas 43% of those who did not buy Black Star Cola one week did buy it the next week. Consider the Markov chain with the two states B and B', where

$$B = \text{a person buys Black Star Cola}$$

$$B' = \text{a person does not buy Black Star Cola.}$$

a. Give the transition matrix P for this Markov chain.

b. Suppose initially that 46% buy Black Star Cola and 54% do not buy Black Star Cola. Compute the state distribution vectors after one, two, and three weeks.

c. Repeat part (b) using the initial state distribution vector $S^{(0)} = [0.5 \quad 0.5]$.

d. Repeat part (b) using the initial state distribution vector $S^{(0)} = [0.2 \quad 0.8]$.

10. A survey is made of shoppers to determine their habits with respect to the two competing supermarkets I and II in Westfield. A person who shops in one of the markets one week will either go to that market again or switch to the other market the following week. The transition probabilities are given in the matrix

$$
\begin{array}{cc}
 & \text{I} \quad \text{II} \\
\begin{array}{c} \text{I} \\ \text{II} \end{array} &
\begin{bmatrix} 0.8 & 0.2 \\ 0.3 & 0.7 \end{bmatrix}.
\end{array}
$$

   a. Suppose the distribution of shoppers between the two markets is initially 55% in market I and 45% in market II. Compute the state distribution vectors after one, two, three, four, and five weeks.

   b. Repeat part (a) using the initial state distribution vector $S^{(0)} = [0.65 \quad 0.35]$.

   c. Repeat part (a) using the initial state distribution vector $S^{(0)} = [0.60 \quad 0.40]$.

11. A mouse is placed in one of the compartments of the maze shown in figure 10.4. A transition takes place when the mouse moves through a door from one compartment to another. For example, the transition probability of moving from compartment A to compartment A is zero. Furthermore, we assume that the mouse will move through any of the doors of a given compartment with equal probability.

   a. Give the transition matrix P for this Markov chain.

   b. Compute P^2 and P^3.

   c. If the mouse is placed in compartment C, what is the probability that it will be in compartment A after two transitions? after three transitions?

   d. If the mouse is placed in compartment D, what is the probability that it will be in compartment D after two transitions? after three transitions?

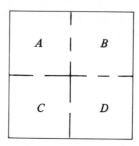

FIGURE 10.4

12. A mouse is placed in one of the four circular compartments of the maze shown in figure 10.5. A transition takes place when the mouse moves through a door from one compartment to another. For example, the transition probability of moving from compartment C to compartment C is zero. Furthermore, we assume that the mouse will move through any of the doors of a given compartment with equal probability.

   a. Give the transition matrix P for this Markov chain.

   b. Compute P^2 and P^3.

   c. If the mouse is placed in compartment B, what is the probability that it will be in compartment C after two transitions? after three transitions?

   d. If the mouse is placed in compartment C, what is the probability that it will be in compartment A after two transitions? after three transitions?

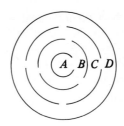

FIGURE 10.5

13. Two urns A and B each contain 2 balls. Two of the 4 balls are blue and 2 are white. At time $t = k (k = 1, 2, 3, \ldots)$ a ball is drawn at random from each urn. The ball drawn from urn A is placed in urn B and the ball drawn from urn B is placed in urn A. The states for this Markov chain are the number of blue balls in urn A.

   a. Give the transition matrix P for this Markov chain.

   b. Compute P^2 and P^3 for the matrix P given in part (a).

   c. Suppose initially urn A contains 1 blue ball. Find $S^{(0)}$, $S^{(1)}$, $S^{(2)}$, and $S^{(3)}$.

14. Two hats A and B contain paper money—three bills in hat A and two bills in hat B. Three of the five are $10 bills and two are $50 bills. At time $t = k (k = 1, 2, 3, \ldots)$ a bill is drawn at random from each hat. The bill drawn from hat A is placed in hat B and the bill drawn from hat B is placed in hat A. The states for this Markov chain are the number of $50 bills in hat A.

   a. Give the transition matrix P for this Markov chain.

   b. Compute P^2 and P^3 for the matrix P given in part (a).

   c. Suppose initially hat A contains no $50 bills. Find $S^{(0)}$, $S^{(1)}$, $S^{(2)}$, and $S^{(3)}$.

15. Refer to the transition matrix given on page 471. Verify the values given for p_{32}, p_{33}, p_{53}, and p_{54}.

16. Refer to the transition matrix given on page 471. Verify the values given for p_{43}, p_{44}, p_{52}, and p_{55}.

17. Shoppers arrive at a check-out counter at random in a department store. The line can hold at most five people at a time. In a two-minute interval at most one shopper arrives and at most one shopper is checked out and leaves. In a given time interval, suppose that the probability that a shopper arrives is $\frac{3}{10}$ and the probability that a shopper is checked out and leaves is $\frac{2}{3}$. Assume that the events of shoppers arriving and shoppers leaving are independent.

   a. Find the transition matrix P for this Markov chain.

   b. Suppose initially, no one is waiting in line. Find $S^{(0)}$, $S^{(1)}$, and $S^{(2)}$.

18. Patients arrive at random at a dentist's office. The waiting room can hold at most five people. In a 15-minute interval at most one person arrives and at most one person is seen by the dentist. In a given time interval, suppose the probability that a person arrives is $\frac{5}{12}$ and the probability that a person is seen by the dentist is $\frac{5}{9}$. Assume that the events of people arriving and people being seen by the dentist are independent.

   a. Find the transition matrix P for this Markov chain.

   b. Suppose initially, no one is in the waiting room. Find $S^{(0)}$, $S^{(1)}$, and $S^{(2)}$.

19. Refer to the transition matrix given on page 473. Suppose that the original population

from which the individual was selected to start the process was composed of 38% genotype AA, 56% genotype Aa, and 6% genotype aa. Find $S^{(1)}$ and $S^{(2)}$.

20. Refer to the transition matrix given on page 473. Suppose that the original population from which the individual was selected to start the process was composed of 26% genotype AA, 32% genotype Aa, and 42% genotype aa. Find $S^{(1)}$ and $S^{(2)}$.

21. Refer to the information given in the inheritance model. Consider the following sequence of matings. We require that the first parent for each mating is of genotype AA. For the first mating, we select a second parent at random from a population in which all genotypes are present. From this mating, we select an offspring at random and mate it with a parent of genotype AA. We continue this process, always mating an offspring selected at random with a parent of genotype AA. Find the transition matrix for this Markov chain.

22. Refer to the information given in the inheritance model. Consider the following sequence of matings. We require that the first parent for each mating be of genotype aa. For the first mating, we select a second parent at random from a population in which all genotypes are present. From this mating, we select an offspring at random and mate it with a parent of genotype aa. We continue this process, always mating an offspring selected at random with a parent of genotype aa. Find the transition matrix for this Markov chain.

23. Refer to the transition matrix found in exercise 21. Suppose that the original population from which the individual was selected to start the process was composed of 15% genotype AA, 69% genotype Aa, and 16% genotype aa. Find $S^{(1)}$ and $S^{(2)}$.

24. Refer to the transition matrix found in exercise 22. Suppose that the original population from which the individual was selected to start the process was composed of 51% genotype AA, 23% genotype Aa, and 26% genotype aa. Find $S^{(1)}$ and $S^{(2)}$.

25. Consider the Markov chain with transition matrix

$$P = \begin{bmatrix} 1 & 0 \\ a & 1-a \end{bmatrix}.$$

Find $P^2, P^3, P^4, \ldots, P^n$.

10.2 Regular Markov Chains and Applications

In the study of Markov chains, one question of interest is what happens to the process as the number of transitions becomes large. Consider again the transition matrix

$$P = \begin{bmatrix} 0.6 & 0.2 & 0.2 \\ 0.1 & 0.7 & 0.2 \\ 0.4 & 0.2 & 0.4 \end{bmatrix}$$

given for the brand-switching situation of example 2 in section 10.1 (page 465). Starting with the initial state distribution vector $S^{(0)} = [0.4 \quad 0.2 \quad 0.4]$, we compute the state distribution vectors for 1, 2, and 3 months of transition. Using theorem 1 (page 468), we obtain the desired state distribution vectors by repeated application of the transition matrix P. We have

$$S^{(1)} = S^{(0)}P = [0.4 \quad 0.2 \quad 0.4] \begin{bmatrix} 0.6 & 0.2 & 0.2 \\ 0.1 & 0.7 & 0.2 \\ 0.4 & 0.2 & 0.4 \end{bmatrix}$$

$$= [0.42 \quad 0.30 \quad 0.28]$$

$$S^{(2)} = S^{(1)}P = [0.42 \quad 0.30 \quad 0.28] \begin{bmatrix} 0.6 & 0.2 & 0.2 \\ 0.1 & 0.7 & 0.2 \\ 0.4 & 0.2 & 0.4 \end{bmatrix}$$

$$= [0.394 \quad 0.350 \quad 0.256]$$

and

$$S^{(3)} = S^{(2)}P = [0.394 \quad 0.350 \quad 0.256] \begin{bmatrix} 0.6 & 0.2 & 0.2 \\ 0.1 & 0.7 & 0.2 \\ 0.4 & 0.2 & 0.4 \end{bmatrix}$$

$$= [0.374 \quad 0.375 \quad 0.251].$$

We can continue applying the transition matrix P to obtain state distribution vectors for additional months of transition. Computations through 30 transitions were carried out on a computer. We see from the computer printout in figure 10.6 that the state distribution vectors are essentially the same after 18 transitions and are close to the vector $[0.35 \quad 0.40 \quad 0.25]$. In fact,

$$[0.35 \quad 0.40 \quad 0.25] \begin{bmatrix} 0.6 & 0.2 & 0.2 \\ 0.1 & 0.7 & 0.2 \\ 0.4 & 0.2 & 0.4 \end{bmatrix} = [0.35 \quad 0.40 \quad 0.25].$$

If there is a state distribution vector that remains unchanged after one transition, it will remain unchanged during further transitions. For example, if S is such that $S = SP$ and if we apply P to S again, we must obtain S. Such a state distribution vector is called a stable state vector.

Definition

■ Consider a Markov chain with an $m \times m$ transition matrix P. A $1 \times m$ matrix $T = [t_1, t_2, \ldots, t_m]$ is called a **stable state vector** for the Markov chain if the entries of T satisfy

a. $0 \le t_1 \le 1, 0 \le t_2 \le 1, \ldots, 0 \le t_m \le 1$

b. $t_1 + t_2 + \cdots + t_m = 1$

and if T satisfies the matrix equation

$$TP = T. \quad ■$$

We saw in our brand-switching example that the state distribution vector $S = [0.35 \quad 0.40 \quad 0.25]$ is a stable state vector since $SP = P$. Note that a stable

K	STATE DISTRIBUTION VECTOR AFTER K TRANSITIONS		
	[FIRST ENTRY	SECOND ENTRY	THIRD ENTRY]
0	.4	.2	.4
1	.42	.3	.28
2	.394	.35	.256
3	.3738	.375	.2512
4	.36226	.3875	.25024
5	.356202	.39375	.250048
6	.353115	.396875	.250009
7	.35156	.398437	.250002
8	.35078	.399218	.25
9	.35039	.399609	.25
10	.350195	.399804	.25
11	.350097	.399902	.25
12	.350048	.39995	.249999
13	.350024	.399975	.249999
14	.350011	.399987	.249999
15	.350005	.399993	.249999
16	.350002	.399996	.249999
17	.350001	.399997	.249999
18	.35	.399998	.249999
19	.349999	.399998	.249999
20	.349999	.399999	.249999
21	.349999	.399999	.249999
22	.349999	.399999	.249999
23	.349999	.399999	.249999
24	.349999	.399998	.249999
25	.349999	.399998	.249999
26	.349999	.399998	.249999
27	.349998	.399998	.249999
28	.349998	.399998	.249999
29	.349998	.399998	.249999
30	.349998	.399998	.249999

FIGURE 10.6

state vector is the long-term effect of repeated applications of the transition matrix. Because of the effects of rounding, we may perceive the existence of a stable state vector in a finite number of transitions. Once we make a conjecture as to the existence of such a vector T, we can verify our conjecture by testing the equation $TP = T$.

Consider again the Markov chain of population transitions described in example 6 in section 10.1 (page 469). The transition matrix is

$$P = \begin{bmatrix} 0.85 & 0.15 \\ 0.05 & 0.95 \end{bmatrix}.$$

Starting with the initial state distribution vector $S^{(0)} = [0.3 \quad 0.7]$, a sequence of state distribution vectors was obtained on a computer. The state distribution vectors for up to 40 transitions were computed using the formula

$$S^{(k+1)} = S^{(k)}P$$

and the results are given in figure 10.7.

```
K                   STATE DISTRIBUTION VECTOR AFTER K TRANSITIONS

                    [FIRST ENTRY    SECOND ENTRY]
0                    .3              .7
1                    .29             .71
2                    .282            .718
3                    .2756           .7244
4                    .27048          .72952
5                    .266384         .733616
6                    .263107         .736892
7                    .260485         .739514
8                    .258388         .741611
9                    .256711         .743288
10                   .255368         .744631
11                   .254295         .745704
12                   .253436         .746563
13                   .252748         .74725
14                   .252199         .7478
15                   .251759         .74824
16                   .251407         .748591
17                   .251125         .748873
18                   .2509           .749098
19                   .25072          .749278
20                   .250576         .749422
21                   .250461         .749537
22                   .250368         .749629
23                   .250294         .749703
24                   .250235         .749762
25                   .250188         .749809
26                   .25015          .749847
27                   .25012          .749877
28                   .250096         .749901
29                   .250077         .749921
30                   .250061         .749936
31                   .250049         .749948
32                   .250039         .749958
33                   .250031         .749966
34                   .250024         .749972
35                   .250019         .749977
36                   .250015         .749981
37                   .250012         .749984
38                   .250009         .749987
39                   .250007         .749989
40                   .250006         .749991
```

FIGURE 10.7

It appears from the computer printout that $T = [0.25 \quad 0.75]$ is a stable state vector for the Markov chain. To establish that this is indeed the case, we need only observe that the entries of T satisfy the definition and that

$$[0.25 \quad 0.75]\begin{bmatrix} 0.85 & 0.15 \\ 0.05 & 0.95 \end{bmatrix} = [0.25 \quad 0.75].$$

Thus T is a stable state vector.

Let us interpret the results for our Markov chain of population transitions. If at the start 30% of the population lives in urban areas and 70% lives in nonurban areas, the distribution of the population will tend to stabilize at 25% in urban areas and 75% in nonurban areas. If the distribution of the population starts with 25% in urban areas and 75% in nonurban areas, this distribution will remain constant.

Note that if we start with a different initial distribution of population, the same stabilization of the population is apparent. For example, starting with the initial state distribution vector $S^{(0)} = [0.2 \quad 0.8]$, the state distribution vectors were computed and the results are given in figure 10.8. We see from the computer printout that the same stable state vector is approached. This result is not unusual. In fact, for a certain type of Markov chain this will always be the case.

K	STATE DISTRIBUTION VECTOR AFTER K TRANSITIONS	
	[FIRST ENTRY	SECOND ENTRY]
0	.2	.8
1	.21	.79
2	.218	.782
3	.2244	.7756
4	.22952	.77048
5	.233616	.766384
6	.236893	.763107
7	.239514	.760485
8	.241611	.758388
9	.243289	.75671
10	.244631	.755368
11	.245705	.754294
12	.246564	.753435
13	.247251	.752748
14	.247801	.752198
15	.24824	.751758
16	.248592	.751406
17	.248874	.751125
18	.249099	.7509
19	.249279	.750719
20	.249423	.750575
21	.249538	.75046
22	.249631	.750368
23	.249704	.750294
24	.249763	.750235
25	.249811	.750188
26	.249848	.75015
27	.249878	.750119
28	.249903	.750095
29	.249922	.750076
30	.249937	.75006
31	.24995	.750048
32	.24996	.750038
33	.249968	.75003
34	.249974	.750023
35	.249979	.750018
36	.249983	.750014
37	.249986	.750011
38	.249989	.750008
39	.249991	.750006
40	.249992	.750004

FIGURE 10.8

Definition

■ A transition matrix P of a Markov chain is **regular** if for some power of P all the entries are positive. A Markov chain is also said to be regular if its transition matrix is regular. ■

The significance of calling a transition matrix "regular" is given in the next theorem. The justification of the theorem requires knowledge of the concept of a limit; therefore, we simply state the theorem.

Theorem 3

■ Any regular Markov chain has a unique stable state vector T. Furthermore, starting from *any* initial state distribution vector, T will be approached through successive transitions. ■

If the transition matrix for a Markov chain has all positive entries, the chain is regular. For example, the Markov chains in our brand-switching example and in our population example are both regular, since the first powers of their respective transition matrices P have all positive entries. Thus the only possibility that a transition matrix is not regular is if at least one of its entries is zero. For example, the matrix

$$P = \begin{bmatrix} 0.5 & 0.5 \\ 0 & 1 \end{bmatrix}$$

is not regular since the $(2, 1)$ entry of P^k, for any k, is zero.

A zero entry in a transition matrix is no guarantee that the matrix is not regular. For example, consider the matrix

$$P = \begin{bmatrix} 0 & 1 \\ 0.5 & 0.5 \end{bmatrix}.$$

Since the entries in

$$P^2 = \begin{bmatrix} 0.5 & 0.5 \\ 0.25 & 0.75 \end{bmatrix}$$

are all positive, P is regular.

Even if a transition matrix is not regular, it may have stable state vectors. For example, consider the matrix

$$P = \begin{bmatrix} 1 & 0 & 0 \\ 0 & 1 & 0 \\ 0 & 0 & 1 \end{bmatrix},$$

which is the 3×3 identity matrix, and suppose it is the transition matrix of a Markov chain. We can interpret from P that for any initial state of the system, the system will remain in that state with probability 1. Therefore the Markov chain has infinitely many stable state vectors.

Instead of starting with an initial state distribution vector and applying successive transitions, a better method for finding the stable state vector $T = [t_1, t_2, \ldots, t_m]$ for a given regular transition matrix P is to obtain the solution to the matrix equation $TP = T$ that satisfies $t_1 + t_2 + \cdots + t_m = 1$.

1. Consider the regular transition matrix

$$P = \begin{bmatrix} 0.85 & 0.15 \\ 0.05 & 0.95 \end{bmatrix}$$

for our population example. Let us find the stable state vector $T = [t_1 \quad t_2]$ by solving the matrix equation

$$[t_1 \quad t_2] \begin{bmatrix} 0.85 & 0.15 \\ 0.05 & 0.95 \end{bmatrix} = [t_1 \quad t_2].$$

Performing the multiplication and equating corresponding entries, we obtain the system of linear equations

$$0.85t_1 + 0.05t_2 = t_1$$

$$0.15t_1 + 0.95t_2 = t_2,$$

which is equivalent to the system

$$-0.15t_1 + 0.05t_2 = 0$$

$$0.15t_1 - 0.05t_2 = 0.$$

We find that this homogeneous system has infinitely many solutions, each of the form $t_2 = 3t_1$. However, the entries of T must also satisfy the condition $t_1 + t_2 = 1$. Therefore we obtain the unique solution $T = [0.25 \quad 0.75]$, which agrees with our previous result.

2. *Machine Performance.* Consider again the computer system discussed at the beginning of section 10.1 (page 463). The transition matrix given there is

$$\begin{array}{cc} & U \quad\ D \\ P = \begin{array}{c} U \\ D \end{array} & \begin{bmatrix} 0.55 & 0.45 \\ 0.65 & 0.35 \end{bmatrix}. \end{array}$$

Recall that the states U and D denote, respectively, that the system is up and operative an entire day and the system is down and inoperative sometime during the day. The transition matrix is regular. Let us find the stable state vector.

To find the stable state vector $T = [t_1 \quad t_2]$, we solve the system of linear equations

$$0.55t_1 + 0.65t_2 = t_1$$

$$0.45t_1 + 0.35t_2 = t_2$$

$$t_1 + \quad t_2 = 1.$$

Solving this system, we obtain the stable state vector $T = [\frac{13}{22} \quad \frac{9}{22}]$.

We can interpret this result as follows. If no attempt is made to improve the system, the system will be down a portion of any given day with probability $\frac{9}{22}$.

3. *Distribution of Inventory.* The Drive-Me Car Rental Company has three locations A, B, and C. A customer can rent a car at any one of these locations and return it to the same location or any of the other two locations. The probabilities

that a customer returns a car to one of the three locations, given the location at which it was rented, appear in the transition matrix

$$P = \begin{matrix} & \begin{matrix} A & B & C \end{matrix} \\ \begin{matrix} A \\ B \\ C \end{matrix} & \begin{bmatrix} 0.35 & 0.30 & 0.35 \\ 0.20 & 0.60 & 0.20 \\ 0 & 0.50 & 0.50 \end{bmatrix} \end{matrix}.$$

Suppose the initial distribution of Drive-Me's fleet of cars is 10% at location A, 40% at location B, and 50% at location C. Let us answer the following two questions:

1. Will the initial distribution remain stable?
2. If not, how should Drive-Me distribute its fleet among the three locations so the distribution will not change?

To answer the first question, we need to observe that

$$[0.10 \quad 0.40 \quad 0.50] \begin{bmatrix} 0.35 & 0.30 & 0.35 \\ 0.20 & 0.60 & 0.20 \\ 0 & 0.50 & 0.50 \end{bmatrix} = [0.115 \quad 0.520 \quad 0.365].$$

Since the system moved away from the distribution $[0.10 \quad 0.40 \quad 0.50]$, the distribution is not a stable state vector.

To answer the second question, we should first observe that P is regular since

$$P^2 = \begin{bmatrix} 0.1825 & 0.4600 & 0.3575 \\ 0.1900 & 0.5200 & 0.2900 \\ 0.1000 & 0.5500 & 0.3500 \end{bmatrix}$$

has all positive entries. Thus there is a distribution of the fleet that will not change; this distribution is the stable state vector $T = [t_1 \quad t_2 \quad t_3]$. We find T by solving the equation

$$[t_1 \quad t_2 \quad t_3] \begin{bmatrix} 0.35 & 0.30 & 0.35 \\ 0.20 & 0.60 & 0.20 \\ 0 & 0.50 & 0.50 \end{bmatrix} = [t_1 \quad t_2 \quad t_3]$$

with the added condition $t_1 + t_2 + t_3 = 1$. Thus we must solve the system with augmented matrix

$$\begin{bmatrix} -0.65 & 0.20 & 0 & | & 0 \\ 0.30 & -0.40 & 0.50 & | & 0 \\ 0.35 & 0.20 & -0.50 & | & 0 \\ 1 & 1 & 1 & | & 1 \end{bmatrix}$$

Performing Gaussian elimination with back-substitution, we arrive at the solution $T = [0.16 \quad 0.52 \quad 0.32]$. Therefore, Drive-Me should place 16% of its fleet at location A, 52% at location B, and 32% at location C.

EXERCISES 10.2

1. Determine which of the following matrices are regular. If the matrix is not regular, explain why.

a. $\begin{bmatrix} 0.54 & 0.46 \\ 0.39 & 0.61 \end{bmatrix}$

b. $\begin{bmatrix} 1 & 0 \\ 0 & 1 \end{bmatrix}$

c. $\begin{bmatrix} 0.2 & 0.3 & 0.5 \\ 0 & 1 & 0 \\ 0.3 & 0.5 & 0.2 \end{bmatrix}$

d. $\begin{bmatrix} 0.23 & 0.18 & 0.59 \\ 0.1 & 0 & 0.9 \\ 0.45 & 0.31 & 0.24 \end{bmatrix}$

2. Determine which of the following matrices are regular. If the matrix is not regular, explain why.

a. $\begin{bmatrix} 1 & 0 & 0 \\ 0 & 1 & 0 \\ 1 & 0 & 0 \end{bmatrix}$

b. $\begin{bmatrix} 0 & 1 & 0 \\ \frac{1}{2} & \frac{1}{2} & 0 \\ \frac{1}{4} & \frac{1}{4} & \frac{1}{2} \end{bmatrix}$

c. $\begin{bmatrix} \frac{1}{5} & \frac{4}{5} \\ \frac{3}{7} & \frac{4}{7} \end{bmatrix}$

d. $\begin{bmatrix} 0.4 & 0.3 & 0.3 \\ 0.1 & 0.6 & 0.3 \\ 0.5 & 0.5 & 0 \end{bmatrix}$

3. In each of the following, verify that the given row matrix is the stable state vector for the given regular transition matrix.

a. $\begin{bmatrix} \frac{1}{3} & \frac{2}{3} \end{bmatrix} \begin{bmatrix} 0 & 1 \\ \frac{1}{2} & \frac{1}{2} \end{bmatrix}$

b. $\begin{bmatrix} 0.384 & 0.616 \end{bmatrix} \begin{bmatrix} 0.23 & 0.77 \\ 0.48 & 0.52 \end{bmatrix}$

c. $\begin{bmatrix} 0.2 & 0.2 & 0.6 \end{bmatrix} \begin{bmatrix} 0.2 & 0.3 & 0.5 \\ 0.5 & 0.4 & 0.1 \\ 0.1 & 0.1 & 0.8 \end{bmatrix}$

4. In each of the following, verify that the given row matrix is the stable state vector for the given regular transition matrix.

a. $\begin{bmatrix} \frac{18}{25} & \frac{7}{25} \end{bmatrix} \begin{bmatrix} \frac{5}{6} & \frac{1}{6} \\ \frac{3}{7} & \frac{4}{7} \end{bmatrix}$

b. $\begin{bmatrix} \frac{2}{11} & \frac{9}{11} \end{bmatrix} \begin{bmatrix} 0.28 & 0.72 \\ 0.16 & 0.84 \end{bmatrix}$

c. $\begin{bmatrix} \frac{1}{3} & \frac{1}{3} & \frac{1}{3} \end{bmatrix} \begin{bmatrix} \frac{1}{10} & \frac{3}{10} & \frac{3}{5} \\ \frac{1}{2} & \frac{1}{2} & 0 \\ \frac{2}{5} & \frac{1}{5} & \frac{2}{5} \end{bmatrix}$

5. Find the stable state vector for each of the following transition matrices.

a. $\begin{bmatrix} \frac{1}{4} & \frac{3}{4} \\ \frac{2}{3} & \frac{1}{3} \end{bmatrix}$

b. $\begin{bmatrix} 0.8 & 0.2 \\ 0.6 & 0.4 \end{bmatrix}$

c. $\begin{bmatrix} 0.12 & 0.88 \\ 0.32 & 0.68 \end{bmatrix}$

d. $\begin{bmatrix} 0.325 & 0.675 \\ 0.593 & 0.407 \end{bmatrix}$

6. Find the stable state vector for each of the following transition matrices.

a. $\begin{bmatrix} \frac{3}{5} & \frac{2}{5} \\ \frac{1}{3} & \frac{2}{3} \end{bmatrix}$

b. $\begin{bmatrix} 0.5 & 0.5 \\ 0.7 & 0.3 \end{bmatrix}$

c. $\begin{bmatrix} 0.26 & 0.74 \\ 0.62 & 0.38 \end{bmatrix}$

d. $\begin{bmatrix} 0.628 & 0.372 \\ 0.457 & 0.543 \end{bmatrix}$

7. Find the stable state vector for each of the following transition matrices.

a. $\begin{bmatrix} \frac{9}{10} & \frac{1}{10} & 0 \\ \frac{2}{10} & \frac{7}{10} & \frac{1}{10} \\ \frac{1}{10} & \frac{2}{10} & \frac{7}{10} \end{bmatrix}$

b. $\begin{bmatrix} 0.2 & 0.7 & 0.1 \\ 0.5 & 0 & 0.5 \\ 0.3 & 0.3 & 0.4 \end{bmatrix}$

c. $\begin{bmatrix} 0.1 & 0.6 & 0.3 \\ 0.4 & 0.5 & 0.1 \\ 0.2 & 0 & 0.8 \end{bmatrix}$

d. $\begin{bmatrix} 0.23 & 0.45 & 0.32 \\ 0.43 & 0.57 & 0 \\ 0.82 & 0.08 & 0.10 \end{bmatrix}$

8. Find the stable state vector for each of the following transition matrices.

a. $\begin{bmatrix} \frac{1}{5} & \frac{2}{5} & \frac{2}{5} \\ \frac{4}{5} & \frac{1}{10} & \frac{1}{10} \\ \frac{2}{5} & 0 & \frac{3}{5} \end{bmatrix}$

b. $\begin{bmatrix} 0.1 & 0.2 & 0.7 \\ 0 & 0.6 & 0.4 \\ 0.5 & 0.1 & 0.4 \end{bmatrix}$

c. $\begin{bmatrix} 0.2 & 0.2 & 0.6 \\ 0.1 & 0.4 & 0.5 \\ 0.6 & 0.1 & 0.3 \end{bmatrix}$

d. $\begin{bmatrix} 0.21 & 0.30 & 0.49 \\ 0.42 & 0.50 & 0.08 \\ 0.75 & 0.10 & 0.15 \end{bmatrix}$

In exercises 9–14, for the referenced exercise, find the stable state vector for the Markov chain and interpret the stable state vector for the system.

9. The system of the rat and the T-maze as described in exercise 5 in section 10.1 (page 474).

10. The market survey described in exercise 10 in section 10.1 (page 476).

11. The survey of cola drinkers as described in exercise 9 in section 10.1 (page 475).

12. The survey of the height of males as described in exercise 6 in section 10.1 (page 475).

13. The Markov chain as described in exercise 13 in section 10.1 (page 477).

14. The Markov chain as described in exercise 14 in section 10.1 (page 477).

15. A mouse is placed in one of the compartments of the maze shown in figure 10.9. A transition takes place when the mouse moves through a door from one compartment to another. We assume that the mouse will move through any of the doors of a given compartment with equal probability.

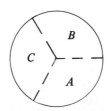

FIGURE 10.9

a. Give the transition matrix P for this Markov chain.

b. Find the stable state vector for this Markov chain.

c. Interpret the stable state vector for the system.

16. A mouse is placed in one of the compartments of the maze shown in figure 10.10. A transition takes place when the mouse moves through a door from one compartment to another. We assume that the mouse will move through any of the doors of a given compartment with equal probability.

a. Give the transition matrix P for this Markov chain.

b. Find the stable state vector for this Markov chain.

c. Interpret the stable state vector for the system.

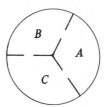

FIGURE 10.10

17. Vigilante Beverage Company is considering the purchase of one of two capping machines A and B. On a monthly basis each machine can be in one of two states:

$$O = \text{the machine is operating properly.}$$

$$N = \text{the machine needs adjustment.}$$

The transition matrices for each machine are the following.

$$
\begin{array}{cc}
 & \begin{array}{cc} O & N \end{array} \\
\begin{array}{c} O \\ N \end{array} & \begin{bmatrix} 0.78 & 0.22 \\ 0.69 & 0.31 \end{bmatrix}
\end{array}
\qquad
\begin{array}{cc}
 & \begin{array}{cc} O & N \end{array} \\
\begin{array}{c} O \\ N \end{array} & \begin{bmatrix} 0.75 & 0.25 \\ 0.82 & 0.18 \end{bmatrix}
\end{array}
$$

Machine A   Machine B

Based on long-term performance, which machine should the company buy?

18. The Horbocker Brewery is considering the purchase of one of two bottling machines A and B. On a monthly basis each machine can be in one of two states:

$$O = \text{the machine is operating properly.}$$

$$N = \text{the machine needs adjustment.}$$

The transition matrices for each machine are the following.

$$
\begin{array}{cc}
 & \begin{array}{cc} O & N \end{array} \\
\begin{array}{c} O \\ N \end{array} & \begin{bmatrix} 0.83 & 0.17 \\ 0.79 & 0.21 \end{bmatrix}
\end{array}
\qquad
\begin{array}{cc}
 & \begin{array}{cc} O & N \end{array} \\
\begin{array}{c} O \\ N \end{array} & \begin{bmatrix} 0.86 & 0.24 \\ 0.71 & 0.29 \end{bmatrix}
\end{array}
$$

Machine A   Machine B

Based on long-term performance, which machine should the company buy?

19. Two urns A and B each contain 3 balls. Two of the 6 balls are blue and 4 are white. At time $t = k\,(k = 1, 2, 3, \ldots)$ a ball is drawn at random from each urn. The ball drawn from urn A is placed in urn B and the ball drawn from urn B is placed in urn A. The states for this Markov chain are the number of blue balls in urn A.

    a. Give the transition matrix for this Markov chain.

    b. Find the stable state vector for this Markov chain.

    c. Interpret the stable state vector for the system.

20. Two urns A and B each contain balls: 4 balls in urn A and 2 balls in urn B. Two of the 6 balls are blue and 4 are red. At time $t = k\ (k = 1, 2, 3, \ldots)$ a ball is drawn at random from each urn. The ball drawn from urn A is placed in urn B and the ball drawn from urn B is placed in urn A. The states for this Markov chain are the number of blue balls in urn A.

    a. Give the transition matrix for this Markov chain.

    b. Find the stable state vector for this Markov chain.

    c. Interpret the stable state vector for the system.

21. Cars arrive at a small comfort station on the Mount Holly Skyway at random. At most two cars are waiting in line. Suppose every 20 minutes at most one car arrives and at most one car leaves. Suppose also that the probability that a car arrives in a 20-minute interval is 0.6 and the probability that a car leaves is 0.5. We are assuming that the events of cars arriving and cars departing are independent.

    a. Give the transition matrix for this Markov chain.

    b. Find the stable state vector for this Markov chain.

    c. Interpret the stable state vector for this system.

22. Tennis players arrive in pairs at the Oakland Tennis Club at random. At most two pairs of players are waiting for a court. Suppose every 30 minutes at most one pair of players arrive and at most one pair of players leave. Suppose also that the probability a pair of players arrive at the club is 0.7 and the probability a pair of players leave the club is 0.4. We are assuming that the events of pairs of players arriving and leaving are independent.

    a. Give the transition matrix for this Markov chain.

    b. Find the stable state vector for this Markov chain.

    c. Interpret the stable state vector for this system.

23. Consider the inheritance model as described in section 10.1 and the transition matrix P given on page 473.

    a. Show that P is regular.

    b. Find the stable state vector for this Markov chain.

    c. Interpret the stable state vector for this system.

24. A particle is located at one of the three points A, B, or C on the following circle (figure 10.11). At time $t = k\ (k = 1, 2, 3, \ldots)$ the particle does one of two things. It moves clockwise to the adjacent point with probability $\frac{2}{5}$ or it moves counterclockwise to the adjacent point with probability $\frac{3}{5}$.

    a. Find the transition matrix P for this Markov chain.

    b. Show that P is regular.

    c. Find the stable state vector for this Markov chain.

    d. Interpret your result of part (c) for this system.

FIGURE 10.11

25. Pinnacle Rental Company has three locations A, B, and C. A customer can rent a truck at any one of these locations and return it to the same location or any of the other two locations. The probability that a truck rented from location A is returned to A is 0.2 and the probability that it is returned to location B is 0.5. The probability that a truck rented from location B is returned to B is 0.7 and the probability that it is returned to C is 0.3. The probability that a truck rented from location C is returned to C is 0.5 and the probability that it is returned to location B is zero.

   a. Give the transition matrix P for this Markov chain.

   b. How should the company distribute its fleet of trucks among the three locations so that the distribution remains stable?

26. Granada Rental Service has three locations A, B, and C located in northern California. A customer can rent a recreational vehicle from any one of these locations and return it to the same location or any of the other two locations. The probability that a vehicle rented from location A is returned to A is 0.5 and the probability that it is returned to location B is 0.4. The probability that a vehicle rented from location B is returned to B is 0.4 and the probability that it is returned to A is 0.6. The probability that a vehicle rented from location C is returned to C is 0.5 and the probability that it is returned to location B is 0.5.

   a. Give the transition matrix P for this Markov chain.

   b. How should the company distribute its fleet of recreational vehicles among the three locations so that the distribution remains stable?

27. A particle is located at one of the three points A, B, or C on the following circle (figure 10.12). At time $t = k$ $(k = 1, 2, 3, \ldots)$ the particle does one of three things. It stays at the same point with probability $\frac{1}{6}$; it moves clockwise to the adjacent point with probability $\frac{2}{9}$; or it moves counterclockwise to the adjacent point with probability $\frac{11}{18}$.

   a. Find the transition matrix P for this Markov chain.

   b. Find the stable state vector. (*Hint*: You may want to use Gaussian elimination or Gauss–Jordan reduction on the system of equations.)

FIGURE 10.12

28. Consider the Markov chain with transition matrix

$$
\begin{bmatrix}
0.1 & 0.2 & 0.2 & 0.5 \\
0.2 & 0 & 0.1 & 0.7 \\
0.3 & 0.4 & 0.1 & 0.2 \\
0.4 & 0.4 & 0 & 0.2
\end{bmatrix}.
$$

Find the stable state vector. (*Hint*:   Use either Gaussian elimination or Gauss–Jordan reduction on the system of equations.)

29. Consider the Markov chain with transition matrix

$$\begin{bmatrix} 0.1 & 0.2 & 0.2 & 0.5 \\ 0.3 & 0.4 & 0.1 & 0.2 \\ 0.2 & 0 & 0.1 & 0.7 \\ 0.4 & 0.4 & 0 & 0.2 \end{bmatrix}.$$

Find the stable state vector. (*Hint*:   Use either Gaussian elimination or Gauss–Jordan reduction on the system of equations.)

30. Consider the Markov chain with transition matrix

$$P = \begin{bmatrix} 1 - a & a \\ b & 1 - b \end{bmatrix},$$

where a and b are not both zero. Show that $[b \quad a]$ is the stable state vector for P.

31. Consider the Markov chain with transition matrix

$$P = \begin{bmatrix} a & b \\ c & d \end{bmatrix}.$$

Let $v = [v_1 \quad v_2]$, where v_1 and v_2 are non-negative numbers with $v_1 + v_2 = 1$. Let $[w_1 \quad w_2] = v \cdot P$. Show that $w_1 + w_2 = 1$.

32. Consider a regular Markov chain with transition matrix

$$\begin{bmatrix} a & b \\ c & d \end{bmatrix}.$$

If $[t_1 \quad t_2]$ is the stable state vector for this Markov chain, show that

$$t_1 = \frac{c}{1 + c - a} \quad \text{and} \quad t_2 = \frac{1 - a}{1 + c - a}.$$

10.3 Matrix Games

In this section we start the study of game theory. We will consider only games in which there are exactly two players, called a **two-person game**. Let us begin by considering the following game played by Ralph and Charlie.

Ralph has two cards, an ace and a two. For a single play of the game, Ralph picks one of the cards and, without showing it to Charlie, puts it face down on the table. Charlie then calls which card it is and the card is turned over. If Charlie is wrong, Ralph receives $1 from Charlie if the card is the ace and $2 if it is the two. If Charlie calls the card correctly, Ralph pays Charlie these amounts.

In this game each player has two possible **moves**, and therefore there are four possible pairs consisting of a card picked by Ralph and a call made by Charlie.

A payment of money, called a **payoff**, is associated with each of the four possible pairs of moves. Ralph's payoffs can be displayed by a 2×2 matrix, as follows:

$$
\begin{array}{cc}
 & \text{Charlie calls} \\
 & \begin{array}{cc} \text{ace} & \text{two} \end{array} \\
\begin{array}{c} \text{Ralph} \\ \text{picks} \end{array} \begin{array}{c} \text{ace} \\ \text{two} \end{array} & \begin{bmatrix} -1 & 1 \\ 2 & -2 \end{bmatrix}
\end{array}
$$

The entry in row 1 and column 1 is -1 since Ralph pays Charlie \$1 if Ralph picks the ace and Charlie calls it correctly.

Since Charlie's payoffs are the negatives (the opposites) of Ralph's payoffs, there is no need to give a similar matrix to describe Charlie's payoffs. A positive payoff for Ralph is a loss to Charlie and a negative payoff for Ralph is a gain to Charlie. A game between two players where one player's gain is the other player's loss is called a **zero-sum** game. The game can now be viewed as a **matrix game**. A single play of the game consists of Ralph picking a row and Charlie picking a column. The entry common to the chosen row and column is Ralph's payoff, a gain from Charlie if positive and a loss to Charlie if negative.

As in our example, any two-person zero-sum game can be modeled as a matrix game. In general, we label the rows of the matrix with the possible moves by the first player, called the **row player R**, and label the columns with the possible moves by the second player, called the **column player C**. The entries of the matrix are the payoffs with respect to the row player R and the matrix is called the **payoff matrix** for the game. A single play of the matrix game consists of player R picking a row and player C picking a column. The entry common to the chosen row and column is the payoff with respect to the row player R.

EXAMPLE 1

Morra. This game consists of two players, where the players show simultaneously 1 or 2 fingers each. As each player shows his fingers, he calls out the total number of fingers that will be shown by both players. If one player calls out the correct total, and his opponent does not, the player receives from his opponent an amount equal to the total number of fingers. If each player calls out the correct total, or if neither calls out the correct total, the play is considered a draw and no payment is made. Let us model this two-person, zero-sum game as a matrix game.

First, each player has four possible moves: show 1, call 2; show 1, call 3; show 2, call 3; and show 2, call 4. These moves can be given by the ordered pairs $(1, 2)$, $(1, 3)$, $(2, 3)$, and $(2, 4)$. The payoff matrix is the following 4×4 matrix; the payoffs are given with respect to player 1.

$$
\begin{array}{cc}
 & \text{player 2} \\
 & \begin{array}{cccc} (1,2) & (1,3) & (2,3) & (2,4) \end{array} \\
\text{player 1} \begin{array}{c} (1,2) \\ (1,3) \\ (2,3) \\ (2,4) \end{array} & \begin{bmatrix} 0 & 2 & -3 & 0 \\ -2 & 0 & 0 & 3 \\ 3 & 0 & 0 & -4 \\ 0 & -3 & 4 & 0 \end{bmatrix}
\end{array}.
$$

The problem for *each* player in a two-person zero-sum game is to determine the **best move**, a move that benefits him or her most. To determine the best move, a player should adopt the following attitude toward an opponent:

1. The opponent has complete knowledge of the possible moves by the player and the associated payoffs.
2. The opponent's interests are in direct conflict with those of the player.
3. The opponent will analyze the game and will pick his or her move intelligently.

Let us now develop a method for determining a best move, if there is one, by analyzing the following matrix game.

player C

$$
\begin{array}{c c}
 & \begin{array}{ccc} c_1 & c_2 & c_3 \end{array} \\
\text{player R} \quad \begin{array}{c} r_1 \\ r_2 \\ r_3 \\ r_4 \end{array} & \left[\begin{array}{ccc} 3 & 4 & ② \\ 6 & 1 & ⊝3 \\ 1 & -② & ① \\ 1 & ⓪ & 1 \end{array}\right].
\end{array}
$$

We first take player R's point of view. (It may appear at first that to achieve the greatest benefit, player R should pick row r_2 since it contains the largest payoff, \$6. However, to obtain this payoff, player C must also pick column c_1. It is not in player C's best interest to pick the first column since player C would always lose money with this choice. Player R must take into account that player C has complete knowledge of the payoff matrix and will choose the move intelligently. Furthermore, player R should assume that player C's interest conflicts directly with his or her own and should only expect the smallest payoff with player C's move. Therefore player R should proceed as follows.

Player R should first determine the smallest entry in each row since it is the smallest payoff that could occur by picking that row. The smallest entry (entries) in a row is called the **row minimum**. We have circled the row minima in the payoff matrix. Thus, by seeking a payoff from among the row minima, player R would be defending against player C's move.

With this first consideration, player R would then obtain the greatest benefit by picking the row containing the largest of the row minima. Therefore **player R's best move** is to pick row r_1 since it contains the **maximum of the row minima**, namely, the payoff \$2. If player R chooses r_1, then regardless of player C's move, player R's payoff will be no less than \$2.

Let us now consider the game from the viewpoint of player C. Player C must adopt the same attitude about his or her opponent as player R did. Player C should assume that player R has complete knowledge of the payoff matrix and should only expect the largest loss from player R's move. Therefore, as in the analysis of the game for player R, player C should first consider the worst that could happen with each of his or her possible moves. Since the payoffs are given with respect to player R, player C should first determine the largest entry (entries) in each column, called the **column maximum**. A column maximum represents the largest possible loss by playing that column. The column maxima are enclosed by squares in the following payoff matrix.

player C

$$
\begin{array}{c}
& \begin{array}{ccc} c_1 & c_2 & c_3 \end{array} \\
\text{player R} \quad
\begin{array}{c} r_1 \\ r_2 \\ r_3 \\ r_4 \end{array}
\left[
\begin{array}{ccc}
3 & 4 & \boxed{2} \\
\boxed{6} & 1 & \boxed{-3} \\
1 & \boxed{-2} & 1 \\
1 & \boxed{0} & 1
\end{array}
\right].
\end{array}
$$

Player C would achieve the greatest benefit by picking the column containing the smallest of the column maxima. Therefore **player C's best move** is to pick column c_3 since it contains the **minimum of the column maxima**, namely, the payoff \$2.

In summary, player R's best move is row r_1 and player C's best move is column c_3. The payoff for these two moves is \$2 to player R. Observe that if player R picks his or her best move and player C does not, player R will win more than \$2. If player C picks his or her best move and player R does not, player C will lose less than \$2; in fact, player C could even win \$3. We should note that the payoff of \$2 is both a row minimum and a column maximum.

Definition

■ An entry in a payoff matrix is called a **saddle point** if it is both a row minimum (an entry less than or equal to all other entries in its row) and a column maximum (an entry greater than or equal to all other entries in its column). ■

The next example illustrates a method for determining saddle points when they exist.

EXAMPLE 2

Consider the game with payoff matrix

player C

$$
\begin{array}{c}
& \begin{array}{cccc} c_1 & c_2 & c_3 & c_4 \end{array} \\
\text{player R} \quad
\begin{array}{c} r_1 \\ r_2 \\ r_3 \\ r_4 \end{array}
\left[
\begin{array}{cccc}
\boxed{-2} & 5 & 3 & 1 \\
1 & \boxed{-2} & 3 & -1 \\
3 & 2 & 5 & \boxed{1} \\
7 & 1 & 2 & \boxed{-2}
\end{array}
\right].
\end{array}
$$

and determine, if possible, any saddle points.

First, we circle the row minima, as we did in the payoff matrix. Next, we check each circled entry to see if it is a column maximum. We observe that 1 is the maximum entry in the fourth column. Thus, 1 in row r_3 and column c_4 is a saddle point for the payoff matrix.

As defined, a saddle point is a row minimum and a column maximum. We observe further that a saddle point is the largest of the row minima and the smallest of the column maxima. Since a saddle point is greater than or equal to any other entry in its column and since, in turn, each entry in that column is greater than or equal to the minimum of its row, the saddle point is the maximum of the row minima. Since a saddle point is less than or equal to any other entry in its row and since, in turn, each entry in that row is less than or equal to the maximum of its column, the saddle point is also the minimum of the column maxima. With this discussion in mind, we have the following method for finding the best moves.

Theorem ■ *Best Moves for a Matrix Game with a Saddle Point:*
If a matrix game has a saddle point v, then the best move for the row player R is to choose the row that contains v. The best move for the column player C is to choose the column that contains v. ■

For a matrix game with a saddle point, the best moves for both players are completely determined. In this case we say that the game is **strictly determined**. Furthermore, if each player chooses his best move, the payoff will be the saddle point and we call the saddle point the **value of the game**. We further observe that if the value is positive, the row player will gain and the column player will lose with each play. If the value is negative, however, the row player will lose and the column player will gain with each play. Therefore we say that the game is **fair** when the value of the game is zero.

A matrix game may have more than one saddle point, as we see in the next example.

EXAMPLE 3

Consider the matrix game with payoff matrix

$$
\begin{array}{c}
 & \text{player C} \\
\text{player R} & \begin{array}{c} & c_1 & c_2 & c_3 \\ r_1 \\ r_2 \\ r_3 \end{array}
\begin{bmatrix} 1 & \overcirc{-2} & 0 \\ \overcirc{2} & 4 & \overcirc{2} \\ \overcirc{-2} & 3 & -1 \end{bmatrix}.
\end{array}
$$

Let us determine saddle points and best moves, if possible.

Following the procedure of example 2, we circle the row minima. We see that two entries of row r_2 are minima for that row and we circle both. Next any circled entry that is the maximum entry in its column is a saddle point. Thus we see that the two circled entries in row r_2 are saddle points. Therefore the game is strictly determined and the best moves are for player R to pick row r_2 and player C to pick either column c_1 or column c_3.

Observe from example 3 that if a matrix game has more than one saddle point, these saddle points are the same number.

EXAMPLE 4

Let us determine whether or not the game played by Ralph and Charlie, described earlier, is strictly determined.

The payoff matrix is repeated here with the row minima circled.

$$
\begin{array}{cc}
 & \text{Charlie calls} \\
 & \begin{array}{cc} \text{ace} & \text{two} \end{array} \\
\begin{array}{c} \text{Ralph} \quad \text{ace} \\ \text{picks} \quad \text{two} \end{array} &
\left[\begin{array}{cc} \text{\textcircled{-1}} & 1 \\ 2 & \text{\textcircled{-2}} \end{array} \right].
\end{array}
$$

We see that there is no row minimum that is also a maximum in its column. Therefore this game has no saddle points and the game is not strictly determined.

Example 4 shows that the calling game played by Ralph and Charlie is not strictly determined, so we cannot apply the methods of this section to the game. In the next section we will investigate matrix games that are not strictly determined.

We close this section with one more example of a conflict situation that can be modeled as a matrix game.

EXAMPLE 5

Two fast-food companies, R and C, plan to expand into a new market. If both companies advertise heavily, it is estimated that R will get 55% and C will get 45% of the market. If both companies advertise lightly, it is estimated that R will get 50% and C will get 50% of the market. If R advertises heavily and C advertises lightly, R will get 75% and C will get 25% of the market. Finally, if R advertises lightly and C advertises heavily, R will get 35% and C will get 65% of the market. Under these conditions, what is the best move for each company?

We can view this conflict situation as a matrix game. Each company has two possible moves, advertise heavily or advertise lightly. Furthermore, we can think of the percentage of the market won by R as a loss to C. Therefore we have the following payoff matrix for this game.

$$
\begin{array}{cc}
 & \text{C} \\
 & \begin{array}{cc} \text{heavy} & \text{light} \end{array} \\
\text{R} \quad \begin{array}{c} \text{heavy} \\ \text{light} \end{array} &
\left[\begin{array}{cc} \text{\textcircled{0.55}} & 0.75 \\ 0.35 & 0.50 \end{array} \right].
\end{array}
$$

We check for saddle points and find that the number 0.55 circled in the payoff matrix is minimum in its row and maximum in its column. Therefore the game is strictly determined and the best move for R is to advertise heavily and the best move for C is also to advertise heavily.

EXERCISES 10.3

For exercises 1–6 the given matrix is the payoff matrix for a two-person zero-sum game. Decide whether or not the game is strictly determined. If it is, give any saddle points.

1.
$$\begin{array}{c} \\ R \end{array} \begin{array}{c} \quad\; c_1 \quad c_2 \\ \begin{array}{c} r_1 \\ r_2 \end{array} \begin{bmatrix} 2 & 3 \\ 4 & -1 \end{bmatrix} \end{array}$$

C

2.
$$\begin{array}{c} \\ R \end{array} \begin{array}{c} \quad\; c_1 \quad c_2 \quad c_3 \\ \begin{array}{c} r_1 \\ r_2 \end{array} \begin{bmatrix} 2 & 1 & -1 \\ 3 & 5 & -2 \end{bmatrix} \end{array}$$

C

3.
$$R \;\; \begin{array}{c} \quad\; c_1 \quad c_2 \quad c_3 \\ \begin{array}{c} r_1 \\ r_2 \end{array} \begin{bmatrix} \frac{1}{2} & \frac{1}{2} & \frac{5}{8} \\ \frac{3}{8} & \frac{1}{4} & -\frac{1}{2} \end{bmatrix} \end{array}$$

C

4.
$$R \;\; \begin{array}{c} \quad\;\; c_1 \quad\; c_2 \quad\; c_3 \quad c_4 \\ \begin{array}{c} r_1 \\ r_2 \\ r_3 \end{array} \begin{bmatrix} -5 & 0 & 3 & -5 \\ -1 & -2 & 3 & -8 \\ 2 & -1 & 0 & -5 \end{bmatrix} \end{array}$$

C

5.
$$R \;\; \begin{array}{c} \quad\;\; c_1 \quad\; c_2 \quad\; c_3 \quad c_4 \\ \begin{array}{c} r_1 \\ r_2 \\ r_3 \\ r_4 \\ r_5 \end{array} \begin{bmatrix} -3 & 1 & -2 & 0 \\ 7 & 6 & -10 & 10 \\ 3 & 4 & -2 & 0 \\ 5 & 7 & -3 & -2 \\ 0 & -6 & -2 & 3 \end{bmatrix} \end{array}$$

C

6.
$$R \;\; \begin{array}{c} \quad\;\; c_1 \quad c_2 \quad c_3 \quad c_4 \\ \begin{array}{c} r_1 \\ r_2 \\ r_3 \\ r_4 \end{array} \begin{bmatrix} 4 & -2 & 0 & -1 \\ -3 & 1 & 2 & 0 \\ 1 & 3 & 0 & 2 \\ 2 & 2 & 4 & 1 \end{bmatrix} \end{array}$$

C

For exercises 7–10 the given matrix is the payoff matrix for a two-person zero-sum game. Each game is strictly determined. Find the best moves for player R and player C and give the value of the game.

7.
$$R \;\; \begin{array}{c} \quad\;\;\; c_1 \quad\;\; c_2 \quad\;\; c_3 \quad\;\; c_4 \\ \begin{array}{c} r_1 \\ r_2 \\ r_3 \end{array} \begin{bmatrix} 0.45 & 0.35 & 0.10 & 0.30 \\ 0.30 & 0.69 & -0.20 & 0.25 \\ 0.00 & -0.75 & 0.09 & -0.15 \end{bmatrix} \end{array}$$

C

8.
$$R \;\; \begin{array}{c} \quad\; c_1 \quad c_2 \quad c_3 \\ \begin{array}{c} r_1 \\ r_2 \\ r_3 \end{array} \begin{bmatrix} \frac{1}{2} & -\frac{1}{4} & -\frac{1}{2} \\ 2 & -\frac{1}{2} & \frac{1}{8} \\ 0 & -\frac{1}{4} & -\frac{1}{8} \end{bmatrix} \end{array}$$

C

9.
$$R \;\; \begin{array}{c} \quad\;\; c_1 \quad\; c_2 \quad\; c_3 \quad c_4 \\ \begin{array}{c} r_1 \\ r_2 \end{array} \begin{bmatrix} 50 & -75 & -25 & 30 \\ -40 & -50 & -45 & 80 \end{bmatrix} \end{array}$$

C

10.
$$R \;\; \begin{array}{c} \quad\;\; c_1 \quad\;\; c_2 \quad\;\; c_3 \quad\;\; c_4 \\ \begin{array}{c} r_1 \\ r_2 \\ r_3 \\ r_4 \end{array} \begin{bmatrix} 0.30 & 0.75 & 0.35 & 0.40 \\ 0.10 & 0.36 & -0.47 & -0.86 \\ 0.35 & 0.66 & 0.35 & 0.85 \\ 0.35 & 0.40 & 0.35 & 0.75 \end{bmatrix} \end{array}$$

C

11. Ron and Cary play the following game. Ron has the 2, 3, and 10 of hearts, and Cary has the 5, 7, and 8 of diamonds. Each picks one of his own cards and they show their selections simultaneously. If the sum of the numbers on the two cards is even, Cary pays Ron that amount in dollars. If the sum of the numbers is odd, Ron pays Cary that amount in dollars.

   a. Give the payoff matrix for this game.

   b. Decide whether or not the game is strictly determined.

   c. If the game is strictly determined, give the best moves for each player.

12. In a game of matching coins, Kim and Tom each show a side of a coin. If both coins show heads, Kim pays Tom $3. If Kim's coin shows a tail and Tom's coin shows a head, Kim pays Tom $4. If Kim's coin shows a head and Tom's coin shows a tail, Tom pays Kim $5. If both coins show tails, Tom pays Kim $8.

   a. Give the payoff matrix for this game.

   b. What are the best moves, if any, for each player?

   c. If the game is strictly determined, find the value of it and decide if the game is fair.

13. Mrs. Black decides to invest in either gold, stocks, or certificates of deposit. She estimates the annual percentage return on each investment corresponding to three possible degrees of inflation. This information is given in the following matrix.

$$
\begin{array}{c}
 \\
 \\
\text{gold} \\
\text{stocks} \\
\text{certificates}
\end{array}
\begin{array}{ccc}
\text{slight} & \text{moderate} & \text{severe} \\
\text{inflation} & \text{inflation} & \text{inflation} \\
\left[\begin{array}{ccc} 5\% & 12\% & 20\% \\ 8\% & 12\% & 9\% \\ 8\% & 8\% & 8\% \end{array}\right]
\end{array}
$$

   a. Considering this situation as a matrix game, what are the best moves for Mrs. Black?

   b. The rates of inflation are 2% when it is slight, 4% when it is moderate, and 9% when it is severe. The percentages given in the preceding matrix were not adjusted for these rates of inflation and therefore do not reflect the real increase in value. For example, if Mrs. Black puts money in certificates, then for any of the three degrees of inflation her dollars increase at the end of the year by 8%. However, since the value of the dollar decreases with increasing inflation, the real returns on this investment are $8 - 2 = 6\%$ for slight inflation, $8 - 4 = 4\%$ for moderate inflation, and $8 - 9 = -1\%$ for severe inflation. Give the payoff matrix of returns adjusted for the rates of inflation and determine now the best moves for Mrs. Black, if any.

14. Two banks, the Farmers' Bank and the First National Bank, each plan to open a new facility in the same area that is not presently served by any bank. To attract customers, each bank offers free checking, national credit cards, or both as services. If Farmers' Bank offers free checking only, it will receive 45%, 65%, or 40% of the business if First National Bank offers free checking only, credit cards only, or both, respectively. If Farmers' Bank offers credit cards only, it will receive 35%, 50%, or 35% of the business if First National Bank offers free checking only, credits cards only, or both, respectively. If Farmers' Bank should offer both services, it will receive 55%, 65%, or 45% according to the option that First National Bank hooses.

   a. Considering this situation as a two-person zero-sum game, give the payoff matrix.

   b. What are the best moves for the two banks?

15. Determine whether or not the game of Morra as described in example 1 (page 492) is strictly determined.

16. Consider the two-person zero-sum game with payoff matrix

$$
\left[\begin{array}{cc} a & b \\ c & d \end{array}\right].
$$

If $a > b$, $d > b$, $d > c$, and $a > c$, what can be said about the game?

17. Consider a strictly determined matrix game with payoff matrix

$$\begin{bmatrix} a & b \\ c & d \end{bmatrix}.$$

Suppose that b and c are both saddle points of the game. Show that $b = c$.

10.4 Two-Person Zero-Sum Games with Mixed Strategies

In this section we will analyze games that are not strictly determined. Before we consider a non-strictly determined game, let us return to the strictly determined game in example 2 in section 10.3 (page 494). Recall that this game has the payoff matrix

player C

$$\begin{array}{c} & \begin{array}{cccc} c_1 & c_2 & c_3 & c_4 \end{array} \\ \text{player R} \begin{array}{c} r_1 \\ r_2 \\ r_3 \\ r_4 \end{array} & \begin{bmatrix} -2 & 5 & 3 & 1 \\ 1 & -2 & 3 & -1 \\ 3 & 2 & 5 & 1 \\ 7 & 1 & 2 & -2 \end{bmatrix}. \end{array}$$

Suppose the game is played several times in succession. Furthermore, suppose that R knows in advance that C is going to choose his or her best move, column c_4, each time the game is played. This additional information about C's move cannot be used to improve R's payoff; the most R can receive is $1. Similarly, C cannot improve his or her payoff even if C knows in advance that R will choose row r_3. These observations about this game are true for any strictly determined game.

With these observations in mind, let us now consider the game with payoff matrix

player C

$$\begin{array}{c} & \begin{array}{cc} c_1 & c_2 \end{array} \\ \text{player R} \begin{array}{c} r_1 \\ r_2 \end{array} & \begin{bmatrix} -3 & 1 \\ 2 & -1 \end{bmatrix}. \end{array}$$

This game is not strictly determined since -3 is the minimum in the first row but not the maximum in its column, and -1 is the minimum in the second row but not the maximum in its column. Therefore the players cannot determine their respective moves by the theorem of the previous section (see page 495). Let us instead analyze the game as follows.

In an attempt to achieve the greatest benefit, suppose player R decides to start with playing row r_2, which contains the largest payoff, $2. Likewise, in an attempt to achieve the greatest benefit, player C decides to start with column c_1, which contains the smallest payoff $-\$3$ to R. Now, suppose the game is played several times in succession and each time player R picks r_2. When player C observes this pattern of play by R, C could improve his or her position by switching to column c_2, resulting in a win of $1 (a payoff of $-\$1$ to player R). As the play continues, if player R observes this switch by player C, then R should switch to row r_1, resulting in a payoff of $1.

For a game that is not strictly determined, we observe that both R and C should change their respective moves as they play. Each player should choose a combination of moves that cannot be anticipated by the other player. Thus each player can choose a move in some random fashion. The problem confronting each player is to decide how frequently a particular move should be played. The frequency at which a move is chosen is determined from a probability. Thus we define a strategy as follows.

Definition

■ A **strategy for player R** is a set of probabilities $P = [p_1 \quad p_2 \quad \cdots \quad p_m]$ where p_i is the probability that row r_i is chosen. A **strategy for player C** is a set of probabilities

$$Q = \begin{bmatrix} q_1 \\ q_2 \\ \vdots \\ q_n \end{bmatrix}$$

where q_j is the probability that column c_j is chosen. ■

Observe that the entries of $P = [p_1 \quad p_2 \quad \cdots \quad p_m]$ represent the probabilities for the complete set of possibilities for player R, so we must have $p_1 + p_2 + \cdots + p_m = 1$ and $p_i \geq 0$ for each $i = 1, 2, \ldots, m$. Likewise, the entries of the strategy Q must satisfy the same properties.

Once a strategy is chosen by each player, these strategies can be evaluated using the concept of **expected payoff**. We illustrate these important concepts in the next examples.

EXAMPLE 1

Consider the game with payoff matrix

player C

$$\begin{array}{c} \\ \text{player R} \end{array} \begin{array}{c} \\ r_1 \\ r_2 \end{array} \begin{array}{cc} c_1 & c_2 \\ \begin{bmatrix} -3 & 1 \\ 2 & -1 \end{bmatrix} \end{array}.$$

Suppose R plays row r_1 50% of the time and row r_2 50% of the time. The actual choice of the row to play can be determined, for example, by the flip of a coin. Suppose that C plays column c_1 60% of the time and column c_2 40% of the time. We now develop a way to evaluate the long-term result of these choices.

We can think of a single play as an experiment. The sample space of all possible outcomes together with the assignment of probabilities is obtained from the probability tree given in figure 10.13.

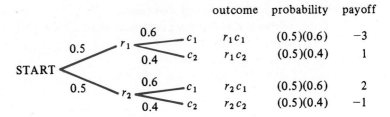

For example, $r_1 c_1$ denotes the outcome that R plays row 1 and C plays column 1. The probability of $r_1 c_1$ is $(0.5)(0.6)$, since playing r_1 and playing c_1 are independent events. The payoff for outcome $r_1 c_1$ is -3, which is obtained from the payoff matrix. The **expected payoff** for the game with respect to the two given methods of play is defined as the expected value

$$E = (-3)(0.5)(0.6) + 1(0.5)(0.4) + 2(0.5)(0.6) + (-1)(0.5)(0.4)$$
$$= -0.3.$$

This means that if the game is played a large number of times, R can expect to lose an average of $0.30 per game.

The next example gives a method that uses matrix multiplication to find the expected payoff for a game with respect to two strategies.

EXAMPLE 2

Consider again the game and methods of play given in example 1. We can denote the strategies of both players by matrices. Let the strategy for R be given by the 1×2 matrix

$$P = [0.5 \quad 0.5]$$

and let the strategy for C be given by the 2×1 matrix

$$Q = \begin{bmatrix} 0.6 \\ 0.4 \end{bmatrix}.$$

Denote the payoff matrix of the game by A. Now, let us compute the quantity PAQ.

$$PAQ = [0.5 \quad 0.5] \begin{bmatrix} -3 & 1 \\ 2 & -1 \end{bmatrix} \begin{bmatrix} 0.6 \\ 0.4 \end{bmatrix}$$

$$= [(-3)(0.5) + 2(0.5) \quad 1(0.5) + (-1)(0.5)] \begin{bmatrix} 0.6 \\ 0.4 \end{bmatrix}$$

$$= (-3)(0.5)(0.6) + 2(0.5)(0.6) + 1(0.5)(0.4) + (-1)(0.5)(0.4)$$

$$= -0.3.$$

Thus we see that PAQ is the expected payoff for the game with respect to the strategies P and Q.

We now give the general form, using matrix multiplication, for the expected payoff of a matrix game.

Theorem ■ *Expected Payoff Formula:*

Consider a two-person zero-sum game with payoff matrix

player C

$$A = \text{player R} \quad \begin{array}{c} \\ r_1 \\ r_2 \\ \vdots \\ r_n \end{array} \begin{array}{ccccc} c_1 & c_2 & \cdots & c_m \\ \begin{bmatrix} a_{11} & a_{12} & \cdots & a_{1m} \\ a_{21} & a_{22} & \cdots & a_{2m} \\ \vdots & \vdots & & \vdots \\ a_{n1} & a_{n2} & \cdots & a_{nm} \end{bmatrix} \end{array}.$$

Let p_i be the probability that R chooses row r_i, $i = 1, \ldots, n$. Set

$$P = [p_1 \quad p_2 \quad \cdots \quad p_n]$$

Let q_j be the probability that C chooses column c_j, $j = 1, \ldots, m$. Set

$$Q = \begin{bmatrix} q_1 \\ q_2 \\ \vdots \\ q_m \end{bmatrix}.$$

Then the **expected payoff** E for the game with respect to the **strategies** P and Q is given by

$$E = PAQ. \quad ■$$

EXAMPLE 3 Consider the matrix game given by the payoff matrix

player C

$$A = \text{player R} \quad \begin{array}{c} \\ r_1 \\ r_2 \\ r_3 \end{array} \begin{array}{ccc} c_1 & c_2 & c_3 \\ \begin{bmatrix} -1 & 5 & 7 \\ 4 & 2 & -6 \\ -3 & 0 & -1 \end{bmatrix} \end{array}.$$

Suppose player R plays row r_1 with a probability of 0.25, row r_2 with a probability of 0.65, and row r_3 with a probability of 0.1. Suppose player C plays column c_1 with a probability of 0.4, column c_2 with a probability of 0.3, and column c_3 with a probability of 0.3. Let us use the expected payoff formula to find the expected payoff for this game with respect to the two given strategies.

We set

$$P = [0.25 \quad 0.65 \quad 0.1] \quad \text{and} \quad Q = \begin{bmatrix} 0.4 \\ 0.3 \\ 0.3 \end{bmatrix}.$$

Then the expected payoff E is given by

$$E = PAQ = [0.25 \quad 0.65 \quad 0.1] \begin{bmatrix} -1 & 5 & 7 \\ 4 & 2 & -6 \\ -3 & 0 & -1 \end{bmatrix} \begin{bmatrix} 0.4 \\ 0.3 \\ 0.3 \end{bmatrix} = 0.91.$$

Thus the long-term average payoff per game is 0.91 units to player R.

The concepts we have been discussing could also be applied to strictly determined games. Recall that if a strictly determined game is played several times, a player cannot improve his or her payoff by making a move different from his or her best move. For example, for the strictly determined game with payoff matrix

$$\begin{array}{c} & \begin{array}{ccc} c_1 & c_2 & c_3 \end{array} \\ \begin{array}{c} r_1 \\ r_2 \\ r_3 \end{array} & \left[\begin{array}{ccc} -1 & 0 & 1 \\ 3 & 4 & ② \\ -2 & 1 & -1 \end{array} \right], \end{array}$$

player R's best move is r_2 and player C's best move is c_3, the row and column of the saddle point 2. Thus player R should play row 2 with probability 1 and all other rows with probability 0. Player C should play column 3 with probability 1 and all other columns with probability 0. Thus player R's and player C's respective strategies are

$$P = \begin{bmatrix} 0 & 1 & 0 \end{bmatrix} \quad \text{and} \quad Q = \begin{bmatrix} 0 \\ 0 \\ 1 \end{bmatrix}.$$

These are called **pure strategies**. If each player uses his or her respective pure strategy, then, using the formula for expected payoff, we have

$$\begin{bmatrix} 0 & 1 & 0 \end{bmatrix} \begin{bmatrix} -1 & 0 & 1 \\ 3 & 4 & 2 \\ -2 & 1 & -1 \end{bmatrix} \begin{bmatrix} 0 \\ 0 \\ 1 \end{bmatrix} = 2,$$

the saddle point of the game.

We realized that for a game that is not strictly determined, each player should not always choose the same move; each player should not use a pure strategy. A strategy in which no move is played with probability 1 is called a **mixed strategy**. The strategies of examples 2 and 3 are mixed strategies.

EXAMPLE 4

Consider the game described in example 2, but this time consider different mixed strategies for the players. Suppose player R plays row r_1 with a probability of 0.3 and plays row r_2 with a probability of 0.7. Suppose player C plays column c_1 with a probability of 0.8 and plays column c_2 with a probability of 0.2. Let us use the formula to find the expected payoff for this game with respect to the two given strategies. We set

$$P = \begin{bmatrix} 0.3 & 0.7 \end{bmatrix} \quad \text{and} \quad Q = \begin{bmatrix} 0.8 \\ 0.2 \end{bmatrix}.$$

Then the expected payoff E is given by

$$E = PAQ = \begin{bmatrix} 0.3 & 0.7 \end{bmatrix} \begin{bmatrix} -3 & 1 \\ 2 & -1 \end{bmatrix} \begin{bmatrix} 0.8 \\ 0.2 \end{bmatrix} = 0.32.$$

Thus the long-term average payoff per game is $0.32 to player R.

Let us compare the results of examples 2 and 4. From example 2, we see that if

$$\text{R uses strategy } [0.5 \quad 0.5] \quad \text{and} \quad \text{C uses strategy } \begin{bmatrix} 0.6 \\ 0.4 \end{bmatrix},$$

the expected payoff is -0.3. From example 3, we see that if

$$\text{R uses strategy } [0.3 \quad 0.7] \quad \text{and} \quad \text{C uses strategy } \begin{bmatrix} 0.8 \\ 0.2 \end{bmatrix},$$

the expected payoff is 0.32.

There are infinitely many mixed strategies each player could use to play the game. However, each player would want to use the strategy that gives the most favorable expected payoff, regardless of the other player's strategy. Player R's objective is to determine a strategy P^* such that the expected payoff is largest, regardless of player C's strategy. Player C's objective is to determine a strategy Q^* such that the expected payoff is smallest, regardless of player R's strategy. The strategies P^* and Q^* are called **optimal mixed strategies**. For any matrix game, optimal strategies P^* and Q^* exist.

The following properties come from the meaning of optimal strategies. If player R uses a strategy P different from the optimal strategy P^*, then the expected payoff PAQ^* is less than the expected payoff P^*AQ^*, that is

$$PAQ^* \leq P^*AQ^*. \tag{4}$$

If player C uses a strategy Q different from the optimal strategy Q^*, then

$$P^*AQ^* \leq P^*AQ. \tag{5}$$

We now consider a method for finding optimal strategies for 2×2 matrix games.

The formulas for finding optimal strategies for a 2×2 matrix game are given next. The formulas can be derived using linear programming; however, the approach is not easy. Therefore we simply state the formulas and show that they are correct by means of an example (see example 7).

Theorem

■ *Optimal Mixed Strategy Formulas:*
Consider a matrix game with payoff matrix

$$A = \begin{bmatrix} a & b \\ c & d \end{bmatrix}.$$

If the game is not strictly determined, then $a + d - b - c \neq 0$. The optimal mixed strategy for row player R is given by $P^* = [p_1 \quad p_2]$ where

$$p_1 = \frac{d - c}{a + d - b - c} \quad \text{and} \quad p_2 = \frac{a - b}{a + d - b - c}. \tag{6}$$

The optimal mixed strategy for column player C is given by

$$Q^* = \begin{bmatrix} q_1 \\ q_2 \end{bmatrix}$$

where

$$q_1 = \frac{d - b}{a + d - b - c} \quad \text{and} \quad q_2 = \frac{a - c}{a + d - b - c}. \tag{7}$$

If both players use their optimal strategies, then the expected payoff E for the game is

$$E = P^*AQ^* = \frac{ad - bc}{a + d - b - c}. \quad \blacksquare \tag{8}$$

EXAMPLE 5

Let us find the optimal strategies for both players for the game described in example 2. Recall that the payoff matrix A is

$$A = \begin{bmatrix} -3 & 1 \\ 2 & -1 \end{bmatrix}.$$

Using the notation of the theorem on optimal strategies, we have $a = -3$, $b = 1$, $c = 2$, and $d = -1$. The optimal mixed strategy for player R is given by formula (6).

$$p_1 = \frac{d - c}{a + d - b - c} = \frac{-1 - 2}{-3 - 1 - 1 - 2} = \frac{3}{7}$$

$$p_2 = \frac{a - b}{a + d - b - c} = \frac{-3 - 1}{-3 - 1 - 1 - 2} = \frac{4}{7}.$$

The optimal mixed strategy for player C is given by formula (7).

$$q_1 = \frac{d - b}{a + d - b - c} = \frac{-1 - 1}{-3 - 1 - 1 - 2} = \frac{2}{7}$$

$$q_2 = \frac{a - c}{a + d - b - c} = \frac{-3 - 2}{-3 - 1 - 1 - 2} = \frac{5}{7}.$$

The expected payoff E for the game, if both players use their optimal strategies, is given by formula (8).

$$E = \frac{ad - bc}{a + d - b - c} = \frac{(-3)(-1) - (1)(2)}{-3 - 1 - 1 - 2}$$

$$= \frac{1}{7}$$

$$= 0.14.$$

Therefore, if both players use their optimal strategies, the long-term average payoff per game is \$0.14 to player R.

Let us compare the results of example 4 with example 5. We saw in example 4 that if player R uses the strategy $P = [0.3 \quad 0.7]$, and player C uses the strategy

$$Q = \begin{bmatrix} 0.8 \\ 0.2 \end{bmatrix},$$

the expected payoff is $0.32, which is larger than the expected payoff $0.14 when each player uses his or her optimal strategy. This may seem to be inconsistent. However, we should realize from inequality (1) that the expected payoff would be smaller if R uses $P \doteq [0.3 \quad 0.7]$ and C uses his optimal strategy

$$Q = \begin{bmatrix} \frac{2}{7} \\ \frac{5}{7} \end{bmatrix}.$$

In particular, for these two strategies we have the expected payoff

$$\begin{bmatrix} \frac{3}{10} & \frac{7}{10} \end{bmatrix} \begin{bmatrix} -3 & 1 \\ 2 & -1 \end{bmatrix} \begin{bmatrix} \frac{2}{7} \\ \frac{5}{7} \end{bmatrix} = -\frac{1}{7}$$

$$= -\$0.14.$$

Since player R should assume that C is an intelligent player and will use his or her optimal strategy, player R should also use his or her optimal strategy.

Note that the p_1 and p_2 given in formula (6) have the property that $p_1 + p_2 = 1$. Therefore, once we have found the value of p_1, p_2 can be found by subtracting p_1 from 1. Similarly, the q_1 and q_2 given in formula (7) are related by $q_1 + q_2 = 1$. For instance, in example 4, $p_1 = \frac{3}{7}$, so $p_2 = 1 - \frac{3}{7} = \frac{4}{7}$.

EXAMPLE 6

A Game with Nature. A farmer must irrigate a crop several times during the growing season. Sometimes the rainfall is below normal between irrigations and sometimes the rainfall is normal.

If the farmer irrigates and the subsequent rainfall is normal, the crop is worth $100 per acre. If the farmer irrigates and the subsequent rainfall is below normal, the crop is worth $150 per acre. If the farmer does not irrigate and the rainfall is normal, the crop is worth $120 per acre. If the farmer does not irrigate and the rainfall is below normal, the crop is worth $70 per acre. What is the farmer's optimal strategy for this game?

To answer this question, we first construct the payoff matrix. In this situation, the farmer's opponent is nature. The farmer has two available strategies:

r_1 = the farmer irrigates.

r_2 = the farmer does not irrigate.

Even though nature is an unthinking opponent, there are two "strategies" available:

c_1 = the rainfall is normal.

c_2 = the rainfall is below normal.

From the given information, the payoff matrix is

nature

		normal rainfall	below normal rainfall
		c_1	c_2
farmer	irrigate r_1	$\begin{bmatrix} 100$	$150 \end{bmatrix}$
	do not irrigate r_2	120	70

This matrix has no saddle points, so the game is not strictly determined. Therefore we use formula (6) to find the farmer's optimal mixed strategy.

Using the notation of the optimal mixed strategy theorem (page 504), we have $a = 100$, $b = 150$, $c = 120$, and $d = 70$. The optimal mixed strategy for the farmer is given by formula (6).

$$p_1 = \frac{d-c}{a+d-b-c} = \frac{70-120}{100+70-150-120} = \frac{1}{2}.$$

Therefore $p_2 = 1 - p_1 = \frac{1}{2}$. Thus the optimal mixed strategy for the farmer is to irrigate 50% of the time.

The expected profit to the farmer is the expected payoff E of the game when the optimal strategy is used. From formula (8), we have

$$E = \frac{ad-bc}{a+d-b-c} = \frac{(100)(70)-(150)(120)}{100+70-150-120} = 110.$$

Thus, in the long run, the farmer can expect a profit of at least $100 per acre regardless of the pattern of rainfall.

The next example is optional. It demonstrates, using linear programming, that the optimal mixed strategy formulas are correct for a specific example.

EXAMPLE 7

A Linear Programming Approach. Consider the 2 × 2 payoff matrix

$$A = \begin{bmatrix} 2 & 3 \\ 4 & 2 \end{bmatrix}.$$

The matrix game is not strictly determined since there is no saddle point. Let the optimal strategies be denoted by

$$P^* = [p_1^* \quad p_2^*] \quad \text{and} \quad Q^* = \begin{bmatrix} q_1^* \\ q_2^* \end{bmatrix}.$$

By formula (7), we have $q_1^* = \frac{1}{3}$ and $q_2^* = \frac{2}{3}$. We will demonstrate here, using linear programming, that these values give the optimal strategy for player C.

Let us denote the expected payoff with respect to the optimal mixed strategies by $v = P^*AQ^*$. From the property of optimal strategies, inequality (4), we have that for any strategy P.

$$PAQ^* \leq v = P^*AQ^*.$$

If player R should always pick row 1, that is, $P = [1 \quad 0]$, we have

$$[1 \quad 0]\begin{bmatrix} 2 & 3 \\ 4 & 2 \end{bmatrix}\begin{bmatrix} q_1^* \\ q_2^* \end{bmatrix} \leq v.$$

If player R should always pick row 2, that is, $P = [0 \quad 1]$, we have

$$[0 \quad 1]\begin{bmatrix} 2 & 3 \\ 4 & 2 \end{bmatrix}\begin{bmatrix} q_1^* \\ q_2^* \end{bmatrix} \leq v.$$

Thus the entries of Q^* must satisfy the inequalities

$$2q_1^* + 3q_2^* \leq v$$
$$4q_1^* + 2q_2^* \leq v.$$

Since the entries of A, P^*, and Q^* are positive, v must be positive. Dividing the inequalities by v, we have the system

$$2\frac{q_1^*}{v} + 3\frac{q_2^*}{v} \leq 1$$

$$4\frac{q_1^*}{v} + 2\frac{q_2^*}{v} \leq 1.$$

Now set

$$x_1 = \frac{q_1^*}{v} \quad \text{and} \quad x_2 = \frac{q_2^*}{v}$$

and

$$x_1 + x_2 = \frac{q_1^* + q_2^*}{v} = \frac{1}{v}.$$

Since player C's objective is to choose Q^* such that the expected payoff is smallest, we must determine q_1^* and q_2^* such that v is minimal. In this case we must determine x_1 and x_2 such that $x_1 + x_2 = 1/v$ is maximal.

Therefore we see that q_1^* and q_2^* can be determined from $q_1^* = vx_1$ and $q_2^* = vx_2$ where x_1 and x_2 are solutions to the linear programming problem:

$$\text{Maximize} \quad z = x_1 + x_2$$
$$\text{subject to} \quad 2x_1 + 3x_2 \leq 1$$
$$4x_1 + 2x_2 \leq 1$$
$$x_1 \geq 0 \quad \text{and} \quad x_2 \geq 0.$$

Transforming the problem to canonical form, we have the initial simplex tableau

$$\begin{array}{c} \\ s_1 \\ s_2 \\ z \end{array} \begin{array}{cc} \begin{array}{ccccc} x_1 & x_2 & s_1 & s_2 & z \end{array} \\ \left[\begin{array}{ccccc|c} 2 & 3 & 1 & 0 & 0 & 1 \\ 4 & 2 & 0 & 1 & 0 & 1 \\ \hline -1 & -1 & 0 & 0 & 1 & 0 \end{array} \right]. \end{array}$$

Applying the simplex method, we derive the final simplex tableau

$$\begin{array}{c} \\ x_2 \\ x_1 \\ z \end{array} \begin{array}{cc} \begin{array}{ccccc} x_1 & x_2 & s_1 & s_2 & z \end{array} \\ \left[\begin{array}{ccccc|c} 0 & 1 & \frac{1}{2} & -\frac{1}{4} & 0 & \frac{1}{4} \\ 1 & 0 & -\frac{1}{4} & \frac{3}{8} & 0 & \frac{1}{8} \\ \hline 0 & 0 & \frac{1}{4} & \frac{1}{8} & 1 & \frac{3}{8} \end{array} \right]. \end{array}$$

From this tableau, we read the solution $x_1 = \frac{1}{8}$, $x_2 = \frac{1}{4}$, and $1/v = x_1 + x_2 = \frac{3}{8}$. Therefore $v = \frac{8}{3}$, $q_1^* = \frac{8}{3} \cdot \frac{1}{8} = \frac{1}{3}$, and $q_2^* = \frac{8}{3} \cdot \frac{1}{4} = \frac{2}{3}$.

EXERCISES 10.4

1. Consider the game with payoff matrix

$$\begin{bmatrix} 6 & 2 & -1 \\ 10 & 5 & 3 \\ 12 & -7 & 44 \end{bmatrix}.$$

Use the expected payoff formula to find the expected payoff for each of the following pairs of strategies.

a.
$$P = [0.6 \quad 0.1 \quad 0.3] \qquad Q = \begin{bmatrix} 0.5 \\ 0.2 \\ 0.3 \end{bmatrix}$$

b.
$$P = [0.26 \quad 0.28 \quad 0.46] \qquad Q = \begin{bmatrix} 0.41 \\ 0.39 \\ 0.20 \end{bmatrix}$$

2. Consider the game with payoff matrix

$$\begin{bmatrix} 20 & 12 \\ 34 & -51 \\ -69 & 47 \end{bmatrix}.$$

Use the expected payoff formula to find the expected payoff for each of the following pairs of strategies.

a. $P = [0.5 \quad 0.4 \quad 0.1] \qquad Q = \begin{bmatrix} 0.42 \\ 0.58 \end{bmatrix}$

b. $P = [0.44 \quad 0.37 \quad 0.19] \qquad Q = \begin{bmatrix} 0.63 \\ 0.37 \end{bmatrix}$

3. Consider the game with payoff matrix

$$A = \begin{bmatrix} -2 & 3 \\ 5 & 1 \end{bmatrix}.$$

a. Find the optimal strategies P^* and Q^* and the corresponding expected payoff P^*AQ^*.

b. Compute P^*AQ for $Q = \begin{bmatrix} \frac{5}{11} \\ \frac{6}{11} \end{bmatrix}$ and show that $P^*AQ \le P^*AQ^*$.

4. Consider the game with payoff matrix

$$A = \begin{bmatrix} 4 & -2 \\ -3 & 1 \end{bmatrix}.$$

a. Find the optimal strategies P^* and Q^* and the corresponding expected payoff P^*AQ^*.

b. Compute PAQ^* for $P = [0.59 \quad 0.41]$ and show that $PAQ^* \le P^*AQ^*$.

For exercises 5–10 find the optimal strategies and the corresponding expected payoff for each of the matrix games.

5. $\begin{bmatrix} 3 & -7 \\ -6 & 8 \end{bmatrix}$

6. $\begin{bmatrix} -45 & 20 \\ 32 & -18 \end{bmatrix}$

7. $\begin{bmatrix} 0.05 & -0.06 \\ -0.51 & 0.12 \end{bmatrix}$

8. $\begin{bmatrix} -0.12 & 0.31 \\ 0.59 & -0.69 \end{bmatrix}$

9. $\begin{bmatrix} 57 & 61 \\ 72 & 42 \end{bmatrix}$

10. $\begin{bmatrix} 120 & 150 \\ 145 & 142 \end{bmatrix}$

11. Juicy Orchards must decide several times during the year whether or not to spray their apple orchards with insecticide. In this situation, Juicy Orchards is competing against nature. The following matrix gives the worth of the apples for the four possibilities.

		nature	
		needs insecticide	does not need insecticide
Juicy Orchards	use insecticide	$200,000	$150,000
	do not use insecticide	−$50,000	$250,000

a. What is Juicy Orchards' optimal strategy?

b. If Juicy Orchards uses their optimal strategy, what is the lowest expected worth of the apples regardless of nature's "strategy?"

12. Mr. Koehler must decide several times during the growing season whether or not to hire a helicopter to spray his soybeans with herbicide. The following payoff matrix gives the worth of the crop for the four possibilities.

		nature	
		needs herbicide	does not need herbicide
Mr. Koehler	use herbicide	$70,000	$60,000
	do not use herbicide	$30,000	$85,000

a. What is Mr. Koehler's optimal strategy?

b. If he uses the optimal strategy, what is the lowest expected worth of the crop, regardless of nature's "strategy?"

13. The maintenance union at State University negotiates a new contract. The union and management can use either aggressive strategies or conciliatory strategies. Suppose the following matrix gives the payoffs corresponding to the four possibilities. The payoffs represent the increase in the hourly wage won for the union members.

management

aggressive   conciliatory

$$\begin{array}{cc} & \text{aggressive} \\ \text{union} & \text{conciliatory} \end{array} \begin{bmatrix} \$4.50 & \$6.00 \\ \$3.00 & \$5.00 \end{bmatrix}$$

Find the optimal strategies for the union and management and determine the corresponding expected payoff.

14. Ruth and Charles play the following game. Ruth writes either the number 10 or 20 on a piece of paper. If Charles guesses the correct number written on the paper, he receives from Ruth that number in dollars. If Charles does not guess the correct number, he pays Ruth that number in dollars.

   a. Give the payoff matrix for this game.

   b. Find the optimal strategies for each player.

   c. Find the expected payoff for the game if both players use their optimal strategies.

15. Consider a matrix game that is not strictly determined with payoff matrix

$$A = \begin{bmatrix} a & b \\ c & d \end{bmatrix}.$$

Show that if each entry of A is increased by 3, the optimal strategies P^* and Q^* remain the same, but the optimal expected value increases by 3.

SUMMARY OF TERMS AND FORMULAS

Markov Chains
   Transition Probability
   Transition Matrix
   State Distribution Vector
   Initial State Distribution Vector

Matrix Games
   Two-Person Game
   Zero-Sum Game
   Matrix Game
   Row Player
   Column Player
   Payoff Matrix
   Best Move
   Row Minimum
   Column Maximum
   Saddle Point
   Strictly Determined Games
   Value of a Game
   Fair Game

Regular Markov Chains
   Stable State Vector

Non-Strictly Determined Games
   Strategy for a Player
   Expected Payoff for a Game
   Pure Strategy
   Mixed Strategy

Formulas

For a non-strictly determined game

$$A = \begin{bmatrix} a & b \\ c & d \end{bmatrix}$$

the optimal mixed strategy $[p_1 \quad p_2]$ for row player R is

$$p_1 = \frac{d-c}{a+d-b-c} \quad \text{and} \quad p_2 = \frac{a-b}{a+d-b-c};$$

the optimal mixed strategy $\begin{bmatrix} q_1 \\ q_2 \end{bmatrix}$ for column player C is

$$q_1 = \frac{d-b}{a+d-b-c} \quad \text{and} \quad q_2 = \frac{a-c}{a+d-b-c};$$

the expected payoff E for the game is

$$E = \frac{ad-bc}{a+d-b-c}.$$

REVIEW EXERCISES (CH. 10)

In exercises 1–4 state whether the matrix can be a transition matrix for a Markov chain. For those that cannot be a transition matrix, give a reason why.

1. $\begin{bmatrix} 0.23 & 0.77 \\ 0.00 & 1.00 \end{bmatrix}$

2. $\begin{bmatrix} \frac{1}{2} & \frac{1}{3} & \frac{1}{6} \\ \frac{1}{5} & \frac{3}{10} & \frac{1}{2} \end{bmatrix}$

3. $\begin{bmatrix} 0.26 & 0.35 & 0.39 \\ 0.81 & 0.09 & 0.10 \\ 0.39 & 0.39 & 0.22 \end{bmatrix}$

4. $\begin{bmatrix} \frac{2}{3} & \frac{5}{3} & -1 \\ 0 & 1 & 0 \end{bmatrix}$

5. Consider the transition matrix

$$P = \begin{bmatrix} 0.39 & 0.51 & 0.10 \\ 0.38 & 0.33 & 0.29 \\ 0.18 & 0.41 & 0.41 \end{bmatrix}.$$

   a. Find P^2.

   b. If $S^{(0)} = [0.3 \quad 0.4 \quad 0.3]$, find $S^{(1)}$ and $S^{(2)}$.

6. Consider the transition matrix

$$
P = \begin{array}{c} \\ A \\ B \\ C \end{array}
\begin{array}{c} \begin{array}{ccc} A & B & C \end{array} \\
\begin{bmatrix} 1 & 0 & 0 \\ 0 & \frac{1}{2} & \frac{1}{2} \\ \frac{1}{3} & \frac{1}{3} & \frac{1}{3} \end{bmatrix} \end{array}.
$$

a. Compute P^2 and P^3.

b. What is the probability that the system moves from state A to state C in one transition?

c. What is the probability that the system moves from state C to state A in two transitions?

d. What is the probability that the system moves from state C to state B in three transitions?

In exercises 7–10 determine which of the matrices are regular. If the matrix is not regular, explain why.

7. $\begin{bmatrix} \frac{1}{2} & \frac{1}{2} \\ 0 & 1 \end{bmatrix}$

8. $\begin{bmatrix} 0.2 & 0.8 \\ 0.2 & 0.8 \end{bmatrix}$

9. $\begin{bmatrix} 0.35 & 0 & 0.65 \\ 0.79 & 0 & 0.21 \\ 0.56 & 0 & 0.44 \end{bmatrix}$

10. $\begin{bmatrix} 0.1 & 0.9 & 0 \\ 0.4 & 0.6 & 0 \\ 0.2 & 0.5 & 0.3 \end{bmatrix}$

In exercises 11–14 find the stable state vector for the given transition matrix.

11. $\begin{bmatrix} 0.5 & 0.5 \\ 0.6 & 0.4 \end{bmatrix}$

12. $\begin{bmatrix} \frac{1}{3} & \frac{2}{3} \\ \frac{3}{5} & \frac{2}{5} \end{bmatrix}$

13. $\begin{bmatrix} 0.1 & 0.2 & 0.7 \\ 0.8 & 0.1 & 0.1 \\ 0.5 & 0 & 0.5 \end{bmatrix}$

14. $\begin{bmatrix} 0.3 & 0.3 & 0.4 \\ 0 & 0.4 & 0.6 \\ 0.2 & 0.6 & 0.2 \end{bmatrix}$

15. During the month of April in the Lehigh Valley, if it rains one day there is a 45% chance of rain the next day. If it is sunny one day, there is a 65% chance that it will be sunny the next day.

a. Write the transition matrix for this Markov chain with two states: rain or sunny.

b. Find the probability that it will rain three days following one on which is was sunny.

16. Two urns A and B contain balls—2 balls in urn A and 4 balls in urn B. Four of the 6 balls are white and 2 are black. At time $t = k$ $(k = 1, 2, 3, \ldots)$ a ball is drawn at random from each urn. The ball drawn from urn A is placed in urn B and the ball drawn from urn B is placed in urn A. The states for this Markov chain are the number of black balls in urn A.

a. Give the transition matrix P for this Markov chain.

b. Show that P is a regular transition matrix.

c. Find the stable state vector for the Markov chain.

d. Interpret the result of part (c) for the system.

17. Trucks arrive at random at a weighing station on an interstate highway. The line can hold at most two trucks. In a 15-minute interval at most one truck arrives and at most one truck is weighed and leaves. In a given time interval, the probability that a truck arrives is $\frac{1}{3}$ and the probability that a truck is weighed and leaves is $\frac{3}{4}$. Assume the events of trucks arriving and trucks leaving are independent.

a. Find the transition matrix P for this Markov chain.

b. Show that P is a regular transition matrix.

    c. Find the stable state vector for the Markov chain.

    d. Interpret the result of part (c) for the system.

For exercises 18–26 the given matrix is the payoff matrix for a two-person zero-sum game. Decide if the game is strictly determined. If it is, find the best moves for each player and give the value of the game.

18. $\begin{bmatrix} 1 & -2 & 3 \\ 0 & 5 & 1 \end{bmatrix}$

19. $\begin{bmatrix} 1 & -1 \\ 2 & -22 \end{bmatrix}$

20. $\begin{bmatrix} 30 & 45 & 2 \\ 4 & 30 & 15 \end{bmatrix}$

21. $\begin{bmatrix} -1 & 0 & 10 & -2 \\ 3 & -1 & -5 & 7 \\ 4 & -7 & -3 & 9 \end{bmatrix}$

22. $\begin{bmatrix} 20 & 25 \\ -1 & 5 \\ 2 & -10 \end{bmatrix}$

23. $\begin{bmatrix} 8 & 7 \\ 2 & 11 \\ 10 & 12 \end{bmatrix}$

24. $\begin{bmatrix} 100 & 250 & -300 \\ 50 & 25 & -200 \end{bmatrix}$

25. $\begin{bmatrix} 0.1 & 0.2 & -0.1 \\ 0.2 & 0.5 & 0.8 \\ -0.2 & 0.6 & -0.9 \end{bmatrix}$

26. $\begin{bmatrix} 0.10 & -0.25 & 0.12 \\ -0.20 & 0.77 & 0.91 \\ 0.45 & 0.33 & 0.82 \end{bmatrix}$

27. Consider the game with payoff matrix

$$\begin{bmatrix} 20 & -21 \\ 12 & 15 \\ 36 & -27 \end{bmatrix}.$$

Find the expected payoff E for each of the following pairs of strategies.

a.

$P = [0.2 \quad 0.5 \quad 0.3] \qquad Q = \begin{bmatrix} 0.56 \\ 0.44 \end{bmatrix}$

b.

$P = [0.29 \quad 0.61 \quad 0.10] \qquad Q = \begin{bmatrix} 0.71 \\ 0.29 \end{bmatrix}$

28. Consider the game with payoff matrix

$$\begin{bmatrix} 123 & 125 & 315 \\ 219 & 371 & 444 \\ 666 & 541 & 379 \end{bmatrix}.$$

Find the expected payoff E for each of the following pairs of strategies.

a.

$P = [0.451 \quad 0.314 \quad 0.235] \qquad Q = \begin{bmatrix} 0.1 \\ 0.6 \\ 0.3 \end{bmatrix}$

b.
$$P = [0.41 \quad 0.34 \quad 0.25] \qquad Q = \begin{bmatrix} 0.012 \\ 0.531 \\ 0.457 \end{bmatrix}$$

For exercises 29–32 find the optimal strategies and the corresponding expected payoff for each of the matrix games.

29. $\begin{bmatrix} 65 & 35 \\ 35 & 45 \end{bmatrix}$

30. $\begin{bmatrix} -27 & 33 \\ 41 & 27 \end{bmatrix}$

31. $\begin{bmatrix} 357 & -481 \\ -492 & 485 \end{bmatrix}$

32. $\begin{bmatrix} 925 & 816 \\ 653 & 977 \end{bmatrix}$

33. At a party, Erica and Rita play the following game. Erica hides either 2 or 5 cigarettes behind her back. Rita guesses the number of cigarettes. If she guesses correctly, Erica gives Rita the cigarettes. If Rita guesses incorrectly, Rita gives Erica three cigarettes.

   a. Give the payoff matrix for this game.

   b. Find the optimal strategies for each player.

   c. Find the value of the game.

34. Robert and Cary play the following game. Both simultaneously hold up either two or three fingers. If the sum of the fingers shown is even, Cary pays Robert that sum in dollars. If the sum of the fingers shown is odd, Robert pays Cary that sum in dollars.

   a. Give the payoff matrix for this game.

   b. Find the optimal strategies for each player.

   c. Find the expected payoff for the game if both players use their optimal strategies.

CHAPTER 11

Computers and BASIC Language Programming

Introduction

Computers are an integral part of our society and our lives. Most companies use computers to prepare customer billings, prepare the payroll, do accounting operations, or control production processes. Not only are computers used to process vast quantities of information, but they also assist researchers in solving problems.

A computer is a very powerful tool, capable of handling large quantities of information and doing lengthy computations very quickly. However, whatever the particular computing problem may be, there is need for human participation in the process. In addition to preparing data for **input** to the computer and interpreting the **output** from the computer, people give the specific instructions to be executed by the computer. If the computer is given incorrect instructions (or incorrect data), we cannot expect the results to be very useful. Thus the menial tasks of information processing are relegated to the computer, but the creative aspects are a human activity.

An important creative activity in using a computer is the design of a procedure for solving a problem. Since the procedure will be carried out by a computer, it must consist of a finite sequence of unambiguous steps that terminate in a finite amount of time. The procedure we design is called an **algorithm**.

Before we can design the steps of an algorithm, we must have a clear statement of the problem. We must know exactly the type of output (answers or information) we seek. We must also have in mind the input needed to obtain the desired output. Of course, the method we use to obtain the output will in part determine the required input.

The set of instructions in the algorithm we create must eventually appear in the language of the computer, called **machine language**. Since it is difficult and tedious to write instructions in machine language, a **compiler** is available to do this. A compiler acts as a translator to change the instructions written in its language into machine language. This is like two persons who attempt to communicate but use different languages. An interpreter who is able to communicate in both languages must be used. One of many compilers, and the one we will be using, is BASIC. BASIC is an acronym for Beginner's All-purpose Symbolic Instruction Code. It was originally developed in the 1960s at Dartmouth College by Professors J. Kemeny and T. Kurtz. Our instructions written in BASIC are called a **source program**. The BASIC compiler translates the source program into a machine language program that can then be executed by the computer. Remember that the algorithm is designed first and then the source program is written. Thus our algorithm encoded in the language of the BASIC compiler is the end product of programming.

We must be very precise when coding our algorithm in BASIC. The BASIC compiler is not able to make judgments. For example, the BASIC compiler can recognize READ in its dictionary of words but it does not recognize REED as a meaningful word. Therefore, if we are imprecise in writing our source program, it will not be translated, and we have committed what is called a **syntax** error.

Even if we correctly encode our algorithm in BASIC and the compiler derives from it a machine language version, when the set of instructions is executed by the computer correct output may not result. In this case we say that there is a **logical** or **execution** error in our program.

When errors arise, either syntax or logical, we want to find them and correct them; that is, we want to **debug** our program. Since a computer will carry out our instructions precisely, any errors that arise were caused by us. Therefore let us begin by learning how to write "bug free" BASIC programs.

11.1 Variables, Constants, and the LET Statement

To learn how to write a source program in BASIC, we need to understand the components of a computer and their functions. One of the computer's components is **memory**. A function of memory is to store data. The data can be put into memory or retrieved from memory. We can think of memory as consisting of individual **cells**. One **constant** is stored in each cell. A constant is either a number such as 7.89, called a **numeric** constant, or a string of characters such as JOE, called a **string** constant. We will consider only numeric constants here and will refer to them only as constants.

A constant is generally written in any one of several forms. For example, we can write the constant "three" as

$$+3, \quad 3, \quad \text{or} \quad 3.0.$$

If a constant is positive, the " + " sign may be used but it is not required. Commas are never used in the representation of numbers. For example, twelve thousand must be written as 12000 and not as 12,000.

Limitations on the number of digits a memory cell can hold limit the number of digits in a constant. For our discussion of BASIC, we will limit a numeric constant to seven decimal digits; however, this will vary depending on the particular computer being used.

EXAMPLE 1

Which of the constants

$$-3.17 \quad 1,345 \quad 0.0003452 \quad -1567845 \quad 78.512385$$

are acceptable in BASIC?

The constants -3.17, 0.0003452, and -1567845 are acceptable. The constant $1,345$ is not acceptable because it contains a comma, and 78.512385 is not acceptable because it contains eight digits.

Certain constants that contain more than seven decimal digits can be written in the program using **scientific** or **exponential notation**. For example, 37580000 contains eight decimal digits and it cannot be written in the program in this form. However, using powers of 10, this constant can be expressed as

$$3758 \times 10^4.$$

In BASIC, we indicate the exponent (power of 10) by E.

$$3758 \times 10^4 \quad \text{is written as} \quad 3758E4$$

Constants less than one and containing more than seven digits may also be expressed using exponential notation. For example,

$$0.000004689 = 0.4689 \times 10^{-5} \quad \text{is written as} \quad 0.4689E - 5.$$

EXAMPLE 2

Express each of the constants

$$121,376,000 \quad \text{and} \quad -0.00000025$$

in BASIC.

The first constant can be written as $121376E3$ and the second can be written as $-0.25E - 6$. Note that the representation is not unique. For example, $121,376,000$ can also be expressed as $1.21376E8$.

To instruct the computer to move constants to and from memory, we use **variables** in our source program. A variable corresponds to one cell in memory. The variable names are chosen by the programmer using the following rule.

Rule 1

■ A **variable name** can be either

1. one of the 26 letters A through Z,
2. a letter followed by one of the digits 0 through 9, or
3. a letter followed by the symbol $. ■

EXAMPLE 3

The variable names

$$A \quad W3 \quad Z \quad T\$$$

are valid in BASIC, whereas the names

$$AB \quad 3T \quad C34 \quad \$W$$

are not acceptable.

A variable name formed with $, such as T$, is used with string data. Since we will consider only numeric constants, variables of the type T$ will not be considered further.

Constants and variables can be combined by arithmetic operations to form an **arithmetic expression**. The operations that we can write in BASIC are given in table 11.1.

TABLE 11.1

Arithmetic operation	BASIC language symbol
exponentiation	** (or ↑)
multiplication	*
division	/
addition	+
subtraction	−

EXAMPLE 4

The following are examples of arithmetic expressions in the BASIC language.

$$5.7 + 3 \qquad (X + Y) ** 2 \qquad 4.5 + (2 * C) \qquad (A + B)/C$$

Note that parentheses may be used freely to form an arithmetic expression. Even though there is usually a priority in the order in which operations are performed, it is better to use parentheses in the expression to avoid confusion. The one rule we must follow when using parentheses is that each left parenthesis must be paired with a right parenthesis. For example, parentheses do not match in the expression.

$$A + ((B − D)/(E * F)$$

This expression when written properly is

$$A + ((B − D)/(E * F))$$

Care must also be taken to avoid having two arithmetic operations next to each other. For example,

$$A + −B$$

is not permissible, whereas the expressions

$$A + (−B) \qquad \text{and} \qquad A − B$$

are proper.

EXAMPLE 5

Write the algebraic expressions

a. $\dfrac{A + B}{3}$ b. $\dfrac{2A − 5B}{C}$ c. $5X^2 + X$ d. $\sqrt{Y + X}$

in BASIC.

The corresponding BASIC language expressions are

a. (A + B)/3

b. ((2 * A) − (5 * B))/C

c. (5 * (X ** 2)) + X

d. (Y + X) ** .5

One way to put a number into memory is to assign the constant to a variable (cell) by the LET statement. For example, the BASIC statement

$$\text{LET A} = 3.5$$

assigns the constant 3.5 to memory cell A.

The LET statement is also used to place in memory the value obtained from an arithmetic expression. For example, the statement

$$\text{LET X} = (3.5 + 8.5)**2$$

is executed by the computer as follows. First, the value of the arithmetic expression to the right of the equal sign is obtained. Next, this number, 144, is placed in X.

The **central processing unit (CPU)** is the component of the computer that performs the arithmetic operations. The CPU also controls the flow of data into and out of memory. Therefore numbers move from memory to the CPU for arithmetic processing and then back to memory. To understand this, consider the statement

$$\text{LET C} = \text{A} + \text{B}$$

To evaluate the arithmetic expression A + B, the CPU retrieves a copy of the numbers appearing in A and B. The contents of these two cells are not destroyed since only a copy of the contents is passed to the CPU. After the CPU performs the addition, the sum is sent back to memory and placed in cell C.

Let us now see how the statement

$$\text{LET A} = \text{A} + \text{B}$$

is carried out. As in our previous example, a copy of the contents of cell A and cell B is sent to the CPU. The addition is performed as before, but this time the result is sent to memory and placed in A. This time, the original content of A is destroyed since it is replaced. For example, if the contents of A and B are 5 and 7, respectively, then after execution of the LET statement, A will contain 12 and B will still contain 7.

From our discussion, we should realize that the equals sign "=" in a LET statement does not have its usual mathematical meaning but takes on the meaning of "place in" or "assign to." The value of the arithmetic expression on the right of the equals sign is placed into the variable (cell) appearing on the left of the sign. Thus only a variable can appear to the left of the equals sign in a LET statement. For example, the assignment statement

$$\text{LET A} = \text{B} + \text{C} - 2$$

is acceptable, but

$$\text{LET A} + 2 = \text{B} + \text{C}$$

is not acceptable since an arithmetic expression, and not a variable, appears to the left of the equals sign.

Rule 2 ■ The general form of the assignment statement is

$$\text{LET variable} = \text{expression}$$

where we use any acceptable BASIC variable name on the left side of the equals sign and any constant, variable or arithmetic expression on the right side of the equals sign. ∎

EXAMPLES

6. Let us write a sequence of LET statements to obtain the value of the expression $3x + 4$ for $x = 5$ and place this value in y.

 Before the value of $3x + 4$ can be obtained, we must place 5 into x. Thus our first statement is LET X = 5. The next statement is LET Y = $(3 * X) + 4$. This second LET statement accomplishes the evaluation of $3x + 4$ and places the value in y. When the sequence of statements

$$\text{LET X} = 5$$
$$\text{LET Y} = (3 * X) + 4$$

is executed, X will contain 5 and Y will contain 19.

 Note that the order in which the statements are executed is important. For example, the sequence

$$\text{LET Y} = (3 * X) + 4$$
$$\text{LET X} = 5$$

will not achieve the desired result.

7. Give the contents of each variable after the following sequence of statements is executed.

$$\text{LET X} = 1$$
$$\text{LET X} = X + 2$$
$$\text{LET Y} = X/2$$
$$\text{LET Z} = Y * X$$

 After the first two LET statements are executed, X will contain 3. The third LET statement causes the contents of X to be divided by 2 and the result, 1.5, placed in Y. Finally, the contents of Y and X are multiplied and this result placed in Z. Since Y contains 1.5 and X contains 3 when the last LET statement is executed, 4.5 will be placed in Z.

EXERCISES 11.1

---

1. State which of the following are acceptable BASIC variable names. For those that are not, give a reason why.

 a. X b. IS c. A13 d. S3

2. State which of the following are acceptable BASIC variable names. For those that are not, give a reason why.

 a. VAR b. Z c. X4 d. 2A

3. State which of the following are valid constants in the BASIC language. For those that are not, give a reason why.

   a. -1.73 b. 1000.007

   c. 12,073   d. $1041.23

   e. $-4.73752E + 11$ f. $0.2735417E - 2$

4. State which of the following are valid constants in the BASIC language. For those that are not, give a reason why.

   a. -0.00754 b. 13400000

   c. 54,781,000   d. $0.234E - 15$

5. For each of the following, give the decimal number representation of the constant.

   a. 1×10^{-3} b. 0.23E6

   c. $1345E - 4$ d. 789×10^2

6. For each of the following, give the decimal number representation of the constant.

   a. 234×10^5 b. 0.287×10^{-4}

   c. $294E + 7$ d. $0.382E - 3$

7. For each of the following, give the number as a BASIC constant using the E notation.

   a. 23×10^{-4} b. 1.7002

   c. 12,736,000   d. 0.000184357

8. For each of the following, give the number as a BASIC constant using the E notation.

   a. 0.0034   b. 739×10^7

   c. 11,000,820   d. 173.25

For exercises 9–18 write the algebraic expression as a BASIC language arithmetic expression.

9. $\dfrac{A + B}{2}$ 10. $AB + AC$ 11. $2x + 3y + 5z$ 12. $\dfrac{3x + z}{4x}$

13. $\sqrt{AB}$ 14. $\sqrt{A^2 + B^2}$

15. $\dfrac{3(x - x_1)^2 + 2(x - x_2)^2}{5}$ 16. $P_1x_1 + P_2x_2 + P_3x_3$

17. $\dfrac{(1 + i)^n - 1}{i}$ 18. $\left(1 + \dfrac{r}{m}\right)^{-tm}$

19. State why each of the following is not acceptable as a BASIC language statement.

   a. LET $A + B = C$ b. LET $AX = A * X$

   c. LET $C = A$ PLUS B   d. LET $Y = A + 2B$

20. State why each of the following is not acceptable as a BASIC language statement.

   a. LET $3 = B$ b. LET $A * B = C$

   c. LET $Z = XY$ d. LET X EQUAL Y

For exercises 21–24 state what the contents of each variable will be if the given sequence of LET statements is executed in the order given.

21. LET X = 2
    LET Y = X + 3
    LET Z = Y * X

22. LET A = 4
    LET A = A ** 2
    LET B = A + 5

23. LET B = 5
    LET C = 7
    LET X = −2
    LET Y = B * (X ** 2) + C * X

24. LET X = 1
    LET Y = X
    LET X = X + 2
    LET Z = Y + X

11.2 Running a Program and More BASIC Statements

In this section we discuss how communication with the computer is accomplished, and then proceed to more BASIC statements. In most large computer systems, BASIC programs are executed in the **time-sharing** mode. In this mode, several users are connected to the same computer at the same time. The computer system gives each user a small fraction of time on a rotating basis. However, the computer operates at such a high speed that each user feels like the sole user of the computer.

A **terminal** is used to communicate with the computer. The keyboard of the terminal looks like a standard typewriter but has additional special keys. In some computer systems, the terminals can be connected to the computer by simply turning the power on at the terminal. In other systems, a special telephone number must be dialed.

Once the terminal is connected to the computer, the computer system will give a command to LOGON. The LOGON command and subsequent responses depend on the system, so the exact details should be checked for the computer system you will be using.

After the LOGON is accepted, we want to request the use of the BASIC system. This is generally accomplished with a command such as

$$\text{EXEC BASIC} \quad \text{or} \quad \text{BASIC}$$

typed at the terminal. As with any line typed at the terminal, we must indicate the completion of the line. A standard method for indicating the completion of an input line is to depress the RETURN key on the terminal (again, this method may be different for your system).

If we have been successful up to this point, the system will give a message at the terminal, such as READY, to indicate that BASIC is ready for our use. We now enter the BASIC statements of our source program. Of course, these statements would have been written prior to our LOGON. We must give a **line number** to each statement in our source program. The line number is typed first, followed by the particular BASIC statement. Any positive integer from 1 to 99999 can generally be used as a line number.

It is always a good idea to number the statements by tens to allow for future modifications. For example, suppose we want the following statements in the indicated order in our source program:

$$LET\ X = 10.2$$
$$LET\ Y = 5.7$$
$$LET\ Z = X * Y$$

but typed the lines

10	LET X = 10.2
20	LET Z = X * Y

Instead of typing these two lines over, to add the missing statement we only have to type the line

15   LET Y = 5.7

and BASIC will insert this line between the other two. The system will always arrange the statements in the order of ascending line numbers. Thus the line number indicates the position and order in which a statement is to be executed.

If we want to **delete** a line in our source program, we need only enter the line number and depress the RETURN key. If we want to change a line, we enter the line number, followed by the new statement. Some systems allow for backspacing on a line and other types of **editing** commands. These should be checked for your system.

Now suppose we have typed a BASIC source program. We wish to have this program translated and executed. The command we give is

RUN

This command is not part of our source program, so it is not given a line number. If there are no syntax errors, the machine language version of our source program as created by BASIC will be executed automatically. The completion of the execution phase is indicated by the system with the message READY.

If the program was executed in the manner we intended, giving us the output we desired, and if we do not wish to do anything further with the program during the session, we may want to **save** the source program in our **library** for future runs or modifications. The manner in which a program is put into the library or retrieved from the library depends on the system.

After we have completed all the work we wanted to do during the LOGON session, we must LOGOFF. This can be accomplished by typing in, for example,

BYE   or   LOGOFF

Let us continue now with more BASIC language statements.

We saw in the previous section how a LET statement can be used to place a number in memory. Another way of assigning values to variables (that is, to memory) is to use a combination of DATA and READ statements. A combination of these BASIC statements is, for example,

DATA −3.5, 0.072, 16, 0.25E − 6
READ W, X, Y, Z

The execution of the READ statement in conjunction with the DATA statement causes -3.5 to be assigned to W, 0.072 to X, 16 to Y, and $0.25E - 6$ to Z. We see that the same assignment is accomplished by the four LET statements

$$LET\ W = -3.5$$
$$LET\ X = 0.072$$
$$LET\ Y = 16$$
$$LET\ Z = 0.25E - 6$$

but this sequence entails more instructions.

To see how READ statements are used in conjunction with DATA statements, let us consider what is called a **data stack**. BASIC translates the source program so that the constants appearing in all the DATA statements of the source program are placed in a stack in the order in which they appear in the program. A **pointer** is set to point to the constant at the top of the stack. Each time a variable is encountered in a READ statement, the constant at the pointer is assigned to the variable and the pointer moves down to the next constant in the stack. If there is not sufficient number of constants in the DATA statements, the program will run out of data for the variables and an error will arise. If there are more constants than variables, no difficulty arises even though there would be constants left in the stack.

EXAMPLE 1

Consider the sequence of statements

```
10    DATA  5
20    READ  X, Y, Z
30    DATA  6, 7, 8
```

and let us describe how they are executed.

The two DATA statements create a data stack containing 5, 6, 7, and 8, in that order. When the READ statement is executed, the constants 5, 6, and 7 are assigned to X, Y, and Z, respectively. The pointer then moves down the stack and points to the constant 8.

Now consider the statements

```
10    DATA  5, 6
20    READ  X, Y, Z
```

The data in the stack will be 5 and 6. Since the READ statement requires three constants for the variables X, Y, and Z and since the stack contains only two, an error will occur because of insufficient data.

If we understand that the DATA statements within the source program create a data stack and that the constants in the stack are copied sequentially from the stack as the READ statements are executed, we see that it does not matter where the DATA statements appear in the source program. However, for readability we should write all DATA statements together.

A third method for putting data in memory is by the use of the

INPUT

statement. When an INPUT statement is encountered, execution of the program pauses and a **prompt** is given at the terminal. Execution of the program resumes after the necessary data are typed in and the RETURN key is depressed. For example, consider the statement

20   INPUT A, B

Execution of this statement causes, for example, the prompt

?

to be displayed at the terminal; whereupon we could type

3, 4

and depress the RETURN key. The program would then resume execution and put 3 into A and 4 into B. Note that commas are typed between the successive constants to be entered.

If, instead of typing 3, 4, we were to enter only 3, the system would give another ? (or give the message ENTER MORE DATA AT LINE 20) since the program is waiting for another constant for B.

The major advantage in using INPUT statements instead of READ and DATA statements is that the data can be entered as the program is running. Therefore partial results can be seen before more data are entered.

A simple question mark caused by the execution of an INPUT statement does not indicate the amount or type of data needed. We can know that only if we know the particular INPUT statement being executed. Therefore it is always a good idea to write our programs with **documentation**. Immediately prior to an INPUT statement we should have a message displayed at the terminal to indicate the needed input. Displaying messages and results is accomplished by the BASIC statement

PRINT

For example, we can cause a message such as ENTER X to be given during program execution by the statement

10   PRINT "ENTER X"

In general, a PRINT statement of the form

PRINT "   "

will cause whatever is between the quotation marks to be given. We can write anything we want between the quotation marks and it will be given as written.

EXAMPLES

2. What will the output be at the terminal when the following statements are executed?

10   PRINT "THIS HAS SPACING"
20   PRINT "THISHASNOSPACING"

The result of execution of line 10 is

THIS HAS SPACING

whereas the result of line 20 is

THISHASNOSPACING

3. What will the output be at the terminal when the following statements are executed?

10   PRINT "ENTER THE VALUE FOR X"

20   INPUT X

Line 10 causes the printing of the message. The result of executing line 20 is a ? printed at the terminal and the program pausing for input. Thus the result we see at the terminal is

ENTER THE VALUE FOR X

?

We also use a PRINT statement to retrieve data from memory and display it at the terminal. To accomplish this, we include with a PRINT statement the particular variable (or variables) that holds the desired data. For example, the statement

PRINT A, B

causes the contents of A and the contents of B to appear on the same line. If we wish to see these two pieces of data on individual lines, we would use two PRINT statements

PRINT A

PRINT B

Each PRINT statement usually creates one line of output; some exceptions to this will be noted.

We can also have a message on the same line with the data being retrieved. For example, the result of execution of the statements

10   DATA -5.2

20   READ A

30   PRINT "A =", A

is the one line

A = -5.2

Note the space between = and -5.2. This space is caused by the comma between " and A in the PRINT statement, line 30. If we wish to have -5.2 closer to the message, we would use a semicolon (;) in place of the comma. The result of the statement

$$30 \qquad \text{PRINT "A ="; A}$$

is $\qquad\qquad$ A = −5.2

To understand the effect of using a comma or a semicolon in a PRINT statement, we need to know how a line of output is structured by a PRINT statement. The explanation we give here will assume that we are using a **hardcopy terminal**. The essentials are no different if we are using a **CRT terminal**.

Each print line has a fixed number of print positions. For our discussion, we will use 60. The 60 print positions are divided into 4 **fields** of 15 each. When a comma is encountered in a PRINT statement, the print bar on the terminal moves right to the next field and begins printing in the leftmost space of that field. When a semicolon is encountered, the print bar does not move to the next field but moves only to the next print position (or possibly moves right 3 print positions to allow for two blanks). A blank line of output can be created by simply using a PRINT statement such as

$$10 \qquad \text{PRINT}$$

EXAMPLE 4 | Give the lines of output created by executing the following sequence of statements.

$$
\begin{aligned}
&10 \qquad \text{DATA } 3.75, -10 \\
&20 \qquad \text{READ A, B} \\
&30 \qquad \text{LET C} = \text{A} + \text{B} \\
&40 \qquad \text{LET D} = \text{A} * \text{B} \\
&50 \qquad \text{PRINT "A ="; A, "B ="; B} \\
&60 \qquad \text{PRINT "THE SUM OF A AND B IS} \qquad \text{"; C} \\
&70 \qquad \text{PRINT} \\
&80 \qquad \text{PRINT "A TIMES B IS", D}
\end{aligned}
$$

The output at the terminal will be

$$
\begin{aligned}
&\text{A} = 3.75 \qquad\qquad \text{B} = -10 \\
&\text{THE SUM OF A AND B IS} \qquad -6.27 \\
&\text{A TIMES B IS} \qquad -37.5
\end{aligned}
$$

Except for one additional statement, the source program of example 4 could be compiled and executed. Any BASIC source program must have an

$$\text{END}$$

statement. The END statement signals to BASIC that the last statement to be translated or executed has been reached. Thus an END statement is written at the end of each source program and has the highest line number used in the source program.

EXAMPLES

5. Describe what the following program does when it is compiled and executed.

```
10      PRINT "ENTER VALUES FOR A, B, C"
20      INPUT A, B, C
30      PRINT "ENTER VALUES FOR X, Y, Z"
40      INPUT X, Y, Z
50      LET P = (A * X) + (B * Y) + (C * Z)
60      PRINT "THE ROW-COLUMN PRODUCT IS   "; P
70      END
```

Through the terminal, the program receives input data for A, B, and C and input data for X, Y, and Z. In lines 50 and 60 in the program the product $AX + BY + CZ$ is computed and printed out. We recall (from chapter 5) that this computer output is the row-column product.

$$[A \quad B \quad C]\begin{bmatrix} X \\ Y \\ Z \end{bmatrix} = AX + BY + CZ$$

Suppose we have already typed in this source program and we have given the RUN command. When the first question mark appears, suppose we enter 3, 0, 1, and when the second question mark appears, we enter $-2, 5, 8$. The lines of print we see at the terminal are as follows:

```
ENTER VALUES FOR A, B, C
?3, 0, 1
ENTER VALUES FOR X, Y, Z
? -2, 5, 8
THE ROW-COLUMN PRODUCT IS     2
```

6. What will the results be when the following source program is compiled and executed?

```
10      DATA 10000, 0.015, 6
20      READ P, I, N
30      LET S = P * ((1 + I) ** N)
40      PRINT "PRINCIPAL", "RATE", "PERIODS"
50      PRINT P, I, N
60      PRINT
70      PRINT "THE AMOUNT IS   "; S
80      END
```

Recall from chapter 7, formula (10) on page 295, that the arithmetic expression in line 30 is the BASIC equivalent for computing the compound amount. Thus the program obtains the compound amount S for the principal P, the rate per period I, and the number of periods N. Values for the variables P, I, and N are obtained through the READ and DATA statements. The output for the data given in line 10 will be

PRINCIPAL	RATE	PERIODS
10000	.015	6

THE AMOUNT IS   10934.42

If we wish to run this program for a different set of data for P, I, and N, we only have to change line 10.

EXERCISES 11.2

For exercises 1 and 2 give DATA and READ statements that will accomplish the same thing as the given LET statements.

1. LET A = 7.2
   LET B = -0.03
   LET C = 6

2. LET X = 4
   LET Y = 5
   LET Z = 3.5

3. Write DATA and READ statements so that when they are executed the contents of the variables X1, X2, and X3 will contain the numbers 72.45, 95, and 10,731.52, respectively.

4. Write DATA and READ statements so that when they are executed the contents of the variables A, B, and C will contain the numbers 0.00007683, 0.23E $-$ 5, and 937.556, respectively.

For exercises 5–8 state what the contents of each variable will be when the given statements are executed. Also state whether or not constants will remain in the data stack or if there will be insufficient data.

5. DATA 3, 5
   DATA $-4, 3$
   READ A, B, C, D

6. DATA 4, 0.01, 3
   READ X
   READ Y

7. DATA 3, $-1, 2$
   DATA 5, 4
   READ X1, X2, X3
   READ Y1, Y2, Y3

8. DATA 3, $-1, 2$
   DATA 5, 4
   READ A, B
   READ C, D, E

For exercises 9 and 10 describe what will occur when the given INPUT statement is executed and the input line

$$3, 4, 5$$

is entered when the question mark appears.

9. INPUT W, X, Y, Z

10. INPUT A, B

For exercises 11 and 12 describe how many question marks will appear at the terminal and how many constants should be entered for each question mark when the given statements are executed.

11. 10   DATA 3, 4
    20   READ A
    30   INPUT X

12. 10   DATA 3, 4, 5
    20   INPUT X, Y, Z
    30   INPUT A

For exercises 13–16 show what the printout will be when the given statements are executed.

13. 10   DATA −0.01, 7.2
    20   READ A, B
    30   PRINT "A IS   "; A
    40   PRINT "B IS   "; B

14. 10   DATA 48.92, 702.58
    20   READ A, B
    30   PRINT "THE NUMBERS ARE"
    40   PRINT A, B

15. 10   DATA 989, 0.02
    20   READ P, R
    30   PRINT "THE PRINCIPAL IS", "THE RATE IS"
    40   PRINT P, R

16. 10   DATA 11, 13
    20   READ X, Y
    30   PRINT "FIRST NUMBER"; X, "SECOND NUMBER"; Y

17. Consider the program given in example 5.
    a. What should the printout at the terminal be if − 7, 4, and − 3 are entered at the first ? and 8, 5, and 6 are entered at the second ? when the program is executed?
    b. Use your computer system to type in the source program of example 5 and run it using the data given in part (a).

18. Consider the program given in example 6.
    a. Suppose the DATA statement is changed so that 500 is assigned to P, 0.02 is assigned to I, and 12 is assigned to N. If the program is executed with these changes, what would the printout be?
    b. Use your computer to type in the source program of example 6 as changed in part (a) and run it.

19. a. Write a program to compute and print the value of the linear form

$$F = 5w + 7x + 8y + 10z + 155$$

where the values for the variables are obtained through an INPUT statement.
    b. Run the program you wrote in part (a), using the values 7.5, 8.2, 9.3, and 6.1 for w, x, y, and z, respectively.

20. a. Write a program to compute and print the row-column product

$$[a_1 \quad a_2 \quad a_3 \quad a_4] \begin{bmatrix} x_1 \\ x_2 \\ x_3 \\ x_4 \end{bmatrix},$$

where the values for the variables are obtained through INPUT statements.

b. Run the program you wrote in part (a) to obtain the value of the row-column product

$$[3 \quad -1 \quad 4 \quad 8] \begin{bmatrix} 9 \\ 11 \\ -3 \\ 7 \end{bmatrix}.$$

21. a. Write a program to compute and print the value of P in the formula

$$P = \frac{S}{(1 + i)^n}$$

where the values for S, i, and n are obtained through an INPUT statement. Your program should also print the values of S, i, and n. (Note from chapter 7 that P is the present value of S.)

b. Run the program you wrote in part (a), using the values $1500 for S, $\frac{5}{12}\%$ for i, and 24 for n.

22. a. Write a program to compute and print the value of R in the formula

$$R = A\frac{i}{1 - (1 + i)^{-n}}$$

where the values for A, i, and n are obtained through an INPUT statement. Your program should also print the values of A, i, and n. (Note from chapter 7 that R is the periodic payment needed to pay a loan A.)

b. Run the program you wrote in part (a), using the values $9000 for A, 1% for i, and 36 for n.

11.3 Control Structures: IF-THEN Statements and FOR-NEXT Loops

A program consists not only of a set of instructions; it also includes an order in which the instructions are to be executed. We consider in this section the order in which instructions are executed, called the **control structure** of the program.

Unless otherwise modified by what are called **control statements**, the control flows through the program in a "one-after-another" fashion; that is, the instructions in a source program will be executed in the order of their line numbers. An example of a program with a one-after-another control structure is

```
10    DATA 0.054, 1.025
20    READ A, B
30    LET X = A + B
40    PRINT X
50    END
```

By using decisions we can design various other control structures in our programs. A decision is written into a program with an IF-THEN statement. An example of this BASIC language statement is

IF X = 5 THEN 60

BASIC translates this statement as follows. The IF condition X = 5 is tested by comparing the contents of X to 5. If the condition is true, program execution **branches** to line 60; otherwise, execution continues to the line immediately following the IF-THEN statement.

The general form of the IF-THEN statement is

IF condition THEN line number

where the condition written after the IF is of the form

expression RELATION expression

The relations that can be used in the IF condition are given in table 11.2.

TABLE 11.2

Relation	BASIC language symbol
equal to	=
greater than	>
greater than or equal to	> =
less than	<
less than or equal to	< =
not equal to	< >

For each expression in the IF condition we use any variable, constant, or arithmetic expression.

EXAMPLE 1

Describe the result of running the following program.

```
10    INPUT X
20    IF X > = 0 THEN 40
30    PRINT "X IS NEGATIVE"
40    END
```

What occurs on a run of the program depends on the data entered when the INPUT X statement is executed. Suppose we enter 3 when ? appears. Since the content of X is positive for this input, and thus the condition X > =0 is true, execution branches to line 40, the program stops, and we do not receive any printout for this input.

If we were to enter −4, however, we would receive the message X IS NEGATIVE before execution stops.

It is disconcerting if there is no printout on a program run, such as for the program in example 1. When there is no output, we cannot be sure whether the program was executed properly or there was some difficulty with the system.

EXAMPLE 2

Consider the following program, which is a modification of the program in example 1. Describe the printout when this program is executed.

```
10      INPUT X
20      IF X > = 0 THEN 40
30      PRINT "X IS NEGATIVE"
40      PRINT "X IS NONNEGATIVE"
50      END
```

For nonnegative input, such as 3, we will receive the message X IS NONNEGATIVE. However, if the input for X is -3, we will not only receive the message X IS NEGATIVE but also the message X IS NONNEGATIVE. This result occurs because the program is executed in one-after-another fashion from line 30.

For the program in example 2, we need a way to branch around line 40 after line 30 is executed. This can be accomplished by inserting between lines 30 and 40 a GO TO statement. The general form of this BASIC language statement is

GO TO line number.

When a GO TO statement is encountered, execution branches to the instruction with the given line number.

EXAMPLE 3

The problem in the program of example 2 is avoided in the following program. Only the message X IS NEGATIVE is printed when the input data are negative.

```
10          INPUT X
20    REM
30    REM     THIS PART TESTS X AND PRINTS A MESSAGE
40          IF X > = 0 THEN 70
50            PRINT "X IS NEGATIVE"
60            GO TO 90
70            PRINT "X IS NONNEGATIVE"
80    REM
90            END
```

Let us see how a GO TO is used with an IF-THEN statement. Consider again the program in example 3. The GO TO (line 60) allows for branching around the "true" instruction (line 70) after the "false" instruction (line 50) is executed. We must take care in how GO TO statements are used in our programs. A GO TO should only be used in conjunction with an IF-THEN, since branching around a set of instructions should only be done when some condition occurs or exists.

Note the use of the BASIC language statement

REM

in the program of example 3. REM statements are not translated by the compiler into a computer instruction. They are used only to insert in the source program remarks and spaces for readability. The program of example 3 consists of the following three parts.

1. Input data (line 10).
2. Test the data and print a message (lines 40–70).
3. End execution (line 90).

We made these units stand out in the source program by inserting spaces (lines 20 and 80) between each of them. A remark on the purpose of lines 40 through 70 was inserted with a REM statement (line 30).

Note a final point on the program of example 3. Lines 50 through 70 are indented under the IF-THEN statement. The indentation is used to show that these lines belong to the IF-THEN statement.

The use of REM statements and indentation in a source program are some aspects of **programming style**. They are not necessary, but they make a source program easier for people to read.

EXAMPLE 4

Let us write a program to enter values for x, y, and z and determine whether or not this triple is a solution to the linear equation $3x - 2y + 5z = 25$. The program should give appropriate output based on the condition that occurs.

To begin our program design, we should first think of the things to be accomplished. The program should consist of the following four parts.

1. Input values for x, y, and z.
2. Compute the value of $3x - 2y + 5z$ for this input.
3. Test this value against 25 and print a message based on the test.
4. End execution of the program.

The first, second, and fourth parts are accomplished by the statements

INPUT X, Y, Z
LET V $= (3 * X) + (-2 * Y) + (5 * Z)$
END

The third part can be accomplished by an IF-THEN structure. The condition of the IF-THEN statement is V $= 25$. If this condition is true, we want to print that the triple is a solution; otherwise, the program is to print that it is not a solution.

A listing of our source program, which includes REM statements and indentation for style, and the printout from two runs of the program, are given in figure 11.1.

```
*LIST

10   REM   THIS PROGRAM TESTS IF A TRIPLE OF INPUT
20   REM   DATA IS A SOLUTION TO 3X - 2Y + 5Z = 25
30   REM
40         PRINT "ENTER VALUES FOR X, Y, AND Z"
50         INPUT X, Y, Z
60   REM
70         LET V = 3*X  - 2*Y + 5*Z
80   REM
90         IF V = 25 THEN 150
100            PRINT
110            PRINT "THE INPUT IS NOT A SOLUTION"
120            PRINT "=== ===== == === = ========="
130            PRINT
140            GO TO 200
150            PRINT
160            PRINT "THE INPUT IS A SOLUTION"
170            PRINT "=== ===== == = ========="
180            PRINT
190  REM
200            END
*RUN


ENTER VALUES FOR X, Y, AND Z
?1,-1,4

THE INPUT IS A SOLUTION
=== ===== == = =========

*RUN
ENTER VALUES FOR X, Y, AND Z
?2,1,2

THE INPUT IS NOT A SOLUTION
=== ===== == === = =========
```

FIGURE 11.1

Often we want to design a program that executes a set of instructions a given number of times. For example, suppose we wish to compute and print the values of the formula

$$S = (1 + 0.015)^n$$

for $n = 1, 2, \ldots, 50$. The repeated execution of a set of instructions is called a **loop** structure. A loop structure can be accomplished in BASIC with the

FOR   and   NEXT

statements. The following is an example of their use.

10   FOR N = 1 TO 50 STEP 1

20   LET S = (1.015) ** N

30   PRINT N, S

40   NEXT N

Line 10 is the beginning of the loop and line 40 is the end of the loop. Lines 20 and 30 form the set of instructions to be repeated for each value of N from 1 to 50. The variable N is called a **counter** or a **control variable** for the loop, since it counts (controls) the number of times execution passes through the body of the loop (lines 20 and 30). The beginning and final values for N and the increment for N are given in the FOR statement. For our example, the beginning value 1, the final value 50, and **step size** 1 for the counter N are given in line 10. If the step size is 1, then STEP 1 need not be included in the FOR statement. For example, FOR N = 1 TO 50 is equivalent to FOR N = 1 TO 50 STEP 1. If the step size is not included in the FOR statement, then the compiler **defaults** to step size 1.

We can write FOR-NEXT loops using any beginning and final values and step sizes for the counter. For example,

$$FOR\ I = 12\ TO\ 6\ STEP\ -2$$

will execute the loop for the value I = 12, 10, 8, and 6. Expressions can be used in the FOR statement as long as their values can be determined when the loop begins. The counter can be used inside the loop but never to the left of the equals sign in an assignment statement. The general rule is that the value of the control variable can never be changed inside the loop.

EXAMPLES

5. Describe the output obtained from executing the following program.

```
10    DATA 11
20    READ K
30    FOR I = 3 TO K STEP 2
40        LET X = I ** 2
50        PRINT I, X
60    NEXT I
70    END
```

The number of times lines 40 and 50 are executed depends on the values of the control variable I in the FOR-NEXT loop. Since the beginning value for I is 3, the final value for I is K = 11, and the step size is 2, the loop will be executed for I = 3, 5, 7, 9, and 11. Therefore the program will create the output

3	9
5	25
7	49
9	81
11	121

6. The source program in figure 11.2 will compute and print the arithmetic average of N numbers in a set of data. The values for N and the numbers must be entered at the terminal during execution. The results of a run of the program are given with the listing.

```
*LIST


10   REM   THIS PROGRAM COMPUTES THE AVERAGE OF N NUMBERS
20   REM
30         PRINT  "HOW MANY NUMBERS ARE THERE"
40         INPUT N
50   REM
60         LET S = 0
70         PRINT  "AT EACH ? ENTER THE NEXT NUMBER"
80         FOR I = 1 TO N
90            INPUT X
100             LET S = S + X
110        NEXT I
120 REM
130        LET A = S/N
140        PRINT
150        PRINT "THE AVERAGE OF THE NUMBERS IS"; A
160        PRINT   "=== ======= == === ======= =="
170 REM
180        END
*RUN


HOW MANY NUMBERS ARE THERE
?5
AT EACH ? ENTER THE NEXT NUMBER
?4
?7
?5
?10
?11


THE AVERAGE OF THE NUMBERS IS 7.4
=== ======= == === ======= ==
```

FIGURE 11.2

EXERCISES 11.3

For exercises 1–4 state what the printout would be if the program were executed.

1.
```
10    DATA 30, 35
20    READ X, Y
30    IF X > Y THEN 60
40        PRINT X
50        GO TO 70
60        PRINT Y
70    END
```

2.
```
10    DATA 15, 42
20    READ X, Y
30    IF X = Y THEN 60
40        PRINT X
50        GO TO 70
60        PRINT Y
70    END
```

3.
```
10    DATA 2, 6, 3
20    READ A, B, C
30    LET D = A * C
40    IF B < = D THEN 60
50        PRINT D
60    END
```

4.
```
10    DATA 4, 11, 5
20    READ A, B, C
30    LET D = A * C
40    IF B < = D Then 60
50        PRINT D
60    END
```

For exercises 5–8 give the complete set of values of the control variable I as determined by the given FOR statement.

5. FOR I = 3 TO 10

6. FOR I = −1 TO 7 STEP 2

7. FOR I = 15 TO 0 STEP −3

8. FOR I = 21 TO 15 STEP −1

For exercises 9–12 give the printout that would result from executing the given program.

9.
```
10    DATA 5, 7, 2, 4
20    LET P = 1
30    FOR I = 1 TO 4
40       READ X
50       LET P = P * X
60    NEXT I
70    PRINT P
80    END
```

10.
```
10    DATA 5, 7, 2, 4
20    LET P = 1
30    FOR I = 1 TO 4
40       READ X
50       PRINT X, P
60       LET P = P * X
70    NEXT I
80    END
```

11.
```
10    DATA 6, −7, 8, 5, −9, 0, 6
20    READ K
30    FOR I = 2 TO K STEP 2
40       READ A, B
50       PRINT B
60    NEXT I
70    PRINT A
80    END
```

12.
```
10    LET K = 3
20    FOR I = 10 TO 5 STEP −1
30       LET K = K + I
40    NEXT I
50    PRINT K
60    END
```

13. a. Suppose the DATA statement and the FOR statement in the program of example 5 (page 537) were changed to

$$\text{DATA } 9 \quad \text{and} \quad \text{FOR I} = -3 \text{ TO K STEP } 3$$

What would the printout be if the program were run with these changes?

b. Make the changes in the source program of example 5 (page 537) as suggested in part (a) and run the program with these changes.

14. a. What would the printout be if we executed the program of example 6 (page 537) and we entered 6 at the first question mark and 2, 4, 5, 7, 12, and 9 at the next question marks?

b. Run the program in example 6 (page 537), using the input given in part (a).

15. Write and run a program to test whether or not a triple (x, y, z) is a solution to the linear equation

$$ax + by + cy = d$$

where the values of $a, b, c,$ and d are obtained through a DATA statement and the values for $x, y,$ and z are entered through an INPUT statement. Use the DATA statement

$$\text{DATA } 1, 4, -3, 15$$

for your program.

16. Write and run a program to test whether or not a triple (x, y, z), obtained through an INPUT statement, is a solution to the system of linear equations

$$x + 4y - 3z = 15$$

$$2x \qquad + 5z = 24.$$

17. a. Write a program to compute the mean of a random variable X with frequency table

value of X	x_1	x_2	x_3	$\cdots$	x_n
frequency	f_1	f_2	f_3	$\cdots$	f

where n, the values of X, and the frequencies are obtained through INPUT statements.

b. Run your program for example 2 of section 9.2 (page 404).

18. a. Write a program to compute the expected value $E(X)$ of a random variable X with probability distribution table

value of X	x_1	x_2	x_3	$\cdots$	x_n
probabilities	P_1	P_2	P_3	$\cdots$	P_n

where n, the values of X, and the probabilities are to be obtained through INPUT statements.

b. Run your program for example 5 of section 9.2 (page 408).

19. a. Write a program to compute and print the values of

$$S = (1 + i)^n$$

where the value for i is entered through an INPUT statement and the values for n are $1, 2, 3, \ldots, 24$.

b. Run the program you wrote in part (a), using $\frac{7}{12}\%$ as input for i, and compare your results to table III of appendix B.

20. Modify the program written in exercise 19 so that the values of S can be obtained during a single run for three different values for i, and n has the values $1, 2, 3, \ldots, 24$ for each value for i.

11.4 Arrays and MAT Statements

The BASIC language provides for the grouping of memory cells to represent a matrix. We call such a collection of cells an **array**. Recall from chapter 5 that a matrix can be a single column (or row) of numbers or several rows and columns. To distinguish between the two, we call an array that represents a column matrix (or a row matrix) a **one-dimensional** array, and an array that represents a matrix with more than one row and column a **two-dimensional** array.

We use the BASIC statement DIM to tell the computer the number of memory cells to be allocated to the array. For example, if we wish to allocate 10 memory cells to a one-dimensional array B, we use the statement

DIM B(10)

The BASIC language statement

DIM C(3, 5)

sets up in memory a two-dimensional array C representing a matrix with 3 rows and

5 columns. The specific dimensions of several arrays can be established with one DIM statement. For example

$$\text{DIM A}(5, 5), X(8)$$

establishes a two-dimensional array A and a one-dimensional array X.

Consider a one-dimensional array B containing 10 cells. We can place data in or retrieve data from these cells by using with the name B a single subscript enclosed within parentheses. For example,

$$\text{LET B}(3) = 0.6$$

places the constant 0.6 in the third location of B.

As with matrices containing more than one row or column, we use two subscripts to gain access to a cell in a two-dimensional array. For example, the statement

$$\text{PRINT A}(1, 2)$$

retrieves and prints the contents of the cell corresponding to the entry in row 1 and column 2 of the matrix.

EXAMPLE 1

Describe what occurs when the following sequence of statements is executed.

```
10    DIM  C(5)
20    DATA  3, −5, 7, −9, 8
30    FOR  I = 1  TO  5
40        READ  C(I)
50    NEXT  I
```

Line 10 creates a one-dimensional array C with 5 cells. The FOR-NEXT loop (lines 30 through 50) causes the constants in the DATA statement (line 20) to be read into the 5 cells of the array C.

The BASIC compiler for some computer systems provides instructions for certain operations on arrays. Use of these instructions can eliminate several lines in the source program. Each of these instructions starts with MAT. For example, the FOR-NEXT loop in example 1 for reading data into the array C can be replaced by the one instruction

$$\text{MAT READ C}$$

MAT PRINT and MAT INPUT statements are also available in BASIC. The first of these instructions is for retrieving and printing the entire array, and the second one is for entering data from the terminal into the entire array. For two-dimensional arrays, MAT PRINT causes the contents of the array to be printed row by row. MAT INPUT and MAT READ cause the data to be placed in the array row by row. The actual form of the input lines entered through the terminal for a MAT INPUT statement depends on the system being used, and you should check this for your system.

EXAMPLE 2

The following program creates an array A with 2 rows and 3 columns, enters data into the entire array, and prints the entire array.

$$
\begin{array}{ll}
10 & \text{DIM A}(2, 3) \\
20 & \text{DATA 2, 5, 6} \\
30 & \text{DATA 8, 9, 1} \\
40 & \text{MAT READ A} \\
50 & \text{MAT PRINT A} \\
60 & \text{END}
\end{array}
$$

The printout for this program will be as follows.

$$
\begin{array}{ccc}
2 & 5 & 6 \\
8 & 9 & 1
\end{array}
$$

If we wish to obtain a more compact printout, we would use a semicolon in the MAT PRINT statement. By changing line 50 to

$$\text{MAT PRINT A;}$$

our printout will be

$$
\begin{array}{ccc}
2 & 5 & 6 \\
8 & 9 & 1
\end{array}
$$

Matrix operations such as adding, subtracting, and multiplying two matrices, finding the transpose of a matrix, and finding the inverse of an invertible matrix were defined and studied in chapter 5. Each of these operations can be accomplished in BASIC with a single MAT instruction. For example, the product of two matrices A and B can be obtained by the one instruction

$$\text{MAT C} = \text{A} * \text{B}$$

Note that if the number of columns of A is not equal to the number of rows of B, then we will receive an error message that the product is not defined.

TABLE 11.3

Statement	Operation
MAT C = A + B	Adds A to B
MAT C = A − B	Subtracts B from A
MAT C = A * B	Multiplies A and B
MAT C = (K) * A	Multiplies A by the scalar K
MAT C = INV(A)	Finds the inverse of A
MAT C = TRN(A)	Finds the transpose of A
MAT C = A	Sets A into C
MAT C = ZER	Sets C equal to the zero matrix
MAT C = IDN	Sets C equal to the identity matrix
MAT READ A	Reads data into A
MAT INPUT A	Enters data from the terminal into A
MAT PRINT A	Prints A

For the remainder of this section examples of programs using MAT statements are given along with sample runs. You will be asked in the exercises to run these programs and also to make minor modifications in them. A list of BASIC language MAT statements is given in table 11.3.

EXAMPLES

3. The program shown in figure 11.3 solves a system of three linear equations in three unknowns using the inverse of the coefficient matrix. The entries of the coefficient matrix and the right-hand sides are to be entered at the terminal during program execution. If the coefficient matrix is not invertible, an appropriate message is printed at the terminal. (See section 5.5.)

```
*LIST

10   REM    THIS PROGRAM SOLVES A SYSTEM OF 3 EQUATIONS IN 3 UNKNOWNS
20   REM
30          DIM A(3,3), B(3), X(3)
40   REM
50          PRINT "ENTER A 3X3 COEFFICIENT MATRIX"
60          MAT INPUT A
70          PRINT
80          PRINT "ENTER THE RIGHT-HAND SIDES OF THE EQUATIONS"
90          MAT INPUT B
100  REM
110         MAT C = INV(A)
120         MAT X = C*B
130  REM
140         PRINT
150         PRINT "THE SOLUTION TO THE SYSTEM IS"
160         PRINT "=== ======== == === ====== =="
170         MAT PRINT X
180         PRINT
190  REM
200         END
*RUN


ENTER A 3X3 COEFFICIENT MATRIX
?1,3,-2,&
?2,-1,10,&
?3,5,-3

ENTER THE RIGHT-HAND SIDES OF THE EQUATIONS
?9,18,22

THE SOLUTION TO THE SYSTEM IS
=== ======== == === ====== ==
 5
 2
 .999999
```

FIGURE 11.3

4. The program shown in figure 11.4 finds the intensity vector for a given 3 × 3 productive input–output matrix and demand vector. The input–output matrix

```
*LIST
  10   REM    THIS PROGRAMS DETERMINES THE INTENSITY VECTOR
  20   REM    FOR A 3X3 PRODUCTIVE INPUT-OUTPUT MATRIX AND
  30   REM    DEMAND VECTOR
  40   REM
  50          DIM A(3,3), I(3,3), D(3), X(3)
  60   REM
  70          PRINT "ENTER A 3X3 PRODUCTIVE INPUT-OUTPUT MATRIX"
  80          MAT INPUT A
  90          PRINT "ENTER A DEMAND VECTOR"
 100          MAT INPUT D
 105 REM
 110          PRINT
 120          PRINT "THE INPUT-OUTPUT MATRIX IS"
 130          PRINT "=== ============ ====== =="
 140          MAT PRINT A;
 150          PRINT
 160          PRINT "THE DEMAND VECTOR IS"
 170          PRINT "=== ====== ====== =="
 180          MAT PRINT D
 185 REM
 190          MAT I = IDN
 200          MAT B = I - A
 210          MAT C = INV(B)
 220          MAT X = C*D
 230 REM
 240          PRINT
 250          PRINT "THE INTENSITY VECTOR IS"
 260          PRINT "=== ========= ====== =="
 270          MAT PRINT X
 280          PRINT
 290 REM
 300          END
*RUN


ENTER A 3X3 PRODUCTIVE INPUT-OUTPUT MATRIX
?.4,.2,.2,.1,.3,.1,.4,.2,.3
ENTER A DEMAND VECTOR
?105,95,70

THE INPUT-OUTPUT MATRIX IS
=== ============ ====== ==
 .4   .2   .2
 .1   .3   .1
 .4   .2   .3

THE DEMAND VECTOR IS
=== ====== ====== ==
 105
  95
  70

THE INTENSITY VECTOR IS
=== ========= ====== ==
 388.249
 247.25
 392.499
```

FIGURE 11.4

and the demand vector are to be entered at the terminal during program execution. (See section 5.6.)

5. The program shown in figure 11.5 finds the line of best fit for a set of N data points, where N ≤ 10. N and the data points are to be entered at the terminal during program execution. (See section 5.4.)

```
*LIST


10   REM   THIS PROGRAM FINDS THE LINE OF BEST FIT
20   REM   FOR UP TO 10 DATA POINTS
30   REM
40         DIM A(10,2), B(10)
50   REM
60         MAT A = ZER
70         MAT B = ZER
80   REM
90         PRINT  "HOW MANY DATA POINTS ARE THERE"
100        INPUT N
110        PRINT "ENTER A DATA POINT (X,Y) AT EACH ?"
120        FOR I = 1 TO N
130           INPUT A(I,2), B(I)
140           LET A(I,1) = 1
150        NEXT I
160 REM
170        MAT C = TRN(A)
180        MAT D = C*A
190        MAT E = INV(D)
200        MAT F = C*B
210        MAT X = E*F
220 REM
230        PRINT
240        PRINT  "THE LINE OF BEST FIT, Y = MX + B, IS"
250        PRINT  "=== ==== == ==== ==== =========== =="
260        PRINT "M = "; X(2), "B = "; X(1)
270        PRINT
280 REM
290        END
*RUN


HOW MANY DATA POINTS ARE THERE
?3
ENTER A DATA POINT (X,Y) AT EACH ?
?30,19.8
?35,19.7
?40,19.3

THE LINE OF BEST FIT, Y = MX + B, IS
=== ==== == ==== ==== =========== ==
M = -.049983    B =   21.3489
```

FIGURE 11.5

EXERCISES 11.4

1. Use your computer to type in the source program of example 3 (page 543) and run it using as input the system of linear equations

$$x - \frac{1}{2}y + \frac{3}{2}z = \frac{1}{2}$$

$$3y - 3z = 6$$

$$\frac{1}{5}z = \frac{4}{5}.$$

2. Use your computer to type in the source program of example 4 (page 543) and run it using as input the input–output matrix A and the demand vector D as follows.

$$A = \begin{bmatrix} 0.4 & 0.1 & 0.1 \\ 0.2 & 0.3 & 0.1 \\ 0.1 & 0.2 & 0.3 \end{bmatrix} \qquad D = \begin{bmatrix} 100 \\ 150 \\ 100 \end{bmatrix}$$

3. a. Change the program in example 3 (page 543) so that it can be used to solve a linear system of four equations in four unknowns.

   b. Use the program of part (a) to solve the system

$$x_1 - 2x_2 + 5x_3 - x_4 = 0$$

$$3x_2 - 6x_3 + 9x_4 = 12$$

$$- x_2 + 4x_3 - 2x_4 = -1$$

$$-3x_1 + 8x_2 - 17x_3 + 11x_4 = 10.$$

4. a. Change the program of example 4 (page 543) so that it can find the intensity vector for a given input–output matrix and demand vector for a two-industry system.

   b. Use the program of part (a) to find the intensity vector for the following input–output matrix A and demand vector D.

$$A = \begin{bmatrix} 0.40 & 0.15 \\ 0.05 & 0.25 \end{bmatrix} \qquad D = \begin{bmatrix} 100 \\ 200 \end{bmatrix}$$

5. To make sure that we entered the data correctly, we should include PRINT statements to print the contents of the variables where the data were placed, as was done for the program in example 4 (page 543). Insert in the source program of example 3 (page 543) appropriate PRINT statements to print the input after it is entered and run your program using the data of example 3.

6. Change the source program of example 5 (page 543) by including appropriate PRINT statements to print the input after it is entered and run your program using the points (1, 11.4), (2, 8.6), (3, 9.2), and (4, 2.1).

7. Below are the essential statements for a program that accepts as input from the terminal two 2 × 3 matrices A and B, computes the sum $A + B$, and prints out this sum.

```
10     DIM A(2, 3), B(2, 3)
20     MAT INPUT A
30     MAT INPUT B
40     MAT C = A + B
50     MAT PRINT C
60     END
```

a. Insert appropriate PRINT statements to give messages during execution that indicate the needed input and identify the output. Also, insert REM statements so that the different units stand out in the printed source program.

b. Use your computer system to type in the source program obtained in part (a) and run it using as data the matrices

$$A = \begin{bmatrix} -1 & 11 & -5 \\ 0.4 & -0.5 & 0.8 \end{bmatrix} \quad \text{and} \quad B = \begin{bmatrix} 3.1 & 5.5 & 0.2 \\ 8 & 9 & 12 \end{bmatrix}.$$

8. a. Write a program to read a 2×3 matrix A and a 3×2 matrix B and compute and print their product. (See chapter 5.)

b. Run your program using as data the matrices

$$A = \begin{bmatrix} -1 & 11 & -5 \\ 0.4 & -0.5 & 0.8 \end{bmatrix} \quad \text{and} \quad B = \begin{bmatrix} 3.1 & 8 \\ 5.5 & 9 \\ 0.2 & 12 \end{bmatrix}.$$

9. Below are the essential statements for a program that reads as data a 3×3 input–output matrix A, and computes the inverse of $(I - A)$ and the intensity vector for each of K demand vectors.

```
10     DIM A(3, 3), D(3), I(3, 3)
20     MAT READ A
30     MAT I = IDN
40     MAT B = I - A
50     MAT C = INV(B)
60     INPUT K
70     FOR I = 1 TO K STEP 1
80     MAT INPUT D
90     MAT X = C * D
100    MAT PRINT X
110    NEXT I
120    END
```

a. Use these statements as the basis of a source program that includes appropriate PRINT statements that indicate the needed input and identify the output. Include REM statements and indentation for style.

b. Run the program of part (a) to obtain the answers to exercise 7 in section 5.6 (page 212).

10. Write a program to read as data a 3×3 matrix A, compute its inverse A^{-1}, and find $A^{-1}B$ for each of K column matrices B obtained as input. (*Hint:* See exercise 9.)

11. Below are the essential statements for a program that accepts as input a 3×3 transition matrix P of a Markov chain and an initial state vector S. It then computes and prints each of K state distribution vectors. (See chapter 10.)

```
10      DIM  P(3, 3),  S(3)
20      MAT INPUT P
30      MAT INPUT S
40      MAT A = TRN(P)
50      INPUT K
60      PRINT 0;  S(1),  S(2),  S(3)
70      FOR I = 1 TO K STEP 1
80      MAT X = A * S
90      PRINT I;  X(1),  X(2),  X(3)
100     MAT S = X
110     NEXT I
120     END
```

a. Use these statements as the basis for a source program that includes appropriate PRINT statements to indicate the needed input and resulting output. Include REM statements and indentation for style.

b. Use your program of part (a) to obtain the state distribution vectors for the transition matrix and initial state vector given in example 3 of section 10.1 (page 466).

12. Rewrite the program in exercise 11 for a 2×2 transition matrix and use it to obtain the answer to exercise 10 of section 10.1 (page 476).

13. Our programs thus far have been written for matrices of fixed dimensions. If we wanted to run the program for matrices of different dimensions, we had to change the DIM statement. However, there is a way to avoid this and to enter the dimensions during execution. Dimensions must be specified in a DIM statement, but these are given as the largest possible size which the program will use. These dimensions can be **trimmed** during execution to the desired size. For example, below is the source program of example 3 (except for the PRINT statements that give messages) as modified for this technique.

```
10      DIM  A(10, 10),  B(10)
20      INPUT N
30      MAT INPUT A(N, N)
40      MAT INPUT B(N)
50      MAT C = INV(A)
60      MAT X = C * B
70      MAT PRINT X
80      END
```

Using the preceding statements as a guide, modify the source program of example 4 to accept an input–output matrix and a demand vector for a system of N industries where N is entered during execution.

SUMMARY OF TERMS

Computer and Programming Terms
  Input, Output
  Central Processing Unit (CPU)
  Memory (Cell)
  Time-sharing
  Terminal (Hardcopy or CRT)
  Algorithm
  Machine Language
  Source Program (Code)
  Compiler
  Syntax Error
  Logical Error
  Debug
  Library
  Data Stack
  Variable
  Constant (Numeric or String)
  Exponential (Scientific) Notation
  Arithmetic Expression
  Line Number
  Fields
  Control Structure
  Control Variable (Counter)
  Array
  MAT Operations
  Variable Name
  Pointer
  Documentation
  Prompt
  Control Statement
  Branches
  Programming Style
  Loop
  Step Size

Commands
  LOGON
  EXEC BASIC
  RUN
  SAVE
  BYE
  LOGOFF

Statements
  LET
  DATA
  READ
  INPUT
  PRINT
  END
  IF-THEN
  FOR-NEXT
  GO TO
  DIM
  REM
  MAT PRINT
  MAT INPUT

REVIEW EXERCISES (CH. 11)

For exercises 1 and 2 write the BASIC arithmetic expression which is equivalent to the given algebraic expression.

1. a. $\dfrac{i}{(1 + i)^{-n} - 1}$

   b. $\sqrt[3]{(A + B)^2}$

2. a. $\dfrac{Ax + By + Cz}{D}$ b. $\dfrac{(x-4)^2 + (y-4)^2}{2}$

For exercises 3–6 state what the contents of each variable would be after the given sequence of lines is executed.

3. 10   LET X = 4
    20   LET Y = 7.2
    30   LET Z = X * Y
    40   LET X = X + 2

4. 10   LET A = 4.3
    20   LET B = 10
    30   LET C = A + B
    40   LET D = A ** 2

5. 10   DATA 4, 25E − 2
    20   DATA 5, 3.2
    30   READ W, X, Y, Z
    40   LET W = W * X
    50   LET Z = Y + Z

6. 10   DATA 5, 3.2, 8.1
    20   DATA 4.1, 5.7
    30   READ N
    40   READ X, Y
    50   LET Z = (X − N) ** 2
    60   LET Z = Z + (Y − N) ** 2

For exercises 7 and 8 give the complete set of values of the counter I as determined by the given FOR statement.

7. FOR I = 2 TO 15 STEP 2

8. FOR I = 9 TO 1 STEP −2

For exercises 9–14 give the printout that would result from execution of the given program.

9. 10   DATA 4
    20   DATA 7.2, 5.4, 4.3, 9.7
    30   READ N
    40   LET S = 0
    50   LET K = 1
    60   IF K > N THEN 100
    70   READ X
    80   LET S = S + X
    90   LET K = K + 1
    100   REM END IF
    110   LET A = S/N
    120   PRINT "A = "; A
    130   END

10. 10   DATA 4, 6
    20   READ X, Y
    30   LET C = 2 * X + 3 * Y
    40   LET D = 3 * X + 2 * Y
    50   IF C < > 26 THEN 90
    60   IF D < > 24 THEN 90
    70   PRINT "YES"
    80   GO TO 110
    90   REM END IF
    100   PRINT "NO"
    110   END

11. 10   DATA 5
    20   DATA 1, 2, 3, 4, 5, 6, 7
    30   READ N
    40   LET Y = 0
    50   FOR I = 1 TO N
    60   READ X
    70   LET Y = Y + (X ** 2)
    80   NEXT I
    90   PRINT "THE RESULT IS"
    100   PRINT, Y
    110   END

```
12. 10        DATA 0.01, 12        13. 10    DIM A(5, 2)
    20        READ I, N                20    DATA 7, 3, 6, 9, 2
    30        PRINT "N", "FACTOR"      30    FOR I = 1 TO 5
    40        PRINT                    40        READ A(I, 1)
    50        FOR K = 1 TO N           50    NEXT I
    60            LET F = (1 + I)**K   60    FOR I = 1 TO 5
    70            PRINT K, F           70        LET K = 6 − I
    80        NEXT K                   80        LET A(I, 2) = A(K, 1)
    90        END                      90    NEXT I
                                      100    MAT PRINT A
                                      110    END
```

```
14. 10    DIM A(5)
    20    DATA 7, 3, 6, 9, 2
    30    MAT READ A
    40    FOR I = 1 TO 5
    50        LET B(I) = A(I)
    60    NEXT I
    70    FOR I = 1 TO 4
    80    FOR J = I + 1 TO 5
    90            LET S = A(1)
   100            LET K = I
   110            IF S < = A(J) THEN 140
   120                LET S = A(J)
   130                LET K = J
   140        NEXT J
   150        LET A(K) = A(I)
   160        LET A(I) = S
   170    NEXT I
   180    FOR I = 1 TO 5
   190        PRINT I, B(I); A(I)
   200    NEXT I
   210    END
```

Appendices

A: Real Numbers

B: Tables

C: Answers to Odd-Numbered Exercises

A Real Numbers

The **real numbers** comprise the rational numbers (positive and negative integers, zero, and ratios of integers such as $\frac{3}{8}$, $\frac{0}{1}$, and $-\frac{5}{2}$, provided the denominator is not zero) and irrational numbers such as $\sqrt{2}$, $-\sqrt[3]{7}$, and π, which cannot be expressed as ratios of integers. The **real number system** comprises the real numbers together with the two operations of addition and multiplication. The real numbers are **closed** with respect to the two operations; that is, for any two numbers a and b there exists a unique real number $a + b$, called the **sum** of a and b, and a unique real number $a \cdot b$, called the **product** of a and b. These operations obey the following properties, where a, b, and c denote any real numbers.

Commutative Laws

$$a + b = b + a$$

$$a \cdot b = b \cdot a$$

That is, the order in which we add or multiply does not matter. ■

Associative Laws

$$a + (b + c) = (a + b) + c$$

$$a \cdot (b \cdot c) = (a \cdot b) \cdot c$$

That is, it does not matter which two numbers we add or multiply first. ■

Distributive Laws

$$a \cdot (b + c) = a \cdot b + a \cdot c$$

$$(a + b) \cdot c = a \cdot c + b \cdot c$$ ■

Identity Elements

■ Zero is called the **additive identity**.

$$a + 0 = a$$

One is called the **multiplicative identity**.

$$a \cdot 1 = a$$ ■

Inverse Elements

■ For every real number a, there exists the real number $-a$, called the **additive inverse** of a (or the **negative** of a) such that $a + (-a) = 0$. For every real number $a \neq 0$, there exists the real number $1/a$, called the **multiplicative inverse** of a (or the **reciprocal** of a) such that $a \cdot (1/a) = 1$. ■

The operation of **subtraction** is defined by $a - b = a + (-b)$; that is, a **minus** b is the sum of a and the additive inverse of b.

The operation of **division** is defined by $a \div b = a \cdot (1/b)$; that is, a **divided** by b is the product of a and the multiplicative inverse of b. Division by 0 is not defined. Rational numbers can be thought of as the quotient of integers. For example, the rational number $\frac{22}{4}$ is the division of 22 by 4 and, therefore, $\frac{22}{4} = 5.5$. We see then that any rational number has a **decimal representation** obtained by division. The decimal representation of any rational number is either **terminating** or **repeating**.

For example, $\frac{17}{5} = 3.4$ and $\frac{7}{400} = 0.0175$ are terminating decimals, whereas $\frac{1}{12} = 0.08333\ldots$ and $\frac{720}{333} = 2.162162\ldots$ are examples of repeating decimals.

Any irrational number also has a decimal representation; however, they are non-terminating and non-repeating. For example, $\sqrt{2} = 1.414213562\ldots$ and $\pi = 3.141592654\ldots$.

For computational purposes, decimal representations are frequently **rounded** to a certain number of decimal places. The rule for rounding is to add 1 to the last place to be kept when the leftmost place to be dropped is 5 or more. For example, $\sqrt{2} = 1.414$ (rounded to 3 places); $\pi = 3.141593$ (rounded to 6 places).

The following additional properties of numbers are useful in computations.

Number Properties ■

$$a \cdot 0 = 0 \qquad (-a)b = -(ab)$$

$$-(-a) = a \qquad a(-b) = -(ab)$$

$$(-1)a = -a \qquad (-a)(-b) = ab$$

$$a \div a = 1 \qquad \frac{-a}{b} = -\frac{a}{b}$$

$$\frac{a}{b} = \frac{ac}{bc} \qquad \frac{a}{-b} = -\frac{a}{b}$$

$$\frac{a}{b} + \frac{c}{b} = \frac{a+c}{b} \qquad \frac{a}{c} - \frac{c}{b} = \frac{a-c}{b}$$

$$\frac{a}{b} \cdot \frac{c}{d} = \frac{ac}{bd} \qquad \frac{a}{b} \div \frac{c}{d} = \frac{a}{b} \cdot \frac{d}{c} \quad ■$$

Relations among real numbers is an important concept. In particular,

$$a > b$$
$$b < a$$

each mean $a - b$ is positive.

The symbols $>$ and $<$ are called **inequality signs**, and expressions such as $a < b$ or $b > a$ are called **inequalities**. Furthermore,

$$a > b \qquad \text{is read as "}a \text{ is greater than } b\text{"}$$

and $\qquad\qquad b < a \qquad$ is read as "b is less than a."

EXAMPLE 1

a. $1 > 0$, since $1 - 0 = 1$ is positive.

b. $5 < 9$, since $9 - 5 = 4$ is positive.

c. $-3 < -1$, since $-1 - (-3) = 2$ is positive.

d. $2 > -4$, since $2 - (-4) = 6$ is positive.

The symbol $a \geq b$ means that a is either greater than or equal to b. Similarly, $a \leq b$ means that a is either less than or equal to b.

The relations $a \le b$ and $b \le c$ can be expressed more concisely by

$$a \le b \le c.$$

It is not proper to write $a \le b \ge c$ for $a \le b$ and $b \ge c$.

Using inequalities we see that

$$a > 0 \text{ means } a \text{ is positive}$$

$$a \ge 0 \text{ means } a \text{ is non-negative}$$

$$a < 0 \text{ means } a \text{ is negative}$$

$$a \le 0 \text{ means } a \text{ is non-positive.}$$

Using the definition of the inequality $<$, we see that

$$\cdots < -3 < -2 < -1 < 0 < 1 < 2 < 3 < \cdots$$

That is, the inequality concept **orders** the real numbers. We can "picture" the real number system geometrically.

There is a **one-to-one correspondence** between the real numbers and points on a line, called a **real number line** (see figure A.1). A real number line is specified by an **origin**, **positive** and **negative directions**, and a **scale**, and is constructed as follows:

1. Pick a point for the origin; this is associated with the number 0.
2. Agree that positive numbers are located to the right of the origin and negative numbers are located to the left of the origin.
3. Pick a point to the right of the origin that we associate with 1; this determines a scale.

FIGURE A.1

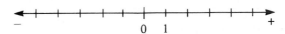

We can easily **graph** a real number x; the graph is the point on the line which corresponds to x. Also, we can determine the real number x associated with a point P, called the **coordinate** of P.

EXAMPLE 2

a. The graph of the real numbers

$$-4, -\frac{3}{2}, \frac{3}{4}, \sqrt{6}, 5$$

is given in figure A.2(a).

b. The graph of the real numbers larger than 3 is given in figure A.2(b).

If a point is included in the graph, we darken it; otherwise, we do not darken it, but draw an open circle.

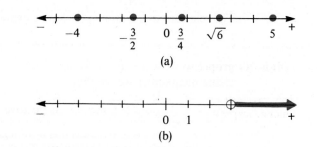

FIGURE A.2

If a and b are the coordinates of the points A and B, respectively, on the number line, then $a < b$ means A lies to the left of B on the number line. We define the **directed distance from A to B** as $b - a$. For example, if A has coordinate -3 and B has coordinate 5, then the directed distance from A to B is $5 - (-3) = 8$.

Exponent notation is used to denote the repeated product of a real number a with itself. We have the following meaning of positive integer exponents:

$$a^1 = a, \quad a^2 = a \cdot a, \quad \ldots, \quad a^n = a \cdot a \cdot a, \ldots, a \quad n \text{ times.}$$

For example,

$$2^5 = 2 \cdot 2 \cdot 2 \cdot 2 \cdot 2 = 32 \quad \text{and} \quad (-5)^3 = (-5)(-5)(-5) = -125.$$

Furthermore, for $a \neq 0$, we have for negative integer and zero exponents

$$a^0 = 1, \quad a^{-1} = \frac{1}{a}, \quad a^{-2} = \frac{1}{a^2} \quad \cdots \quad a^{-n} = \frac{1}{a^n}.$$

For example,

$$\left(\frac{1}{2}\right)^0 = 1 \quad \text{and} \quad 3^{-2} = \frac{1}{3^2} = \frac{1}{9}.$$

0^0 is undefined.

We have the following three rules of exponents that are useful.

$$a^n \cdot a^m = a^{n+m}; \quad (-2)^2(-2)^3 = (-2)^{2+3} = (-2)^5 = -32$$

$$\frac{a^n}{a^m} = a^{n-m}; \quad \frac{6^{10}}{6^8} = 6^{10-8} = 6^2 = 36$$

$$(a^n)^m = a^{n \cdot m}; \quad (5^2)^3 = 5^{2 \cdot 3} = 5^6 = 15{,}625.$$

TABLE I   *Binomial Random Variable X*

The entries in the table give the probabilities $Pr(X = x)$ where: n = number of trials of the binomial experiment, x = number of successes on the n trials, and p = probability of a success on a single trial.

N	X	0.10	0.20	0.25	0.30	0.40	0.50
1	0	0.9000	0.8000	0.7500	0.7000	0.6000	0.5000
	1	0.1000	0.2000	0.2500	0.3000	0.4000	0.5000
2	0	0.8100	0.6400	0.5625	0.4900	0.3600	0.2500
	1	0.1800	0.3200	0.3750	0.4200	0.4800	0.5000
	2	0.0100	0.0400	0.0625	0.0900	0.1600	0.2500
3	0	0.7290	0.5120	0.4219	0.3430	0.2160	0.1250
	1	0.2430	0.3840	0.4219	0.4410	0.4320	0.3750
	2	0.0270	0.0960	0.1406	0.1890	0.2880	0.3750
	3	0.0010	0.0080	0.0156	0.0270	0.0640	0.1250
4	0	0.6561	0.4096	0.3164	0.2401	0.1296	0.0625
	1	0.2916	0.4096	0.4219	0.4116	0.3456	0.2500
	2	0.0486	0.1536	0.2109	0.2646	0.3456	0.3750
	3	0.0036	0.0256	0.0469	0.0756	0.1536	0.2500
	4	0.0001	0.0016	0.0039	0.0081	0.0256	0.0625
5	0	0.5905	0.3277	0.2373	0.1681	0.0778	0.0312
	1	0.3280	0.4096	0.3955	0.3602	0.2592	0.1562
	2	0.0729	0.2048	0.2637	0.3087	0.3456	0.3125
	3	0.0081	0.0512	0.0879	0.1323	0.2304	0.3125
	4	0.0005	0.0064	0.0146	0.0283	0.0768	0.1562
	5	0.0000	0.0003	0.0010	0.0024	0.0102	0.0312
6	0	0.5314	0.2621	0.1780	0.1176	0.0467	0.0156
	1	0.3543	0.3932	0.3560	0.3025	0.1866	0.0937
	2	0.0984	0.2458	0.2966	0.3241	0.3110	0.2344
	3	0.0146	0.0819	0.1318	0.1852	0.2765	0.3125
	4	0.0012	0.0154	0.0330	0.0595	0.1382	0.2344
	5	0.0001	0.0015	0.0044	0.0102	0.0369	0.0937
	6	0.0000	0.0001	0.0002	0.0007	0.0041	0.0156
7	0	0.4783	0.2097	0.1335	0.0824	0.0280	0.0078
	1	0.3720	0.3670	0.3115	0.2471	0.1306	0.0547
	2	0.1240	0.2753	0.3115	0.3177	0.2613	0.1641
	3	0.0230	0.1147	0.1730	0.2269	0.2903	0.2734
	4	0.0026	0.0287	0.0577	0.0972	0.1935	0.2734
	5	0.0002	0.0043	0.0115	0.0250	0.0774	0.1641
	6	0.0000	0.0004	0.0013	0.0036	0.0172	0.0547
	7	0.0000	0.0000	0.0001	0.0002	0.0016	0.0078

TABLE I (*continued*)

N	X	0.10	0.20	0.25	0.30	0.40	0.50
					P		
8	0	0.4305	0.1678	0.1001	0.0576	0.0168	0.0039
	1	0.3826	0.3355	0.2670	0.1977	0.0896	0.0312
	2	0.1488	0.2936	0.3115	0.2965	0.2090	0.1094
	3	0.0331	0.1468	0.2076	0.2541	0.2787	0.2187
	4	0.0046	0.0459	0.0865	0.1361	0.2322	0.2734
	5	0.0004	0.0092	0.0231	0.0467	0.1239	0.2187
	6	0.0000	0.0011	0.0038	0.0100	0.0413	0.1094
	7	0.0000	0.0001	0.0004	0.0012	0.0079	0.0312
	8	0.0000	0.0000	0.0000	0.0001	0.0007	0.0039
9	0	0.3874	0.1342	0.0751	0.0404	0.0101	0.0020
	1	0.3874	0.3020	0.2253	0.1556	0.0605	0.0176
	2	0.1722	0.3020	0.3003	0.2668	0.1612	0.0703
	3	0.0446	0.1762	0.2336	0.2668	0.2508	0.1641
	4	0.0074	0.0661	0.1168	0.1715	0.2508	0.2461
	5	0.0008	0.0165	0.0389	0.0735	0.1672	0.2461
	6	0.0001	0.0028	0.0087	0.0210	0.0743	0.1641
	7	0.0000	0.0003	0.0012	0.0039	0.0212	0.0703
	8	0.0000	0.0000	0.0001	0.0004	0.0035	0.0176
	9	0.0000	0.0000	0.0000	0.0000	0.0003	0.0020
10	0	0.3487	0.1074	0.0563	0.0282	0.0060	0.0010
	1	0.3874	0.2684	0.1877	0.1211	0.0403	0.0098
	2	0.1937	0.3020	0.2816	0.2335	0.1209	0.0439
	3	0.0574	0.2013	0.2503	0.2668	0.2150	0.1172
	4	0.0112	0.0881	0.1460	0.2001	0.2508	0.2051
	5	0.0015	0.0264	0.0584	0.1029	0.2007	0.2461
	6	0.0001	0.0055	0.0162	0.0368	0.1115	0.2051
	7	0.0000	0.0008	0.0031	0.0090	0.0425	0.1172
	8	0.0000	0.0001	0.0004	0.0014	0.0106	0.0439
	9	0.0000	0.0000	0.0000	0.0001	0.0016	0.0098
	10	0.0000	0.0000	0.0000	0.0000	0.0001	0.0010
11	0	0.3138	0.0859	0.0422	0.0198	0.0036	0.0005
	1	0.3835	0.2362	0.1549	0.0932	0.0266	0.0054
	2	0.2131	0.2953	0.2581	0.1998	0.0887	0.0269
	3	0.0710	0.2215	0.2581	0.2568	0.1774	0.0806
	4	0.0158	0.1107	0.1721	0.2201	0.2365	0.1611
	5	0.0025	0.0388	0.0803	0.1321	0.2207	0.2256
	6	0.0003	0.0097	0.0268	0.0566	0.1471	0.2256
	7	0.0000	0.0017	0.0064	0.0173	0.0701	0.1611
	8	0.0000	0.0002	0.0011	0.0037	0.0234	0.0806
	9	0.0000	0.0000	0.0001	0.0005	0.0052	0.0269
	10	0.0000	0.0000	0.0000	0.0000	0.0007	0.0054
	11	0.0000	0.0000	0.0000	0.0000	0.0000	0.0005
12	0	0.2824	0.0687	0.0317	0.0138	0.0022	0.0002
	1	0.3766	0.2062	0.1267	0.0712	0.0174	0.0029
	2	0.2301	0.2835	0.2323	0.1678	0.0639	0.0161
	3	0.0852	0.2362	0.2581	0.2397	0.1419	0.0537
	4	0.0213	0.1329	0.1936	0.2311	0.2128	0.1208
	5	0.0038	0.0532	0.1032	0.1585	0.2270	0.1934
	6	0.0005	0.0155	0.0401	0.0792	0.1766	0.2256
	7	0.0000	0.0033	0.0115	0.0291	0.1009	0.1934
	8	0.0000	0.0005	0.0024	0.0078	0.0420	0.1208
	9	0.0000	0.0001	0.0004	0.0015	0.0125	0.0537
	10	0.0000	0.0000	0.0000	0.0002	0.0025	0.0161
	11	0.0000	0.0000	0.0000	0.0000	0.0003	0.0029
	12	0.0000	0.0000	0.0000	0.0000	0.0000	0.0002

TABLE I Binomial Random Variable *X* / **561**

TABLE I (*continued*)

N	X	P					
		0.10	0.20	0.25	0.30	0.40	0.50
13	0	0.2542	0.0550	0.0238	0.0097	0.0013	0.0001
	1	0.3672	0.1787	0.1029	0.0540	0.0113	0.0016
	2	0.2448	0.2680	0.2059	0.1388	0.0453	0.0095
	3	0.0997	0.2457	0.2517	0.2181	0.1107	0.0349
	4	0.0277	0.1535	0.2097	0.2337	0.1845	0.0873
	5	0.0055	0.0691	0.1258	0.1803	0.2214	0.1571
	6	0.0008	0.0230	0.0559	0.1030	0.1968	0.2095
	7	0.0001	0.0058	0.0186	0.0442	0.1312	0.2095
	8	0.0000	0.0011	0.0047	0.0142	0.0656	0.1571
	9	0.0000	0.0001	0.0009	0.0034	0.0243	0.0873
	10	0.0000	0.0000	0.0001	0.0006	0.0065	0.0349
	11	0.0000	0.0000	0.0000	0.0001	0.0012	0.0095
	12	0.0000	0.0000	0.0000	0.0000	0.0001	0.0016
	13	0.0000	0.0000	0.0000	0.0000	0.0000	0.0001
14	0	0.2288	0.0440	0.0178	0.0068	0.0008	0.0001
	1	0.3559	0.1539	0.0832	0.0407	0.0073	0.0009
	2	0.2570	0.2501	0.1802	0.1134	0.0317	0.0056
	3	0.1142	0.2501	0.2402	0.1943	0.0845	0.0222
	4	0.0349	0.1720	0.2202	0.2290	0.1549	0.0611
	5	0.0078	0.0860	0.1468	0.1963	0.2066	0.1222
	6	0.0013	0.0322	0.0734	0.1262	0.2066	0.1833
	7	0.0002	0.0092	0.0280	0.0618	0.1574	0.2095
	8	0.0000	0.0020	0.0082	0.0232	0.0918	0.1833
	9	0.0000	0.0003	0.0018	0.0066	0.0408	0.1222
	10	0.0000	0.0000	0.0003	0.0014	0.0136	0.0611
	11	0.0000	0.0000	0.0000	0.0002	0.0033	0.0222
	12	0.0000	0.0000	0.0000	0.0000	0.0005	0.0056
	13	0.0000	0.0000	0.0000	0.0000	0.0001	0.0009
	14	0.0000	0.0000	0.0000	0.0000	0.0000	0.0001
15	0	0.2059	0.0352	0.0134	0.0047	0.0005	0.0000
	1	0.3432	0.1319	0.0668	0.0305	0.0047	0.0005
	2	0.2669	0.2309	0.1559	0.0916	0.0219	0.0032
	3	0.1285	0.2501	0.2252	0.1700	0.0634	0.0139
	4	0.0428	0.1876	0.2252	0.2186	0.1268	0.0417
	5	0.0105	0.1032	0.1651	0.2061	0.1859	0.0916
	6	0.0019	0.0430	0.0917	0.1472	0.2066	0.1527
	7	0.0003	0.0138	0.0393	0.0811	0.1771	0.1964
	8	0.0000	0.0035	0.0131	0.0348	0.1181	0.1964
	9	0.0000	0.0007	0.0034	0.0116	0.0612	0.1527
	10	0.0000	0.0001	0.0007	0.0030	0.0245	0.0916
	11	0.0000	0.0000	0.0001	0.0006	0.0074	0.0417
	12	0.0000	0.0000	0.0000	0.0001	0.0016	0.0139
	13	0.0000	0.0000	0.0000	0.0000	0.0003	0.0032
	14	0.0000	0.0000	0.0000	0.0000	0.0000	0.0005
	15	0.0000	0.0000	0.0000	0.0000	0.0000	0.0000

TABLE I (*continued*)

N	X	P 0.10	0.20	0.25	0.30	0.40	0.50
16	0	0.1853	0.0281	0.0100	0.0033	0.0003	0.0000
	1	0.3294	0.1126	0.0535	0.0228	0.0030	0.0002
	2	0.2745	0.2111	0.1336	0.0732	0.0150	0.0018
	3	0.1423	0.2463	0.2079	0.1465	0.0468	0.0085
	4	0.0514	0.2001	0.2252	0.2040	0.1014	0.0278
	5	0.0137	0.1201	0.1802	0.2099	0.1623	0.0667
	6	0.0028	0.0550	0.1101	0.1649	0.1983	0.1222
	7	0.0004	0.0197	0.0524	0.1010	0.1889	0.1746
	8	0.0001	0.0055	0.0197	0.0487	0.1417	0.1964
	9	0.0000	0.0012	0.0058	0.0185	0.0840	0.1746
	10	0.0000	0.0002	0.0014	0.0056	0.0392	0.1222
	11	0.0000	0.0000	0.0002	0.0013	0.0142	0.0667
	12	0.0000	0.0000	0.0000	0.0002	0.0040	0.0278
	13	0.0000	0.0000	0.0000	0.0000	0.0008	0.0085
	14	0.0000	0.0000	0.0000	0.0000	0.0001	0.0018
	15	0.0000	0.0000	0.0000	0.0000	0.0000	0.0002
	16	0.0000	0.0000	0.0000	0.0000	0.0000	0.0000
17	0	0.1668	0.0225	0.0075	0.0023	0.0002	0.0000
	1	0.3150	0.0957	0.0426	0.0169	0.0019	0.0001
	2	0.2800	0.1914	0.1136	0.0581	0.0102	0.0010
	3	0.1556	0.2393	0.1893	0.1245	0.0341	0.0052
	4	0.0605	0.2093	0.2209	0.1868	0.0796	0.0182
	5	0.0175	0.1361	0.1914	0.2081	0.1379	0.0472
	6	0.0039	0.0680	0.1276	0.1784	0.1839	0.0944
	7	0.0007	0.0267	0.0668	0.1201	0.1927	0.1484
	8	0.0001	0.0084	0.0279	0.0644	0.1606	0.1855
	9	0.0000	0.0021	0.0093	0.0276	0.1070	0.1855
	10	0.0000	0.0004	0.0025	0.0095	0.0571	0.1484
	11	0.0000	0.0001	0.0005	0.0026	0.0242	0.0944
	12	0.0000	0.0000	0.0001	0.0006	0.0081	0.0472
	13	0.0000	0.0000	0.0000	0.0001	0.0021	0.0182
	14	0.0000	0.0000	0.0000	0.0000	0.0004	0.0052
	15	0.0000	0.0000	0.0000	0.0000	0.0001	0.0010
	16	0.0000	0.0000	0.0000	0.0000	0.0000	0.0001
	17	0.0000	0.0000	0.0000	0.0000	0.0000	0.0000
18	0	0.1501	0.0180	0.0056	0.0016	0.0001	0.0000
	1	0.3002	0.0811	0.0338	0.0126	0.0012	0.0001
	2	0.2835	0.1723	0.0958	0.0458	0.0069	0.0006
	3	0.1680	0.2297	0.1704	0.1046	0.0246	0.0031
	4	0.0700	0.2153	0.2130	0.1681	0.0614	0.0117
	5	0.0218	0.1507	0.1988	0.2017	0.1146	0.0327
	6	0.0052	0.0816	0.1436	0.1873	0.1655	0.0708
	7	0.0010	0.0350	0.0820	0.1376	0.1892	0.1214
	8	0.0002	0.0120	0.0376	0.0811	0.1734	0.1669
	9	0.0000	0.0033	0.0139	0.0386	0.1284	0.1855
	10	0.0000	0.0008	0.0042	0.0149	0.0771	0.1669
	11	0.0000	0.0001	0.0010	0.0046	0.0374	0.1214
	12	0.0000	0.0000	0.0002	0.0012	0.0145	0.0708
	13	0.0000	0.0000	0.0000	0.0002	0.0045	0.0327
	14	0.0000	0.0000	0.0000	0.0000	0.0011	0.0117
	15	0.0000	0.0000	0.0000	0.0000	0.0002	0.0031
	16	0.0000	0.0000	0.0000	0.0000	0.0000	0.0006
	17	0.0000	0.0000	0.0000	0.0000	0.0000	0.0001
	18	0.0000	0.0000	0.0000	0.0000	0.0000	0.0000

TABLE I Binomial Random Variable *X* / **563**

TABLE I (*continued*)

N	X	\multicolumn{6}{c}{P}					
		0.10	0.20	0.25	0.30	0.40	0.50
19	0	0.1351	0.0144	0.0042	0.0011	0.0001	0.0000
	1	0.2852	0.0685	0.0268	0.0093	0.0008	0.0000
	2	0.2852	0.1540	0.0803	0.0358	0.0046	0.0003
	3	0.1796	0.2182	0.1517	0.0869	0.0175	0.0018
	4	0.0798	0.2182	0.2023	0.1491	0.0467	0.0074
	5	0.0266	0.1636	0.2023	0.1916	0.0933	0.0222
	6	0.0069	0.0955	0.1574	0.1916	0.1451	0.0518
	7	0.0014	0.0443	0.0974	0.1525	0.1797	0.0961
	8	0.0002	0.0166	0.0487	0.0981	0.1797	0.1442
	9	0.0000	0.0051	0.0198	0.0514	0.1464	0.1762
	10	0.0000	0.0013	0.0066	0.0220	0.0976	0.1762
	11	0.0000	0.0003	0.0018	0.0077	0.0532	0.1442
	12	0.0000	0.0000	0.0004	0.0022	0.0237	0.0961
	13	0.0000	0.0000	0.0001	0.0005	0.0085	0.0518
	14	0.0000	0.0000	0.0000	0.0001	0.0024	0.0222
	15	0.0000	0.0000	0.0000	0.0000	0.0005	0.0074
	16	0.0000	0.0000	0.0000	0.0000	0.0001	0.0018
	17	0.0000	0.0000	0.0000	0.0000	0.0000	0.0003
	18	0.0000	0.0000	0.0000	0.0000	0.0000	0.0000
	19	0.0000	0.0000	0.0000	0.0000	0.0000	0.0000
20	0	0.1216	0.0115	0.0032	0.0008	0.0000	0.0000
	1	0.2702	0.0576	0.0211	0.0068	0.0005	0.0000
	2	0.2852	0.1369	0.0669	0.0278	0.0031	0.0002
	3	0.1901	0.2054	0.1339	0.0716	0.0123	0.0011
	4	0.0898	0.2182	0.1897	0.1304	0.0350	0.0046
	5	0.0319	0.1746	0.2023	0.1789	0.0746	0.0148
	6	0.0089	0.1091	0.1686	0.1916	0.1244	0.0370
	7	0.0020	0.0545	0.1124	0.1643	0.1659	0.0739
	8	0.0004	0.0222	0.0609	0.1144	0.1797	0.1201
	9	0.0001	0.0074	0.0271	0.0654	0.1597	0.1602
	10	0.0000	0.0020	0.0099	0.0308	0.1171	0.1762
	11	0.0000	0.0005	0.0030	0.0120	0.0710	0.1602
	12	0.0000	0.0001	0.0008	0.0039	0.0355	0.1201
	13	0.0000	0.0000	0.0002	0.0010	0.0146	0.0739
	14	0.0000	0.0000	0.0000	0.0002	0.0049	0.0370
	15	0.0000	0.0000	0.0000	0.0000	0.0013	0.0148
	16	0.0000	0.0000	0.0000	0.0000	0.0003	0.0046
	17	0.0000	0.0000	0.0000	0.0000	0.0000	0.0011
	18	0.0000	0.0000	0.0000	0.0000	0.0000	0.0002
	19	0.0000	0.0000	0.0000	0.0000	0.0000	0.0000
	20	0.0000	0.0000	0.0000	0.0000	0.0000	0.0000

TABLE II *Standard Normal Curve*

The entries in the table give the probability

$Pr(0 < Z < z)$.

area of shaded region from 0 to z

Z	0	1	2	3	4	5	6	7	8	9
0.0	0.0000	0.0040	0.0080	0.0120	0.0160	0.0199	0.0239	0.0279	0.0319	0.0359
0.1	0.0398	0.0438	0.0478	0.0517	0.0557	0.0596	0.0636	0.0675	0.0714	0.0753
0.2	0.0793	0.0832	0.0871	0.0910	0.0948	0.0987	0.1026	0.1064	0.1103	0.1141
0.3	0.1179	0.1217	0.1255	0.1293	0.1331	0.1368	0.1406	0.1443	0.1480	0.1517
0.4	0.1554	0.1591	0.1628	0.1664	0.1700	0.1736	0.1772	0.1808	0.1844	0.1879
0.5	0.1915	0.1950	0.1985	0.2019	0.2054	0.2088	0.2123	0.2157	0.2190	0.2224
0.6	0.2257	0.2291	0.2324	0.2357	0.2389	0.2422	0.2454	0.2486	0.2517	0.2549
0.7	0.2580	0.2611	0.2642	0.2673	0.2703	0.2734	0.2764	0.2793	0.2823	0.2852
0.8	0.2881	0.2910	0.2939	0.2967	0.2995	0.3023	0.3051	0.3078	0.3106	0.3133
0.9	0.3159	0.3186	0.3212	0.3238	0.3264	0.3289	0.3315	0.3340	0.3364	0.3389
1.0	0.3413	0.3438	0.3461	0.3485	0.3508	0.3531	0.3554	0.3577	0.3599	0.3621
1.1	0.3643	0.3665	0.3686	0.3708	0.3729	0.3749	0.3770	0.3790	0.3810	0.3830
1.2	0.3849	0.3869	0.3888	0.3907	0.3925	0.3944	0.3962	0.3980	0.3997	0.4015
1.3	0.4032	0.4049	0.4066	0.4082	0.4099	0.4115	0.4131	0.4147	0.4162	0.4177
1.4	0.4192	0.4207	0.4222	0.4236	0.4251	0.4265	0.4279	0.4292	0.4306	0.4319
1.5	0.4332	0.4345	0.4357	0.4370	0.4382	0.4394	0.4406	0.4418	0.4429	0.4441
1.6	0.4452	0.4463	0.4474	0.4485	0.4495	0.4505	0.4515	0.4525	0.4535	0.4545
1.7	0.4554	0.4564	0.4573	0.4582	0.4591	0.4599	0.4608	0.4616	0.4625	0.4633
1.8	0.4641	0.4649	0.4656	0.4664	0.4671	0.4678	0.4686	0.4693	0.4699	0.4706
1.9	0.4713	0.4719	0.4726	0.4732	0.4738	0.4744	0.4750	0.4756	0.4762	0.4767
2.0	0.4773	0.4778	0.4783	0.4788	0.4793	0.4798	0.4803	0.4808	0.4812	0.4817
2.1	0.4821	0.4826	0.4830	0.4834	0.4838	0.4842	0.4846	0.4850	0.4854	0.4857
2.2	0.4861	0.4865	0.4868	0.4871	0.4875	0.4878	0.4881	0.4884	0.4887	0.4890
2.3	0.4893	0.4896	0.4898	0.4901	0.4904	0.4906	0.4909	0.4911	0.4913	0.4916
2.4	0.4918	0.4920	0.4922	0.4925	0.4927	0.4929	0.4931	0.4932	0.4934	0.4936
2.5	0.4938	0.4940	0.4941	0.4943	0.4945	0.4946	0.4948	0.4949	0.4951	0.4952
2.6	0.4953	0.4955	0.4956	0.4957	0.4959	0.4960	0.4961	0.4962	0.4963	0.4964
2.7	0.4965	0.4966	0.4967	0.4968	0.4969	0.4970	0.4971	0.4972	0.4973	0.4974
2.8	0.4975	0.4975	0.4976	0.4977	0.4978	0.4978	0.4979	0.4980	0.4980	0.4981
2.9	0.4981	0.4982	0.4983	0.4983	0.4984	0.4984	0.4985	0.4985	0.4986	0.4986

TABLE III   *Compound Interest and Annuity*

Accumulation Factor: $(1 + i)^n$; Amount of \$1 Per Period: $s_{\overline{n}|i} = [(1 + i)^n - 1]/i$; Periodic Deposit Needed to Yield \$1: $1/s_{\overline{n}|i} = i/[(1 + i)^n - 1]$; Present Value of \$1: $(1 + i)^{-n}$; Present Value of \$1 Per Period: $a_{\overline{n}|i} = [1 - (1 + i)^{-n}]/i$; Periodic Payment Needed to Pay \$1: $1/a_{\overline{n}|i} = i/[1 - (1 + i)^{-n}]$.

$i = \frac{5}{12}\% = 0.004166667$

Number of periods n	Accumulation factor $(1 + i)^n$	Amount of \$1 per period $s_{\overline{n}\|i}$	Periodic deposit needed to yield \$1 $1/s_{\overline{n}\|i}$	Present value of \$1 $(1 + i)^{-n}$	Present value of \$1 per period $a_{\overline{n}\|i}$	Periodic payment needed to pay \$1 $1/a_{\overline{n}\|i}$	Number of periods n
1	1.00416667	1.00000000	1.00000000	0.99585062	0.99585062	1.00416667	1
2	1.00835069	2.00416667	0.49896050	0.99171846	1.98756908	0.50312717	2
3	1.01255216	3.01251736	0.33194829	0.98760345	2.97517253	0.33611496	3
4	1.01677112	4.02506952	0.24844291	0.98350551	3.95867804	0.25260958	4
5	1.02100767	5.04184064	0.19834026	0.97942457	4.93810261	0.20250625	5
6	1.02526187	6.06284831	0.16493898	0.97536057	5.91346318	0.16910564	6
7	1.02953379	7.08811018	0.14108133	0.97131343	6.88477661	0.14524800	7
8	1.03382352	8.11764397	0.12318845	0.96728308	7.85205970	0.12735512	8
9	1.03813111	9.15146749	0.10927209	0.96326946	8.81532916	0.11343876	9
10	1.04245666	10.18959860	0.09813929	0.95927249	9.77460165	0.10230596	10
11	1.04680023	11.23205526	0.08903090	0.95529211	10.72989376	0.09319757	11
12	1.05116190	12.27885549	0.08144082	0.95132824	11.68122200	0.08560748	12
13	1.05554174	13.33001739	0.07501866	0.94738082	12.62860283	0.07918532	13
14	1.05993983	14.38555913	0.06951416	0.94344978	13.57205261	0.07368082	14
15	1.06435625	15.44549896	0.06474378	0.93953505	14.51158766	0.06891045	15
16	1.06879106	16.50985520	0.06056988	0.93563457	15.44722422	0.06473655	16
17	1.07324436	17.57864627	0.05688720	0.93175426	16.37897848	0.06105387	17
18	1.07771621	18.65189063	0.05361387	0.92788806	17.30686654	0.05778053	18
19	1.08220670	19.72960684	0.05068525	0.92403790	18.23090443	0.05485191	19
20	1.08671589	20.81181353	0.04804963	0.92020372	19.15110815	0.05221630	20
21	1.09124387	21.89852942	0.04566517	0.91638544	20.06749359	0.04983183	21
22	1.09579072	22.98977330	0.04349760	0.91258301	20.98007661	0.04766427	22
23	1.10035652	24.08556402	0.04151865	0.90879636	21.88887297	0.04568531	23
24	1.10494134	25.18592053	0.03970472	0.90502542	22.79389839	0.04387139	24
25	1.10954526	26.29086187	0.03803603	0.90127013	23.69516853	0.04220270	25
26	1.11416836	27.40040713	0.03649581	0.89753042	24.59269895	0.04066247	26
27	1.11881073	28.51457549	0.03506978	0.89380623	25.48650517	0.03923645	27
28	1.12347244	29.63338622	0.03374572	0.89009749	26.37660266	0.03791239	28
29	1.12815358	30.75685866	0.03251307	0.88640414	27.26300680	0.03667974	29
30	1.13285422	31.88501224	0.03136270	0.88272611	28.14573291	0.03552936	30
31	1.13757444	33.01786646	0.03028663	0.87906335	29.02479626	0.03445330	31
32	1.14231434	34.15544090	0.02927791	0.87541578	29.90021205	0.03344458	32
33	1.14707398	35.29775524	0.02833041	0.87178335	30.77199540	0.03249708	33
34	1.15185346	36.44482922	0.02743873	0.86816599	31.64016139	0.03160540	34
35	1.15665284	37.59668268	0.02659809	0.86456365	32.50472504	0.03076476	35
36	1.16147223	38.75333552	0.02580423	0.86097624	33.36570128	0.02997090	36
37	1.16631170	39.91480775	0.02505336	0.85740373	34.22310501	0.02922003	37
38	1.17117133	41.08111945	0.02434208	0.85384604	35.07695105	0.02850875	38
39	1.17605121	42.25229078	0.02366736	0.85030311	35.92725416	0.02783402	39
40	1.18095142	43.42834199	0.02302644	0.84677488	36.77402904	0.02719310	40
41	1.18587206	44.60929342	0.02241685	0.84326129	37.61729033	0.02658352	41
42	1.19081319	45.79516547	0.02183637	0.83976228	38.45705261	0.02600303	42
43	1.19577491	46.98597866	0.02128295	0.83627779	39.29333040	0.02544961	43
44	1.20075731	48.18175357	0.02075474	0.83280776	40.12613816	0.02492141	44
45	1.20576046	49.38251088	0.02025008	0.82935312	40.95549028	0.02441675	45
46	1.21078446	50.58827134	0.01976743	0.82591083	41.78140111	0.02393409	46
47	1.21582940	51.79905581	0.01930537	0.82248381	42.60388492	0.02347204	47
48	1.22089536	53.01488563	0.01886263	0.81907102	43.42295594	0.02302929	48
49	1.22598242	54.23578056	0.01843801	0.81567238	44.23862832	0.02260468	49
50	1.23109068	55.46176298	0.01803044	0.81228785	45.05091617	0.02219711	50

TABLE III (*continued*)

$i = \frac{6\%}{12} = 0.005$

Number of periods n	Accumulation factor $(1+i)^n$	Amount of $1 per period $s_{\overline{n}\mid}$	Periodic deposit needed to yield $1 $1/s_{\overline{n}\mid}$	Present value of $1 $(1+i)^{-n}$	Present value of $1 per period $a_{\overline{n}\mid}$	Periodic payment needed to pay $1 $1/a_{\overline{n}\mid}$	Number of periods n
1	1.00500000	1.00000000	1.00000000	0.99502488	0.99502488	1.00500000	1
2	1.01002500	2.00500000	0.49875312	0.99007450	1.98509938	0.50375312	2
3	1.01507513	3.01502500	0.33167221	0.98514876	2.97024814	0.33667221	3
4	1.02015050	4.03010013	0.24813279	0.98024752	3.95049566	0.25313279	4
5	1.02525125	5.05025063	0.19800997	0.97537067	4.92586633	0.20300997	5
6	1.03037751	6.07550188	0.16459546	0.97051808	5.89638441	0.16959546	6
7	1.03552940	7.10587939	0.14072854	0.96568963	6.86207404	0.14572854	7
8	1.04070704	8.14140879	0.12282886	0.96088520	7.82295924	0.12782886	8
9	1.04591058	9.18211583	0.10890736	0.95610468	8.77906392	0.11390736	9
10	1.05114013	10.22802641	0.09777057	0.95134794	9.73041186	0.10277057	10
11	1.05639583	11.27916654	0.08865903	0.94661487	10.67702673	0.09365903	11
12	1.06167781	12.33556237	0.08106643	0.94190534	11.61893207	0.08606643	12
13	1.06698620	13.39724018	0.07464224	0.93721924	12.55615131	0.07964224	13
14	1.07232113	14.46422639	0.06913609	0.93255646	13.48870777	0.07413609	14
15	1.07768274	15.53654752	0.06436436	0.92791688	14.41662465	0.06936436	15
16	1.08307115	16.61423026	0.06018937	0.92330037	15.33992502	0.06518937	16
17	1.08848651	17.69730141	0.05650579	0.91870684	16.25863186	0.06150579	17
18	1.09392894	18.78578791	0.05323173	0.91413616	17.17276802	0.05823173	18
19	1.09939858	19.87971685	0.05030253	0.90958822	18.08235624	0.05530253	19
20	1.10489558	20.97911544	0.04766645	0.90506290	18.98741915	0.05266645	20
21	1.11042006	22.08401101	0.04528163	0.90056010	19.88797925	0.05028163	21
22	1.11597216	23.19443107	0.04311380	0.89607971	20.78405896	0.04811380	22
23	1.12155202	24.31040322	0.04113465	0.89162160	21.67568055	0.04613465	23
24	1.12715978	25.43195524	0.03932061	0.88718567	22.56286622	0.04432061	24
25	1.13279558	26.55911502	0.03765186	0.88277181	23.44563803	0.04265186	25
26	1.13845955	27.69191059	0.03611163	0.87837991	24.32401794	0.04111163	26
27	1.14415185	28.83037015	0.03468565	0.87400986	25.19802780	0.03968565	27
28	1.14987261	29.97452200	0.03336167	0.86966155	26.06768936	0.03836167	28
29	1.15562197	31.12439461	0.03212914	0.86533488	26.93302423	0.03712914	29
30	1.16140008	32.28001658	0.03097892	0.86102973	27.79405397	0.03597892	30
31	1.16720708	33.44141666	0.02990304	0.85674600	28.65079997	0.03490304	31
32	1.17304312	34.60862375	0.02889453	0.85248358	29.50328355	0.03389453	32
33	1.17890833	35.78166686	0.02794727	0.84824237	30.35152592	0.03294727	33
34	1.18480288	36.96057520	0.02705586	0.84402226	31.19554818	0.03205586	34
35	1.19072689	38.14537807	0.02621550	0.83982314	32.03537132	0.03121550	35
36	1.19668052	39.33610496	0.02542194	0.83564492	32.87101624	0.03042194	36
37	1.20266393	40.53278549	0.02467139	0.83148748	33.70250372	0.02967139	37
38	1.20867725	41.73544942	0.02396045	0.82735073	34.52985445	0.02896045	38
39	1.21472063	42.94412666	0.02328607	0.82323455	35.35308900	0.02828607	39
40	1.22079424	44.15884730	0.02264552	0.81913886	36.17222786	0.02764552	40
41	1.22689821	45.37964153	0.02203631	0.81506354	36.98729141	0.02703631	41
42	1.23303270	46.60653974	0.02145622	0.81100850	37.79829991	0.02645622	42
43	1.23919786	47.83957244	0.02090320	0.80697363	38.60527354	0.02590320	43
44	1.24539385	49.07877030	0.02037541	0.80295884	39.40823238	0.02537541	44
45	1.25162082	50.32416415	0.01987117	0.79896402	40.20719640	0.02487117	45
46	1.25787892	51.57578497	0.01938894	0.79499907	41.00218547	0.02438894	46
47	1.26416832	52.83366390	0.01892733	0.79103390	41.79321937	0.02392733	47
48	1.27048916	54.09783222	0.01848503	0.78709841	42.58031778	0.02348503	48
49	1.27684161	55.36832138	0.01806087	0.78318250	43.36350028	0.02306087	49
50	1.28322581	56.64516299	0.01765376	0.77928607	44.14278635	0.02265376	50

TABLE III Compound Interest and Annuity / 567

TABLE III (*continued*)

$i = \frac{7}{12}\% = 0.005833333$

Number of periods n	Accumulation factor $(1+i)^n$	Amount of $1 per period $s_{\overline{n}\|i}$	Periodic deposit needed to yield $1 $1/s_{\overline{n}\|i}$	Present value of $1 $(1+i)^{-n}$	Present value of $1 per period $a_{\overline{n}\|i}$	Periodic payment needed to pay $1 $1/a_{\overline{n}\|i}$	Number of periods n
1	1.00583333	1.00000000	1.00000000	0.99420050	0.99420050	1.00583333	1
2	1.01170069	2.00583333	0.49854591	0.98845463	1.98265513	0.50437924	2
3	1.01760228	3.01753403	0.33139643	0.98270220	2.96535732	0.33722976	3
4	1.02353830	4.03516631	0.24782310	0.97700301	3.94234034	0.25365644	4
5	1.02950894	5.05867460	0.19768024	0.97133688	4.91367722	0.20351357	5
6	1.03551440	6.08818354	0.16425260	0.96570361	5.87938083	0.17008594	6
7	1.04155490	7.12369794	0.14037653	0.96010301	6.83948384	0.14620986	7
8	1.04763064	8.16525285	0.12247018	0.95453489	7.79401874	0.12830352	8
9	1.05374182	9.21288349	0.10854365	0.94899906	8.74301780	0.11437698	9
10	1.05988865	10.26662531	0.09740299	0.94347534	9.68651314	0.10323632	10
11	1.06607133	11.32651396	0.08828842	0.93802354	10.62453667	0.09412175	11
12	1.07229008	12.39258529	0.08069341	0.93258847	11.55712014	0.08652675	12
13	1.07854511	13.46487537	0.07426730	0.92717495	12.48429509	0.08010064	13
14	1.08483662	14.54342048	0.06875962	0.92179779	13.40609288	0.07459295	14
15	1.09116483	15.62825710	0.06398666	0.91645182	14.32254470	0.06982000	15
16	1.09752996	16.71942193	0.05981068	0.91113686	15.23368156	0.06564401	16
17	1.10393222	17.81695189	0.05612632	0.90585272	16.13953427	0.06195966	17
18	1.11037182	18.92088411	0.05285165	0.90059922	17.04013350	0.05868499	18
19	1.11684899	20.03125593	0.04992198	0.89537619	17.93550969	0.05575532	19
20	1.12336395	21.14810493	0.04728556	0.89018346	18.82569315	0.05311889	20
21	1.12991690	22.27146887	0.04490050	0.88502084	19.71071398	0.05073383	21
22	1.13650808	23.40138577	0.04273251	0.87988815	20.59060213	0.04856585	22
23	1.14313771	24.53789386	0.04075329	0.87478524	21.46538738	0.04658663	23
24	1.14980602	25.68103157	0.03893225	0.86971192	22.33509930	0.04477258	24
25	1.15651322	26.83083759	0.03727055	0.86466802	23.19976732	0.04310388	25
26	1.16325955	27.98735081	0.03573043	0.85965338	24.05942070	0.04156376	26
27	1.17004523	29.15061036	0.03430460	0.85466782	24.91408852	0.04013793	27
28	1.17687049	30.32065558	0.03298082	0.84971117	25.76379968	0.03881415	28
29	1.18373557	31.49752607	0.03174853	0.84478327	26.60858295	0.03758186	29
30	1.19064069	32.68126164	0.03059857	0.83988394	27.44846689	0.03643191	30
31	1.19758610	33.87190233	0.02952299	0.83501303	28.28347993	0.03535633	31
32	1.20457202	35.06948843	0.02851482	0.83017037	29.11365030	0.03434815	32
33	1.21159869	36.27406045	0.02756791	0.82535580	29.93900610	0.03340124	33
34	1.21866634	37.48565913	0.02667687	0.82056914	30.75957524	0.03251020	34
35	1.22577523	38.70432548	0.02583691	0.81581025	31.57538549	0.03167024	35
36	1.23292559	39.93010071	0.02504376	0.81107896	32.38646445	0.03087710	36
37	1.24011765	41.16302630	0.02429365	0.80637510	33.19283955	0.03012698	37
38	1.24735167	42.40314395	0.02358316	0.80169853	33.99453808	0.02941649	38
39	1.25462789	43.65049562	0.02290925	0.79704907	34.79158716	0.02874258	39
40	1.26194655	44.90512352	0.02226917	0.79242659	35.58401374	0.02810251	40
41	1.26930791	46.16707007	0.02166046	0.78783091	36.37184465	0.02749379	41
42	1.27671220	47.43637798	0.02108087	0.78326188	37.15510653	0.02691420	42
43	1.28415969	48.71309018	0.02052836	0.77871935	37.93382588	0.02636170	43
44	1.29165062	49.99724988	0.02000110	0.77420316	38.70802904	0.02583443	44
45	1.29918525	51.28890050	0.01949740	0.76971317	39.47774221	0.02533073	45
46	1.30676383	52.58808575	0.01901571	0.76524922	40.24299143	0.02484905	46
47	1.31438662	53.89484959	0.01855465	0.76081115	41.00380258	0.02438798	47
48	1.32205388	55.20923621	0.01811291	0.75639883	41.76020141	0.02394624	48
49	1.32976586	56.53129009	0.01768932	0.75201209	42.51221349	0.02352265	49
50	1.33752283	57.86105595	0.01728278	0.74765079	43.25986428	0.02311612	50

TABLE III (*continued*)

$i = \frac{8}{12}\% = 0.006666667$

Number of periods n	Accumulation factor $(1+i)^n$	Amount of $1 per period $s_{\overline{n}|i}$	Periodic deposit needed to yield $1 $1/s_{\overline{n}|i}$	Present value of $1 $(1+i)^{-n}$	Present value of $1 per period $a_{\overline{n}|i}$	Periodic payment needed to pay $1 $1/a_{\overline{n}|i}$	Number of periods n
1	1.00666667	1.00000000	1.00000000	0.99337748	0.99337748	1.00666667	1
2	1.01337778	2.00666667	0.49833887	0.98679882	1.98017631	0.50500554	2
3	1.02013363	3.02004444	0.33112095	0.98026373	2.96044004	0.33778762	3
4	1.02693452	4.04017807	0.24751384	0.97377192	3.93421196	0.25418051	4
5	1.03378075	5.06711259	0.19735105	0.96732310	4.90153506	0.20401772	5
6	1.04067262	6.10089335	0.16391042	0.96091699	5.86245205	0.17057709	6
7	1.04761044	7.14156597	0.14002531	0.95455330	6.81700535	0.14669198	7
8	1.05459451	8.18917641	0.12211240	0.94823175	7.76523710	0.12877907	8
9	1.06162514	9.24377092	0.10818096	0.94195207	8.70718917	0.11484763	9
10	1.06870264	10.30539606	0.09703654	0.93571398	9.64290315	0.10370321	10
11	1.07582732	11.37409870	0.08791905	0.92951720	10.57242035	0.09458572	11
12	1.08299951	12.44992602	0.08032176	0.92336145	11.49578180	0.08698843	12
13	1.09021950	13.53292553	0.07389385	0.91724648	12.41302828	0.08056052	13
14	1.09748763	14.62314503	0.06838474	0.91117200	13.32420028	0.07505141	14
15	1.10480422	15.72063266	0.06361067	0.90513775	14.22933802	0.07027734	15
16	1.11216958	16.82543688	0.05943382	0.89914346	15.12848148	0.06610049	16
17	1.11958404	17.93760646	0.05574880	0.89318886	16.02167035	0.06241546	17
18	1.12704794	19.05719051	0.05247363	0.88727371	16.90894405	0.05914030	18
19	1.13456159	20.18423844	0.04954361	0.88139772	17.79034177	0.05621027	19
20	1.14212533	21.31880003	0.04690696	0.87556065	18.66590242	0.05357362	20
21	1.14973950	22.46092536	0.04452176	0.86976224	19.53566466	0.05118843	21
22	1.15740443	23.61066487	0.04235374	0.86400222	20.39966688	0.04902041	22
23	1.16512046	24.76806930	0.04037456	0.85828035	21.25794723	0.04704123	23
24	1.17288793	25.93318976	0.03856062	0.85259638	22.11054361	0.04522729	24
25	1.18070718	27.10607769	0.03689210	0.84695004	22.95749365	0.04355876	25
26	1.18857857	28.28678488	0.03535220	0.84134110	23.79883475	0.04201886	26
27	1.19650242	29.47536344	0.03392664	0.83576931	24.63460406	0.04059331	27
28	1.20447911	30.67186587	0.03260317	0.83023441	25.46483847	0.03926983	28
29	1.21250897	31.87634497	0.03137123	0.82473617	26.28957464	0.03803789	29
30	1.22059236	33.08885394	0.03022166	0.81927434	27.10884898	0.03688832	30
31	1.22872964	34.30944630	0.02914649	0.81384868	27.92269766	0.03581316	31
32	1.23692117	35.53817594	0.02813875	0.80845896	28.73115662	0.03480542	32
33	1.24516731	36.77509711	0.02719231	0.80310492	29.53426154	0.03385898	33
34	1.25346843	38.02026443	0.02630176	0.79778635	30.33204789	0.03296843	34
35	1.26182489	39.27373286	0.02546231	0.79250299	31.12455088	0.03212898	35
36	1.27023705	40.53555774	0.02466970	0.78725463	31.91180551	0.03133637	36
37	1.27870530	41.80579479	0.02392013	0.78204102	32.69384653	0.03058680	37
38	1.28723000	43.08450009	0.02321020	0.77686194	33.47070848	0.02987687	38
39	1.29581153	44.37173009	0.02253687	0.77171716	34.24242564	0.02920354	39
40	1.30445028	45.66754163	0.02189739	0.76660645	35.00903209	0.02856406	40
41	1.31314661	46.97199191	0.02128928	0.76152959	35.77056168	0.02795595	41
42	1.32190092	48.28513852	0.02071031	0.75648635	36.52704803	0.02737697	42
43	1.33071360	49.60703944	0.02015843	0.75147650	37.27852453	0.02682510	43
44	1.33958502	50.93775304	0.01963180	0.74649984	38.02502437	0.02629847	44
45	1.34851559	52.27733806	0.01912875	0.74155613	38.76658050	0.02579541	45
46	1.35750569	53.62585365	0.01864772	0.73664516	39.50322566	0.02531439	46
47	1.36655573	54.98335934	0.01818732	0.73176672	40.23499238	0.02485399	47
48	1.37566610	56.34991507	0.01774626	0.72692058	40.96191296	0.02441292	48
49	1.38483721	57.72558117	0.01732334	0.72210654	41.68401949	0.02399001	49
50	1.39406946	59.11041837	0.01691749	0.71732437	42.40134387	0.02358416	50

TABLE III Compound Interest and Annuity / 569

TABLE III (*continued*)

$i = \frac{9}{12}\% = 0.0075$

Number of periods n	Accumulation factor $(1+i)^n$	Amount of $1 per period $s_{\overline{n}|i}$	Periodic deposit needed to yield $1 $1/s_{\overline{n}|i}$	Present value of $1 $(1+i)^{-n}$	Present value of $1 per period $a_{\overline{n}|i}$	Periodic payment needed to pay $1 $1/a_{\overline{n}|i}$	Number of periods n
1	1.00750000	1.00000000	1.00000000	0.99255583	0.99255583	1.00750000	1
2	1.01505625	2.00750000	0.49813200	0.98516708	1.97772291	0.50563200	2
3	1.02266917	3.02255625	0.33084579	0.97783333	2.95555624	0.33834579	3
4	1.03033919	4.04522542	0.24720501	0.97055417	3.92611041	0.25470501	4
5	1.03806673	5.07556461	0.19702242	0.96332920	4.88943961	0.20452242	5
6	1.04585224	6.11363135	0.16356891	0.95615802	5.84559763	0.17106891	6
7	1.05369613	7.15948358	0.13967488	0.94904022	6.79463785	0.14717488	7
8	1.06159885	8.21317971	0.12175552	0.94197540	7.73661325	0.12925552	8
9	1.06956084	9.27477856	0.10781929	0.93496318	8.67157642	0.11531929	9
10	1.07758255	10.34433940	0.09667123	0.92800315	9.59957958	0.10417123	10
11	1.08566441	11.42192194	0.08755094	0.92109494	10.52067452	0.09505094	11
12	1.09380690	12.50758636	0.07995148	0.91423815	11.43491267	0.08745148	12
13	1.10201045	13.60139325	0.07352188	0.90743241	12.34234508	0.08102188	13
14	1.11027553	14.70340370	0.06801146	0.90067733	13.24302242	0.07551146	14
15	1.11860259	15.81367923	0.06323639	0.89397254	14.13699495	0.07073639	15
16	1.12699211	16.93228183	0.05905879	0.88731766	15.02431261	0.06655879	16
17	1.13544455	18.05927394	0.05537321	0.88071231	15.90502492	0.06287321	17
18	1.14396039	19.19471849	0.05209766	0.87415614	16.77918107	0.05959766	18
19	1.15254009	20.33867888	0.04916740	0.86764878	17.64682984	0.05666740	19
20	1.16118414	21.49121897	0.04653063	0.86118985	18.50801969	0.05403063	20
21	1.16989302	22.65240312	0.04414543	0.85477901	19.36279870	0.05164543	21
22	1.17866722	23.82229614	0.04197748	0.84841589	20.21121459	0.04947748	22
23	1.18750723	25.00096336	0.03999846	0.84210014	21.05331473	0.04749846	23
24	1.19641353	26.18847059	0.03818474	0.83583140	21.88914614	0.04568474	24
25	1.20538663	27.38488412	0.03651650	0.82960933	22.71875547	0.04401650	25
26	1.21442703	28.59027075	0.03497693	0.82343358	23.54218905	0.04247693	26
27	1.22353523	29.80469778	0.03355176	0.81730380	24.35949286	0.04105176	27
28	1.23271175	31.02823301	0.03222871	0.81121966	25.17071251	0.03972871	28
29	1.24195709	32.26094476	0.03099723	0.80518080	25.97589331	0.03849723	29
30	1.25127176	33.50290184	0.02984816	0.79918690	26.77508021	0.03734816	30
31	1.26065630	34.75417361	0.02877352	0.79323762	27.56831783	0.03627352	31
32	1.27011122	36.01482991	0.02776634	0.78733262	28.35565045	0.03526634	32
33	1.27963706	37.28494113	0.02682048	0.78147158	29.13712203	0.03432048	33
34	1.28923434	38.56457819	0.02593053	0.77565418	29.91277621	0.03343053	34
35	1.29890359	39.85381253	0.02509170	0.76988008	30.68265629	0.03259170	35
36	1.30864537	41.15271612	0.02429973	0.76414896	31.44680525	0.03179973	36
37	1.31846021	42.46136149	0.02355082	0.75846051	32.20526576	0.03105082	37
38	1.32834866	43.77982170	0.02284157	0.75281440	32.95808016	0.03034157	38
39	1.33831128	45.10817037	0.02216893	0.74721032	33.70529048	0.02966893	39
40	1.34834861	46.44648164	0.02153016	0.74164796	34.44693844	0.02903016	40
41	1.35846123	47.79483026	0.02092276	0.73612701	35.18306545	0.02842276	41
42	1.36864969	49.15329148	0.02034452	0.73064716	35.91371260	0.02784452	42
43	1.37891456	50.52194117	0.01979338	0.72520809	36.63892070	0.02729338	43
44	1.38925642	51.90085573	0.01926751	0.71980952	37.35873022	0.02676751	44
45	1.39967584	53.29011215	0.01876521	0.71445114	38.07318136	0.02626521	45
46	1.41017341	54.68978799	0.01828495	0.70913264	38.78231401	0.02578495	46
47	1.42074971	56.09966140	0.01782532	0.70385374	39.48616775	0.02532532	47
48	1.43140533	57.52071111	0.01738504	0.69861414	40.18478189	0.02488504	48
49	1.44214087	58.95211644	0.01696292	0.69341353	40.87819542	0.02446292	49
50	1.45295693	60.39425732	0.01655787	0.68825165	41.56644707	0.02405787	50

TABLE III (*continued*)

$i = \frac{10\%}{12} = 0.008333333$

Number of periods n	Accumulation factor $(1+i)^n$	Amount of $1 per period $s_{\overline{n}|i}$	Periodic deposit needed to yield $1 $1/s_{\overline{n}|i}$	Present value of $1 $(1+i)^{-n}$	Present value of $1 per period $a_{\overline{n}|i}$	Periodic payment needed to pay $1 $1/a_{\overline{n}|i}$	Number of periods n
1	1.00833333	1.00000000	1.00000000	0.99173554	0.99173554	1.00833333	1
2	1.01673611	2.00833333	0.49792531	0.98353938	1.97527491	0.50625864	2
3	1.02520891	3.02506944	0.33057092	0.97541095	2.95068586	0.33890426	3
4	1.03375232	4.05027836	0.24689661	0.96734970	3.91803557	0.25522994	4
5	1.04236692	5.08403068	0.19669433	0.95953508	4.87739065	0.20502766	5
6	1.05105331	6.12639760	0.16322806	0.95142652	5.82881717	0.17156139	6
7	1.05981209	7.17745091	0.13932523	0.94356349	6.77238066	0.14765856	7
8	1.06864386	8.23726300	0.12139995	0.93576545	7.70814611	0.12973288	8
9	1.07754922	9.30590686	0.10745863	0.92803185	8.63617796	0.11579196	9
10	1.08652880	10.38345608	0.09630705	0.92036217	9.55654013	0.10464038	10
11	1.09558321	11.46998489	0.08718407	0.91275587	10.46929600	0.09551741	11
12	1.10471307	12.55556809	0.07958255	0.90521243	11.37450843	0.08791589	12
13	1.11391901	13.67028116	0.07315138	0.89773134	12.27223976	0.08148472	13
14	1.12320167	14.78420017	0.06763978	0.89031207	13.16255183	0.07597311	14
15	1.13256168	15.90740184	0.06286382	0.88295412	14.04550595	0.07119715	15
16	1.14199970	17.03996352	0.05868557	0.87565698	14.92116292	0.06701890	16
17	1.15151636	18.18196322	0.05499956	0.86842014	15.78958306	0.06333289	17
18	1.16111233	19.33347958	0.05172375	0.86124312	16.65082618	0.06005708	18
19	1.17078827	20.49459191	0.04879336	0.85412540	17.50495158	0.05712669	19
20	1.18054483	21.66538017	0.04615659	0.84706652	18.35201810	0.05448992	20
21	1.19038271	22.84592501	0.04377148	0.84006597	19.19208406	0.05210482	21
22	1.20030256	24.03630772	0.04160373	0.83312327	20.02520734	0.04993706	22
23	1.21030509	25.23661028	0.03962497	0.82623796	20.85144529	0.04795831	23
24	1.22039096	26.44691537	0.03781159	0.81940954	21.67085483	0.04614493	24
25	1.23056089	27.66730633	0.03614374	0.81263756	22.48349240	0.04447708	25
26	1.24081556	28.89786721	0.03460463	0.80592155	23.28941395	0.04293796	26
27	1.25115569	30.13868277	0.03317995	0.79926104	24.08867499	0.04151328	27
28	1.26158199	31.38983846	0.03185744	0.79265558	24.88133057	0.04019078	28
29	1.27209517	32.65142045	0.03062654	0.78610471	25.66743527	0.03895987	29
30	1.28269596	33.92351562	0.02947808	0.77960797	26.44704325	0.03781141	30
31	1.29338510	35.20621158	0.02840408	0.77316493	27.22020818	0.03673741	31
32	1.30416331	36.49959668	0.02739756	0.76677514	27.98698332	0.03573090	32
33	1.31503133	37.80375999	0.02645240	0.76043815	28.74742147	0.03478553	33
34	1.32598993	39.11879132	0.02556316	0.75415354	29.50157501	0.03389650	34
35	1.33703984	40.44478125	0.02472507	0.74792087	30.24949588	0.03305840	35
36	1.34818184	41.78182109	0.02393385	0.74173970	30.99123559	0.03226719	36
37	1.35941669	43.13000293	0.02318572	0.73560962	31.72684521	0.03151905	37
38	1.37074516	44.48941962	0.02247725	0.72953020	32.45637541	0.03081059	38
39	1.38216804	45.86016479	0.02180542	0.72350103	33.17987644	0.03013875	39
40	1.39368611	47.24233283	0.02116746	0.71752168	33.89739813	0.02950079	40
41	1.40530016	48.63601893	0.02056049	0.71159175	34.60898988	0.02889423	41
42	1.41701099	50.04131909	0.01998349	0.70571083	35.31470070	0.02831682	42
43	1.42881942	51.45833008	0.01943320	0.69987851	36.01457921	0.02776653	43
44	1.44072625	52.88714950	0.01890818	0.69409439	36.70867360	0.02724152	44
45	1.45273230	54.32787575	0.01840676	0.68835807	37.39703167	0.02674009	45
46	1.46483840	55.78060805	0.01792738	0.68266916	38.07970083	0.02626071	46
47	1.47704539	57.24544645	0.01746864	0.67702727	38.75672809	0.02580197	47
48	1.48935410	58.72249183	0.01702925	0.67143200	39.42816009	0.02536258	48
49	1.50176538	60.21184593	0.01660803	0.66588297	40.09404307	0.02494136	49
50	1.51428009	61.71361131	0.01620388	0.66037981	40.75442288	0.02453721	50

TABLE III Compound Interest and Annuity / 571

TABLE III (*continued*)

$i = \frac{11}{12}\% = 0.00916667$

Number of periods n	Accumulation factor $(1+i)^n$	Amount of $1 per period $s_{\overline{n}|i}$	Periodic deposit needed to yield $1 $1/s_{\overline{n}|i}$	Present value of $1 $(1+i)^{-n}$	Present value of $1 per period $a_{\overline{n}|i}$	Periodic payment needed to pay $1 $1/a_{\overline{n}|i}$	Number of periods n
1	1.00916667	1.00000000	1.00000000	0.99091660	0.99091660	1.00916667	1
2	1.01841736	2.00916667	0.49771637	0.98191570	1.97283230	0.50688546	2
3	1.02775285	3.02758403	0.33029637	0.97299657	2.94582887	0.33946303	3
4	1.03717392	4.05533688	0.24658864	0.96415845	3.90998732	0.25575531	4
5	1.04668135	5.09251080	0.19636679	0.95540061	4.86538793	0.20553346	5
6	1.05627593	6.13919215	0.16288788	0.94672232	5.81211025	0.17205455	6
7	1.06595846	7.19546808	0.13897637	0.93812286	6.75023312	0.14814303	7
8	1.07572974	8.26142654	0.12104447	0.92960152	7.67983463	0.13021114	8
9	1.08559060	9.33715628	0.10709899	0.92115757	8.60099220	0.11626566	9
10	1.09554185	10.42274688	0.09594400	0.91279033	9.51378253	0.10511066	10
11	1.10558431	11.51828873	0.08681845	0.90449909	10.41828162	0.09598512	11
12	1.11571884	12.62387304	0.07921499	0.89628316	11.31456477	0.08838166	12
13	1.12594626	13.73959188	0.07278237	0.88814186	12.20270663	0.08194903	13
14	1.13626743	14.86553813	0.06726968	0.88007451	13.08278113	0.07643635	14
15	1.14668322	16.00180557	0.06249295	0.87208044	13.95486157	0.07165961	15
16	1.15719448	17.14848879	0.05831418	0.86415898	14.81902055	0.06748084	16
17	1.16780210	18.30568327	0.05462784	0.85630947	15.67533002	0.06379451	17
18	1.17850695	19.47348536	0.05135188	0.84853127	16.52386129	0.06051854	18
19	1.18930993	20.65199231	0.04842148	0.84082372	17.36468502	0.05758815	19
20	1.20021194	21.84130224	0.04578482	0.83318618	18.19787120	0.05495148	20
21	1.21121388	23.04151418	0.04339993	0.82561802	19.02348921	0.05256659	21
22	1.22231667	24.25272806	0.04123247	0.81811860	19.84160781	0.05039914	22
23	1.23352124	25.47504473	0.03925410	0.81068729	20.65229510	0.04842077	23
24	1.24482852	26.70856598	0.03744117	0.80332350	21.45561860	0.04660784	24
25	1.25623945	27.95339450	0.03577383	0.79602659	22.25164518	0.04494050	25
26	1.26775498	29.20963395	0.03423528	0.78879596	23.04044114	0.04340195	26
27	1.27937607	30.47738892	0.03281121	0.78163101	23.82207214	0.04197788	27
28	1.29110368	31.75676499	0.03148935	0.77453114	24.59660328	0.04065602	28
29	1.30293880	33.04786867	0.03025914	0.76749576	25.36409904	0.03942580	29
30	1.31488240	34.35080746	0.02911140	0.76052429	26.12462333	0.03827806	30
31	1.32693549	35.66568987	0.02803815	0.75361614	26.87823946	0.03720482	31
32	1.33909907	36.99262536	0.02703241	0.74677074	27.62501020	0.03619908	32
33	1.35137414	38.33172442	0.02608805	0.73998752	28.36499773	0.03525472	33
34	1.36376174	39.68309856	0.02519965	0.73326592	29.09826364	0.03436631	34
35	1.37626289	41.04686030	0.02436240	0.72660537	29.82486901	0.03352907	35
36	1.38887863	42.42312319	0.02357205	0.72000532	30.54487433	0.03273872	36
37	1.40161002	43.81200182	0.02282480	0.71346522	31.25833755	0.03199146	37
38	1.41445811	45.21361183	0.02211723	0.70698453	31.96532408	0.03128390	38
39	1.42742397	46.62806994	0.02144631	0.70056270	32.66588678	0.03061298	39
40	1.44050869	48.05549391	0.02080928	0.69419921	33.36008599	0.02997594	40
41	1.45371336	49.49600261	0.02020365	0.68789352	34.04797952	0.02937032	41
42	1.46703906	50.94971597	0.01962719	0.68164511	34.72962462	0.02879386	42
43	1.48048692	52.41675503	0.01907787	0.67545345	35.40507807	0.02824454	43
44	1.49405805	53.89724195	0.01855383	0.66931804	36.07439611	0.02772049	44
45	1.50775358	55.39130000	0.01805338	0.66323835	36.73763446	0.02722004	45
46	1.52157466	56.89905358	0.01757498	0.65721389	37.39484835	0.02674165	46
47	1.53552243	58.42062824	0.01711724	0.65124415	38.04609250	0.02628391	47
48	1.54959805	59.95615067	0.01667886	0.64532864	38.69142114	0.02584552	48
49	1.56380270	61.50574872	0.01625864	0.63946686	39.33088800	0.02542531	49
50	1.57813755	63.06975141	0.01585551	0.63365833	39.96454633	0.02502218	50

TABLE III (continued)

$i = 1\% = 0.01$

Number of periods n	Accumulation factor $(1+i)^n$	Amount of $1 per period $s_{\overline{n}|i}$	Periodic deposit needed to yield $1 $1/s_{\overline{n}|i}$	Present value of $1 $(1+i)^{-n}$	Present value of $1 per period $a_{\overline{n}|i}$	Periodic payment needed to pay $1 $1/a_{\overline{n}|i}$	Number of periods n
1	1.01000000	1.00000000	1.00000000	0.99009901	0.99009901	1.01000000	1
2	1.02010000	2.01000000	0.49751244	0.98029605	1.97039506	0.50751244	2
3	1.03030100	3.03010000	0.33002211	0.97059015	2.94098521	0.34002211	3
4	1.04060401	4.06040100	0.24628109	0.96098034	3.90196555	0.25628109	4
5	1.05101005	5.10100501	0.19603980	0.95146569	4.85343124	0.20603980	5
6	1.06152015	6.15201506	0.16254837	0.94204524	5.79547647	0.17254837	6
7	1.07213535	7.21353521	0.13862828	0.93271805	6.72819453	0.14862828	7
8	1.08285671	8.28567056	0.12069029	0.92348322	7.65167775	0.13069029	8
9	1.09368527	9.36852727	0.10674036	0.91433982	8.56601758	0.11674036	9
10	1.10462213	10.46221254	0.09558208	0.90528695	9.47130453	0.10558208	10
11	1.11566835	11.56683467	0.08645408	0.89632372	10.36762825	0.09645408	11
12	1.12682503	12.68250301	0.07888479	0.88744923	11.25507747	0.08888479	12
13	1.13809328	13.80932804	0.07241482	0.87866260	12.13374007	0.08241482	13
14	1.14947421	14.94742132	0.06690117	0.86996297	13.00370304	0.07690117	14
15	1.16096896	16.09689554	0.06212378	0.86134947	13.86505252	0.07212378	15
16	1.17257864	17.25786449	0.05794460	0.85282126	14.71787378	0.06794460	16
17	1.18430443	18.43044314	0.05425806	0.84437749	15.56225127	0.06425806	17
18	1.19614748	19.61474757	0.05098205	0.83601731	16.39826858	0.06098205	18
19	1.20810895	20.81089504	0.04805175	0.82773997	17.22600855	0.05805175	19
20	1.22019004	22.01900399	0.04541531	0.81954447	18.04555297	0.05541531	20
21	1.23239194	23.23919403	0.04303075	0.81143017	18.85698313	0.05303075	21
22	1.24471586	24.47158586	0.04086372	0.80339621	19.66037934	0.05086372	22
23	1.25716302	25.71630183	0.03888584	0.79544179	20.45582113	0.04888584	23
24	1.26973465	26.97346485	0.03707347	0.78756613	21.24338726	0.04707347	24
25	1.28243200	28.24319950	0.03540675	0.77976844	22.02315570	0.04540675	25
26	1.29525631	29.52563150	0.03386888	0.77204796	22.79520366	0.04386888	26
27	1.30820888	30.82088781	0.03244553	0.76440392	23.55960759	0.04244553	27
28	1.32129097	32.12909669	0.03112444	0.75683557	24.31644316	0.04112444	28
29	1.33450388	33.45038766	0.02989502	0.74934215	25.06578530	0.03989502	29
30	1.34784892	34.78489153	0.02874811	0.74192292	25.80770822	0.03874811	30
31	1.36132740	36.13274045	0.02767573	0.73457715	26.54228537	0.03767573	31
32	1.37494068	37.49406785	0.02667089	0.72730411	27.26958947	0.03667089	32
33	1.38869009	38.86900853	0.02572744	0.72010307	27.98969255	0.03572744	33
34	1.40257699	40.25769862	0.02483997	0.71297334	28.70266589	0.03483997	34
35	1.41660276	41.66027560	0.02400368	0.70591420	29.40858009	0.03400368	35
36	1.43076878	43.07687836	0.02321431	0.69892495	30.10750504	0.03321431	36
37	1.44507647	44.50764714	0.02246805	0.69200490	30.79950994	0.03246805	37
38	1.45952724	45.95272361	0.02176150	0.68515337	31.48466330	0.03176150	38
39	1.47412251	47.41225085	0.02109160	0.67836967	32.16303298	0.03109160	39
40	1.48886373	48.88637336	0.02045560	0.67165314	32.83468611	0.03045560	40
41	1.50375237	50.37523709	0.01985102	0.66500311	33.49968922	0.02985102	41
42	1.51878989	51.87898946	0.01927563	0.65841892	34.15810814	0.02927563	42
43	1.53397779	53.39777936	0.01872737	0.65189992	34.81000806	0.02872737	43
44	1.54931757	54.93175715	0.01820441	0.64544546	35.45545352	0.02820441	44
45	1.56481075	56.48107472	0.01770505	0.63905492	36.09450844	0.02770505	45
46	1.58045885	58.04588547	0.01722775	0.63272764	36.72723608	0.02722775	46
47	1.59626344	59.62634432	0.01677111	0.62646301	37.35369909	0.02677111	47
48	1.61222608	61.22260777	0.01633384	0.62026041	37.97395949	0.02633384	48
49	1.62834834	62.83483385	0.01591474	0.61411921	38.58807871	0.02591474	49
50	1.64463182	64.46318218	0.01551273	0.60803882	39.19611753	0.02551273	50

TABLE III Compound Interest and Annuity / 573

TABLE III (*continued*)

$i = 1\frac{1}{4}\% = 0.0125$

Number of periods n	Accumulation factor $(1+i)^n$	Amount of $1 per period $s_{\overline{n}\|i}$	Periodic deposit needed to yield $1 $1/s_{\overline{n}\|i}$	Present value of $1 $(1+i)^{-n}$	Present value of $1 per period $a_{\overline{n}\|i}$	Periodic payment needed to pay $1 $1/a_{\overline{n}\|i}$	Number of periods n
1	1.01250000	1.00000000	1.00000000	0.98765432	0.9876543	1.01250000	1
2	1.02515625	2.01250000	0.49689441	0.97546106	1.96311538	0.50939441	2
3	1.03797070	3.03765625	0.32920117	0.96341833	2.92653371	0.34170117	3
4	1.05094534	4.07562695	0.24536102	0.95152428	3.87805798	0.25786102	4
5	1.06408215	5.12657229	0.19506211	0.93977706	4.81783504	0.20756211	5
6	1.07738318	6.19065444	0.16153381	0.92817488	5.74600992	0.17403381	6
7	1.09085047	7.26803762	0.13758872	0.91671593	6.66272585	0.15008872	7
8	1.10448610	8.35888809	0.11963314	0.90539845	7.56812429	0.13213314	8
9	1.11829218	9.46337420	0.10567055	0.89422069	8.46234498	0.11817055	9
10	1.13227083	10.58166637	0.09450307	0.88318093	9.34552591	0.10700307	10
11	1.14642422	11.71393720	0.08536839	0.87227746	10.21780337	0.09786839	11
12	1.16075452	12.86036142	0.07775831	0.86150860	11.07931197	0.09025831	12
13	1.17526395	14.02111594	0.07132100	0.85087269	11.93018466	0.08382100	13
14	1.18995475	15.19637988	0.06580515	0.84036809	12.77055275	0.07830515	14
15	1.20482918	16.38633463	0.06102646	0.82999318	13.60054592	0.07352646	15
16	1.21988955	17.59116382	0.05684672	0.81974635	14.42029227	0.06934672	16
17	1.23513817	18.81105336	0.05316023	0.80962602	15.22991829	0.06566023	17
18	1.25057739	20.04619153	0.04988479	0.79963064	16.02954893	0.06238479	18
19	1.26620961	21.29676893	0.04695548	0.78975866	16.81930759	0.05945548	19
20	1.28203723	22.56297854	0.04432039	0.78000855	17.59931613	0.05682039	20
21	1.29806270	23.84501577	0.04193749	0.77037881	18.36969495	0.05443749	21
22	1.31428848	25.14307847	0.03977238	0.76086796	19.13056291	0.05227238	22
23	1.33071709	26.45736695	0.03779666	0.75147453	19.88203744	0.05029666	23
24	1.34735105	27.78808403	0.03598665	0.74219707	20.62423451	0.04848665	24
25	1.36419294	29.13543508	0.03432247	0.73303414	21.35726865	0.04682247	25
26	1.38124535	30.49962802	0.03278729	0.72398434	22.08125299	0.04528729	26
27	1.39851092	31.88087337	0.03136677	0.71504626	22.79629925	0.04386677	27
28	1.41599230	33.27938429	0.03004863	0.70621853	23.50251778	0.04254863	28
29	1.43369221	34.69537659	0.02882228	0.69749978	24.20001756	0.04132228	29
30	1.45161336	36.12906880	0.02767854	0.68888867	24.88890623	0.04017854	30
31	1.46975853	37.58068216	0.02660942	0.68038387	25.56929010	0.03910942	31
32	1.48813051	39.05044069	0.02560791	0.67198407	26.24127418	0.03810791	32
33	1.50673120	40.53857120	0.02466786	0.66368797	26.90496215	0.03716786	33
34	1.52556629	42.04530334	0.02378387	0.65549429	27.56045644	0.03628387	34
35	1.54463587	43.57086963	0.02295111	0.64740177	28.20785822	0.03545111	35
36	1.56394382	45.11550550	0.02216533	0.63940916	28.84726737	0.03466533	36
37	1.58349312	46.67944932	0.02142270	0.63151522	29.47878259	0.03392270	37
38	1.60328678	48.26294243	0.02071983	0.62371873	30.10250133	0.03321983	38
39	1.62332787	49.86622921	0.02005365	0.61601850	30.71851983	0.03255365	39
40	1.64361946	51.48955708	0.01942141	0.60841334	31.32693316	0.03192141	40
41	1.66416471	53.13317654	0.01882063	0.60090206	31.92783522	0.03132063	41
42	1.68496677	54.79734125	0.01824906	0.59348352	32.52131874	0.03074906	42
43	1.70602885	56.48230801	0.01770466	0.58615656	33.10747530	0.03020466	43
44	1.72735421	58.18833687	0.01718557	0.57892006	33.68639536	0.02968557	44
45	1.74894614	59.91569108	0.01669012	0.57177290	34.25816825	0.02919012	45
46	1.77080797	61.64463721	0.01621675	0.56471397	34.82288222	0.02871675	46
47	1.79294306	63.43545518	0.01576406	0.55774219	35.38062442	0.02826406	47
48	1.81535485	65.22838824	0.01533075	0.55085649	35.93148091	0.02783075	48
49	1.83804679	67.04374310	0.01491563	0.54405579	36.47553670	0.02741563	49
50	1.86102237	68.88178989	0.01451763	0.53733905	37.01287575	0.02701763	50

TABLE III *(continued)*

$i = 1\frac{1}{2}\% = 0.015$

Number of periods n	Accumulation factor $(1+i)^n$	Amount of \$1 per period $s_{\overline{n}\mid i}$	Periodic deposit needed to yield \$1 $1/s_{\overline{n}\mid i}$	Present value of \$1 $(1+i)^{-n}$	Present value of \$1 per period $a_{\overline{n}\mid i}$	Periodic payment needed to pay \$1 $1/a_{\overline{n}\mid i}$	Number of periods n
1	1.01500000	1.00000000	1.00000000	0.98522167	0.98522167	1.01500000	1
2	1.03022500	2.01500000	0.49627792	0.97066175	1.95588342	0.51127792	2
3	1.04567838	3.04522500	0.32838296	0.95631699	2.91220042	0.34338296	3
4	1.06136355	4.09090338	0.24444479	0.94218423	3.85438465	0.25944479	4
5	1.07728400	5.15226693	0.19408932	0.92826033	4.78264497	0.20908932	5
6	1.09344326	6.22955093	0.16052521	0.91454219	5.69718717	0.17552521	6
7	1.10984491	7.32299419	0.13655616	0.90102679	6.59821396	0.15155616	7
8	1.12649259	8.43283911	0.11858402	0.88771112	7.48592508	0.13358402	8
9	1.14338998	9.55933169	0.10460982	0.87459224	8.36051732	0.11960982	9
10	1.16054083	10.70272167	0.09343418	0.86164723	9.22218455	0.10843418	10
11	1.17794894	11.86326249	0.08429384	0.84893323	10.07111779	0.09929384	11
12	1.19561817	13.04121143	0.07667999	0.83638742	10.90750521	0.09167999	12
13	1.21355244	14.23682960	0.07024036	0.82402702	11.73153222	0.08524036	13
14	1.23175573	15.45038205	0.06472332	0.81184928	12.54338150	0.07972332	14
15	1.25023207	16.68213778	0.05994436	0.79985150	13.34323301	0.07494436	15
16	1.26898555	17.93236984	0.05576508	0.78803104	14.13126405	0.07076508	16
17	1.28802033	19.20135539	0.05207966	0.77638526	14.90764931	0.06707966	17
18	1.30734064	20.48937572	0.04880578	0.76491159	15.67256089	0.06380578	18
19	1.32695075	21.79671636	0.04587847	0.75360747	16.42616837	0.06087847	19
20	1.34685501	23.12366710	0.04324574	0.74247042	17.16863879	0.05824574	20
21	1.36705783	24.47052211	0.04086550	0.73149795	17.90013673	0.05586550	21
22	1.38756370	25.83757994	0.03870332	0.72068763	18.62082437	0.05370332	22
23	1.40837715	27.22514364	0.03673075	0.71003708	19.33086145	0.05173075	23
24	1.42950281	28.63352080	0.03492410	0.69954396	20.03040537	0.04992410	24
25	1.45094535	30.06302361	0.03326345	0.68920583	20.71961120	0.04826345	25
26	1.47270953	31.51396896	0.03173196	0.67902052	21.39863172	0.04673196	26
27	1.49480018	32.98667850	0.03031527	0.66898574	22.06761746	0.04531527	27
28	1.51722218	34.48147867	0.02900108	0.65909925	22.72671671	0.04400108	28
29	1.53998051	35.99870085	0.02777878	0.64935887	23.37607558	0.04277878	29
30	1.56308022	37.53868137	0.02663919	0.63976243	24.01583801	0.04163919	30
31	1.58652642	39.10176159	0.02557430	0.63030781	24.64614582	0.04057430	31
32	1.61032432	40.68828801	0.02457710	0.62099292	25.26713874	0.03957710	32
33	1.63447918	42.29861233	0.02364144	0.61181568	25.87895442	0.03864144	33
34	1.65899637	43.93309152	0.02276189	0.60277407	26.48172849	0.03776189	34
35	1.68388132	45.59208789	0.02193363	0.59386608	27.07559458	0.03693363	35
36	1.70913954	47.27596921	0.02115240	0.58508974	27.66068431	0.03615240	36
37	1.73477663	48.98510874	0.02041437	0.57644309	28.23712740	0.03541437	37
38	1.76079828	50.71988538	0.01971613	0.56792423	28.80505163	0.03471613	38
39	1.78721025	52.48068366	0.01905463	0.55953126	29.36458288	0.03405463	39
40	1.81401841	54.26789391	0.01842710	0.55126232	29.91584520	0.03342710	40
41	1.84122868	56.08191232	0.01783106	0.54311559	30.45896079	0.03283106	41
42	1.86884712	57.92314100	0.01726426	0.53508925	30.99405004	0.03226426	42
43	1.89687982	59.79198812	0.01672465	0.52718153	31.52123157	0.03172465	43
44	1.92533302	61.68886794	0.01621038	0.51939067	32.04062223	0.03121038	44
45	1.95421301	63.61420096	0.01571976	0.51171494	32.55233718	0.03071976	45
46	1.98352621	65.56841398	0.01525125	0.50415265	33.05648983	0.03025125	46
47	2.01327910	67.55194018	0.01480342	0.49670212	33.55319195	0.02980342	47
48	2.04347829	69.56521929	0.01437500	0.48936170	34.04255365	0.02937500	48
49	2.07413046	71.60869758	0.01396478	0.48212975	34.52468339	0.02896478	49
50	2.10524242	73.68282804	0.01357168	0.47500468	34.99968807	0.02857168	50

TABLE III Compound Interest and Annuity / 575

TABLE III (*continued*)

$i = 1\frac{3}{4}\% = 0.0175$

Number of periods n	Accumulation factor $(1+i)^n$	Amount of $1 per period $s_{\overline{n}|}$	Periodic deposit needed to yield $1 $1/s_{\overline{n}|}$	Present value of $1 $(1+i)^{-n}$	Present value of $1 per period $a_{\overline{n}|}$	Periodic payment needed to pay $1 $1/a_{\overline{n}|}$	Number of periods n
1	1.01750000	1.00000000	1.00000000	0.98280098	0.98280098	1.01750000	1
2	1.03530625	2.01750000	0.49566295	0.96589777	1.94869875	0.51316295	2
3	1.05342411	3.05280625	0.32756746	0.94928528	2.89798403	0.34506746	3
4	1.07185903	4.10623036	0.24353237	0.93295851	3.83094254	0.26103237	4
5	1.09061656	5.17808939	0.19312142	0.91691254	4.74785508	0.21062142	5
6	1.10970235	6.26870596	0.15952256	0.90114254	5.64899762	0.17702256	6
7	1.12912215	7.37840831	0.13553059	0.88564378	6.53464139	0.15303059	7
8	1.14888178	8.50753045	0.11754292	0.87041157	7.40505297	0.13504292	8
9	1.16898721	9.65641224	0.10355813	0.85541135	8.26049432	0.12105813	9
10	1.18944449	10.82539945	0.09237534	0.84072860	9.10122291	0.10987534	10
11	1.21025977	12.01484394	0.08323038	0.82626889	9.92749181	0.10073038	11
12	1.23143931	13.22510371	0.07561377	0.81205788	10.73954969	0.09311377	12
13	1.25298950	14.45654303	0.06917283	0.79809128	11.53764097	0.08667283	13
14	1.27491682	15.70953253	0.06365562	0.78436490	12.32200587	0.08115562	14
15	1.29722786	16.98444935	0.05887739	0.77087459	13.09288046	0.07637739	15
16	1.31992935	18.28167721	0.05469958	0.75761631	13.85049677	0.07219958	16
17	1.34302811	19.60160656	0.05101623	0.74458605	14.59508282	0.06851623	17
18	1.36653111	20.94463468	0.04774492	0.73177990	15.32686272	0.06524492	18
19	1.39044540	22.31116578	0.04482061	0.71919401	16.04605673	0.06232061	19
20	1.41477820	23.70161119	0.04219122	0.70682458	16.75288130	0.05969122	20
21	1.43953681	25.11638938	0.03981464	0.69466789	17.44754919	0.05731464	21
22	1.46472871	26.55592620	0.03765638	0.68272028	18.13026948	0.05515638	22
23	1.49036146	28.02065490	0.03588796	0.67097811	18.80124764	0.05338796	23
24	1.51644279	29.51101637	0.03388565	0.65943800	19.46068565	0.05138565	24
25	1.54298054	31.02745915	0.03222952	0.64809632	20.10878196	0.04972952	25
26	1.56998269	32.57043969	0.03070269	0.63694970	20.74573166	0.04820269	26
27	1.59745739	34.14042238	0.02929079	0.62599479	21.37172644	0.04679079	27
28	1.62541290	35.73787977	0.02798151	0.61522829	21.98695474	0.04548151	28
29	1.65385762	37.36329267	0.02676424	0.60464697	22.59160171	0.04426424	29
30	1.68280013	39.01715029	0.02562975	0.59424764	23.18584934	0.04312975	30
31	1.71224913	40.69995042	0.02457005	0.58402716	23.76987650	0.04207005	31
32	1.74221349	42.41219955	0.02357812	0.57398247	24.34385897	0.04107812	32
33	1.77270223	44.15441305	0.02264779	0.56411053	24.90796951	0.04014779	33
34	1.80372452	45.92711527	0.02177363	0.55440839	25.46237789	0.03927363	34
35	1.83528970	47.73083979	0.02095082	0.54487311	26.00725100	0.03845082	35
36	1.86740727	49.56612949	0.02017507	0.53550183	26.54275283	0.03767507	36
37	1.90008689	51.43353675	0.01944257	0.52629172	27.06904455	0.03694257	37
38	1.93333841	53.33362365	0.01874990	0.51724002	27.58628457	0.03624990	38
39	1.96717184	55.26696206	0.01809399	0.50834400	28.09462857	0.03559399	39
40	2.00159734	57.23413390	0.01747209	0.49960098	28.59422955	0.03497209	40
41	2.03662530	59.23573124	0.01688170	0.49100834	29.08523789	0.03438170	41
42	2.07226624	61.27235654	0.01632057	0.48256348	29.56780136	0.03382057	42
43	2.10853090	63.34462278	0.01578666	0.47426386	30.04206522	0.03328666	43
44	2.14543019	65.45315367	0.01527810	0.46610699	30.50817221	0.03277810	44
45	2.18297522	67.59858386	0.01479321	0.45809040	30.96626261	0.03229321	45
46	2.22117728	69.78155908	0.01433043	0.45021170	31.41647431	0.03183043	46
47	2.26004789	72.00273637	0.01388836	0.44246850	31.85894281	0.03138836	47
48	2.29959872	74.26278425	0.01346569	0.43485848	32.29380129	0.03096569	48
49	2.33984170	76.56238298	0.01306124	0.42737934	32.72118063	0.03056124	49
50	2.38078893	78.90222468	0.01267391	0.42002883	33.14120946	0.03017391	50

TABLE III *(continued)*

$i = 2\% = 0.02$

Number of periods n	Accumulation factor $(1+i)^n$	Amount of $1 per period $s_{\overline{n}|i}$	Periodic deposit needed to yield $1 $1/s_{\overline{n}|i}$	Present value of $1 $(1+i)^{-n}$	Present value of $1 per period $a_{\overline{n}|i}$	Periodic payment needed to pay $1 $1/a_{\overline{n}|i}$	Number of periods n
1	1.02000000	1.00000000	1.00000000	0.98039216	0.98039216	1.02000000	1
2	1.04040000	2.02000000	0.49504950	0.96116878	1.94156094	0.51504950	2
3	1.06120800	3.06040000	0.32675467	0.94232233	2.88388327	0.34675467	3
4	1.08243216	4.12160800	0.24262375	0.92384543	3.80772870	0.26262375	4
5	1.10408080	5.20404016	0.19215839	0.90573081	4.71345951	0.21215839	5
6	1.12616242	6.30812096	0.15852581	0.88797138	5.60143089	0.17852581	6
7	1.14868567	7.43428338	0.13451196	0.87056018	6.47199107	0.15451196	7
8	1.17165938	8.58296905	0.11650980	0.85349037	7.32548144	0.13650980	8
9	1.19509257	9.75462843	0.10251544	0.83675527	8.16223671	0.12251544	9
10	1.21899442	10.94972100	0.09132653	0.82034830	8.98258501	0.11132653	10
11	1.24337431	12.16871542	0.08217794	0.80426304	9.78684805	0.10217794	11
12	1.26824179	13.41208973	0.07455960	0.78849318	10.57534122	0.09455960	12
13	1.29360663	14.68033152	0.06811835	0.77303253	11.34837375	0.08811835	13
14	1.31947876	15.97393815	0.06260197	0.75787502	12.10624877	0.08260197	14
15	1.34586834	17.29341692	0.05782547	0.74301473	12.84926350	0.07782547	15
16	1.37278571	18.63928525	0.05365013	0.72844581	13.57770931	0.07365013	16
17	1.40024142	20.01207096	0.04996984	0.71414256	14.29187188	0.06996984	17
18	1.42824625	21.41231238	0.04670210	0.70015937	14.99203125	0.06670210	18
19	1.45681117	22.84055863	0.04378177	0.68643076	15.67846201	0.06378177	19
20	1.48594740	24.29736980	0.04115672	0.67297133	16.35143334	0.06115672	20
21	1.51566634	25.78331719	0.03878477	0.65977582	17.01120916	0.05878477	21
22	1.54597967	27.29898354	0.03663141	0.64683904	17.65804820	0.05663140	22
23	1.57689926	28.84496321	0.03466810	0.63415592	18.29220412	0.05466810	23
24	1.60843725	30.42186247	0.03287110	0.62172149	18.91392560	0.05287110	24
25	1.64060599	32.03029972	0.03122044	0.60953087	19.52345647	0.05122044	25
26	1.67341811	33.67090572	0.02969923	0.59757928	20.12103576	0.04969923	26
27	1.70688648	35.34432383	0.02829309	0.58586204	20.70689780	0.04829309	27
28	1.74102421	37.05121031	0.02698967	0.57437455	21.28127236	0.04698967	28
29	1.77584469	38.79223451	0.02577836	0.56311231	21.84438466	0.04577836	29
30	1.81136158	40.56807921	0.02464992	0.55207089	22.39645555	0.04464992	30
31	1.84758882	42.37944079	0.02359635	0.54124597	22.93770152	0.04359635	31
32	1.88454059	44.22702961	0.02261061	0.53063330	23.46833482	0.04261061	32
33	1.92223140	46.11157020	0.02168653	0.52022873	23.98856355	0.04168653	33
34	1.96067603	48.03380160	0.02081867	0.51002817	24.49859172	0.04081867	34
35	1.99988955	49.99447763	0.02000221	0.50002761	24.99861933	0.04000221	35
36	2.03988734	51.99436719	0.01923285	0.49022315	25.48884248	0.03923285	36
37	2.08068509	54.03425453	0.01850678	0.48061093	25.96945341	0.03850678	37
38	2.12229879	56.11493962	0.01782057	0.47118719	26.44064060	0.03782057	38
39	2.16474477	58.23723841	0.01717114	0.46194822	26.90258883	0.03717114	39
40	2.20803966	60.40198318	0.01655575	0.45289042	27.35547924	0.03655575	40
41	2.25220046	62.61002284	0.01597188	0.44401021	27.79948945	0.03597188	41
42	2.29724447	64.86222330	0.01541729	0.43530413	28.23479358	0.03541729	42
43	2.34318936	67.15946777	0.01488993	0.42678875	28.66156233	0.03488993	43
44	2.39005314	69.50265712	0.01438794	0.41840074	29.07996307	0.03438794	44
45	2.43785421	71.89271027	0.01390962	0.41019680	29.49015987	0.03390962	45
46	2.48661129	74.33056447	0.01345342	0.40215373	29.89231360	0.03345342	46
47	2.53634352	76.81717576	0.01301792	0.39426836	30.28658196	0.03301792	47
48	2.58707039	79.35351927	0.01260184	0.38653761	30.67311957	0.03260184	48
49	2.63881179	81.94058966	0.01220396	0.37895844	31.05207801	0.03220396	49
50	2.69158803	84.57940145	0.01182321	0.37152788	31.42360589	0.03182321	50

TABLE III Compound Interest and Annuity / **577**

TABLE III (*continued*)

$i = 3\% = 0.03$

Number of periods n	Accumulation factor $(1 + i)^n$	Amount of $1 per period $s_{\overline{n}|i}$	Periodic deposit needed to yield $1 $\dfrac{1}{s_{\overline{n}|i}}$	Present value of $1 $(1 + i)^{-n}$	Present value of $1 per period $a_{\overline{n}|i}$	Periodic payment needed to pay $1 $\dfrac{1}{a_{\overline{n}|i}}$	Number of periods n
1	1.03000000	1.00000000	1.00000000	0.97087379	0.97087379	1.03000000	1
2	1.06090000	2.03000000	0.49261084	0.94259591	1.91346970	0.52261084	2
3	1.09272700	3.09090000	0.32353036	0.91514166	2.82861135	0.35353036	3
4	1.12550881	4.18362700	0.23902700	0.88848705	3.71709840	0.26902705	4
5	1.15927407	5.30913581	0.18835457	0.86260878	4.57970719	0.21835457	5
6	1.19405230	6.46840988	0.15459750	0.83748426	5.41719144	0.18459750	6
7	1.22987387	7.66246218	0.13050635	0.81309151	6.23028296	0.16050635	7
8	1.26677008	8.89233605	0.11245639	0.78940923	7.01969219	0.14245639	8
9	1.30477318	10.15910613	0.09843386	0.76641673	7.78610892	0.12843386	9
10	1.34391638	11.46387931	0.08723051	0.74409391	8.53020284	0.11723051	10
11	1.38423387	12.80779569	0.07807745	0.72242128	9.25262411	0.10807745	11
12	1.42576089	14.19202956	0.07046209	0.70137988	9.95400399	0.10046209	12
13	1.46853371	15.61779045	0.06402954	0.68095134	10.63495533	0.09402954	13
14	1.51258972	17.08632416	0.05852634	0.66111781	11.29607314	0.08852634	14
15	1.55796742	18.59891389	0.05376658	0.64186195	11.93793509	0.08376658	15
16	1.60470644	20.15688130	0.04961085	0.62316694	12.56110203	0.07961085	16
17	1.65284763	21.76158774	0.04595253	0.60501645	13.16611847	0.07595253	17
18	1.70243306	23.41443537	0.04270870	0.58739461	13.75351308	0.07270870	18
19	1.75350605	25.11686844	0.03981388	0.57028603	14.32379911	0.06981388	19
20	1.80611123	26.87037449	0.03721571	0.55367575	14.87747486	0.06721571	20
21	1.86029457	28.67648572	0.03487178	0.53754928	15.41502414	0.06487178	21
22	1.91610341	30.53678030	0.03274739	0.52189250	15.93691664	0.06274739	22
23	1.97358651	32.45288370	0.03081390	0.50669175	16.44360839	0.06081390	23
24	2.03279411	34.42647022	0.02904742	0.49193374	16.93554212	0.05904742	24
25	2.09377793	36.45926432	0.02742787	0.47760557	17.41314769	0.05742787	25
26	2.15659127	38.55304225	0.02593829	0.46369473	17.87684242	0.05593829	26
27	2.22128901	40.70963352	0.02456421	0.45018906	18.32703147	0.05456421	27
28	2.28792768	42.93092252	0.02329323	0.43707675	18.76410823	0.05329323	28
29	2.35656551	45.21885020	0.02211467	0.42434636	19.18845459	0.05211467	29
30	2.42726247	47.57541571	0.02101926	0.41198676	19.60044135	0.05101926	30
31	2.50008035	50.00267818	0.01999893	0.39998715	20.00042849	0.04999893	31
32	2.57508276	52.50275852	0.01904662	0.38833703	20.38876553	0.04904662	32
33	2.65233524	55.07784128	0.01815612	0.37702625	20.76579178	0.04815612	33
34	2.73190530	57.73017652	0.01732196	0.36604490	21.13183668	0.04732196	34
35	2.81386245	60.46208181	0.01653929	0.35538340	21.48722007	0.04653929	35
36	2.89827833	63.27594427	0.01580379	0.34503243	21.83225250	0.04580379	36
37	2.98522668	66.17422259	0.01511162	0.33498244	22.16723544	0.04511162	37
38	3.07478348	69.15944927	0.01445954	0.32522615	22.49246159	0.04445954	38
39	3.16702698	72.23423275	0.01384385	0.31575355	22.80821513	0.04384385	39
40	3.26203779	75.40125973	0.01326238	0.30655684	23.11477197	0.04326238	40
41	3.35989893	78.66329753	0.01271241	0.29762800	23.41239997	0.04271241	41
42	3.46069589	82.02319645	0.01219167	0.28895922	23.70135920	0.04219167	42
43	3.56451677	85.48389234	0.01169811	0.28054294	23.98190213	0.04169811	43
44	3.67145227	89.04840911	0.01122985	0.27237178	24.25427392	0.04122985	44
45	3.78159584	92.71986139	0.01078518	0.26443862	24.51871254	0.04078518	45
46	3.89504372	96.50145723	0.01036254	0.25673653	24.77544907	0.04036254	46
47	4.01189503	100.39650095	0.00996051	0.24925876	25.02470783	0.03996051	47
48	4.13225188	104.40839598	0.00957777	0.24199880	25.26670664	0.03957777	48
49	4.25621944	108.54064785	0.00921314	0.23495029	25.50165693	0.03921314	49
50	4.38390602	112.79686729	0.00886549	0.22810708	25.72976401	0.03886549	50

TABLE III (continued)

$i = 4\% = 0.04$

Number of periods n	Accumulation factor $(1+i)^n$	Amount of $1 per period $s_{\overline{n}|i}$	Periodic deposit needed to yield $1 $1/s_{\overline{n}|i}$	Present value of $1 $(1+i)^{-n}$	Present value of $1 per period $a_{\overline{n}|i}$	Periodic payment needed to pay $1 $1/a_{\overline{n}|i}$	Number of periods n
1	1.04000000	1.00000000	1.00000000	0.96153846	0.96153846	1.04000000	1
2	1.08160000	2.04000000	0.49019608	0.92455621	1.88609467	0.53019608	2
3	1.12486400	3.12160000	0.32034854	0.88899636	2.77509103	0.36034854	3
4	1.16985856	4.24646400	0.23549005	0.85480419	3.62989522	0.27549005	4
5	1.21665290	5.41632256	0.18462711	0.82192711	4.45182233	0.22462711	5
6	1.26531902	6.63297546	0.15076190	0.79031453	5.24213686	0.19076190	6
7	1.31593178	7.89829448	0.12660961	0.75997781	6.00205467	0.16660961	7
8	1.36856905	9.21422626	0.10852783	0.73069021	6.73274487	0.14852783	8
9	1.42331181	10.58279531	0.09449299	0.70258674	7.43533161	0.13449299	9
10	1.48024428	12.00610712	0.08329094	0.67556417	8.11089578	0.12329094	10
11	1.53945406	13.48635141	0.07414904	0.64958093	8.76047671	0.11414904	11
12	1.60103222	15.02580546	0.06655217	0.62459705	9.38507376	0.10655217	12
13	1.66507351	16.62683768	0.06014373	0.60057409	9.98564785	0.10014373	13
14	1.73167645	18.29191119	0.05466897	0.57747508	10.56312293	0.09466897	14
15	1.80094351	20.02358764	0.04994110	0.55526450	11.11838743	0.08994110	15
16	1.87298125	21.82453114	0.04582000	0.53390818	11.65229561	0.08582000	16
17	1.94790050	23.69751239	0.04219852	0.51337325	12.16566885	0.08219852	17
18	2.02581652	25.64541288	0.03899333	0.49362812	12.65929697	0.07899333	18
19	2.10684918	27.67122940	0.03613862	0.47464242	13.13393940	0.07613862	19
20	2.19112314	29.77807858	0.03358175	0.45638695	13.59032634	0.07358175	20
21	2.27876807	31.96920172	0.03128011	0.43883360	14.02915995	0.07128011	21
22	2.36991879	34.24796979	0.02919881	0.42195539	14.45111533	0.06919881	22
23	2.46471554	36.61788858	0.02730906	0.40572633	14.85684167	0.06730906	23
24	2.56330416	39.08260412	0.02558683	0.39012147	15.24696314	0.06558683	24
25	2.66583633	41.64590829	0.02401196	0.37511680	15.62207994	0.06401196	25
26	2.77246978	44.31174462	0.02256738	0.36068923	15.98276918	0.06256738	26
27	2.88336858	47.08421440	0.02123854	0.34681657	16.32958575	0.06123854	27
28	2.99870332	49.96758298	0.02001298	0.33347747	16.66306322	0.06001298	28
29	3.11865145	52.96628630	0.01887993	0.32065141	16.98371463	0.05887993	29
30	3.24339751	56.08493775	0.01783010	0.30831867	17.29203330	0.05783010	30
31	3.37313341	59.32833526	0.01685535	0.29646026	17.58849356	0.05685535	31
32	3.50805875	62.70146867	0.01594859	0.28505794	17.87355150	0.05594859	32
33	3.64838110	66.20952742	0.01510357	0.27409417	18.14764567	0.05510357	33
34	3.79431634	69.85790851	0.01431477	0.26355209	18.41119776	0.05431477	34
35	3.94608899	73.65222486	0.01357732	0.25341547	18.66461323	0.05357732	35
36	4.10393255	77.59831385	0.01288688	0.24366872	18.90828195	0.05288688	36
37	4.26808986	81.70224640	0.01223957	0.23429685	19.14257880	0.05223957	37
38	4.43881345	85.97033626	0.01163192	0.22528543	19.36786423	0.05163192	38
39	4.61636599	90.40914971	0.01106083	0.21662061	19.58448484	0.05106083	39
40	4.80102063	95.02551570	0.01052349	0.20828904	19.79277388	0.05052349	40
41	4.99306145	99.82653633	0.01001738	0.20027793	19.99305181	0.05001738	41
42	5.19278391	104.81959778	0.00954020	0.19257493	20.18562674	0.04954020	42
43	5.40049527	110.01238169	0.00908989	0.18516820	20.37079494	0.04908989	43
44	5.61651508	115.41287696	0.00866454	0.17804635	20.54884129	0.04866454	44
45	5.84117568	121.02939204	0.00826246	0.17119841	20.72003970	0.04826246	45
46	6.07482271	126.87056772	0.00788205	0.16461386	20.88465356	0.04788205	46
47	6.31781562	132.94539043	0.00752189	0.15828256	21.04293612	0.04752189	47
48	6.57052824	139.26320604	0.00718065	0.15219476	21.19513088	0.04718065	48
49	6.83334937	145.83373429	0.00685712	0.14634112	21.34147200	0.04685712	49
50	7.10668335	152.66708366	0.00655020	0.14071262	21.48218462	0.04655020	50

TABLE III Compound Interest and Annuity / 579

TABLE III (*continued*)

$i = 5\% = 0.05$

Number of periods n	Accumulation factor $(1+i)^n$	Amount of $1 per period $s_{\overline{n}|}$	Periodic deposit needed to yield $1 $1/s_{\overline{n}|}$	Present value of $1 $(1+i)^{-n}$	Present value of $1 per period $a_{\overline{n}|}$	Periodic payment needed to pay $1 $1/a_{\overline{n}|}$	Number of periods n
1	1.05000000	1.00000000	1.00000000	0.95238095	0.95238095	1.05000000	1
2	1.10250000	2.05000000	0.48780488	0.90702948	1.85941043	0.53780488	2
3	1.15762500	3.15250000	0.31720856	0.86383760	2.72324803	0.36720856	3
4	1.21550625	4.31012500	0.23201183	0.82270247	3.54595050	0.28201183	4
5	1.27628156	5.52563125	0.18097480	0.78352617	4.32947667	0.23097480	5
6	1.34009564	6.80191281	0.14701747	0.74621540	5.07569207	0.19701747	6
7	1.40710042	8.14200845	0.12281982	0.71068133	5.78637340	0.17281982	7
8	1.47745544	9.54910888	0.10472181	0.67683936	6.46321276	0.15472181	8
9	1.55132822	11.02656432	0.09069008	0.64460892	7.10782168	0.14069008	9
10	1.62889463	12.57789254	0.07950457	0.61391325	7.72173493	0.12950457	10
11	1.71033936	14.20678716	0.07038889	0.58467929	8.30641422	0.12038889	11
12	1.79585633	15.91712652	0.06282541	0.55683742	8.86325164	0.11282541	12
13	1.88564914	17.71298285	0.05645577	0.53032135	9.39357299	0.10645577	13
14	1.97993160	19.59863199	0.05102397	0.50506795	9.89864094	0.10102397	14
15	2.07892818	21.57856359	0.04634229	0.48101710	10.37965804	0.09634229	15
16	2.18287459	23.65749177	0.04226991	0.45811152	10.83776956	0.09226991	16
17	2.29201832	25.84036636	0.03869914	0.43629669	11.27406625	0.08869914	17
18	2.40661923	28.13238467	0.03554622	0.41552065	11.68958690	0.08554622	18
19	2.52695020	30.53900391	0.03274501	0.39573396	12.08532086	0.08274501	19
20	2.65329771	33.06595410	0.03024259	0.37688948	12.46221034	0.08024259	20
21	2.78596259	35.71925181	0.02799611	0.35894236	12.82115271	0.07799611	21
22	2.92526072	38.50521440	0.02597051	0.34184987	13.16300258	0.07597051	22
23	3.07152376	41.43047512	0.02413682	0.32557131	13.48857388	0.07413682	23
24	3.22509994	44.50199887	0.02247090	0.31006791	13.79864179	0.07247090	24
25	3.38635494	47.72709882	0.02095246	0.29530277	14.09394457	0.07095246	25
26	3.55567269	51.11345376	0.01956432	0.28124073	14.37518530	0.06956432	26
27	3.73345632	54.66912645	0.01829186	0.26784832	14.64303362	0.06829186	27
28	3.92013914	58.40258277	0.01712253	0.25507364	14.89812726	0.06712253	28
29	4.11613560	62.32271191	0.01604551	0.24294632	15.14107358	0.06604551	29
30	4.32194238	66.43884750	0.01505144	0.23137745	15.37245103	0.06505144	30
31	4.53803949	70.76078988	0.01413212	0.22035947	15.59281050	0.06413212	31
32	4.76494147	75.29882937	0.01328042	0.20986617	15.80267667	0.06328042	32
33	5.00318854	80.06377084	0.01249004	0.19987254	16.00254921	0.06249004	33
34	5.25334797	85.06695938	0.01175545	0.19035480	16.19240401	0.06175545	34
35	5.51601537	90.32030735	0.01107171	0.18129029	16.37419429	0.06107171	35
36	5.79181614	95.83632272	0.01043446	0.17265741	16.54685171	0.06043446	36
37	6.08140694	101.62813886	0.00983979	0.16443563	16.71128734	0.05983979	37
38	6.38547729	107.70954580	0.00928423	0.15660536	16.86789271	0.05928423	38
39	6.70475115	114.09502309	0.00876462	0.14914797	17.01704067	0.05876462	39
40	7.03998871	120.79977424	0.00827816	0.14204568	17.15908635	0.05827816	40
41	7.39198815	127.83976295	0.00782229	0.13528160	17.29436796	0.05782229	41
42	7.76158756	135.23175110	0.00739471	0.12883962	17.42320758	0.05739471	42
43	8.14966693	142.99333866	0.00699333	0.12270440	17.54591198	0.05699333	43
44	8.55715028	151.14300559	0.00661625	0.11686133	17.66277331	0.05661625	44
45	8.98500779	159.70015587	0.00626173	0.11129651	17.77406982	0.05626173	45
46	9.43425818	168.68516366	0.00592820	0.10599668	17.88006650	0.05592820	46
47	9.90597109	178.11942185	0.00561421	0.10094921	17.98101571	0.05561421	47
48	10.40126965	188.02539294	0.00531843	0.09614211	18.07715782	0.05531843	48
49	10.92133313	198.42666259	0.00503965	0.09156391	18.16872173	0.05503965	49
50	11.46739979	209.34799572	0.00477674	0.08720373	18.25592546	0.05477674	50

C Answers to Odd-Numbered Exercises

CHAPTER 1

Exercises 1.1, page 12

1. not equivalent   3. equivalent   5. equivalent   7. not equivalent   9. $x = 4$ 11. $x = \dfrac{32}{3}$

13. $x = -8$ 15. $x = -1$ 17. $x = \dfrac{1}{7}$ 19. $x = 5$ 21. $x = \dfrac{6}{5}$ 23. $x = -5$ 25. 60 pizzas

27. invest $4800 in the income fund and $3200 in the growth fund

29. false   31. true   33. false

35.

37.

39.

41.

43. not equivalent   45. not equivalent   47. equivalent

49. $x \geq -5$

51. $x \geq \dfrac{7}{2}$

53. $x < 8$

55. $x < -2$

57. $y \geq 0$ 59. $x \leq 0$ 61. $x \geq 10$ 63. $y \geq 1$ 65. $x \leq y$ 67. $y \geq x$

Exercises 1.2, page 25

For exercises 1, 3, and 5, see the following figure.

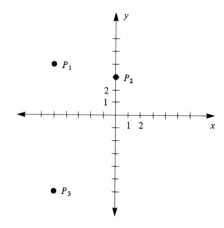

7. -9 9. 0   11. $(10, -9)$ 13. $(0, -6)$ 15. on or to the left of the y-axis   17. on the y-axis

580

19. the origin or on the negative y-axis   21. in quadrant II

23. a. $(0,4)$ b. $\left(\dfrac{12}{5},0\right)$ c. $(-3,9)$ d. $\left(-\dfrac{21}{5},11\right)$ 25. a. $\left(0,\dfrac{3}{2}\right)$ b. $\left(16,-\dfrac{33}{2}\right)$ c. $\left(\dfrac{1}{9},\dfrac{11}{8}\right)$ d. $\left(\dfrac{4}{3},0\right)$

27. slope $= 3$

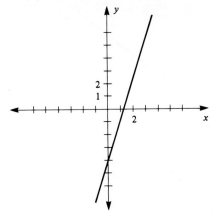

29. slope $= \dfrac{7}{2}$

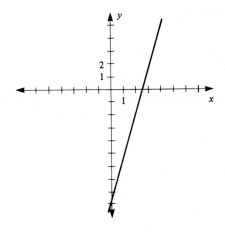

31. slope $= -\dfrac{9}{8}$

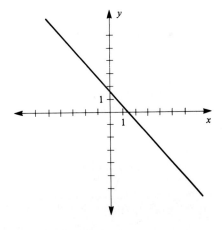

33. $m = 1$ 35. undefined   37. $m = -\dfrac{5}{8}$ 39. undefined   41. 0

43. a.

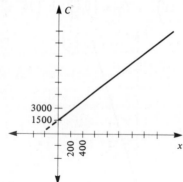

b. $m = 5.80$ c. \$73000

Exercises 1.3, page 30

1. $y = 7x - 33$ 3. $y = -2x + 35$ 5. $y = x + 4$ 7. $y = -6$ 9. $y = -5x$

11. $y = -3x - 12$ 13. $y = 3.7x + 7.5$ 15. $y = -x + 2$ 17. $y = -2x + 38$ 19. $y = -\dfrac{8}{5}x + \dfrac{44}{5}$

21. $y = \dfrac{2}{3}x - 4$ 23. $y = -\dfrac{3}{2}x + 6$ 25. $x = -4$

Exercises 1.4, page 41

1. The sentences in parts (a), (c), and (e) are statements, whereas those in parts (b), (d), and (f) are not.
3. a. Jesse is smart and is a lawyer.
   b. Jesse is either smart or is a lawyer.
   c. Jesse is smart but is not a lawyer.
   d. Jesse is neither a lawyer nor is smart.
   e. It is not true that Jesse is smart and is a lawyer.
   f. It is not true that Jesse is not smart or is a lawyer.
   g. If Jesse is smart, then he is a lawyer.
   h. If Jesse is a lawyer, then he is smart.
   i. If Jesse is not a lawyer, then he is not smart.
   j. If Jesse is smart, then he is both smart and a lawyer.
   k. Jesse is smart if and only if he is not a lawyer.
   l. It is not true that, if Jesse is not a lawyer, then he is smart.
5. a. $p \wedge q$ b. $\sim q$ c. $p \vee q$ d. $p \rightarrow q$ e. $\sim p \wedge \sim q$ f. $\sim(p \vee q)$ g. $\sim q \rightarrow \sim p$ h. $p \leftrightarrow q$
7. a. Ted is angry, tired, and not playing tennis.
   b. Ted is either angry, tired, or playing tennis.
   c. If Ted is neither angry nor tired, then he is not playing tennis.
   d. If Ted is playing tennis, then he is either angry or tired.
   e. Ted is angry and tired if and only if he is playing tennis.
9. a. $p \wedge q \wedge \sim r$ b. $p \wedge \sim q \wedge \sim r$ c. $(p \wedge q) \rightarrow r$ d. $(p \wedge \sim q) \rightarrow \sim r$
11. converse: If John is tired, then he is a hard worker.
    contrapositive: If John is not tired, then he is not a hard worker.
13. converse: If I am either rich or famous, then I work hard.
    contrapositive: If I am neither rich nor famous, then I do not work hard.
15. a. false b. true c. true d. true e. true f. false

17. a.

p	q	~p	~q	~p ∧ ~q
T	T	F	F	F
T	F	F	T	F
F	T	T	F	F
F	F	T	T	T

b.

p	q	~p	~p ∨ q
T	T	F	T
T	F	F	F
F	T	T	T
F	F	T	T

c.

p	q	~(p ∨ q)	
T	T	F	T
T	F	F	T
F	T	F	T
F	F	T	F

d.

p	q	~(p → q)	
T	T	F	T
T	F	T	F
F	T	F	T
F	F	F	T

e.

p	q	p ↔ ~q	
T	T	F	F
T	F	T	T
F	T	T	F
F	F	F	T

f.

p	q	(q ∧ p) → ~p		
T	T	T	F	F
T	F	F	T	F
F	T	F	T	T
F	F	F	T	T

19. a.

p	q	r	(p ∧ q) → r	
T	T	T	T	T
T	T	F	T	F
T	F	T	F	T
T	F	F	F	T
F	T	T	F	T
F	T	F	F	T
F	F	T	F	T
F	F	F	F	T

b.

p	q	r	(p ∧ q) → (r ∨ p)		
T	T	T	T	T	T
T	T	F	T	T	T
T	F	T	F	T	T
T	F	F	F	T	T
F	T	T	F	T	T
F	T	F	F	T	F
F	F	T	F	T	T
F	F	F	F	T	F

c.

p	q	r	(r ∧ ~q) ↔ ~p			
T	T	T	F	F	T	F
T	T	F	F	F	T	F
T	F	T	T	T	F	F
T	F	F	F	T	T	F
F	T	T	F	F	F	T
F	T	F	F	F	F	T
F	F	T	T	T	T	T
F	F	F	F	T	F	T

21. a.

p	q	(p ∨ q) ↔ (q ∨ p)		
T	T	T	T	T
T	F	T	T	T
F	T	T	T	T
F	F	F	T	F

b.

p	~ ~p ↔ p		
T	T	F	T
F	F	T	T

c.

p	q	(p → q) ↔ (~q → ~p)				
T	T	T	T	F	T	F
T	F	F	T	T	F	F
F	T	T	T	F	T	T
F	F	T	T	T	T	T

d.

p	q	~(p → q) → p		
T	T	F	T	T
T	F	T	F	T
F	T	F	T	T
F	F	F	T	T

e.

p	q	r	[(p → q) ∧ (q → r)] → (p → r)				
T	T	T	T	T	T	T	T
T	T	F	T	F	F	T	F
T	F	T	F	F	T	T	T
T	F	F	F	F	T	T	F
F	T	T	T	T	T	T	T
F	T	F	T	F	F	T	T
F	F	T	T	T	T	T	T
F	F	F	T	T	T	T	T

23.

p	q	~(p → q) ↔ p ∧ ~q				
T	T	F	T	T	F	F
T	F	T	F	T	T	T
F	T	F	T	T	F	F
F	F	F	T	T	F	T

25. not valid   27. not valid   29. valid   31. valid   33. not valid   35. valid   37. not valid
39. valid   41. valid   43. not valid   45. valid

47.

p	*q*	*p* ⊻ *q*
T	T	F
T	F	T
F	T	T
F	F	F

Review Exercises 1.1, page 47

1. $x = 6$ 3. $x = 2$

5. $x < 2$    7. $x \geq -3$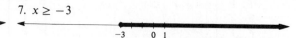

9. $x \leq 150$ 11. $x \geq 150$

Review Exercises 1.2, page 48

For exercises 1 and 3, see the following figure.

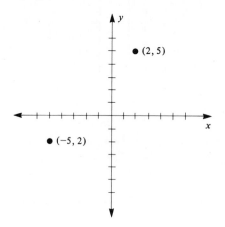

5. $m = 2$ 7. $m = -\dfrac{6}{11}$ 9. the abscissa is negative   11. $x < 0$ and $y < 0$ 13. the rise is negative

15. a. $(0, 4)$ b. $\left(\dfrac{21}{2}, 1\right)$ c. $\left(-1, \dfrac{30}{7}\right)$ d. $\left(\dfrac{7}{2}, -3\right)$

17. $m = \dfrac{3}{2}; b = 0$ 19. $m = 3; b = -\dfrac{1}{2}$

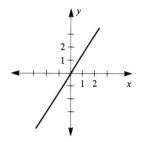   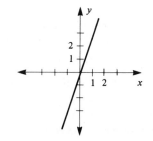

Review Exercises 1.3, page 49

1. $y = -7x - 3$　　3. $y = 15$　　5. $y = -4x + 3$　　7. $y = 5x + 9$　　9. $y = 2x$　　11. $y = -\dfrac{5}{4}x + 5$

Review Exercises 1.4, page 49

1. a. If Bob does the homework regularly, then he will get an A on the next test.
　b. Bob does not get an A on the next test, but he does the homework regularly.
　c. If Bob gets an A on the next test, then he does not do the homework regularly.
3. a. false　b. false　c. false

5.

p	q	r	$[p \wedge (q \vee r)]$		$\leftrightarrow$	$[(p \wedge q) \vee (p \wedge r)]$		
T	T	T	T	T	T	T	T	T
T	T	F	T	T	T	T	T	F
T	F	T	T	T	T	F	T	T
T	F	F	F	F	T	F	F	F
F	T	T	F	T	T	F	F	F
F	T	F	F	T	T	F	F	F
F	F	T	F	T	T	F	F	F
F	F	F	F	F	T	F	F	F

7. valid　　9. valid　　11. valid

CHAPTER 2

Exercises 2.1, page 56

1. Let C = grams of carbohydrates and p = ounces of potatoes; $C = 17p$
3. Let T = income tax and I = gross income; $T = 0.024I$
5. Let L = total length and t = length of tail; $L = \frac{29}{4}t$, $t \geq 30$
7. Let S = sale price and l = list price; $S = 0.80l$; $S = \$120$ when $l = \$150$
9. Let I = monthly income and x = number of dictionaries sold; $I = 2x + 300$; $I = \$550$ when $x = 125$
11. Let d = the distance away Chicago and t = time in hours; $d = 90 - 50t$; $d = 65$ miles when $t = \frac{1}{2}$ hour
13. Let V = value of comic book and t = time in years ($t = 0$ means 1981); $V = 195t + 675$; $V = \$1845$ in 1987.
15. Let P = mosquito population and t = time in minutes; $P = 100,000 - 10t$
17. a. Let F = taxi fare and t = time in minutes; $F = 0.15t + 1.00$
　b. 15 cents
19. a. Let F = percentage of fat in diet and t = time in years; $F = -\frac{6}{5}t + 42$
　b. Let P = percentage of protein in diet and t = time in years; $P = 12$
21. a. $D = 48t + 12$
　b. $d = 54t$
　c. Let D' = the distance between Steve and Gary; $D' = -6t + 12$
　d. Gary will catch Steve when $t = 2$ hours; that is, at 10:15 AM
23. Let P = the assembly line worker's daily pay and x = the number of items assembled
　a. $P = \begin{cases} 80, & \text{when } x \leq 90 \\ 0.50(x - 90) + 80, & \text{when } x > 90 \end{cases}$
　b. $P = \$80$ when $x = 75$
　c. $P = \$95$ when $x = 120$
25. The apparent temperature is not a linear function of wind velocity.

Exercises 2.2, page 63

1. $V = -1500t + 26,500$ 3. $V = -95.50t + 778.50$; $V = \$492$ when $t = 3$

5. a. $C = 0.75x + 660$ b. $R = 1.85x$ c. $P = 1.10x - 660$ 7. a. $C = 27x + 15,000$ b. $P = 60x - 15,000$

9. $C = \begin{cases} 2x + 296, & \text{when } x \le 20 \\ 1.50(x - 20) + 336, & \text{when } x > 20 \end{cases}$

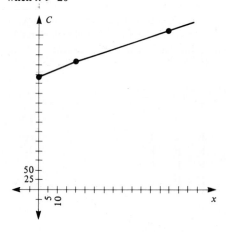

11. $p = -\dfrac{1}{8}q + \dfrac{65}{8}$

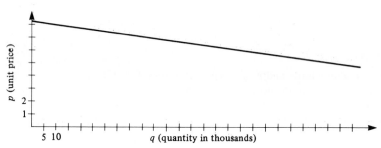

13. $p = -0.1125q + 6.525$

15. $p = 0.001q + 1.05$

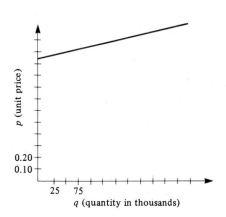

17. $p = 0.09q + 3.15$

Exercises 2.3, page 69

1. a. It is correct.   b. It is correct.   c. It is incorrect.   d. It is incorrect.

3. $x = 4$ and $y = \dfrac{3}{2}$ 5. $x = 1$ and $y = 3$

7. The solution set is empty.   9. The solution set has infinitely many solutions.

11. $x = 5$ and $y = 18$ 13. $x = 7.5$ and $y = 2$ 15. $x = \dfrac{12}{7}$ and $y = \dfrac{12}{7}$ 17. The solution set is empty.

19. $x = 11.3$ and $y = 8.4$ 21. $x = 16$ and $y = 21$ 23. $x = 150$ and $y = 170$

Exercises 2.4, page 78

1. The break-even point is $x = 600$.

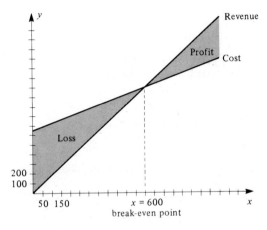

3. The break-even point is $x = 350$.
5. The break-even point is $x = 36$.
7. Equilibrium quantity = 8 thousand and the equilibrium price = \$9

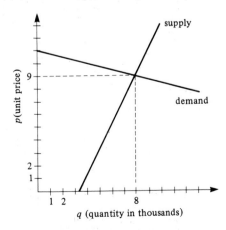

9. The equilibrium quantity = 4000 and the equilibrium price = \$6.00
11. a. $C = 110x + 90y + 2550$ b. $C = \$41,250$

13. Let I = total interest, x = amount in savings account, and y = amount in certificate of deposit $I =$
$0.005x + 0.0075y$
  b. $I = \$15$

15. Let x = the number of unfinished tables to be made

  y = the number of unfinished bookcases to be made

  and solve

$$12x + 28y = 840$$
$$2x + 3y = 115.$$

Solution: Allegheny Cabinets can make 35 unfinished tables and 15 unfinished bookcases in a week.

17. Let x = the number of acres to be planted in barley

  y = the number of acres to be planted in wheat

  and solve

$$x + y = 375$$
$$110x + 80y = 35,100.$$

Solution: Crestfield Farms should plant 170 acres in barley and 205 acres in wheat.

19. Let x = the tons of the lesser grade stainless steel to be made

  y = the tons of the better grade stainless steel to be made

  and solve

$$0.05x + 0.15y = 4.7$$
$$0.95x + 0.85y = 48.3.$$

Solution: Beth Metals can make 32.5 tons of the lesser grade stainless steel and 20.5 tons of the better grade stainless steel.

21. Let x = units of rice in the diet

  y = units of chicken breasts in the diet

  and solve

$$350x + 86y = 1070$$
$$3x + 19y = 32.$$

Solution: The diet should consist of 2.75 units of rice and 1.25 units of chicken breasts.

23. Let x = the amount invested at 6%

  y = the amount invested at 9%

  and solve

$$x + y = 15,000$$
$$0.06x + 0.09y = 1,200.$$

Solution: The individual should invest $5000 at 6% and $10,000 at 9%.

25. Let x = the pounds of ingredient A used in the fertilizer

  y = the pounds of ingredient B used in the fertilizer

  and solve

$$0.40x + 0.20y = 18$$
$$0.30x + 0.50y = 24.$$

Solution: The Deerfield Company should use 30 pounds of each ingredient per bag of the fertilizer.

27. Let $x =$ the number of undergraduate students at the university

$y =$ the number of graduate students at the university

and solve

$$\frac{1}{20}x + \frac{1}{10}y = \quad 320$$

$$700x + 900y = 4,255,000.$$

Solution: There should be 5500 undergraduate and 450 graduate students at the university.

29. Let $x =$ the amount invested at 13%

$y =$ the amount invested at 8%

The maximum amount that the individual can invest at 8% is found from solving the system of linear equations

$$x + \quad y = \quad 1,000$$

$$0.13x + 0.08y = 10,000.$$

Solution: The maximum amount that can be invested at 8% is $6000 in which case, $4000 would be invested at 13%.

31. Let $x =$ the number of bars of gold held by "Us" at the beginning of a trade period

$y =$ the number of bars of gold held by "Them" at the beginning of a trade period

and solve

$$\frac{3}{4}x + \frac{3}{5}y + 1700 = 2x$$

$$\frac{1}{4}x + \frac{2}{5}y + 400 = 2y.$$

Solution: "Us" should hold 1600 bars of gold and "Them" should hold 500 bars of gold at the beginning of a trade period.

Review Exercises (Chapter 2), page 83

1. Let $S =$ sale price and $l =$ list price; $S = 0.85$
3. Let $d =$ distance in miles and $t =$ time in hours; $d = 32 - 52t$
5. $V = -2150t + 35,800$ 7. $p = -0.175q + 7.9$ 9. $x = 25$ and $y = 20$
11. Let $x =$ tons of High-Protein made and $y =$ tons of Regular made

  a. tons of beef by-products $= \frac{5}{8}x + \frac{1}{3}y$ b. tons of grain $= \frac{3}{8}x + \frac{2}{3}y$

13. Let $x =$ the amount invested at 14%

$y =$ the amount invested at 9%

and solve

$$x + \quad y = 10,000$$

$$0.14x + 0.09y = \quad 1250.$$

Solution: The individual should invest $7000 at 14% and $3000 at 9%.

CHAPTER 3

Exercises 3.1, page 95

1.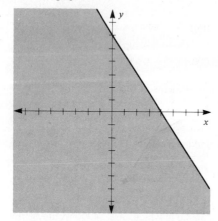

3.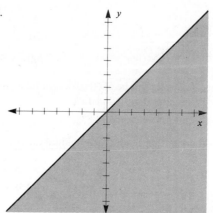

5.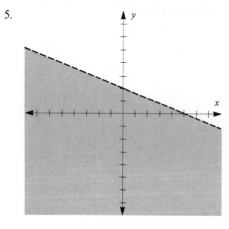

7.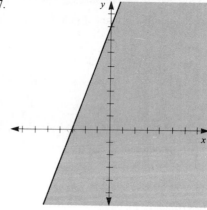

9.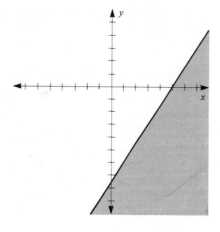

11.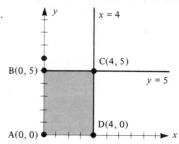

13.

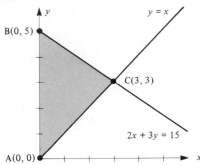

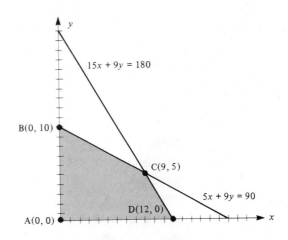

17.

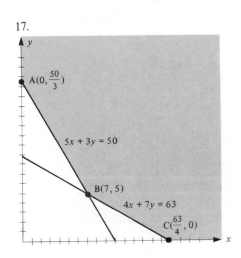

19.

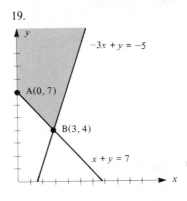

21.

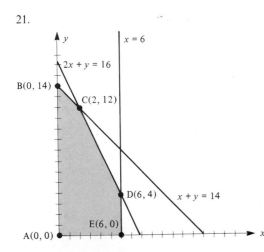

23.

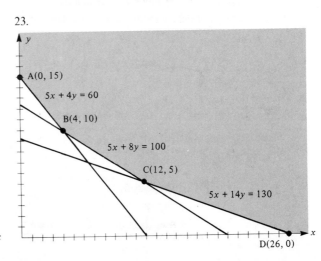

25.

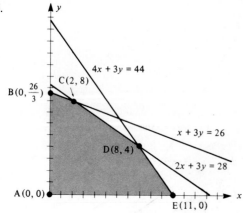

27.

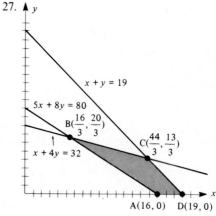

29. The graphs in exercises 11, 13, 15, 21, 25, and 27 are bounded.

31. a. $4.00x + 5.40y \leq 1026$

$$y \leq \frac{2}{3}x$$

$x \geq 0$ and $y \geq 0$

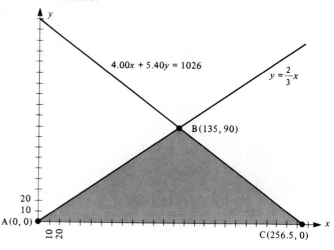

33. Let x = dozens of utility models made in a day

y = dozens of deluxe models made in a day

a. $2x + 3y \leq 9$

$3x + y \leq 10$

$x \geq 0$ and $y \geq 0$

b.

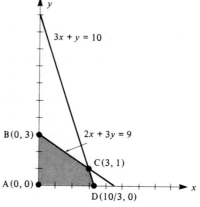

35. Let x = number of pounds of medium-density plastic made

    y = number of pounds of high-density plastic made

    $0.5x + 0.7y \leq 2100$

    $0.5x + 0.3y \leq 1500$

    $x \geq 0$ and $y \geq 0$

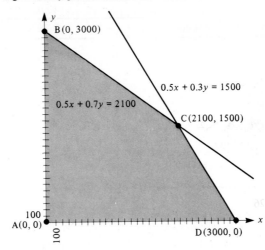

37. Let x = number of pounds of medium-density plastic made

    y = number of pounds of high-density plastic made

    $0.5x + 0.7y \leq 2100$

    $0.5x + 0.3y \leq 1500$

    $0.1x + 0.1y \leq 340$

    $x \geq 0$ and $y \geq 0$

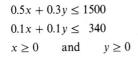

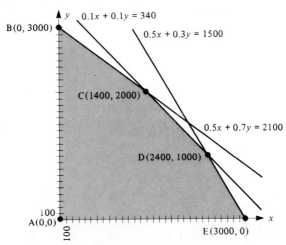

Exercises 3.2, page 110

1.

Vertex	System of equations	Coordinates	Value of		
			$z = x + 3y$	$z = 2x + 3y$	$z = 5x + 3y$
A	$x = 0$ $y = 0$	$(0, 0)$	0	0	0
B	$x = 0$ $4x + 7y = 63$	$(0, 9)$	27	27	27
C	$4x + 7y = 63$ $5x + 3y = 50$	$(7, 5)$	22	29	50
D	$y = 0$ $5x + 3y = 50$	$(10, 0)$	10	20	50

a. z is maximum at $x = 0$ and $y = 9$
b. z is maximum at $x = 7$ and $y = 5$
c. z is maximum at all points on the line $5x + 3y = 50$ between the points $(7, 5)$ and $(10, 0)$, inclusively

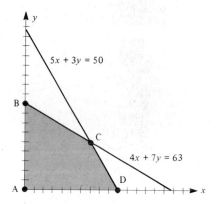

3.

Vertex	System of equations	Coordinates	Value of		
			$z = 3x + 5y$	$z = 4x + 5y$	$z = 12x + 5y$
A	$x = 0$ $y = 0$	$(0, 0)$	0	0	0
B	$x = 0$ $x + 2y = 24$	$(0, 12)$	60	60	60
C	$x + 2y = 24$ $4x + 5y = 72$	$(8, 8)$	64	72	136
D	$4x + 5y = 72$ $2x + y = 30$	$(13, 4)$	59	72	176
E	$y = 0$ $2x + y = 30$	$(15, 0)$	45	60	180

a. z is maximum at $x = 8$ and $y = 8$

b. z is maximum at all points on the line $4x + 5y = 72$ between $(8, 8)$ and $(13, 4)$, inclusively

c. z is maximum at $x = 15$ and $y = 0$

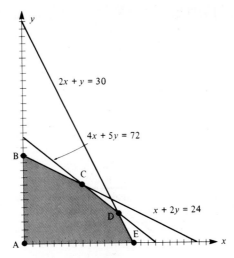

5.

Vertex	System of equations	Coordinates	Value of $z = x + y$
A	$x = 0$ $y = 0$	$(0, 0)$	0
B	$x = 3$ $7x - 3y = 0$	$(3, 7)$	10
C	$x = 3$ $y = 0$	$(3, 0)$	3

The maximum value of z is 10 occurring at $x = 3$ and $y = 7$.

7.

Vertex	System of equations	Coordinates	Value of $z = 3x + y$
A	$x = 0$ $y = 0$	$(0, 0)$	0
B	$x = 0$ $x + 3y = 24$	$(0, 8)$	8
C	$x + 3y = 24$ $7x + 2y = 35$	$(3, 7)$	16
D	$7x + 2y = 35$ $y = 0$	$(5, 0)$	15

The maximum value of z is 16 occurring at $x = 3$ and $y = 7$.

9.

Vertex	System of equations	Coordinates	Value of $z = 7x + 5y$
A	$x = 0$ $5x + 3y = 30$	$(0, 10)$	50
B	$5x + 3y = 30$ $x + y = 8$	$(3, 5)$	46
C	$x + y = 8$ $y = 0$	$(8, 0)$	56

The minimum value of z is 46 occurring at $x = 3$ and $y = 5$.

11.

Vertex	System of equations	Coordinates	Value of $z = 5x + 7y$
A	$x = 0$ $y = 0$	$(0, 0)$	0
B	$x = 0$ $y = 8$	$(0, 8)$	56
C	$y = 8$ $3x + 4y = 40$	$\left(\frac{8}{3}, 8\right)$	$69\frac{1}{3}$
D	$3x + 4y = 40$ $x = 2y$	$(8, 4)$	68

The maximum value of z is $69\frac{1}{3}$ occurring at $x = \frac{8}{3}$ and $y = 8$.

13.

Vertex	System of equations	Coordinates	Value of $z = 7x + 12y$
A	$x = 0$ $2x + 3y = 16$	$\left(0, \frac{16}{3}\right)$	64
B	$x = 0$ $y = 10$	$(0, 10)$	120
C	$x = 5$ $y = 10$	$(5, 10)$	155
D	$x = 5$ $2x + 3y = 16$	$(5, 2)$	59

The minimum value of z is 59 occurring at $x = 5$ and $y = 2$.

15.

Vertex	System of equations	Coordinates	Value of $z = \frac{1}{3}x + y$
A	$x = 0$ $-x + 12y = 24$	$(0, 2)$	2
B	$x = 0$ $2x + 9y = 99$	$(0, 11)$	11
C	$2x + 9y = 99$ $4x - 3y = 9$	$(9, 9)$	12
D	$4x - 3y = 9$ $-x + 12y = 24$	$\left(4, 2\frac{1}{3}\right)$	$3\frac{2}{3}$

The maximum value of z is 12 occurring at $x = 9$ and $y = 9$.

17.

Vertex	System of equations	Coordinates	Value of $z = 3x + y$
A	$x = 0$ $7x + 3y = 52$	$\left(0, 17\frac{1}{3}\right)$	$17\frac{1}{3}$
B	$7x + 3y = 52$ $5y + 4y = 52$	$(4, 8)$	20
C	$5x + 4y = 52$ $3x + 5y = 39$	$(8, 3)$	27
D	$3x + 5y = 39$ $y = 0$	$(13, 0)$	39

The minimum value of z is $17\frac{1}{3}$ occurring at $x = 0$ and $y = 17\frac{1}{3}$.

19.

Vertex	System of equations	Coordinates	Value of $z = 3x - 5y$
A	$x = 0$ $5x + 7y = 35$	$(0, 5)$	-25
B	$x = 0$ $-5x + 3y = 24$	$(0, 8)$	-40
C	$-5x + 3y = 24$ $3x + 4y = 61$	$(3, 13)$	-56

(*continued on p. 598*)

(continued from p. 597)

Vertex	System of equations	Coordinates	Value of $z = 3x - 5y$
D	$3x + 4y = 61$ $x = 7$	$(7, 10)$	-29
E	$x = 7$ $y = 0$ $5x + 7y = 35$	$(7, 0)$	21

The minimum value of z is -56 occurring at $x = 3$ and $y = 13$.

21.

Vertex	System of equations	Coordinates	Value of $z = \frac{1}{2}x + 5y$
A	$x = 0$ $x + y = 1$	$(0, 1)$	5
B	$x = 0$ $x + 3y = 9$	$(0, 3)$	15
C	$x + 3y = 9$ $x + y = 5$	$(3, 2)$	$11\frac{1}{2}$
D	$x + y = 5$ $y = 0$	$(5, 0)$	$2\frac{1}{2}$
E	$x + y = 1$ $y = 0$	$(1, 0)$	$\frac{1}{2}$

The maximum value of z is 15 occurring at $x = 0$ and $y = 3$, whereas the minimum value of z is $\frac{1}{2}$ occurring at $x = 1$ and $y = 0$.

23. Let x = the number of regular hours to be scheduled

   y = the number of overtime hours to be scheduled

Problem:

$$\text{Maximize} \quad z = 19x + 27y$$
$$\text{subject to} \quad 4.00x + 5.40y \leq 1026$$
$$y \leq \frac{2}{3}x$$
$$x \geq 0 \quad \text{and} \quad y \geq 0.$$

Solution: Linda should schedule 135 regular hours and 90 overtime hours to achieve the maximum amount of $4995.

25. Let x = dozens of utility models made in a day

   y = dozens of deluxe models made in a day

Problem:

$$\text{Maximize} \quad z = 666x + 855y$$
$$\text{subject to} \quad 2x + 3y \leq 9$$
$$3x + y \leq 10$$
$$x \geq 0 \quad \text{and} \quad y \geq 0.$$

Solution: The Tennessee Chainsaw Company should make 3 dozen utility models and 1 dozen deluxe models to achieve a maximum daily profit of $2853.

27. Let x = number of pounds of medium-density plastic made

y = number of pounds of high-density plastic made

Problem:

$$\text{Maximize} \quad z = 52x + 70y$$
$$\text{subject to} \quad 0.5x + 0.7y \leq 2100$$
$$0.5x + 0.3y \leq 1500$$
$$0.1x + 0.1y \leq 340$$
$$x \geq 0 \quad \text{and} \quad y \geq 0.$$

Solution: Petro Chemicals should make 1400 pounds of medium-density plastic and 2000 pounds of high-density plastic to achieve the maximum profit of $212,800.

29. Let x = number of ceramic bowls to be made

y = number of ceramic plates to be made

Problem:

$$\text{Maximize} \quad z = 1.75x + 1.00y$$
$$\text{subject to} \quad 0.20x + 0.10y \leq 72$$
$$0.05x + 0.05y \leq 26$$
$$x \geq 0 \quad \text{and} \quad y \geq 0.$$

Solution: Pottery Makers should make 200 ceramic bowls and 320 ceramic plates to achieve its maximum profit of $670.

31. Let x = pounds of ingredient A used per bag

y = pounds of ingredient B used per bag

Problem:

$$\text{Minimize} \quad z = 0.30x + 0.40y$$
$$\text{subject to} \quad 0.70x + 0.50y \geq 60$$
$$0.30x + 0.50y \geq 30$$
$$x \geq 0 \quad \text{and} \quad y \geq 0.$$

Solution: Pendrake Lawn Fertilizer should contain 75 pounds of ingredient A and 15 pounds of ingredient B per bag to achieve the minimum cost of $28.50 per bag.

33. Let x = units of steak in the meal

y = units of salad in the meal

Problem:

$$\text{Minimize} \quad z = 1.20x + 0.20y$$
$$\text{subject to} \quad 200x + 20y \geq 500$$
$$3x + 4y \geq 24$$
$$2x + y \geq 11$$
$$x \geq 0 \quad \text{and} \quad y \geq 0.$$

Solution: For the meal Erica should plan $1\frac{3}{4}$ units of steak and $7\frac{1}{2}$ units of salad to achieve the minimal cost of $3.60.

35. Let x = tons of newsprint to be produced in a day

  y = tons of mando stock to be produced in a day

Problem:

$$\text{Maximize} \quad z = 1500x + 1800y$$
$$\text{subject to} \quad x \geq 8$$
$$y \geq 5$$
$$x + y \leq 30$$
$$y \leq 2x$$
$$x \geq 0 \quad \text{and} \quad y \geq 0.$$

Solution: Satellite Paper Company should produce 10 tons of newsprint and 20 tons of mando stock to achieve the maximum profit of $51,000.

Review Exercises (Chapter 3), page 116

1.
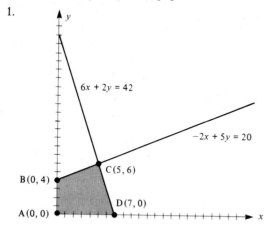

3.
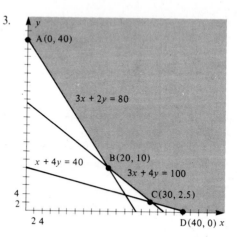

5. The maximum value of z is 17.5 occurring at $x = 70$ and $y = 0$.
7. The maximum value of z is 20 occurring at $x = 0$ and $y = 5$.
9. The maximum value of z is 101 occurring at $x = 25$ and $y = 13$.
11. The minimum value of z is 4.5 occurring at $x = 3$ and $y = 0$.
13. Let x = crates of apple cider to be made

  y = crates of vinegar to be made

Problem:

$$\begin{aligned} \text{Maximize} \quad & z = 300x + 250y \\ \text{subject to} \quad & 2x + 5y \le 9 \\ & 3x + 4y \le 10 \\ & y \ge \frac{1}{2} \\ & x \ge 0 \quad \text{and} \quad y \ge 0. \end{aligned}$$

Solution: Juicy Orchards should make $2\frac{2}{3}$ crates of apple cider and $\frac{1}{2}$ crate of vinegar to achieve the maximum profit of $925.

15. Let x = grams of ingredient I used in each bottle

y = grams of ingredient II used in each bottle

Problem:

$$\begin{aligned} \text{Minimize} \quad & z = 2x + 2.5y \\ \text{subject to} \quad & 0.10x + 0.20y \ge 12 \\ & 0.30x + 0.10y \ge 18 \\ & y \le 60 \\ & x \ge 0 \quad \text{and} \quad y \ge 0. \end{aligned}$$

Solution: Feline Pet Food Company should use 48 grams of ingredient I and 36 grams of ingredient II in each bottle to achieve the minimal cost of $1.86.

CHAPTER 4

Exercises 4.1, page 123

1. a. $C = 4x + 7y + 12z$ b. $R = 6x + 11y + 15z$ c. $P = 2x + 4y + 3z$ d. $C = \$1591$, $R = \$2185$, $P = \$594$
3. a. $TC = 350x_1 + 86x_2 + 85x_3 + 123x_4$ b. 4745 calories
5. Let x_i = the number of bicycles made at factory i, $i = 1, 2, 3, 4$.
 a. $TC = 105x_1 + 90x_2 + 102x_3 + 114x_4 + 5065$ b. $70,927
7. Let x_1 = the number of 10″ pizzas, x_2 = the number of 13″ pizzas, and x_3 = the number of 15″ pizzas.
 a. $C = 0.60x_1 + 1.20x_2 + 1.85x_3$ b. $R = 2.50x_1 + 4.00x_2 + 5.25x_3$ c. $P = 1.90x_1 + 2.80x_2 + 3.40x_3$
9. a. a solution b. not a solution c. a solution d. a solution

13. $x = 4$ and $y = 3$ 15. $x = -\frac{5}{2}$, $y = 6$, and $z = 4$ 17. $x = -2$, $y = -1$, and $z = 3$

19. $x = 3$, $y = -2$, and $z = -1$ 21. $x = \frac{1}{2}$, $y = \frac{1}{4}$, and $z = -\frac{1}{2}$

Exercises 4.2, page 137

1. a. $\left[\begin{array}{ccc|c} -1 & 0 & 6 & 0 \\ 4 & 5 & -2 & 1 \\ 2 & -6 & 0 & -3 \end{array}\right]$ b. $\left[\begin{array}{cccc|c} 8 & -1 & 3 & 1 & 2 \\ 0 & 3 & -1 & 2 & -3 \\ 1 & 4 & -2 & -1 & 1 \end{array}\right]$

3. a. $\begin{aligned} x - y &= 4 \\ y &= 2 \end{aligned}$ b. $\begin{aligned} x + 2y - z &= 3 \\ 2x \quad\;\; + 5z &= -7 \\ -3y + 4z &= 2 \end{aligned}$

7. $x = 6$ and $y = 2$ 9. $x = -\dfrac{5}{2}, \quad y = 6, \quad z = 4$ 11. no solutions   13. $x = 1, \quad y = -2, \quad z = 0$

15. $x_1 = -14, \quad x_2 = -45, \quad x_3 = -8, \quad x_4 = -11$ 17. $x_1 = -\dfrac{1}{2}, \quad x_2 = 1, \quad x_3 = 0, \quad x_4 = 1, \quad x_5 = -2$

19. a. $T_1 = \dfrac{1}{2}x + \dfrac{3}{4}y + z$

$T_2 = \dfrac{1}{2}x + \dfrac{1}{2}y + \dfrac{1}{2}z$

$T_3 = \dfrac{1}{4}x + \dfrac{1}{2}y + z$

b. Make 8 boxes of lift rods, 8 boxes of slip nuts, and 4 boxes of trap plugs.
21. Plant 30 acres in wheat, 215.75 in soybeans, and 129.25 in corn.   23. $a = 3$

Exercises 4.3, page 153

1. a. yes b. no c. yes d. no   3. a. no b. yes c. yes d. no
5. a. $x_1 = 0, \quad x_2 = 6, \quad x_3 = -8, \quad x_4 = 0$ b. $x_1 = 6 - 3t, \quad x_2 = s, \quad x_3 = -7 + 2t, \quad x_4 = t$
7. $x = -\dfrac{9}{4}t - \dfrac{1}{2}, \quad y = -\dfrac{1}{4}t - \dfrac{3}{2}, \quad z = t$ 9. $x = 1, \quad y = -2, \quad z = -3$

11. $x = \dfrac{1}{2}, \quad y = \dfrac{3}{2}, \quad z = -\dfrac{1}{2}$ 13. $x_1 = 2, \quad x_2 = 0, \quad x_3 = -3, \quad x_4 = \dfrac{1}{2}$

15. $x = -\dfrac{1}{3}, \quad y = \dfrac{1}{3}, \quad z = \dfrac{1}{7}$ 17. no solution   19. $x = \dfrac{1}{2}, \quad y = \dfrac{3}{2}, \quad z = -\dfrac{1}{2}$

21. $x = -\dfrac{5}{8}t + 2, \quad y = \dfrac{1}{8}t + 1, \quad z = t$ 23. $x = 2t + \dfrac{5}{2}, \quad y = t, \quad z = -\dfrac{13}{2}$

25. $x_1 = 3t + 3, \quad x_2 = 2t + 1, \quad x_3 = t$ 27. $x_1 = 2s + \dfrac{1}{2}t - 1, \quad x_2 = s, \quad x_3 = \dfrac{1}{2}t - 1, \quad x_4 = t$

29. $x_1 = 3s - t, \quad x_2 = -\dfrac{8}{9}s + \dfrac{1}{3}t + \dfrac{2}{3}, \quad x_3 = -\dfrac{8}{9}s + \dfrac{7}{3}t + \dfrac{5}{3}, \quad x_4 = s, \quad x_5 = t$

31. $x_1 = \dfrac{5}{13}t + \dfrac{1}{5}, \quad x_2 = -\dfrac{6}{13}t, \quad x_3 = -\dfrac{4}{13}t - \dfrac{1}{5}, \quad x_4 = t$

33. $x_1 = 2t - 2, \quad x_2 = t, \quad x_3 = -2, \quad x_4 = 1, \quad x_5 = -1$
35. Let x = the number of plaques to be made
y = the number of trophy bases to be made
z = the number of picture frames to be made
and solve

$$\frac{1}{4}x + \frac{1}{2}y + \frac{1}{4}z = 12$$

$$\frac{1}{4}x + \frac{1}{2}y + \frac{1}{2}z = 15$$

$$\frac{1}{8}x + \frac{1}{2}y + \frac{1}{4}z = 10.$$

Solution: White Birch Enterprises should make 16 plaques, 10 trophy bases, and 12 picture frames.

37. Let $x =$ the number of cases of Lady Jane cigars to be made per day

      $y =$ the number of cases of Slim cigars to be made per day

      $z =$ the number of cases of Concord cigars to be made per day

and solve

$$x + 0.2y + 0.4z = 14$$
$$0.4x + 0.2y + 0.4z = 11.6$$
$$x + 0.8y + 0.2z = 16.$$

Solution: The Imperial Cigar Company should make 4 cases of Lady Janes, 10 cases of Slims, and 20 cases of Concords per day.

39. Let $x =$ the pounds of type A food to be used in the 100-pound mixture

      $y =$ the pounds of type B food to be used in the 100-pound mixture

      $z =$ the pounds of type C food to be used in the 100-pound mixture

and solve

$$x + y + z = 100$$
$$25x + 30y + 15z = 2475$$
$$200x + 100y + 150z = 14{,}250.$$

Solution: The 100-pound mixture should contain 30 pounds of type A food, 45 pounds of type B food, and 25 pounds of type C food.

41. Let $A_i =$ the number of pleasure boats shipped from factory A to warehouse i, $i = 1, 2$

      $B_i =$ the number of pleasure boats shipped from factory B to warehouse i, $i = 1, 2$

and solve the system of linear equations

$$A_1 + A_2 \qquad\qquad = 400$$
$$B_1 + B_2 = 200$$
$$A_1 \qquad + B_1 \qquad = 500$$
$$A_2 \qquad + B_2 = 100.$$

Solution: The shipping plan is described by

$$A_1 = t + 300, \qquad A_2 = -t + 100$$
$$B_1 = -t + 200, \quad \text{and} \quad B_2 = t,$$

where t is an integer and $0 \le t \le 100$.

43. Let $x_1 =$ the amount invested at 7%

      $x_2 =$ the amount invested at 8%

      $x_3 =$ the amount invested at 9.5%

      $x_4 =$ the amount invested at 10.5%

and solve the system of linear equations

$$x_1 + x_2 + x_3 + x_4 = 30{,}000$$
$$x_1 + x_2 - x_3 - x_4 = 0$$
$$0.7x_1 + 0.8x_2 + 0.095x_3 + 0.105x_4 = 2500.$$

Solution: The amounts invested at each percentage are described by

$$x_1 = t + 12{,}500, \qquad\qquad x_2 = 2500 - t,$$
$$x_3 = 15{,}000 - t, \quad \text{and} \quad x_4 = t,$$

where t is any real number and $0 \le t \le 2500$.

45. To determine a, b, c, and d, solve the system of linear equations

$$
\begin{aligned}
a + b + c + d &= 8.0 \\
3b + 4c + d &= 13.8 \\
3a \quad\quad + 2c + d &= 17.1 \\
a + 2b \quad\quad + d &= 5.9.
\end{aligned}
$$

Solution: $a = 3.5$, $b = 0.6$, $c = 2.7$,   and $d = 1.2$. Therefore,
$D = 3.5x + 0.6y + 2.7z + 1.2$

47. a.

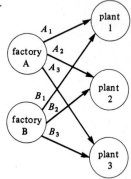

   b. Solve the system of linear equations

$$
\begin{aligned}
A_1 + A_2 + A_3 &= 120 \\
B_1 + B_2 + B_3 &= 400 \\
A_1 \quad\quad + B_1 \quad\quad &= 200 \\
A_2 \quad\quad + B_2 \quad &= 150 \\
A_3 \quad\quad + B_3 &= 170.
\end{aligned}
$$

Solution: A schedule is described by the equations

$$A_1 = s + t - 200, \quad A_2 = -s + 150, \quad A_3 = -t + 170,$$
$$B_1 = -s - t + 400, \quad B_2 = s, \quad \text{and} \quad B_3 = t,$$

where s and t are integers satisfying

$$0 \le s \le 150, \quad 0 \le t \le 170, \quad \text{and} \quad s + t \ge 200.$$

49. If $a = -4$, then the system will have infinitely many solutions.

Review Exercises (Chapter 4), page 158

1. Let x, y, and z represent, respectively, the pounds of type A, B, and C meat products made and let T represent the total cost.
   a. $T = 0.22x + 0.43y + 0.31z$ b. $T = \$156.20$

3. a. not a solution   b. a solution   c. not a solution

5. $x_1 = 2t - 2$, $x_2 = 6t - 3$, $x_3 = -t + 1$, $x_4 = t$

7. $x_1 = -3t - 1$, $x_2 = -2t - \dfrac{5}{2}$, $x_3 = -\dfrac{15}{4}t - \dfrac{41}{4}$, $x_4 = \dfrac{1}{4}t + \dfrac{7}{4}$, $x_5 = t$

9. The system has no solutions.

11. Let x, y, and z represent, respectively, the number of 1-acre, 2-acre, and 3-acre lots and solve the system of linear equations

$$\begin{aligned} x + \quad y + \quad z &= \quad 240 \\ 1500x + 2500y + 3000z &= 255{,}000 \end{aligned}$$

Solution: The number of lots of each size that can be developed are described by the equations

$$x = 3t - 180, \quad y = -3t + 210, \quad \text{and} \quad z = t,$$

where t is any integer and $0 \leq t \leq 70$.

13. Let x, y, and z represent, respectively, the pounds of ingredients A, B, and C used in a 50-pound bag of lawn fertilizer and solve

$$\begin{aligned} 0.30x + 0.60y + 0.30z &= 22.2 \\ 0.60x + 0.30y + 0.30z &= 18.6 \\ x + \quad y + \quad z &= 50. \end{aligned}$$

Solution: A 50-pound of lawn fertilizer should contain 12 pounds of ingredient A, 24 pounds of ingredient B, and 14 pounds of ingredient C.

CHAPTER 5

Exercises 5.1, page 166

1. $\begin{bmatrix} 3 & 1 & -2 \\ 2 & 8 & 4 \end{bmatrix}$ 3. not defined   5. $\begin{bmatrix} 5 & 0 & -4 \\ 5 & 13 & 4 \end{bmatrix}$ 7. $\begin{bmatrix} 3 & 2 & 9 \\ 1 & 2 & 4 \\ 3 & 6 & 0 \end{bmatrix}$

9. $\begin{bmatrix} 1 & 2 & \frac{5}{2} \\ 3 & 1 & \frac{1}{2} \\ \frac{3}{2} & 1 & \frac{1}{2} \end{bmatrix}$ 11. $\begin{bmatrix} -1 & -6 & -1 \\ -11 & -2 & 2 \\ -3 & 2 & -2 \end{bmatrix}$ 13. $-A = \begin{bmatrix} 1 & 2 & 0 & -5 \\ -2 & 0 & -7 & 9 \\ -3 & 1 & -4 & -8 \end{bmatrix}$ 19. $\begin{bmatrix} 2 \\ -3 \\ 4 \\ 9 \end{bmatrix}$

21. $\begin{bmatrix} 5 & -6 & 7 \end{bmatrix}$ 23. $\begin{bmatrix} 1 & 3 & 2 \\ 0 & 0 & 0 \\ -2 & 4 & -5 \end{bmatrix}$ 25.

	Men	Women	Children
Kutztown	15	15	20
Reading	20	40	16
Allentown	30	25	17

27.

	Model			
	1	2	3	4
Fischer	2	0	4	5
Computer World	0	3	7	0

29.

	Men	Women	Children
Kutztown	120	120	160
Reading	160	320	128
Allentown	240	200	136

Exercises 5.2, page 172

1. -10 3. not defined   5. -10 7. -46 9.

$$[160 \quad 100 \quad 60 \quad 80]\begin{bmatrix} 9.75 \\ 9.20 \\ 8.75 \\ 12.20 \end{bmatrix} = \$3981$$

11. $\begin{bmatrix} 2 & 3 \\ 4 & 2 \\ 3 & 5 \end{bmatrix}\begin{bmatrix} 2.5 \\ 3 \end{bmatrix} = \begin{bmatrix} 14 \\ 16 \\ 22.5 \end{bmatrix}$ 13. $[2 \quad 3 \quad -4]\begin{bmatrix} x \\ y \\ z \end{bmatrix} = 7$ 15. $\begin{bmatrix} 2 & 3 & -4 \\ 1 & 0 & 2 \\ 0 & 4 & -1 \end{bmatrix}\begin{bmatrix} x \\ y \\ z \end{bmatrix} = \begin{bmatrix} 7 \\ 5 \\ 8 \end{bmatrix}$

17. $\begin{bmatrix} 2 & 3 & 0 & -1 \\ 0 & 1 & 5 & 0 \end{bmatrix}\begin{bmatrix} x_1 \\ x_2 \\ x_3 \\ x_4 \end{bmatrix} = \begin{bmatrix} 7 \\ 8 \end{bmatrix}$

Exercises 5.3, page 180

1. a. 2×3 b. not defined   c. 4×4 d. 3×3 3. $\begin{bmatrix} 2 & 3 \\ 1 & 8 \\ -3 & 5 \\ 0 & 3 \end{bmatrix}$ 5. not defined

7. $\begin{bmatrix} 3 & -6 & 12 \\ -20 & 0 & 12 \end{bmatrix}$ 9. $\begin{bmatrix} 2 & 4 & 5 \\ 6 & 2 & 1 \\ 8 & 6 & 6 \end{bmatrix}$ 17. $7\begin{bmatrix} 1 & -2 \\ -3 & 4 \end{bmatrix} = \begin{bmatrix} 7 & -14 \\ -21 & 28 \end{bmatrix}$

19. $\begin{bmatrix} 3 & -1 \\ -5 & 4 \end{bmatrix}\begin{bmatrix} -2 & 3 \\ 4 & 1 \end{bmatrix} = \begin{bmatrix} -10 & 8 \\ 26 & -11 \end{bmatrix}$ 21. $\begin{bmatrix} 1 & -1 \\ 2 & 3 \\ 0 & 4 \end{bmatrix}\begin{bmatrix} 7 & 2 \\ 14 & 4 \end{bmatrix} = \begin{bmatrix} -7 & -2 \\ 56 & 16 \\ 56 & 16 \end{bmatrix}$

23. $\begin{bmatrix} 8 & 6 & 4 \\ 4 & 0 & 7 \\ 9 & 3 & 5 \end{bmatrix}\begin{bmatrix} 2.5 & 35 \\ 4 & 50 \\ 3.5 & 70 \end{bmatrix} = \begin{array}{l} \text{eaten 1} \\ \text{at 2} \\ \text{meal 3} \end{array}\overset{\text{Protein \quad Calories}}{\begin{bmatrix} 58 & 860 \\ 34.5 & 630 \\ 52 & 815 \end{bmatrix}}$

25. a.

	Quart	Pint	Half-pint
Plant I	1500	1500	2000
Plant II	2000	4000	1500
Plant III	3000	2500	1000

b.

	Plant I	Plant II	Plant III
Week 1	35	30	40
Week 2	40	60	30

c.

$$\begin{bmatrix} 35 & 30 & 40 \\ 40 & 60 & 30 \end{bmatrix}\begin{bmatrix} 1500 & 1500 & 2000 \\ 2000 & 4000 & 1500 \\ 3000 & 2500 & 1000 \end{bmatrix} = \begin{array}{l} \text{Week 1} \\ \text{Week 2} \end{array}\overset{\text{Quart \qquad Pint \qquad Half-pint}}{\begin{bmatrix} 232,500 & 275,500 & 155,000 \\ 270,000 & 375,000 & 200,000 \end{bmatrix}}$$

27. $[15.00 \quad 18.00 \quad 12.50]\begin{bmatrix} 0 & 13 & 11 & 8 & 5 \\ 9 & 0 & 4 & 7 & 10 \\ 4 & 2 & 10 & 11 & 0 \end{bmatrix} = \overset{\text{Mon. \quad Tues. \quad Wed. \quad Thurs. \quad Fri.}}{[212.00 \quad 220.00 \quad 362.00 \quad 383.50 \quad 255.00]}$

Exercises 5.4, page 190

1. $y = 0.5x + 0.5$; $E^t = [0.5 \quad -1 \quad 0.5]$; none of the data points lies on the line.
3. $y = 1.2x + 1.2$; $E^t = [0 \quad 0 \quad 0]$; all of the data points lie on the line.
5. $y = 0.11x + 0.16$; $E^t = [0.06 \quad -0.13 \quad 0.08 \quad -0.01]$; none of the data points lie on the line.
7. $y = 0.4x + 3.8$; $E^t = [1.5 \quad -0.1 \quad -0.6 \quad -2.5 \quad 1.7]$; none of the data points lie on the line.
9. a. $S = 1.57t + 21.16$ b. $S = 29.01$ in 1985 and $S = 30.58$ in 1986
11. a. $P = -0.032t + 1.004$
    b. $P = .84$ when $t = 5$. Therefore, in the fifth year, a dollar would purchase only 84% of what it could purchase in the base year.
13. b. $y = 0.7314x - 142.6$

Exercises 5.5, page 199

1. a. $[3 \quad 4]$ b. $[-2 \quad 6]$ c. $[5 \quad 0]$ d. $\begin{bmatrix} 3 & 4 \\ -2 & 6 \\ 5 & 0 \end{bmatrix}$ 3. $\begin{bmatrix} 9 & -5 \\ -7 & 4 \end{bmatrix}$

5. $\begin{bmatrix} -1 & -\frac{1}{2} \\ -3 & -1 \end{bmatrix}$ 7. $\begin{bmatrix} \frac{1}{2} & 0 & 0 \\ 0 & -\frac{1}{3} & 0 \\ 0 & 0 & \frac{1}{4} \end{bmatrix}$ 9. $\begin{bmatrix} -16 & 4 & -13 \\ 5 & -1 & 4 \\ -12 & 3 & -10 \end{bmatrix}$ 11. $\begin{bmatrix} -\frac{2}{3} & -\frac{1}{6} & \frac{1}{3} \\ \frac{2}{3} & -\frac{1}{3} & -\frac{1}{3} \\ 1 & -\frac{1}{6} & -\frac{1}{3} \end{bmatrix}$

13. The inverse does not exist.   15. $\begin{bmatrix} \frac{1}{2} & -\frac{3}{2} & -\frac{19}{6} & \frac{39}{2} \\ 0 & 1 & \frac{5}{3} & -12 \\ 0 & 0 & -\frac{1}{3} & 2 \\ 0 & 0 & 0 & 1 \end{bmatrix}$ 17. $\begin{bmatrix} 2 & 7 & -4 & 6 \\ -\frac{1}{3} & -\frac{5}{3} & 1 & -\frac{4}{3} \\ -1 & -3 & 2 & -3 \\ -\frac{17}{3} & -\frac{61}{3} & 12 & -\frac{56}{3} \end{bmatrix}$

19. $\begin{bmatrix} 5 & -4 & -1 & -4 \\ 1 & -1 & 0 & -1 \\ -\frac{1}{2} & \frac{1}{2} & \frac{1}{2} & 1 \\ -4 & 3 & 0 & 1 \end{bmatrix}$ 21. $A^{-1}B = \begin{bmatrix} -7 \\ 6 \end{bmatrix}$ 23. $A^{-1}B = \begin{bmatrix} 1 \\ 0 \\ 1 \end{bmatrix}$

25. a. $\begin{bmatrix} 19 \\ 13 \\ 4 \end{bmatrix}$ b. $\begin{bmatrix} 9 \\ 6 \\ 1 \end{bmatrix}$ c. $\begin{bmatrix} -6 \\ -7 \\ -2 \end{bmatrix}$ d. $\begin{bmatrix} 0 \\ 0 \\ 0 \end{bmatrix}$

Exercises 5.6, page 210

1. a. \$0.20 b. \$0.60   c.

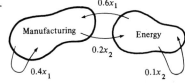

d. $x_1 = 0.4x_1 + 0.2x_2 + d_1$
   $x_2 = 0.6x_1 + 0.1x_2 + d_2$

e. $\begin{bmatrix} \frac{2550}{7} \\ \frac{2400}{7} \end{bmatrix}$; $\begin{bmatrix} \frac{4250}{7} \\ \frac{4000}{7} \end{bmatrix}$

3. $\begin{bmatrix} 91.67 \\ 131.25 \end{bmatrix}$; $\begin{bmatrix} 183.33 \\ 262.50 \end{bmatrix}$; $\begin{bmatrix} 45.833 \\ 65.625 \end{bmatrix}$ 5. $\begin{bmatrix} 320 \\ 600 \end{bmatrix}$; $\begin{bmatrix} 460 \\ 800 \end{bmatrix}$; $\begin{bmatrix} 180 \\ 400 \end{bmatrix}$

7. $\begin{bmatrix} 178.125 \\ 355.000 \\ 268.750 \end{bmatrix}$; $\begin{bmatrix} 375 \\ 1000 \\ 750 \end{bmatrix}$; $\begin{bmatrix} 328.125 \\ 895.000 \\ 568.750 \end{bmatrix}$ 9. a. $(I - A)^{-1} = \begin{bmatrix} 16 & 4 \\ 12 & 4 \end{bmatrix}$ b. $\begin{bmatrix} 5400 \\ 4200 \end{bmatrix}$; $\begin{bmatrix} 2320 \\ 1840 \end{bmatrix}$; $\begin{bmatrix} 1640 \\ 1280 \end{bmatrix}$

11. a.

$$\begin{array}{cc} & \text{A} \quad \text{M} \\ \begin{array}{c} \text{A} \\ \text{M} \end{array} & \begin{bmatrix} 0.25 & 0.05 \\ 0.15 & 0.40 \end{bmatrix} \end{array}$$

b. $\begin{bmatrix} 31.6 \\ 174.6 \end{bmatrix}$

13. a.

$$\begin{array}{cccc} & \text{S} & \text{C} & \text{F} \\ \begin{array}{c} \text{S} \\ \text{C} \\ \text{F} \end{array} & \begin{bmatrix} 0.15 & 0.25 & 0.55 \\ 0.10 & 0.30 & 0.35 \\ 0.10 & 0.20 & 0.30 \end{bmatrix} \end{array}$$

b. \$2100 worth of shelter, \$1300 worth of clothing, and \$1200 worth of food

Review Exercises (Chapter 5), page 217

1. $\begin{bmatrix} 6 & 1 & -8 & 7 \\ 0 & 5 & 4 & 0 \\ 6 & -6 & -1 & 16 \end{bmatrix}$ 3. -0.1 5. not defined   7. $\begin{bmatrix} -1 \\ 12 \\ -11 \end{bmatrix}$

9. $\begin{bmatrix} -8 & 6 & 0 & -13 \\ 35 & -33 & -13 & 48 \\ -10 & 12 & 18 & -15 \end{bmatrix}$ 11. $\begin{bmatrix} 1.2 & 9.3 & 0.5 \\ -1.3 & -39.2 & -7.5 \\ -8.6 & 19.5 & 3.7 \end{bmatrix}$ 13. $\begin{bmatrix} 0.125 & 0.5 & 0.75 \\ 0.500 & -1.0 & -1.00 \\ -0.375 & 0.5 & 0.75 \end{bmatrix}$

15. $\begin{bmatrix} -213 & -36 & -4 & 18 \\ -425 & -70 & -10 & 40 \\ -49 & -8 & -2 & 4 \\ 120 & 20 & 0 & -10 \end{bmatrix}$ 17. a. $\begin{bmatrix} 6 \\ 4 \\ -2 \end{bmatrix}$ b. $\begin{bmatrix} -0.375 \\ -0.500 \\ 0.125 \end{bmatrix}$ c. $\begin{bmatrix} -0.5 \\ 1.0 \\ 1.0 \end{bmatrix}$

19. a.

$$\begin{array}{c} \quad\quad\quad \text{Hand} \\ \quad\quad \text{Calculators} \quad \text{Microcomputers} \\ \begin{array}{c} \text{Plant I} \\ \text{Plant II} \end{array} \begin{bmatrix} 465 & 25 \\ 364 & 57 \end{bmatrix} \end{array}$$

b.

$$\begin{array}{c} \quad\quad\quad \text{Hand} \\ \quad\quad \text{Calculators} \quad \text{Microcomputers} \\ \begin{array}{c} \text{Plant I} \\ \text{Plant II} \end{array} \begin{bmatrix} 824 & 49 \\ 765 & 105 \end{bmatrix} \end{array}$$

21. $y = -0.82x + 17.18$; $E^t = [-0.02 \quad -0.04 \quad 0.1 \quad -0.08]$ 23. $\begin{bmatrix} 1545 \\ 1085 \\ 2050 \end{bmatrix}$

CHAPTER 6

Exercises 6.1, page 229

1. Find a solution to

$$\begin{array}{rcl} 3x + 4y + s_1 & = & 26 \\ 4x + 3y \quad\quad + s_2 & = & 23 \\ -5x - 2y \quad\quad\quad\quad + z & = & 0 \end{array}$$

for which $x \geq 0$, $y \geq 0$, $s_1 \geq 0$, $s_2 \geq 0$, and z is maximum.

3. Find a solution to

$$\begin{array}{rcl} x + 3y + s_1 & = & 26 \\ 4x + 3y \quad\quad + s_2 & = & 44 \\ 2x + 3y \quad\quad\quad\quad + s_3 & = & 28 \\ -4x - 5y \quad\quad\quad\quad\quad\quad + z & = & 0 \end{array}$$

for which $x \geq 0$, $y \geq 0$, $s_i \geq 0$, $i = 1, 2, 3$, and z is maximum.

5. Find a solution to

$$
\begin{aligned}
5x_1 + 3x_2 \quad\quad\quad + s_1 \quad\quad\quad\quad\quad &= 30 \\
4x_1 + x_2 + 7x_3 \quad\quad + s_2 \quad\quad\quad &= 20 \\
5x_2 + 8x_3 \quad\quad\quad + s_3 \quad &= 32 \\
-1.5x_1 - 0.75x_2 - 0.50x_3 \quad\quad\quad\quad\quad + z &= 0
\end{aligned}
$$

for which $x_i \geq 0$, $s_i \geq 0$, $i = 1, 2, 3$, and z is maximum.

7. The problem is in canonical form.

9. The problem is not in canonical form since each of the first two equations do not contain a variable with coefficient 1 which appears only in that equation.

11. Basic solution is $x = y = 0$, $s_1 = 6$, $s_2 = 8$, and $z = 0$. The basic variables are s_1, s_2, and z. The solution is feasible.

13. Basic solution is $x_1 = 0$, $x_2 = 5$, $x_3 = 6$, $s_1 = s_2 = 0$, and $z = 24$. The basic variables are x_2, x_3, and z. The solution is feasible.

15. a. $x = 8$, $s_2 = 4$, $z = 40$, and $y = s_1 = 0$.
    b. $x = 9$, $s_1 = -3$, $z = 45$, and $y = s_2 = 0$.
    c. The solution of part (a) is feasible whereas the one of part (b) is not.

17. a. $x_1 = 6$, $s_2 = 4$, $s_3 = 40$, $z = 42$, and $x_2 = x_3 = s_1 = 0$.
    b. $x_1 = 7$, $s_1 = -5$, $s_3 = 40$, $z = 49$, and $x_2 = x_3 = s_2 = 0$.
    c. The solution in part (a) is feasible whereas the one of part (b) is not.

19. a. The basic solution is $y = 4$, $s_1 = 32$, $z = 32$, and $x = s_2 = 0$, which is feasible.
    b. The basic solution is $y = 12$, $s_2 = -112$, $z = 96$, and $x = s_1 = 0$, but this solution is not feasible.
    c. s_2 should be changed to a non-basic variable when changing y to a basic variable.

21. a.

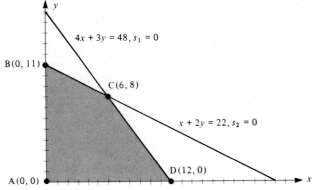

b.

Basic Solution	Corresponding to Vertex
$s_1 = 48$, $s_2 = 22$, $z = 0$ $x = y = 0$	A
$y = 11$, $s_1 = 15$, $z = 11$ $x = s_2 = 0$	B
$x = 6$, $y = 8$, $z = 14$ $s_1 = s_2 = 0$	C
$x = 12$, $s_2 = 10$, $z = 12$ $y = s_1 = 0$	D

Exercises 6.2, page 246

1. Canonical form:
   Find a solution to

$$
\begin{aligned}
5x + 3y + s_1 &&&&= 41 \\
x + y &+ s_2 &&&= 11 \\
2x + y &&+ s_3 &&= 18 \\
-3x - 2y &&&+ z &= 0
\end{aligned}
$$

for which $x \geq 0$, $y \geq 0$, $s_i \geq 0$, $i = 1, 2, 3$, and z is maximum.
Initial Simplex Tableau:

$$
\begin{array}{c}
\\ s_1 \\ s_2 \\ s_3 \\ \\ z
\end{array}
\begin{array}{c}
\begin{array}{cccccc}
x & y & s_1 & s_2 & s_3 & z
\end{array} \\
\left[
\begin{array}{cccccc|c}
5 & 3 & 1 & 0 & 0 & 0 & 41 \\
1 & 1 & 0 & 1 & 0 & 0 & 11 \\
2 & 1 & 0 & 0 & 1 & 0 & 18 \\
\hline
-3 & -2 & 0 & 0 & 0 & 1 & 0
\end{array}
\right]
\end{array}
$$

3. The basic feasible solution is $s_1 = 24$, $s_2 = 42$, $z = 0$, and $x_1 = x_2 = 0$, which is not optimal. The entering variable is x_2 and the departing variable is s_1. The new simplex tableau is

$$
\begin{array}{c}
\\ x_2 \\ s_2 \\ \\ z
\end{array}
\begin{array}{c}
\begin{array}{ccccc}
x_1 & x_2 & s_1 & s_2 & z
\end{array} \\
\left[
\begin{array}{ccccc|c}
\frac{1}{3} & 1 & \frac{1}{3} & 0 & 0 & 8 \\
3 & 0 & -1 & 1 & 0 & 18 \\
\hline
-\frac{2}{3} & 0 & \frac{4}{3} & 0 & 1 & 32
\end{array}
\right]
\end{array}
$$

with the new basic feasible solution $x_2 = 8$, $s_2 = 18$, $z = 32$, and $x_1 = s_1 = 0$.

5. The basic feasible solution is $x_2 = 15$, $x_4 = 17$, $s_1 = 9$, $z = 35$, and $x_1 = x_3 = s_2 = s_3 = 0$, which is not optimal. The entering variable is x_1 and the departing variable can be either s_1 or x_2; we choose s_1 as the departing variable. The new simplex tableau is

$$
\begin{array}{c}
\\ x_1 \\ x_4 \\ x_2 \\ \\ z
\end{array}
\begin{array}{c}
\begin{array}{cccccccc}
x_1 & x_2 & x_3 & x_4 & s_1 & s_2 & s_3 & z
\end{array} \\
\left[
\begin{array}{cccccccc|c}
1 & 0 & -\frac{2}{3} & 0 & \frac{1}{3} & \frac{1}{3} & -\frac{1}{3} & 0 & 3 \\
0 & 0 & -\frac{2}{3} & 1 & \frac{4}{3} & \frac{7}{3} & \frac{2}{3} & 0 & 29 \\
0 & 1 & \frac{7}{3} & 0 & -\frac{5}{3} & -\frac{2}{3} & \frac{8}{3} & 0 & 0 \\
\hline
0 & 0 & -\frac{32}{3} & 0 & \frac{7}{3} & \frac{22}{3} & \frac{17}{3} & 1 & 56
\end{array}
\right]
\end{array}
$$

with the new basic feasible solution $x_1 = 3$, $x_2 = 0$, $x_4 = 29$, $z = 56$, and $x_3 = s_1 = s_2 = s_3 = 0$.

7. Initial Simplex Tableau:

$$
\begin{array}{c}
\\ s_1 \\ s_2 \\ \\ z
\end{array}
\begin{array}{c}
\begin{array}{cccccc}
x_1 & x_2 & x_3 & s_1 & s_2 & z
\end{array} \\
\left[
\begin{array}{cccccc|c}
5 & 0 & 2 & 1 & 0 & 0 & 21 \\
2 & 3 & 0 & 0 & 1 & 0 & 11 \\
\hline
-7 & -10 & -8 & 0 & 0 & 1 & 0
\end{array}
\right]
\end{array}
$$

Initial Basic Feasible Solution:

$$
s_1 = 21, \; s_2 = 11, \; z = 0, \text{ and } x_1 = x_2 = x_3 = 0
$$

Solution to Problem: $x_1 = 0$, $x_2 = 3\frac{2}{3}$, $x_3 = 10\frac{1}{2}$, $z = 120\frac{2}{3}$

9. The optimal solution is $x = 0$, $y = 9$, and $z = 18$.
11. The optimal solution is $x = 4$, $y = 7$, and $z = 29$.
13. The optimal solution is $x_1 = 40$, $x_2 = 20$, $x_3 = 0$, and $z = 560$.
15. The optimal solution is $x_1 = x_2 = 0$, $x_3 = 6$, and $z = 12$.
17. The optimal solution is $x_1 = x_2 = x_3 = 0$, $x_4 = 7.5$, and $z = 45$.
19. The optimal solution is $x_1 = x_2 = x_5 = 0$, $x_3 = 3$, $x_4 = 6$, and $z = 45$.
21. a. The graph of the region of feasible solutions is given in the following figure. Since the region is unbounded and the problem is to maximize $z = 4x + 5y$, the problem has an unbounded solution.

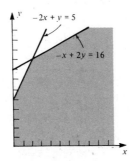

b. When the simplex algorithm is used, the tableau given below is obtained; we see from the s_1 column in the tableau that the problem has an unbounded solution.

$$
\begin{array}{c}
 \\
y \\
x \\
\\
z
\end{array}
\begin{array}{ccccc}
x & y & s_1 & s_2 & z \\
\left[\begin{array}{ccccc|c}
0 & 1 & -\frac{1}{3} & \frac{2}{3} & 0 & 9 \\
1 & 0 & -\frac{2}{3} & \frac{1}{3} & 0 & 2 \\
\hline
0 & 0 & -\frac{13}{3} & \frac{14}{3} & 1 & 53
\end{array}\right]
\end{array}
$$

23. Let x_1 = dozens of Licorice Delights to be made
    x_2 = dozens of Centre Surprises to be made
    x_3 = dozens of All-Day Suckers to be made

Problem:

$$
\begin{aligned}
\text{Maximize} \quad & z = 2x_1 + 3x_2 + 4x_3 \\
\text{subject to} \quad & 0.1x_1 + 0.1x_2 + 0.2x_3 \le 10 \\
& 0.4x_1 + 0.2x_2 + 0.2x_3 \le 12 \\
& x_i \ge 0, \quad i = 1, 2, 3.
\end{aligned}
$$

Solution: The Spangle Confection Company should make 20 dozen of Centre Surprise, 40 dozen of All-Day Suckers, and no Licorice Delights, in which case it will achieve a maximum profit of $220.

25. Let x_1 = number of cartons of Zapman games to be made
    x_2 = number of cartons of Snakepit games to be made
    x_3 = number of cartons of Monkey Kong games to be made

Problem:

$$\text{Maximize} \quad z = 100x_1 + 120x_2 + 110x_3$$
$$\text{subject to} \quad 0.75x_1 + \quad x_2 + 2.25x_3 \leq 10$$
$$0.50x_1 + \quad x_2 + \quad x_3 \leq 8$$
$$x_1 + 0.50x_2 + 0.25x_3 \leq 7$$
$$x_i \geq 0, \quad i = 1, 2, 3.$$

Solution: Starmont Products should make $4\frac{1}{4}$ cartons of Zapman games, $5\frac{1}{8}$ cartons of Snakepit games, and $\frac{3}{4}$ carton of Monkey Kong games per day, in which case it will achieve a maximum profit of $1122.50 per day.

27. Let $x_1 =$ batches of Extra-Strength to be made

   $x_2 =$ batches of Regular to be made

   $x_3 =$ batches of Placebo Plus to be made

Problem:

$$\text{Maximize} \quad z = 2000x_1 + 3000x_2 + 1500x_3$$
$$\text{subject to} \quad 0.5x_1 + \quad x_2 + 0.5x_3 \leq 6.5$$
$$0.5x_1 + \quad x_2 + 0.25x_3 \leq 6$$
$$1.5x_1 + 0.5x_2 + \quad x_3 \leq 5$$
$$x_i \geq 0, \quad i = 1, 2, 3.$$

Solution: The Grendal Corporation should make 0.2 batch of Extra-Strength, 5.4 batches of Regular, and 2 batches of Placebo Plus painkillers each day, in which case its profit will be $19,600 and will be maximum.

29. Let $x_1 =$ cases of Lady Jane to be made

   $x_2 =$ cases of El Macho to be made

   $x_3 =$ cases of Slim to be made

   $x_4 =$ cases of Mascot to be made

Problem:

$$\text{Maximize} \quad z = 30x_1 + 30x_2 + 40x_3 + 10x_4$$
$$\text{subject to} \quad 0.5x_1 + \quad 2x_2 + \quad x_3 + \quad x_4 \leq 17$$
$$0.25x_1 + \quad x_2 + \quad x_3 + \quad x_4 \leq 9$$
$$0.5x_1 + \quad x_2 + 2x_3 + 2x_4 \leq 19$$
$$x_1 + 0.5x_2 + \quad x_3 + 2x_4 \leq 12$$
$$x_i \geq 0, \quad i = 1, 2, 3, 4.$$

Solution: The Imperial Cigar Company should make each day 8 cases of Lady Jane cigars, 6 cases of El Macho cigars, 1 case of Slim cigars, and none of the Mascot cigars, in which case it will achieve the maximum profit of $460.

Exercises 6.3, page 262

1. Minimize $\quad Z = CX$

   subject to $\quad AX \geq B$

   $\quad\quad\quad\quad X \geq 0$

   where $C = [4 \quad 5]$, $A = \begin{bmatrix} 5 & 9 \\ 15 & 9 \end{bmatrix}$, $B = \begin{bmatrix} 90 \\ 180 \end{bmatrix}$, and $X = \begin{bmatrix} x \\ y \end{bmatrix}$.

3. Maximize $z = B'Y$
   subject to $A'Y \leq C'$
               $Y \geq 0$

   where $B' = [2 \quad 3]$, $A' = \begin{bmatrix} -4 & 3 \\ 14 & 6 \end{bmatrix}$, $C' = \begin{bmatrix} 1 \\ 35 \end{bmatrix}$, and $Y = \begin{bmatrix} x \\ y \end{bmatrix}$.

5. Maximize $z = 90x + 180y$
   subject to $5x + 15y \leq 4$
               $9x + 9y \leq 5$
               $x \geq 0$ and $y \geq 0.$

7. Minimize $Z = x + 35y$
   subject to $-4x + 14y \geq 2$
               $3x + 6y \geq 3$
               $x \geq 0$ and $y \geq 0.$

9. The optimal solution is $x = 9$, $y = 5$, and $Z = 61$.

11. The optimal solution is $x = 1$, $y = 1$, and $Z = 18$.

13. The optimal solution is $x = 0$, $y = 17\frac{1}{3}$, and $Z = 17\frac{1}{3}$.

15. The optimal solution is $x_1 = 0$, $x_2 = 8$, $x_3 = 0$, and $Z = 8$.

17. The optimal solution is $x_1 = 6$, $x_2 = 0$, $x_3 = 4$, and $Z = 86$.

19. The optimal solution is $x_1 = 2.25$, $x_2 = 0.5$, $x_3 = 0$, $x_4 = 1.25$, and $Z = 52.5$.

21. The optimal value for Z is \$3.75. There is not a unique diet that yields this minimal cost. One possible diet is 7.5 units of type B food and none of the other two. Another diet that gives the minimal cost is $3\frac{2}{3}$ units of type A food, $\frac{1}{6}$ unit of type B food, and none of type C food.

23. Let x = kilograms of ingredient A used per bag

      y = kilograms of ingredient B used per bag

   Problem:

            Minimize $Z = 0.50x + 0.75y$
            subject to $0.25x + 0.35y \geq 50$
                           $0.30x + 0.20y \geq 25$
                           $0.10x + 0.15y \geq 20$
                           $x \geq 0$ and $y \geq 0.$

   Solution: The minimal cost is \$1.00 per bag which Mayfair Feed Company can achieve by using 200 kilograms of ingredient A and none of ingredient B.

25. The minimal amount of waste is 80 inches which can be achieved by cutting 10 boards under option 1 and 20 boards under option 3 and not cutting any boards under option 2. Another cutting plan that will achieve the minimal waste of 80 inches is to cut 20 boards under option 2, 10 boards under option 3, and none under option 1.

27.

	Number of 4" rolls	Number of 6" rolls	Number of $8\frac{1}{2}$" rolls	Amount of waste
option 1	3	0	0	3"
option 2	2	1	0	1"
option 3	1	0	1	2.5"
option 4	0	2	0	3"
option 5	0	1	1	0.5"

The minimal waste is 53 inches which can be achieved by cutting 25 rolls under option 2 and 56 rolls under option 5.

29. Let x_1 = the number of hours to operate the Bristol plant

   x_2 = the number of hours to operate the Lewistown plant

   x_3 = the number of hours to operate the Acton plant

   Problem:

$$\text{Minimize} \quad Z = 200x_1 + 125x_2 + 150x_3$$
$$\text{subject to} \quad 100x_1 \qquad\quad + 50x_3 \geq 2500$$
$$50x_1 + 75x_2 + 80x_3 \geq 1800$$
$$x_i \geq 0, \quad i = 1, 2, 3.$$

   Solution: Sun Tile Company can achieve a minimal cost of $5500 by operating the Bristol plant for 20 hours and the Acton plant for 10 hours and not use the Lewistown plant to produce the desired number of tiles.

Exercises 6.4, page 277

1. The optimal value of z is 32 which occurs at $x = 6$ and $y = 7$.
3. The optimal value of z is 66 which occurs at $x = 3$ and $y = 0$.
5. The optimal value of z is 109 which occurs at $x_1 = 4$, $x_2 = 7$, and $x_3 = 9$.
7. The problem has no feasible solutions.
9. Let x_1 = tons of beef shipped from Bradford to Topps

   x_2 = tons of beef shipped from Bradford to Marketville

   x_3 = tons of beef shipped from Lewis Run to Topps

   x_4 = tons of beef shipped from Lewis Run to Marketville

   Problem:

$$\text{Minimize} \quad z = 60x_1 + 50x_2 + 30x_3 + 40x_4$$
$$\text{subject to} \quad x_1 + x_2 \qquad\qquad \leq 9$$
$$x_3 + x_4 \leq 13$$
$$x_1 \qquad + x_3 \qquad \geq 10$$
$$x_2 \qquad + x_4 \geq 12$$
$$x_i \geq 0, \quad i = 1, 2, 3, 4.$$

   The minimal cost to Bradford Beef is $870 which is obtained by shipping 9 tons from Bradford to Marketville, 10 tons from Lewis Run to Topps, and 3 tons from Lewis Run to Marketville.

11. Let x_1 = the number of single-sided disks from CompuDisk

   x_2 = the number of double-sided disks from CompuDisk

   x_3 = the number of single-sided disks from PCDisk

   x_4 = the number of double-sided disks from PCDisk

   Problem:

$$\text{Minimize} \quad z = 3x_1 + 3.5x_2 + 3x_3 + 4x_4$$
$$\text{subject to} \quad x_1 \qquad + x_3 \qquad \geq 1500$$
$$x_2 \qquad + x_4 \geq 1200$$
$$x_1 \qquad\qquad \leq 900$$
$$x_1 + x_2 \qquad\qquad \leq 1300$$
$$x_4 \leq 1100$$
$$x_3 + x_4 \leq 1700$$
$$x_i \geq 0, \quad i = 1, 2, 3, 4.$$

Solution: The minimal cost to Micro Company is $8700 which is achieved by purchasing 100 single-sided disks from CompuDisk along with 1400 of these from PCDisk while purchasing all of the double-sided disks (1200) from CompuDisk.

13. Let x_1 = pounds of ingredient I used in each bag

x_2 = pounds of ingredient II used in each bag

x_2 = pounds of ingredient III used in each bag

Problem:

$$\text{Minimize} \quad z = 0.50x_1 + 0.30x_2 + 0.30x_3$$
$$\text{subject to} \quad 0.40x_1 + 0.20x_2 + 0.30x_3 \geq 12$$
$$0.45x_1 + 0.30x_2 + 0.15x_3 \geq 15$$
$$0.15x_1 + 0.50x_2 + 0.30x_3 \geq 9.75$$
$$x_1 + x_2 \geq x_3$$
$$x_i \geq 0, \quad i = 1, 2, 3.$$

Solution: To achieve the minimal cost of $16 per bag, Derby Feed Company would need to use 20 pounds of ingredient I, 20 pounds of ingredient II, but none of the third ingredient in each bag.

15. Let x_1 = the amount invested at 9%

x_2 = the amount invested at 15%

x_3 = the amount invested at 20%.

Problem:

$$\text{Maximize} \quad z = 0.09x_1 + 0.15x_2 + 0.20x_3$$
$$\text{subject to} \quad x_1 + x_2 + x_3 \leq 20{,}000$$
$$x_2 + x_3 \leq x_1$$
$$x_3 \leq 5000$$
$$x_1 \geq 10{,}000$$
$$x_i \geq 0, \quad i = 1, 2, 3.$$

Solution: The optimal return for the investor is $2650 which is achieved by investing $10,000 in the low-risk investment and investing $5000 in each of the other two investments.

Review Exercises (Chapter 6), page 280

1. Find a solution to

$$2x_1 + x_2 + 6x_3 + s_1 \qquad\qquad = 20$$
$$2x_1 + 5x_2 + 7x_3 \qquad + s_2 \qquad = 22$$
$$-8x_1 - 6x_2 - 5x_3 \qquad\qquad + z = 0$$

for which $x_i \geq 0$, $i = 1, 2, 3$, and z is maximum.

3. a. The basic solution is $x_1 = 5$, $s_1 = 5$, $s_3 = 10$, $z = 35$, and $x_2 = x_3 = s_2 = 0$, which is feasible.

b. The basic solution is $x_1 = \frac{15}{2}$, $s_1 = -12$, $s_2 = -\frac{5}{2}$, $z = \frac{105}{2}$, and $x_2 = x_3 = s_3 = 0$; this solution is not feasible.

5. The basic feasible solution is $s_1 = \frac{5}{2}$, $s_2 = \frac{13}{2}$, $z = 0$, and $x_1 = x_2 = x_3 = 0$, but it is not optimal. The entering variable is x_3 and the departing variable is s_1. The new basic feasible solution is $x_3 = 10$, $s_2 = \frac{3}{2}$, $z = 100$, and $x_1 = x_2 = s_1 = 0$.

7. The maximum value for z is 81 and it occurs at $x_1 = 9.75$, $x_2 = 0.5$, and $x_3 = 0$.

9. The optimal value for z is 16 which occurs at $x = 3$ and $y = 7$.

11. The maximum value of z is 211 which occurs at $x_2 = 13$, $x_3 = 11$, and $x_1 = x_4 = 0$.

13. The optimal solution is $x = 8.6$, $y = 21.1$, and $z = 433.8$.

15. The problem has an unbounded solution.

17. Let x_1 = the number of ranch houses to be built

 x_2 = the number of split-level houses to be built

 x_3 = the number of modern houses to be built

 x_4 = the number of Cape Cod houses to be built

Problem:

$$\text{Maximize} \quad z = 5000x_1 + 7500x_2 + 7000x_3 + 6000x_4$$
$$\text{subject to} \quad x_1 + x_2 + x_3 + x_4 \le 80$$
$$x_1 + 2x_2 + 2x_3 + x_4 \le 130$$
$$3x_1 + 6x_2 + 5x_3 + 4x_4 \le 387$$
$$x_i \ge 0, \quad i = 1, 2, 3, 4.$$

Solution: The maximum profit is $538,500 obtained by building 17 split-levels, 33 modern houses, and 30 Cape Cods, but no ranch houses.

CHAPTER 7

Exercises 7.1, page 291

1. $P = \$600, A = \$632, I = \$32$ 3. $P = \$5000, A = \$5020.68, I = \$20.68$ 5. $I = \$224, A = \$1624$
7. $I = \$180, A = \$1680$ 9. $I = \$104.83, A = \$954.83$ 11. $I = \$50.75, A = \$1500.75$
13. $r = 10\%$ 15. $r = 12.5\%$ 17. $r = 9.2\%$ 19. $P = \$961.54$ 21. $P = \$733.50$
23. $P = \$1072.13$ 25. $\$12.50$ 27. $\$620.16$ 29. 18% 31. $\$13,495$ 33. $\$25,082.87$
35. $D = \$150, P = \$850$ 37. $D = \$245, P = \$4755$ 39. $\$1764.71$
41. The simple interest rate of 10.5% is slightly better.

Exercises 7.2, page 303

1. $S = \$8682.19$ 3. $S = \$4163.89$ 5. $S = \$17,125.53$ 7. $S = \$3524.58$ 9. $S = \$5898.75$
11. $r_E = 6.14\%$ 13. $r_E = 8.3\%$ 15. $r_E = 10.38\%$ 17. $r_E = 8.33\%$
19. 7.3% interest converted annually is better   21. 9.8% interest converted continuously is better
23. $P = \$4319.19$ 25. $P = \$6217.21$ 27. $P = \$6969.60$ 29. $P = \$1802.86$ 31. $\$1561.06$
33. $\$18,715.31$ 35. $\$1526.56$ 37. $\$134,487.66$ 39. $\$47,640.82$

Exercises 7.3, page 312

1. $S = \$7814.13$ 3. $S = \$2618.85$ 5. $S = \$3488.50$ 7. $R = \$614.14$ 9. $R = \$498.85$
11. $R = \$191.07$ 13. $A = \$12,701.66$ 15. $A = \$6449.70$ 17. $A = \$25,770.70$ 19. $R = \$253.07$
21. $R = \$131.18$ 23. $R = \$269.05$ 25. $B = \$3698.65$ 27. $B = \$9930.95$ 29. $R = \$526.58$

Payment number	Size of payment	Portion to interest	Portion to Loan	Outstanding balance
1	526.58	45.00	481.58	2518.42
2	526.58	37.78	488.80	2029.62
3	526.58	30.45	496.13	1533.49
4	526.58	23.01	503.57	1029.92
5	526.58	15.45	511.13	518.79
6	526.58	7.79	518.79	—

31. $\$263.28$ 33. $\$895$ 35. $\$21,710.86$ 37. $\$86.93$ 39. $\$3967.42$ 41. $\$11,356.10$

Review Exercises (Chapter 7), page 317

1. $r = 9\%$ 3. $r = 12.5\%$ 5. \$157.35   7. \$9862.49   9. $D = \$280, P = \$3220$
11. $S = \$15,687.78$ 13. $S = \$2252.99$ 15. $r_E = 7.23\%$ 17. \$4508.21   19. \$6483.30
21. \$260.49   23. a. \$259.99   b. \$4167.52

CHAPTER 8

Exercises 8.1, page 328

1. a. sample space b. not a sample space c. sample space

3. a. $S = \{\text{blue, purple}\}$ b. $Pr(\text{blue}) = \dfrac{1}{5}, Pr(\text{purple}) = \dfrac{4}{5}$

5. a. $S = \{\text{soda, beer, iced tea, none of the three}\}$

  b. $Pr(\text{soda}) = \dfrac{26}{75}, Pr(\text{beer}) = \dfrac{13}{50}$

    $Pr(\text{iced tea}) = \dfrac{9}{50}, Pr(\text{none of the three}) = \dfrac{1}{30}$

  c. $\dfrac{39}{50}$

7. a. $S = \{0, 1, 2, 3, 4 \text{ or more}\}$

  b. $Pr(0) = \dfrac{2}{25}, Pr(1) = \dfrac{3}{8}, Pr(2) = \dfrac{11}{40}$

    $Pr(3) = \dfrac{9}{100}, Pr(4 \text{ or more}) = \dfrac{1}{10}$

  c. $Pr(2 \text{ or more}) = \dfrac{93}{200}$ d. $Pr(\text{at most } 3) = \dfrac{9}{10}$ 9. $\dfrac{2}{3}$

11. a. $S = \{1, 2, 3, 4\}$ b. $Pr(1) = Pr(2) = Pr(3) = Pr(4) = \dfrac{1}{4}$ c. $\dfrac{1}{2}$

13. a. $Pr(e_1) = \dfrac{3}{16}, Pr(e_2) = \dfrac{1}{16}, Pr(e_3) = \dfrac{3}{8}, Pr(e_4) = \dfrac{3}{8}$ b. $Pr(E) = \dfrac{9}{19}$

  c. An urn contains 16 balls, 3 are of color e_1, 1 is of color e_2, 6 are of color e_3, and 6 are of color e_4. A ball is drawn at random from the urn and its color observed.
15. Erica will do more studying.

17. a. $S = \{\text{defective, not defective}\}$ b. $Pr(\text{defective}) = \dfrac{13}{300}, Pr(\text{not defective}) = \dfrac{287}{300}$

19. $\dfrac{5}{18}, \dfrac{13}{18}$ 21. a. $S = \{1H, 1T, 2H, 2T, 3H, 3T, 4H, 4T, 5H, 5T, 6H, 6T\}$
             b. $S = \{T, HT, HHT, HHH\}$

Exercises 8.2, page 337

1. a. The person selected makes more than \$30,000 a year and is a Republican.
  b. The person selected either makes more than \$30,000 a year or is a Republican.
  c. The person selected is not a Republican.

d. The person selected makes more than $30,000 a year but is not a Republican.

e. The person selected makes more than $30,000 a year and is not a Republican.

f. The person selected does not make more than $30,000 a year but is not a Republican.

g. It is not the case that the person selected either makes more than $30,000 a year or is a Republican.

h. The person selected neither makes more than $30,000 a year nor is a Republican.

i. The person selected does not make more than $30,000 a year and is a Republican.

j. The person selected is a Republican who makes more than $30,000 a year.

3. a. $E \cup F$ b. $E \cap F$ c. $E \cup F$ d. $E' \cap F'$

5. It is not the case that the student selected either smokes cigarettes or drinks beer means that the student neither smokes cigarettes nor drinks beer.

It is not the case that the student selected smokes cigarettes and drinks beer means that the student either does not smoke cigarettes or does not drink beer.

7. No, a student may own both a car and a bicycle.

9. a. 0.53   b. $\dfrac{1}{12}$ 11. a. 0.70   b. $\dfrac{2}{21}$ 13. 0.73   15. a. $\dfrac{3}{5}$ b. $\dfrac{1}{5}$ c. $\dfrac{27}{35}$ 17. a. $\dfrac{5}{7}$ b. 0.277

19. a. 0.75   b. 0.23   21. a. $\dfrac{1}{15}$ b. $\dfrac{2}{5}$ 23. $\dfrac{1}{3}$ 25. 0.2   27. 0.42   31. 0.35   33. 0.366

35. a. $E_1 \cap E_2 \cap E_3$ b. $E'_1 \cap E'_2 \cap E'_3$ c. $(E_1 \cap E'_2 \cap E'_3) \cup (E'_1 \cap E_2 \cap E'_3) \cup (E'_1 \cap E'_2 \cap E_3)$

Exercises 8.3, page 350

1. 70   3   500   5. a. 18,954   b. 13,104   7. 4, 8, 16, 2^n 9. a. 24   b. 720

11.

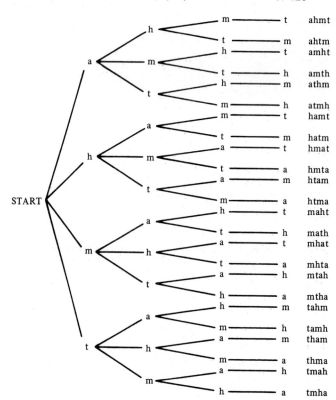

13.

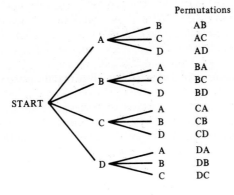

15.

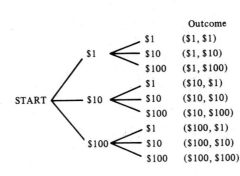

17.

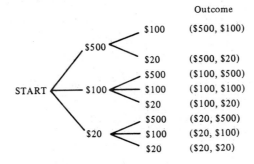

19. a. 1 b. 120 c. 362,880 d. 20 e. 840 f. 33,390,720

21. a. 360 b. 1296   23. 110   25. 120   27. $\dfrac{1}{840}$ 29. $1 - \dfrac{365 \cdot 364 \cdot 363 \cdot 362}{365^4} = 0.0164$

Exercises 8.4, page 356

1. a. 3 b. 21 c. 1 d. 56
3. $\{A, B, C\}\{A, B, D\}\{A, B, E\}\{A, C, D\}\{A, C, E\}\{A, D, E\}\{B, C, D\}\{B, C, E\}\{B, D, E\}\{C, D, E\}\,C[5, 3] = 10$
5. $C[28, 4] = 20{,}475$ 7. $C[6, 3]C[5, 2]C[6, 1] = 1200$ 9. $C[40, 6] = 2{,}763{,}633{,}600$
11. $C[2, 1]C[6, 2]C[7, 2] = 630$ 13. $C[7, 5]C[11, 5] = C[7, 2]C[11, 5] = 9702$ 15. $C[50, 3] = 19{,}600$

17. $P[50, 3] = 117{,}600$ 19. $\dfrac{C[6, 5]}{2^6} = \dfrac{3}{32}$ 21. $1 - \dfrac{C[6, 6]}{2^6} = \dfrac{31}{32}$ 23. $\dfrac{1}{120}$

25. a. $\dfrac{1}{64}$ b. $\dfrac{3}{32}$ c. $\dfrac{7}{64}$ 27. $\dfrac{1}{120}$ 29. $\dfrac{C[13, 2]C[13, 3]}{C[52, 5]} = 0.0086$ 31. $\dfrac{C[3, 1]C[9, 5]}{C[12, 6]} = \dfrac{9}{22}$

33. $\dfrac{C[8,4] + C[8,5]}{C[15,5]} = 0.042$ 35. $\dfrac{5}{9}$

Exercises 8.5, page 367

1. a. The probability that the person selected likes to hunt, given that the person is in favor of stricter gun control laws.
   b. The probability that the person selected is in favor of stricter gun control laws, given that the person likes to hunt.
   c. The probability that the person selected does not like to hunt, given that the person is in favor of stricter gun control laws.
   d. The probability that the person selected is not in favor of stricter gun control laws, given that the person does not like to hunt.

3. a. $Pr(E|F)$ b. $Pr(E'|F')$ c. $Pr(F'|E)$ 5. a. $\dfrac{1}{7}$ b. $\dfrac{3}{7}$ c. 0   d. $\dfrac{6}{7}$ 7. $\dfrac{3}{10}$

9. a. $\dfrac{12}{37}$ b. $\dfrac{12}{39}$ c. $\dfrac{25}{37}$ d. $\dfrac{3}{7}$ e. $\dfrac{4}{7}$ f. $\dfrac{36}{61}$ 11. 0.45   13. $\dfrac{4}{17}$ 15. a. $\dfrac{1}{3}$ b. $\dfrac{1}{2}$

17. $\dfrac{4}{5}$ 19. 0.75   21. 0.637   23. 0.075   25. 0.438   27. No   29. $\dfrac{2}{5}$ 31. 0.1287

33. a. 0.113   b. 0.435   c. 0.274   35. a. $\dfrac{4}{9}$ b. $\dfrac{1}{9}$ c. $\dfrac{4}{9}$ 37. a. 0.4096   b. 0.0016

39. a. 0.16807   b. 0.3087   c. 0.96922   43. a. $\dfrac{5}{7}$ b. $\dfrac{2}{7}$ c. $\dfrac{4}{7}$ 47. $\dfrac{1}{5525}$

Exercises 8.6, page 378

1.

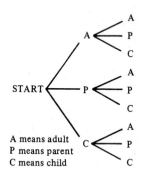

3. a.

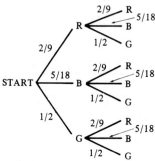

b. $\frac{1}{4}$ c. $\frac{14}{81}$ d. $\frac{4}{81}$

5. a.

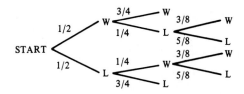

b. $\frac{4}{35}$ c. $\frac{1}{7}$

7. a.

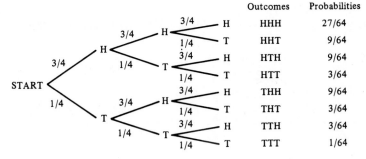

b. $\frac{15}{32}$ c. $\frac{1}{4}$

9. (a), (b), and (c).

		Outcomes	Probabilities
		HHH	27/64
		HHT	9/64
		HTH	9/64
		HTT	3/64
		THH	9/64
		THT	3/64
		TTH	3/64
		TTT	1/64

d. Pr(at least 2 heads) $= \frac{27}{32}$ 11. a. $\frac{1}{8}$ b. $\frac{3}{8}$ c. $\frac{1}{8}$ 13. 0.82   15. a. 0.7125   b. 0.069

17. a.

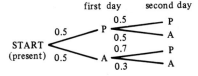

Answer: $(0.5)(0.5) + (0.5)(0.3) = 0.40$

b.

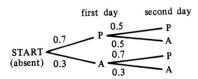

Answer: $(0.7)(0.5) + (0.3)(0.3) = 0.44$ 19. $\frac{3}{8}$

Exercises 8.7, page 387

1. a. $\dfrac{27}{44}$ b. $\dfrac{11}{27}$ c. $\dfrac{16}{27}$ 3. a. 0.8282 b. 0.5110 c. 0.4890 5. a. $\dfrac{7}{12}$ b. $\dfrac{2}{7}$ c. $\dfrac{5}{7}$

7. a. 0.00011 b. 0.72727 9. a. 0.81 b. 0.19 11. a. 0.9383 b. 0.0617
13. a. 0.5781 b. 0.2705 c. 0.2520 d. 0.4774

Review Exercises (Chapter 8), page 390

1. a. $S = \{R, W, G, B\}$

   b. $Pr(R) = \dfrac{2}{13}$.

   $Pr(W) = \dfrac{6}{13}, Pr(G) = \dfrac{1}{13}, Pr(B) = \dfrac{4}{13}$

   c. $\dfrac{10}{13}$

3. 0.47 5. a. 0.524 b. $\dfrac{17}{21}$ 7. a. 24 b. 6 c. 21 d. 1770 9. a. $\dfrac{5}{32}$ b. $\dfrac{5}{8}$ c. $\dfrac{31}{32}$

11. 0.4835 13. a. $\dfrac{27}{59}$ b. $\dfrac{32}{59}$ c. $\dfrac{32}{63}$ d. $\dfrac{31}{41}$ 15. a. 0.1160 b. 0.3364 c. 0.8840

17. a.   b. $\dfrac{3}{5}$

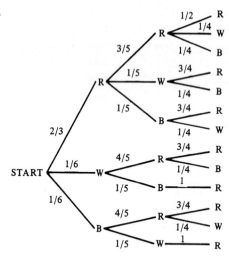

19. a. 0.0016 b. 0.1536 21. a. $\dfrac{13}{24}$ b. $\dfrac{9}{13}$ 25. a. 0.2941 b. 0.3529 c. 0.3529 d. 0.6471

CHAPTER 9

Exercises 9.1, page 400

1. a. discrete b. continuous c. discrete d. continuous e. continuous f. discrete

3. a.

value of X	0	1	2
$Pr(X = x)$	$\frac{1}{5}$	$\frac{3}{5}$	$\frac{1}{5}$

b.

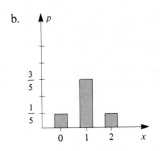

5. a.

value of X	0	1	2
$Pr(X = x)$	$\frac{4}{9}$	$\frac{4}{9}$	$\frac{1}{9}$

b.

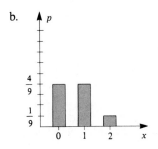

7. a.

value of X	0	1	2
$Pr(X = x)$	0.09	0.42	0.49

b.

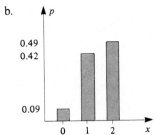

9.

value of X	-1	1	2	3
$Pr(X = x)$	0.2	0.1	0.4	0.3

11.

value of X	0	1	2	3
$Pr(X = x)$	0.614125	0.325125	0.057375	0.003375

13.

value of X	0	6	9	12	15	18
$Pr(X = x)$	$\frac{1}{9}$	$\frac{2}{9}$	$\frac{2}{9}$	$\frac{1}{9}$	$\frac{2}{9}$	$\frac{1}{9}$

15. $k = \dfrac{8}{23}$

Exercises 9.2, page 410

1. \$126.42   3. 19.28   5. $x = 3.12$ 7. $E(X) = 12.11$ 9. $E(X) = 1$ 11. $E(X) = 1.6$
13. $E(X) = 0.45$ 15. $E(X) = 10$ 17. $E(X) = 4.60$, \$4.60   19. \$620   21. 45,500
23. The professor should write a textbook.   25. a. \$120,000 b. \$142,500 c. yes   27. 1.168
29. \$22   31. a. $E(Y) = 15.5$ b. $E(Y) = 10E(X)$

33. $E(mX) = p_1 m x_1 + p_2 m x_2 + \cdots + p_n m x_n$
$ = m(p_1 x_1 + p_2 x_2 + \cdots + p_n x_n)$
$ = mE(X)$

35. $\left(\dfrac{1}{5}\right)(\$60) = \$12$

Exercises 9.3, page 420

1. a. $\dfrac{27}{64}$ b. 0.2388 c. 0.1525 d. 0.2355   3. a. 0.0283 b. 0.0033 c. 0.1070

   d. 0.9700 e. 0.9998 f. 0.2436   5. a. 0.3037 b. 0.9620 c. 0.7333 d. 0.2667

7. a. 0.2054 b. 0.0000 c. 0.8042 d. 0.5886   9. 0.0006   11. a. 0.0981 b. 0.9857 c. 0.5277

13. a. 0.1348 b. 0.9942 c. 0.0203   15. a. 0.0688 b. 0.6064 c. 0.9916

17. a. 0.8672 b. 0.0000 c. 0.7164   19. a. 0.2968 b. 0.4846 c. 0.9847

21. a. 0.5982 b. 0.2272 c. 0.2272

23. a.

value of X	0	1	2	3
$Pr(X = x)$	q^3	$3pq^2$	$3p^2q$	p^3

  b. $E(X) = q^3(0) + 3pq^2(1) + 3p^2q(2) + p^3(3)$
$$= 3p(q^2 + 2qp + p^2)$$
$$= 3p(q + p)^2$$
$$= 3p, \quad \text{since } q + p = 1$$

Exercises 9.4, page 427

1. $s^2 = 1880$, $s = 43.37$ 3. $s^2 = 2.40$, $s = 1.55$ 5. $\mu = 6.25$, $\sigma^2 = 332.19$, $\sigma = 18.23$

7. $\mu = 99.668$, $\sigma^2 = 51{,}422$, $\sigma = 226.76$ 9. $\sigma^2 = \dfrac{1}{5}$, $\sigma = \dfrac{1}{\sqrt{5}}$ 11. $\sigma^2 = 2.04$, $\sigma = 1.43$

13. $\sigma^2 = 0.3825$, $\sigma = 0.6185$ 15. $\sigma^2 = 28$, $\sigma = 5.29$

17. Let σ^2 be the variance of X. The variance of $X + b$ is

$$p_1(x_1 + b)^2 + p_2(x_2 + b)^2 + \cdots + p_n(x_n + b)^2 - (E(X) + b)^2$$
$$= p_1(x_1^2 + 2bx_1 + b^2) + p_2(x_2^2 + 2bx_2 + b^2) + \cdots + p_n(x_n^2 + 2bx_n + b^2) - (E(X)^2 + 2bE(X) + b^2)$$
$$= (p_1x_1^2 + p_2x_2^2 + \cdots + p_nx_n^2 - E(X)^2) + 2b(p_1x_1 + p_2x_2 + \cdots + p_nx_n)$$
$$\quad + b^2(p_1 + p_2 + \cdots + p_n) - 2bE(X) - b^2$$
$$= \sigma^2 + 2bE(X) + b^2 - 2bE(X) - b^2$$
$$= \sigma^2$$

Exercises 9.5, page 440

1. a. 0.4744 b. 0.0197 c. 0.5832 d. 0.2148 e. 0.1539 f. 0.1599 g. 0.0085 h. 0.5978

3. a. 0.0918 b. 0.5030 c. 0.2186 d. 0.1100   5. a. 0.8133 b. 0.8888 c. 0.7021 d. 0.1032

7. a. 0.0217 b. 0.2920 c. 0.8186   9. a. 0.0401 b. 0.1056 c. 0.0546

11. a. 0.1867 b. 0.5222 c. 0.0764   13. a. 0.1335 b. 0.6070 c. 0.1867   15. 87.7

Exercises 9.6, page 448

1. a. $\mu = 375$, $\sigma = 12.25$ b. $\mu = 274.01$, $\sigma = 11.35$ c. $\mu = 64.056$, $\sigma = 5.82$ d. $\mu = 3169.049$, $\sigma = 27.06$

3. a.

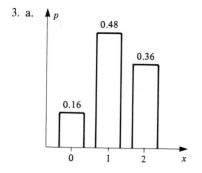

   b.

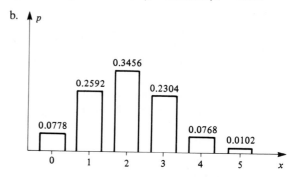

c.

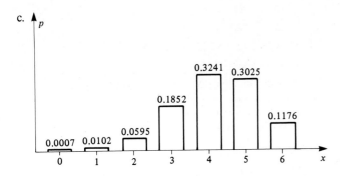

5. a. 0.0694   b. 0.1075   c. 0.7503   d. 0.1091   7. a. 0.1093   b. 0.3520   c. 0.2679   d. 0.0000
9. a. 0.0150   b. 0.4380   c. 0.0197   11. 0.7764   13. a. 0.6700   b. 0.3183   c. 0.0000
15. a. 0.7995   b. 0.0192   c. 0.0139   17. $p = 0.75, q = 0.25$

Exercises 9.7, page 457

1. yes   3. yes   5. no   7. yes   9. no   11. no   13. yes

15.

One-tailed test for $\alpha = 0.01$

Review Exercises (Chapter 9), page 460

3.

value of X	0	1	2	3
$Pr(X = x)$	$\frac{4}{33}$	$\frac{5}{11}$	$\frac{4}{11}$	$\frac{2}{33}$

5. a. $\mu = 63.81$ b. $\sigma^2 = 32{,}494$ c. $\sigma = 180.26$ 7. 0.50, no   9. a. 0.2252   b. 0.3670   c. 0.0577
11. a. 0.1696   b. 0.0189   13. a. 0.0475   b. 0.8533   c. 0.0475   15. a. 0.1357   b. 0.4871   c. 0.2285
17. a. 0.7995   b. 0.0192   c. 0.0139   19. yes

CHAPTER 10

Exercises 10.1, page 473

1. Parts (a) and (d) can be transition matrices. Part (b) cannot be a transition matrix since the sum of the numbers in the first row is not equal to one. Part (c) cannot be a transition matrix since the sum of the numbers in the second row is not equal to one.

3. a.

$$P^2 = \begin{array}{c} \\ A \\ B \\ C \end{array} \begin{array}{ccc} A & B & C \\ \begin{bmatrix} 0.3466 & 0.3405 & 0.3129 \\ 0.2496 & 0.4375 & 0.3129 \\ 0.2496 & 0.3405 & 0.4099 \end{bmatrix} \end{array}$$

b.

$$P^3 = \begin{array}{c} \\ A \\ B \\ C \end{array} \begin{array}{ccc} A & B & C \\ \begin{bmatrix} 0.29907 & 0.36378 & 0.33715 \\ 0.28355 & 0.35893 & 0.35752 \\ 0.25057 & 0.40743 & 0.34200 \end{bmatrix} \end{array}$$

c. 0.4375   d. 0.33715

5. a.
$$P = \begin{array}{c} \\ R \\ L \end{array} \begin{array}{cc} R & L \\ \begin{bmatrix} \frac{3}{5} & \frac{2}{5} \\ \frac{7}{10} & \frac{3}{10} \end{bmatrix} \end{array}$$
b.
$$P^2 = \begin{array}{c} \\ R \\ L \end{array} \begin{array}{cc} R & L \\ \begin{bmatrix} \frac{16}{25} & \frac{9}{25} \\ \frac{63}{100} & \frac{37}{100} \end{bmatrix} \end{array}$$
c. $\dfrac{16}{25}$ d. $\dfrac{37}{100}$

7. a.
$$P = \begin{array}{c} \\ C \\ N \end{array} \begin{array}{cc} C & N \\ \begin{bmatrix} 0.72 & 0.28 \\ 0.033 & 0.67 \end{bmatrix} \end{array}$$

where the states are college degree (C) and no college degree (N).

b.
$$P^2 = \begin{array}{c} \\ C \\ N \end{array} \begin{array}{cc} C & N \\ \begin{bmatrix} 0.6108 & 0.3892 \\ 0.4587 & 0.5413 \end{bmatrix} \end{array} \qquad P^3 = \begin{array}{c} \\ C \\ N \end{array} \begin{array}{cc} C & N \\ \begin{bmatrix} 0.568212 & 0.431788 \\ 0.508893 & 0.491107 \end{bmatrix} \end{array}$$

c. 0.4587   d. 0.431788   e. 0.491107

9. a.
$$P = \begin{array}{c} \\ B \\ B' \end{array} \begin{array}{cc} B & B' \\ \begin{bmatrix} 0.58 & 0.42 \\ 0.43 & 0.57 \end{bmatrix} \end{array}$$

b. $S^{(1)} = [0.499 \quad 0.501]$
$S^{(2)} = [0.50485 \quad 0.49515]$
$S^{(3)} = [0.5057275 \quad 0.4942725]$

c. $S^{(1)} = [0.505 \quad 0.495]$
$S^{(2)} = [0.50575 \quad 0.49425]$
$S^{(3)} = [0.5058625 \quad 0.4941375]$

d. $S^{(1)} = [0.46 \quad 0.54]$
$S^{(2)} = [0.499 \quad 0.501]$
$S^{(3)} = [0.50485 \quad 0.49515]$

11. a.
$$P = \begin{array}{c} \\ a \\ b \\ c \\ d \end{array} \begin{array}{cccc} a & b & c & d \\ \begin{bmatrix} 0 & \frac{2}{3} & \frac{1}{3} & 0 \\ \frac{1}{2} & 0 & 0 & \frac{1}{2} \\ \frac{1}{2} & 0 & 0 & \frac{1}{2} \\ 0 & \frac{2}{3} & \frac{1}{3} & 0 \end{bmatrix} \end{array}$$

b.
$$P^2 = \begin{array}{c} \\ a \\ b \\ c \\ d \end{array} \begin{array}{cccc} a & b & c & d \\ \begin{bmatrix} \frac{1}{2} & 0 & 0 & \frac{1}{2} \\ 0 & \frac{2}{3} & \frac{1}{3} & 0 \\ 0 & \frac{2}{3} & \frac{1}{3} & 0 \\ \frac{1}{2} & 0 & 0 & \frac{1}{2} \end{bmatrix} \end{array}$$

$$P^3 = \begin{array}{c} \\ a \\ b \\ c \\ d \end{array} \begin{array}{cccc} a & b & c & d \\ \begin{bmatrix} 0 & \frac{2}{3} & \frac{1}{3} & 0 \\ \frac{1}{2} & 0 & 0 & \frac{1}{2} \\ \frac{1}{2} & 0 & 0 & \frac{1}{2} \\ 0 & \frac{2}{3} & \frac{1}{3} & 0 \end{bmatrix} \end{array}$$

c. $0, \dfrac{1}{2}$ d. $\dfrac{1}{2}, 0$

13. a.
$$P = \begin{array}{c} \\ 0 \\ 1 \\ 2 \end{array} \begin{array}{ccc} 0 & 1 & 2 \\ \begin{bmatrix} 0 & 1 & 0 \\ \frac{1}{4} & \frac{1}{2} & \frac{1}{4} \\ 0 & 1 & 0 \end{bmatrix} \end{array}$$

b.

$$P^2 = \begin{array}{c} \\ 0 \\ 1 \\ 2 \end{array} \begin{array}{ccc} 0 & 1 & 2 \\ \left[\begin{array}{ccc} \frac{1}{4} & \frac{1}{2} & \frac{1}{4} \\ \frac{1}{8} & \frac{3}{4} & \frac{1}{8} \\ \frac{1}{4} & \frac{1}{2} & \frac{1}{4} \end{array}\right] \end{array}$$

$$P^3 = \begin{array}{c} \\ 0 \\ 1 \\ 2 \end{array} \begin{array}{ccc} 0 & 1 & 2 \\ \left[\begin{array}{ccc} \frac{1}{8} & \frac{3}{4} & \frac{1}{8} \\ \frac{3}{16} & \frac{5}{8} & \frac{3}{16} \\ \frac{1}{8} & \frac{3}{4} & \frac{1}{8} \end{array}\right] \end{array}$$

c. $S^{(0)} = [0 \quad 1 \quad 0]$

$S^{(1)} = [\frac{1}{4} \quad \frac{1}{2} \quad \frac{1}{4}]$

$S^{(2)} = [\frac{1}{8} \quad \frac{3}{4} \quad \frac{1}{8}]$

$S^{(3)} = [\frac{3}{16} \quad \frac{5}{8} \quad \frac{3}{16}]$

17. a.

$$P = \begin{array}{c} \\ 0 \\ 1 \\ 2 \\ 3 \\ 4 \\ 5 \end{array} \begin{array}{cccccc} 0 & 1 & 2 & 3 & 4 & 5 \\ \left[\begin{array}{cccccc} \frac{7}{10} & \frac{3}{10} & 0 & 0 & 0 & 0 \\ \frac{7}{15} & \frac{13}{30} & \frac{1}{10} & 0 & 0 & 0 \\ 0 & \frac{7}{15} & \frac{13}{30} & \frac{1}{10} & 0 & 0 \\ 0 & 0 & \frac{7}{15} & \frac{13}{30} & \frac{1}{10} & 0 \\ 0 & 0 & 0 & \frac{7}{15} & \frac{13}{30} & \frac{1}{10} \\ 0 & 0 & 0 & 0 & \frac{2}{3} & \frac{1}{3} \end{array}\right] \end{array}$$

b. $S^{(0)} = [1 \quad 0 \quad 0 \quad 0 \quad 0 \quad 0]$

$S^{(1)} = [\frac{7}{10} \quad \frac{3}{10} \quad 0 \quad 0 \quad 0 \quad 0]$

$S^{(2)} = [\frac{63}{100} \quad \frac{34}{100} \quad \frac{3}{100} \quad 0 \quad 0 \quad 0]$

19. a. $S^{(1)} = [\frac{33}{100} \quad \frac{1}{2} \quad \frac{17}{100}]$

$S^{(2)} = [\frac{29}{100} \quad \frac{1}{2} \quad \frac{21}{100}]$

21.

$$\text{AA cross} \quad \begin{array}{c} \\ \text{AA} \\ \text{Aa} \\ \text{aa} \end{array} \begin{array}{ccc} \text{A} & \text{Aa} & \text{aa} \\ \left[\begin{array}{ccc} 1 & 0 & 0 \\ \frac{1}{2} & \frac{1}{2} & 0 \\ 0 & 1 & 0 \end{array}\right] \end{array}$$

23. $S^{(1)} = [0.495 \quad 0.505 \quad 0]$

$S^{(2)} = [0.7475 \quad 0.2525 \quad 0]$

25. $P^2 = \begin{bmatrix} 1 & 0 \\ a + a(1-a) & (1-a)^2 \end{bmatrix}$

$P^3 = \begin{bmatrix} 1 & 0 \\ a + a(1-a) + a(1-a)^2 & (1-a)^3 \end{bmatrix}$

$P^4 = \begin{bmatrix} 1 & 0 \\ a + a(1-a) + a(1-a)^2 + a(1-a)^3 & (1-a)^4 \end{bmatrix}$

$P^n = \begin{bmatrix} 1 & 0 \\ a + a(1-a) + a(1-a)^2 + \cdots + a(1-a)^{n-1} & (1-a)^n \end{bmatrix}$

Exercises 10.2, page 486

1. a. regular   b. not regular   c. not regular   d. regular
5. a. $[\frac{8}{17} \quad \frac{9}{17}]$ b. $[\frac{3}{4} \quad \frac{1}{4}]$ c. $[\frac{4}{15} \quad \frac{11}{15}]$ d. $[\frac{593}{1268} \quad \frac{675}{1268}]$
7. a. $[\frac{7}{11} \quad \frac{3}{11} \quad \frac{1}{11}]$ b. $[\frac{1}{3} \quad \frac{1}{3} \quad \frac{1}{3}]$ c. $[\frac{10}{43} \quad \frac{12}{43} \quad \frac{21}{43}]$ d. $[\frac{645}{1592} \quad \frac{2153}{4776} \quad \frac{86}{597}]$
9. a. $[\frac{7}{11} \quad \frac{4}{11}]$ b. In the long run, the rat will turn right with probability $\frac{7}{11}$ and left with probability $\frac{4}{11}$.
11. $[\frac{43}{85} \quad \frac{42}{85}]$ In the long run, 51% will buy Black Star Cola and 49% will not buy Black Star Cola.

13. $[\frac{1}{6} \quad \frac{2}{3} \quad \frac{1}{6}]$ In the long run, the probability that there are no blue balls in urn A is $\frac{1}{6}$, the probability that there is one blue ball in urn A is $\frac{2}{3}$, and the probability that there are two blue balls in urn A is $\frac{1}{6}$.

15. a.

$$P = \begin{array}{c} \\ A \\ B \\ C \end{array} \begin{array}{ccc} A & B & C \\ \left[\begin{array}{ccc} 0 & \frac{1}{2} & \frac{1}{2} \\ \frac{2}{3} & 0 & \frac{1}{3} \\ \frac{2}{3} & \frac{1}{3} & 0 \end{array}\right] \end{array}$$

   b. $[\frac{2}{5} \quad \frac{3}{10} \quad \frac{3}{10}]$
   c. In the long run, the mouse will be in compartment A with probability $\frac{2}{5}$, in compartment B with probability $\frac{3}{10}$, and in compartment C with probability $\frac{3}{10}$.

17. The company should buy machine B.

19. a.

$$P = \begin{array}{c} \\ 0 \\ 1 \\ 2 \end{array} \begin{array}{ccc} 0 & 1 & 2 \\ \left[\begin{array}{ccc} \frac{1}{3} & \frac{2}{3} & 0 \\ \frac{2}{9} & \frac{5}{9} & \frac{2}{9} \\ 0 & \frac{2}{3} & \frac{1}{3} \end{array}\right] \end{array}$$

   b. $[\frac{1}{5} \quad \frac{3}{5} \quad \frac{1}{5}]$
   c. In the long run, the probability that there are no blue balls in urn A is $\frac{1}{5}$, the probability that there is one blue ball in urn A is $\frac{3}{5}$, and the probability that there are two blue balls in urn A is $\frac{1}{5}$.

21. a.

$$P = \begin{array}{c} \\ 0 \\ 1 \\ 2 \end{array} \begin{array}{ccc} 0 & 1 & 2 \\ \left[\begin{array}{ccc} 0.4 & 0.6 & 0 \\ 0.2 & 0.5 & 0.3 \\ 0 & 0.5 & 0.5 \end{array}\right] \end{array}$$

   b. $[\frac{5}{29} \quad \frac{15}{29} \quad \frac{9}{29}]$
   c. In the long run, the probability that no cars are waiting in line is $\frac{5}{29}$, that one car is waiting in line is $\frac{15}{29}$, and that two cars are waiting in line is $\frac{9}{29}$.

23. b. $[\frac{1}{4} \quad \frac{1}{2} \quad \frac{1}{4}]$
   c. In the long run, the population will be composed of 25% genotype AA, 50% genotype Aa, and 25% genotype aa.

25. a.

$$P = \begin{array}{c} \\ A \\ B \\ C \end{array} \begin{array}{ccc} A & B & C \\ \left[\begin{array}{ccc} 0.2 & 0.5 & 0.3 \\ 0 & 0.7 & 0.3 \\ 0.5 & 0 & 0.5 \end{array}\right] \end{array}$$

   b. The trucks should be distributed as follows: 23% at location A, 39% at location B, and 38% at location C.

27. a.

$$P = \begin{array}{c} \\ A \\ B \\ C \end{array} \begin{array}{ccc} A & B & C \\ \left[\begin{array}{ccc} \frac{1}{6} & \frac{2}{9} & \frac{11}{18} \\ \frac{11}{18} & \frac{1}{6} & \frac{2}{9} \\ \frac{2}{9} & \frac{11}{18} & \frac{1}{6} \end{array}\right] \end{array}$$

   b. $[\frac{1}{3} \quad \frac{1}{3} \quad \frac{1}{3}]$

29. $[\frac{83}{308} \quad \frac{95}{308} \quad \frac{29}{308} \quad \frac{101}{308}]$

31. $w_1 = av_1 + cv_2$ and $w_2 = bv_1 + dv_2$.

Therefore,

$$w_1 + w_2 = (av_1 + cv_2) + (bv_1 + dv_2)$$
$$= (a + b)v_1 + (c + d)v_2$$
$$= v_1 + v_2$$
$$= 1$$

Exercises 10.3, page 496

1. The game is not strictly determined.
3. The game is strictly determined with value $\frac{1}{2}$.
5. The game is strictly determined with value -2.
7. r_1 and c_3 with value 0.10.
9. r_2 and c_2 with value -50.
11. a.

$$\begin{array}{c} & & \text{Cary} \\ & & 5 \quad\ 7 \quad\ 8 \\ \text{Ron} \begin{array}{c} 2 \\ 3 \\ 10 \end{array} & \left[\begin{array}{rrr} -7 & -9 & 10 \\ 8 & 10 & -11 \\ -15 & -17 & 18 \end{array}\right] \end{array}$$

  b. The game is not strictly determined.
13. a. Mrs. Black should invest in either stocks or certificates.

  b. $\begin{bmatrix} 3\% & 8\% & 11\% \\ 6\% & 8\% & 0\% \\ 6\% & 4\% & -1\% \end{bmatrix}$

  The game is not strictly determined.
15. The game of Morra is not strictly determined.
17. Since b is a saddle point, $b \le a$ and $b \ge d$. Since c is a saddle point, $c \le d$ and $c \ge a$. Now, $b \le a$ and $a \le c$ imply that $b \le c$, and $b \ge d$ and $d \ge c$ imply that $b \ge c$. Therefore, $b = c$.

Exercises 10.4, page 509

1. a. 7.89 b. 7.7078

3. a. $P^* = [\frac{4}{9} \ \frac{5}{9}]$, $Q^* = [\frac{2}{9} \ \frac{7}{9}]$, $E = \dfrac{17}{9}$ b. $P^*AQ = \frac{17}{9} \le \frac{17}{9} = P^*AQ^*$

5. $P^* = [\frac{7}{12} \ \frac{5}{12}]$, $Q^* = [\frac{5}{8} \ \frac{3}{8}]$, $E = -\dfrac{3}{4}$

7. $P^* = [\frac{63}{74} \ \frac{11}{74}]$, $Q^* = [\frac{9}{37} \ \frac{28}{37}]$, $E = -0.0332$

9. $P^* = [\frac{15}{17} \ \frac{2}{17}]$, $Q^* = [\frac{19}{34} \ \frac{15}{34}]$, $E = \dfrac{999}{17}$

11. a. Juicy Orchards should spray $\frac{6}{7}$ of the time and should not spray $\frac{1}{7}$ of the time. b. $164,285.71
13. Both the union and the university should use aggressive strategies, in which case the expected payoff is $4.50.
15. Let $P^* = [p_1 \quad p_2]$ and $Q^* = [q_1 \quad q_2]$ be the row and column strategies, respectively, for the game

$$\begin{bmatrix} a & b \\ c & d \end{bmatrix}$$

Let $U^* = [u_1 \quad u_2]$ and $V^* = [v_1 \quad v_2]$ be the row and column strategies, respectively, for the game

$$\begin{bmatrix} a + 3 & b + 3 \\ c + 3 & d + 3 \end{bmatrix}$$

Then,

$$u_1 = \frac{d + 3 - (c + 3)}{a + 3 + d + 3 - (b + 3) - (c + 3)}$$

$$= \frac{d - c}{a + d - b - c} = p_1$$

Therefore, $U^* = P^*$. Similarly, $V^* = Q^*$.

Now, the optimal expected value of the second game is

$$\frac{(a + 3)(d + 3) - (b + 3)(c + 3)}{a + 3 + d + 3 - (b + 3) - (c + 3)}$$

$$= \frac{ad - bc}{a + d - b - c} + 3$$

Review Exercises (Chapter 10), page 512

1. It can be a transition matrix.
3. It can be a transition matrix.

5. a.
$$P^2 = \begin{bmatrix} 0.3639 & 0.4082 & 0.2279 \\ 0.3258 & 0.4216 & 0.2526 \\ 0.2998 & 0.3952 & 0.3050 \end{bmatrix}$$

  b. $S^{(1)} = [0.323 \quad 0.408 \quad 0.269]$
  $S^{(2)} = [0.32943 \quad 0.40966 \quad 0.26091]$

7. not regular   9. not regular   11. $[\frac{6}{11} \quad \frac{5}{11}]$ 13. $[\frac{3}{8} \quad \frac{1}{12} \quad \frac{13}{24}]$

15. a.   rain sunny   b. 0.3885
$$\begin{array}{c} \text{rain} \\ \text{sunny} \end{array} \begin{bmatrix} 0.45 & 0.55 \\ 0.35 & 0.65 \end{bmatrix}$$

17. a.   0   1   2
$$\begin{array}{c} 0 \\ 1 \\ 2 \end{array} \begin{bmatrix} \frac{2}{3} & \frac{1}{3} & 0 \\ \frac{1}{2} & \frac{5}{12} & \frac{1}{12} \\ 0 & \frac{3}{4} & \frac{1}{4} \end{bmatrix}$$

  b. $\frac{27}{47}$ $\frac{18}{47}$ $\frac{2}{47}$
  c. In the long run, the probability that no trucks are waiting in line is $\frac{27}{47}$, the probability that one truck is waiting in line is $\frac{18}{47}$, and the probability that two trucks are waiting in line is $\frac{2}{47}$.
19. The game is strictly determined. The row player should use row one and the column player should use column two. The value of the game is -1.
21. The game is not strictly determined.
23. The game is strictly determined. The row player should use row three and the column player should use column one. The value of the game is 10.
25. The game is strictly determined. The row player should use row two and the column player should use column one. The value of the game is 0.2.

27. a. 9.536 b. 11.9756   29. $P^* = [\frac{1}{4} \quad \frac{3}{4}]$, $Q^* = [\frac{1}{4} \quad \frac{3}{4}]$, $E = 42.5$

31. $P^* = [\frac{977}{1815} \quad \frac{838}{1815}]$, $Q^* = [\frac{966}{1815} \quad \frac{849}{1815}]$, $E = -34.99$

33. a.
$$\begin{array}{cc} & \text{Rita} \\ & \begin{array}{cc} 2 & 5 \end{array} \\ \text{Erica} \begin{array}{c} 2 \\ 5 \end{array} & \begin{bmatrix} -2 & 3 \\ 3 & -5 \end{bmatrix} \end{array}$$
b. $P^* = [\frac{8}{13} \quad \frac{5}{13}]$, $Q^* = [\frac{8}{13} \quad \frac{5}{13}]$ c. $E = -\dfrac{1}{13}$

CHAPTER 11

Exercises 11.1, page 521

1. a. acceptable b. not acceptable c. not acceptable d. acceptable

3. a. acceptable b. acceptable c. not acceptable d. not acceptable e. acceptable f. not acceptable

5. a. 0.001 b. 230,000 c. 0.1345 d. 78,900

7. a. 23E − 4 b. 17002E − 4 c. 12736E3 d. 184357E − 9   9. (A + B)/2

11. $(2 * X) + (3 * Y) + (5 * Z)$ 13. $(A * B) ** 0.5$ 15. $(3 * ((X - X1) ** 2) + 2 * ((X - X2) ** 2))/5$

17. $(((1 + I) ** N) - 1)/I$

19. a. A + B is not a variable

    b. AX is not an acceptable variable

    c. A PLUS B is not an arithmetic expression

    d. A + 2B is not an acceptable arithmetic expression

21. X = 2, Y = 5, Z = 10   23. B = 5, C = 7, X = −2, Y = 6

Exercises 11.2, page 530

1. DATA 7.2, −0.03, 6 READ A, B, C   3. DATA 72.45, 95, 10731.52 READ X1, X2, X3

5. A = 3, B = 5, C = −4, D = 3   7. insufficient data

9. Another ? will appear since there are not enough data in the input line.

11. One ? will appear and one constant is needed for X.

13. A IS −0.01

    B IS   7.2

15. THE PRINCIPAL IS   THE RATE IS

    989   0.02

17. a. THE ROW-COLUMN PRODUCT IS   − 54

19. a. 10   DATA 5, 7, 8, 10, 155

      20   READ A, B, C, D, E

      30   PRINT "ENTER VALUES FOR W, X, Y, Z"

      40   INPUT W, X, Y, Z

      50   LET F = $(A * W) + (B * X) + (C * Y) + (D * Z) + E$

      60   PRINT "THE VALUE OF F IS   "; F

      70   END

21. a. 10   PRINT "ENTER VALUES FOR S, I, N"

      20   INPUT S, I, N

      30   LET P = $S/((1 + I) ** N)$

      40   PRINT

      50   PRINT "FOR S =   "; S, "I =   "; I, "N =   "; N

      60   PRINT "   P =   "; P

      70   END

Exercises 11.3, page 538

1. 30   3. no printout   5. 3, 4, 5, 6, 7, 8, 9, 10   7. 15, 12, 9, 6, 3, 0   9. 280   11. 8
                                                                                        −9
                                                                                        6
                                                                                        0

13. a. −3   9
        0   0
        3   9
        6   36
        9   81

```
15. 10          DATA 1, 4, −3, 15
    20          READ A, B, C, D
    30          PRINT "ENTER VALUES FOR X, Y, Z"
    40          INPUT X, Y, Z
    50   REM
    60          LET T = (A * X) + (B * Y) + (C * Z)
    70   REM
    80          IF T = D THEN 110
    90          PRINT "("; X; Y; Z;") IS NOT A SOLUTION"
    100         GO TO 130
    110         PRINT "("; X; Y; Z; ") IS A SOLUTION"
    120  REM
    130         END
17. a. 10       PRINT "ENTER NUMBER OF VALUES FOR X"
    20          INPUT N
    30   REM
    40          PRINT "AT EACH   ?, ENTER NEXT VALUE AND FREQUENCY"
    50   REM
    60          LET S = 0
    70          LET K = 0
    80          FOR I = 1 TO N
    90             INPUT X, F
    100            LET S = S + (X * F)
    110            LET K = K + F
    120         NEXT I
    130  REM
    140         LET M = S/K
    150         PRINT "THE MEAN OF X IS   "; M
    160  REM
    170         END
19. a. 10       PRINT "ENTER THE RATE PER PERIOD"
    20          INPUT I
    30   REM
    40          PRINT "PERIOD", "AMOUNT OF $1"
    50   REM
    60          FOR K = 1 TO 24
    70             LET S = (1 + I)**K
    80             PRINT K, S
    90          NEXT K
```

```
100   REM
110          END
```

Note: This program can be made more efficient by changing the manner in which S is obtained.

Exercises 11.4, page 546

3. a. Change only the dimension statement to

```
10          DIM A(4, 4), B(4), X(4)
```

5. Insert, for example, the following lines.

```
92          PRINT "THE COEFFICIENT MATRIX IS"
94          MAT PRINT A
96          PRINT
98          PRINT "THE RIGHT-HAND SIDES ARE"
99          MAT PRINT B
```

7. Insert, for example, the following lines

```
12   REM
15          PRINT "ENTER AN 2 × 3 MATRIX A"
25          PRINT "ENTER AN 2 × 3 MATRIX B"
35   REM
45   REM
47          PRINT
49          PRINT "THE SUM A + B IS"
55   REM
```

9. Insert, for example, the following lines.

```
12   REM
15   REM "ENTER A 3 × 3 INPUT-OUTPUT MATRIX A"
25   REM
55   REM
57          PRINT "HOW MANY DEMAND VECTORS ARE THERE?"
65   REM
75          PRINT "ENTER THE NEXT DEMAND VECTOR"
95          PRINT "THE INTENSITY VECTOR IS"
97          PRINT
115  REM
```

Also, indent lines 80 through 100.

11. a. Insert, for example, the following lines.

```
15   REM
17          PRINT "ENTER A 3 × 3 TRANSITION MATRIX"
25          PRINT "ENTER AN INITIAL STATE VECTOR"
```

35	REM
45	REM
47	PRINT "HOW MANY TRANSITIONS DO YOU WANT?"
55	REM
57	PRINT "K"; "ENTRIES OF STATE DISTRIBUTION VECTOR"
65	REM
115	REM

Also, indent lines 80 through 100.

13. Except for PRINT statements, the essential changes that need to be made are as follows:
   1. change line 10 to

$$5 \qquad \text{DIM } A(10, 10), I(10, 10), D(10), X(10)$$

   2. insert between lines 60 and 70 the statement

$$\text{INPUT N}$$

   3. change line 80 to

$$80 \qquad \text{MAT INPUT } A(N, N)$$

   4. change line 100 to

$$100 \qquad \text{MAT INPUT } D(N)$$

   5. change line 190 to

$$190 \qquad \text{MAT I} = \text{IDN}(N, N)$$

Review Exercises (Chapter 11), page 549

1. a. $I/(((1 + I) ** (-N)) - 1)$ b. $(A + B) ** (2/3)$ 3. $X = 6, Y = 7.2, Z = 28.8$
5. $W = 1, X = 0.25, Y = 5, Z = 8.2$ 7. 2, 4, 6, 8, 10, 12, 14   9. $A = 6.65$
11. THE RESULT IS

	55
13. 7	2
3	9
6	6
9	3
2	7

Index

Abscissa 16
Accumulation factor 296
Addition of matrices 162
Additive identity
  matrix 165
  real numbers 555
Additive inverse
  of a matrix 165
  for a real number 555
Algorithm 516
Alternate hypothesis 450, 452
Amortization schedule 310
Amount
  annuity 305
  bank discount 288
  compound interest 295
  simple interest 284
Annuity 304
  amount 305
  amount of $1 per period 306
  certain 305
  contingent 305
  due 305
  future value 305
  ordinary 305

payment period 305
periodic deposit 307
periodic payment 304, 309
present value 308
present value of $1 per
  period 308
term 305
Argument 40
  conclusion of 40
  hypothesis of 40
  invalid 40
  valid 40
Arithmetic average 403
Arithmetic expression 519
Array 540
  one-dimensional 540
  two-dimensional 540
Artificial variable 267
Assignment of probabilities
  321, 373
Associative property
  addition of matrices 165
  addition of real numbers 555
  multiplication of matrices
    178

multiplication of real number
  555
Augmented matrix 128, 129

Back substitution 122
Bank discount 284, 288
  amount 288
  proceeds 288
  rate 288
Bar graph 398
BASIC 517
Basic feasible solution 219
  initial 223
Basic solution 219, 223
Basic variable 147, 223
Bayes's formula 385, 387
Behavioral pattern 464
Bernoulli trials 415
Best fit, line of 185
Best moves 493, 495
Binomial experiment 415
Binomial random variable 415
  table of 559
Boundary line 88

Bounded convex polygonal set
95
Branches 533
Break-even point 71

Canonical form 221, 269
Cartesian coordinate system 15
Cartesian product 341, 342
Center of symmetry 431
Central processing unit (CPU)
520
Central tendency, measure of
403
Certain event 320
Coefficient
matrix 171
of variables 120
Column matrix 161
Column maximum 494
Column player 492
best move for 494
Combination 352
counting 353
Commutative property
addition of matrices 165
addition of real numbers 555
multiplication of real numbers
555
Complement of a set 332
Compiler 516
Compound interest 284, 293
amount 295
amount of $1 296
accumulation factor 296
continuous compounding
299
conversion period 293
discount factor 300
effective rate 299
frequency of conversion 293
interest rate 293
interest period 293
nominal rate 293
present value 300
present value of $1 300
Compound statements 33
Conclusion 40

Conditional probability 360
Conjunction 33
truth table for 35
Consistent system of linear
equations 66, 121
Constant 517
numeric 517
string 517
Constraints 99, 220
Continuous compounding 299
Continuous random variable
396
Contrapositive 37
Control
statement 532
structure 532
variable 537
Converse 37
Conversion rule 436
Convex set 90
Convex polygonal set 90
boundary line 88
bounded 95
unbounded 95
vertices 90
Coordinate axes 15
scale 15
x 15
y 15
Coordinate plane 15
Coordinate system
Cartesian 15
rectangular 15
Coordinate
of a real number 557
x 16
y 16
Coordinates of a point 16
abscissa 16
ordinate 16
Cost
fixed 60
total 60
variable 60
Counting
combinations 353
permutations 346, 349
principle of 343, 344
CRT terminal 528

Data points 19
Data stack 525
Debug 517
Decimal representation 555
repeating 555
terminating 555
Default 537
Degeneracy 244
Degenerate solution 228
Demand equation 61
Demand vector 206
De Morgan's laws 336
Departing variable 229, 233
criterion for selecting 238
Dependent events 363
Dependent variable 52, 75, 119
Depreciation 59
straight-line 59
sum-of-the-years-digits 60
Deterministic model 51
Deviation
from the mean 423
standard 425, 426
Diagonal matrix 192
Diet problem 108, 260
Difference
of matrices 165
of sets 332
Directed distance 558
Discount factor 300
Discount rate 288
Discrete random variable 396
finite 396
infinite 396
Disjoint sets 333
Disjunction 33
truth table for 36
Dispersion, measure of 423
Distribution curve 430
normal 431
Distribution vector 176
Distributive property
for matrices 179
for real numbers 555
Division of real numbers 555
Documentation 526
Drug company claim 419
Dual problem 251
Duality 251, 256

Editing 524
Effective rate of interest 299
Elimination method 66
Empty event 320
Entering variable 229, 233
  criterion for selecting 236
Entry of a matrix 128
Equal matrices 162
Equilibrium
  market 72
  price 73
  quantity 73
Equally likely 341
Equation
  demand 61
  equivalent 8
  linear 18
  normal 186
  solution of 8
  supply 62
Equiprobable 341
Equivalence 33
  truth table for 38
Equivalent
  equations 8
  inequalities 11
  rates of interest 289
  systems of linear equations
    67, 122
Error vector 185
  size of 185
Exact interest year 285
Exclusive OR 36
Expansion rule 383, 386
Expected payoff 500, 501, 502
Expected value 405
Experiment 319
  binomial 415
  urn-like 375
Exponent 558
Exponential notation 517
Event 320
  certain 320
  empty 320
  impossible 320
  probability of 326
Events
  dependent 363
  independent 363

mutually exclusive 333
pair-wise mutually exclusive
  386
probability of 326

Factorial 346
Fair game 409, 495
Feasible solution 87, 99, 222
  basic 219
  initial basic 223
Fixed cost 60
Focal date 287
Frequency 403
Frequency table 404
Function 52
  graph of 53
  linear 52
Future value 283, 305

Game 491
  fair 409, 495
  matrix 492
  mixed strategies 503
  moves 491
  optimal mixed strategy 504
  pure strategies 503
  saddle point for 494
  strategy 500
  strictly determined 495
  two-person 491
  value of 495
  zero-sum 492
Game with nature 506
Gaussian elimination 131
Gauss-Jordan reduction 140,
  142, 144
Genetics 418, 471
Graphical method 103
Graph
  bar 398
  inequality 11
  linear equation 18
  linear function 53
  linear inequality 88
  real numbers 557
  systems of linear inequalities
    87

Half-plane 88
Hardcopy terminal 528
Histogram 443
Homogeneous system of linear
  equations 149
Hypothesis
  alternate 450, 452
  null 450, 452
  of an argument 40
  testing 452

Identity matrix 193
Identity for real numbers 555
Implication 33
  contrapositive of 37
  converse of 37
  truth table for 37
Impossible event 320
Inclusive OR 37
Inconsistent system of linear
  equations 66, 121
Independent events 363
Independent trials 414
Independent variable 52, 75,
  119
Inequalities
  equivalent 11
  graph of 88
  matrix 251
  rules of 11
  system of 86
Inequality signs 556
Input 516
Input-output analysis 204
  demand vector 206
  goods 204
  industries 204
  intensity vector 206
  matrix 206
Integer programming 110
Intensity vector 206
Interest 283
  compound 284, 293
  effective rate of 299
  nominal rate 293
  period 293
  rate of 284, 293
  simple 284
  tables of 565

Intersection of sets   332
Invalid argument   40
Inventory, distribution of   484
Inverse of a matrix   193
   procedure for finding   196
Invertible matrix   193
Inverse of a real number   555
Investment problem   109, 150

Law of large numbers   324
Leading one   141, 146
Level of significance   451, 452
Life insurance premium   410
Library   524
Life of a manufactured item
   438
Line numbers   523
Line of best fit   185
   error vector   185
   normal equations   186
Linear equations   18, 120
   graph of   18
   point-slope form   28
   slope-intercept form   29
   solution of   120
   system of   65, 121
Linear form   75, 119
Linear function   52
   graph of   53
Linear inequality in two
      variables   86
   graph of   88
   system of   86
Linear programming problem
   in canonical form   221
   in two variables   99
   standard maximization type
      of   100, 101, 220
   standard minimization type
      of   101, 102, 250, 251
Loan balance   310
Logical connectives   33
   conjunction   33
   disjunction   33
   equivalence   33
   implication   33
   negation   33
Logical error   517
Logically equivalent   40
Loop   536

Machine language   516
Machine performance   484
Market equilibrium   72
Markov chain   462, 463, 464
   regular   483
   stable state vector   479
Mathematical model   51
Matrices
   addition of   162
   difference of   165
   equal   162
   multiplication of   174
   row-column product of   169
   row equivalent   130
   row operations on   130
   substraction of   165
   sum of   162
Matrix   161
   addition   162
   augmented   128, 129
   coefficient   171
   column   161
   diagonal   192
   entry of   128
   equation   172
   game   429
   identity   193
   inequality   251
   input-output   206
   inverse of   193
   invertible   193
   main diagonal of   192
   payoff   492
   product   174
   productive   207
   reduced row-echelon form
      141
   row   161
   row-echelon form   141
   size of   128
   square   192
   substraction   165
   transition   176, 464
   transpose of   166
   upper triangular   198
   zero   164
Mean   403, 404, 406
Measure
   of central tendency   403
   of dispersion   423
Mixed strategies   503

   optimal   504
Mixture problem   133
Model
   deterministic   51
   mathematical   51
   probabilistic   51
Memory (cells)   517
Moves   491
Multiplication
   by a scalar   165
   of matrices   174
Mutually exclusive events   333

Natural conditions   86
Negation   33
   truth table for   35
Nominal rate of interest   293
Non-basic variable   223
Non-negative conditions   222
Non-trivial solution   149
Normal curve   431
   center of symmetry   431
   standard   432
Normal equations   186
Normal random variable   431
   probabilities of   433
Null hypothesis   450
Number properties   556
Numeric constant   517

Objective function   99, 222
One-tailed test   453
Operations on equations   122
Opinion polls   419
Optimal mixed strategy   504
Optimal
   solution   99, 222
   value   99, 222
Ordered pair   14
Ordered subset   352
Ordinary annuity   305
Ordinary interest year   285
Ordinate   16
Origin   15, 557
Outcomes   319
   equally likely   341
Output   516
Outstanding principle   310

Parameter 147
Parametric equation 147
Partition of a sample space
    383, 386
Pair-wise mutually exclusive
    386
Payoff 396, 492
    expected 500, 501, 502
Payoff matrix 492
Payment
    period 305
    periodic 304, 309
Permutation 346, 348, 352
    counting 346, 349
Phase I problem 267
Phase II problem 267
Pivot 142, 238
Pivot operations 238
Plot 15
Pointer 525
Point-slope form 28
Population transition 469
Present value 283, 300, 308
Present value of $1 300
Present value of $1 per period
    308
Price vector 169
Primal problem 251
Principal 284
    outstanding 310
Principle of counting 343, 344
Probabilistic model 51
Probability 321
    assignment of 321
    conditional 360
    distribution 398
    model 319
    of an event 326
    subjective 325
    transition 464
    tree 373
Proceeds 288
Production problem 134, 244
Product rule of events 362,
    364, 365
Productive matrix 207
Profit 61
Programming style 535
Prompt 526
Propositional variable 33
Pure strategies 503

Quadrants 16
Queuing theory 469

Random variable 395
    binomial 415
    continuous 396
    discrete 396
    finite 396
    infinite 396
    normal 431
    standard normal 432
    value of 395
Real number line 557
Real number system 555
Reduced row-echelon form
    141
Regular Markov chain 483
Relative frequency 323, 324
Reliability of a production
    process 456
Rent 305
Revenue 60
Rise 22
Row equivalent matrices 130
Row matrix 161
Row minimum 493
Row operation 130
Row player 492
    best move for 493
Row-column product 169
Row-echelon form 141
Rules of
    equalities 8
    inequalities 11
Run 22

Saddle point 494
Salvage value 59
Sample space 319
    partition of 383, 386
Scalar 165
Scale 15, 557
Scientific notation 517
Selecting from a machine's
    production 438
Sequence of Bernoulli trials
    415

Set
    complement of 332
    number of elements in 341
    solution 11, 66
    universal 331
Sets
    difference 332
    disjoint 333
    intersection of 332
    union of 332
    Venn diagram of 332
Shipment problem 151
Simple interest 284
    amount 284
    rate of 284
Simplex algorithm 240
Simplex method 219, 269
Simplex tableau 233
Sinking fund 307
Slack variable 220
Slope 21
    ratio of rise to run 22
Slope-intercept form 29
Solution
    basic 219, 223
    basic feasible 219, 223
    degenerate 228
    feasible 87, 99, 222
    non-trivial 149
    of a matrix equation 172
    optimal 99, 222
    set 11, 66
    trivial 149
    to a linear equation 120
    to a system of linear equa-
        tions 65, 121
    to a system of linear inequal-
        ities 86
    to an equation 8
    unbounded 107
Solving a linear programming
    problem
    by duality 256
    by the graphical method 103
    by the simplex algorithm 240
    by the simplex method 269
Solving a system of linear equa-
    tions
    by elimination method 66
    by Gauss-Jordan reduction
        140, 142

by Gaussian elimination 131
by substitution method 67
Source program 517
Square
  matrix 192
  system 146
Stable state vector 479
Standard deviation 425, 426
Standard maximization
    problem 100, 101, 220
  method for solving 103,
    240, 269
Standard minimization problem
    101, 102, 250, 251
  method for solving 103,
    256, 269
Standard normal curve 432
  table for 564
Standard normal random
    variable 432
State distribution vector 465
  initial 465
Statements 32
  compound 33
  logically equivalent 40
States of Markov chains 464
States of stochastic process 383
Statistic, test 450, 452
Step size 537
Stochastic process 372
Stock cutting problem 261
Straight-line depreciation 59
Strategy 500, 502, 503
  mixed 503
  pure 503
Strictly determined game 495
String constant 517
Subscripted variable 119
Subset 332
  ordered 352
  unordered 352
Substraction
  of matrices 165
  of real numbers 555
Substitution method 67
Supply equation 62
Supply problem 276
Surplus 266
Surplus variable 266
Syntax 517

System of linear equations 65,
    121
  consistent 66, 121
  equivalent 67, 122
  homogeneous 149
  inconsistent 66, 121
  solution of 65, 121
  square 146
  triangular form 122
System of linear inequalities 86
  graph of 87, 90
  solution of 86

Tables
  binomial random variable
    559
  interest 565
  standard normal curve 564
Tautology 39
Terminal 523
  CRT 523
  hardcopy 523
Test statistic 450, 452
Test taking by guessing 419
Time diagram 287
Time-sharing 523
Total cost 60
Transition matrix 176, 464
Transition probability 464
Transportation problem 272
Transpose of a matrix 166
Tree diagram 343
  branch of 343
  path in 343
Triangular form 122
Trivial solution 149
Truth table 35
  conjunction 35
  disjunction 36
  equivalence 38
  implication 37
  negation 35
Two-person game 491
Two-phase method 265, 269
Two-tailed test 454

Unbounded
  convex polygonal set 95
  solution 107
Union of sets 332
Universal set 331
Unknowns 120
Unordered subset 352
Upper triangular matrix 198
Urn-like experiment 375
Useful life 59

Valid argument 40
Value of a game 495
Values of a random variable
    395
Variable
  artificial 267
  basic 147, 223
  coefficient of 120
  control 537
  cost 60
  departing 229, 233
  dependent 52, 75, 119
  entering 229, 233
  independent 52, 75, 119
  name 518
  non-basic 233
  propositional 33
  random 395
  slack 220
  subscripted 119
  surplus 266
Variance 424, 426
Vector
  demand 206
  distribution 176
  error 185
  intensity 206
  price 169
  stable state 479
  state distribution 465
Venn diagram 332
Vertices 90
Voter preference 455

x-axis   15

y-axis   15
y-intercept   29

zero matrix   164
zero-sum game   492